Magnetic, Ferroelectric, and Multiferroic Metal Oxides

Metal Oxides Series

Magnetic, Ferroelectric, and Multiferroic Metal Oxides

Edited by

Biljana D. Stojanovic
Institute for Multidisciplinary Research, University of Belgrade, Belgrade, Serbia

Series Editor

Ghenadii Korotcenkov

ELSEVIER

Elsevier
Radarweg 29, PO Box 211, 1000 AE Amsterdam, Netherlands
The Boulevard, Langford Lane, Kidlington, Oxford OX5 1GB, United Kingdom
50 Hampshire Street, 5th Floor, Cambridge, MA 02139, United States

British Library Cataloguing-in-Publication Data
A catalogue record for this book is available from the British Library

Library of Congress Cataloging-in-Publication Data
A catalog record for this book is available from the Library of Congress

ISBN: 978-0-12-811180-2

For Information on all Elsevier publications visit our website at https://www.elsevier.com/books-and-journals

Publisher: Matthew Deans
Acquisition Editor: Kayla Dos Santos
Editorial Project Manager: Andrae Akeh
Production Project Manager: Omer Mukthar
Cover Designer: Miles Hitchen

Typeset by MPS Limited, Chennai, India

Contents

List of contributors

Luis A.O. Araujo Federal University of São Carlos (UFSCar), São Carlos, Brazil

Juras Banys Vilnius University, Vilnius, Lithuania

Antoine Barbier Service de Physique de l'Etat Condensé, SPEC, CEA, CNRS, UMR 3680, Université Paris-Saclay, CEA Saclay, 91191 Gif-sur-Yvette Cedex, France

Wagner B. Bastos Institute of Chemistry, São Paulo State University (Unesp), Araraquara, Brazil

Andreja Bencan Jozef Stefan Institute, Ljubljana, Slovenia

Nelu Blaž University of Novi Sad, Novi Sad, Serbia

Jelena D. Bobic Institute for Multidisciplinary Research, University of Belgrade, Belgrade, Serbia

Mario Cilense São Paulo State University (Unesp), Institute of Chemistry, Araraquara, Brazil

Cristina E. Ciomaga Faculty of Physics, Dielectrics, Ferroelectrics & Multiferroics Group, Al. I. Cuza Univ. of Iasi, 11 Carol I Bv., 700506, Iasi, Romania

Sasa T. Cirkovic University of Belgrade, Belgrade, Serbia

Johan A. Cortés São Paulo State University (Unesp), School of Engineering, Guaratinguet́, Brazil

Zeljka Cvejić Department of Physics, Faculty of Sciences, University of Novi Sad, Novi Sad, Serbia

Mirjana Damnjanović University of Novi Sad, Novi Sad, Serbia

Drago M. Djordjevic University of Belgrade, Belgrade, Serbia

Idalci Cruvinel dos Reis Federal Institute of Education, Science and Technology of Goiano, Rio Verde, GO, Brazil

Adis S. Dzunuzovic Institute for Multidisciplinary Research, University of Belgrade, Belgrade, Serbia

Antonio Feteira Christian Doppler Lab for Advanced Ferroic Oxides, Materials and Engineering Research Institute, Sheffield Hallam University, Sheffield, United Kingdom

Carlos A. Fortulan University of São Paulo (USP), São Carlos, Brazil

Cesar R. Foschini São Paulo State University (Unesp), School of Engineering, Bauru, Brazil

Amador M. Gonzalez CEMDATIC, ETS Ingeniería y Sistemas de Telecomunicación, Campus Sur UPM, c/ Nikola Tesla s/n. 28031-Madrid, Spain

Yuslín González-Abreu University of Havana, Havana, Cuba

Robertas Grigalaitis Vilnius University, Vilnius, Lithuania

José de los Santos Guerra Federal University of Uberlândia, Uberlândia, MG, Brazil

Juan Hinojosa Technical University of Cartagena, Murcia, Spain

Adelina-Carmen Ianculescu "Politehnica" University of Bucharest, Bucharest, Romania

Nikola I. Ilic Institute for Multidisciplinary Research, University of Belgrade, Belgrade, Serbia

Maksim Ivanov Vilnius University, Vilnius, Lithuania

Ram S. Katiyar University of Puerto Rico, Puerto Rico, United States

Evgeniya Khomyakova Jozef Stefan Institute, Ljubljana, Slovenia

Milica Kisić University of Novi Sad, Novi Sad, Serbia

Lalitha Kodumudi Venkataraman Technische Universität Darmstadt, Darmstadt, Germany

Ling Bing Kong Nanyang Technological University, Nanyang, Singapore; Bengbu University, Bengbu, Anhui, P. R. China

Jurij Koruza Technische Universität Darmstadt, Darmstadt, Germany

Ashok Kumar CSIR-National Physical Laboratory, New Delhi, India

Shalini Kumari University of Puerto Rico, Puerto Rico, United States; West Virginia University, Morgantown, WV, United States

Danjela Kuščer Jožef Stefan Institute, Ljubljana, Slovenia

Sean Li The University of New South Wales, Sydney, NSW, Australia

Lie Liu General Test Systems Inc., Shenzhen, Guangdong, P.R. China

Elson Longo Federal University of São Carlos (UFSCar), São Carlos, Brazil

Glauco M.M.M. Lustosa Institute of Chemistry, São Paulo State University (Unesp), Araraquara, Brazil

Jan Macutkevic Vilnius University, Vilnius, Lithuania

Barbara Malič Jožef Stefan Institute, Ljubljana, Slovenia

Dusanka S. Mandic MADU Clinic, Belgrade, Serbia

Andrea Marić University of Novi Sad, Novi Sad, Serbia

Félix Lorenzo Martínez-Viviente Technical University of Cartagena, Murcia, Spain

Marija Milanović Department of Materials Engineering, Faculty of Technology Novi Sad, University of Novi Sad, Novi Sad, Serbia

Liliana Mitoseriu Faculty of Physics, Dielectrics, Ferroelectrics & Multiferroics Group, Al. I. Cuza Univ. of Iasi, 11 Carol I Bv., 700506, Iasi, Romania

Emilio Morán Departamento de Química Inorgánica. Facultad de Ciencias Químicas. Universidad Complutense de Madrid. 28040-Madrid, Spain

Francisco Moura Federal University of Itajubá (Unifei), Itabira, Brazil

Carlos O. Paiva-Santos São Paulo State University (Unesp), Institute of Chemistry, Araraquara, Brazil

Lorena Pardo Instituto de Ciencia de Materiales de Madrid (ICMM). CSIC. c/ Sor Juana Inés de la Cruz, 3. Cantoblanco. 28049-Madrid, Spain

Cristina Pascual-Gonzalez Christian Doppler Lab for Advanced Ferroic Oxides, Materials and Engineering Research Institute, Sheffield Hallam University, Sheffield, United Kingdom

Aimé Peláiz-Barranco University of Havana, Havana, Cuba

Leinig A. Perazolli Institute of Chemistry, São Paulo State University (Unesp), Araraquara, Brazil

Ioana Pintilie National Institute of Materials Physics, Magurele, Romania

Lucian Pintilie National Institute of Materials Physics, Magurele, Romania

Dhiren K. Pradhan University of Puerto Rico, Puerto Rico, United States; Carnegie Institution for Science, Washington, DC, United States

Goran Radosavljević Vienna University of Technology, Vienna, Austria

Srdjan Rakić Department of Physics, Faculty of Sciences, University of Novi Sad, Novi Sad, Serbia

Miguel A. Ramirez São Paulo State University (Unesp), School of Engineering, Guaratinguetá, Brazil

Armando Reyes-Montero Instituto de Investigaciones en Materiales, Universidad Nacional Autónoma de México, Circuito Exterior S/N, A.P. 70-360, México, CDMX, México

Leandro S.R. Rocha São Paulo State University (Unesp), School of Engineering, Guaratinguetá, Brazil

Tadej Rojac Jozef Stefan Institute, Ljubljana, Slovenia

Giorgio Schileo Department of Materials Science and Engineering, The University of Sheffield, Sheffield, United Kingdom

Alexandre Z. Simões São Paulo State University (Unesp), School of Engineering, Guaratinguetá, Brazil

Vladimir V. Srdić Department of Materials Engineering, Faculty of Technology Novi Sad, University of Novi Sad, Novi Sad, Serbia

Catalina-Andreea Stanciu "Politehnica" University of Bucharest, Bucharest, Romania; National Institute for Lasers, Plasma and Radiation Physics, Magurele, Romania

Biljana D. Stojanovic Institute for Multidisciplinary Research, University of Belgrade, Belgrade, Serbia

Goran Stojanović Department of Microelectronics, Faculty of Technical Sciences, University of Novi Sad, Novi Sad, Serbia

Ronald Tararam CEITEC S.A Semiconductors, Porto Alegre, Brazil

Guilhermina F. Teixeira Institute of Chemistry, São Paulo State University (Unesp), Araraquara, Brazil

Hana Ursic Jozef Stefan Institute, Ljubljana, Slovenia

Mirjana M. Vijatovic Petrovic Institute for Multidisciplinary Research, University of Belgrade, Belgrade, Serbia

María E. Villafuerte-Castrejón Instituto de Investigaciones en Materiales, Universidad Nacional Autónoma de México, Circuito Exterior S/N, A.P. 70-360, México, CDMX, México

Rodrigo Vivar-Ocampo Instituto de Investigaciones en Materiales, Universidad Nacional Autónoma de México, Circuito Exterior S/N, A.P. 70-360, México, CDMX, México

Marko Vrabelj Jožef Stefan Institute, Ljubljana, Slovenia

Julian Walker The Pennsylvania State University, State College, PA, United States

Chuanhu Wang Bengbu University, Bengbu, Anhui, P. R. China

Zhihong Yang Nanjing University of Aeronautics and Astronautics, Nanjing, P.R. China

Maria A. Zaghete Institute of Chemistry, São Paulo State University (Unesp), Araraquara, Brazil

Sonia M. Zanetti Institute of Chemistry, São Paulo State University (Unesp), Araraquara, Brazil

Tianshu Zhang Anhui Target Advanced Ceramics Technology Co. Ltd., Hefei, Anhui, P.R. China

Ljiljana Živanov University of Novi Sad, Novi Sad, Serbia

About the series editor

Ghenadii Korotcenkov received his PhD in Physics and Technology of Semiconductor Materials and Devices from Technical University of Moldova in 1976 and Doctor of Science in Physics of Semiconductors and Dielectrics from Academy of Science of Moldova in 1990 (Highest Qualification Committee of the USSR, Moscow). He has more than 40 years of experience as a teacher and scientific researcher. Long time, he was a leader of gas sensor group and a manager of various national and international scientific and engineering projects carried out in the Laboratory of Micro- and Optoelectronics, Technical University of Moldova. He received financial support from International Foundations and Programs such as CRDF, MRDA, ICTP, INTAS, INCO-COPERNICUS, COST, and NATO. In 2007–2008, he was an invited scientist in Korea Institute of Energy Research (Daejeon). Since 2008, Dr. Korotcenkov is a research professor in the School of Materials Science and Engineering at Gwangju Institute of Science and Technology (GIST) in Korea.

Specialists from Former Soviet Union know Dr. Korotcenkov's research results in the field of study of Schottky barriers, MOS structures, native oxides, and photoreceivers on the basis of III–Vs compounds such as InP, GaP, AlGaAs, and InGaAs. His present scientific interests since 1995 include material sciences, focusing on metal-oxide film deposition and characterization, surface science, and design of thin-film gas sensors and thermoelectric convertors. He is the author or editor of 35 books and special issues, including 11 volume of *Chemical Sensors* series published by Momentum Press (USA), 15 volume of *Chemical Sensors* series published by Harbin Institute of Technology Press (China), 3 volume of *Porous Silicon: From Formation to Application* issue published by CRC Press (USA), and 2 volume *Handbook of Gas Sensor Materials* published by Springer (USA). Dr. Korotcenkov is author and coauthor of more than 550 scientific publications including 20 review papers, 35 book chapters, more than 250 articles published in peer-reviewed scientific journals and more than 50 refereed papers published in international conference proceedings (h-factor = 36 (Scopus) and $h = 42$ (Google scholar citation)). He is a holder of 17 patents. He presented more than 200 reports on the National and International conferences, including 15 invited talks. He was also co-organizer of several international conferences. His research activities are honored by Award of the Supreme Council of Science and Advanced Technology of the Republic of Moldova (2004), The Prize of the Presidents of Ukrainian, Belarus and Moldovan Academies of Sciences (2003), Senior Research Excellence Award of Technical University of Moldova (2001, 2003, 2005), National Youth Prize of the Republic of Moldova in the field of science and technology (1980), among others. He also had Fellowship from International Research Exchange Board (IREX, USA, 1998), Brain Korea 21 Program (Korea, 2008–2012), and Brain Pool Program (Korea, 2015–2017).

About the editor

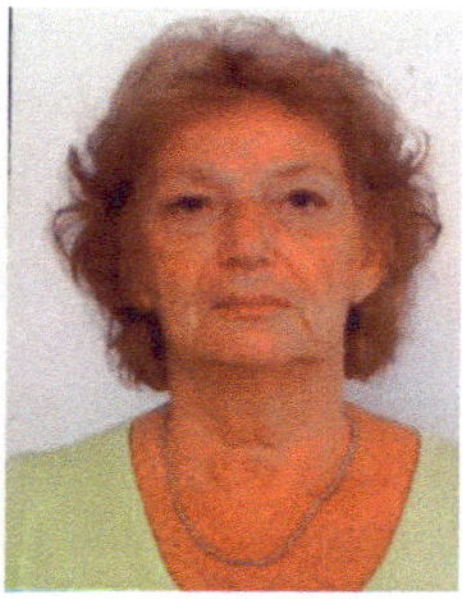

BILJANA D. STOJANOVIC, full member Academy of the Engineering Sciences-ESS, Serbia from 2004, full professor at University of Belgrade, Serbia (retired), research professor and scientist emeritus. She was born on 31st May 1940 in Zagreb, Croatia, to Dušan Oreščanin and mother Djurdjica Oreščanin, née Hadzic. She received her B.Sc. at the Faculty of Technology and Metallurgy-TMF, Belgrade in 1964. And her Masters at the Faculty for Electronic Engineering, University of Nis in 1974. She completed her PHD. at the same faculty in 1977. She was employed in "Electronic Industry"-EI Nis (Belgrade, 1964–1970) in Military Technical Institute (VTI KoV). Belgrade, from 1970–1990, at Faculty of Chemistry, University of Pristine during 1993, and from 1991–2008 in the Centre for the Multidisciplinary Studies University of Belgrade, (the Institute for Multidisciplinary Research.) Biljana was elected for the Assistant Research Professor at the TMF in 1985 and for the Associate professor in 1990. She became full Research Professor in 1994 and Full Professor in 1966 at the Faculty of Technology and Metallurgy, Belgrade. She was elected for the Scientist Emeritus, according to the Law for scientific research, in 2008. She received the Military award with Silver swords for the special research contribution (1986), Gold award for the best research results (1989), Golden Award for the best patent disclosure at the First Technical Exhibition, Belgrade (2003) and several other medals and awards.

In the frame of the **teaching activities and international cooperation**, Professor Stojanovic has lectured on a number of subjects such as "Chemical Technology," "Materials Science", "Sintered Electronic Materials", "High temperature processing", "Ceramic Materials" and "Ferroelectric Materials". She was mentor for 8 diploma works (Yugoslavia, Serbia, Brazil, France), 16 master theses (Yugoslavia, Serbia, France) and 20 PhD theses (Yugoslavia, Serbia, Brazil, Slovenia, Spain). She participated in several Commissions for the defense of the doctoral thesis. She was engaged as the visiting professor 1999–2006 and 2013–2017 at UNESP-Institute for Chemistry, San Paolo, Araraquara, Brazil. Besides this, she gave lectures on Technical faculty University of Mar del Plata, Argentina (2000), University Carlos III, Madrid, Spain (2001), Faculty for Industrial Ceramics, Limoge, France (2002), University "Tor Vergata", Roma, Italy (2003, 2007), Keio University, Jokohama, Japan (2004) and the International School for Postgraduate Education, IJS, Ljubljana, Slovenia (2015). According to

Brazilian scientific and technological development programs (CAPES) "Science without borders", she teaches and manages research cooperation projects (Brazil-Serbia).

Regarding **scientific-research activities**, she is engaged in the field of materials for the application in the electronics with special attention to synthesis, structure, properties and application of nanostructured ceramic materials. She published 3 books, (in English), 2 monographs (in Serbian and with translation in English), 8 book chapters, and has edited 2 books (in English). She has published more than 280 articles in renowned international journals from the SCI list, more than 150 abstracts or shorter articles in conference proceedings, and has presented about 20 plenary or key-note lectures at the international conferences. Her articles were cited app. 2800 times. She is a member of several scientific boards related to International conferences. She is permanent reviewer in respectable scientific journals (Ceram. Inter., JALCOM, Mat. Chem. Phys., PAC, etc) and a member of several scientific societies. She participated in several projects of the Serbian Ministry for education, science, and technological development, projects from the programs of European Union-cycles FP6 and 7 and other European or worldwide projects (COST, thematic Latino-American projects, Spanish, French, Japanese and with few production organizations).

Engineering activity she is responsible for app. 70 original technical designs and 5 patents. She is a member of several international professional associations (Serbian Chemical Society, ETRAN, SSCC, European Society ECeRS, Brazilian Materials Society-MAB, Latino-American Materials Association-CMAA, International thermoanalytical Association-ICTA, Balkan Society of Chemistry, Glass and Ceramics, etc).

In **management activities** she was engaged as national COST coordinator of Serbia in the European scientific program COST, period 2006–2014, and as a representative of Serbia in the Committee of Senior Officials-CSO COST, Brussels, Belgium. She is a permanent member of the several boards of various scientific conferences. She is currently a deputy secretary in the Technology, Metallurgy and Material Engineering Sciences Department AESS and president of the Interdepartmental Committee for International Cooperation AESS (ICIC).

Biljana has two children, Aleksandra (born 1966) and son Milos (born 1978) and has a grandson Nemanja (born 1994). Her hobbies include; classical music, theater, cinema and traveling.

Preface to the series

The field of synthesis, study, and application of metal oxides is one of the most rapidly progressing areas of science and technology. Metal oxides are one of the most ubiquitous compound groups on earth, with a large variety of chemical compositions, atomic structures, and crystalline shapes. In addition, metal oxides are known to possess unique functionalities that are absent or inferior in other solid materials. In particular, metal oxides represent an assorted and appealing class of materials, properties of which exhibit a full spectrum of electronic properties, from insulating to semiconducting, metallic, and superconducting. Moreover, almost all the known effects including superconductivity, thermoelectric effects, photoelectrical effects, luminescence, and magnetism can be observed in metal oxides. Therefore, metal oxides have emerged as an important class of multifunctional materials with a rich collection of properties, which have great potential for numerous device applications. Specific properties of the metal oxides, such as the wide variety of materials with different electrophysical, optical, and chemical characteristics; their high thermal and temporal stability; and their ability to function in harsh environments, make metal oxides very suitable materials for designing transparent electrodes, high-mobility transistors, gas sensors, actuators, acoustical transducers, photovoltaic and photonic devices, photo- and heterogeneous catalysts, solid-state coolers, high-frequency and micromechanical devices, energy harvesting and storage devices, nonvolatile memories, and many others in the electronics, energy, and health sectors. In these devices metal oxides can be successfully used as sensing or active layers, substrates, electrodes, promoters, structure modifiers, membranes, and fibers; that is, they can be used as active and passive components.

Among other advantages of metal oxides are the low fabrication cost and robustness in practical applications. Furthermore, the metal oxides can be prepared in various forms such as ceramics, thick films, and thin film. For thin-film deposition, techniques that are compatible with standard microelectronic technology can be used. This last factor is very important for large-scale production, because the microelectronic approach promotes low cost for mass production, offers the possibility of manufacturing devices on a chip, and guarantees good reproducibility. Various metal oxide nanostructures, including nanowires, nanotubes, nanofibers, core-shell structures, and hollow nanostructures, also can be synthesized. The field of metal-oxide nanostructured morphologies (e.g., nanowires, nanorods, nanotubes, etc.) has become one of the most active research areas within the nanoscience community.

The ability to create a variety of metal oxide–based composites and the ability to synthesize various multicomponent compounds significantly expand the range of properties that metal oxide–based materials can have, making metal oxides a truly versatile multifunctional material for widespread use. Small changes in their chemical composition and atomic structure can be accompanied by a spectacular variation in properties and behavior of metal oxides. Even now, advances in synthesizing and characterizing techniques are revealing numerous new functions of metal oxides.

Taking into account the importance of metal oxides for progress in microelectronics, optoelectronics, photonics, energy conversion, sensors, and catalysis, a large number of various books devoted to this class of materials have been published. However, some books from this list are too general, some are collections of various original works without any generalizations, and others were published many years ago. But during the past decade great progress has been made on the synthesis as well as on the structural, physical, and chemical characterization and application of metal oxides in various devices, and a large number of papers have been published on metal oxides. In addition, until now many important topics related to the study and application of metal oxides have not been discussed. To remedy this situation, we decided to generalize and systematize the results of research in this direction and to publish a series of books devoted to metal oxides.

The proposed book series "Metal Oxides" is the first devoted to consideration of metal oxides only. We believe that combining books on metal oxides in a series could help readers when searching for required information on the subject. In particular, we plan that the books from our series, which have a clear specialization by content, will provide interdisciplinary discussion for various oxide materials with a wide range of topics, from material synthesis and deposition to characterizations, processing, and then to device fabrications and applications. This book series is prepared by a team of highly qualified experts, which guarantees its high quality.

I hope that our books will be both useful and comfortable to use. I also hope that readers will consider this "Metal Oxides" series like an encyclopedia of metal oxides that helps them understand the present status of metal oxides, estimate the role of multifunctional metal oxides in the design of advanced devices, and then, based on observed knowledge, formulate new goals for further research.

The intended audience of the "Metal Oxides" book series is scientists and researchers, working or planning to work in the field of materials related to metal oxides; that is, scientists and researchers whose activities are related to electronics, optoelectronics, energy, catalysis, sensors, electrical engineering, ceramics, biomedical designs, and so on. I believe that this book series will also be interesting for practicing engineers or project managers in industries and national laboratories who would like to design metal oxide–based devices, but don't know how to do it, and who would like to know how to select optimal metal oxides for specific applications. With many references to the vast resource of recently published literature on the subject, this book series will serve as a significant and insightful source of valuable information, providing scientists and engineers with new insights for understanding and improving existing metal oxide–based devices and for designing new metal oxide–based materials with new and unexpected properties.

This book series should be very helpful for university students, postdoctorates, and professors. The structure of these books offers a basis for courses in the field of material sciences, chemical engineering, electronics, electrical engineering, optoelectronics, energy technologies, environmental control, and many others. Graduate students could also find the book series to be very useful in their research and understanding features of metal oxide synthesis, study, and application of this multifunctional material in various devices. We are sure that all of them will find information useful for their activity.

Finally, I thank all contributing authors and book editors who have been involved in the creation of these books. I am thankful that they agreed to participate in this project and for their efforts in the preparation of these books. Without their participation, this project would have not been possible. I also express my gratitude to Elsevier for giving us the opportunity to publish this series. I especially thank the entire editorial team at Elsevier for their patience during the development of this project and for encouraging us during the various stages of preparation.

Ghenadii Korotcenkov

Preface

During the past decades one of the main goals within the current technologies, processing, and devices, is increasing the speed, integration, and miniaturization of electronic components. Therefore current research is focused on the study of new materials and functionalities, and new devices with more complex design. With their promising properties, magnetic, ferroic, and multiferroic materials could likely offer new types of devices and functionality, which has motivated a lot of current research activity in the area of ferroelectric, magnetic, and multiferroic materials.

Despite the growing number of books dedicated to materials and devices with ferroic or magnetic properties, there are only few that tracked theory, structure, processing, properties, functionalities, and applications from the viewpoint of chemists, materials science, physicists, and medicine in one book. The book, *Magnetic, Ferroelectric, and Multiferroic Metal Oxides*, appeals to readers with different degrees of background knowledge in the areas discussed. Our book is an attempt to provide the opportunity to acquire knowledge in the theory, materials, devices, design, and application of magnetic metal oxides, ferromagnetics, ferroelectrics, and multiferroics. Because a book of this size could not cover all topics, we have chosen to concentrate on the basic theoretical approach, structures, synthesis, properties, and applications of magnetic metal oxides, primarily ferrite-based ferromagnetics, most known ferroelectric materials, and multiferroics, either single or composites.

Part I is dedicated to *ferroelectric metal oxides* and is organized in three sections and 14 chapters. The first section addresses the fundamental problems in ferroelectrics (4 chapters); the second addresses the synthesis of ferroelectric materials in bulk or film forms (5 chapters); and the third is dedicated to ferroelectric metal oxide applications (5 chapters). The first chapter presents a general view of ferroelectrics: origin of ferroelectricity in metal oxide ferroelectrics and ferroelectric properties. The following chapters introduce the perovskite and Aurivillius types of ferroelectric metal oxides, lead-free perovskite ferroelectrics, and perovskite layer-structured ferroelectrics (see chapter 2: Perovskite and Aurivillius: Types of ferroelectric metal oxides, chapter 3: Lead-free perovskite ferroelectrics, and chapter 4: Perovskite layer-structured ferroelectrics). In the next section, problems of synthesis and deposition of ferroelectric films, more precisely a review of methods for powder-based processing, chemical synthesis, and epitaxial growth methods for preparation of ferroelectric ceramics and thick and thin films are given in Chapter 5, Review of methods for powder-based processing and Chapter 6, Chemical synthesis and epitaxial growth methods for the preparation of ferroelectric ceramics and thin films. Also, nanosize ferroelectrics—preparation, properties, and applications, especially nanosize $BaTiO_3$-based systems and ecological lead-free ferroelectrics—are considered in Chapter 7, Nanosized ferroelectrics: Preparation, properties, and applications, Chapter 8, Nanosized $BaTiO_3$-based systems, and Chapter 9, Ecological, lead-free ferroelectrics.

Section 3 addresses ferroelectric metal oxide applications such as compositionally-graded ferroelectric ceramics and multilayers for electronic and sensing applications (see chapter 10: Compositionally-graded ferroelectric ceramics and multilayers for electronic and sensing applications), relaxor ferroelectrics (see chapter 11: Review of the most common relaxor ferroelectrics and their applications), tunable ferroelectrics for frequency agile microwave and THz devices (see chapter 12: Tunable ferroelectrics for frequency agile microwave and THz devices), piezoelectric energy harvest device based on quartz as power generator (see chapter 13: Piezoelectric energy harvesting device based on quartz as a power generator), and nonvolatile memories (see chapter 14: Nonvolatile memories).

Part II of the book contains three sections considering *magnetic and multiferroic metal oxides*, as follows. Section 4 addresses magnetic oxides: ferromagnetics, antiferromagnetics, and ferrimagnetics. Section 5 addresses fundamentals of multiferroics, and Session 6 deals with properties and applications of multiferroic metal oxides. Introductory chapters in this part of the book are Chapter 15, Theory of ferrimagnetism and ferrimagnetic metal oxides, related to the theory of ferrimagnetism and ferrimagnetic metal oxides; and Chapter 16, Metal oxide structure, crystal chemistry, and magnetic properties, addressing metal oxide structure, crystal chemistry, and magnetic properties. Review of methods for the preparation of magnetic metal oxides is given in Chapter 17, Review of methods for the preparation of magnetic metal oxides, and progress in ferrite-based composites for microwave absorbing applications is discussed in Chapter 18, Ferrite-based composites for microwave absorbing applications. Applications of ferromagnetic materials as soft ferrites and in biomedical applications are considered in Chapter 19, Soft ferrite applications and Chapter 20, Biomedical applications.

Many attempts have been made to develop processes and techniques that would yield to quality and reproducibility of powders and products of multiferroic materials as single multiferroics or composites. Due to the unique combination of magnetic and ferroelectric properties, recent research has focused on magnetic and ferroelectric materials ranging from more simple devices to more specific devices like sensors used in integrated circuits or as energy storage devices. In this book, Section 5 is dedicated to ferroelectric perovskite−spinel ferrite ceramics in Chapter 21, Ferroelectric perovskite−spinel ferrite ceramics and to single-phase, composite, and laminate multiferroics in Chapter 22, Single-phase, composite, and laminate multiferroics. The last session is devoted to properties and application of multiferroics starting from single and heterostructure multiferroic thin films in Chapter 23, Single and heterostructure multiferroic thin films, $BiFeO_3$ ceramics and thick films in Chapter 24, $BiFeO_3$ ceramics and thick films: Processing issues and electromechanical properties, complex transition metal oxides as single multiferroics in Chapter 25, Properties of single multiferroics: Complex transition metal oxides, and $BaTiO_3$-ferrites as composite multiferroics in Chapter 26, Bulk composite multiferroics: $BaTiO_3$-ferrites.

Besides efforts to improve the performance of electronic devices that are based on hard and brittle substrates, a new challenge in electronics is recently the ability to survive the mechanical deformation of material. Flexible electronics, which can be bent, rolled, and stretched into arbitrary shapes, would significantly expand the application of modern electronic devices in the areas of computation, communications, displays, sensing, energy harvesting, and storage devices. There have been tremendous efforts to develop electronic devices based on mechanically flexible substrates. Several examples dedicated to this issue are presented in Chapter 27, Complex composites: Polymer matrix-ferroics or multiferroics. Ferroelectric, ferromagnetic, and multiferroic heterostructures for possible applications as tunnel junctions are presented in Chapter 28, Ferroelectric, ferromagnetic, and multiferroic heterostructures for possible applications as tunnel junctions.

Last, but certainly not least, I would like to point out that the realization of the book would not be possible without the influence of many people, colleagues, and friends. Of course, this book is the compilation of input from a number of my professional colleagues. It was often quite difficult to find and convince the authors from various countries, universities, and institutions to write chapters for this book. The best scientists are always very engaged and sometimes they cannot find time to write the manuscripts, even if they like and strongly support the subject. Consequently, it was necessary to have a lot of patience and make a big effort to find the appropriate authors who were able to respect the deadlines. Sometimes it seemed that everything was impossible and that nothing could be done on time. Such facts made me feel a little frustrated and almost desperate from time to time. However, I always received the gentle support of the main editor, Prof. Ghenadii Korotchenkov, and the project manager, Mr. Andrae Akeh, and I am extremely grateful to them. I would also like to acknowledge the Elsevier publishing team: Praveen Swapna and Chandran Sujithkumar (ELS-CHN), and Heather Cain and Kayla Dos Santos (ELS-CMA) for their help during preparation of the book.

I also wish to express my special gratitude to my young collaborators who worked with great dedication behind the scenes, Nikola I. Ilic and Adis S. Dzunuzovic. Their hard work, expressed interest, patience, enthusiasm, and assistance in preparing the materials for the book helped me a lot. It is also my pleasure to thank my colleagues for their useful discussions, suggestions, and support. Finally, I would like to extend my gratitude to my family and friends for their patience, tolerance, and moral support through the extended period of time when I was very engaged.

Biljana D. Stojanovic
The Book Editor

Introduction to ferroics and multiferroics: Essential background

Biljana D. Stojanovic
University of Belgrade, Belgrade, Serbia

During recent years, labeled by the integration and miniaturization of electronic components, the study of new materials and their functionalities became very widespread, and one of the main goals is to be integrated within the current technologies, processing, and devices. Due to a unique combination of properties, the research has been focusing on magnetic and ferroelectric materials ranging from giant devices like electrical transformers to tiny devices like sensors used in integrated circuits or as a storage device. The present research is gradually becoming more dedicated to multiferroic materials because of their promising properties and applications. Furthermore, the properties of these materials are likely to offer new kinds of devices and functionality, which have motivated a lot of current research activity in the area of ferroelectric, magnetic, and multiferroic materials.

In recent years, scientific interest for *magnetic metal oxides* has again increased, and various methods have been developed to synthesize these materials. Many attempts have been made to develop processes and techniques that would yield to quality and reproducibility of powders and products. Magnetic oxide powders have found various applications in photography, inks in high-speed printing, catalysis, and especially in medicine. Magnetic nanomaterials have gained great interest in hard works sensors, catalysts, permanent magnets, recording heads, microwave absorbers, magnetic drug delivery, and more.

To the class of magnetics belong *ferrimagnetic* oxides, which are strongly magnetic and insulating if they contain only ferric ions. The hexagonal ferrites were the first low-cost magnetic materials that could be manufactured into magnets of any desired shape. The microwave application of ferrites or the drug and gene delivery will continue to impact significantly on the practice of electronics and biomedicine. Magnetic drug-targeting will undoubtedly dramatically improve the therapeutic potential of many water-insoluble and unstable drugs. Further, the development of drugs able to target selected cells or organs within the body will also deeply improve the benefits in vivo of biomedical applications of nanomagnet metal oxide materials. Recently, the development of novel iron oxide magnetic nanoparticles for the drug delivery systems allows the remarkable feature of ionically bound pharmaceutical drugs, because the active low molecular weight substances can desorb from the carriers after a defined time period and hence diffuse from the vascular wall into the tissue. The following chapters will present the details and a hard working review of methods for the preparation of metal oxide magnetic materials, as well as most important issues related to magnetic oxide structures and properties, a theoretical approach to ferromagnetism, and examples of actual applications of magnetic materials.

A number of classical monographs served as a comprehensive introduction into the field of *ferroelectrics*, but a recent challenge to the low-dimensional structures indicates advances in the synthesis and fabrication of micro- and nanoscale ferroelectric structures with a significant divergence of the bulk properties. The history of ferroelectrics is long, as has been described in the excellent review by Busch [1]. Ferroelectricity was first discovered by Valasek in 1920. He found that the polarization of sodium potassium tartarate tetrahydrate ($NaKC_4H_4O_6 \cdot 4H_2O$), better known as Rochelle salt, could be reversed by the application of an external electric field. Many distinguished scientists were involved in its early development, including Brewster (who studied pyroelectricity), Curie (who discovered piezoelectricity), Boltzmann, Pockels, and Debye. After 1949, the ferroelectric research field emerged due to the discovery of ferroelectricity in barium titanate ($BaTiO_3$) ceramics [2], which belong to the biggest class of ferroelectrics: perovskite oxides with the general formula ABO_3. The fact that hydrogen is not present in ferroelectric $BaTiO_3$ indicated that the first assumption about hydrogen bonds being essential for ferroelectricity is not correct. More than 700 ferroelectrics have been found to date, from the water-soluble crystals to oxides, polymers, ceramics, and even liquid crystals, most of them not hydrogen-bounded or oxide. The prefix *ferro*, meaning iron, was chosen to describe the property because of a resemblance with already known ferromagnetic properties despite the fact that most of the ferroelectric materials do not contain iron.

Ferroelectric materials exhibit a wide range of functional properties, including switchable polarization, piezoelectricity, high nonlinear optical activity, pyroelectricity, nonlinear dielectric behavior, among others. They are applicable for capacitors; sensors such as accelerometers, microphones, and hydrophones; actuators and resonant wave devices, such as radio-frequency filters and microwave phase filters; nonvolatile memories; infrared detectors; electric-motor overload protection; circuits; optical switches, such as laser Q-switches; optical shutters and integrated optical (photonic) devices, for laser frequency doubling and optical mixing; and more. The extensive investigations of *multiferroics* has gained much attention with advance of novel methodologies for the development of nanoferrites, expanding their use in other areas such as targeted drug delivery, heterogeneous catalysis, repulsion-suspension for use in levitated railway system, ferro-fluids, magnetic refrigeration antennas, and so on.

The internal electric dipoles of a ferroelectric material are coupled to the material lattice so any changes in the lattice structure would change the strength of the dipoles (in other words, a change in the spontaneous polarization). The resulting change in the spontaneous polarization also impacts the surface charge. The mechanical force and temperature are some of the stimuli that impact the lattice dimensions of a material. The generation of a surface charge in a material subjected to external stress is called piezoelectricity. A change in the spontaneous polarization of a material in response to a change in temperature is called pyroelectricity. Generally, 230 space groups divided in 32 crystalline classes can be found in crystals. There are 21 noncentrosymmetric classes, and 20 of them are piezoelectric. Ten of the piezoelectric classes have a spontaneous electric polarization that varies with the temperature, therefore they are pyroelectric. Among pyroelectric materials, some are also ferroelectric (Fig. 1).

Ferroelectric materials demonstrate a spontaneous nonzero polarization even when the applied electric field, E, is zero. The distinguishing feature of ferroelectrics is that the spontaneous polarization can be reversed by a suitably strong applied electric field in the opposite direction; the polarization is therefore dependent not only on the current electric field but also on its history, yielding a hysteresis loop. Fig. 2 shows a schematic plot of polarization, P, versus E and the

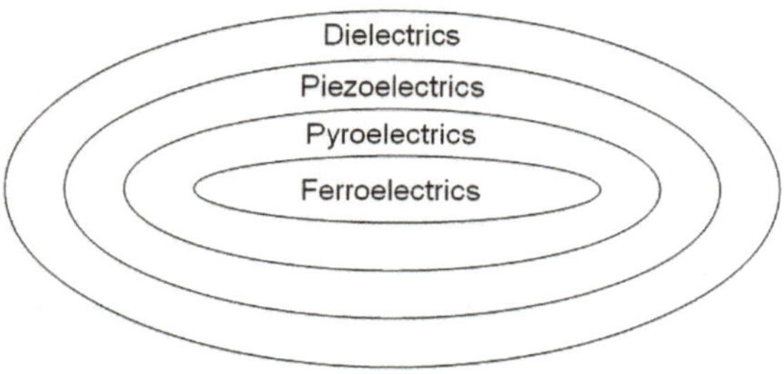

Figure 1 Venn diagram of dielectric materials.

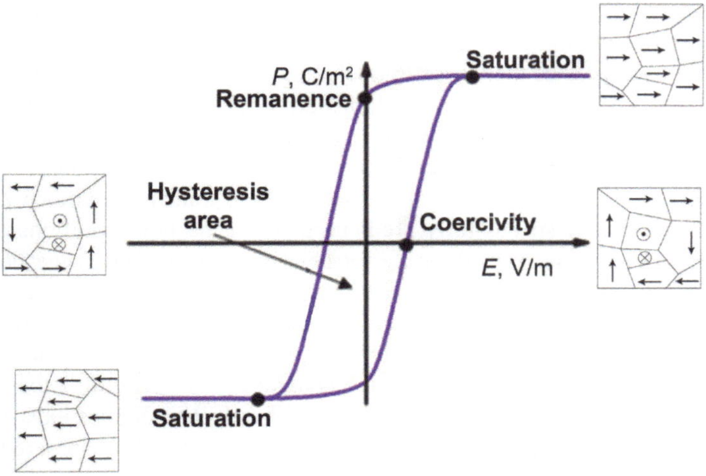

Figure 2 Typical ferroelectric hysteresis loop with schematically presented domain orientation at different parts of the loop.

ferroelectric domains orientation. Increasing the field from zero, the overall polarization in the crystal increases as the polarizations in the different dipolar regions are aligning. Eventually, it reaches a saturation P_S point where the further increase in P is due to the relative permittivity of the material. The gradient of the P/E curve for a linear dielectric is equal to its permittivity.

Typically, materials demonstrate ferroelectricity only below a certain phase transition temperature, called the Curie temperature, T_C. Above this temperature, the spontaneous polarization vanishes, and the ferroelectric crystal transforms into the paraelectric state. This ferroelectric-to-paraelectric phase transition is a characteristic of most ferroelectrics with several exceptions. Some ferroelectric materials melt or decompose before T_C is reached; the polymeric ferroelectric poly(vinylidene fluoride), PVDF, is one such example. For most ferroelectrics, known as *proper ferroelectrics*, the relative permittivity increases with temperature, reaching a peak at T_C and decreases above T_C. *Improper* or *extrinsic ferroelectrics* do not possess such peak in permittivity and the anomaly is present. However, high values of relative permittivity can be reached at T_C, leading to use of ferroelectric materials as high-value capacitors. The phase transition at T_C can either be continuous (second-order transition) or discontinuous (first-order transition). In either case, the dependence of permittivity (ε) on temperature above T_C can be described by $\varepsilon = C/(T - T_0)$. The parameter T_0 is the Curie–Weiss temperature, which is only equal to T_C for a second-order phase transition, and C is the Curie–Weiss constant [3]. There is actually a region of coexistence of ferroelectric and paraelectric phases where both phases exist together at the same temperature. In this case, the Curie temperature is defined as the temperature at which the ferroelectric and paraelectric phases have the same Gibbs free energies. The permittivity rises as the temperature decreases, with a peak at the transition, where there is a discontinuous step down. First-order phase transitions tend to show thermal hysteresis in the transitions; the transition occurs at a higher temperature, which means it approaches from the low-temperature side and not from the high-temperature side.

There are two types of theoretical models that describe the ferroelectric phenomenon. The first is the phenomenological or thermodynamic theory first proposed by Devonshire [4] where the Gibbs free energy of the ferroelectric system is described in terms of a power series in the spontaneous polarization-P_S. The theory is similar to the generalized phenomenological theory of phase transitions known as Landau theory [3]. The theory successfully predicts many of the behavioral characteristics of ferroelectric materials, including the ferroelectric hysteresis and the behavior of the permittivity near T_C related to macroscopic properties. However, the theory does not explain the microscopic origins of ferroelectricity. There are two types of microscopic models of ferroelectric behavior concerning that issue. The first one is the *order–disorder model* of ferroelectric behavior and the second is the *displacive model*. According to the order–disorder model, the electric dipoles exist within the structure in the paraelectric phase above T_C, but are thermally disordered between two or more states and the average polarization is zero. There are several materials that act like that. Potassium dihydrogen phosphate is paraelectric, tetragonal (42 *m*) above 123 K and ferroelectric, orthorhombic (*mm*2) below that temperature.

The polarization appears along the tetragonal *c*-axis. Another example is sodium nitrite ($NaNO_2$), in which the polar groups are NO_2^- ions, which become ordered below $T_C = 163°C$ with dipoles all pointing along the *b*-axis of the crystal structure.

The displacive model for ferroelectrics is based on the soft-mode theory considering that the dipoles appear due to the cooperative motion of ions, but there are no dipoles in the structure above T_C. Examples of such ferroelectrics include the perovskite-structured oxides such as $BaTiO_3$. In practice, the distinction between order–disorder and displacive ferroelectrics becomes rather fuzzy, as some ferroelectrics that are considered displacive may exhibit significant disordered cation displacement above T_C, while some order–disorder ferroelectrics show soft-mode behavior [3,5,6].

Ferroelectric materials might be inorganic and organic, but the metal oxides are most known and most used. Among them the ferroelectrics with perovskite crystal structure with the general chemical formula ABO_3 are by far the most technologically important class of ferroelectric materials. Probably the most interesting perovskites from the application point of view of are $BaTiO_3$, $PbTiO_3$, $Pb(Zr_{0.53}Ti_{0.47})O_3$ (PZT), and $KNbO_3$. Some perovskites show nonferroelectric phase transitions. The $SrTiO_3$ exhibits a phase transition with linked rotations, or tilts, of the TiO_6 octahedra about the cubic [100] direction. Tilting of the octahedra is a common feature of the phase transitions that occur in perovskites and can lead to very complex series of phase transitions, as has been observed for $NaNbO_3$. There is another very important class of oxides, named complex perovskites, where the A-sites or, more commonly, B-sites in the structure are occupied by ions of different valence in a fixed molar ratio. Doping certain ferroelectric ceramics with electron donors (e.g., $BaTiO_3$ with La^{3+}) can render them semiconducting. Heating these ceramics through their Curie temperature causes a very large, reversible increase in resistivity (by several orders of magnitude in some cases) over a narrow range of temperature. This large positive temperature coefficient of resistance is widely exploited in electric-motor overload-protection devices and self-stabilizing ceramic heating elements. Another idea of recent interest is the ferroelectric tunnel junction, in which a contact made up by nanometer-thick ferroelectric film is placed between metal electrodes.

It was pointed out that the ferroelectrics might have various applications. The nonlinear nature of ferroelectric materials can be used to make capacitors with high-value of permittivity or tunable capacitance. Usually, ferroelectric tunable capacitors are small in size compared to dielectric (nontunable) capacitors of similar capacitance. Ferroelectric capacitors have an interesting application in the medical ultrasound machines where the capacitors generate the energy further transforming the ultrasound ping to an image of the internal organs of a human body. Another purpose is for high-quality infrared cameras where the infrared image is projected onto a two-dimensional array of ferroelectric capacitors capable of detecting temperature differences as small as a millionth of a degree Celsius. The ability of ferroelectrics to store information via the spontaneous polarization has made them a useful candidate for the application in computers. The growth of ferroelectric thin films onto silicon at relatively low temperatures and the replacement of the dielectric layer in dynamic random-access memories (DRAMs) make their use a reality; besides DRAM devices, thin films have not yet reached wide commercialization.

The thickness of the ferroelectric layer may be small enough to allow tunneling of electrons. The piezoelectric and interface effects as well as the depolarization field may lead to a giant electroresistance switching effect. During recent years, the study of microelectromechanical systems (MEMS) has shown significant opportunities for miniaturized mechanical devices based on thin-film materials and silicon technology. In most of the structures applied in MEMS, the piezoelectric film is part of a composite structure, i.e., the piezoelectric film is clamped to another elastic body [7]. The spontaneous polarization of ferroelectric materials implies a hysteresis effect, which can be used as a memory function such as ferroelectric random access memories (FERAMs) for computers and memory cards. For such use, the most common are thin films of ferroelectric materials, but great attention must be paid to the interfaces, electrodes, and sample quality for devices to function reliably. As the ferroelectric materials are also piezoelectric and pyroelectric, those properties might be combined for sensor applications such as fire sensors, sonar, vibration sensors, and even fuel injectors on diesel engines. Many of these previously mentioned will be discussed in detail in the following chapters of this book.

Ferroelectrics have undergone a minor renaissance in the past 20 years with the development of high-quality thin (<300 nm) oxide films, capable of performing switching at the 5 V standard logic level for silicon transistors. In a well-known article, Scott [8] presented his viewpoint of what the most promising lines of investigation would be over the next decade. He indicated that the general field of integrated ferroelectrics is not limited to memory devices entailing polarization reversal but includes electrically controlled tunnel junctions, which are much more demanding in their thickness requirements (<7 nm); electrocaloric cooling devices; THz emitters; resistive RAMs; photochromics; domain nanoelectronics; flexible polymeric ferroelectrics; and photovoltaics. Scott pointed out that in the area of memory applications they may still be competitive with phase-change memories, magnetic RAMs, and FLASH, and that ferroelectrically gated memory looked promising. In general, future trends and future devices will undoubtedly be nanoscale ferroelectrics (nanocells, nanotubes, nanowires) [9] as well as thin films, operating with low voltages (<5 V). Emphasis will still be on oxides—"oxide electronics." It is interesting to note that today there are more than 500 laboratories worldwide working with ferroelectric films and searching for new ferroelectric materials, properties, and applications [8].

The evolution of passive magnetoelectronic components open the field of the next generation of devices, which combine memory and logic functions and promise to set new standards in future information technology and nanoelectronics [10]. These perspectives require a new class of multifunctional materials and structures whose properties can be manipulated by several independent stimuli by affecting physical degrees of freedom set by the order parameters. The field of *multiferroic* (MF) and *magnetoelectric* (ME) materials has recently received a great deal of attention. The researchers have been looking for ways to couple magnetic and ferroelectric ordering within a material for decades. The development of MF and ME materials possessing coupled magnetic and electrical order parameters started with pioneering work in Russia in the late 1950s and continued intensively in the 1960s. There have been very few MFs existing in nature or synthesized in the laboratory. The first MF material discovered was nickel iodine boracite ($Ni_3B_2O_{13}I$). This discovery was followed by the synthesis of several MF boracite compounds, all of which have complex structures with many atoms per formula unit and more than one formula unit per unit cell. However, the phenomena of ME/MF materials have mostly remained in the theoretical domain owing to the fact that in single-phase materials such couplings are rare and weak. Since the end of the 1990s, significant progress has been made in synthesizing and characterizing various oxides, which show the MF properties at high temperatures, e.g., Bi-based compounds, namely $BiMnO_3$ or $BiFeO_3$. Recently, the situation has changed dramatically for at least two reasons: first, advances in materials fabrication have made it possible to manufacture these materials in structures of lower dimensionality, such as thin films or nanowires, or in compound structures such as laminates, epitaxial-layered heterostructures, or complex composites including flexible structures [11].

MF materials belong to a special class of materials that exhibit an ME effect, and in recent years they have been drawing significant attention. ME effect is a product property and it needs biphasic surrounding for its exhibition. MF materials exhibit two or more ferroic orders: magnetic (ferromagnetic (FM), antiferromagnetic (AFM), or ferrimagnetic), ferroelectric (FE), ferroelastic, or ferrotoroidic (Fig. 3) [12]. ME materials, on the other hand, exhibit coupling between the electric and magnetic degrees of freedom, so an electric (magnetic) polarization can be induced in them by a magnetic (electric) field. ME coupling in composite MFs is larger than in single-phase MF materials. As ME/MFs allow the possibility of switching the magnetization with electric field, they have the opportunity for information storage applications offering the miniaturization of magnetic random access memory (MRAM), where the write operation will require magnetic fields or large currents [13]. Another possibility is the development of memory bits with multiple stable states [14], or mixed memory and logic functions [15].

The ME effect appears when electric polarization occurs under the influence of an external magnetic field and magnetization occurs under the influence of an electric field. Macroscopically, the interaction of the electric (E) and magnetic (H) fields and the stress tensor (s) with the material can be described within the Landau theory of phase transitions. This allows an electric field to be used to control the magnetic properties of a material and a magnetic field to be used to control the electric properties. Traditionally, one distinguishes linear, quadratic, and higher order ME effects, but more recently the term "magnetoelectric effect" is often used to describe any form of cross-correlation between magnetic and (di)electric properties. It is important to point out that not all magnetic ferroelectrics exhibit a linear ME effect (in the original sense) and that not all materials that exhibit a linear ME effect are simultaneously multiferroic. A MF that is ferromagnetic and ferroelectric is liable to display large linear ME effects and the ME effect in a single-phase crystal is traditionally described in Landau theory by writing the free energy, F, of the

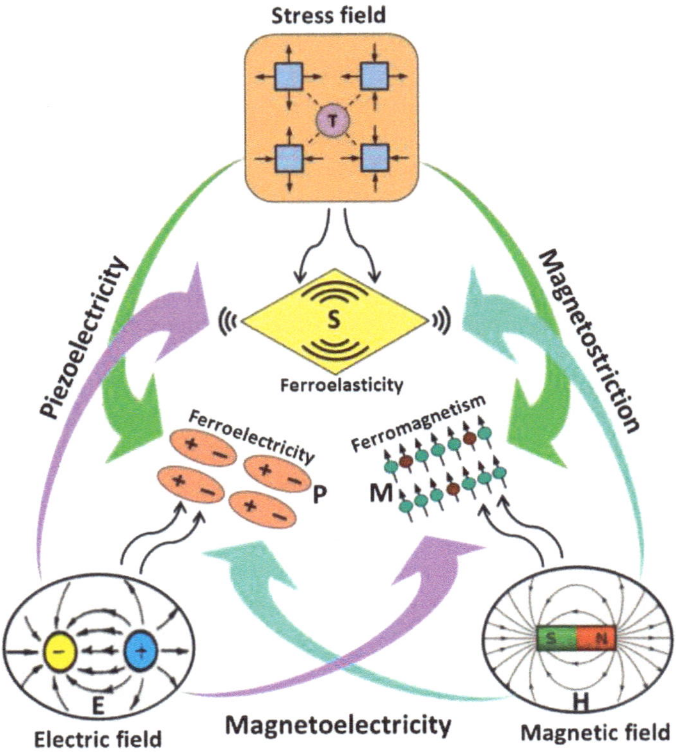

Figure 3 Schematic illustration of magnetic-elastic-electric couplings in multiferroic materials. *M*, magnetization; *S*, mechanical strain; *P*, dielectric polarization.
Source: Adapted from [12], *Open Access.*

system in terms of an applied magnetic field *H*. If the value of the ME effect is high (hundreds of times higher than normal), it is called a giant ME effect. Neither ferroelectric nor ferromagnetic phases manifest the ME effect but composites of these two have remarkable ME effect. This follows because FE and FM materials often (but not always) possess a large permittivity and permeability, respectively, but a large permittivity is not a prerequisite for a material to be ferroelectric (or vice versa). Similarly, ferromagnets do not necessarily possess large permeability. For example, the ferroelectric KNO_3 possesses a small permittivity $\varepsilon \sim 25$ near its Curie temperature of 120°C, whereas paraelectric $SrTiO_3$ exhibits permeability $\varepsilon \sim 50{,}000$ at low temperatures. Therefore, large ME coupling is not restricted to MF materials [16].

Research into single-phase MEs and MFs took off during the latter half of the twentieth century. In 1957, the linear ME coupling coefficient-α was predicted to occur in Cr_2O_3. In the 1960s, α was experimentally observed to be nonzero below the antiferromagnetic Néel temperature of 307 K, near which it peaked to a value $\alpha < 4.1$ psm^{-1}. This work on Cr_2O_3 and other AFM crystals, such as Gd_2CuO_4,

Sm_2CuO_4, $KNiPO_4$, $LiCoPO_4$, and $BiFeO_3$, is well summarized by Schmid [17]. As ME coupling is determined by the structure and magnetic symmetry of a crystal, small modifications might alter, eliminate, or allow ME effects. Source of magnetically induced ferroelectricity might be spiral spin ordering, or collinear spin ordering where ferroelectricity is induced with competing FM and AFM interactions, such as in the orthorhombic RMn_2O_5 with Mn^{3+} ions ($S = 2$) in oxygen pyramids and Mn^{4+} ions ($S = 3/2$) in oxygen octahedra. The nearest-neighbor magnetic coupling in the loop is AFM, favoring antiparallel alignment of neighboring spins. However, because of the odd number of spins in the loop, ordered spins cannot be antiparallel to each other at all bonds, which gives rise to frustration and favors more complex magnetic structures [18].

The application of single-phase MFs in data storage lies in the possibility of reversing the magnetization by applying an electric field or the polarization by applying a magnetic field. Recently, ME switching has been observed in orthorhombic manganites, $REMnO_3$ or $REMn_2O_5$, where RE is a rare-earth element. These are antiferromagnets that display improper (weak) ferroelectricity. The most well-known single-phase MF is $BiFeO_3$, which is a commensurate ferroelectric and an incommensurate antiferromagnet at room temperature. Promising MF behavior was reported in the thin epitaxial $BiFeO_3$ films but the large magnetization only appeared in the deoxygenated films followed with high electrical conductivity [16]. Bulk $BiFeO_3$ exhibits G-type AFM order, where the magnetic moment of each Fe cation is antiparallel to that of its nearest neighbors. The calculated magnetocrystalline anisotropy predicts that a preferred orientation of the Fe magnetic moments is perpendicular to the polar [111] direction. A small canting of the Fe magnetic moment leads to a net magnetic moment of about 0.05 μ_B per Fe atom. This weak ferromagnetism raises a question of whether the small magnetization is coupled with the electric polarization, so that it can be manipulated by applied electric fields [16].

An alternative strategy to obtain ME effects is to introduce indirect coupling via strain between two materials such as a ferroelectric and a ferromagnet. Each phase may then be independently optimized for room temperature performance where strain coupling requires close contact between a piezomagnetic (or magnetostrictive) material and a piezoelectric (or electrostrictive) material. This can be achieved in the form of composite, laminate, or epitaxial multilayers. Thin-film geometry naturally leads to the idea of composite MF materials, such as layered heterostructures of FE and FM materials. The advantage of composites is that fundamental limitations on the strength of the ME coupling no longer apply. However, the problem of how to couple the ferroic orders in the different phases arises. The ferroelectricity in MF materials is generally caused by a different mechanism than in regular FE materials such as $BaTiO_3$, $PbTiO_3$, or $KNbO_3$, which all contain transition metal cations with a formal d^0 configuration. Among others, the first principles calculations have been used to clarify the possible mechanisms for ferroelectricity that are compatible with the simultaneous presence of magnetic order, as well as to investigate possible mechanisms for coupling between the electric polarization and the magnetic order, and, finally, to design the new MF and ME materials [19]. In general, two types of ME coupling can be expected: (1) direct, owing to electronic effects at the interface (limited to the interface); and (2) indirect, mediated by strain (bulk effect) [10]. The coupling constant depends on the frequency of the *ac* applied magnetic field, and such MF structures could thus find applications, e.g., in microwave frequency transducers [15].

Boomgaard and Born [20] had given some requirements of good ME effect in composites, summarized as follows: (1) two phases should be in equilibrium and no chemical reaction should occur between the constituent phases, (2) magnetostriction coefficient of ferromagnetic phase and piezoelectric coefficient of ferroelectric phase should be high, (3) resistivity of the constituent phases should be high to avoid the leakage of the charges developed in ferroelectric phase, and (4) proper poling strategy should be adopted to get a higher ME signal. These MF composites are being extensively studied due to various practical applications, such as state logic memory devices, ME RAMs, sensors, microwave devices (transducers, filters, phase shifters, oscillators), spintronic devices, and so on. [21]. The extensive investigations of MFs has gained much attention with advance of novel methodologies for the development of nanoferrites, expanding their use in other areas, such as targeted drug delivery, heterogeneous catalysis, repulsion-suspension for use in levitated railway systems, ferro-fluids, and magnetic refrigeration antennas [22].

The selection of *ferroelectric* and *ferromagnetic* materials for the preparation of MF materials depends on the different factors and application that we need. Ferroelectric materials used as a component of a MF are usually those that possess perovskite structure. The perovskites have the general formula ABO_3, where A (larger cation) is a monovalent, divalent, or trivalent metal and B (smaller cation) is a pentavalent, tetravalent, or trivalent element [23]. From this family, the most important ferroelectric materials are barium titanate, barium zirconium titanate, and lead zirconate titanate. The most crucial and desired properties of one ferroelectric material as a constituent of MF composite are high dielectric constant, low dielectric and piezoelectric losses, and high piezoelectric coefficients. Further substantial properties of ferroelectrics that should be considered are Curie temperature (T_C), remnant polarization (P_R), coercive field (E_C), and so on. On the other hand, ferrites are mostly used as a magnetic component of MF composite. Ferrites are mixed metal oxides with iron (III) oxide as a main component, with the general formula MFe_2O_4 (where M is a divalent cation like Fe, Ni, Zn, Co, Mn, Mg). They crystallize in the three crystal types: spinel (cubic crystal structure), garnet (cubic crystal structure), and magnetoplumbite (hexagonal crystal structure). The most commonly used ferrites are nickel-ferrite (NF), nickel zinc ferrite, and nickel-cobalt-copper ferrite (NCCF). The relevant properties of one magnetic material in composites are magnetic permeability (μ), remnant magnetization (M_R), coercive magnetic field (H_C), magnetostriction (λ_{ij}), piezomagnetic coefficient (q_{ij}), and Curie and Néel temperatures (T_C and T_N) [21].

The field of FE and MF materials has a relatively long history; many of those materials were previously studied, but with discovery of their new function, the application recently has been rediscovered. Due to substantial advancements in experimental synthesis and characterization techniques on one side, and with new theoretical approaches on the other side, substantial progress in understanding these materials has been achieved during recent years. Nevertheless, many open questions related to MFs still remain, such as how to achieve large polarization, large magnetization, and strong ME coupling above room temperature; and what mechanisms for coupling between magnetic and ferroelectric properties do exist in these materials. Due to that, the modeling and deep theoretical approach based on the computational physics to FE an ME/MF effects, as well as the material science research based on physical phenomena and chemical structure, are of essential importance for achieving the new applications and novel devices. To search for new ferroelectrics and MFs, the user point of view is undoubtedly very important [24]. Some of the outstanding issues in the area of MFs, not only related to theoretical approach but also in the field of processing, properties, and applications, will be considered in the following chapters of this book.

References

[1] G. Busch, Early history of ferroelectricity, Ferroelectrics 74 (1987) 267–284.

[2] R. Gray, USA Patent no. 2 486 560, 1949.

[3] M. Lines, A. Glass, Principles and Applications of Ferroelectric Materials, Clarendon, Oxford, 1977.

[4] A. Devonshire, Theory of barium titanate, Phil. Mag. 40 (1949) 1040–1063.

[5] R. Whatmore, A. Glazer, Structural phase transitions in lead zirconate, J. Phys. C Solid State Phys. 12 (1979) 1505.

[6] B. Noheda, J. Gonzalo, A. Caballero, C. Moure, D. Cox, G. Sirane, New features of the morphotropic phase boundary in the Pb $(Zr_{1-x}Ti_x)O_3$ system, Ferroelectrics 237 (2000) 237–244.

[7] J. Muralt, Ferroelectric thin films for micro-sensors and actuators: a review, J. Micromech. Microeng. 10 (2000) 136–146.

[8] J. Scott, Prospects for Ferroelectrics: 2012–2022, Hindawi Publishing Corporation, ISRN Mater. Sci. (2013), 24 pages.

[9] A. Gruverman, A. Kholkin, Nanoscale ferroelectrics: processing, characterization, Rep. Prog. Phys. 69 (2006) 2443–2474.

[10] J. Velev, S. Jaswali, E. Tsymbal, Multi-ferroic and magnetoelectric materials and interfaces, Phil. Trans. R. Soc. A 369 (2011) 3069–3097.

[11] W. Prellier, M. Singh, P. Murugavel, The single-phase multiferroic oxides: from bulk to thin film, J. Phys. Condens. Matter. 17 (2005) R803–R832.

[12] H. Palneedi, V. Annapureddy, S. Priya, J. Ryu, Status and perspectives of multiferroic magnetoelectric composite materials and application, Actuators 5 (2016) 9.

[13] Y. Chu, L. Martin, M. Holcomb, R. Ramesh, Controlling magnetism with multiferroics, Mater. Today 10 (2007) 16–23.

[14] M. Bibes, A. Barthelemy, Towards a magnetoelectric memory, Nat. Mater. 7 (2008) 425–426.

[15] M. Gajek, M. Bibes, S. Fusil, K. Bouzehouane, J. Fontcuberta, A. Barthelemy, et al., Tunnel junctions with multiferroic barriers, Nat. Mater. 6 (2007) 296–302.

[16] W. Eerenstein, N. Mathur, J.F. Scott, Multiferroic and magnetoelectric materials, Nature 442 (2006) 759–765.

[17] H. Schmid, Introduction to the proceedings of the 2nd international conference on magnetoelectric interaction phenomena in crystals, MEIPIC-2, Ferroelectrics 161 (1994) 1–28.
[18] J. Velev, S. Jaswal, E. Tsymbal, Multi-ferroic and magnetoelectric materials and interfaces, Phil. Trans. R Soc. 369 (2011) 3069–3097.
[19] S. Picozzi, C. Ederer, First principles studies of multiferroic materials, J. Phys. Condens. Matter. 21 (2009) 303201.
[20] R. Devan, B. Chougule, Magnetic properties and dielectric behavior in ferrite/ferroelectric particulate composites, Physica B 393 (2007) 161–166.
[21] S. Upadhyay, V. Reddy, N. Lakshmi, Study of $(1-x)$ $BaTiO_3$–x $Ni_{0.5}Zn_{0.5}Fe_2O_4$ ($x = 5$, 10 and 15%) magneto-electric ceramic composites, J. Asian Ceram. Soc. 1 (2013) 346–350.
[22] R. Grigalaitis, M. Vijatović Petrović, J. Bobić, A. Dzunuzovic, R. Sobiestianskas, A. Brilingas, et al., Dielectric and magnetic properties of $BaTiO_3$ –$NiFe_2O_4$ multiferroic composites, Ceram. Int. 40 (2013) 6165–6170.
[23] M. Vijatović, J. Bobić, B. Stojanović, History and Challenges of Barium Titanate: Part I, Sci. Sinter. 40 (2) (2008) 155–165.
[24] J. Scott, Searching for new ferroelectrics and multiferroics: A user's point of view, Computational materials. <www.nature.com/npjcompumats>.

Part I

Ferroelectric Metal Oxides

Section I

Ferroelectrics: Fundamentals

General view of ferroelectrics: Origin of ferroelectricity in metal oxide ferroelectrics and ferroelectric properties

1

Maksim Ivanov, Jan Macutkevic, Robertas Grigalaitis and Juras Banys
Vilnius University, Vilnius, Lithuania

1.1 Introduction

A highly peculiar effect was experimentally observed by Valasek in Rochelle salt in 1921 [1]. He proved that there is a loop in polarization versus electric field, which is analogous to magnetization versus magnetic field in ferromagnets. This also meant that the material is spontaneously polarized even in the absence of any external electric fields. If the spontaneous polarization of a material can be reoriented by an external electric field, then, in an analogy to ferromagnets, we speak about ferroelectricity. Thus, it is not the existence of the spontaneous polarization alone, but its "switchability" by an external field that defines a ferroelectric material. After the Valasek work a lot of other ferroelectric materials were discovered. A significant part of these materials are oxides (i.e., materials containing oxygen).

A sketch of a ferroelectric hysteresis loop is presented in Fig. 1.1. E_C is the coercive field, which is required to decrease the polarization to 0. P_S corresponds to spontaneous polarization, and P_R is the remanent polarization, which is observed under zero electric field. The latter is smaller than the former due to partial backswitching observable in real life ferroelectrics.

Unfortunately, the switchability is well defined only in an engineering sense. Some materials exhibit spontaneous polarization, but it is impossible to change its direction. These are known as pyroelectrics. All ferroelectrics are pyroelectrics, and all pyroelectrics are piezoelectrics. All of these crystalline materials cannot have inversion symmetry, so it is possible to classify them according to their crystal structure. However, the property of switchability can only be checked by a direct experiment, making it difficult to foresee the ferroelectricity theoretically.

Despite these problems, ferroelectric materials can act as very good memory cells, energy harvesters, and electromechanical transducers. This makes them highly interesting for researchers and engineers.

The aim of the chapter is to introduce the readers to the very basic properties of ferroelectric materials. Theoretical treatment as well as some typical examples of experimental results on crystals, ceramics, and thin films will be presented in the following sections.

Magnetic, Ferroelectric, and Multiferroic Metal Oxides. DOI: http://dx.doi.org/10.1016/B978-0-12-811180-2.00001-3

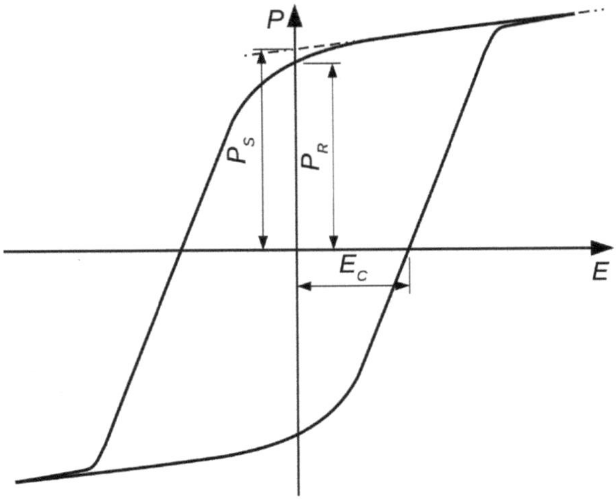

Figure 1.1 A sketch of a ferroelectric hysteresis loop.

1.2 Macroscopic phenomenological theory of ferroelectric phase transitions

According to the Landau–Ginzburg–Devonshire theory, the phase transition can be characterized by an order parameter (η), which appears close to the phase transition temperature [2,3]. The order parameter characterizes the displacement of atoms from its position in high temperature phase (for displacive phase transitions) or the degree of ordering (for order-disorder phase transitions). The order parameter is proportional to the spontaneous polarization in the case of proper ferroelectric phase transitions. From a thermodynamic standpoint, the phase transition must be analyzed via thermodynamic potentials. The free energy is the function of choice to describe real-world experiments. Close to the phase transition temperature the free energy may be written as an expansion of powers of the order parameter, with odd powers omitted for symmetry reasons:

$$F(\sigma, T, \eta) = F_0(\sigma, T) + \frac{1}{2}A_2(\sigma, T)\eta^2 + \frac{1}{4}A_4(\sigma, T)\eta^4 + \frac{1}{6}A_6(\sigma, T)\eta^6 \ldots \quad (1.1)$$

where A_2, A_4, A_6 are temperature- and pressure-dependent coefficients. Stable states are characterized by minima of the free energy with the necessary and sufficient conditions:

$$\frac{\partial F}{\partial \eta} = \eta(A_2 + A_4\eta^2 + A_6\eta^4) = 0 \quad (1.2)$$

and

$$\frac{\partial^2 F}{\partial \eta^2} = A_2 + 3A_4\eta^2 + 5A_6\eta^4 > 0. \quad (1.3)$$

In ferroelectric phase the crystal will be stable if $F(\sigma, T, \eta)$ has a minimum at $\eta \neq 0$, while the trivial minimum for $\eta = 0$ corresponds to the paraelectric phase. Two cases can be distinguished: (1) $A_4 > 0$ (then $A_6 = 0$ is possible for simplicity), which corresponds to a phase transition of the second order; and (2) $A_4 < 0 \Rightarrow A_6 > 0$, which is the case of a phase transition of the first order.

For a ferroelectric phase transition the order parameter is the polarization ($\eta = P$) and:

$$\frac{\partial^2 F}{\partial P^2} = \frac{1}{\varepsilon - \varepsilon_0} \approx \frac{1}{\varepsilon}. \quad (1.4)$$

Here ε is the *absolute* dielectric permittivity of the material and ε_0 is the dielectric permittivity of vacuum. Although all coefficients in Eq. (1.1) are temperature-dependent, the coefficient A_2 can be approximated as a linear temperature function close to the phase transition temperature; that is, $A_2 = (T - \Theta)/C$ (where Θ is the Curie–Weiss temperature and C is the Curie–Weiss constant) and other coefficients can be assumed temperature independent. In such cases close to the phase transition temperature in the paraelectric phase ($\eta = 0$), the inverse dielectric permittivity can be expressed as (according to Eqs. (1.3) and (1.4)):

$$\frac{1}{\varepsilon} = \frac{1}{C}(T - \Theta) \quad (1.5)$$

In the case of second-order ferroelectric phase transition the Curie–Weiss temperature is equal to ferroelectric phase transition temperature ($\Theta = T_C$) and, to satisfy Eq. (1.2), the spontaneous polarization P_s depends on the temperature as a square root law:

$$P_s = \sqrt{\frac{T_C - T}{CA_4}}. \quad (1.6)$$

For $T > T_C$, a minimum is found for $P = 0$. At $T = T_C$, this minimum shifts continuously to finite values of the polarization. The temperature dependence of the permittivity in the ferroelectric phase is obtained by inserting Eq. (1.6) into Eq. (1.4):

$$\left.\frac{1}{\varepsilon}\right|_{T<T_c} = 2\frac{T_C - T}{C}. \quad (1.7)$$

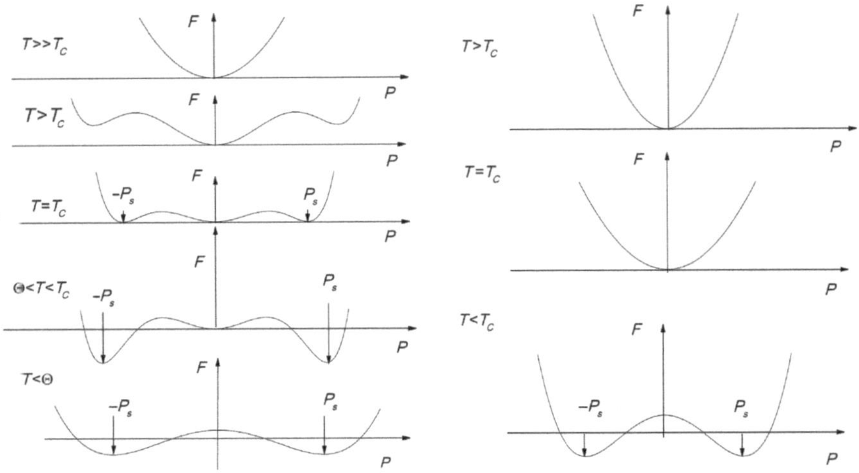

Figure 1.2 Schematic evolution of the free energy for first-order phase transition (left) and second-order phase transition (right).

If the expansion coefficients are chosen as $A_4 < 0$ and $A_6 > 0$ (the first-order ferroelectric phase transition), the stable states will again be found from Eq. (1.2):

$$P_s^2 = \frac{1}{2A_6}\left(|A_4| + \sqrt{A_4^2 - 4C^{-1}(T-\Theta)A_6}\right). \tag{1.8}$$

Inserting Eq. (1.8) into Eq. (1.1) results in the free energy as a function of polarization and of temperature. At high temperatures the free energy assumes a parabolic shape with a minimum corresponding to a stable paraelectric phase (Fig. 1.2). On cooling, secondary minima at finite polarizations become visible. Their energy level at the beginning, however, is higher than that at $P = 0$. In this regime, the paraelectric phase is stable and the ferroelectric phase is metastable. Lowering the temperature further, at $T = T_C$, all three minima of the free energy are at the same level. Below T_C, the finite spontaneous polarization is favored. In the temperature regime between T_C and Θ, the paraelectric phase coexists with the ferroelectric phase with the paraelectric phase being metastable. Somewhere on cooling through this regime, the first-order phase transition to the ferroelectric state will occur with a corresponding jump of the spontaneous polarization from zero to a finite value. The permittivity in the ferroelectric phase is given by:

$$\left.\frac{1}{\varepsilon}\right|_{T<T_c} = \frac{T-\Theta}{C} + \frac{3A_4}{2A_6}\left(|A_4| + \sqrt{A_4^2 - 4C^{-1}(T-\Theta)A_6}\right) + \frac{1.25}{A_6}\left(|A_4| + \sqrt{A_4^2 - 4C^{-1}(T-\Theta)A_6}\right)^2. \tag{1.9}$$

For ferroelectric phase transitions from cubic phase $O_h = \mathrm{m3m}$ all three axes are equivalent, therefore the order parameter should have three components. In such cases the free energy is:

$$F(\sigma, T, \eta_1, \eta_2, \eta_3) = F_0(\sigma, T) + \frac{1}{2C}(T - T_c)\left(\eta_1^2 + \eta_2^2 - \eta_3^2\right) + \\ + \frac{1}{4}A_4\left(\eta_1^2 + \eta_2^2 + \eta_3^2\right)^2 + \frac{1}{2}A_4'\left(\eta_1^2 \cdot \eta_2^2 + \eta_1^2 \cdot \eta_3^2 + \eta_2^2\eta_3^2\right) + \ldots. \tag{1.10}$$

Stable states can be obtained from relations:

$$\frac{\partial F}{\partial \eta_1} = 0, \quad \frac{\partial F}{\partial \eta_2} = 0, \quad \frac{\partial F}{\partial \eta_3} = 0. \tag{1.11}$$

The solutions of Eq. (1.11) are:

$$\eta_{01} = \eta_{02} = \eta_{03} = 0 \tag{1.12}$$

for cubic $O_h = \mathrm{m3m}$ phase;

$$\eta_{01} = \eta_{02} = 0, \quad \eta_{03} \neq 0,$$

or

$$\eta_{01} = \eta_{03} = 0, \quad \eta_{02} \neq 0, \tag{1.13}$$

or

$$\eta_{02} = \eta_{03} = 0, \quad \eta_{01} \neq 0,$$

for tetragonal $C_{4v} = \mathrm{4mm}$ phase, while solutions

$$\begin{aligned} \eta_{01} = \eta_{02} \neq 0, \quad \eta_{03} = 0, \\ \eta_{01} = \eta_{03} \neq 0, \quad \eta_{02} = 0, \\ \eta_{02} = \eta_{03} \neq 0, \quad \eta_{01} = 0 \end{aligned} \tag{1.14}$$

for orthorhombic $C_{2v} = \mathrm{mm2}$ phase and $\eta_{01} = \eta_{02} = \eta_{03} \neq 0$ in the case of rhombohedral phase $C_{3v} = \mathrm{3m}$.

Therefore three different phase transitions are possible starting from the parent cubic phase $O_h = \mathrm{m3m}$. The stability of different phases is determined by coefficients in Eq. 1.10. For example, $BaTiO_3$ crystals have three polar phases: $O_h \leftrightarrow C_{4v} \leftrightarrow C_{2v} \leftrightarrow C_{3v}$.

Moreover, the phase will be stable if:

$$\frac{\partial^2 F(\sigma, T, \eta_1, \eta_2, \eta_3)}{\partial \eta_{1,2,3}^2} > 0. \tag{1.15}$$

Despite the fact that the expression of the free energy is more complicated when the order parameter is three-dimensional (Eq. 1.10) in comparison with the case of the one-dimensional order parameter (Eq. 1.1), the Curie–Weiss law is always valid for the temperature dependence of the dielectric permittivity. This is due to the fact that the dielectric permittivity is strongly temperature-dependent close to the phase transition temperature, while the inverse dielectric permittivity (Eq. 1.4) is the weaker temperature function and can be effectively approximated by the Taylor series, with the linear part corresponding to the Curie–Weiss law. So, close to any ferroelectric phase transition some temperature interval can be found where the Curie–Weiss law is valid (Fig. 1.3). The ratio of Curie–Weiss constants in paraelectric and ferroelectric phases is close to 2 for the second-order phase transition and higher than 2 for the first-order phase transition. A big difference between values of the Curie–Weiss temperatures obtained from the fits in the paraelectric and ferroelectric phases is observed due to coexistence of the two phases in some temperature range (Fig. 1.3). According to the Landau theory of the second-order phase transition the Curie–Weiss temperature should be the same in both paraelectric and ferroelectric phases and it indicates the phase transition temperature.

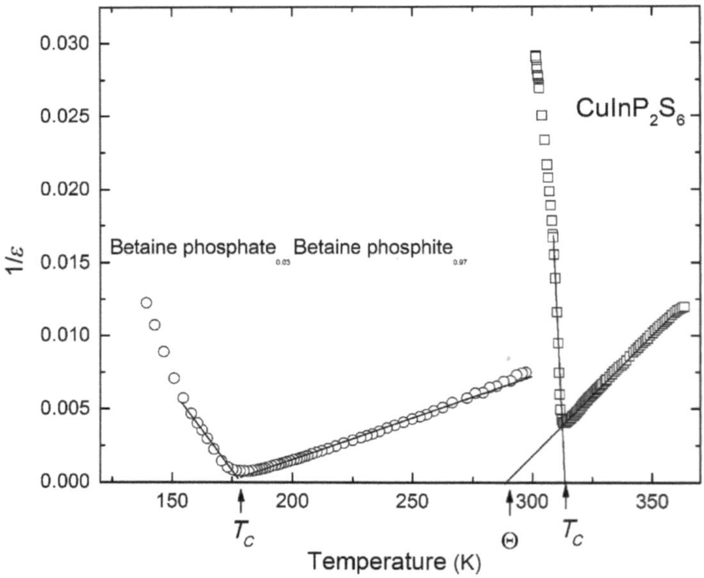

Figure 1.3 Temperature dependence of the static dielectric permittivity of $BP_{0.03}BPI_{0.97}$ (second-order phase transition) and $CuInP_2S_6$ (first-order phase transition) ferroelectric crystals.
Source: Based on [4] J. Banys, J. Macutkevic, A. Brilingas, J. Grigas, C. Klimm, G. Völkel, Broadband dielectric spectroscopy of betaine phosphate0.03 betaine phosphite0.97 crystals in the vicinity of the ferroelectric phase transitions, Phase Transit. 78 (2005) 869–881, DOI:10.1080/01411590500335584; [5] J. Banys, J. Macutkevic, V. Samulionis, A. Brilingas, Y. Vysochanskii, Dielectric and ultrasonic investigation of phase transition in CuInP2S6 crystals, Phase Transit. 77 (2004) 345–358, DOI:10.1080/01411590410001667608.

However, for real crystals the dielectric permittivity remains finite even at ferroelectric phase transition temperature and very close to it. Therefore, for real crystals the Curie–Weiss law is not valid at $T = T_C$ and some uncertainty of determination of T_C always exists in both paraelectric and ferroelectric phases.

1.3 Microscopic theory of ferroelectrics: the mean field

Very often ferroelectric phase transitions are classified as order-disorder or displacive. The displacive phase transitions are characterized by shift of some ions from its position in centrosymmetric phase, while in the case of order-disorder phase transitions the ferroelectric order appears due to the ordering of dipoles in the same orientation. The first group of phase transitions is related to instability in the vibration of atoms, while the second one is related to orientation of dipoles.

In contrast to the macroscopic Landau–Ginzburg–Devonshire model, which classifies the ferroelectric phase transitions only according to different symmetry types, the phase transition type, the phase transition temperature, and so on, the microscopic models try to find the relation between microscopic parameters, such as the interaction energy between ferroelectrically active ions and macroscopic parameters of materials; for example, the temperature of phase transition. Ferroelectric materials with an order-disorder phase transition are most often described by the three-dimensional Ising model with the following Hamiltonian:

$$H = -\frac{1}{2}\sum_{R,R'} J_{R,R'}\sigma_R\sigma_{R'} - E_z\mu_z\sum_R \sigma_R, \tag{1.16}$$

where $J_{R,R'}$ is the interaction between particles in positions R and R' and $E_z\mu_z\sum_R \sigma_R$ is the interaction with an external electric field. σ_R is the variable reflecting the unit dipoles, which may have value of 1 or -1, corresponding to two opposite orientations.

The statistic sum Z is:

$$Z = \sum_{\sigma_{R1}=\mp 1} \cdots \sum_{\sigma_{RN}=\mp 1} \exp\left[-\frac{U(\sigma_{R1,\dots}\sigma_{RN})}{kT}\right] \tag{1.17}$$

and the free energy:

$$F = -kT \ln Z. \tag{1.18}$$

The order-disorder ferroelectric phase transitions can be characterized by an average dipole parameter of the form:

$$<\sigma_R> = s = \frac{N_1 - N_2}{N_1 + N_2} = \frac{N_1 - N_2}{N}, \tag{1.19}$$

where N_1 and N_2 are the numbers of ferroelectrically active ions in two different positions, say 1 or 2, respectively; N is the total number of ferroelectrically active ions. For $N_1 = N_2$ the system is disordered and no spontaneous polarization occurs. The order in crystal appears for $N_1 \neq N_2$ and then the nonzero mean dipolar moment is:

$$<\mu_z> = \frac{N_1 - N_2}{N}\mu_z = \mu_z s, \tag{1.20}$$

and the spontaneous polarization is:

$$P_c = \frac{\mu_z}{v_{cell}}s. \tag{1.21}$$

Here, μ_z is the dipolar moment of the ordering entities, and v_{cell} is the volume of the unit cell. The mean field theory is used most frequently as an approximation of microscopic theory [6]. According to this theory the interaction of some particle with all other particles is replaced by the effective mean field, which is independent of position of the particle and its closest environment. The accuracy of this approximation is good when the interaction between particles is a very weak function of distance. According to this approximation the Hamiltonian is:

$$H = \frac{J_0 N s^2}{2} - s\sum_R J_{R,R'}\sigma_R - E_z\mu_z\sum_R \sigma_R = \frac{J_0 N s^2}{2} - (sJ_0 + E_z\mu_z)\sum_R \sigma_R \tag{1.22}$$

The free energy is:

$$F = -kT \ln Z = \frac{J_0 N s^2}{2} - NkT \ln 2\mathrm{ch}\left(\frac{J_0 s + E_z\mu_z}{kT}\right). \tag{1.23}$$

The crystal will be stable if:

$$\frac{\partial F}{\partial s} = J_0 N s - \frac{NJ_0\left(2\mathrm{sh}\frac{J_0 s + E_z\mu_z}{kT}\right)}{2\mathrm{ch}\frac{J_0 s + E_z\mu_z}{kT}} = 0 \tag{1.24}$$

or

$$s = \mathrm{th}\frac{J_0 s + E_z\mu_z}{kT}. \tag{1.25}$$

In the system described by Eq. (1.25), the ferroelectric phase transition occurs, because for small s values:

$$\text{th}\frac{sJ_0}{kT} = \frac{sJ_0}{kT} - \frac{1}{3}\left(\frac{sJ_0}{kT}\right)^3 + . \tag{1.26}$$

or

$$s^2 = \frac{3k^3T^3}{J_0^3}\left(\frac{J_0}{kT} - 1\right) \tag{1.27}$$

That is, $s^2 = 3(T_C - T)/T_C$, where $T_C = J_0/k$ and it was assumed that close to T_C, $T = T_C = J_0/k$.

According to Eq. (1.21) the spontaneous polarization will be:

$$P = \frac{\mu_z}{v_{cell}}\sqrt{3\frac{(T_C - T)}{T_C}}. \tag{1.28}$$

This is the same temperature dependence as for the second-order ferroelectric phase transitions in Landau theory (Eq. (1.6)). Therefore the mean field theory is equivalent to Landau theory and Eq. (1.23) for free energy is equivalent to Eq. (1.1). Coefficients A_2, A_4, and A_6 can be obtained as a function of T, N, μ_z, J_0.

The dielectric permittivity in mean field theory:

$$\varepsilon - \varepsilon_0 = \frac{\partial P}{\partial E} = \frac{\mu_z}{v_{cell}}\frac{\partial s}{\partial E}. \tag{1.29}$$

according to Eq. (1.25):

$$\frac{\partial s}{\partial E} = \frac{\mu_z}{kTch^2\left[\frac{J_0 s + \mu_z E}{kT}\right] - J_0}, \tag{1.30}$$

or

$$\varepsilon = \varepsilon_0 + \frac{\mu_z^2}{v_{cell}kTch^2\left[\frac{J_0 s + \mu_z E}{kT}\right] - J_0}. \tag{1.31}$$

In the paraelectric phase close to the phase transition temperature, Eq. (1.31) transforms into:

$$\varepsilon = \varepsilon_0 + \frac{\mu_z^2}{J_0 v_{cell}}\frac{T_C}{T - T_C}, \tag{1.32}$$

which is the Curie–Weiss law with $C = \frac{\mu_z^2}{J_0 v_{cell}} T_C$. For further evaluation it is necessary to know J_0, which determines the main interactions in the system. In the case of dipole–dipole interactions the mean local electric field for each dipole is:

$$E = E_z + s\frac{J_0}{\mu} = E_z + \frac{J_0 v_{cell}}{\mu_z^2} P_C = E_z + \frac{1}{\varepsilon_0}\beta P_C \tag{1.33}$$

where E is the local electric field, E_z is the mean macroscopic field, and $\frac{\varepsilon_0 J_0 v_{cell}}{\mu_z^2} = \beta$ is the Lorentz depolarization factor, which is determined by local anisotropy; for example, for cubic crystals $\beta = 1/3$ and $C/T_C = 3\varepsilon_0$. In the case of relative dielectric permittivity the Curie–Weiss constant is $C_R = C/\varepsilon_0$ and $C_R/T_C = 3$. Indeed it was observed experimentally that for order-disorder ferroelectrics $C_R \approx 10T_C$ (so there is a slight discrepancy), while for displacive ferroelectrics, $C_R >> 10T_C$.

We have to keep in mind that in the case of ferroelectrics, Curie–Weiss law is valid only in the case of one of the components of dielectric permittivity, which originates from some ordering dipoles, driving the ferroelectric phase transition. For example, this could be the contribution of the unstable (soft) mode to the total dielectric permittivity. However, other contributions might be insignificant at low frequencies and are often neglected. Still, these multiple processes can be distinguished from dynamic studies of ferroelectrics.

1.4 Dynamic properties of ferroelectrics: theory

The atoms of all materials vibrate at nonzero temperatures around their equilibrium position. The lattice waves (phonons) propagate through the material and can be characterized by the wavelength (λ) and the frequency (ω) [7]. The characteristic wave vector $K = 2\pi/\lambda$ can be reduced to the first Brillouin zone of the reciprocal lattice, $0 \le K \le \pi/a$, where a is the lattice constant. The acoustic phonon is the coherent vibration of atoms as a linear chain. For every lattice there are three modes with different frequencies belonging to one longitudinal (LA) and two transverse (TA) branches of the acoustic phonons. Optical phonons are vibrations of the sublattices against each other, when more than one type of atoms are present in the elementary cell. A vibration of the atoms perpendicular to the propagation corresponds to a transverse wave (TO), a vibration in the direction of the propagation corresponds to a longitudinal wave (LO). The interaction of electromagnetic waves is possible only with optic phonons when $K \approx 0$, when impulse and energy conservation laws are valid.

The structure of the material will be stable if frequencies of all phonons are positive. This suggests that close to a phase transition temperature the frequency of a particular phonon is not stable. A linear relation (Cochran's law) between ω^2_{TO} at the zone center and T is found, suggesting that the temperature dependence of the optic mode frequency relates to the phase transition [8]:

$$\omega_{TO}^2 = A_{cr}|T - T_C| \tag{1.34}$$

in accordance with the Lyddane–Sachs–Teller relation:

$$\frac{\varepsilon(T,0)}{\varepsilon_\infty} = \frac{\omega_{LO}^2}{\omega_{TO}^2(T)}. \tag{1.35}$$

Here, ε_∞ is the contribution of all the higher modes and electronic polarization, ω_{LO} is the longitudinal optic mode frequency, and $\varepsilon(T, 0)$ is the contribution of the particular TO mode to the static dielectric permittivity. Substitution of Eq. (1.34) into Eq. (1.35) gives the Curie–Weiss law. This means that Cochran's law is valid in both paraelectric and ferroelectric phases, but the A_{cr} constant decreases twice during the second-order paraelectric to ferroelectric phase transition.

The frequency dependence of the complex dielectric permittivity caused by a polar phonon vibration can be described by the Lorentz oscillator:

$$\varepsilon(\omega)^* = \varepsilon_\infty + \frac{(\varepsilon(0) - \varepsilon_\infty)\omega_{TO}^2}{\omega_{TO}^2 - \omega^2 + i\gamma\omega}. \tag{1.36}$$

In the case of order-disorder ferroelectrics the soft mode is strongly overdamped (i.e., the damping of the mode is much higher than its frequency, $\gamma >> \omega_{\mathrm{TO}}$) and Eq. (1.36) transforms into the simple Debye equation:

$$\varepsilon^*(\omega) = \varepsilon_\infty + \frac{\Delta\varepsilon}{1 + i\omega\tau}, \tag{1.37}$$

where $\tau = \gamma/{\omega^2}_{\mathrm{TO}}$ and $\Delta\varepsilon = \varepsilon(0) - \varepsilon_\infty$.

If the damping is temperature independent (which is a relatively good approximation), then the Cochran's law transforms into a linear temperature dependence of the frequency of the relaxational mode $\nu = 1/\tau = {\omega^2}_{\mathrm{TO}}/\gamma$, which is in good agreement with Curie–Weiss law (Eq. (1.5)):

$$\nu = \nu_0|T - T_C| \tag{1.38}$$

Again, the ν_0 constant changes its value by a factor of two during the second-order phase transition, as it is proportional to A_{cr} in Eq. (1.34).

Until now we have assumed that the order parameter is time-independent. If the material is affected by time-dependent external fields—for example, an electrical pulse—Eq. (1.2) should be generalized into the Landau–Khalatnikov equation [9]:

$$L\frac{\partial\eta}{\partial t} = -\frac{\partial F}{\partial\eta}. \tag{1.39}$$

which describes the relaxation of a fluctuation in order parameter to an equilibrium without an external field. Here, L is a proportionality coefficient. In the case of equilibrium, the right side of Eq. (1.39) is equal to zero. However, for small fluctuations ($\delta\eta = \eta - \eta_0$ and $\delta\eta << \eta_0$) it can be approximated:

$$L\frac{\partial(\delta\eta)}{\partial t} = -\frac{\partial^2 F}{\partial\eta^2}\delta\eta \tag{1.40}$$

In such case the relaxation time to the equilibrium is equal to:

$$\tau = L/\frac{\partial^2 F}{\partial \eta^2}. \tag{1.41}$$

According to Eqs. (1.4–1.7), in the case of second-order phase transitions the relaxation time $\tau \sim L/(T - T_C)$ for $T > T_C$ and $\tau \sim L/(2(T_C - T))$ for $T < T_C$. If the order parameter is proportional to the polarization (the case of proper ferroelectrics), then this relaxation time describes the relaxation time of dielectric permittivity, and thus Eq. (1.38) is obtained from Landau–Khalatnikov theory.

It can be shown that in the limit of high frequencies ($\omega\tau >> 1$), if both relaxation frequency ν and $\Delta\varepsilon$ vary in accordance with the Curie–Weiss law (Eqs. (1.5) and (1.38)), the real part of the Debye equation (Eq. (1.37)) transforms into $\varepsilon'(\omega) \sim |T - T_C|$. This means that at high frequencies there is a minimum in the real part of dielectric permittivity close to the phase transition temperature instead of the maximum according to Eq. (1.5). The existence of the minimum is related to the critical slowing down [10]. The Eyring equation is often used to describe the temperature dependence of the relaxation time close to the order-disorder ferroelectric phase transition [11]:

$$\tau = \frac{h}{kT} \times \exp\left(\frac{U}{kT}\right), \tag{1.42}$$

which in fact is a modified Arrhenius law. Another possibility to describe the temperature dependence of the relaxation time is to combine the Curie–Weiss and the Arrhenius law:

$$\tau = \frac{C_R}{T - T_C} \times \exp\left(\frac{U}{kT}\right). \tag{1.43}$$

In that case the exponential factor describes discrepancies between the phenomenological Landau theory close to the phase transition temperature and the experimental results, which appear to be due to critical fluctuations [12].

In the case of the presence of several types of dipoles in the same material with close relaxation times or defects, which locally alter the interactions between dipolar moments, the relaxational dispersion broadens and/or becomes asymmetric in comparison with Debye relaxation (Eq. (1.37)). Thus other empirical formulas are introduced to reflect this behavior. The most general one at the moment is the Havriliak–Negami formula [13]:

$$\varepsilon^*(\omega) = \varepsilon_\infty + \frac{\Delta\varepsilon}{\left(1 + (i\omega\tau_{NH})^{1-\alpha}\right)^\gamma}. \tag{1.44}$$

Here additional parameters α and γ are introduced to describe the broadening and the asymmetric shape of the dielectric dispersion, respectively. Other parameters are analogous to the Debye formula (1.37). The other approach to get the characteristic relaxation times of the order–disorder-type systems is to calculate the distribution of relaxation times as a weighted sum of single Debye-type relaxations:

$$\begin{aligned} \varepsilon'(\omega) &= \varepsilon_\infty + \Delta\varepsilon \int_{-\infty}^{+\infty} d(\lg\tau) \frac{f(\tau)}{1+(\omega\tau)^2} \\ \varepsilon''(\omega) &= \Delta\varepsilon \int_{-\infty}^{+\infty} d(\lg\tau) \frac{f(\tau)(\omega\tau)}{1+(\omega\tau)^2} \end{aligned} \quad . \tag{1.45}$$

From this set of equations a distribution of relaxation times $f(\tau)$, which corresponds to the probability to find a relaxator with particular relaxation time, can be calculated. Since various distributions of relaxation times can give the same $\varepsilon^*(\omega)$ values, Eq. (1.45) is known as an ill-posed problem and can be solved only by involving regularization methods. It was shown that by choosing the appropriate regularization parameters, the correct distribution of the relaxation times can be obtained even for highly inhomogeneous systems [14].

Phonon dynamics and the resonance-type dispersions in microwaves, caused by domain wall vibrations, are described by the superposition of damped harmonic oscillators:

$$\varepsilon^*(\omega) = \varepsilon_\infty + \sum_{j=1}^{n} \frac{\Delta\varepsilon_j \omega_{TOj}^2}{\omega_{TOj}^2 - \omega^2 + i\omega\gamma_{TOj}^2}. \tag{1.46}$$

Here $\Delta\varepsilon_j$ is the dielectric strength of the j-th mode, and ω_{TOj} and γ_{TOj} are the frequency and damping of the j-th transverse oscillator, respectively. From a general summation rule it follows that the sum of product $\Delta\varepsilon_j\omega_{TOj}^2$ should be stable; that is, $\sum_j \Delta\varepsilon_j\omega_{TOj}^2 = const.$ If all oscillators are uncoupled, each term should be temperature independent. This means that the decrease in ω_{TOj} should be accompanied by the increase in $\Delta\varepsilon_j$. In the case of the soft mode the oscillator strength typically increases on cooling, and to preserve the constant sum of $\Delta\varepsilon_j\omega_{TOj}^2$ this mode should receive some input from other higher-frequency modes (i.e., these modes become coupled).

In the case there is an appreciable splitting in transverse and longitudinal frequencies of polar phonons, a generalized-oscillator model with the factorized form of the complex permittivity is typically used [15]:

$$\varepsilon^*(\omega) = \varepsilon_\infty \prod_{j=1}^{n} \frac{\omega_{LOj}^2 - \omega^2 + i\omega\gamma_{LOj}^2}{\omega_{TOj}^2 - \omega^2 + i\omega\gamma_{TOj}^2}. \tag{1.47}$$

Here ω_{LOj} and γ_{LOj} are frequency and damping of the j-th longitudinal oscillator, respectively. Other parameters are the same as in Eq. (1.46).

1.5 Raman, infrared, and dielectric spectroscopy of ferroelectrics

Investigation of the soft vibrational modes is the main route to understanding the microscopic mechanisms responsible for the appearance of phase transitions in ferroelectrics. The lattice dynamics of bulk single crystals and ceramics can be studied using neutron scattering, infrared (IR) absorption or reflection spectroscopy, and Raman scattering techniques. On the other hand, the dielectric spectroscopy can provide useful information about the collective processes (e.g., domain dynamics) in ferroelectrics. Thus, complementing these methods, we can obtain a unified picture about the polarization dynamics in ferroelectric materials starting from optical frequencies down to radio and microwave ones.

In Raman scattering, the incident laser light (visible or UV) interacts with optical (typically long wavelength) phonons of the material being studied, which results in the change in the wavelength of the scattered photon. If this interaction transfers some energy to the interacting media, the resulting photon wavelength becomes larger. This is called the Stokes scattering. In the case of antiStokes scattering the energy is transferred from the material to the photon, which results in the decrease of its wavelength. Since the Raman effect can be understood as the distortion of the laser light induced dipole moment by the vibrations of atoms in the lattice, the selection rules permit the interaction with those modes, which are accompanied with the change of the polarizability.

In IR spectroscopy the interaction between incident IR radiation and media being studied occurs via the absorption process rather than the scattering one. The IR energy covering a range of frequencies is directed onto the sample. Absorption occurs where the frequency of the incident radiation matches that of a vibration so that the molecule is promoted to a vibrational excited state. Thus, IR spectroscopy measures transitions between molecular vibrational energy levels. In order for particular mode to be IR active, it must be accompanied by the change in the dipolar moment of the molecule.

It is well known that each system, comprised of N atoms, has $3N$ degrees of freedom. The number of vibrational modes can be calculated omitting three translational and three rotational (except for linear molecules, which have only two of them) ones. Thus, triatomic asymmetric molecules will have $3N-6=3$ vibrational modes. Not all of these vibrations are active in IR and Raman spectra simultaneously. Raman spectroscopy is best at symmetric vibrations of nonpolar groups while IR is best at the asymmetric vibrations of polar groups. For ferroelectric materials, in centrosymmetric phase, group theory permits only those vibrations that are active either in IR or Raman spectrum. In the ferroelectric phase, where there is no inversion center, modes can be active in both kinds of spectra (there could be situations when certain modes are silent; i.e., they are inactive in both Raman and IR). So, these two spectroscopies can be considered as complementary techniques providing a unified picture of vibrational states of materials under study. Since the laser beam in Raman spectrometer can be focused to a relatively small surface (few microns), this micro-Raman spectroscopic technique is suitable to examine individual domains of single crystals, ceramic grains, nanoparticles, and so on. On the other hand, IR spectroscopy, coupled with the time-domain terahertz spectroscopy, can provide the complex dielectric function of the material under test.

In the case of perovskites, which probably have the simplest structure in terms of ferroelectricity, the group analysis predicts three IR active modes in the paraelectric Pm3m phase. One additional mode (corresponding to B′−B″ vibrations) occurs for the ordered $A(B'_{1/2}B''_{1/2})O_3$ $Fm\overline{3}m$ phase [16]. Other more complex structures, like tetragonal tungsten-bronze [17] or Aurivillius [18], have many more modes even in the paraelectric phase. The site occupancy disorder further increases the number of allowed modes via the increase in number of effective atoms in the unit cell. The possibility of the detection of forbidden modes (i.e., modes that should not present in this particular symmetry) appears.

One of the most interesting points studying ferroelectric materials with the help of IR and Raman spectroscopies is the so-called soft-mode behavior. The soft-mode concept was introduced to describe the mechanism of the ferroelectric structural phase transitions. This concept is based on the assumption that the material becomes unstable against a particular normal vibration of the lattice. This unstable phonon, whose frequency decreases as the temperature is approaching the phase transition temperature T_C (i.e., the phonon softens), becomes zero at T_C meaning that the atomic positions (or the correspondent vibration) become frozen at this temperature and produce a structure of different symmetry with a macroscopic dipole moment. Thus, the soft phonon must be both polar (IR active) and of long wavelength [19]. Since only transverse optical modes satisfy this condition, they are the main source of the increase of the static dielectric permittivity at and below the phase transition point and the soft-mode frequency satisfies Cochran's law (Eq. (1.34)). However, if the phase transition takes place at low temperatures where quantum fluctuations are significant, the soft-mode frequency is better described by the Barrett formula:

$$\omega_{TO} \sim \sqrt{A\left[\left(\frac{T_1}{2}\right)\coth\left(\frac{T_1}{2T}\right) - T_0\right]}. \tag{1.48}$$

Here T_0 is the critical temperature, indicating the tendency to hypothetical lattice instability at negative temperatures; T_1 is the temperature below which the quantum fluctuations start to play a role. This concept is clearly illustrated in Fig. 1.4.

The dielectric dispersion below phonon frequencies can be successfully analyzed with the help of the dielectric spectroscopy. It is a powerful tool revealing collective dynamics of phase transitions in ferroelectric materials. The complex dielectric permittivity $\varepsilon^*(\omega) = \varepsilon'(\omega) - i\varepsilon''(\omega)$ can be measured in extremely broad frequency range from 10^{-6} to 10^{11} Hz. Different techniques should be used to cover it fully. At frequencies from 10^{-6} to 10^{7} Hz lumped circuits methods can be applied to extract the dielectric permittivity from the measured data typically represented as a

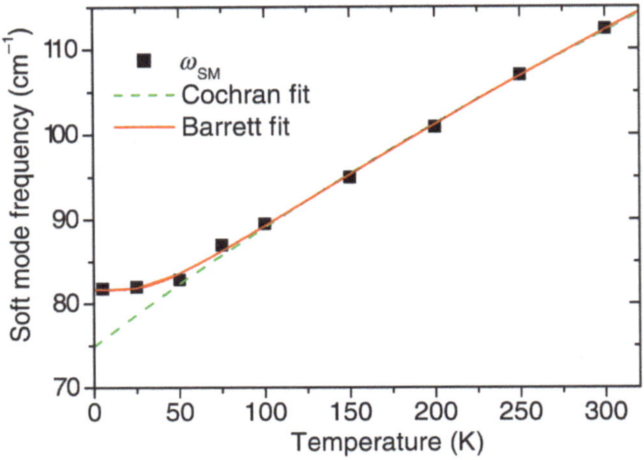

Figure 1.4 Temperature dependence of the soft mode frequency fitted with Cochran and Barrett formula for quantum paraelectric $EuTiO_3$ ceramics.
Source: Adapted from [20] S. Kamba, D. Nuzhnyy, P. Vaněk, M. Savinov, K. Knížek, Z. Shen, et al., Magnetodielectric effect and optic soft mode behaviour in quantum paraelectric $EuTiO_3$ ceramics, EPL Europhys. Lett. 80 (2007) 27002, DOI:10.1209/0295-5075/80/27002.

parallel *RC* circuit where the capacitance *C* can be related to the real part of the dielectric permittivity and the resistance *R* reflects dielectric losses. Two types of instrumentation can be employed to measure the dielectric data, one operating in frequency domain, another in time domain. While the first one allows the direct calculation of the dielectric permittivity applying simple plain capacitance model, the second one requires the Fourier transformation to recalculate $\varepsilon(t)$ to $\varepsilon^*(\omega)$. A comprehensive overview of these methods can be found in [21].

At higher frequencies, where parasitic capacitances highly impact measurement results and the possibility of standing waves even in connection cabling occurs, the measurements in various waveguide lines or resonators are used. In the frequency range from 10^6 to 10^{10} Hz a coaxial transmission line connected to a network analyzer is often employed. Here the propagating TEM wave in the line partially reflects and partially transmits through the sample, mounted in the center of it. The dielectric permittivity for low capacitance samples can be calculated either from complex reflection or transmission data according to the flat capacitor method. For high dielectric permittivity materials standing waves can be formed even in the sample, thus the flat capacitor method fails and it is important to take into account the inhomogeneous distribution of the electric field inside the sample [22]. In the frequency range between 10^{10} and 10^{11} Hz rectangular waveguide systems replace the coaxial lines. Here the sample can fully cover the waveguide if its dielectric permittivity is sufficiently low or a needle-shaped sample can be placed in the center of the waveguide's wide wall parallel to the electric vector of the TE_{10} waveguide mode. More detailed description of these methods can be found in [10].

Typically, the dielectric permittivity in ferroelectrics is most extensively studied close to phase transition temperature, where the dielectric permittivity is strongly temperature dependent. In the case of the order-disorder ferroelectrics the dielectric dispersion occurs solely in radio- and microwave frequency range and can be successfully analyzed by the dielectric spectroscopy tools, while the investigations of displacive type ferroelectrics are based mostly on soft mode analysis by (quasi)optical techniques like Raman, IR, and Terahertz time-domain spectroscopy. In only some cases (typically in materials with high dielectric permittivity values) the soft mode frequency goes down below 10^{11} Hz and can be visible in microwaves.

As discussed earlier, the phenomenon of critical slowing down is often observed in ferroelectrics described by the order–disorder-type dynamics. When the probing frequency exceeds the inverse characteristic relaxation time of the dipolar moments, a minimum in the temperature dependence of the dielectric permittivity is observed as shown in Fig. 1.5. This phenomenon is connected with the decrease in the relaxation frequency of the dipoles in accordance with the linear law (Eq. (1.38)).

It is energetically unfavorable for a ferroelectric crystal of finite size to stay as a single domain in the ferroelectric phase. Thus it breaks into domains with different polar orientations to minimize the depolarizing field at its faces. A number of publications show significant contribution of ferroelectric domains to dielectric and (electro)mechanical properties either in the case of single crystals [24–28], ceramics [29–34], or thin films [35–37]. This very same contribution is shown in Fig. 1.5 as the hatched area and is only visible in the ferroelectric state, as ferroelectric domains are absent in the paraelectric phase. The data is numerous, as well as the theoretical explanations. However, the basic idea stands on movement of the ferroelectric domain walls.

As we can see from Fig. 1.6, the poly-domain (twinned) crystal has zero net polarization. However, under external electric field the energetically favorable domains would grow at the expense of others (this can be interpreted as domain wall movement). As a result the net polarization becomes nonzero, and the crystal deforms. The deformation under external electric field is extrinsically observed as the reverse piezoelectric effect; that is, the domains contribute to the total electromechanical coupling. On the other hand, the dielectric permittivity can be determined in accordance with Eq. (1.4). In the case of domain wall movement the partial derivative is nonzero. The ferroelectric domains contribute to the total dielectric permittivity proportionally to the volume density of the domain walls.

The domain contribution is highly important, thus it becomes interesting to engineer the domain structure to enhance the dielectric permittivity [26–28,38]. One of the ways to control domain width and hence their density in the case of crystal slabs is to control the thickness of the crystal, which was first shown by Kittel for magnetic domains, but is applicable for ferroelectrics, too [39]:

$$w = \sqrt{\frac{\sigma}{U} d}. \tag{1.49}$$

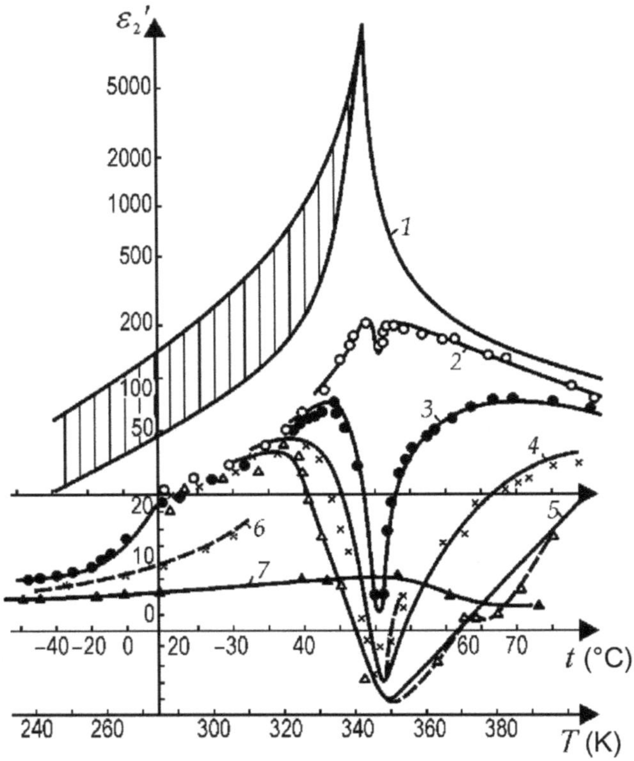

Figure 1.5 Temperature dependence of the real part of relative dielectric permittivity of TGS crystal at different frequencies: 1-1 kHz, 2-1 GHz, 3-16 GHz, 4-26 GHz, 5-37 GHz, 6-80 GHz, 7-250 GHz.
Source: © 2011 [23] Y. Poplavko, Y. Prokopenko, V. Molchanov, V. Kazmirenko, Ferroelectrics study at microwaves, in: Ferroelectrics - Characterization and Modeling, InTech, 2011, pp. 203–226, under CC BY-NC-SA 3.0 license.

Here, d is the crystal thickness, U is the volume energy density of the domains, and σ is the surface energy density of the domain wall. However, all domain walls have a certain thickness, which depends on the material under consideration, and this influences the domain width at small thicknesses d, as in the case of thin films [37].

Interestingly, Kittel's law is observed in the case of ceramics, too [29,40,41]. And this is precisely the reason for increase of the dielectric permittivity of ceramics upon decrease of grain size as depicted in Fig. 1.7. However, every contribution has a certain time scale. As the domain walls move, they emit many acoustic waves [42]. In the case of ceramics, the domain walls move in a correlated fashion due to the external field, and the distance between the walls is quite constant in the same crystallite. This means that at some frequencies the domain wall motion is enhanced, while it is impeded at the others due to destructive interference from

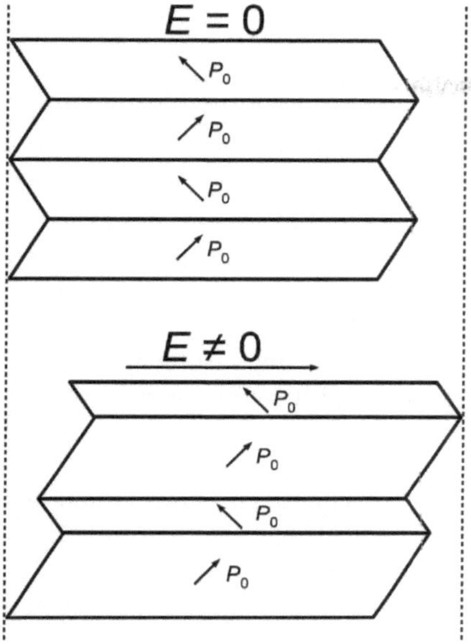

Figure 1.6 Domain wall motion under external electric field. P_0 denotes spontaneous polarization.

neighboring walls via acoustic waves. However, the resulting resonance-type system might be quite overdamped due to effective emission of the waves and the dielectric relaxation-like dispersion is visible around the fundamental frequency only [31,34,41,43–45].

Even though ferroelectric ceramics and thin films are likely to have quite small domains, crystals allow the use of the anisotropy of the crystal lattice and corresponding properties, and the direction of the crystal cut can be relatively easily controlled. Possible directions of the spontaneous polarization are limited by the crystal structure. As a result it is possible to pole crystals in a frustrative way; that is, when two or more energetically favorable polarization directions are allowed. Such domain engineering results in a high density of domain walls, leading to high dielectric permittivity and a highly enhanced piezoelectric effect [26–28,33,46]. Similar ideas can be applied to thin epitaxial films, but mechanical clamping to the substrate and nonideality of metal contacts (if any) must also be considered [47,48].

1.6 Other spectroscopic techniques

To study dynamical local properties of ferroelectrics undergoing phase transitions, other experimental methods need to be used. Some of the most exploited are magnetic resonance methods, which can provide comprehensive local information about dynamical structural changes in the lattice. Magnetic methods include nuclear magnetic resonance (NMR), nuclear quadrupole resonance (NQR), electron paramagnetic resonance (EPR), electron-nuclear double resonance (ENDOR), and others. The information about the distribution and dynamics of the local electric fields at the nuclear or electronic subsystems can be obtained. The main requirement is that the nuclei of the atom should have a nonzero spin (for NMR) or the electronic system should have an unpaired electron (for EPR).

The main advantage of NMR measurements is the ability to detect the change of the redistribution of the electron density at different time scales and convert it to the change of different relaxation times while, for example, in X-ray diffraction this dynamical disorder will be transformed into spatial disorder with probabilistic distribution.

Usually an onset of structural ordering results in a change in the number of resonance lines because there is a distinct set of energy levels for each group of inequivalent nuclei. In addition, phase transitions can involve the shift (chemical shift) or the change in the line width of the resonance. Also, the reorientational dynamics of polar entities affect nuclear spin-lattice relaxation, which measures the rate of return of the arbitrary nuclear spin magnetization to the equilibrium state along an external magnetic field after its perturbation by an electromagnetic pulse. In general, the spin-lattice relaxation depends on the spectral density of fluctuations of local order parameter (i.e., local polarization). For the Debye-type relaxation, the expression for the spectral density is proportional to the dielectric losses [49]. In the oxide group, ferroelectric lead-containing materials comprise a big and technologically important subgroup. The examination of ^{207}Pb nuclear system in $PbTiO_3$ typically shows remarkable and weakly temperature-dependent but significantly anisotropic chemical shifts [50]. Investigations of the relaxational properties in $PbTiO_3$ and $PbZr_{1-x}Ti_xO_3$ powders revealed that the spin-lattice relaxation time varies between several seconds and longer than an hour at different temperatures, while the spin-spin relaxation time is nearly temperature independent [51]. The NQR studies of the ^{47}Ti and ^{49}Ti in lead-free textbook ferroelectric $BaTiO_3$ revealed a local symmetry different from the cubic one in its paraelectric state.

The development of observed satellite background into well-defined satellite lines below phase transition temperature was attributed by authors to the confirmation of mixed displacive and order-disorder type of phase transition [52]. It is needed to point out that NMR is quite successfully employed to examine the local structure of various disordered ferroelectric materials (ferroelectric relaxors) like lead magnesium niobate. Here the experimentally obtained ^{207}Pb spectrum was decomposed into anisotropic and isotropic parts corresponding to the polar nano clusters and spherical glassy matrix contributions, respectively [53].

EPR technique is widely used to determine the electronic states of paramagnetic centers in various solid solutions. Since not many ferroelectric oxides have these in their original configuration, in most cases they are artificially introduced as dopants in small quantities to prevent the interaction with each other and broadening of the EPR lines. In perovskite ferroelectrics EPR is intensively used to determine the presence and dynamics of oxygen vacancies and other defects that play a key role in various important physical processes such as aging, electrical conductivity, and so on [54–56]. Also, the nature of the evolution of the local crystalline structure during phase transitions can be clarified [55,57].

1.7 The size and mechanical strain effect in ferroelectric ceramics and thin films

The characteristic size of the ferroelectric structure can be as important as the lattice structure of the bulk material. However, the size effects should be primarily (but not exclusively) related to the surface effects as well as strain, which can be either induced by the environment or the surface itself.

The notion of the so-called size effect encompasses several phenomena; for example, the reduction of dielectric permittivity, spontaneous polarization, Curie temperature, piezoelectric properties with the decrease of size of the nanostructured materials [58–60]. But it must be stressed that the effect is not as straightforward, and the dielectric permittivity can increase as the characteristic size is decreased [29,61,62].

As Fig. 1.7 shows, roughly three regions can be distinguished in the dependence of the dielectric permittivity on grain size in the case of ceramics. In the case of big grains (more than 5 μm in the case of $BaTiO_3$) the dielectric properties are dominated by the bulk material. At intermediate grain sizes (700 nm–5 μm in the case of BTO) the permittivity increases as the ceramic grain size is decreased. This is primarily due to contribution of the ferroelectric domains [29,40], as was discussed in Section 1.5. The third region is at grain sizes smaller than 700 nm, where the dielectric permittivity increases with the grain size.

According to Kittel's law the domain size (domain wall density) must monotonically decrease (increase) with diminishing of the characteristic crystal size (slab

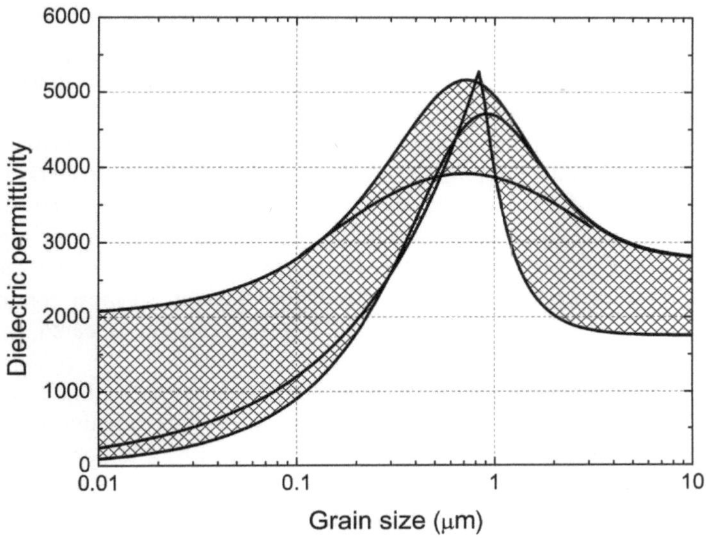

Figure 1.7 A sketch of grain-size dependence of dielectric permittivity of $BaTiO_3$ at room temperature. The hatched area represents possible values of the dielectric permittivity. The solid lines represent different approximations based on [61,62] and references therein.

thickness or grain diameter) and the dielectric permittivity would diverge around zero. However, as we can see from Fig. 1.7, it is not the case. The reason for that is interfacial effects, also known as grain boundary dilution.

The experimental probing field usually has a quite large wavelength compared to the grain size. In such case the externally observed electrical properties of a non-uniform multiphase system with sharp boundaries between the phases can be described by the effective medium approximation. There are quite a few papers describing various geometries [63–68], however only a few of them are suitable for single-phase ceramics [68–71]. Brick layer model (also known as brick wall model) is one of the most usually applied, as it is quite simple mathematically and easy to justify from the physical point of view. It assumes that the ceramics are made of cuboid grains separated by thin grain boundaries. If the current is flowing through the series-connected grain boundary and the grain itself, and current through the grain boundary parallel to the grain is negligible, then the externally observed effective dielectric permittivity (or conductivity) can be expressed as:

$$\frac{D + d_{gb}}{\varepsilon_{eff}} = \frac{D}{\varepsilon_{bulk}} + \frac{R}{\varepsilon_{gb}}. \tag{1.50}$$

Here, ε_{bulk} is dielectric permittivity of the bulk crystal (or in the center of the grain), ε_{gb} is permittivity of the grain boundary, d_{gb} is thickness of the grain boundary, D is the diameter of the grain. The bulk permittivity of ferroelectrics is relatively high, unlike in the case of grain boundary. The latter could consist of secondary phases, or strained nonferroelectric material. In any case, the thickness of the boundary can be expected to be independent of the grain size and to be around 1–10 nm depending on different evaluation approaches [72–74]. So in the case of diminishing grain size, the effective permittivity would approach a much smaller permittivity of the grain boundary, hence the decrease in permittivity in $BaTiO_3$ below 700 nm as shown in Fig. 1.7.

Such an approach is quite simplistic, but helps to understand the main physics behind it. When studied in detail, we must consider many more effects; for example, no grain is cuboid (they are actually irregular). No grain boundaries are created equal—they might have a distribution of properties, and they may or may not be more conductive than the bulk crystal. No sharp boundaries between the phases might exist. The porosity must also be taken into account, so most ceramics must be analyzed as a three-phase composite. The topic is quite complicated, hence a lot of research is conducted in this field.

The strain imposed by substrates is highly important for thin films, especially in the case of epitaxial single crystal films. Either positive or negative biaxial strain means that an anisotropic distortion is imposed on the crystal lattice, which is quite reminiscent of typical ferroelectric phase transformations. As a result, ferroelectric phase transitions might happen at a higher temperature that in bulk material [75,76] or, as in the case of $SrTiO_3$, it may be induced to happen in a normally nonferroelectric material [77]. It was experimentally shown [78] that heteroepitaxially grown $BaTiO_3$ thin films with a compressive biaxial strain can markedly enhance ferroelectric properties: temperature of the ferroelectric phase transition may be shifted by as much as 500°C and the spontaneous polarization at room temperature can reach 70 μC/cm^2, compared to 26 μC/cm^2 in bulk.

The strain may control other properties, thus making it another way of tailoring a material's properties. For example, it was shown in [79] that the mobility of charge carriers can be controlled by strain, making it important for electronics applications.

Speaking of electronics, ferroelectric materials usually should be considered as wide bandgap semiconductors [80]. In this case the interface with other materials might have different properties than the bulk, in particular due to space charge near the interface [81]. This space charge along with orientation of the substrate can become one of the most important factors for the formation of domain walls [82]. On the other hand, a similar space charge might be formed across the interface in the neighboring material (unless the material is an absolute vacuum or a perfect metal). This would give rise to a depolarizing field, which, again, can be used to alter properties of the film, including the domain structure [48].

A plethora of other interesting phenomena was briefly discussed in a recent review by Rappe [83]. A curious reader can find additional information there and in its references.

1.8 Summary

This chapter was conceived as a very brief introduction to the physics of ferroelectrics, trying to keep it as simple as possible. The very basic theoretical background and some of the experimental facts are presented and illustrated. Even though the origin of the ferroelectricity is in collective instability, which creates a net polarization even in the absence of external electric field, it manifests itself in other properties, such as pyroelectricity, piezoelectricity, NMR, and EPR. Ferroelectric domains play a very important role in determining the externally observable properties, which are coupled to electric field. At the nanoscale, interfacial effects are the most influential, leading to the possibility of tuning properties of materials, structures, and devices. Phenomena, closely related to ferroelectricity, are mentioned and referred to different papers, enabling nonexperts in ferroelectricity to read and understand the rest of this book.

References

[1] J. Valasek, Piezo-electric and allied phenomena in Rochelle salt, Phys. Rev. 17 (1921) 475–481. Available from: http://dx.doi.org/10.1103/PhysRev.17.475.

[2] L.D. Landau, Theory of phase transformations, I, Phys. Z Sowjetunion 11 (1937) 26.

[3] L.D. Landau, Theory of phase transformations, II, Phys. Z Sowjetunion 11 (1937) 545.

[4] J. Banys, J. Macutkevic, A. Brilingas, J. Grigas, C. Klimm, G. Völkel, Broadband dielectric spectroscopy of betaine $phosphate_{0.03}$ betaine $phosphite_{0.97}$ crystals in the vicinity of the ferroelectric phase transitions, Phase Transit. 78 (2005) 869–881. Available from: http://dx.doi.org/10.1080/01411590500335584.

[5] J. Banys, J. Macutkevic, V. Samulionis, A. Brilingas, Y. Vysochanskii, Dielectric and ultrasonic investigation of phase transition in $CuInP_2S_6$ crystals, Phase Transit. 77 (2004) 345–358. Available from: http://dx.doi.org/10.1080/01411590410001667608.

[6] J.A. Gonzalo, Effective Field Approach to Phase Transitions and Some Applications to Ferroelectrics, World Scientific, Singapore, Hackensack, NJ, 2006.

[7] L. Brillouin, Wave Propagation in Periodic Structures: Electric Filters and Crystal Lattices, second ed., with corrections and additions. Dover Publications, Mineola, NY, 2003

[8] W. Cochran, Crystal stability and the theory of ferroelectricity, Phys. Rev. Lett. 3 (1959) 412–414. Available from: http://dx.doi.org/10.1103/PhysRevLett.3.412.

[9] L.D. Landau, I.M. Khalatnikov, On the anomalous absorption of a sound near to points of phase transition of the second kind, Dokl. Akad. Nauk. SSSR (1954) 469.

[10] J. Grigas, Microwave Dielectric Spectroscopy of Ferroelectrics and Related Materials, first ed., CRC Press, Amsterdam, the Netherlands, 1996.

[11] A. Piecha, G. Bator, R. Jakubas, Critical slowing down of low-frequency dielectric relaxation in ferroelectric $(C_3N_2H_5)_5Bi_2Cl_{11}$, J. Phys. Condens. Matter 17 (2005) L411. Available from: http://dx.doi.org/10.1088/0953-8984/17/41/L03.

[12] A. Dziaugys, J. Banys, J. Macutkevic, R. Sobiestianskas, Y. Vysochanskii, Dipolar glass phase in ferrielectrics: $CuInP_2S_6$ and $Ag_{0.1}Cu_{0.9}InP_2S_6$ crystals, Phys. Status Solidi A 207 (2010) 1960–1967. Available from: http://dx.doi.org/10.1002/pssa.200925346.

[13] S. Havriliak, S. Negami, A complex plane analysis of α-dispersions in some polymer systems, J. Polym. Sci. Part C Polym. Symp. 14 (1966) 99–117. Available from: http://dx.doi.org/10.1002/polc.5070140111.

[14] J. Banys, S. Lapinskas, A. Kajokas, A. Matulis, C. Klimm, G. Völkel, et al., Dynamic dielectric susceptibility of the betaine phosphate (0.15) betaine phosphite (0.85) dipolar glass, Phys. Rev. B 66 (2002) 144113. Available from: http://dx.doi.org/10.1103/PhysRevB.66.144113.

[15] K.J. Button (Ed.), Infrared and Millimeter Waves V8: Electromagnetic Waves in Matter, Elsevier, the Netherlands, 1983.

[16] M.E. Lines, A.M. Glass, Principles and Applications of Ferroelectrics and Related Materials, OUP Oxford, 1977.

[17] E. Buixaderas, M. Savinov, M. Kempa, S. Veljko, S. Kamba, J. Petzelt, et al., Infrared and dielectric spectroscopy of the relaxor ferroelectric Sr 0.61 Ba 0.39 Nb 2 O 6, J. Phys. Condens. Matter 17 (2005) 653. Available from: http://dx.doi.org/10.1088/0953-8984/17/4/008.

[18] M.P. Moret, R. Zallen, R.E. Newnham, P.C. Joshi, S.B. Desu, Infrared activity in the Aurivillius layered ferroelectric $SrBi_2Ta_2O_9$, Phys. Rev. B 57 (1998) 5715.

[19] R. Blinc, B. Zeks, Soft Modes in Ferroelectrics and Antiferroelectrics, American Elsevier Pub. Co, Amsterdam, NY, 1974.

[20] S. Kamba, D. Nuzhnyy, P. Vaněk, M. Savinov, K. Knížek, Z. Shen, et al., Magnetodielectric effect and optic soft mode behaviour in quantum paraelectric $EuTiO_3$ ceramics, EPL Europhys. Lett. 80 (2007) 27002. Available from: http://dx.doi.org/10.1209/0295-5075/80/27002.

[21] F. Kremer, A. Schönhals (Eds.), Broadband Dielectric Spectroscopy, first ed., Springer-Verlag, Berlin Heidelberg, 2003, http://www.springer.com/us/book/9783540434078 (accessed 30.11.16).

[22] J. Banys, S. Lapinskas, S. Rudys, S. Greicius, R. Grigalaitis, High frequency measurements of ferroelecrics and related materials in coaxial line, Ferroelectrics 414 (2011) 64–69. Available from: http://dx.doi.org/10.1080/00150193.2011.577308.

[23] Y. Poplavko, Y. Prokopenko, V. Molchanov, V. Kazmirenko, Ferroelectrics study at microwaves, Ferroelectrics - Characterization and Modeling, InTech (2011) 203–226, http://dx.doi.org/10.5772/16709 (accessed 12.04.17).

[24] E. Nakamura, Anomalous dielectric behavior of KH_2PO_4 type crystals in the ferroelectric phase, Ferroelectrics 135 (1992) 237–247. Available from: http://dx.doi.org/10.1080/00150199208230027.

[25] Y.N. Huang, X. Li, Y. Ding, Y.N. Wang, H.M. Shen, Z.F. Zhang, et al., Domain freezing in potassium dihydrogen phosphate, triglycine sulfate, and CuAlZnNi, Phys. Rev. B 55 (1997) 16159–16167. Available from: http://dx.doi.org/10.1103/PhysRevB.55.16159.

[26] Y.L. Wang, A.K. Tagantsev, D. Damjanovic, N. Setter, Giant domain wall contribution to the dielectric susceptibility in $BaTiO_3$ single crystals, Appl. Phys. Lett. 91 (2007) 062905. Available from: http://dx.doi.org/10.1063/1.2751135.

[27] S. Wada, K. Muraoka, H. Kakemoto, T. Tsurumi, H. Kumagai, Enhanced piezoelectric properties of potassium niobate single crystals with fine engineered domain configurations, Mater. Sci. Eng. B 120 (2005) 186–189. Available from: http://dx.doi.org/10.1016/j.mseb.2005.02.033.

[28] C. Xu, Q. Li, Q. Yan, N. Luo, Y. Zhang, X. Chu, Domain structure evolutions during the poling process for [011]-oriented PMN–xPT crystals across the MPB region, J. Am. Ceram. Soc. (2016), n/a-n/a, http://dx.doi.org/10.1111/jace.14226.

[29] G. Arlt, D. Hennings, G. de With, Dielectric properties of fine-grained barium titanate ceramics, J. Appl. Phys. 58 (1985) 1619–1625. Available from: http://dx.doi.org/10.1063/1.336051.

[30] R. Theissmann, L.A. Schmitt, J. Kling, R. Schierholz, K.A. Schönau, H. Fuess, et al., Nanodomains in morphotropic lead zirconate titanate ceramics: on the origin of the strong piezoelectric effect, J. Appl. Phys. 102 (2007) 024111. Available from: http://dx.doi.org/10.1063/1.2753569.

[31] E. Buixaderas, V. Bovtun, M. Kempa, M. Savinov, D. Nuzhnyy, F. Kadlec, et al., Broadband dielectric response and grain-size effect in $K_{0.5}Na_{0.5}NbO_3$ ceramics, J. Appl. Phys. 107 (2010) 014111. Available from: http://dx.doi.org/10.1063/1.3273490.

[32] L. Jin, V. Porokhonskyy, D. Damjanovic, Domain wall contributions in $Pb(Zr,Ti)O_3$ ceramics at morphotropic phase boundary: a study of dielectric dispersion, Appl. Phys. Lett. 96 (2010) 242902. Available from: http://dx.doi.org/10.1063/1.3455328.

[33] M. Davis, D. Damjanovic, D. Hayem, N. Setter, Domain engineering of the transverse piezoelectric coefficient in perovskite ferroelectrics, J. Appl. Phys. 98 (2005) 014102. Available from: http://dx.doi.org/10.1063/1.1929091.

[34] T. Hoshina, Y. Kigoshi, S. Hatta, T. Teranishi, H. Takeda, T. Tsurumi, Size effect and domain-wall contribution of barium titanate ceramics, Ferroelectrics 402 (2010) 29–36. Available from: http://dx.doi.org/10.1080/00150191003697179.

[35] T. Tybell, P. Paruch, T. Giamarchi, J.-M. Triscone, Domain wall creep in epitaxial ferroelectric $Pb(Zr_{0.2}Ti_{0.8})O_3$ thin films, Phys. Rev. Lett. 89 (2002) 097601. Available from: http://dx.doi.org/10.1103/PhysRevLett.89.097601.

[36] A.S. Sidorkin, P. Saint-Gregoire, L.P. Nesterenko, Fatigue phenomena in thin ferroelectric films stimulated by repeated switching of the polarization, J. Adv. Dielectr. 03 (2013) 1350002. Available from: http://dx.doi.org/10.1142/S2010135X13500021.

[37] G. Catalan, J. Seidel, R. Ramesh, J.F. Scott, Domain wall nanoelectronics, Rev. Mod. Phys. 84 (2012) 119–156. Available from: http://dx.doi.org/10.1103/RevModPhys.84.119.

[38] B.L. Cheng, M. Gabbay, M. Maglione, Y. Jorand, G. Fantozzi, Domain walls motions in barium titanate ceramics, J. Phys. IV 06 (1996), C8-647-C8-650, http://dx.doi.org/10.1051/jp4:19968139.

[39] C. Kittel, Theory of the structure of ferromagnetic domains in films and small particles, Phys Rev. 70 (1946) 965–971. Available from: http://dx.doi.org/10.1103/PhysRev.70.965.

[40] G. Arlt, Twinning in ferroelectric and ferroelastic ceramics: stress relief, J. Mater. Sci. 25 (1990) 2655–2666. Available from: http://dx.doi.org/10.1007/BF00584864.

[41] V. Porokhonskyy, L. Jin, D. Damjanovic, Separation of piezoelectric grain resonance and domain wall dispersion in $Pb(Zr,Ti)O_3$ ceramics, Appl. Phys. Lett. 94 (2009) 212906. Available from: http://dx.doi.org/10.1063/1.3147166.

[42] R.R. Newton, A.J. Ahearn, K.G. McKay, Observation of the ferro-electric Barkhausen effect in barium titanate, Phys. Rev. 75 (1949) 103–106. Available from: http://dx.doi.org/10.1103/PhysRev.75.103.

[43] G. Arlt, N.A. Pertsev, Force constant and effective mass of 90° domain walls in ferroelectric ceramics, J. Appl. Phys. 70 (1991) 2283–2289. Available from: http://dx.doi.org/10.1063/1.349421.

[44] G. Arlt, U. Böttger, S. Witte, Emission of GHz shear waves by ferroelastic domain walls in ferroelectrics, Appl. Phys. Lett. 63 (1993) 602–604. Available from: http://dx.doi.org/10.1063/1.109962.

[45] V. Porokhonskyy, S. Kamba, A. Pashkin, M. Savinov, J. Petzelt, R.E. Eitel, et al., Broadband dielectric spectroscopy of $(1-x)BiScO_3-xPbTiO_3$ piezoelectrics, Appl. Phys. Lett. 83 (2003) 1605–1607. Available from: http://dx.doi.org/10.1063/1.1604945.

[46] S. Wada, K. Yako, K. Yokoo, H. Kakemoto, T. Tsurumi, Domain wall engineering in barium titanate single crystals for enhanced piezoelectric properties, Ferroelectrics 334 (2006) 17–27. Available from: http://dx.doi.org/10.1080/00150190600689647.

[47] A.P. Levanyuk, Phase transitions and domain structures in thin ferroelectric films, Phase Transit. 0 (2015) 1–10. Available from: http://dx.doi.org/10.1080/01411594.2014.986732.

[48] C. Lichtensteiger, S. Fernandez-Pena, C. Weymann, P. Zubko, J.-M. Triscone, Tuning of the depolarization field and nanodomain structure in ferroelectric thin films, Nano Lett. (2014). Available from: http://dx.doi.org/10.1021/nl404734z.

[49] A. Abragam, The Principles of Nuclear Magnetism, Clarendon Press, Oxford, 1961.

[50] D.A. Bussian, G.S. Harbison, Variable temperature 207Pb NMR of $PbTiO_3$, Solid State Commun 115 (2000) 95–98, http://dx.doi.org/10.1016/S0038-1098(00)00128-9.

[51] L.-S. Bouchard, A.O. Sushkov, D. Budker, J.J. Ford, A.S. Lipton, Nuclear-spin relaxation of Pb_{207} in ferroelectric powders, Phys. Rev. A 77 (2008) 022102. Available from: http://dx.doi.org/10.1103/PhysRevA.77.022102.

[52] B. Zalar, V.V. Laguta, R. Blinc, NMR evidence for the coexistence of order-disorder and displacive components in barium titanate, Phys. Rev. Lett. 90 (2003) 037601. Available from: http://dx.doi.org/10.1103/PhysRevLett.90.037601.

[53] R. Blinc, V. Laguta, B. Zalar, Field cooled and zero field cooled Pb_{207} NMR and the local structure of relaxor $PbMg_{1/3}Nb_{2/3}O_3$, Phys. Rev. Lett. 91 (2003) 247601. Available from: http://dx.doi.org/10.1103/PhysRevLett.91.247601.
[54] D.V. Azamat, A. Dejneka, J. Lancok, V.A. Trepakov, L. Jastrabik, A.G. Badalyan, Electron paramagnetic resonance studies of manganese centers in $SrTiO_3$: non-Kramers Mn^{3+} ions and spin-spin coupled Mn^{4+} dimers, J. Appl. Phys. 111 (2012) 104119. Available from: http://dx.doi.org/10.1063/1.4723653.
[55] V.V. Laguta, M.D. Glinchuk, I.P. Bykov, Y.L. Maksimenko, J. Rosa, L. Jastrabík, Impurity centers in $PbTiO_3$ single crystals: an electron-spin-resonance analysis, Phys. Rev. B 54 (1996) 12353–12360. Available from: http://dx.doi.org/10.1103/PhysRevB.54.12353.
[56] W.L. Warren, K. Vanheusden, D. Dimos, G.E. Pike, B.A. Tuttle, Oxygen vacancy motion in perovskite oxides, J. Am. Ceram. Soc. 79 (1996) 536–538. Available from: http://dx.doi.org/10.1111/j.1151-2916.1996.tb08162.x.
[57] W.L. Warren, B.A. Tuttle, E.C. Rong, G.J. Gerardi, E.H. Poindexter, Electron paramagnetic resonance investigation of acceptor centers in $Pb(Zr,Ti)O_3$ ceramics, J. Am. Ceram. Soc. 80 (1997) 680–684. Available from: http://dx.doi.org/10.1111/j.1151-2916.1997.tb02884.x.
[58] C.A. Randall, N. Kim, J.-P. Kucera, W. Cao, T.R. Shrout, Intrinsic and extrinsic size effects in fine-grained morphotropic-phase-boundary lead zirconate titanate ceramics, J. Am. Ceram. Soc. 81 (1998) 677–688. Available from: http://dx.doi.org/10.1111/j.1151-2916.1998.tb02389.x.
[59] S. Lin, T. Lü, C. Jin, X. Wang, Size effect on the dielectric properties of $BaTiO_3$ nanoceramics in a modified Ginsburg-Landau-Devonshire thermodynamic theory, Phys. Rev. B 74 (2006) 134115. Available from: http://dx.doi.org/10.1103/PhysRevB.74.134115.
[60] A.Y. Emelyanov, N.A. Pertsev, S. Hoffmann-Eifert, U. Böttger, R. Waser, Grain-boundary effect on the Curie-Weiss law of ferroelectric ceramics and polycrystalline thin films: calculation by the method of effective medium, J. Electroceramics 9 (2002) 5–16. Available from: http://dx.doi.org/10.1023/A:1021665300233.
[61] M.T. Buscaglia, M. Viviani, V. Buscaglia, L. Mitoseriu, A. Testino, P. Nanni, et al., High dielectric constant and frozen macroscopic polarization in dense nanocrystalline $BaTiO_3$ ceramics, Phys. Rev. B 73 (2006) 064114. Available from: http://dx.doi.org/10.1103/PhysRevB.73.064114.
[62] A.V. Polotai, A.V. Ragulya, C.A. Randall, Preparation and size effect in pure nanocrystalline barium titanate ceramics, Ferroelectrics 288 (2003) 93–102. Available from: http://dx.doi.org/10.1080/00150190390211972.
[63] Z. Hashin, S. Shtrikman, A variational approach to the theory of the effective magnetic permeability of multiphase materials, J. Appl. Phys. 33 (1962) 3125–3131. Available from: http://dx.doi.org/10.1063/1.1728579.
[64] S. Kirkpatrick, Percolation and conduction, Rev. Mod. Phys. 45 (1973) 574–588. Available from: http://dx.doi.org/10.1103/RevModPhys.45.574.
[65] C.E. Ciomaga, C.S. Olariu, L. Padurariu, A.V. Sandu, C. Galassi, L. Mitoseriu, Low field permittivity of ferroelectric-ferrite ceramic composites: experiment and modeling, J. Appl. Phys. 112 (2012) 094103. Available from: http://dx.doi.org/10.1063/1.4764037.
[66] D. Schmidt, M. Schubert, Anisotropic Bruggeman effective medium approaches for slanted columnar thin films, J. Appl. Phys. 114 (2013) 083510. Available from: http://dx.doi.org/10.1063/1.4819240.

[67] A. Sakanas, R. Grigalaitis, M. Ivanov, J. Banys, L. Mitoseriu, V. Buscaglia, et al., The alternative expression of Lichtenecker's logarithmic mixture formula and its application to the broadband dielectric spectroscopy of $BaTiO_3$-$Ni_{0.5}Zn_{0.5}Fe_2O_4$ composites, Ferroelectrics 479 (2015) 90–97. Available from: http://dx.doi.org/10.1080/00150193.2015.1031618.

[68] J. Petzelt, I. Rychetsky, D. Nuzhnyy, Dynamic ferroelectric–like softening due to the conduction in disordered and inhomogeneous systems: giant permittivity phenomena, Ferroelectrics 426 (2012) 171–193. Available from: http://dx.doi.org/10.1080/00150193.2012.671732.

[69] N.J. Kidner, N.H. Perry, T.O. Mason, E.J. Garboczi, The brick layer model revisited: introducing the nano-grain composite model, J. Am. Ceram. Soc. 91 (2008) 1733–1746. Available from: http://dx.doi.org/10.1111/j.1551-2916.2008.02445.x.

[70] I. Rychetsky, J. Petzelt, Dielectric spectra of grainy high-permittivity materials, Ferroelectrics 303 (2004) 137–140. Available from: http://dx.doi.org/10.1080/00150190490453072.

[71] D.S. McLachlan, Analytical functions for the DC and AC conductivity of conductor-insulator composites, J. Electroceramics 5 (2000) 93–110. Available from: http://dx.doi.org/10.1023/A:1009954017351.

[72] P. Papet, J.P. Dougherty, T.R. Shrout, Particle and grain size effects on the dielectric behavior of the relaxor ferroelectric $Pb(Mg_{1/3}Nb_{2/3})O_3$, J. Mater. Res. 5 (1990) 2902–2909. Available from: http://dx.doi.org/10.1557/JMR.1990.2902.

[73] J.M. Kiat, C. Bogicevic, P. Gemeiner, A. Al-Zein, F. Karolak, N. Guiblin, et al., Structural investigation of strontium titanate nanoparticles and the core-shell model, Phys. Rev. B 87 (2013) 024106. Available from: http://dx.doi.org/10.1103/PhysRevB.87.024106.

[74] T. Hoshina, S. Wada, Y. Kuroiwa, T. Tsurumi, Composite structure and size effect of barium titanate nanoparticles, Appl. Phys. Lett. 93 (2008) 192914. Available from: http://dx.doi.org/10.1063/1.3027067.

[75] N.A. Pertsev, A.G. Zembilgotov, A.K. Tagantsev, Effect of mechanical boundary conditions on phase diagrams of epitaxial ferroelectric thin films, Phys. Rev. Lett. 80 (1998) 1988–1991. Available from: http://dx.doi.org/10.1103/PhysRevLett.80.1988.

[76] S. Bagdzevicius, R. Mackeviciute, M. Ivanov, B. Fraygola, C.S. Sandu, N. Setter, et al., Internal electrical and strain fields influence on the electrical tunability of epitaxial $Ba_{0.7}Sr_{0.3}TiO_3$ thin films, Appl. Phys. Lett. 108 (2016) 132901. Available from: http://dx.doi.org/10.1063/1.4944997.

[77] N.A. Pertsev, A.K. Tagantsev, N. Setter, Phase transitions and strain-induced ferroelectricity in $SrTiO_3$ epitaxial thin films, Phys. Rev. B 61 (2000) R825–R829. Available from: http://dx.doi.org/10.1103/PhysRevB.61.R825.

[78] K.J. Choi, Enhancement of ferroelectricity in strained $BaTiO_3$ thin films, Science 306 (2004) 1005–1009. Available from: http://dx.doi.org/10.1126/science.1103218.

[79] R. Mackevičiūtė, Š. Bagdzevičius, M. Ivanov, B. Fraygola, R. Grigalaitis, N. Setter, et al., Strain engineering of electrical conductivity in epitaxial thin $Ba_{0.7}Sr_{0.3}TiO_3$ film heterostructures, Lith. J. Phys. 56 (2016) 173–181. Available from: http://dx.doi.org/10.3952/physics.v56i3.3366.

[80] I. Grinberg, D.V. West, M. Torres, G. Gou, D.M. Stein, L. Wu, et al., Perovskite oxides for visible-light-absorbing ferroelectric and photovoltaic materials, Nature 503 (2013) 509–512. Available from: http://dx.doi.org/10.1038/nature12622.

[81] W. Känzig, Space charge layer near the surface of a ferroelectric, Phys. Rev. 98 (1955) 549–550. Available from: http://dx.doi.org/10.1103/PhysRev.98.549.

[82] A. Crassous, T. Sluka, C.S. Sandu, N. Setter, Thickness dependence of domain-wall patterns in $BiFeO_3$ thin films, Ferroelectrics 480 (2015) 41–48. Available from: http://dx.doi.org/10.1080/00150193.2015.1012414.
[83] L.W. Martin, A.M. Rappe, Thin-film ferroelectric materials and their applications, Nat. Rev. Mater 2 (2016) 16087. Available from: http://dx.doi.org/10.1038/natrevmats.2016.87.

Perovskite and Aurivillius: Types of ferroelectric metal oxides

2

Mirjana M. Vijatovic Petrovic and Jelena D. Bobic
Institute for Multidisciplinary Research, University of Belgrade, Belgrade, Serbia

2.1 Introduction

Crystal structures can be divided into 32 classes, or point groups, according to the number of rotational axes and reflection planes they exhibit that leave the crystal structure unchanged. These 32 classes are divided into seven crystal systems: cubic, hexagonal, tetragonal, orthorombic, trigonal, monoclinic, and triclinic. Twenty-one of the thirty-two crystal classes do not possess a center of symmetry (which is the main characteristic for being a piezoelectric), from which twenty are piezoelectric. Any material develops a dielectric polarization when an electric field is applied, but a substance with such a natural charge separation even in the absence of a field is called a polar material. Piezoelectric materials are polarized by the applied pressure, and in pyroelectrics spontaneous polarization occurs. Under normal circumstances, even polar materials do not display a net dipole moment. As a consequence, there are no electric dipole equivalents of bar magnets because the intrinsic dipole moment is neutralized by a free electric charge that builds up on the surface by internal conduction or from the ambient atmosphere. The possibility of inorganic crystals being polar (pyroelectric or piezoelectric) is strictly a function of their point group symmetry [1]. Ferroelectric materials are the subgroup of spontaneous polarized pyroelectrics, possessing the spontaneous polarized dipoles that can be reversed by the applied electric field.

Ferroelectric oxides could be classified into four structure types: perovskite, pyrochlore, tungsten-bronzes, and bismuth layer-structured compounds (BLSF, so-called Aurivillius). The perovskite and layered perovskite structures can incorporate ions of a variety of sizes and charges allowing subtle distortions that can ease the bond strains created by size mismatch. The perovskite structure has shown great flexibility of composition, incorporating nearly every member of the periodic table (Fig. 2.1). The basic criterion for the structure to be stable is that the valences of the ions should balance and that the ionic radii meet the Goldschmidt criteria [2]. The factor that predicts the stability of the structure is a tolerance factor defined by a Goldschmidt equation:

$$R_A + R_O = t\sqrt{2}(R_B + R_O), \tag{2.1}$$

where t is a tolerance factor and R_A, R_B, and R_O are ionic radii of A, B, and oxygen ions, respectively. In the case of ideal cubic structure, the tolerance factor is 1.

Magnetic, Ferroelectric, and Multiferroic Metal Oxides. DOI: http://dx.doi.org/10.1016/B978-0-12-811180-2.00002-5

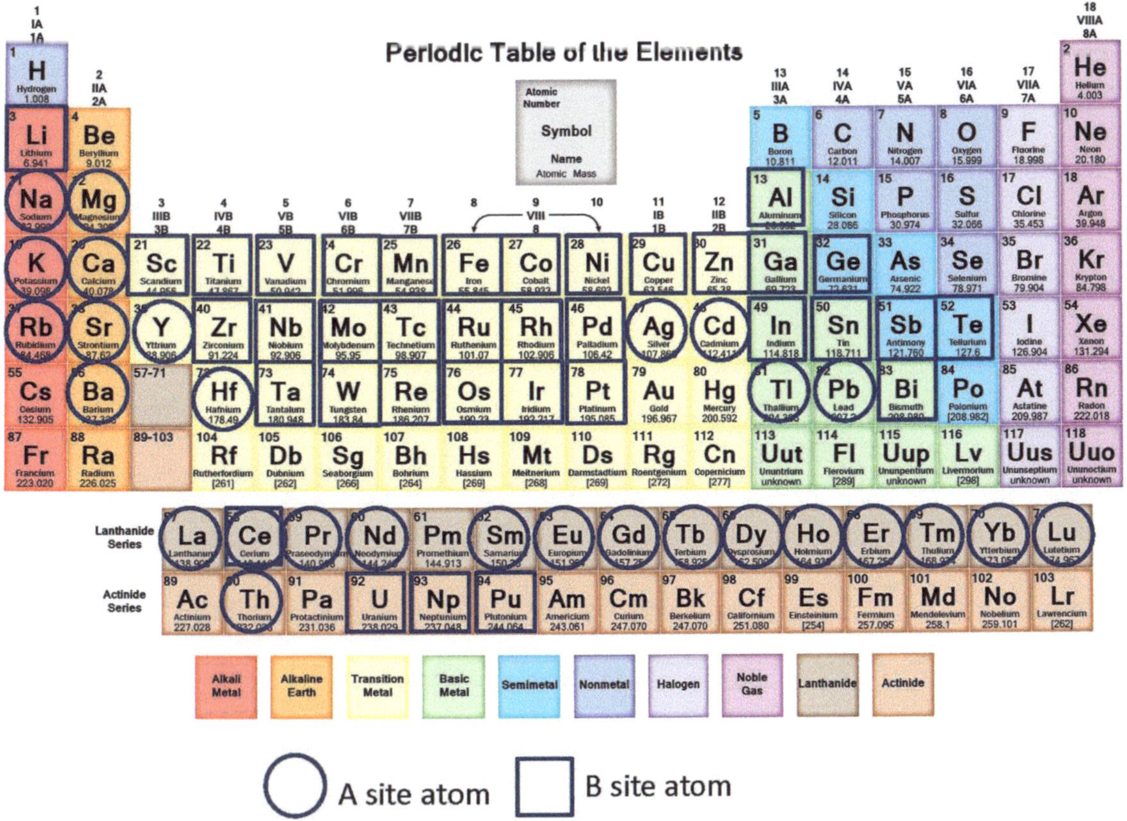

Figure 2.1 The periodic table of elements with marked elements found in the perovskite structure.

Lower values of t will lower the symmetry of the crystal structure. The more the tolerance factor value is closer to 1, the structure is more like a cubic. The structure is stable if the value of t is between 0.85 and 1.06; perovskites with $t < 1$ are ferroelectrics [3]. This factor can be applied to the Aurivillius structure in which the stability region is smaller ($0.81 < t < 0.89$), being determined not only by the stability of the perovskite layer but also by the stability of the bismuth oxide layer. Since perovskites are not truly ionic compounds and since the t values also depend on what values are taken for the ionic radii, the tolerance factor is only a rough estimate [4].

In this chapter perovskite and Aurivillius structure will be presented as well as the most representative compounds with those structures, such as $BaTiO_3$ and $Bi_4Ti_3O_{12}$. General overview of possible dopants and their influence on structure, microstructure, and some properties of these compounds are discussed.

2.2 Perovskite structure

The perovskite structure takes its name from the mineral perovskite, $CaTiO_3$, which was first found in the Ural Mountains in 1839, and named after Lev Perovski who was the founder of the Russian Geographical Society. A perovskite structure is any compound that has the same crystallographic structure as the perovskite mineral and generic formula ABX_3. The general crystal structure is a primitive cube, with the A-larger cation in the corner, the B-smaller cation in the middle of the cube, and the anion, commonly oxygen, in the center of the face edges, where A is a monovalent, divalent, or trivalent metal and B a pentavalent, tetravalent, or trivalent element, respectively [5].

Geometrical packing of ions in the lattice is a very important factor determining the structure type. The perovskite structure can be considered a three-dimensional framework of BO_6 octahedra, but it can also be regarded as a cubic close-packed arrangement of A and O ions, with the B ions filling the interstitial positions (Fig. 2.2) [7]. The coordination number of cation A is 12 and for cation B is 6.

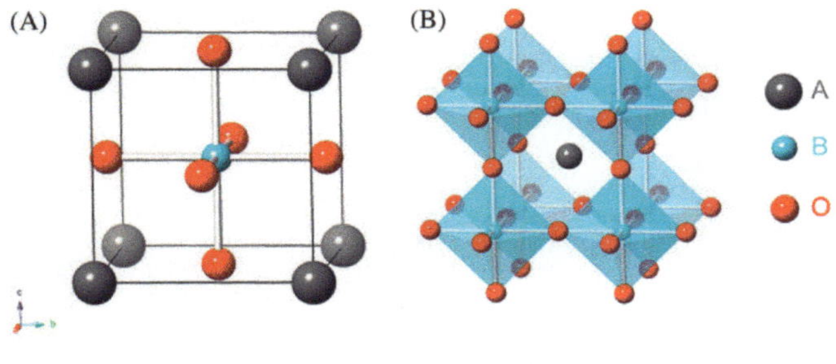

Figure 2.2 The unit cell of cubic barium titanate presented in two different ways: (A) cubic arrangement of A and O ions with B ions in the interstitial position, (B) as a three-dimensional structure consisting of BO_6 octahedrons.
Reprinted from [6] H.Du, X. Li, Z. Xu, D. Chu, Electric double-layered transistors: a review of recent progress, J Mat. Sci., 50 (2015) 5641–5673, Copyright (2015) with permission from Springer.

As it is presented in the periodic table of elements in the introduction section, it can be noticed that a large number of elements are found to be part of the perovskite family of compounds. The most investigated are titanates, then galates, cerates, zirconates, stanates, and the rare-earth perovskites. Since a lot of materials belong to this group of compounds a variety of applications are found. Materials with perovskite structure, their properties, and potential application are presented in Table 2.1.

One of the most interesting and the most investigated perovskites is barium titanate, $BaTiO_3$ (BT). Its ferroelectric properties are connected with a series of three structural phase transitions. The Curie point T_c of BT was found at 120°C. Above 120°C, the original cubic cell is stable up to 1460°C with the space group Pm-3m. Above this temperature a hexagonal structure is stable [11]. When the temperature is below the Curie point, crystallographic changes in BT occur; first, at about 120°C a ferroelectric transition between the cubic, paraelectric, and ferroelectric phase of tetragonal structure (space group P4mm) takes place. At 5°C, the transition to a phase of the orthorhombic structure (with Amm2 space group) goes on and at −90°C goes to the low temperature phase having a trigonal structure (space group-R3m) [1,12]. Fig. 2.3 illustrates crystallographic changes of $BaTiO_3$ [2]. All of these ferroelectric phases, presented in Fig. 2.4, have a spontaneous polarization based on a significant extent on movement of the Ti atom in the O_6 octahedra in the following way (using pseudocubic notation). At the Curie point Ti ions are all in equilibrium positions in the center of their octahedra, but with a decrease of the temperature, Ti ions jump between energetically favorable positions out of the octahedron center, as shown in Fig. 2.4. These changes are related to structural distortion, lengthening or shortening of the bonds, so crystallographic dimensions of the BT lattice change with temperature [13]. A lot of papers indicate that the mechanisms responsible for these phase transitions in $BaTiO_3$ are rather complex. The variation of $BaTiO_3$ lattice parameters at different cooling temperatures is showing three phase transition areas. The exact transition temperatures and amount of distortion are a function of cooling rate, residual stresses, and purity of the material [8]. It can also depend on

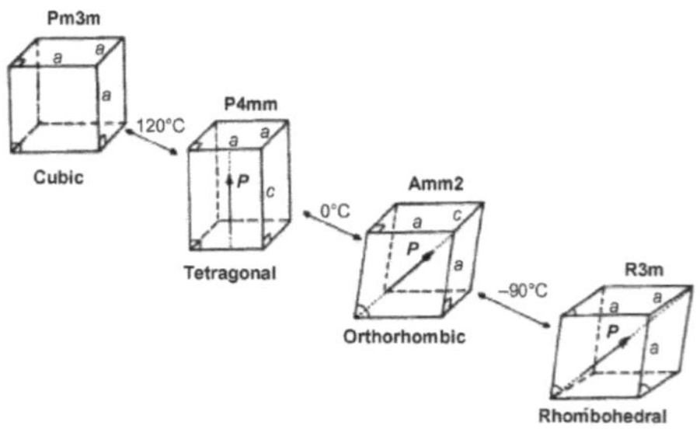

Figure 2.3 Four structural phases of barium titanate with appropriate space groups. Reprinted from A.J. Moulson, J.M. Herbert, Electroceramics, John Wiley & Sons Ltd, 2003, Copyright (2003) with permission from John Wiley and Sons.

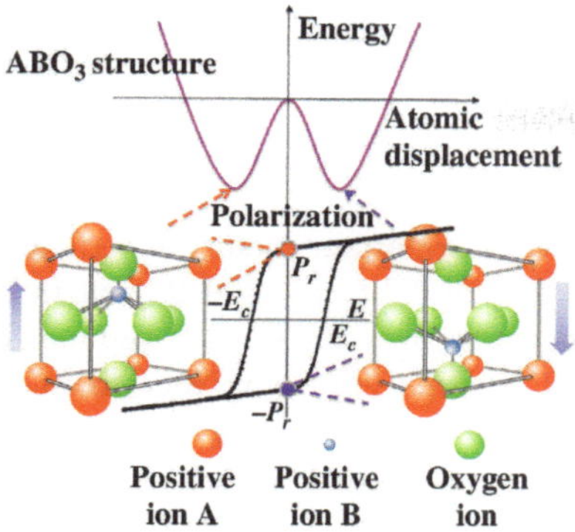

Figure 2.4 Illustration of unit cell structure of typical ferroelectric perovskite and how spontaneous polarization is formed when the temperature drops below Curie point at ferroelectric phase.
Reproduced from [12] F. Yang, G. Hu, B. Xu, W. Wu, C. Yang, H. Wu, Statistical mechanical origin of hysteresis in ferroelectrics, J. Appl. Phys., 112 (2012) 034113, with the permission of AIP Publishing.

synthesis methods, sintering temperature, grain size, number and type of defects and impurities, and so on. The size ratio of the *c*-axis to *a*-axis for pure $BaTiO_3$ is approximately 1.01 at room temperature, and therefore, the spontaneous strain for $BaTiO_3$ is approximately 1% at room temperature.

2.2.1 Substitutions in the barium titanate lattice

Perovskite crystal structure is very flexible and it has the capabilty to host ions with higher and lesser oxidation states at a particular site in the lattice. Ordering of these ions in the perovskite crystal lattice affects the symmetry of the structure, microstructure development, and therefore the mechanical and electrical properties of the obtained materials. By doping, the material's electrical properties can be significantly modified. Pure BT is an electrical insulator at room temperature. The presence of dopants in the BT lattice could cause lowering of the material's electrical resistivity, and dopant concentration is a very important factor when doping is involved. It can lead to the appearance of semiconductivity in the material, while reaching a critical concentration turns the material into insulator again. The positive temperature coefficient of resistivity (PTCR) effect is a consequence of grain boundary phenomena and it is typical for semiconducting BT. As it was mentioned, lots of added dopants can induce the PTCR effect in the material [14,15].

Based on available literature data, some of the possible conclusions regarding substitutions in the BT lattice can be presented:

1. Ba^{2+} ions can usually be substituted by monovalent, divalent, and trivalent cations with ion radii 1–1.4 Å (Ag^{+}, Sr^{2+}, Pb^{2+},Ca^{2+}, Cd^{2+}, Sb^{3+}, La^{3+}, Y^{3+}, Pr^{3+}, Sm^{3+}, Ce^{3+}, Bi^{3+}, etc.).

2. Ti^{4+} ions can be replaced by trivalent, tetravalent, and pentavalent cations in with the ionic radii 0.45–0.75 Å (Co^{3+}, Ga^{3+}, W^{3+}, Mn^{3+}, Mo^{3+}, Sn^{4+}, Hf^{4+}, Zr^{4+}, Ce^{4+}, Th^{4+}, V^{5+}, Ta^{5+}, Nb^{5+}, Bi^{5+}, Sb^{5+}, Mo^{6+}, W^{6+}, Mn^{6+}, etc.).
3. O^{2-} ions are very hard for the substitution in comparison with Ba and Ti ions, but partial substitution with Cl^-, F^-, and S^{2-} is possible.

The ion's radii are a very important factor as well as the concentration. It is quite important to know the solubility of used dopant in the BT lattice. It means that there is a critical concentration when A- or B-site ions cannot be further substituted by dopant ions and dopants are being located in the interstitial positions, at the grain boundaries, and so on, influencing the electrical properties of obtained materials [16].

From the literature data, in the pure BT powders obtained via different kinds of methods the cubic structure is usually found. If BT is doped with a very small amount of dopants, usually less than 0.1 mol%, crystal changes cannot be noticed from the standard X-ray diffraction (XRD) analysis. To obtain dense BT ceramics, pressed samples are being sintered at high temperatures, most frequently between 1250°C and 1400°C. In the sintering process, the physical and chemical changes occur in the materials and influence the density and microstructure development. Defect structure developed during this process determines not only the electronic but the mass transport. Sometimes it can significantly reduce or increase the diffuseness coefficient that influences the sintering mechanism [16–18]. Powder traits such as particle size, particle size distribution, shape of the particles, agglomeration factor, dopants concentration as well as the atmosphere, temperature, sintering time, and rate [18] have a significant influence on the sintering process. In order to produce fine-grained microstructure it is necessary for the starting powder to have small particles with narrow size distribution, to be nonagglomerated, and to have high purity. When material is being doped, careful selection of the dopant is required. Favorable are dopants that can influence the increase of ions defects, which can increase the density and grain growth inhibition. The atmosphere can be of great importance since the reaction between the atmosphere, dopants, and primary particles can lead to elimination of the remaining porosity. Controlled sintering temperature rate affects the densification, defects concentration, and grain growth kinetics.

In the study of BT powders, pure and doped with La and Sb (0.01, 0.03, and 0.05 mol%, each dopant separately), XRD analysis has shown the formation of cubic crystal structure with space group Pm-3m. Such a small amount of dopants cannot be clearly noticed in the XRD diagrams but the calculated crystallite size of Sb-doped samples was ~17 nm and was lower than in pure and La-doped BT 20–24 nm. This can be connected with ionic radii of Sb, which is smaller than La and causes shortening of the bonds in the lattice. Available research data have shown that the solubility of donor elements in the crystal lattice of BT is generally quite high and reaches the values of 10 at% [19,20]. The authors working on the dopants incorporation in the crystal lattice of BT have the same conclusion. The addition of dopants leads to a change of lattice parameters due to overall elongation and shrinkage of the BT unit cell. With the dopant concentration increase, the tetragonality c/a, the decrease is in agreement with literature data found for BT doped with different dopants [21–27]. The splitting of reflections and appearance of new ones are the indication of the symmetry change. This is especially evident from the splitting of the [002] and [200], [102] and [201], and [112] and [211] diffraction peaks at around $2\Theta = 45°$, $51°$, and $56°$ onto two peaks and it is indicative of the fact that materials possess tetragonal structure. The irregularities in the ions packing in the crystal lattice can cause the change into a lower symmetry. Therefore, incorporation of dopants into the BT crystal lattice can cause the emergence of irregularities in the lattice and affect the symmetry change. In the homogeneous and fine-grained materials, the appearance of mechanical stress in the grains leads to a stabilization of the cubic structure [21]. Buscaglia et al. [19] connected the lowering of the symmetry of XRD peaks with the existence of two structurally different regions, cubic and tetragonal, in the doped BT grains. Thus, it is a connection between the dopant's incorporation in the crystal lattice and the transition of the material from the paraelectric to ferroelectric phase.

Wu et al. [28] connected the symmetry lowering with grain size increase. They have been studying pure BT ceramics with different grain sizes and showed that in the ceramics with the largest grains the tetragonality was the highest and it decreased when the concentration of smaller grains in the sample increased. The grain size effect can also be valid in the doped samples, since particular dopants caused the inhibition of grain growth. Therefore, the addition of dopants can cause the lowering of grain size and as a consequence, the lowering of the symmetry occurs and the structure changes from tetragonal to pseudocubic.

High-density nanostructured ceramics is very hard to obtain due to many factors that can influence the process of ceramics preparation. The highest impact results from the chosen synthesis method, the degree of agglomeration and sintering regime (temperature, time, and atmosphere). Dopant incorporation also has a significant effect on densification and microstructure of BT. The optimization of the sintering process is very important. If the sintering temperature is too high the secondary phases can appear in the material and nonhomogeneous microstructure can be formed. When sintering time is varied, a big difference between ceramics grain size, grain shape, homogeneity, porosity, and such can be noticed. For the ceramics doped with La and Sb and sintered for 2, 4, and 8 h, it was found that sintering for 8 h enabled the microstructure development and good densification of the materials. All ceramics possessed the polygonal grains but with differences in grain size and in number of pores. On the other hand, the porosity of the samples doped with La is quite high in comparison with ceramics doped with Sb. Doping with Sm has an effect similar to La doping. Some dopants, such as Sb, also have a much stronger inhibition effect in comparison with others. Sb-doped ceramics consist of much smaller grains and also bigger grains surrounded by smaller grains. The change of dopant concentration hardly influences the discontinuous grain growth until it comes to a certain concentration of dopants, after which the grain growth is being blocked [15]. A lot of different dopants are grain growth inhibitors—Nb, Dy, Sc, Ni, Zr [24,25,29]—but there are also dopants that promote grain growth as in the case of W [27]. The influence on density is quite different. The density of ceramics doped with Sb, Nb, or Ta are even higher than for pure BT obtained by the same method [22,24], in comparison with La, Sm, and Zr, in which the density is lower and porosity highly developed [26,29]. It can be shown that deliberate substitutions of ions in the BT lattice with a particular kind of dopant could enable tailoring of crystal structure and microstructure and obtaining the materials with the desired properties (Table 2.1).

Table 2.1 Materials with perovskite structure and their properties and application [8–10]

Material	Properties	Application
$BaTiO_3$	Ferroelectricity, piezoelectricity, high dielectric constant	Multilayer ceramic capacitor, PTCR resistor, embedded capacitance
$(Ba,Sr)TiO_3$	Nonlinear dielectric properties	Tunable microwave devices, pyrodetector
$PbTiO_3$	Pyroelectricity, piezoelectricity	Acoustic transducer, pyrodetector
$Pb(Zr,Ti)O_3$	Dielectric, ferroelectricity, piezoelectricity, electrooptic	Piezoelectric transducers and actuators, ferroelectric memories, acoustic wave device, substrate waveguide device
$(Pb,La)(Zr,Ti)O_3$	Pyroelectricity, electrooptic, Transparent ferroelectric	Pyrodetector, waveguide device, optoelectronic devices
$Bi_4Ti_3O_{12}$	Ferroelectricity, high dielectric constant	High-temperature actuators
$LiNbO_3$	Piezoelectricity	Pyrodetector, surface acoustic wave device
$K(Ta,Nb)O_3$	Piezoelectricity, electrooptic	Lead-free piezoceramics
$PbMg_{1/3}Nb_{1/3}O_3$	Relaxor ferroelectrics	Capacitors, actuators, memory
$BiFeO_3$	Magnetoelectric coupling, high Curie temperature	Magnetic field detectors, memories
$LaGaO_3$	Oxide–ion conduction	Electrolyte in solid oxide fuel cells (P-SOFCs)
$BaCeO_3$, $SrCeO_3$	Proton conduction	Electrolyte in protonic solid oxide fuel cells (P-SOFCs), hydrogen sensor
$LaAlO_3$, $YAlO_3$	Host materials for rare-earth luminescent ions	Lasers, substrates for epitaxial film deposition

2.3 Aurivillius type of ferroelectric metal oxides

2.3.1 Crystal structure of the Aurivillius type of compounds

Aurivillius oxides are a family of materials first synthesized by Bengt Aurivillius in 1949 [30]. The Aurivillius phases may be described structurally as intergrowths of fluorite-like $(Bi_2O_2)^{2+}$ layers with perovskite $(A_{n-1}B_nO_{3n+1})^{2-}$ layers ($n = 1$, 2, 3, 4), where n is a number of perovskite slabs (oxygen octahedrons). The 12-coordinate perovskite-like A-site is typically occupied by large cations (from 1.34 to 1.61 Å) such as Na^+, K^+, Ca^{2+}, Sr^{2+}, Ba^{2+}, Pb^{2+}, Bi^{3+}, or Ln^{3+} and the 6-coordinate perovskite-like B-site by smaller cations (from 0.59 to 0.65 Å) such as Fe^{3+}, Cr^{3+}, Ti^{4+}, Nb^{5+}, or W^6. Depending on the number of pseudoperovskite units between $(Bi_2O_2)^{2+}$ layers, Aurivillius compounds can be classified into different groups. The compounds with an odd number of perovskite layers crystallized in *B2cb* structure and even-layered members prefer $A2_1ma$ crystal structure. The third group contains the structures with two different values of n by unit cell unit or intergrowth compounds. This type of compound requires introducing a new parameter, m, so the general formula can be presented as: $(Bi_2O_2)^{2+}(A_{n-1}B_nO_{3n+3})^{2-}(Bi_2O_2)^{2+}(A_{m-1}B_mO_{3m+3})^2$ [31]. The tolerance factor introduced by Goldschmidt (Eq. 2.1) can be also applied to the Aurivillius structure, but the stability of the region is smaller. It is determined not only by the stability of the perovskite layer, but also by the need to fit the same lateral dimensions in the $(Bi_2O_2)^{2+}$ layer and the adjacent pseudoperovskite layers [32]. As already mentioned, the tolerance factor of compounds with perovskite structure varies between 0.77 and 1.01; however, the value of t in the layered structures is reduced to the interval 0.81−0.89 with t value decreasing with n [4].

Typical examples of Aurivillius compounds with a different number of perovskite units are Bi_2WO_6 ($n = 1$), $Bi_2SrTa_2O_9$ ($n = 2$), $Bi_4Ti_3O_{12}$ ($n = 3$), $SrBi_4Ti_4O_{15}$ ($n = 4$), and $Sr_2Bi_4Ti_5O_{18}$ ($n = 5$). Schematic drawings of these structures are presented in Fig. 2.5 [33]. Within this huge number of different compounds, bismuth titanate ($Bi_4Ti_3O_{12}$−BIT) is one of particular interest as the first and most

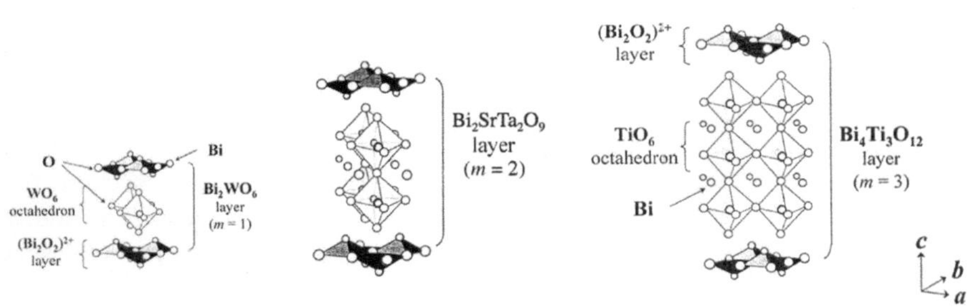

Figure 2.5 Schematic drawings of Bi_2WO_6 ($n = 1$), $Bi_2SrTa_2O_9$ ($n = 2$) and $Bi_4Ti_3O_{12}$ ($n = 3$) structure.
Reprinted from S. Luoa, Y. Noguchia, M. Miyayama, T. Kudo, Rietveld analysis and dielectric properties of Bi2WO6Bi4Ti3O12 ferroelectric system, Mater. Res. Bull. 36 (2001) 531−540, Copyright (2001) with permission from Elsevier.

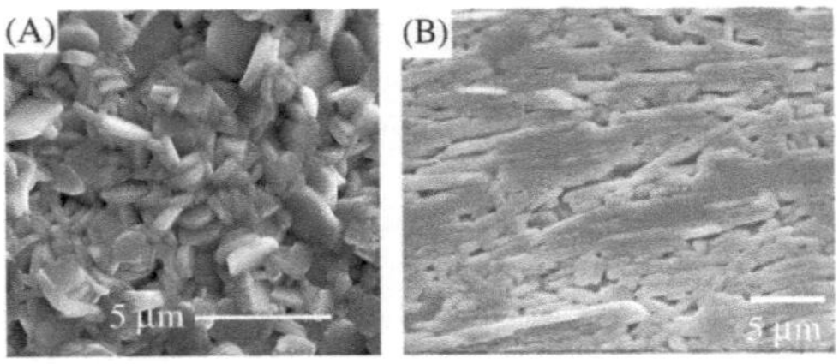

Figure 2.6 Plate-like grains of Aurivillius family of compounds: (A) randomly arranged, (B) textured grains.
Reprinted from Y. Sakuma, T. Kimura, Mechanisms of texture development in bismuth layer-structured ferroelectrics prepared by templated grain growth, J. Electroceramics, 13 (2004) 537–541, Copyright (2004) with permission from Springer.

investigated member of this large family. $Bi_4Ti_3O_{12}$ presents monoclinic symmetry at a temperature below the Curie temperature that allows two different possible directions of polarization. However, the structure of BIT is usually considered as orthorhombic, although the choice of the axes does not follow the conventional $c < a < b$ assumed in the orthorhombic system. The other compounds belonging to Aurivillius family exhibit orthorhombic or tetragonal symmetry [34,35].

A typical characteristic of all BLSFs is plate-like formation of the grains (Fig 2.6A). The layered structure is described by a high value of the *c*-parameter in comparison with parameters *a* and *b* of the orthorhombic cell and leads to a high anisotropy of the crystallographic structure, which is directly connected with the high anisotropy of the electric, piezoelectric, and ferroelectric properties [8,34]. Fig. 2.6B shows textured ceramics with a high degree of grain orientation [36]. These ceramics have higher remnant polarization and d_{33} than randomly oriented ceramics, but at the same time it has fewer disadvantages. First, the piezoelectric properties are not homogeneous in any polarization direction and second, specific manufacturing methods are needed to obtain this kind of ceramics. A main undesired feature of texture ceramics is that polarization and electrical conductivity are maximized in the same *ab* direction.

There are no such a number of reports about adjusting the electrical properties of Aurivillius compounds by controlling the microstructure. It is known that electrical conductivity increases with the increase the size of the platelets, so it is important for large anisotropic grain growth to be strongly controlled [32,37,38]. Thus, the desired characteristics of the starting powder become a critical issue. It is important to use appropriate thermal treatment during the preparation process because high calcinated Aurivillius powders possess a lamellar morphology, which makes it difficult to produce dense ceramics since a high degree of packing of the particles is not easy by the normal cold-pressing.

2.3.2 Substitution in Aurivillius type of structure

A feature of the Aurivillius compounds, similar to the case of classical perovskites, resides in their compositional flexibility of the perovskite block, which allows the incorporation of various cations. Doping-effects investigation of the structure and properties of BLSF compounds have been made by many authors. The electron configuration of ions in the unit cell and polarizability of the cations ***A*** and B play an important role. The ions in perovskite A position can be substituted with alkaline, alkaline earth and rare-earth ions being the largest cations with greater solubility limits. The B position could also accept ions, but with ionic radii between 0.58 and 0.65 Å such as W^{6+}, Nb^{5+}, Fe^{3+}, V^{5+}, and Zr^{4+}.

In the cases when cations ***A*** are different (like in the case of $BaBi_2Nb_2O_9$, $BaBi_4Ti_4O_{15}$, etc.), the concept of an average ionic radius is applicable if the sizes of these ions do not considerably differ. There are different opinions about substitution of Bi^{3+} ions by other ions in the bismuth-oxygen layers. According to investigation of Subbarao et al. [39], Bi^{3+} ions in $[Bi_2O_2]^{2+}$ layers cannot be replaced without destroying the structure. Tellier et al. [40] demonstrated that although Ba is partially disordered into the Bi_2O_2 layers in $BaBi_4Ti_4O_{15}$, the Ca in $CaBi_4Ti_4O_{15}$ is not. Ismaialzet's investigation showed that in the bismuth-oxygen layer only one of the two Bi^{3+} ions can be substituted, and only by trivalent cations [41]. However, many authors agree that in structural refinements of $n = 1$, 2, and 3 Aurivillius structures the M-site in the $(M_2O_2)^{2+}$ layer can be occupied by Pb^{2+}, and also by cations without a lone pair, such as Ba^{2+}, Sr^{2+}, and Ca^{2+} [42–46].

It is believed that the chemical stability of the perovskite layer against oxygen vacancies is drastically improved with isovalent A substitution of $Bi^{3+}$for La^{3+}. Opinions are divided regarding La^{3+} substitution at the Bi^{3+} site. Some authors claim that La ions incorporate just into the perovskite lattice at the place of Bi^{3+} but not to the $(Bi_2O_2)^{2+}$ layers [47–49]. Others consider that La cations occupy the A-site in the perovksite layers as well as the Bi^{3+} sites in the bismuth oxide layer [50–52]. That was confirmed by Raman spectroscopy (especially for higher concentration of dopands) by Osada and coworkers. Therefore, a partial cation mixing can occur at M- and A-sites, which is referred to as the *cation disorder*. Bi and La have the same valence (+3) but slightly different ionic radii (0.103 Å for Bi^{3+} and 0.106 Å for La^{3+}). Some authors in earlier research showed that La substitution in $Bi_{4-x}La_xTi_3O_{12}$ (BLT) changes from an orthorhombic phase ($x < 0.75$) to a tetragonal phase for $x > 0.75$ [53,54], while recent investigation showed just orthorhombic distortion that is attributed to the smaller displacements of the constituent ions from the corresponding positions [52,55]. This distortion results in a decrease of Curie temperature.

Donor doping with ions such as V^{5+}, Nb^{5+}, and W^{6+} could also effectively enhance the ferroelectric properties of BIT and other Aurivillius compounds by inducing the distortion of oxygen octahedral and reducing the oxygen vacancy concentration [56–59]. B-site substitution in BIT does not change the lattice symmetry since the radius of the dopant is not significantly different from the Ti^{4+} ion although the small invariance of the crystal structure is expected [60]. The most effective dopant is niobium ($Bi_4Ti_{2.96}Nb_{0.04}O_{12}$), which decreases the conductivity in three orders of magnitude, although this composition with the lowest conductivity did not possess the highest piezoelectric response [61]. Decreasing the conductivity in BIT is a necessary but not sufficient condition for achieving a useful piezoelectric response.

In all Aurivillius bismuth-based compounds the loss of Bi_2O_3 during sintering is a potential problem resulting in nonstoichiometry that leads to undesirable conductivity effects, which tend to degrade the performance at high temperature. Often, the sintering is performed in sealed alumina crucibles in order to avoid bismuth loss. Also, introducing a slight excess of Bi_2O_3 is found to act as a sintering aid in suppressing the bismuth and oxygen vacancies and stabilizing the bismuth layer-structured phase in BLSFs. SEM analyses confirm that the increase in Bi excess favored the grain growth in the ceramic. For different Aurivillius compounds the various amounts of bismuth aid, together with appropriate sintering temperature, is optimum for obtaining the best properties and performance.

2.4 Summary

The present review has shown a general overview of the progress in the studies on the ferroelectric ceramics of perovskite and the Aurivillius family of compounds.

Over the years, perovskite and bismuth layer-structure ferroelectrics have attracted considerable interest because of their technological importance. A lot of studies based on the doping effect and improvement of these ferroelectric materials properties have been reported. When material is being doped, careful selection of the dopant is required. The ion radii is a very important factor as well as its concentration. Based on available literature data, some of the possible conclusions regarding substitutions in the BT, bismuth titanate, and barium bismuth titanate lattice were presented. Defect structure developed during this process determines not only the electronic but the mass transport. Powder traits such as particle size, particle size distribution, shape of the particles, agglomeration factor, dopant concentration, as well as the atmosphere, temperature, sintering time, and rate have a significant influence on the sintering process. The optimization of this process in highly important since the loss of bismuth, oxygen, and such could potentially cause a problem resulting in nonstoichiometry that leads to undesirable conductivity effects, which tend to degrade the performance of materials. On the other hand, the presence of dopants in the lattice can also cause the change of the material's electrical resistivity and reaching the critical concentration of dopants enables the opposite effect.

References

[1] R.C. Buchanan, Ceramic Materials for Electronics, University of Ilinois at Urbana-Champaign, Urbana, Ilinois, 1986.
[2] A.J. Moulson, J.M. Herbert, Electroceramics, John Wiley & Sons Ltd, 2003.
[3] S. Kasap, P. Capper, Ferroelectric Materials, Springer Handbook of Electronic and Photonic Materials, Springer, 2006.
[4] T. Jardiel, A.C. Caballero, M. Villegas, Aurivillius ceramics: $Bi_4Ti_3O_{12}$-based piezoelectrics, J. Ceram. Soc. Jpn. 116 (2008) 511−518.

[5] F. Jona, G. Shirane, Ferroelectric Crystals, Dover Publications, Inc, New York, NY, 1993.
[6] H. Du, X. Li, Z. Xu, D. Chu, Electric double-layered transistors: a review of recent progress, J Mat. Sci. 50 (2015) 5641–5673.
[7] M.M. Vijatović, J.D. Bobić, B.D. Stojanović, History and challenges of barium titanate: Part I, Sci. Sintering 40 (2) (2008) 155–165.
[8] S. Pradhan, G.S. Roy, Study the crystal structure and phase transition of $BaTiO_3$_a perovskite, Researcher 5 (2013) 63–67.
[9] E. Caetano Camilo de Souza, R. Muccillo, Properties and application of perovskite proton conductors, Mat. Res. 13 (2010) 385–394.
[10] Y. Mao, H. Zhau, S. Wong, Synthesis, properties and application of perovskite-phase metal oxide nanostructures, Mater. Matters 5 (2010) 50.
[11] W.-S. Cho, Structural evolution and of $BaTiO_3$ nanoparticles synthesized from polymeric precursors, J. Phys. Chem. Solids 59 (1998) 659.
[12] F. Yang, G. Hu, B. Xu, W. Wu, C. Yang, H. Wu, Statistical mechanical origin of hysteresis in ferroelectrics, J. Appl. Phys. 112 (2012) 034113.
[13] A. Koelzynski, K. Tkacz-Smiech, From the molecular picture to the band structure of cubic and tetragonal barium titanate, Ferroelectrics 314 (2005) 123.
[14] S. Chatterjee, B. Stojanovic, H. Maiti, Effect of additives and powder preparation techniques on PTCR properties of barium titanate, Mater. Chem. Phys. 78 (2003) 702–710.
[15] S. Urek, M. Drofenik, D. Makovec, Sintering and properties of highly donor doped barium titanate ceramics, J. Mater. Sci 35 (2000) 895–901.
[16] C. Barry, M. Grant Norton, Ceramics Materials, Science and Engineering, Springer, New York, NY, 2007.
[17] V.V. Srdić, Processing of New Ceramics Materials, University of Novi Sad, Faculty of Technology, 2004.
[18] M.M. Ristić, Basics of Materials Science, Naučna knjiga, Belgrade, 1977.
[19] M.T. Buscaglia, V. Buscaglia, M. Viviani, P. Nanni, M. Hanuskova, Influence of foreign ions on the crystal structure of $BaTiO_3$, J. Euro. Ceram. Soc. 20 (2000) 1997–2007.
[20] D. Makovec, Z. Samardzija, M. Drofenik, Solid solubility of holmium, yttrium, and dysprosium in $BaTiO_3$, J. Am. Ceram. Soc 87 (7) (2004) 1324–1329.
[21] E. Brzozowski, M.S. Castro, Influence of Nb^{5+} and Sb^{3+} dopants on the defect profile, PTCR effect and GBBL characteristics of $BaTiO_3$ ceramics, J. Eur. Ceram. Soc. 24 (2004) 2499–2507.
[22] M.M. VijatovicPetrovic, J.D. Bobic, T. Ramoska, J. Banys, B.D. Stojanovic, Antimony doping effect on barium titanate structure and electrical properties, Ceram. Int. 37 (2011) (2011) 2669–2677.
[23] M. Ganguly, S.K. Rout, W.S. Woo, C.W. Ahn, I.W. Kim, Characterization of A-site deficient samarium doped barium titanate, Physica B 411 (2013) 26–34.
[24] M.M. Vijatovic Petrovic, J.D. Bobic, R. Grigalaitis, N.I. Ilic, A.S. Dzunuzovic, V. Jankausikaite, et al., Donor-acceptor joint effect in barium titanate systems, Ceram. Int. 41 (2015) 11365–11371.
[25] M.C. Ferrarelli, C.C. Tan, D.C. Sinclair, Ferroelectric, electrical and structural properties of Dy and Sc co-doped $BaTiO_3$, J. Mater. Chem. 21 (2011) 6292.
[26] P. Kumar, S. Singh, J.K. Juneja, C. Prakash, K.K. Raina, Ferroelectric properties of substituted barium titanate ceramics, Phys. B 404 (2009) 1752–1756.
[27] S. Devi, A.K. Jha, Phase transitions and electrical characteristics of tungsten substituted barium titanate, Phys. B 404 (2009) 4290–4294.

[28] L. Wu, M.C. Chure, K.K. Wu, W.C. Chang, M.J. Yang, W.K. Liu, et al., Dielectric properties of barium titanate ceramics with different materials powder size, Ceram. Int. 35 (2009) 957–960.
[29] S. Shoudhury, S. Akter, M.J. Rahman, H. Bhuiyan, S.N. Rahman, N. Khatun, et al., Dielectric and electrical properties of zirconium doped barium titanate perovskite, J. Bangladesh Acad. Sci. 32 (2008) 221–229.
[30] B. Aurivillius, Mixed bismuth oxides with layer lattices, II structure of $Bi_4Ti_3O_{12}$, Ark. Kemi 1 (1949) 499–512.
[31] T. Kikuchi, A. Watanabe, K. Uchida, A family of mixed-layer type bismuth compounds, Mater. Res. Bull. 12 (1997) 299–304.
[32] E.C. Subbarao, Systematics of bismuth layer compounds, Ferroelectrics 12 (1996) 33–41.
[33] S. Luoa, Y. Noguchia, M. Miyayama, T. Kudo, Rietveld analysis and dielectric properties of $Bi_2WO_6Bi_4Ti_3O_{12}$ ferroelectric system, Mater. Res. Bull. 36 (2001) 531–540.
[34] R.E. Newnham, R.W. Wolf, J.E. Dorrian, Ferroelectric bismuth titanates, Mater. Res. Bull. 6 (5) (1971) 1029–1033.
[35] B.J. Kennedy, Y. Kubota, B.A. Hunter, Ismunandar, K. Kato, Structural phase transition in the layered bismuth oxide $BaBi_4Ti_4O_{15}$, Solid State Commun 126 (2003) 653–658.
[36] Y. Sakuma, T. Kimura, Mechanisms of texture development in bismuth layer-structured ferroelectrics prepared by templated grain growth, J. Electroceramics 13 (2004) 537–541.
[37] H. Chen, B. Shen, J. Xu, J. Zhai, The grain size-dependent electrical properties of $Bi_4Ti_3O_{12}$ piezoelectric ceramics, J. Alloy. Compd. 551 (2013) 92–97.
[38] J.D. Bobić, M.M. Vijatović Petrović, N.I. Ilić, E. Palaimiene, R. Grigalaitis, C.O. Paiva-Santos, et al., Lead-free $BaBi_4Ti_4O_{15}$ ceramics: effect of synthesis methods on phase formation and electrical properties, Ceram. Int. 41 (2015) 309–316.
[39] E.C. Subbarao, Crystal chemistry of mixed bismuth oxides with layer-type structure, J. Am. Ceram. Soc. 45 (1962) 166–169.
[40] J. Tellier, Ph Boullay, M. Manier, D. Mercurio, A comparative study of the Aurivillius phase ferroelectrics $CaBi_4Ti_4O_{15}$ and $BaBi_4Ti_4O_{15}$, J. Solid State Chem. 177 (2004) 1829–1837.
[41] I.H. Ismailzade, Crystal chemistry of perovskite-like layer-type ferroelectrics, J. Phys. Colloques 33 (1972), C2-237-C2-239.
[42] G. Nalini, T.N. Guru Row, Structure determination at room temperature and phase transition studies above T_c in $ABi_4Ti_4O_{15}$ (A = Ba, Sr or Pb), B, Mater. Sci. 25 (2002) 275–281.
[43] B.J. Kennedy, Q. Zhou, Ismunandar, Y. Kubota, K. Kato, Cation disorder and phase transition in the four-layer ferroelectric Aurivillius phase $ABi_4Ti_4O_{15}$ (A = Ca, Sr, Ba, Pb), J. Solid State Chem. 181 (2008) 1377–1386.
[44] P. Millan, A. Castro, J.B. Torrance, The first doping of lead 2 + into the bismuth oxide layers of the Aurivillius oxides, Mater. Res. Bull. 28 (1993) 117–122.
[45] T. Rentschler, Substitution of lead into the bismuth oxide layers of the n = 2- and n = 3-Aurivillius phases, Mater. Res. Bull. 32 (3) (1997) 351–369.
[46] Ismunandar, T. Kamiyama, A. Hoshikawa, Q. Zhou, B.J. Kennedy, Y. Kubota, et al., Structural studies offive layer Aurivillius oxides: $A_2Bi_4Ti_5O_{18}$ (A = Ca, Sr, Ba, and Pb), J. Solid State Chem. 177 (2004) 4188–4196.
[47] M. Takahashi, Y. Noguchi, M. Miyayama, Estimation of ionic and hole conductivity in bismuth titanate polycrystals at high temperature, Solid State Ionics 172 (2004) 325–329.

[48] S. Rachna, S. Bhattacharyya, S.M. Gupta, Correlating structure, dielectric and impedance studies with lanthanum-ion substitution in bismuth titanate, M, Sci. Eng. B-adv 175 (2010) 207–212.
[49] A.Z. Simoes, C. Quinelato, A. Ries, B.D. Stojanovic, E. Longo, J.A. Varela, Preparation of lanthanum doped $Bi_4Ti_4O_{12}$ ceramics by the polymeric precursors method, Mater. Chem. Phys. 98 (2006) 481–485.
[50] M. Osada, M. Tada, M. Kakihana, T. Watanabe, H. Funakubo, Cation distribution and structural instability in $Bi_{4-x}La_xTi_3O_{12}$, Jpn. J. Appl. Phys. 40 (2001) 5572–5575.
[51] X.B. Chen, R. Hui, J. Zhu, W.P. Lu, X.Y. Mao, Relaxor properties of lanthanum-doped bismuth layer-structured ferroelectrics, J. Appl. Phys. 96 (10) (2004) 5697–5700.
[52] S.J. Kim, C. Moriyoshi, S. Kimura, Y. Kuroiwa, K. Kato, M. Takata, et al., Direct observation of oxygen stabilization in layered ferroelectric, $Bi_{3.25}La_{0.75}Ti_3O_{12}$, Appl. Phys. Lett. 91 (2007) 062913.
[53] M.S. Shimazu, J. Tanaka, K. Uramatsu, M. Sukioka, Phase transition in the family $La_xBi_{4-x}Ti_3O_{12}$: in relation to lattice symmetry and distortion, J. Solid State Chem. 35 (1980) 402–406.
[54] T. Takenaka, K. Sakata, Electrical properties of grain-oriented ferroelectric ceramics in some lanthanum-modified layer-structured oxides, Ferroelectrics 38 (1981) 769.
[55] C. Moriyoshi, S. Kimura, Y. Uroiwa, S.J. Kim, Y. Noguchi, Charge density distribution in $Bi_4Ti_3O_{12}$ and $Bi_{3.75}La_{0.75}Ti_3O_{12}$ in the paraelectric phase, J. Korean Soc 55 (2009) 794–798.
[56] J.S. Kim, I.V. Kim, Role of the doped Nb ion in ferroelectric bismuth titanate and bismuth lanthanum titanate ceramics: different dielectric behaviors, J. Korean Phys. Soc. 46 (1) (2005) 143–146.
[57] L. Zhang, R. Chu, S. Zhao, G. Li, Q. Yin, Microstructure and electrical properties of niobium doped $Bi_4Ti_3O_{12}$ layer-structured piezoelectric ceramics, Mater. Sci. Eng. B 116 (2005) 99–103.
[58] J.D. Bobic, M.M. VijatovicPetrovic, J. Banys, B.D. Stojanovic, Electrical properties of niobium doped barium bismuth-titanate ceramics, Mater. Res. Bull. 47 (2012) 1874–1880.
[59] C. Jin, C. Du, Jun Zhu, J. He, X. Mao, X. Chen, Ferroelectric and dielectric properties ofniobium-doped $SrBi_4Ti_4O_{15}$ ceramics, J. Phys. D: Appl. Phys 39 (2006) 2415–2418.
[60] W. Li, J. Gu, C. Song, D. Su, J. Zhu, B-site doping effect on ferroelectric property of bismuth titante ceramic, J. Appl. Phys. 98 (2005) 114104.
[61] H.S. Shulman, M. Testorf, D. Damijanovic, N. Setter, Microstructure, electrical conductivity and piezoelectric properties of bismuth titanate, J. Am. Ceram. Soc. 79 (12) (1996) 3124–3128.

Lead-free perovskite ferroelectrics

3

Jurij Koruza[1], Lalitha Kodumudi Venkataraman[1] and Barbara Malič[2]
[1]Technische Universität Darmstadt, Darmstadt, Germany
[2]Jožef Stefan Institute, Ljubljana, Slovenia

3.1 Introduction

The ferroelectrics utilized in piezoelectric applications for decades have been based on lead-containing materials, predominantly $Pb(Zr,Ti)O_3$ (PZT) and its derivatives. This can be explained by the excellent piezoelectric properties, the flexibility of the compositions, the well-understood processing, and the low costs [1,2]. However, lead represents a serious hazard to humans and the environment, especially during the production and recycling of these materials [3]. Increased environmental awareness over the past 20 years has resulted in several governmental regulations worldwide that are putting increasingly stringent requirements on the use of materials in electronic devices [4–7]. Consequently, this period witnessed considerable scientific efforts to develop new ferroelectric materials with an improved electromechanical response to be used in sensors, actuators, and transducers. While many new compositions were discovered, some developed in the 1950s and 1960s were also revived.

The most important lead-free ferroelectrics crystallize in the perovskite structure and can be roughly divided into three main groups: alkaline niobates, alkali-bismuth titanates, and barium-titanate-based compositions (Fig. 3.1). Although many of them possess excellent properties under specific conditions, sometimes even outperforming the state-of-the-art PZT, they typically lack the flexibility of having good properties over a broad temperature range. For example, barium-titanate-based materials have exceptionally high piezoelectric coefficients at room temperature, but depolarize at 50–120°C. On the other hand, alkaline niobates have very high Curie temperatures, but often inferior functional properties when compared to PZT. Nevertheless, many lead-free materials were successfully used to prepare prototype devices and the first products are starting to enter the market [8].

This chapter highlights the most important structural and functional characteristics of the three main groups of lead-free perovskite ferroelectrics, while more extensive reviews have been published elsewhere [8–13]. The processing of these materials is described in Chapter 5, Review of Methods for Powder-Based Processing. A further, recently emerging group of ferroelectrics are the $BiFeO_3$-based materials [14], which are reviewed in Chapter 24, $BiFeO_3$ ceramics and thick films: processing issues and electromechanical properties.

Magnetic, Ferroelectric, and Multiferroic Metal Oxides. DOI: http://dx.doi.org/10.1016/B978-0-12-811180-2.00003-7

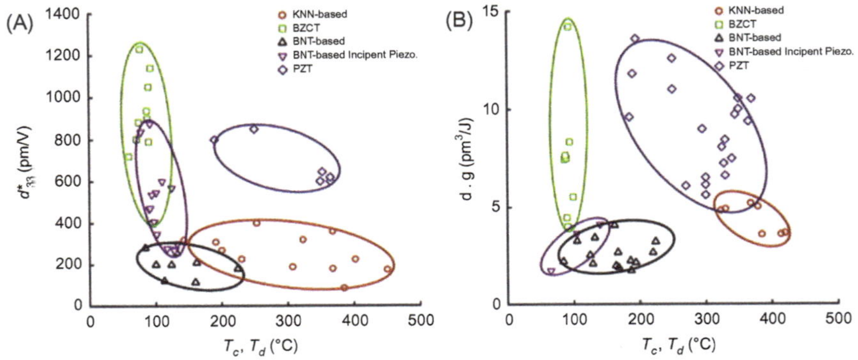

Figure 3.1 Piezoelectric properties of ferroelectric ceramics: (A) large-signal piezoelectric coefficient (normalized strain), important for actuation applications, and (B) product of the piezoelectric charge and voltage coefficients, the figure of merit for nonresonant sensor applications. *KNN*, (K,Na)NbO_3-based; *BZCT*, (Ba,Ca)(Ti,Zr)O_3-based; *BNT*, $(Na_{1/2}Bi_{1/2})TiO_3$-based; and *PZT*, Pb(Zr,Ti)O_3-based compositions. T_C and T_d mark the Curie and depolarization temperatures, respectively.
Reprinted from J. Rödel, K.G. Webber, R. Dittmer, W. Jo, M. Kimura, D. Damjanovic, Transferring lead-free piezoelectric ceramics into application, J. Eur. Ceram. Soc. 35 (6) (2015) 1659–1681, with permission from Elsevier.

3.2 Alkaline niobates

Sodium potassium niobate $(K_{1-x}Na_x)NbO_3$ has been the most widely studied lead-free piezoelectric material since early 2000. This research was summarized in a number of review papers [8–10,13,15–19]. The first studies of the $KNbO_3$–$NaNbO_3$ pseudobinary phase equilibria were reported in the late 1960s and early 1970s [20,21]. The binaries $KNbO_3$ and $NaNbO_3$ exhibit complete solid solubility (see Fig. 3.2). $NaNbO_3$ crystallizes in an orthorhombic syngony (*Pbma*) at room temperature. Besides being a prototype antiferroelectric [22,23], $NaNbO_3$ is also known for the highest number of phase transitions among the perovskite oxides [21,24–27]. At room temperature, $KNbO_3$ is ferroelectric and exhibits orthorhombic (*Amm*2) symmetry; the same is true for all $(K_{1-x}Na_x)NbO_3$ solid solutions with at least 2 mole% $KNbO_3$. The solid solution with a K/Na molar ratio 1/1, $K_{0.5}Na_{0.5}NbO_3$ or KNN, has been the most thoroughly investigated composition for piezoelectric applications. It is rhombohedral ($R3c$) at low temperatures, orthorhombic between ≈ −160°C and ≈ 200°C, tetragonal ($P4mm$) up to ≈ 400°C (Curie temperature; T_C), and cubic ($Pm\overline{3}m$) at higher temperatures [28]. Alternatively, a monoclinic structure (*Pm*) has been proposed for the room-temperature phase. While the orthorhombic assignment assumes a multiple unit cell ($a \approx c \approx 4\sqrt{2}$ Å and $b \approx 4$ Å), the monoclinic assumes a simple perovskite unit cell ($a' \approx b' \approx c' \approx 4$ Å, $\beta \approx 90.3°$) [29]. The solidus temperature is at 1140°C. Early

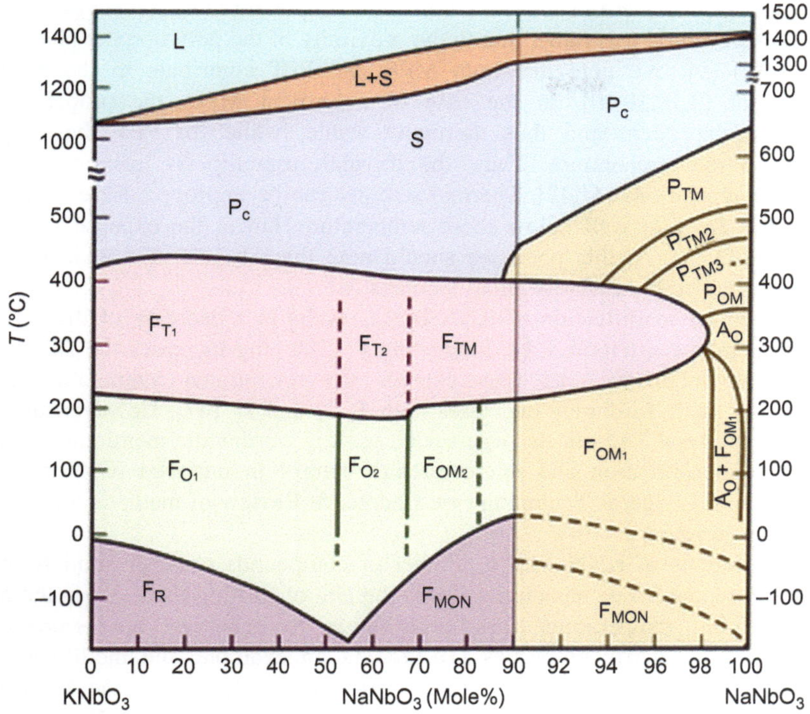

Figure 3.2 Pseudobinary $KNbO_3$–$NaNbO_3$ phase diagram [17,21].
Reprinted from J.F. Li, K. Wang, F.Y. Zhu, L.Q. Cheng, F.Z. Yao, (K,Na)NbO_3-based lead-free piezoceramics: fundamental aspects, processing technologies, and remaining challenges, J. Am. Ceram. Soc. 96 (12) (2013) 3677–3696, with permission from John Wiley & Sons.

studies proposed a morphotropic phase boundary (MPB) between two orthorhombic phases at $x = 0.475$ [20], but later its existence could not be confirmed [29,30].

The piezoelectric d_{33} coefficient of KNN obtained by atmosphere sintering is 80 pC/N [31], while processing optimization and/or changing of the K/Na molar ratio resulted in a few 10 pC/N higher d_{33} values and coupling coefficients k_p of about 0.38 [32–35]. The highest values of $d_{33} = 150–160$ pC/N and $k_p = 0.45$ were obtained by hot-pressing or spark-plasma sintering [36–38].

Clearly, the piezoelectric properties of pure KNN cannot compete with those of PZT, as was already recognized a few decades earlier [21]. However, in early 2000, legislation [4–7] and environmental awareness stimulated extensive research on modified KNN compositions. In 2004, the breakthrough Toyota paper reported that the d_{33} coefficient of $(K_{0.44}Na_{0.52}Li_{0.04})(Nb_{0.86}Ta_{0.10}Sb_{0.04})O_3$ (KNLNTS or LF4) could exceed 300 pC/N, and with texturing, even 400 pC/N, although at the expense of decreasing the T_C to 253°C [39]. The enhancement of the piezoelectric response was first explained as being due to the formation of a MPB between the orthorhombic (O) and tetragonal (T) phases. It was later recognized that the modifiers contributed to a decrease of the O–T transition temperature (T_{O-T}), suggesting that the d_{33} enhancement was rather due to the proximity of the polymorphic phase transition (PPT) [9]. We note that both MPB and PPT contribute to the property enhancement [9,10,19,40]. In the case of a vertical MPB, the properties are composition-dependent and thus thermally stable, while for PPT they depend strongly on the temperature. Thus, the thermal instability is inherent to PPT-designed compositions [41,42]. Thermal stability can be improved, for example, by shifting the T_{O-T} to well below room temperature, but at the expense of poorer properties [43,44]. At this point we should note that Sb is also a toxic element, albeit being added in relatively small amounts.

The chemical modification of KNN by Li results in a decrease of the T_{O-T} to room temperature at 6 mole% Li. Interestingly, Li-doping increases the T_C, in contrast to the modifiers, and thus extends the operational temperature range [10,17,41,45–47]. Ta-doping decreases both T_{O-T} and T_C [47]. However, the high cost and scarcity of Ta limit its large-scale use. It is worthwhile mentioning that the problem of densification and microstructure control in undoped KNN could be overcome with Li and/or Ta doping (see Chapter 5: Review of methods for powder-based processing).

Solid solutions of KNN with a number of compounds or other solid solutions have been prepared and characterized with the aim of shifting the T_{O-T} in the proximity of room temperature, thus contributing to enhanced but temperature-dependent piezoelectric properties [10,17]. As an example, the modification of KNN by $LiSbO_3$ decreased the T_C from $\approx$400°C to 370°C and the T_{O-T} from $\approx$200°C to 35°C, which resulted in an enhanced d_{33} coefficient of 265 pC/N, but also reduced the temperature stability [48]. The addition of $CaTiO_3$ effectively decreased the T_{O-T} transition temperature of $LiSbO_3$-modified KNN, thus resulting in a stable piezoelectric response (d_{33} of 210 pC/N upon addition of 1 wt% $CaTiO_3$) in the temperature range −50–200°C [43]. Similarly, the addition of $CaZrO_3$ to Li- and Ta-modified KNN resulted in a high d_{33} coefficient of about 300 pC/N and stable large-signal properties between room temperature and 175°C [44].

An alternative proposition to form a MPB in KNN was to stabilize the low-temperature rhombohedral (R) phase at room temperature by a chemical modification with other alkaline-earth perovskites [49,50]. Based on such reasoning, a thermally stable and large piezoelectric response was reported for $0.92(Na_{0.5}K_{0.5})NbO_3-0.02(Bi_{1/2}Li_{1/2})TiO_3-0.06BaZrO_3$ with d_{33} values of 252 pC/N and 170 pC/N at 30°C and 200°C, respectively [51]. Wang et al. reported a d_{33} of 490 pC/N for the MPB composition $0.96(K_{0.48}Na_{0.52})(Nb_{0.95}Sb_{0.05})O_3-0.04Bi_{0.5}(Na_{0.82}K_{0.18})_{0.5}ZrO_3$ [52]. We note that the stoichiometry and chemical homogeneity in such multicomponent systems, which contain several volatile species (K, Na, Bi), can be difficult to attain and control, and thus the reproducibility of the properties could be a challenge.

Various sintering aids have been introduced to obtain KNN materials with a high relative density, the correct stoichiometry, and a homogeneous microstructure to overcome inherent problems related to sintering and densification (see Chapter 5, Review of Methods for Powder-Based Processing). Copper [53–56] and zinc compounds [57,58], and borate [59] and germanate [60] glassy phases have been employed as liquid-phase sintering aids. Among them, copper-based additives, such as CuO [54], $K_4CuNb_8O_{23}$ [53,55], and $K_{5.4}Cu_{1.3}Ta_{10}O_{29}$ (KCT) [56], have been studied most thoroughly, since, in addition to promoting densification, they also contribute to piezoelectric hardening. The mechanical quality factor (Q_m) increased from 90 in KNN to 1300 upon the addition of 0.38 mole% KCT [56]. A further modification with CuO resulted in Q_m values between 2600 and 3000 [61,62] as a consequence of the increased concentration of defect dipoles in the material [63].

3.3 Alkaline bismuth titanates

The unique piezoelectric properties of lead-based systems are largely attributed to the $6s^2$ lone-pair on the Pb^{2+} ion and its high polarizability. As the Bi^{3+} ion possesses a similar electronic structure [64], the alkaline bismuth titanates were expected to exhibit comparable properties [65]. This section describes the two most widely investigated alkaline bismuth titanate compounds, $K_{1/2}Bi_{1/2}TiO_3$ (KBT) and $Na_{1/2}Bi_{1/2}TiO_3$ (NBT), and their derivatives [8−12]. We note that although the notation $Bi_{1/2}Na_{1/2}TiO_3$ (BNT) is often used in the literature, the IUPAC recommendations dictate the use of the form $Na_{1/2}Bi_{1/2}TiO_3$ (NBT) due to the higher electronegativity of Bi.

Pure NBT exhibits a rhombohedral symmetry at room temperature with a remanent polarization (P_r) of 38 μC/cm², a d_{33} of 91 pC/N, and a coupling coefficient (k_{33}) of 0.47 [66,67]. The rhombohedral phase transforms to an intermediate tetragonal ($P4bm$) phase at 300°C and a cubic ($Pm\bar{3}m$) phase at 540°C [68]. The rhombohedral ($R3c$) symmetry has long been under debate and the occurrence of local disordered regions has been reported [69−72]. A non-rhombohedral, low-symmetry (monoclinic) average structure has been evidenced from high-resolution X-ray diffraction [73,74], transmission electron microscopy (TEM) [75], and optical birefringence microscopy [76]. A first-principles study in conjunction with electric-field-dependent diffraction confirmed the stability of the rhombohedral symmetry, while the monoclinic (Cc) distortion was attributed to local structural and strain heterogeneities [77].

The electromechanical properties of NBT are softened by codoping with K^+ and Li^+, resulting in $d_{33} = 231$ pC/N, $k_p = 0.41$, $P_r = 40.2$ μC/cm², and $E_c = 2.47$ kV/mm [78]. Akin to PZT, B-site-substituted Fe and Mn are shown to result in hardening, while A-site-substituted La softens the electromechanical properties [79,80]. Low dopant concentrations of Fe and Mn have been shown to increase the depolarization temperature, T_d, by ~20−30°C [81]. The bulk properties of NBT are sensitive to A-site nonstoichiometry: compositions with the starting ratio Bi/Na > 1 are insulators, while Bi/Na ≤ 1 materials are predominantly oxide ion conductors [82]. The main drawbacks of pure NBT are its high E_c of 7.3 kV/mm [83] and its relatively low piezoelectric properties. Several solid solutions based on NBT with improved properties have been developed over the past two decades. The most important ones are briefly described next.

The $(1-x)Na_{1/2}Bi_{1/2}TiO_3-xBaTiO_3$ (NBT-BT) is a solid solution with a MPB at $x = 0.06 \sim 0.07$ (Fig. 3.3A) [83]. The MPB composition has a $d_{33} = 125$ pC/N and a $k_{33} = 0.55$. At $T_d \approx 107°C$, the depolarization fields suppress the long-range ferroelectric order achieved upon poling. Ma et al. demonstrated that these compositions undergo an electric-field-induced phase transition and reported a field-dependent phase diagram (Fig. 3.3B) [84]. The majority of the investigations into phase transitions were carried out for compositions around $x = 0.06$, which exhibits the highest dielectric and piezoelectric response [85–89], while reports on the tetragonal end of the diagram are comparatively scarce [90–93].

One of the most interesting chemical modifications of the NBT-BT system, the ternary solid solution with $K_{0.5}Na_{0.5}NbO_3$ (KNN) [94,95], exhibits a high unipolar strain of 0.45% at 8 kV/mm. This was correlated to an electric-field-induced reversible transformation from the nonpolar (*P4bm*) to the ferroelectric rhombohedral (*R3c*) phase [96]. The drawback is the large field required to realize such a macroscopic strain.

This problem was reduced by developing 3-0 composites of relaxor NBT-BT-KNN (matrix) and ferroelectric NBT-7BT (inclusion), where a reduction of the threshold field was demonstrated [97]. The optimum response of the composites involves polarization and strain-coupling mechanisms, as demonstrated using 2-2 composites with layers parallel (polarization-coupled) and perpendicular (strain-coupled) to the electrodes [98].

Relaxor/semiconductor composites address the problem of the thermal depolarization of NBT-BT. The increase in T_d in the 3-0 composites of NBT-6BT with ZnO inclusions was demonstrated by Zhang et al., wherein the charges from ZnO are suggested to form a local field that stabilizes the poled state of the NBT-6BT [99].

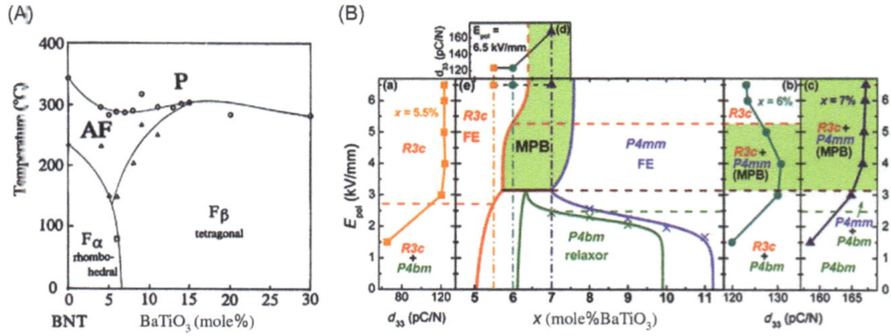

Figure 3.3 (A) Zero-field $Na_{1/2}Bi_{1/2}TiO_3-BaTiO_3$ phase diagram. (B) Field-dependent $Na_{1/2}Bi_{1/2}TiO_3-BaTiO_3$ phase diagram.
For (A) From T. Takenaka, K.-i. Maruyama, K. Sakata, $(Bi_{1/2}Na_{1/2})TiO_3$-$BaTiO_3$ system for lead-free piezoelectric ceramics, Jpn. J. Appl. Phys. 30 (9S) (1991) 2236. Copyright 1991 by The Japan Society of Applied Physics; For (B) Reprinted with permission from C. Ma, H. Guo, S.P. Beckman, X. Tan, Creation and destruction of morphotropic phase boundaries through electrical poling: a case study of lead-free $(Bi_{1/2}Na_{1/2})TiO_3$-$BaTiO_3$ piezoelectrics, Phys. Rev. Lett. 109 (10) (2012) 107602. Copyright 2012 by the American Physical Society.

Riemer et al. [100] recently reported a ~40–60°C increase in T_d and revealed the importance of the residual thermal stresses. The residual stresses stabilize the ferroelectric phase in the virgin state of the samples, in accordance with the stress-induced relaxor-ferroelectric transformation [101]. Moreover, the composites exhibit piezoelectric hardening with a two-fold increase of Q_m [102].

$K_{1/2}Bi_{1/2}TiO_3$ (KBT) is tetragonal (*P4mm*) at room temperature and has a relatively high T_C of 380°C [103,104]. The $(1-x)Na_{1/2}Bi_{1/2}TiO_3-xK_{1/2}Bi_{1/2}TiO_3$ solid solution exhibits promising electromechanical properties; for example, $d_{33} = 192$ pC/N and $k_p = 0.32$ for $x = 0.22$ [105]. The enhanced electromechanical properties are consistently reported for compositions with $0.15 < x < 0.2$ [104,106–111], but discrepancies exist regarding the exact range of the MPB. For $x = 0.2$, the application of an electric field results in the transformation of the rhombohedral (*R*3*c*) phase to a mixture of tetragonal and rhombohedral phases [112]. Also, rapidly increasing fields result in partial transformation from the rhombohedral to the tetragonal phase, whereas lower rates inhibit the transformation due to the internal stresses caused by the domain switching [113]. Using combined diffraction and TEM studies, it was demonstrated that the electric-field-induced and stress-induced transitions largely depend on the initial structural state; while compressive stress results in the formation of textured rhombohedral domains, an electric field resulted in textured rhombohedral and tetragonal structures [114]. The ternary solid solution $xNa_{1/2}Bi_{1/2}TiO_3-yBi_{1/2}Li_{1/2}TiO_3-zK_{1/2}Bi_{1/2}TiO_3$ (BNLKT) with Mn addition was demonstrated to exhibit high-power characteristics superior to hard PZT [115].

Sakata et al. [116]. were the first to report on the $(1-x)Na_{1/2}Bi_{1/2}TiO_3-xSrTiO_3$ (NBT-ST) solid solution with $0 \leq x \leq 0.5$. The transition temperatures decrease with increase of the ST content. This was confirmed using temperature-dependent Raman and dielectric studies [117]. A MPB was reported between the rhombohedral ferroelectric and pseudocubic (tetragonal) paraelectric phases in the range $0.26 \leq x \leq 0.28$, with $x = 0.28$ exhibiting the maximum large-signal d_{33}^* (normalized strain) of 488 pm/V [118]. A $d_{33}^* \sim 600$ pm/V at moderate fields of 4 kV/mm was reported for $x = 0.25$ [119]. The large strain is attributed to the field-induced relaxor-to-ferroelectric phase transition, also observed in other solid solutions modified by ST [120,121]. A correlation between the microstructure and the electrical properties was established for $x = 0.25$ [122,123]. Sintering was shown to result in a heterogeneous microstructure consisting of two solid solutions, a NBT-rich core and a ST-rich shell.

3.4 Barium titanate-based piezoelectrics

$BaTiO_3$ (BT) was the first oxide ferroelectric, discovered during World War II, and even today, remains one of the most widely utilized and studied ferroelectric and dielectric materials [2]. Details about pure BT are given in Chapter 2, Perovskite and Aurivillius Type of Ferroelectric Metal Oxides and Chapter 8, Nanosize

$BaTiO_3$-Based Systems, while this section reviews the BT-based solid solutions developed as potential lead-free piezoelectrics. Although some of these compositions were already investigated for dielectric applications in the 1950s [124], extensive research of their electromechanical properties only started in 2009, when Liu and Ren reported exceptional piezoelectric coefficients in Ca- and Zr-modified BT [125].

The most widely studied compositions are the $(Ba_xCa_{1-x})(Zr_{1-y}Ti_y)O_3$ solid solutions, in particular the ones lying at the $(1\text{-}x)Ba(Zr_{0.2}Ti_{0.8})O_3-x(Ba_{0.7}Ca_{0.3})TiO_3$ isopleths of the $BaTiO_3-BaZrO_3-CaTiO_3$ ternary phase diagram. The latter are hereafter abbreviated as BZT-xBCT, whereby x marks the mole% of the $(Ba_{0.7}Ca_{0.3})TiO_3$ solid solution. Several pseudobinary BZT-xBCT phase diagrams were reported (Fig. 3.4) [125–128]. Common features include the presence of a rhombohedral (R; *R*3*m*) phase on the BZT-rich side and a tetragonal (T; *P*4*mm*) phase on the BCT-rich side. The intermediate region was initially proposed to be a MPB, converging into a triple or tricritical point at 57°C and $x = 0.32$. However, its evident temperature dependence tends to suggest a PPT-like nature. Moreover, later studies revealed the presence of another PPT and thus an intermediate region between the R and T phases, which is considered a mixed R + T region [126,129] or a bridging orthorhombic phase (O; *Amm*2) [127,130,131]. The triple-point region is difficult to resolve, nevertheless some theoretical and experimental observations indicate the presence of a quadruple point [127], two triple points [132,133], or a field-dependent line of triple points [134]. At elevated temperatures, the material transforms into the cubic (C; $Pm\bar{3}m$) structure, whereby the T_C spans from 61°C for BZT-0.3BCT to 113°C for BZT-0.9BCT.

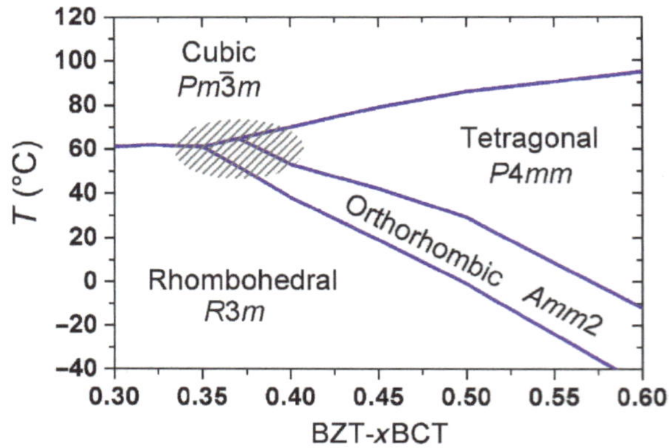

Figure 3.4 Pseudobinary phase diagram of the $(1-x)Ba(Ti_{0.8}Zr_{0.2})O_3-x(Ba_{0.7}Ca_{0.3})TiO_3$ system. The dashed area marks the phase-convergence region that could not be experimentally determined.
Modified after D.S. Keeble, F. Benabdallah, P.A. Thomas, M. Maglione, J. Kreisel, Revised structural phase diagram of $(Ba_{0.7}Ca_{0.3}TiO_3)$-$(BaZr_{0.2}Ti_{0.8}O_3)$, Appl. Phys. Lett. 102 (9) (2013) 092903; M. Acosta, N. Novak, W. Jo, J. Rödel, Relationship between electromechanical properties and phase diagram in the $Ba(Zr_{0.2}Ti_{0.8})O_3$-$x(Ba_{0.7}Ca_{0.3})TiO_3$ lead-free piezoceramic, Acta Mater. 80 (2014) 48–55.

The best room-temperature electromechanical properties were reported for the BZT-0.5BCT composition: a small-signal d_{33} of 560−620 pC/N, a large-signal $d_{33}{}^*$ of 1140 pm/V at 0.5 kV/mm, a k_p of 0.53, a P_r of 14.5 μC/cm^2, and an E_c of 0.168 kV/mm [125,135]. It should be noted that these properties are strongly influenced by the grain size, whereby the best results are obtained for sizes above 10 μm [136]. The strain response could be further enhanced by compressive stresses of 8−10 MPa, which increase the room-temperature large-signal $d_{33}{}^*$ by over 30% to 1500 pm/V (at 0.5 kV/mm) [137]. In addition, the reported blocking force/stress exceeds the ones reported for PZT [138], which is an important characteristic for actuator applications [139]. Due to the presence of two PPTs and the relatively low T_C, the electromechanical properties of the BZT-*x*BCT compositions are very susceptible to temperature changes. This remains one of the main limitations of this system. Interestingly, the highest d_{33} is obtained at the O−T PPT, while a high $d_{33}{}^*$ is obtained along both PPTs [128,132].

The exceptionally good properties of the BZT-*x*BCT compositions were initially related to a facilitated polarization rotation and an isotropic free energy due to the presence of triple and tricritical points [125]. The importance of elastic softening along the T_{O-T} [132] and the strong contribution of electrostriction [140] were also reported. On the other hand, studies using in situ X-ray diffraction [141,142] and Rayleigh analyses [143] pointed out the dominance of the extrinsic contributions to the macroscopic strain. The ratio of the intrinsic and extrinsic contributions to the strain strongly depends on the composition, temperature, and applied electric field [134].

The fatigue associated with the electromechanical properties upon bipolar cycling was higher than for other ferroelectrics [144], while the fatigue during unipolar cycling was comparable [145]. The Young's modulus determined from the ferroelastic curves was 165−179 GPa [138], while a value of 105 GPa was obtained for the 0.385BCT composition using the resonance method [146]. The fracture toughness was reported to be lower than for PZT [146].

Doping of the BZT-*x*BCT system was investigated primarily to reduce the high sintering temperatures and further improve the functional properties. The former is described in Chapter 5, Review of Methods for Powder-Based Processing. The piezoelectric properties could be increased by doping with Y [147,148], Cu [149], Zn [150], B [151], Bi [152], Ce [153], and Pr [154]. Furthermore, hardening of the piezoelectric properties and Q_m in the range of 550 were reported for Cr [155], Mn [156], and Co [157]. At this point it should be noted that many of these dopants also resulted in increased densification; therefore, the exact origin of the improved properties is not always clear. In addition to BZT-*x*BCT, promising electromechanical properties were also reported for other BT-based solid solutions: $(Ba,Ca)(Sn,Ti)O_3$ [158], $(Ba,Ca)(Hf,Ti)O_3$ [159], and $(Ba,Sr)(Zr,Ti)O_3$ [160].

3.5 Conclusions

After more than 15 years of intensive research, the lead-free perovskite ferroelectrics have seen considerable improvements, resulting in the first prototypes and products. Unlike PZT, which can be tailored for a range of applications, a single lead-free material cannot be used in all applications. Instead, specific compositions with a carefully designed microstructure could replace or even outperform PZT in individual devices. While future environmental regulations are expected to aid the transfer of lead-free piezoelectrics to applications, much remains to be done by the scientific community. In particular, a better understanding of the physical mechanisms and comprehensive investigations of the processing conditions, yielding homogeneous and reproducible results, are needed. The use of lead-free materials in devices will also be governed by the price of the raw materials and the compatibility with low-cost base-metal electrodes. Finally, the application of these materials on an industrial scale will require knowledge about the so-called secondary properties, such as the mechanical properties, blocking force, fatigue behavior, or frequency dependence of the material's response.

Acknowledgments

This work was supported by the Deutsche Forschungs gemeinschaft (Grant No. KO5100/1-1). L.K.V. acknowledges the support of the Alexander von Humboldt foundation. B.M. acknowledges the support of the Slovenian Research Agency under the program P2-0105.

References

[1] K. Uchino, Ferroelectric Devices, second ed., CRC Press, Boca Raton, FL, 2009.
[2] G.H. Haertling, Ferroelectric ceramics: history and technology, J. Am. Ceram. Soc. 82 (4) (1999) 797–818.
[3] P. Lyn, Lead toxicity, a review of the literature. Part I: Exposure, evaluation, and treatment, Altern. Med. Rev. 11 (1) (2006) 1–22.
[4] Solid Waste: Hazardous Electronic Waste. U.S. California Senate Bill No. 50, 2004.
[5] Measures for the Administration on Pollution Control of Electronic Information Products, Ministry of Information Industry China Order No. 39 2006.
[6] Regulation (EC) No. 1907/2006 concerning the Registration, Evaluation, Authorisation and Restriction of Chemicals (REACH), European Parliament and the European Council, Official Journal of the European Union 2006.
[7] Directive 2011/65/EU on the restriction of the use of certain hazardous substances in electrical and electronic equipment, European Parliament and the Council, Official Journal of the EU 2011.
[8] J. Rödel, K.G. Webber, R. Dittmer, W. Jo, M. Kimura, D. Damjanovic, Transferring lead-free piezoelectric ceramics into application, J. Eur. Ceram. Soc. 35 (6) (2015) 1659–1681.
[9] T.R. Shrout, S.J. Zhang, Lead-free piezoelectric ceramics: alternatives for PZT? J. Electroceram. 19 (1) (2007) 113–126.
[10] J. Rödel, W. Jo, K.T.P. Seifert, E.M. Anton, T. Granzow, D. Damjanovic, Perspective on the development of lead-free piezoceramics, J. Am. Ceram. Soc. 92 (6) (2009) 1153–1177.

[11] S.O. Leontsev, R.E. Eitel, Progress in engineering high strain lead-free piezoelectric ceramics, Sci. Technol. Adv. Mat. 11 (4) (2010) 044302.
[12] V.V. Shvartsman, D.C. Lupascu, Lead-free relaxor ferroelectrics, J. Am. Ceram. Soc. 95 (1) (2011) 1–26.
[13] S. Priya, S. Nahm, Lead-Free Piezoelectrics, Springer, New York, 2012.
[14] T. Rojac, A. Bencan, B. Malic, G. Tutuncu, J.L. Jones, J.E. Daniels, et al., $BiFeO_3$ ceramics: processing, electrical, and electromechanical properties, J. Am. Ceram. Soc. 97 (7) (2014) 1993–2011.
[15] B. Malic, A. Bencan, T. Rojac, M. Kosec, Lead-free piezoelectrics based on alkaline niobates: synthesis, sintering and microstructure, Acta. Chim. Slov. 55 (4) (2008) 719–726.
[16] P.K. Panda, Review: environmental friendly lead-free piezoelectric materials, J. Mater. Sci. 44 (19) (2009) 5049–5062.
[17] J.F. Li, K. Wang, F.Y. Zhu, L.Q. Cheng, F.Z. Yao, $(K,Na)NbO_3$-based lead-free piezoceramics: fundamental aspects, processing technologies, and remaining challenges, J. Am. Ceram. Soc. 96 (12) (2013) 3677–3696.
[18] B. Malič, J. Koruza, J. Hreščak, J. Bernard, K. Wang, J.G. Fisher, et al., Sintering of lead-free piezoelectric sodium potassium niobate ceramics, Materials 8 (12) (2015) 8117–8146.
[19] J.G. Wu, D.Q. Xiao, J.G. Zhu, Potassium-sodium niobate lead-free piezoelectric materials: past, present, and future of phase boundaries, Chem. Rev. 115 (7) (2015) 2559–2595.
[20] V.J. Tennery, K.W. Hang, Thermal and x-ray diffraction studies of $NaNbO_3$-$KNbO_3$ system, J. Appl. Phys. 39 (10) (1968) 4749–4753.
[21] B. Jaffe, W.R.J. Cook, H. Jaffe, Piezoelectric ceramics, in: first ed., J.P. Roberts, P. Popper (Eds.), Non-Metallic Solids, vol. 3, Academic Press, London, New York, 1971.
[22] P. Vousden, The structure of ferroelectric sodium niobate at room temperature, Acta Crystallogr. 4 (6) (1951) 545–551.
[23] L.E. Cross, B.J. Nicholson, The optical and electrical properties of single crystals of sodium niobate, Philos. Mag. 46 (376) (1955) 453–466.
[24] H.D. Megaw, 7 phases of sodium niobate, Ferroelectrics 7 (1-4) (1974) 87–89.
[25] I.P. Raevskii, L.A. Reznichenko, V.G. Smotrakov, V.V. Eremkin, M.A. Malitskaya, E.M. Kuznetsova, et al., A new phase transition in sodium niobate, Tech. Phys. Lett. 26 (8) (2000) 744–746.
[26] Y. Shiratori, A. Magrez, J. Dornseiffer, F.H. Haegel, C. Pithan, R. Waser, Polymorphism in micro-, submicro-, and nanocrystalline $NaNbO_3$, J. Phys. Chem. B 109 (43) (2005) 20122–20130.
[27] J. Koruza, J. Tellier, B. Malič, V. Bobnar, M. Kosec, Phase transitions of sodium niobate powder and ceramics, prepared by solid state synthesis, J. Appl. Phys. 108 (11) (2010) 113509.
[28] G. Shirane, H. Danner, A. Pavlovic, R. Pepinsky, Phase transitions in ferroelectric $KNbO_3$, Phys. Rev. 93 (4) (1954) 672–673.
[29] J. Tellier, B. Malic, B. Dkhil, D. Jenko, J. Cilensek, M. Kosec, Crystal structure and phase transitions of sodium potassium niobate perovskites, Solid State Sci. 11 (2) (2009) 320–324.
[30] D.W. Baker, P.A. Thomas, N. Zhang, A.M. Glazer, A comprehensive study of the phase diagram of $K_xNa_{1-x}NbO_3$, Appl. Phys. Lett. 95 (9) (2009) 091903.
[31] L. Egerton, D.M. Dillon, Piezoelectric and dielectric properties of ceramics in the system potassium-sodium niobate, J. Am. Ceram. Soc. 42 (9) (1959) 438–442.

[32] M. Fukada, T. Saito, H. Kume, T. Wada, Fabrication of lead-free piezoelectric $(Na_{0.5}K_{0.5})NbO_3$ ceramics by a modified solid-state reaction method, IEEE T. Ultrason. Ferr. 55 (5) (2008) 988–993.
[33] Y.H. Lee, J.H. Cho, B.I. Kim, D.K. Choi, Piezoelectric properties and densification based on control of volatile mass of potassium and sodium in $(K_{0.5}Na_{0.5})NbO_3$ ceramics, Jpn. J. Appl. Phys. 47 (6) (2008) 4620–4622.
[34] B.P. Zhang, L.M. Zhang, J.F. Li, X.N. Ding, H.L. Zhang, Effect of sintering temperature on electrical properties of $Na_{0.5}K_{0.5}NbO_3$ lead-free piezoelectric ceramics prepared by normal sintering, Ferroelectrics 358 (2007) 188–195.
[35] R.Z. Zuo, J. Rödel, R.Z. Chen, L.T. Li, Sintering and electrical properties of lead-free $Na_{0.5}K_{0.5}NbO_3$ piezoelectric ceramics, J. Am. Ceram. Soc. 89 (6) (2006) 2010–2015.
[36] R.E. Jaeger, L. Egerton, Hot pressing of potassium-sodium niobates, J. Am. Ceram. Soc. 45 (5) (1962) 209–213.
[37] G.H. Haertling, Properties of hot-pressed ferroelectric alkali niobate ceramics, J. Am. Ceram. Soc. 50 (6) (1967) 329–330.
[38] J.F. Li, K. Wang, B.P. Zhang, L.M. Zhang, Ferroelectric and piezoelectric properties of fine-grained $Na_{0.5}K_{0.5}NbO_3$ lead-free piezoelectric ceramics prepared by spark plasma sintering, J. Am. Ceram. Soc. 89 (2) (2006) 706–709.
[39] Y. Saito, H. Takao, T. Tani, T. Nonoyama, K. Takatori, T. Homma, et al., Lead-free piezoceramics, Nature 432 (7013) (2004) 84–87.
[40] D. Damjanovic, Contributions to the piezoelectric effect in ferroelectric single crystals and ceramics, J. Am. Ceram. Soc. 88 (10) (2005) 2663–2676.
[41] E. Hollenstein, D. Damjanovic, N. Setter, Temperature stability of the piezoelectric properties of Li-modified KNN ceramics, J. Eur. Ceram. Soc. 27 (13-15) (2007) 4093–4097.
[42] E.K. Akdogan, K. Kerman, M. Abazari, A. Safari, Origin of high piezoelectric activity in ferroelectric $(K_{0.44}Na_{0.52}Li_{0.04})(Nb_{0.84}Ta_{0.1}Sb_{0.06})O_3$ ceramics, Appl. Phys. Lett. 92 (11) (2008).
[43] S.J. Zhang, R. Xia, T.R. Shrout, Modified $(K_{0.5}Na_{0.5})NbO_3$ based lead-free piezoelectrics with broad temperature usage range, Appl. Phys. Lett. 91 (13) (2007).
[44] K. Wang, F.Z. Yao, W. Jo, D. Gobeljic, V.V. Shvartsman, D.C. Lupascu, et al., Temperature-insensitive $(K,Na)NbO_3$-based lead-free piezoactuator ceramics, Adv. Funct. Mater. 23 (33) (2013) 4079–4086.
[45] Y.P. Guo, K. Kakimoto, H. Ohsato, Phase transitional behavior and piezoelectric properties of $(Na_{0.5}K_{0.5})NbO_3$-$LiNbO_3$ ceramics, Appl. Phys. Lett. 85 (18) (2004) 4121–4123.
[46] K. Higashide, K.I. Kakimoto, H. Ohsato, Temperature dependence on the piezoelectric property of $(1-x)(Na_{0.5}K_{0.5})NbO_3$-$xLiNbO_3$ ceramics, J. Eur. Ceram. Soc. 27 (13-15) (2007) 4107–4110.
[47] E. Hollenstein, M. Davis, D. Damjanovic, N. Setter, Piezoelectric properties of Li- and Ta-modified $(K_{0.5}Na_{0.5})NbO_3$ ceramics, Appl. Phys. Lett. 87 (18) (2005) 182905.
[48] S.J. Zhang, R. Xia, T.R. Shrout, G.Z. Zang, J.F. Wang, Characterization of lead free $(K_{0.5}Na_{0.5})NbO_3$-$LiSbO_3$ piezoceramic, Solid State Commun. 141 (12) (2007) 675–679.
[49] R.P. Wang, R.J. Xie, K. Hanada, K. Matsusaki, H. Bando, M. Itoh, Phase diagram and enhanced piezoelectricity in the strontium titanate doped potassium-sodium niobate solid solution, Phys. Status Solidi. A 202 (6) (2005) R57–R59.
[50] R.P. Wang, H. Bando, T. Katsumata, Y. Inaguma, H. Taniguchi, M. Itoh, Tuning the orthorhombic-rhombohedral phase transition temperature in sodium potassium niobate by incorporating barium zirconate, Phys Status Solidi-R 3 (5) (2009) 142–144.

[51] R.P. Wang, K. Wang, F.Z. Yao, J.F. Li, F.H. Schader, K.G. Webber, et al., Temperature stability of lead-free niobate piezoceramics with engineered morphotropic phase boundary, J. Am. Ceram. Soc. 98 (7) (2015) 2177–2182.
[52] X.P. Wang, J.G. Wu, D.Q. Xiao, J.G. Zhu, X.J. Cheng, T. Zheng, et al., Giant piezoelectricity in potassium-sodium niobate lead-free ceramics, J. Am. Chem. Soc. 136 (7) (2014) 2905–2910.
[53] M. Matsubara, T. Yamaguchi, K. Kikuta, S. Hirano, Sinterability and piezoelectric properties of $(K,Na)NbO_3$ ceramics with novel sintering aid, Jpn. J. Appl. Phys. 43 (10) (2004) 7159–7163.
[54] M. Matsubara, T. Yamaguchi, W. Sakamoto, K. Kikuta, T. Yogo, S. Hirano, Processing and piezoelectric properties of lead-free $(K,Na)(Nb,Ta)O_3$ ceramics, J. Am. Ceram. Soc. 88 (5) (2005) 1190–1196.
[55] M. Matsubara, T. Yamaguchi, K. Kikuta, S. Hirano, Synthesis and characterization of $(K_{0.5}Na_{0.5})(Nb_{0.7}Ta_{0.3})O_3$ piezoelectric ceramics sintered with sintering aid $K_{5.4}Cu_{1.3}Ta_{10}O_{29}$, Jpn, J. Appl. Phys. 44 (9a) (2005) 6618–6623.
[56] M. Matsubara, T. Yamaguchi, K. Kikuta, S. Hirano, Sintering and piezoelectric properties of potassium sodium niobate ceramics with newly developed sintering aid, Jpn. J. Appl. Phys. 44 (1a) (2005) 258–263.
[57] S.H. Park, C.W. Ahn, S. Nahm, J.S. Song, Microstructure and piezoelectric properties of ZnO-added $(Na_{0.5}K_{0.5})NbO_3$ ceramics, Jpn. J. Appl. Phys. 43 (8b) (2004) L1072–L1074.
[58] J. Ryu, J.J. Choi, B.D. Hahn, D.S. Park, W.H. Yoon, K.Y. Kim, Sintering and piezoelectric properties of KNN ceramics doped with KZT, IEEE T. Ultrason. Ferr. 54 (12) (2007) 2510–2515.
[59] K.P. Chen, F.L. Zhang, J.Q. Zhou, X.W. Zhang, C.W. Li, L.N. An, Effect of borax addition on sintering and electrical properties of $(K_{0.5}Na_{0.5})NbO_3$ lead-free piezoceramics, Ceram. Int. 41 (8) (2015) 10232–10236.
[60] J. Bernard, A. Bencan, T. Rojac, J. Holc, B. Malic, M. Kosec, Low-temperature sintering of $K_{0.5}Na_{0.5}NbO_3$ ceramics, J. Am. Ceram. Soc. 91 (7) (2008) 2409–2411.
[61] B.C. Park, I.K. Hong, H.D. Jang, V.D.N. Tran, W.P. Tai, J.S. Lee, Highly enhanced mechanical quality factor in lead-free $(K_{0.5}Na_{0.5})NbO_3$ piezoelectric ceramics by co-doping with $K_{5.4}Cu_{1.3}Ta_{10}O_{29}$ and CuO, Mater. Lett. 64 (14) (2010) 1577–1579.
[62] H.-S. Han, J. Koruza, E.A. Patterson, J. Schultheiß, E. Erdem, W. Jo, et al., Hardening behavior and highly enhanced mechanical quality factor in $(K_{0.5}Na_{.5})NbO_3$-based ceramics, J. Eur. Ceram. Soc. 37 (5) (2017) 2083–2089.
[63] R.A. Eichel, E. Erünal, P. Jakes, S. Körbel, C. Elsässer, H. Kungl, et al., Interactions of defect complexes and domain walls in CuO-doped ferroelectric $(K,Na)NbO_3$, Appl. Phys. Lett. 102 (24) (2013) 242908.
[64] D. Schütz, M. Deluca, W. Krauss, A. Feteira, T. Jackson, K. Reichmann, Lone-pair-induced covalency as the cause of temperature- and field-induced instabilities in bismuth sodium titanate, Adv. Funct. Mater. 22 (11) (2012) 2285–2294.
[65] P. Baettig, C.F. Schelle, R. LeSar, U.V. Waghmare, N.A. Spaldin, Theoretical prediction of new high-performance lead-free piezoelectrics, Chem. Mater. 17 (6) (2005) 1376–1380.
[66] G.A. Smolenskii, V.A. Isupov, A.I. Agranovskaya, N.N. Krainik, New ferroelectrics of complex composition, Sov. Phys.-Sol. State 2 (11) (1961) 2651–2654.
[67] T. Takenaka, H. Nagata, Current status and prospects of lead-free piezoelectric ceramics, J. Eur. Ceram. Soc. 25 (12) (2005) 2693–2700.
[68] G.O. Jones, P.A. Thomas, Investigation of the structure and phase transitions in the novel A-site substituted distorted perovskite compound $Na_{0.5}Bi_{0.5}TiO_3$, Acta Crystallogr. B 58 (2) (2002) 168–178.

[69] A.M. Balagurov, E.Y. Koroleva, A.A. Naberezhnov, V.P. Sakhnenko, B.N. Savenko, N.V. Ter-Oganessian, et al., The rhombohedral phase with incommensurate modulation in $Na_{1/2}Bi_{1/2}TiO_3$, Phase Transit. 79 (1-2) (2006) 163–173.

[70] D.S. Keeble, E.R. Barney, D.A. Keen, M.G. Tucker, J. Kreisel, P.A. Thomas, Bifurcated polarization rotation in bismuth-based piezoelectrics, Adv. Funct. Mater. 23 (2) (2013) 185–190.

[71] E. Aksel, J.S. Forrester, J.C. Nino, K. Page, D.P. Shoemaker, J.L. Jones, Local atomic structure deviation from average structure of $Na_{0.5}Bi_{0.5}TiO_3$: combined x-ray and neutron total scattering study, Phys. Rev. B 87 (10) (2013) 104113.

[72] J. Kreisel, P. Bouvier, B. Dkhil, P.A. Thomas, A.M. Glazer, T.R. Welberry, et al., High-pressure x-ray scattering of oxides with a nanoscale local structure: application to $Na_{1/2}Bi_{1/2}TiO_3$, Phys. Rev. B 68 (1) (2003) 014113.

[73] S. Gorfman, P.A. Thomas, Evidence for a non-rhombohedral average structure in the lead-free piezoelectric material $Na_{0.5}Bi_{0.5}TiO_3$, J. Appl. Crystallogr. 43 (6) (2010) 1409–1414.

[74] E. Aksel, J.S. Forrester, J.L. Jones, P.A. Thomas, K. Page, M.R. Suchomel, Monoclinic crystal structure of polycrystalline $Na_{0.5}Bi_{0.5}TiO_3$, Appl. Phys. Lett. 98 (15) (2011) 152901.

[75] I. Levin, I.M. Reaney, Nano- and mesoscale structure of $Na_{1\backslash 2}Bi_{1\backslash 2}TiO_3$: a TEM perspective, Adv. Funct. Mater. 22 (16) (2012) 3445–3452.

[76] S. Gorfman, A.M. Glazer, Y. Noguchi, M. Miyayama, H. Luo, P.A. Thomas, Observation of a low-symmetry phase in $Na_{0.5}Bi_{0.5}TiO_3$ crystals by optical birefringence microscopy, J. Appl. Crystallogr. 45 (3) (2012) 444–452.

[77] B.N. Rao, R. Datta, S.S. Chandrashekaran, D.K. Mishra, V. Sathe, A. Senyshyn, et al., Local structural disorder and its influence on the average global structure and polar properties in $Na_{0.5}Bi_{0.5}TiO_3$, Phys. Rev. B 88 (22) (2013) 224103.

[78] D. Lin, D. Xiao, J. Zhu, P. Yu, Piezoelectric and ferroelectric properties of $[Bi_{0.5}(Na_{1-x-y}K_xLi_y)_{0.5}]TiO_3$ lead-free piezoelectric ceramics, Appl. Phys. Lett. 88 (6) (2006) 062901.

[79] E. Aksel, H. Foronda, K.A. Calhoun, J.L. Jones, S. Schaab, T. Granzow, Processing and properties of $Na_{0.5}Bi_{0.5}TiO_3$ piezoelectric ceramics modified with La, Mn and Fe, Funct. Mater. Lett. 3 (1) (2010) 45–48.

[80] E. Aksel, E. Erdem, P. Jakes, J.L. Jones, R.-A. Eichel, Defect structure and materials "hardening" in Fe_2O_3-doped $[Bi_{0.5}Na_{0.5}]TiO_3$ ferroelectrics, Appl. Phys. Lett. 97 (1) (2010) 012903.

[81] M. Davies, E. Aksel, J.L. Jones, Enhanced high-temperature piezoelectric coefficients and thermal stability of Fe- and Mn-substituted $Na_{0.5}Bi_{0.5}TiO_3$ ceramics, J. Am. Ceram. Soc. 94 (5) (2011) 1314–1316.

[82] M. Li, H. Zhang, S.N. Cook, L. Li, J.A. Kilner, I.M. Reaney, et al., Dramatic influence of A-site nonstoichiometry on the electrical conductivity and conduction mechanisms in the perovskite oxide $Na_{0.5}Bi_{0.5}TiO_3$, Chem. Mater. 27 (2) (2015) 629–634.

[83] T. Takenaka, K.-i Maruyama, K. Sakata, $Bi_{1/2}Na_{1/2})TiO_3$-$BaTiO_3$ system for lead-free piezoelectric ceramics, Jpn. J. Appl. Phys. 30 (9S) (1991) 2236.

[84] C. Ma, H. Guo, S.P. Beckman, X. Tan, Creation and destruction of morphotropic phase boundaries through electrical poling: a case study of lead-free $(Bi_{1/2}Na_{1/2})TiO_3$-$BaTiO_3$ piezoelectrics, Phys. Rev. Lett. 109 (10) (2012) 107602.

[85] F. Cordero, F. Craciun, F. Trequattrini, E. Mercadelli, C. Galassi, Phase transitions and phase diagram of the ferroelectric perovskite$(Na_{0.5}Bi_{0.5})_{1-x}Ba_xTiO_3$ by anelastic and dielectric measurements, Phys. Rev. B 81 (14) (2010) 144124.

[86] R. Garg, B.N. Rao, A. Senyshyn, P.S.R. Krishna, R. Ranjan, Lead-free piezoelectric system $(Na_{0.5}Bi_{0.5})TiO_3$-$BaTiO_3$: equilibrium structures and irreversible structural transformations driven by electric field and mechanical impact, Phys. Rev. B 88 (1) (2013) 014103.

[87] R. Ranjan, A. Dviwedi, Structure and dielectric properties of $(Na_{0.50}Bi_{0.50})_{1-x}Ba_xTiO_3$: $0 \leq x \leq 0.10$, Solid State Comm. 135 (6) (2005) 394–399.

[88] B.N. Rao, M. Avdeev, B. Kennedy, R. Ranjan, Phase boundary at $x = 0.03$ and its anomalous influence on the structure and properties in the lead-free piezoelectric $(1-x)Na_{1/2}Bi_{1/2}TiO_3-(x)BaTiO_3$, Phys. Rev. B 92 (21) (2015) 214107.

[89] W. Jo, S. Schaab, E. Sapper, L.A. Schmitt, H.-J. Kleebe, A.J. Bell, et al., On the phase identity and its thermal evolution of lead free $(Bi_{1/2}Na_{1/2})TiO_3$-6 mol% $BaTiO_3$, J. Appl. Phys. 110 (7) (2011) 074106.

[90] G. Picht, J. Töpfer, E. Hennig, Structural properties of $(Bi_{0.5}Na_{0.5})_{1-x}Ba_xTiO_3$ lead-free piezoelectric ceramics, J. Eur. Ceram. Soc. 30 (16) (2010) 3445–3453.

[91] M. Deluca, G. Picht, M.J. Hoffmann, A. Rechtenbach, J. Töpfer, F.H. Schader, et al., Chemical and structural effects on the high-temperature mechanical behavior of $(1-x)(Na_{1/2}Bi_{1/2})TiO_3$-$xBaTiO_3$ ceramics, J. Appl. Phys. 117 (13) (2015) 134110.

[92] K. Datta, K. Roleder, P.A. Thomas, Enhanced tetragonality in lead-free piezoelectric $(1-x)BaTiO_3-xNa_{1/2}Bi_{1/2}TiO_3$ solid solutions where $x = 0.05-0.40$, J. Appl. Phys. 106 (12) (2009) 123512.

[93] K. Datta, P.A. Thomas, K. Roleder, Anomalous phase transitions of lead-free piezoelectric $xNa_{0.5}Bi_{0.5}TiO_3-(1-x)BaTiO_3$ solid solutions with enhanced phase transition temperatures, Phys. Rev. B 82 (22) (2010) 224105.

[94] S.-T. Zhang, A.B. Kounga, E. Aulbach, H. Ehrenberg, J. Rödel, Giant strain in lead-free piezoceramics $Bi_{0.5}Na_{0.5}TiO_3-BaTiO_3-K_{0.5}Na_{0.5}NbO_3$ system, Appl. Phys. Lett. 91 (11) (2007) 112906.

[95] W. Jo, R. Dittmer, M. Acosta, J. Zang, C. Groh, E. Sapper, et al., Giant electric-field-induced strains in lead-free ceramics for actuator applications – status and perspective, J. Electroceram. 29 (1) (2012) 71–93.

[96] M. Hinterstein, M. Knapp, M. Hölzel, W. Jo, A. Cervellino, H. Ehrenberg, et al., Field-induced phase transition in $Bi_{1/2}Na_{1/2}TiO_3$-based lead-free piezoelectric ceramics, J. Appl. Crystallogr. 43 (6) (2010) 1314–1321.

[97] C. Groh, D.J. Franzbach, W. Jo, K.G. Webber, J. Kling, L.A. Schmitt, et al., Relaxor/ferroelectric composites: a solution in the quest for practically viable lead-free incipient piezoceramics, Adv. Funct. Mater. 24 (3) (2014) 356–362.

[98] H. Zhang, C. Groh, Q. Zhang, W. Jo, K.G. Webber, J. Rödel, Large strain in relaxor/ferroelectric composite lead-free piezoceramics, Adv. Electr. Mater. 1 (6) (2015) 1500018.

[99] J. Zhang, Z. Pan, F.F. Guo, W.C. Liu, H. Ning, Y.B. Chen, et al., Semiconductor/relaxor 0-3 type composites without thermal depolarization in $Bi_{0.5}Na_{0.5}TiO_3$-based lead-free piezoceramics, Nat. Commun. 6 (2015) 6615.

[100] L. Riemer, K.V. Lalitha, X. Jiang, N. Liu, C. Dietz, R. Stark, et al., Stress-induced phase transition in lead-free relaxor ferroelectric composites, Acta Mater. 136 (2017) 271–280.

[101] F.H. Schader, Z. Wang, M. Hinterstein, J.E. Daniels, K.G. Webber, Stress-modulated relaxor-to-ferroelectric transition in lead-free$(Na_{1/2}Bi_{1/2})TiO_3-BaTiO_3$ ferroelectrics, Phys. Rev. B 93 (13) (2016) 134111.

[102] K.V. Lalitha, L. Riemer, J. Koruza, J. Rödel, Hardening of electromechanical properties in piezoceramics using a composite approach, Appl. Phys. Lett. 111 (2) (2017) 022905.

[103] C.F. Buhrer, Some properties of bismuth perovskites, J. Chem. Phys. 36 (3) (1962) 798–803.

[104] V.A. Isupov, Ferroelectric $Na_{0.5}Bi_{0.5}TiO_3$ and $K_{0.5}Bi_{0.5}TiO_3$ perovskites and their solid solutions, Ferroelectrics 315 (1) (2005) 123–147.

[105] Y.-R. Zhang, J.-F. Li, B.-P. Zhang, Enhancing electrical properties in NBT-KBT lead-free piezoelectric ceramics by optimizing sintering temperature, J. Am. Ceram. Soc. 91 (8) (2008) 2716–2719.

[106] J. Kreisel, A.M. Glazer, G. Jones, P.A. Thomas, L. Abello, G. Lucazeau, An x-ray diffraction and Raman spectroscopy investigation of A-site substituted perovskite compounds: the $(Na_{1-x}K_x)_{0.5}Bi_{0.5}TiO_3$ $(0 \leq x \leq 1)$ solid solution, J. Phys.-Condens. Mat 12 (14) (2000) 3267–3280.

[107] Y. Hiruma, K. Yoshii, H. Nagata, T. Takenaka, Investigation of phase transition temperatures on $(Bi_{1/2}Na_{1/2})TiO_3$-$(Bi_{1/2}K_{1/2})TiO_3$ and $(Bi_{1/2}Na_{1/2})TiO_3$-$BaTiO_3$ lead-free piezoelectric ceramics by electrical measurements, Ferroelectrics 346 (1) (2007) 114–119.

[108] S. Zhao, G. Li, A. Ding, T. Wang, Q. Yin, Ferroelectric and piezoelectric properties of $(Na, K)_{0.5}Bi_{0.5}TiO_3$ lead free ceramics, J. Phys. D Appl. Phys. 39 (10) (2006) 2277–2281.

[109] M. Otoničar, S.D. Škapin, M. Spreitzer, D. Suvorov, Compositional range and electrical properties of the morphotropic phase boundary in the $Na_{0.5}Bi_{0.5}TiO_3$–$K_{0.5}Bi_{0.5}TiO_3$ system, J. Eur. Ceram. Soc. 30 (4) (2010) 971–979.

[110] I. Levin, I.M. Reaney, E.M. Anton, W. Jo, J. Rödel, J. Pokorny, et al., Local structure, pseudosymmetry, and phase transitions in $Na_{1/2}Bi_{1/2}TiO_3$–$K_{1/2}Bi_{1/2}TiO_3$ ceramics, Phys. Rev. B 87 (2) (2013) 024113.

[111] T.H. Dinh, J.K. Kang, J.S. Lee, N.H. Khansur, J. Daniels, H.Y. Lee, et al., Nanoscale ferroelectric/relaxor composites: origin of large strain in lead-free Bi-based incipient piezoelectric ceramics, J. Eur. Ceram. Soc. 36 (14) (2016) 3401–3407.

[112] A.J. Royles, A.J. Bell, A.P. Jephcoat, A.K. Kleppe, S.J. Milne, T.P. Comyn, Electric-field-induced phase switching in the lead free piezoelectric potassium sodium bismuth titanate, Appl. Phys. Lett. 97 (13) (2010) 132909.

[113] A.J. Royles, A.J. Bell, J.E. Daniels, S.J. Milne, T.P. Comyn, Observation of a time-dependent structural phase transition in potassium sodium bismuth titanate, Appl. Phys. Lett. 98 (18) (2011) 182904.

[114] M. Otonicar, J. Park, M. Logar, G. Esteves, J.L. Jones, B. Jancar, External-field-induced crystal structure and domain texture in $(1-x)Na_{0.5}Bi_{0.5}TiO_3$–$xK_{0.5}Bi_{0.5}TiO_3$ piezoceramics, Acta Mater. 127 (1) (2017) 319–331.

[115] Y. Hiruma, Y. Watanabe, H. Nagata, T. Takenaka, Piezoelectric properties of $(Bi_{1/2}Na_{1/2})TiO_3$-based solid solution for lead-free high-power applications, Jpn. J. Appl. Phys. 47 (9) (2008) 7659–7663.

[116] K. Sakata, Y. Masuda, Ferroelectric and antiferroelectric properties of $(Na_{0.5}Bi_{0.5})TiO_3$-$SrTiO_3$ solid solution ceramics, Ferroelectrics 7 (1) (1974) 347–349.

[117] D. Rout, K.-S. Moon, S.-J.L. Kang, I.W. Kim, Dielectric and Raman scattering studies of phase transitions in the $(100-x)Na_{0.5}Bi_{0.5}TiO_3$–$xSrTiO_3$ system, J. Appl. Phys. 108 (8) (2010) 084102.

[118] Y. Hiruma, Y. Imai, Y. Watanabe, H. Nagata, T. Takenaka, Large electrostrain near the phase transition temperature of $(Bi_{0.5}Na_{0.5})TiO_3$–$SrTiO_3$ ferroelectric ceramics, Appl. Phys. Lett. 92 (26) (2008) 262904.

[119] M. Acosta, W. Jo, J. Rödel, D.C. Lupascu, Temperature- and frequency-dependent properties of the $0.75Bi_{1/2}Na_{1/2}TiO_3$-$0.25SrTiO_3$ lead-free incipient piezoceramic, J. Am. Ceram. Soc. 97 (6) (2014) 1937–1943.
[120] F. Wang, M. Xu, Y. Tang, T. Wang, W. Shi, C.M. Leung, Large strain response in the ternary $Bi_{0.5}Na_{0.5}TiO_3$-$BaTiO_3$-$SrTiO_3$ solid solutions, J. Am. Ceram. Soc. 95 (6) (2012) 1955–1959.
[121] K. Wang, A. Hussain, W. Jo, J. Rödel, Temperature-dependent properties of $(Bi_{1/2}Na_{1/2})TiO_3$-$(Bi_{1/2}K_{1/2})TiO_3$-$SrTiO_3$ lead-free piezoceramics, J. Am. Ceram. Soc. 95 (7) (2012) 2241–2247.
[122] J. Koruza, V. Rojas, L. Molina-Luna, U. Kunz, M. Duerrschnabel, H.-J. Kleebe, et al., Formation of the core–shell microstructure in lead-free $Bi_{1/2}Na_{1/2}TiO_3$-$SrTiO_3$ piezoceramics and its influence on the electromechanical properties, J. Eur. Ceram. Soc. 36 (4) (2016) 1009–1016.
[123] M. Acosta, L.A. Schmitt, L. Molina-Luna, M.C. Scherrer, M. Brilz, K.G. Webber, et al., Core-shell lead-free piezoelectric ceramics: current status and advanced characterization of the $Bi_{1/2}Na_{1/2}TiO_3$-$SrTiO_3$ system, J. Am. Ceram. Soc. 98 (11) (2015) 3405–3422.
[124] M. McQuarrie, F.W. Behnke, Structural and dielectric studies in the system (Ba,Ca)(Ti,Zr)O_3, J. Am. Ceram. Soc. 37 (11) (1954) 539–543.
[125] W.F. Liu, X.B. Ren, Large piezoelectric effect in Pb-free ceramics, Phys. Rev. Lett. 103 (25) (2009) 257602.
[126] M.C. Ehmke, S.N. Ehrlich, J.E. Blendell, K.J. Bowman, Phase coexistence and ferroelastic texture in high strain (1-x)$Ba(Zr_{0.2}Ti_{0.8})O_3$-x$(Ba_{0.7}Ca_{0.3})TiO_3$ piezoceramics, J. Appl. Phys. 111 (12) (2012) 124110.
[127] D.S. Keeble, F. Benabdallah, P.A. Thomas, M. Maglione, J. Kreisel, Revised structural phase diagram of $(Ba_{0.7}Ca_{0.3}TiO_3)$-$(BaZr_{0.2}Ti_{0.8}O_3)$, Appl. Phys. Lett. 102 (9) (2013) 092903.
[128] M. Acosta, N. Novak, W. Jo, J. Rödel, Relationship between electromechanical properties and phase diagram in the $Ba(Zr_{0.2}Ti_{0.8})O_3$-x$(Ba_{0.7}Ca_{0.3})TiO_3$ lead-free piezoceramic, Acta Mater. 80 (2014) 48–55.
[129] A.B. Haugen, J.S. Forrester, D. Damjanovic, B.Z. Li, K.J. Bowman, J.L. Jones, Structure and phase transitions in $0.5(Ba_{0.7}Ca_{0.3}TiO_3)$-$0.5(BaZr_{0.2}Ti_{0.8}O_3)$ from -100 °C to 150 °C, J. Appl. Phys. 113 (1) (2013) 014103.
[130] Y. Tian, L.L. Wei, X.L. Chao, Z.H. Liu, Z.P. Yang, Phase transition behavior and large piezoelectricity near the morphotropic phase boundary of lead-free $(Ba_{0.85}Ca_{0.15})(Zr_{0.1}Ti_{0.9})O_3$ ceramics, J. Am. Ceram. Soc. 96 (2) (2013) 496–502.
[131] L. Zhang, M. Zhang, L. Wang, C. Zhou, Z. Zhang, Y.G. Yao, et al., Phase transitions and the piezoelectricity around morphotropic phase boundary in $Ba(Zr_{0.2}Ti_{0.8})O_3$-x$(Ba_{0.7}Ca_{0.3})TiO_3$ lead-free solid solution, Appl. Phys. Lett. 105 (16) (2014) 162908.
[132] M. Acosta, N. Khakpash, T. Someya, N. Novak, W. Jo, H. Nagata, et al., Origin of the large piezoelectric activity in (1-x)$Ba(Zr_{0.2}Ti_{0.8})O_3$-x$(Ba_{0.7}Ca_{0.3})TiO_3$ ceramics, Phys. Rev. B 91 (10) (2015) 104108.
[133] A.A. Heitmann, G.A. Rossetti, Thermodynamics of ferroelectric solid solutions with morphotropic phase boundaries, J. Am. Ceram. Soc. 97 (6) (2014) 1661–1685.
[134] M. Acosta, N. Novak, G.A. Rossetti, J. Rödel, Mechanisms of electromechanical response in (1-x)$Ba(Zr_{0.2}Ti_{0.8})O_3$-x$(Ba_{0.7}Ca_{0.3})TiO_3$ ceramics, Appl. Phys. Lett. 107 (14) (2015) 142906.
[135] P. Wang, Y.X. Li, Y.Q. Lu, Enhanced piezoelectric properties of $(Ba_{0.85}Ca_{0.15})(Ti_{0.9}Zr_{0.1})O_3$ lead-free ceramics by optimizing calcination and sintering temperature, J. Eur. Ceram. Soc. 31 (11) (2011) 2005–2012.

[136] J.G. Hao, W.F. Bai, W. Li, J.W. Zhai, Correlation between the microstructure and electrical properties in high-performance $(Ba_{0.85}Ca_{0.15})(Zr_{0.1}Ti_{0.9})O_3$ lead-free piezoelectric ceramics, J. Am. Ceram. Soc. 95 (6) (2012) 1998–2006.

[137] M.C. Ehmke, F.H. Schader, K.G. Webber, J. Rödel, J.E. Blendell, K.J. Bowman, Stress, temperature and electric field effects in the lead-free $(Ba,Ca)(Ti,Zr)O_3$ piezoelectric system, Acta Mater. 78 (2014) 37–45.

[138] D.R.J. Brandt, M. Acosta, J. Koruza, K.G. Webber, Mechanical constitutive behavior and exceptional blocking force of lead-free BZT-xBCT piezoceramics, J. Appl. Phys. 115 (20) (2014) 204107.

[139] K.G. Webber, D.J. Franzbach, J. Koruza, Determination of the true operational range of a piezoelectric actuator, J. Am. Ceram. Soc. 97 (9) (2014) 2842–2849.

[140] F. Li, L. Jin, R.P. Guo, High electrostrictive coefficient Q_{33} in lead-free $Ba(Zr_{0.2}Ti_{0.8})O_3$-$x(Ba_{0.7}Ca_{0.3})TiO_3$ piezoelectric ceramics, Appl. Phys. Lett. 105 (23) (2014) 232903.

[141] G. Tutuncu, B.Z. Li, K. Bowman, J.L. Jones, Domain wall motion and electromechanical strain in lead-free piezoelectrics: Insight from the model system $(1-x)Ba(Zr_{0.2}Ti_{0.8})O_3$-$x(Ba_{0.7}Ca_{0.3})TiO_3$ using in situ high-energy X-ray diffraction during application of electric fields, J. Appl. Phys. 115 (14) (2014) 144104.

[142] M.C. Ehmke, N.H. Khansur, J.E. Daniels, J.E. Blendell, K.J. Bowman, Resolving structural contributions to the electric-field-induced strain in lead-free $(1-x)Ba(Zr_{0.2}Ti_{0.8})O_3$ - $x(Ba_{0.7}Ca_{0.3})TiO_3$ piezoceramics, Acta Mater. 66 (2014) 340–348.

[143] J.H. Gao, X.H. Hu, L. Zhang, F. Li, L.X. Zhang, Y. Wang, et al., Major contributor to the large piezoelectric response in $(1-x)Ba(Zr_{0.2}Ti_{0.8})O_3$ - $x(Ba_{0.7}Ca_{0.3})TiO_3$ ceramics: domain wall motion, Appl. Phys. Lett. 104 (25) (2014) 252909.

[144] Y.C. Zhang, J. Glaum, M.C. Ehmke, J.E. Blendell, K.J. Bowman, M.J. Hoffman, High bipolar fatigue resistance of BCTZ lead-free piezoelectric ceramics, J. Am. Ceram. Soc. 99 (1) (2016) 174–182.

[145] Y.C. Zhang, J. Glaum, M.C. Ehmke, K.J. Bowman, J.E. Blendell, M.J. Hoffman, Unipolar fatigue behavior of BCTZ lead-free piezoelectric ceramics, J. Am. Ceram. Soc. 99 (4) (2016) 1287–1293.

[146] M. Vögler, M. Acosta, D.R.J. Brandt, L. Molina-Luna, K.G. Webber, Temperature-dependent R-curve behavior of the lead-free ferroelectric $0.615Ba(Zr_{0.2}Ti_{0.8})O_3$-$0.385(Ba_{0.7}Ca_{0.3})TiO_3$ ceramic, Eng. Fract. Mech. 144 (2015) 68–77.

[147] X.F. Wang, P.F. Liang, L.L. Wei, X.L. Chao, Z.P. Yang, Phase evolution and enhanced electrical properties of $(Ba_{0.85}Ca_{0.15-x}Y_x)(Zr_{0.1}Ti_{0.9})O_3$ lead-free ceramics, J. Mater. Sci. 26 (7) (2015) 5217–5225.

[148] H.R. Liu, Q. Li, Y.Y. Li, N.N. Luo, J. Shim, J.H. Gao, et al., Structure evolution and electrical properties of Y^{3+}-doped $Ba_{1-x}Ca_xZr_{0.07}Ti_{0.93}O_3$ ceramics, J. Am. Ceram. Soc. 97 (7) (2014) 2076–2081.

[149] J. Li, X.J. Sun, X.S. Zhang, Q. Chen, Z.H. Peng, P. Yu, Synthesis and characterization of sol-gel derived $Ba_{0.85}Ca_{0.15}Ti_{0.9}Zr_{0.1}O_3$-xCuO ceramics, Phys. Status Solidi. A 210 (3) (2013) 533–537.

[150] J.G. Wu, D.Q. Xiao, W.J. Wu, Q. Chen, J.G. Zhu, Z.C. Yang, et al., Role of room-temperature phase transition in the electrical properties of $(Ba,Ca)(Ti,Zr)O_3$ ceramics, Scripta Mater. 65 (9) (2011) 771–774.

[151] S.H. Shin, J.D. Han, J. Yoo, Piezoelectric and dielectric properties of B_2O_3-added $(Ba_{0.85}Ca_{0.15})(Ti_{0.915}Zr_{0.085})O_3$ ceramics sintered at low temperature, Mater. Lett. 154 (2015) 120–123.

[152] R. Hayati, M.A. Bahrevar, T. Ebadzadeh, V. Rojas, N. Novak, J. Koruza, Effects of Bi_2O_3 additive on sintering process and dielectric, ferroelectric, and piezoelectric

properties of $(Ba_{0.85}Ca_{0.15})(Zr_{0.1}Ti_{0.9})O_3$ lead-free piezoceramics, J. Eur. Ceram. Soc. 36 (14) (2016) 3391–3400.
[153] E. Chandrakala, J.P. Praveen, A. Kumar, A.R. James, D. Das, Strain-induced structural phase transition and its effect on piezoelectric properties of (BZT-BCT)-(CeO_2) ceramics, J. Am. Ceram. Soc. 99 (11) (2016) 3659–3669.
[154] C. Han, J.G. Wu, C.H. Pu, S. Qiao, B. Wu, J.G. Zhu, et al., High piezoelectric coefficient of Pr_2O_3-doped $Ba_{0.85}Ca_{0.15}Ti_{0.90}Zr_{0.10}O_3$ ceramics, Ceram. Int. 38 (8) (2012) 6359–6363.
[155] X. Xia, X.P. Jiang, C. Chen, X.G. Jiang, N. Tu, Y.J. Chen, Effects of Cr_2O_3 doping on the microstructure and electrical properties of $(Ba,Ca)(Zr,Ti)O_3$ lead-free ceramics, Front, Mater. Sci. 10 (2) (2016) 203–210.
[156] X.F. Wang, P.F. Liang, X.L. Chao, Z.P. Yang, Dielectric properties and impedance spectroscopy of $MnCO_3$-modified $(Ba_{0.85}Ca_{0.15})(Zr_{0.1}Ti_{0.9})O_3$ lead-free ceramics, J. Am. Ceram. Soc. 98 (5) (2015) 1506–1514.
[157] H.J. Sun, Y. Zhang, X.F. Liu, Y. Liu, S.S. Guo, W. Chen, Effects of cobalt and sintering temperature on electrical properties of $Ba_{0.98}Ca_{0.02}Zr_{0.02}Ti_{0.98}O_3$ lead-free ceramics, J. Mater. Sci. 25 (9) (2014) 3962–3966.
[158] D.Z. Xue, Y.M. Zhou, H.X. Bao, J.H. Gao, C. Zhou, X.B. Ren, Large piezoelectric effect in Pb-free $Ba(Ti,Sn)O_3$-x$(Ba,Ca)TiO_3$ ceramics, Appl. Phys. Lett. 99 (12) (2011) 122901.
[159] C. Zhou, W.F. Liu, D.Z. Xue, X.B. Ren, H.X. Bao, J.H. Gao, et al., Triple-point-type morphotropic phase boundary based large piezoelectric Pb-free material-$Ba(Ti_{0.8}Hf_{0.2})O_3$-$(Ba_{0.7}Ca_{0.3})TiO_3$, Appl. Phys. Lett. 100 (22) (2012) 222910.
[160] D.Y. Liang, X.H. Zhu, Y. Zhang, W. Shi, J.L. Zhu, Large piezoelectric effect in (1-x)$Ba(Zr_{0.15}Ti_{0.85})O_3$-x$(Ba_{0.8}Sr_{0.2})TiO_3$ lead-free ceramics, Ceram. Int. 41 (6) (2015) 8261–8266.

Perovskite layer-structured ferroelectrics

4

Aimé Peláiz-Barranco[1], José de los Santos Guerra[2], Yuslín González-Abreu[1] and Idalci Cruvinel dos Reis[3]
[1]University of Havana, Havana, Cuba, [2]Federal University of Uberlândia, Uberlândia, MG, Brazil, [3]Federal Institute of Education, Science and Technology of Goiano, Rio Verde, GO, Brazil

4.1 General overview

The investigation of the physical properties of materials, viewing the optimization as well as the reduction in their dimension, to be used in several application fields has been the focus of most of the research during the last few decades. In this way, dielectric systems have been conventionally used as host-matrix containing numerous compounds and many different dopants [1−3]. Those materials reveal peculiar characteristics, which make them potential candidates for practical applications. For instance, high dielectric permittivity, low dielectric losses, and high mechanical resistance can be mentioned as some of the required characteristics for application in specific electronic devices when compared with other kinds of materials [4]. In particular, ferroelectric systems, well known as classical dielectric materials, have shown additional important characteristics, such as the electric field modulation of the macroscopic polarization, which was revealed as the starting point for the development of a new class of materials with a promissory future, from technological and scientific points of view [1]. The application fields extend in a very broad temperature range and can be delimited by a critical point, named the Curie temperature (T_C). Below T_C the materials are characterized by a spontaneous polarization, which determines the ferroelectric phase, whereas above T_C the spontaneous polarization disappears, rendering the system in a nonpolar state, the so-called paraelectric phase [4]. Some of the most important applications of the ferroelectric systems are related to their use in random access memories, piezoelectric transducers, pyroelectric sensors, and others [2,5], which are determined by the enhancement of a property. During the last years intensive efforts have been focused on the optimization of both synthesis and processing methods of these materials, not only in ceramics, but also as thin films, in order to improve their characteristics for their integration in electronic devices [1,5]. In this way, a wide variety of ferroelectric systems have been obtained containing pure and/or modified compounds based on $Pb(Zr,Ti)O_3$ (PZT), $Pb(Mg_{1/3}Nb_{2/3})O_3$ (PMN), $Pb(Zn_{1/3}Nb_{2/3})O_3$ (PZN), and others, with excellent physical properties [6,7].

Magnetic, Ferroelectric, and Multiferroic Metal Oxides. DOI: http://dx.doi.org/10.1016/B978-0-12-811180-2.00004-9

However, there has been an increasing interest in recent years in the reduction of the use of toxic and hazardous elements in the fabrication process of materials, considering environment preservation. In this way, a different class of systems—lead-free ferroelectrics—are now being considered by the scientific community as potentially attractive and suitable alternatives for numerous applications, substituting for lead-containing compounds. Nowadays, there has been a continuous interest on bismuth layer-structured ferroelectrics (BLSFs) because of their potential use for nonvolatile memories and high temperature piezoelectrics applications [2,8]. These compounds, commonly known as the Aurivillius family, possess a complex crystal structure where bismuth layers ($[Bi_2O_2]^{2+}$) alternate with perovskite blocks (BO_6 octahedral) [4,9]. They can be represented by the chemical formula $A_{m-1}Bi_2B_mO_{3m+3}$, m being the number of perovskite blocks. The A- and B-sites can be occupied by different elements, so that it is possible to obtain numerous compounds for this family. On the other hand, it is possible to find a cation-site mixing among specific atoms positions, in particular between the Bi^{3+} cation and the A-site of the perovskite blocks, which contribute to the enhancement of their physical properties [10].

From a fundamental point of view, the highly anisotropic crystal structure of these materials, where the c-axis is larger than a and b, results in a very strong anisotropy in both electrical and optical properties of such compounds. It has been reported that the ferroelectric characteristics occur predominantly in the a–b plane of the perovskite blocks, leading to a high coercive electric field and reduced piezoelectric properties [11]. In fact, the major contribution to the ferroelectric polarization can arise from the displacement of the A-site cations along the a-axis. It is important to point out that most of the applications of the Aurivillius compounds are strongly limited by some undesirable characteristics, such as low remnant polarization, high dielectric losses (in some cases), and relatively high electrical conductivity, which acts mostly along the polarization plane [12,13]. To overcome such difficulties, the doping and codoping with elements of different natures, in both A- and B-sites, lead to a significant improvement of the physical properties. This topic has been in fact the motivation of recent studies on BLSFs, providing interesting and promising new results [14–19], when compared to the investigation of the physical properties in classic ferroelectrics. Figs. 4.1A and B summarize the publications on classic ferroelectric systems and bismuth-layer-structured perovskites, respectively, reported from 1945 up to 2016, according to the Web of Science database. The sudden increase on the investigation of BLSF compounds from the 1990s to today reveals their importance and priority concerning the awareness for the development of environmentally friendly lead-free materials. The aim of the present chapter is to present the recent advances on the physical properties of multilayer BLSF compounds. In particular, the current status of their structural, dielectric, electrical, piezoelectric, and pyroelectric characteristics is presented and discussed, viewing the potentiality for practical applications.

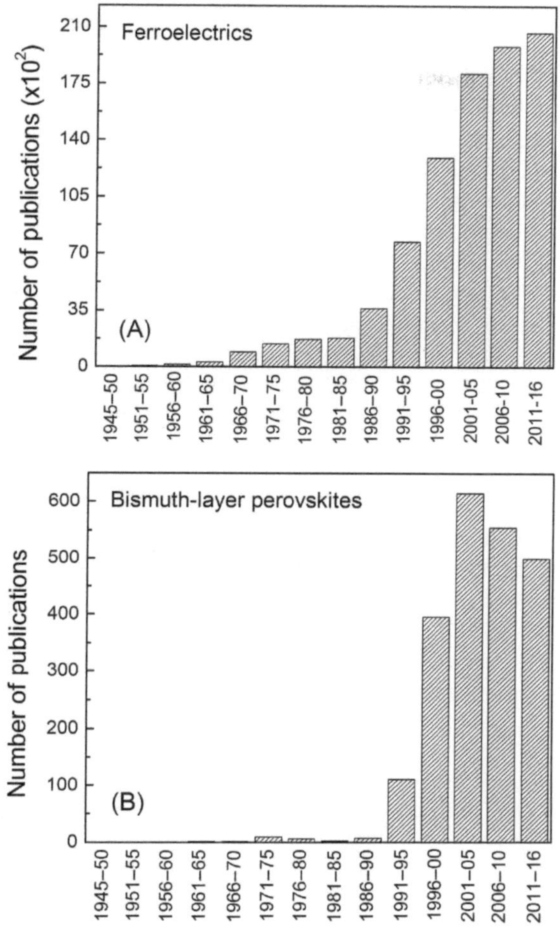

Figure 4.1 Publications on ferroelectric systems (A) and bismuth-later-structured perovskites ferroelectrics (B) reported from 1945 to 2016, according to the Web of Science database.

4.2 Physical properties

4.2.1 Single and bilayer bismuth layer-structured ferroelectrics

The structural studies in Aurivillius materials have shown important changes in the symmetry from the incorporation of different elements in the A-sites of the structure or with the increase of the number of perovskite blocks [10,14,20–23]. By using room-temperature X-ray diffraction technique, Wolfe et al. reported on the most studied single-layered Aurivillius material, Bi_2WO_6, the same orthorhombic structure (with *B2cb* space group) as that reported for an iso-structural Bi_2MoO_6 system [24]. Raman spectroscopy studies have shown light difference between the spectra of the Bi_2MoO_6 and Bi_2WO_6 systems in agreement with the X-ray diffraction results. The results reveal that the effect of substitution of Mo^{6+} by W^{6+} is small since the B-site is a center of symmetry and the size of the BO_6 octahedral is relatively independent of the cationic substitution [25].

On the other hand, similar studies for bilayered Aurivillius systems have shown different scenarios due to the existence of an additional perovskite block, or the substitution of the element in the A-site [10,14,20–23,25–30]. The structural properties results, which were reported by Blake et al. for the $ABi_2Nb_2O_9$ (A = Ba, Sr, Ca) system, showed an increase of the structural distortion with the decrease of the A-site cation size. The samples containing Sr^{2+} and Ca^{2+} cations showed an orthorhombic (*A2₁am* space group) distortion, whereas the sample containing Ba^{2+} revealed a tetragonal symmetry (*I4/mmm* space group). An additional study carried out by Ismunandar et al. in $SrBi_2NbO_9$ and $BaBi_2NbO_9$ samples revealed a noticeable disorder for the Sr^{2+} and Ba^{2+} cations [22]. The degree of the disorder has been reported to be higher for Ba^{2+} doping compositions [22]. The inhomogeneous distribution of cations provides the structural disorder and could be reflected in the distortion of the BO_6 octahedral [14]. From this point of view, several works have shown a different approach to explore the structure of bilayered Aurivillius systems [25–31]. Extensive research on the structural properties have been carried out in $Sr_{1-x}Ba_xBi_2Nb_2O_9$ (SBBN-*x*) [26–28], $Sr_{1-3x/2}La_xBi_2Nb_2O_9$ (SLBN-*x*), and $Sr_{1-3x/2}La_xBi_2Ta_2O_9$ (SLBT-*x*) [31] systems. The X-ray diffraction at room temperature and the refined data by using Rietveld refinement showed an orthorhombic distortion ($A2_1am$ space group). Fig. 4.2 shows the doping cation concentration (*x*)

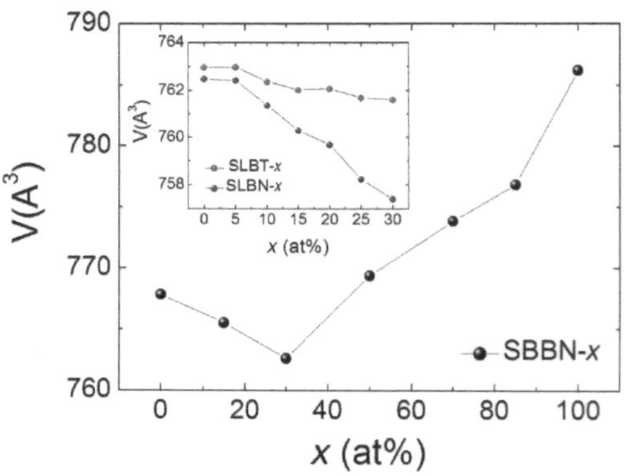

Figure 4.2 Doping cation (La^{3+} and Ba^{2+}) concentration (*x*) dependence of the unit-cell volume (V) for SLBT-*x*, SLBN-*x*, and SBBN-*x* ceramic systems.
Source: Data taken from Y.G. Abreu, Obtención y caracterización de materiales ferroeléctricos libres de plomo, PhD Thesis, Universidad de la Habana, La Habana, Cuba, 2014; I. Cruvinel dos Reis, Estudo das propriedades físicas de sistemas ferroelétricos com estruturas tipo Aurivillius, PhD Thesis, Universidade Estadual Paulista – UNESP "Júlio de Mesquita Filho", Ilha Solteira, São Paulo, Brazil, 2015.

dependence of the unit-cell volume (V). As can be seen in the inset of Fig. 4.2, the SLBT-*x* and SLBN-*x* systems show a reduction of the unit-cell volume with the increase of La^{3+} content, suggesting a higher incorporation of this element in the structure through the substitution of the Sr^{2+} cation. Room-temperature Raman spectroscopy results have also suggested the substitution of Sr^{2+} by La^{3+}ions for the SLBT-*x* and SLBN-*x* samples [31]. For the SLBN-*x* system, the possibility of a lower incorporation of La^{3+} in the $[Bi_2O_2]$ layers, replacing the Bi^{3+}, has been also considered [31].

On the other hand, the SBBN-*x* system showed a quite different behavior of the unit-cell volume with the increase of the doping element (Ba^{2+}) concentration [28]. The volume revealed to decreases for Ba^{2+} concentration from 0 to 30 at% and then increases for higher concentrations. The Raman spectroscopy measurements [32] and the refined structural results showed a cation mixing (Ba^{2+}, Sr^{2+}, Bi^{3+}) between the A- and bismuth-sites in $[Bi_2O_2]^{2-}$ layers. For $x \leq 30$ at%, an increase of the Bi^{3+}occupation at A-sites has been obtained, while for $x \geq 50$ at% such occupation tends to decrease. The cations mixing between the A- and bismuth-sites occurs in order to preserve the stability between the $[Bi_2O_2]^{2-}$ layers and the perovskite blocks. Additional studies on the structural properties in bilayered Aurivillius systems have shown a tetragonal structure (*I4/mmm* space group) at high temperatures for most of such compounds [29,33]. A particular behavior has been observed for $Sr_{0.85}Bi_{2.1}Ta_2O_9$ by using Rietveld refinement. Subsequence structural transitions from orthorhombic ($A2_1am$ space group) to orthorhombic (*Amam* space group) at 375°C and from orthorhombic (*Amam* space group) to tetragonal (*I4/mmm* space group) for temperatures above 550°C have been reported [33].

The main contribution to the polarization in Aurivillius systems is related to the displacement of the A-site cations. For odd *m* values, the highest polarization values have been found to occur in the *a*–*b* plane, whereas lower polarization values are obtained along the *c* axes; for even *m* values, the polarization is limited to the *a*–*b* plane due the presence of a mirror symmetry element perpendicular to the *c* axes [13,29].

Several studies have been carried out concerning the polarization values for single- and bilayered Aurivillius systems [13,24,29,33]. A spontaneous polarization value around $P_s = 42\ \mu C/cm^2$ has been reported for Bi_2WO_6 [24]. For bilayered materials some reported values are found to be around 5.84 $\mu C/cm^2$ and 8.0 $\mu C/cm^2$, for the $SrBi_2Ta_2O_9$ and $Sr_{0.5}Ba_{0.5}Bi_2Ta_2O_9$ systems, respectively [33]. For pure ($Na_{0.5}Bi_{2.5}TaO_9$) and doped $(Na,Bi)_{0.47}(Li,Ce)_{0.03}Bi_2Ta_{1.97}Sc_{0.03}O_{8.97}$ ceramics P_s values around 24.1 and 23.9 $\mu C/cm^2$, respectively, have been reported from the crystallographic analyses [29]. The low P_s values for some bilayered Aurivillius material have been associated with a high electrical conductivity in the polarization plane [13]. Several reported values for the remnant polarization (P_r) in ceramic and thin films are shown in Tables 4.1 and 4.2 [28–31,34–36]. The SLBN-*x* and SLBT-*x* ceramics showed an enhancement of the ferroelectricity with the increase of the lanthanum concentration, which has been associated with the increase of the grain size [31]. The increase of the grain-size allows the mobility and orientation of the

Table 4.1 Remnant polarization at room temperature for some ceramic systems

Ceramics	P_r (μC/cm^2)	Ceramics	P_r (μC/cm^2)	Ceramics	P_r (μC/cm^2)
SBN [28]	19	SBN [31]	0.13	SBT [31]	0.03
SBBN-15 [28]	34	SLBN-5 [31]	0.70	SLBT-5 [31]	2.13
SBBN-30 [28]	13	SLBN-10 [31]	1.24	SLBT-10 [31]	4.51
SBBN-50 [28]	8	SLBN-15 [31]	1.15	SLBT-15 [31]	6.63
SBBN-70 [28]	5	SLBN-20 [31]	1.54	SLBT-20 [31]	5.97
SBBN-85 [28]	4	SLBN-25 [31]	0.88	SLBT-25 [31]	5.82
BBN [28]	4	SBLN-30 [31]	0.53	SLBT-30 [31]	5.24
$Sr_{0.6}(BiNa)_{0.2}$BN [35]	3.3	$SrBi_2Ta_2O_9$ [31]	5.89	$SrBi_2(Ta_{0.925}W_{0.075})_2O_9$ [36]	6.90

Table 4.2 Remnant polarization at room temperature for some thin films

Thin films	P_r (μC/cm^2)
$Bi_2VO_{5.5}$ [34]	1.08
$Bi_{1.9}Eu_{0.1}VO_{5.5}$ [34]	1.36
$Bi_{1.8}Eu_{0.2}VO_{5.5}$ [34]	0.96

domains. The decreases of P_r for the higher lanthanum compositions could be associated with the incorporation of this doping element in the $[Bi_2O_2]^{2+}$ layers [31]. Figure 4.3 shows the hysteresis loops for a SLBT-x system at room temperature and 1 Hz.

The barium-doped SBBN-x samples have shown a tendency to decrease the P_r values with the increase of the Ba^{2+} cation [28,37]. Fig. 4.4 shows the hysteresis loops obtained at room temperature for the SBBN-30 and SBBN-70 compositions, as examples of the ferroelectric behavior observed for all the compositions. The samples with $x \leq 30$ at% show hysteresis loops typical for normal ferroelectrics, whereas the samples with $x \geq 50$ at% shows relaxor-like hysteresis loops [28,37]. This behavior has been associated with the inhomogeneous distribution of the Ba^{2+} ions between A- and bismuth-sites [28,38].

The piezoelectric response also has been studied for single- and bilayered Aurivillius systems [38–42]. The Bi_2WO_6 single crystal has been reported as a high-temperature piezoelectric material because of its high Curie temperature and high piezoelectric coefficients $d_{33} = 40$ pC/N [39]. On the other hand, bilayered

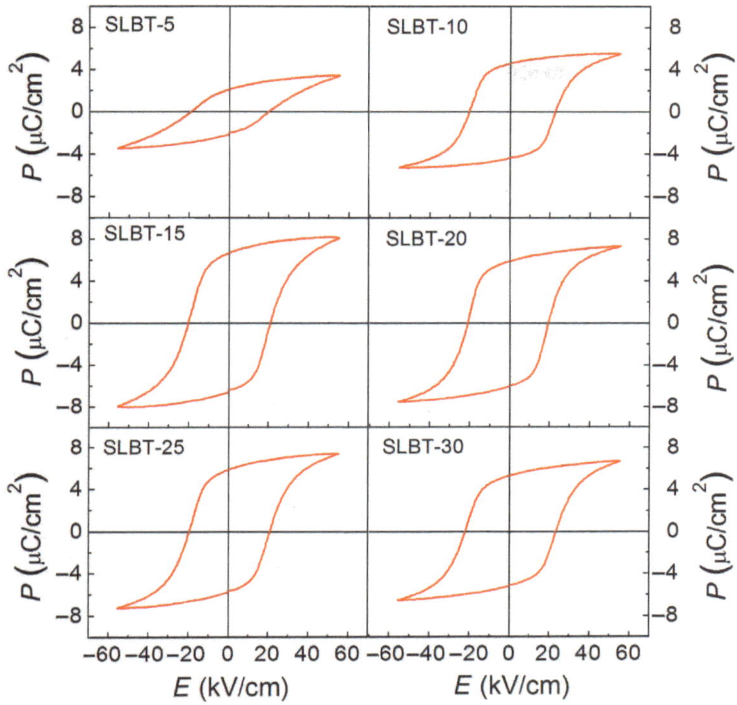

Figure 4.3 Hysteresis loops, at room temperature and 1 Hz, for SLBT-x ceramic samples [31].

Aurivillius materials are well known by their low piezoelectric activity and electromechanical coupling factors, because the rotation of the spontaneous polarization is limited to the $a-b$ plane [40]. However, for some of these materials the doping by using different elements has been reported to improve the piezoelectric activity [38,40–42]. For instance, $Bi_3Ti_{0.8}Nb_{1.2}O_9$ [40], $(Bi_3TiNb_2O_9)_{0.6}(SrBi_2Nb_2O_9)_{0.4}$ [41], and $Bi_{3-x}Nd_xNbTiO_9$ [42] ceramic systems have shown high piezoelectric coefficient (d_{33}) values around 7, 9.5, and 11 pC/N, respectively, which were revealed to be higher than that reported for Bi_3TiNbO_9 (around 1–6 pC/N) [40–42]. For the $(CaBi_2Nb_2O_9)_{1-x}(BaBi_2Nb_2O_9)_x$ ceramic system, the d_{33} value has been enhanced by considering the Ba^{2+} doping on the A-sites ($x \leq 0.3$ at%) showing values between 6 and 11 pC/N [40,41]. Similar results have been reported for SBBN-x ceramics, where the d_{33} values show a variation from 14 to 36 pC/N, for Ba^{2+} concentrations between 0 and 30 at% [38].

The dielectric behavior and the ferroelectric-paraelectric phase transitions analysis in single- and bilayered Aurivillius systems showed a strong variation of the transition temperature (T_m) due mainly to the substitutions in the A-site of the structure, whereas the substitutions in the B-site have been reported to be accompanied

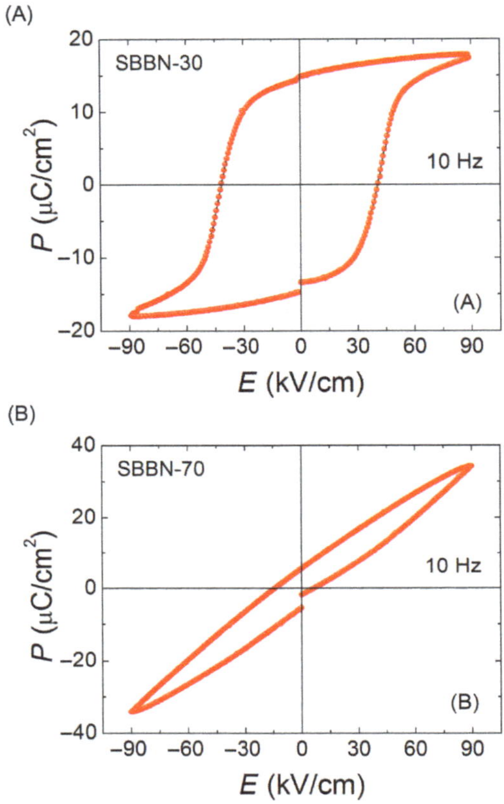

Figure 4.4 Hysteresis loops, at room temperature and 10 Hz, for the SBBN-30 (A) and SBBN-70 (B) compositions [28].

by slight changes in the transition temperature [28,33,37,38,43–49]. Table 4.3 shows the reported T_m values for some of these systems.

A strong dispersion of the dielectric parameters with frequency for $Ba_2Bi_2Nb_2O_9$ (BBN) and $Ba_2Bi_2Ta_2O_9$ (BBT) ceramics, reported by Miranda et al., has been associated with a relaxor ferroelectric behavior [47]. The frequency dispersion degree (ΔT) between 1 kHz and 1 MHz at 100°C for the BBN sample was higher than that for $Pb(Mg_{1/3}Nb_{2/3})O_3$ (PMN), a classical relaxor ferroelectric system [48]. The results for the freezing temperature and the activation energy values [47], which were obtained by using the Vogel-Fulcher relation, have also suggested that the dynamic for the polar nanoregions is different to that observed for PMN and $Pb_{1-3x/2}La_x(Zr_{0.60}Ti_{0.40})O_3$ (PLZT) systems [48].

The influence of the composition on T_m also has been studied in $Sr_{1-x}Ba_xBi_2Nb_2O_9$, $Sr_{x-1}Ca_xBi_2Nb_2O_9$, and $SrBi_2(V_xNb_{1-x})_2O_9$ ceramic systems [49]. The increase of the Ca^{2+} concentration leads to an increase of T_m, whereas the increase of the Ba^{2+} concentration reduces T_m. This different behavior has been

Table 4.3 The T_m values for some mono- and bi-layered Aurivillius ceramic systems

System	T_m (°C)
Bi_2WO_6 [33]	690 and 960
$SrBi_2Nb_2O_9$ [33]	418/440
$SrBi_2Nb_2O_9$ [28]	365
$SrBi_2Nb_2O_9$ [31]	341
$BaBi_2Nb_2O_9$ [33]	200
$BaBi_2Nb_2O_9$ [28]	160
$BaBi_2Nb_2O_9$ [31]	210
$SrBi_2Ta_2O_9$ [31]	302

explained considering the difference between the ionic radii for the corresponding elements occupying the A-sites. On the other hand, for V^{5+} occupying the B-sites, the influence of the concentration on T_m was even lower. The results showed the strong influence of the elements occupying the A-sites on the large-range interactions, which are responsible for the ferroelectricity [49]. Additional studies on $Sr_{1-x}Ba_xBi_2Nb_2O_9$ ceramics have shown the influence of Ba^{2+} on T_m, revealing a transition from normal to relaxor behavior with the increase of the Ba^{2+} content [50]. For $x \leq 60$ at%, T_m was found to decrease with the increase of the Ba^{2+} concentration and then increase for higher Ba^{2+} content. This behavior has been associated with the incorporation of Ba^{2+}cation in the Bi^{3+}-sites [50]. For $ABi_2Nb_2O_9$ and $ABi_2Ta_2O_9$ ceramics, with $A = Sr^{2+}$, Ba^{2+}, and $Sr_{0.5}Ba_{0.5}Bi_2Ta_2O_9$ system, it has been reported that the inclusion of Ba^{2+} provides a relaxor ferroelectric behavior [33]. Furthermore, it has been shown that the T_m values decrease with the increase of the Ba^{2+} concentration, whereas T_m increases with the increase of the Sr^{2+} content [33]. On the other hand, it has been reported to be an important influence of the Ba^{2+} concentration on the dielectric permittivity values (real component, ε') for $Ca_{1-x}Ba_xBi_2Nb_2O_9$ ceramic compounds [43]. The increase of ε' with the increase of the Ba^{2+} content has been explained, considering the growing of the polarization related to the larger ionic radius of the Ba^{2+} cation [43].

Fig. 4.5 shows the Ba^{2+} concentration dependence of T_m and the corresponding dielectric permittivity value (ε_m') for the $Sr_{1-x}Ba_xBi_2Nb_2O_9$ ceramic system at 50 kHz. Both T_m and ε_m' parameters show a tendency to increase for $x \leq 30$ at%. These compositions have also shown a normal ferroelectric-paraelectric phase transition in the temperature dependence of the dielectric permittivity [28,38]. For compositions with $x \geq 50$ at%, both T_m and ε_m' parameters decrease and typical characteristics of relaxor ferroelectric behavior have been reported [28,38]. These results have been explained considering the results obtained from refined structural data [28,38]. The cation-site mixing among atomic positions (i.e., Ba^{2+}, Sr^{2+}, and Bi^{3+} ions occupy A-sites and Bi^{3+}-sites at the layers) provides an inhomogeneous distribution of these elements in the structure. This compositional disorder affects the long-range polar interactions.

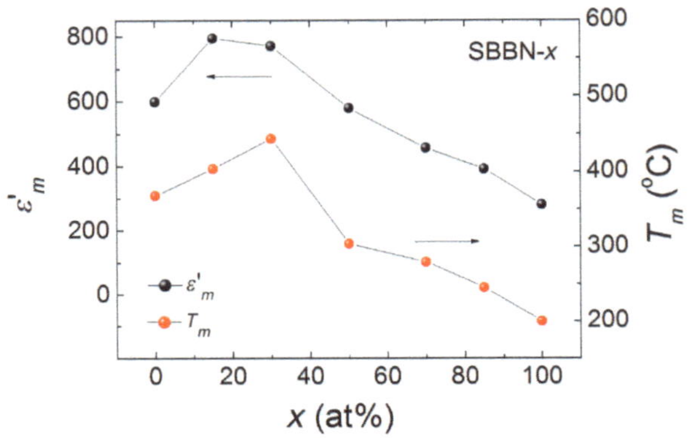

Figure 4.5 Ba^{2+} concentration dependence of T_m and the corresponding dielectric permittivity value (ε_m') for the $Sr_{1-x}Ba_xBi_2Nb_2O_9$ ceramic system at 50 kHz. *Source*: Data taken from Y.G. Abreu, Obtención y caracterización de materiales ferroeléctricos libres de plomo, PhD Thesis, Universidad de la Habana, La Habana, Cuba, 2014.

In addition, the dielectric analysis of bilayered systems doping with rare-earth elements has shown that the doping concentration affects the T_m value [44,45]. For Bi_3TiNbO_9 ceramics, the doping with Ce^{3+} ions gradually reduces the T_m value with the increase of the Ce^{3+} concentration up to 0.7 wt% [44]. Similar behavior has been reported for $SrBi_{2-x}Ho_xTa_2O_9$ ceramics when the Ho^{3+} content increases [45]. Fig. 4.6 shows the La^{3+} concentration dependence of T_m for $Sr_{1-3x/2}La_xBi_2Nb_2O_9$ (SLBN-x) ceramics [31]. It can be seen that T_m decreases with the increase of the La^{3+} content, which was expected considering the incorporation of the La^{3+}cation at the A-site, replacing the Sr^{2+}cation [31].

4.2.2 Multilayer bismuth layer-structured ferroelectrics (m ≥ 3)

4.2.2.1 m = 3

Bismuth titanate ($Bi_4Ti_3O_{12}$), the classic three-layer perovskite, exhibits a high Curie temperature of 675°C, excellent fatigue endurance properties, but low piezoelectric activity because of the high electrical conductivity [51]. The spontaneous polarization has been reported for single crystals as 50 and 4 μC/cm^2 along the *a*- and *c*-axes, respectively [52]. However, for ceramics and thin films only low remnant polarization values have been achieved [8,53]. These results have been associated with a high leakage current and the domain pinning, which is caused by defects. Numerous studies have been carried out in order to resolve that; B-site substitutions by using donor dopants such as V^{5+}, Nb^{5+}, Ta^{5+}, and others have been evaluated [54,55]. A good piezoelectric coefficient d_{33} value of 22.8 pC/N has been reported for $Bi_4Ti_{2.9}W_{0.05}Nb_{0.05}O_{12}$ ceramics, with high remnant polarization and Curie temperature values [56]. It has been reported that W/Nb doping improves the electrical, piezoelectric, and dielectric properties. Recently studies

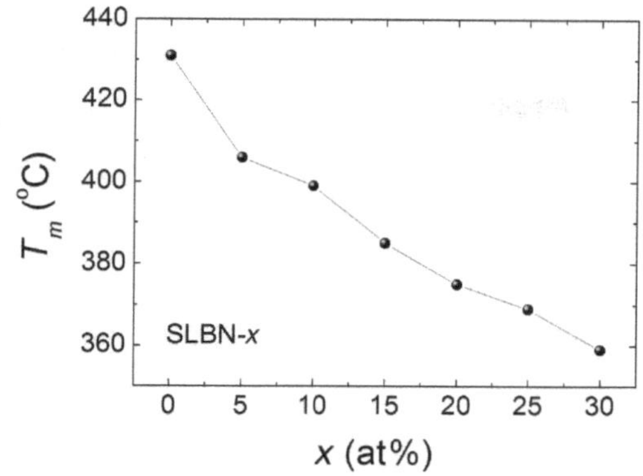

Figure 4.6 La^{3+} concentration dependence of T_m for the $Sr_{1-3x/2}La_xBi_2Nb_2O_9$ (SLBN-*x*) ceramic system.
Source: Data taken from I. Cruvinel dos Reis, Estudo das propriedades físicas de sistemas ferroelétricos com estruturas tipo Aurivillius, PhD Thesis, Universidade Estadual Paulista – UNESP "Júlio de Mesquita Filho", Ilha Solteira, São Paulo, Brazil, 2015.

on $Bi_4Ti_{2.95}(B_{1/3}Nb_{2/3})_{0.05}O_{12}$ [B = Mg^{2+}, Zn^{2+}, Cu^{2+}] ceramics have shown an important decreasing of the electrical conductivity for B = Mg^{2+} because of the enhanced structural distortions and the decrease of the oxygen vacancies concentration [57]. P_r and d_{33} values of 12.9 μC/cm^2 and 20 pC/N were also reported. For $Bi_{3.15}Nd_{0.85}Ti_3O_{12}$ thin films, which were obtained by sol-gel, high remnant polarization has been reported ($2P_r$ = 103 μC/cm^2) [58,59].

The first structural studies on $Bi_4Ti_3O_{12}$ were made by Aurivillius, reporting an orthorhombic structure [28]. Some X-ray refinements on this system have showed a monoclinic structure and the polarization direction in the *ac* plane, indicating a break of the orthorhombic symmetry. A monoclinic symmetry allows two different possible directions for the polarization; the main polarization direction is along the *a*-axis and there is another slow polarization along the *c*-axis. Both polarizations can be independently switched. The component on the *c*-axis has been attributed to the reorientation of the lone electron pairs associated with the Bi^{3+} ions at the A-sites of the perovskite blocks [2]. The origin of the ferroelectricity has been proposed [60] to be associated with (1) the displacement along the *a*-axis of the cations in B-sites with respect to its octahedral surroundings (main contribution to the spontaneous polarization); and (2) the Bi-O connecting between bismuth layer and the perovskite component, which provides the rotation of the octahedrons (upswing the antiparallel displacements normal to the polar axis). It also has been reported that the grain shape of the Aurivillius phase ferroelectric is different in nano- and micrometer scale [61]. The grain morphology has been studied for the $Bi_{3.15}Nd_{0.85}Ti_3O_{12}$ system considering ultrafine powders and bulk ceramics [61]. The ultrafine powders with high purity were prepared by a hydrolysis method and the bulk ceramics were sintered at different temperatures (1023, 1123, and 1273 K). The nanoparticles, which were obtained by hydrolysis, showed equiaxed shapes and formed agglomerates. The bulk ceramic sample, which was sintered at 1023 K, preserved the nanoparticles and their agglomerates. The sintered sample at 1123 K showed a limited grain growth; the initial agglomerates disappeared to form large equiaxed grains. The ceramic sample, which was sintered at 1273 K, showed an anisotropic grain growth with faceted platelet grains (2011 ± 30 nm of length and 307 ± 30 nm of thickness). Yan and colleagues [40] showed that ceramics with two, three, and four layers and orthorhombic structure have better resistance to thermal depoling up to temperatures close to their Curie temperature than that of the monoclinic structure. There is an important reduction in d_{33} for the monoclinic structure well before its corresponding transition temperature. The monoclinic structure generates more non-180° ferroelectric domain structures than those the orthorhombic phase, whose thermal instability has a significant contribution to the thermal depoling behavior of the system. It also has been analyzed that fatigue depends on orientation in thin films [62]. It depends on the competition of charge trapping at domain walls and charge detrapping. The *a*-axis-oriented $Bi_{3.15}Nd_{0.85}Ti_3O_{12}$ thin film has been reported to have the largest remnant polarization and the highest leakage current; the *c*-axis-oriented film exhibits the smallest remnant polarization and the lowest leakage current. This intrinsic anisotropic ferroelectric behavior must be controlled in order to provide its potential for practical applications.

The dielectric behavior has also shown interesting results for modified $Bi_4Ti_3O_{12}$. A normal and displacive-type ferroelectric with a transition temperature of 903 K has been reported for $Bi_4Ti_2Nb_{0.5}Fe_{0.5}O_{12}$ ceramics [63]. The dielectric spectra for $Bi_{2.5}Gd_{1.5}Ti_3O_{12}$ ceramics have shown a complex dielectric behavior with a transition temperature around 468 K. A high ionic conductivity above that temperature has been associated with oxygen vacancies migration [64]. For intergrowth $BaBi_2Nb_2O_9{-}Bi_4Ti_3O_{12}$ ceramics, a relaxor ferroelectric behavior with the highest transition temperature value (909 K) than those of all the known relaxor ferroelectric materials has been reported [65]. The influence of simultaneous doping on A- and B-sites has been also studied for $Bi_4Ti_3O_{12}$ [66]. The dielectric analysis of $Bi_{4-x}Ba_xTi_{3-x}Nb_xO_{12}$ ($0 \leq x \leq 1.4$) has shown a normal ferroelectric-paraelectric transition for $x \leq 0.4$. For $x \geq 0.6$, a typical relaxor behavior was reported [66]. The frequency dispersion degree (ΔT) and the diffuseness (γ) of the phase transition also increased with the doping concentration (Fig. 4.7).

The electrical conductivity in the $Bi_4Ti_3O_{12}$ system is anisotropic between *a*–*b*-plane and *c*-axis. For textured ceramics it has been reported one order of magnitude between 500°C and 700°C, but the spontaneous polarization was maximized too. The doping at B-sites has provided a decrease of the electrical conductivity; the Nb^{5-} ions showed the best results [2]. Two different mechanisms for the electrical conductivity also have been reported: electronic *p*-type conductivity along the *c*-axis and an ionic mechanism due to oxygen vacancies in the *a*–*b*-plane [2].

The temperature dependence of the pyroelectric coefficient on niobium and thallium modified bismuth titanate has been also studied by Hou and colleagues [67]. A pyroelectric coefficient of 12 $\mu C/m^2K$ has been reported at room temperature, which is better than that for pure bismuth titanate ceramics [68]. Several works

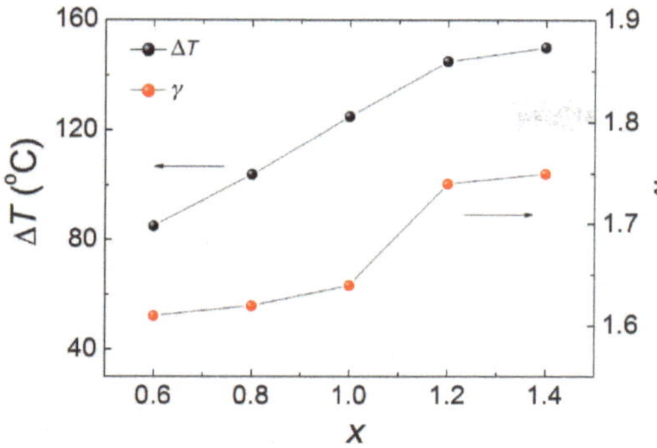

Figure 4.7 Composition dependence of the frequency dispersion degree (ΔT) and the diffuseness (γ) of the phase transition for $Bi_{4-x}Ba_xTi_{3-x}Nb_xO_{12}$ ceramics ($x \geq 0.6$). *Source*: Data taken from D. Ben Jennet, P. Marchet, M. El Maaoui, J.P. Mercurio, From ferroelectric to relaxor behaviour in the Aurivillius-type Bi4-xBaxTi3-xNbxO12 ($0 \leq x \leq 1.4$) solid solutions, Mat. Lett. 59 (2005) 376–382.

have shown research concerning the nanosized powder preparation of bismuth titanate, which has allowed obtaining ceramics with excellent pyroelectric properties for applications as detectors [68–72].

In addition to $Bi_4Ti_3O_{12}$ three-layer perovskite, there is another that deserves to be mentioned. $KSrBi_2Ta_3O_{12}$ thin films [73] have been developed in order to improve the remnant polarization of $SrBi_2Ta_2O_9$, which has been extensively studied for nonvolatile random access memories. This two-layer perovskite exhibits excellent fatigue-free behavior but small remnant polarization [74]. The $SrBi_2Ta_2O_9$ system has two TaO_6 octahedra between close bismuth layers, which have been reported to make no contribution to polarization along the *c*-axis in thin films [75]. The inclusion of $KTaO_3$ items into the adjacent bismuth layers provides three TaO_6 octahedra and a remnant polarization value along the *c*-axis. A higher remnant polarization value has been reported ($2P_r = 15.6\ \mu C/cm^2$) than that for $SrBi_2Ta_2O_9$ thin films, keeping the excellent fatigue-free properties [73].

4.2.2.2 m = *4*

The $BaBi_4Ti_4O_{15}$ system has been reported as a relaxor ferroelectric with diffuse transition due to positional disorder and compositional fluctuations, where the cation at A-sites plays an important role in the dielectric behavior [76,77]. The La^{3+} substitution for Bi^{3+} ions has showed that the diffuseness and relaxor behavior increase with increasing lanthanum concentration [78]. It has been attributed to the compositional fluctuation induced by La^{3+}, partly in perovskite and partly in $(Bi_2O_2)^{2+}$ layers. The freezing temperature and the dc-conductivity also decreased with the lanthanum content. The last result has confirmed the decrease in the oxygen vacancy concentration with La^{3+} substitution. Fig. 4.8 shows the composition dependence of the frequency dispersion degree (ΔT) and the freezing temperature (T_F) for the

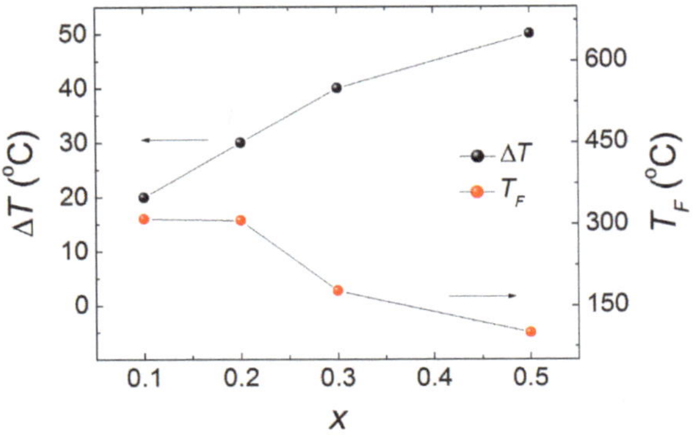

Figure 4.8 Composition dependence of the frequency dispersion degree (ΔT) and the freezing temperature (T_F) for $BaBi_{4-x}La_xTi_4O_{15}$ ceramic system.
Source: Data taken from A. Chakrabarti, J. Bera, Effect of La-substitution on the structure and dielectric properties of BaBi4Ti4O15 ceramics, J. All. Comp. 505 (2010) 668–674.

$BaBi_{4-x}La_xTi_4O_{15}$ ceramic system [78]. On the other hand, the Sm^{3+} doping has shown similar results for the dielectric properties [79].

The lead-free $Na_{0.5}Bi_{4.5}Ti_4O_{15}$ system exhibits high values of mechanical quality factor and ferroelectric-paraelectric transition temperature (655°C), being very interesting for ferroelectric and piezoelectric applications [80,81]. However, obtaining it is not easy because of the high volatility of Na^{2+} and Bi^{3+} ions, which provides secondary phase formation, especially in thin films. In order to resolve that situation, rare earths have been used as doping [82,83]. Tb^{3+} ion doping has been shown to be the best value for remnant polarization (20 $\mu C/cm^2$) [83]. Also, simultaneous cerium and lithium doping has enhanced the piezoelectric properties; a d_{33} value of 30 pC/N has been reported. The results have been associated with the lattice distortion and the bismuth vacancies [81].

$CaBi_4Ti_4O_{15}$ also has been studied for its potential for high-temperature piezoelectric applications, but its low d_{33} value and high electric conductivity are important disadvantages for those purposes [84]. Several studies have been carried out to improve piezoelectric and ferroelectric properties. The effect of Li^+, K^+, and Ce^{3+} doped $CaBi_4Ti_4O_{15}$ layered-structured ceramics has showed that simultaneous substitution by using lithium-cerium or potassium-cerium can improve the electrical resistivity at high temperatures and also the d_{33} piezoelectric coefficient of the modified ceramics [85]. Recently, the Co^{3+} doping at B-sites has been shown to provide better ferroelectric and piezoelectric properties [86].

A four-layer Aurivillius-type structure also has been reported for cerium-doped $K_{0.5}La_{0.5}Bi_4Ti_4O_{15}$ ceramics showing a higher value of d_{33} piezoelectric coefficient than modified potassium lanthanum bismuth titanate ceramics [87]. The composition with the higher d_{33} value shows the higher electromechanical coupling and the lower dielectric losses. The results have been explained considering that the incorporation of the cerium increases the resistivity, making the poling easier, and then

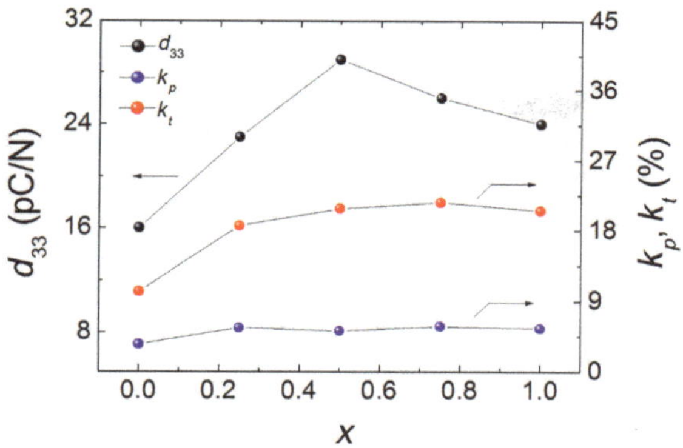

Figure 4.9 Cerium concentration dependence of d_{33}, k_p, and k_t piezoelectric parameters for $Na_{0.5}La_{0.5}Bi_4Ti_4O_{15} + x$ wt% CeO_2 ceramic system.
Source: Data taken from Ch.-M. Wang, L. Zhao, J.-F. Wang, L.-M. Zheng, J. Du, M.-L. Zhao, et al., Cerium-modified Aurivillius-type sodium lanthanum bismuth titanate with enhanced piezoactivities, Mater. Sci. Eng. B 163 (2009) 179–183.

provides better piezoelectric properties. Similar results have been found for $Na_{0.25}K_{0.25}Bi_{4.5}Ti_4O_{15}$ and $Na_{0.5}La_{0.5}Bi_4Ti_4O_{15}$ ceramics [88,89]. Figure 4.9 shows the cerium concentration dependence of d_{33}, k_p, and k_t piezoelectric parameters for $Na_{0.5}La_{0.5}Bi_4Ti_4O_{15} + x$ wt% CeO_2 ceramic system [89].

4.2.2.3 m ≥ 5

Several structural studies have been carried out on five-layer perovskite systems [90,91]. For $A_2Bi_4Ti_5O_{18}$ ($A = Ca^{2+}$, Sr^{2+}, Ba^{2+}, Pb^{2+}), an orthorhombic structure at room temperature (B2eb as space group) has been proposed [91]; another report has suggested that $Ba_2Bi_4Ti_5O_{18}$ shows a tetragonal structure (I4mm as space group) [90]. More recently, for $Pb_2Bi_4Ti_5O_{18}$, which shows a relaxor ferroelectric behavior [92], a detailed structural analysis by using neutron diffraction has shown a particular phase transition sequence F2mm→I4mm→I4/mmm on heating [90].

On the other hand, a $Sr_2Bi_4Ti_5O_{18}$ ceramic system has been reported to have better piezoelectric response ($d_{33} = 26$ pC/N) than that of $Ca_2Bi_4Ti_5O_{18}$ ($d_{33} = 8.83$ pC/N), but lower Curie temperature [93,94]. The dielectric and piezoelectric studies on $(Sr_{1-x}Ca_x)_2Bi_4Ti_5O_{18}$ ($0 \leq x \leq 1$) ceramic systems showed that $SrCaBi_4Ti_5O_{18}$ possesses the higher ferroelectric-paraelectric phase transition temperature and the best piezoelectric coefficient, $d_{33} = 26$ pC/N [95]. More recently it has been shown that $Sr_2Bi_4Ti_5O_{18}$ is a normal ferroelectric whereas $Ba_2Bi_4Ti_5O_{18}$ and $Pb_2Bi_4Ti_5O_{18}$ are typical relaxor ferroelectrics [96]. A $Pb_2Bi_4Ti_5O_{18}$ system exhibits the higher remnant polarization, which has been associated with lower electrical conductivity and higher polarizability of Pb^{2+} ions.

It is known that layer perovskites exhibit piezoelectric properties lower than typical lead-containing perovskites because the ferroelectric displacements of the constituent ions are restricted in the $a-b$ plane. In order to improve the ferroelectric and piezoelectric behavior, mixed layer intergrowth Aurivillius phases have been developed from common layer perovskites ($1 \leq m \leq 4$). These systems ($m \geq 5$) are obtained from one-half the unit cell of one usual phase and one-half of another one along the *c*-axis. Piezoelectric analyses have been carried out on intergrowth $BaBi_2Nb_2O_9-Bi_4Ti_3O_{12}$ ($BaBi_6Ti_3Nb_2O_{21}$) Aurivillius ferroelectric ceramics [65]. A piezoelectric response has been observed even in the paraelectric phase. The results have indicated that there is spontaneous polarization above the transition temperature, which is typical for relaxor ferroelectric behavior. Other studies on $Bi_{7-x}La_xTi_4NbO_{21}$ ($0 \leq x \leq 1.75$) ceramics have showed an increase for d_{33} piezoelectric coefficient until $x = 0.75$ ($d_{33} = 16.6$ pC/N) and a maximum value for remnant polarization was obtained for $x = 0.50$ ($2P_r = 24.4\ \mu C/cm^2$) [97]. The d_{33} values are higher than those for other materials from the Aurivillius family, showing a promising material for piezoelectric applications [97].

Thin films of $Sr_4Bi_4Ti_7O_{24}$ showed a remnant polarization of $P_r = 5.3\ \mu C/cm^2$ and a Curie temperature of 324 K [98]. Also, a relaxor ferroelectric behavior has been evaluated, which is not typical for $Sr_{n-1}Bi_2Ti_nO_{3m+3}$ Aurivillius systems. On the other hand for $SrBi_8Ti_7O_{27}$ a higher remnant polarization ($2P_r = 30\ \mu C/cm^2$) than that of $Bi_4Ti_3O_{12}$ and $SrBi_4Ti_4O_{15}$ systems has been reported [99]. Note that these three-layer and four-layer Aurivillius phases, respectively, are used to obtain the $SrBi_8Ti_7O_{27}$ intergrowth BLSF. The refined atomic data has shown that the bismuth layers for $SrBi_8Ti_7O_{27}$ are different than those for $Bi_4Ti_3O_{12}$ and $SrBi_4Ti_4O_{15}$. It has been found that there are two kinds of Bi^{3+} ions; one of them is displacing with respect to the other one along the polarization direction. This nontypical situation has been discussed as the main cause for the better ferroelectric polarization. Rare-earth doping on that system has also improved the remnant polarization values [100,101]; the substitution of Bi^{3+} by lanthanide ions contributes to stabilize the oxygen in the structure. A remnant polarization value of $13.5\ \mu C/cm^2$ and also a fatigue-free behavior have been reported for La-doping $SrBi_8Ti_7O_{27}$ thin films [100]. For Nd-doping in ceramic systems, $2P_r$ value reaches $33.2\ \mu C/cm^2$ in $SrBi_{7.50}Nd_{0.50}Ti_7O_{27}$ [101]. A maximum value for d_{33} (14 pC/N) is also obtained for this composition.

Acknowledgements

The authors wish to thank to the ICTP, Trieste-Italy, for financial support of Latin-American Network of Ferroelectric Materials (NET-43). Thanks to CNPq, CAPES, and FAPEMIG agencies from Brazil.

References

[1] Y. Xu, Ferroelectric Materials and Their Applications, Elsevier Science, Amsterdam, North-Holland, 1991.

[2] T. Jardiel, A.C. Caballero, M. Villegas, Aurivillius ceramics: $Bi_4Ti_3O_{12}$ – based piezoelectrics, J. Cer. Soc. Jap. 116 (2008) 511–518.

[3] R.D. Richtmyer, Dielectric resonators, J. Appl. Phys. 10 (1939) 391−398.

[4] T. Takenaka, K. Sakata, Grain orientation effects on electrical properties of bismuth layer-structured ferroelectric $Pb_{(1-x)}(NaCe)_{x/2}Bi_4Ti_4O_{15}$ solid solution, J. Appl. Phys. 55 (1984) 1092−1099.

[5] K. Uchino, Ferroelectric Devices, Marcel Dekker Inc, London, 2000.

[6] A.K. Tagantsev, V.O. Sherman, K.F. Astafiev, J. Venkatesh, N.R. Setter, Ferroelectric materials for microwave tunable applications, J. Electro-ceramics 11 (2003) 5−66.

[7] J. de los Santos Guerra, A. Peláiz Barranco, F. Calderón Piñar, D. Garcia, J.A. Eiras, Substituição do Pb + 2 por La + 3 na rede cristalina do sistema ferroelétrico PZT (53/47): influência nas propriedades estruturais, microestruturais e elétricas, Cerâmica 51 (317) (2005) 19−23.

[8] X. Du, I.W. Chen, Ferroelectric thin films of bismuth-containing layered perovskites: Part I, $Bi_4Ti_3O_{12}$, J. Am. Ceram. Soc. 81 (1998) 3253−3259.

[9] E.C. Subbarao, Crystal chemistry of mixed bismuth oxides with layer-type structure, J. Am. Cer. Soc. 45 (1962) 166−169.

[10] S.M. Blake, M.J. Falconer, M. McCreedy, P. Lightfoot, Cation disorder in ferroelectric Aurivillius phases of the type $Bi_2ANb_2O_9$ (A = Ba, Cr, Ca), J. Mater. Chem. 7 (1997) 1609−1613.

[11] R.L. Withers, J.G. Thompson, A.D. Rae, The crystal chemistry underlying ferroelectricity in $Bi_4Ti_3O_{12}$, Bi_3TiNbO_9, and Bi_2WO_6, J. Solid State Chem. 94 (1991) 404−417.

[12] N. Zhong, S. Okamura, K. Uchiyama, T. Shiosaki, Single-ionized-oxygen-vacancy related dielectric relaxation in $Bi_{3.25}La_{0.75}Ti_3O_{12}$ ferroelectric, Appl. Phys. Lett. 87 (2005) 252901.

[13] L. Lazcano, Propiedades ferroeléctricas de materiales cerámicos con estructura Aurivillius de composición $Pb_xBi_4Ti_{3+x}O_{12+3x}$, x = 0,1,2,3, PhD Thesis, Universidad Autónoma de Madrid, Spain, 2000.

[14] B. Wachsmuth, E. Zschech, N.W. Thomas, S.G. Brodie, S.J. Gurman, S. Baker, et al., Structure model of aurivillius compounds, Phys. Sta. Sol. A 135 (1993) 59−71.

[15] B.J.Kennedy Ismunandar, Structure of $ABi_2Nb_2O_9$ (A = Sr, Ba): refinement of powder neutron diffraction, Data, J. Sol. Sta. Chem. 126 (1996) 136−141.

[16] J.M. Perez-Mato, M. Aroyo, A. García, P. Blaha, K. Schwarz, J. Schweifer, et al., Competing structural instabilities in the ferroelectric Aurivillius compound $SrBi_2Ta_2O_9$, Phys. Rev. B 70 (2004) 214111.

[17] G. Zhen Liu, C. Wang, H. Shuang, H. Bin Lu, Raman scattering study of La-doped $SrBi_2Nb_2O_9$, J. Phys. D: Appl. Phys. 40 (2007) 7817−7820.

[18] D. Kajewski, Z. Ujma, K. Szot, M. Pawełczyk, Dielectric properties and phase transition in $SrBi_2Nb_2O_9$−$SrBi_2Ta_2O_9$ solid solution, Cer. Inter. 35 (2009) 2351−2355.

[19] A. Peláiz-Barranco, Y. González-Abreu, Ferroelectric ceramic materials of the Aurivillius family, J. Adv. Diel. 3 (2013) 1330003.

[20] G.A. Smolenski, V.A. Isupov, I. Agranovskaya, Ferroelectrics of the oxygen-octahedral type with layered structure, Sov. Phys. Sol. Sta. 3 (1961) 651−655.

[21] E.C. Subbarao, A family of ferroelectric bismuth compounds, J. Phys. Chem. Sol. 26 (1962) 665−676.

[22] B.J.Kennedy Ismunandar, Effect of temperature on cation disorder in $ABi_2Nb_2O_9$ (A = Sr, Ba), J. Mater. Chem. 9 (1999) 541−544.

[23] R.E. Newnham, R.W. Wolfe, R.S. Horsey, F.A. Diaz-Colon, Crystal structure of $(Sr,Ba)Bi_2Ta_2O_9$, Mater. Res. Bull. 8 (1973) 1183−1195.

[24] R.W. Wolfe, R.E. Newnahm, M.I. Kay, Sol. Sta. Comm. 7 (1969) 1797−1801.

[25] P.R. Graves, G. Hua, S. Myhra, J.G. Thompson, The Raman modes of the Aurivillius phases: temperature and polarization dependence, J. Sol. Sta. Chem. 144 (1995) 112–122.
[26] Y. González Abreu, Relajación dieléctrica y conductividad eléctrica en cerámicas ferroeléctricas tipo perovskita. MSc Thesis, Universidad de la Habana, La Habana, Cuba, 2010.
[27] A. Peláiz-Barranco, Y. González-Abreu, Dielectric relaxation mechanisms in relaxor bi-layered perovskites, Ferroelectrics 426 (2012) 122–131.
[28] Y.G. Abreu, Obtención y caracterización de materiales ferroeléctricos libres de plomo, PhD Thesis, Universidad de la Habana, La Habana, Cuba, 2014.
[29] Ch Long, H. Fan, M. Li, A ferroelectric polarization contribution from defect dipoles in acceptor Aurivillius oxide, $(Na,Bi)_{0.47}(Li,Ce)_{0.03}Bi_2Ta_{1.97}Sc_{0.03}O_{8.97}$, Appl. Phys. Lett. 103 (2013) 192908.
[30] Y. Sun, Zh Li, H. Zhang, Ch Yu, G. Viola, Sh Fu, et al., Lead free Bi_3TaTiO_9 ferroelectric ceramics with high Curie point, Mat. Lett. 175 (2016) 79–81.
[31] I. Cruvinel dos Reis, Estudo das propriedades físicas de sistemas ferroelétricos com estruturas tipo Aurivillius, PhD Thesis, Universidade Estadual Paulista – UNESP "Júlio de Mesquita Filho", Ilha Solteira, São Paulo, Brazil, 2015.
[32] Y. González-Abreu, A. Peláiz-Barranco, Y. Gagou, J. Belhadi, P. Saint-Grégoire, Vibrational analysis on two-layer Aurivillius phase $Sr_{1-x}Ba_xBi_2Nb_2O_9$ using Raman spectroscopy, Vib. Spect. 77 (2015) 1–4.
[33] D. Nuzhnyy, S. Kamba, P. Kužel, S. Veljko, V. Bovtun, M. Savinov, et al., Dynamics of the phase transitions in bi-layered ferroelectrics with Aurivillius structure: dielectric response in the terahertz spectral range, Phys. Rev. B 74 (2006) 134105–134107.
[34] W. Xie, L. Su, T. Sa, J. Fu, W. Hu, N. Qin, et al., Strong red emission and electrical properties of highly c-axis oriented Eu^{3+}-doped bismuth vanadate ferroelectric thin films, Mat. Chem. Phys. 148 (2014) 854–858.
[35] P. Fang, Z. Xi, W. Long, X. Li, Ferroelectric, dielectric and piezoelectric properties of $Sr_{0.6}(BiNa)_{0.2}Bi_2Nb_2O_9$ ceramics, Mat. Res. Bull. 48 (2013) 4907–4910.
[36] S. Swain, P. Kumar, R. Bilash Choudhary, Electrical and ferroelectric studies of the 2-layered $SrBi_2Ta_2O_9$ based ceramics, Phys. B 477 (2015) 56–63.
[37] A. Peláiz-Barranco, Y. González-Abreu, J. de los Santos Guerra, J. Wang, T. Yang, P. Saint-Grégoire, Ferroelectric Materials - Synthesis and Characterization, InTech, 2015 (Chapter 4).
[38] Y. González-Abreu, A. Peláiz-Barranco, J.D.S. Guerra, Y. Gagou, P. Saint-Grégoire, From normal ferroelectric transition to relaxor behavior in Aurivillius ferroelectric ceramics, J. Mater. Sci. 49 (2014) 7437–7444.
[39] H. Takeda, T. Nishida, S. Okamura, T. Shiosaki, Crystal growth of bismuth tungstate Bi_2WO_6 by slow cooling method using borate fluxes, J. Eur. Ceram. Soc. 25 (2005) 2731–2733.
[40] H. Yan, H. Zhang, M.J. Reece, Thermal depoling of high Curie point Aurivillius phase ferroelectric ceramics, Appl. Phys. Lett. 87 (2005) 082911.
[41] H. Zhang, H. Yan, M.J. Reece, Microstructure and electrical properties of Aurivillius phase $(CaBi_2Nb_2O_9)_{1-x}(BaBi_2Nb_2O_9)_x(CaBi_2Nb_2O_9)_{1-x}(BaBi_2Nb_2O_9)_x$ solid solution, J. Appl. Phys. 108 (2010) 014109.
[42] H. Zhang, H. Yan, M.J. Reece, The effect of Nd substitution on the electrical properties of Bi_3NbTiO_9 Aurivillius phase ceramics, J. Appl. Phys. 106 (2009) 044106.

[43] V.D. Phadtare, V.R. Puri, Studies on electrical and dielectric properties of co-precipitated Aurivillius phase $Ca_{1-x}Ba_xBi_2Nb_2O_9$ ceramics, Cer. Inter. 42 (2016) 8581–8586.
[44] Q. Wanga, Ch Wanga, J. Wanga, Sh Zhangb, High performance Aurivillius-type bismuth titanate niobate (Bi_3TiNbO_9) piezoelectric ceramics for high temperature applications, Cer. Inter. 42 (2016) 6993–7000.
[45] A. K. Jha Sugandha, Effect of holmium substitution on electrical properties of strontium bismuth tantalate ferroelectric ceramics, Cer. Inter. 39 (2013) 9397–9403.
[46] K. Shi, L. Peng, M. Li, Zh Zhou, K. Jiang, J. Zhang, et al., Structural distortion, phonon behavior and electronic transition of Aurivillius layered ferroelectric $CaBi_2Nb_{2-x}W_xO_9$ ceramics, J. All. Comp. 653 (2015) 168–174.
[47] C. Miranda, M.E.V. Costa, M. Avdeev, A.L. Kholkin, J.L. Baptista, Relaxor properties of Ba-based layered perovskites, J. Eur. Ceram. Soc. 21 (2001) 1303–1306.
[48] D. Viehland, S.J. Jang, L. Eric Cross, Freezing of the polarization fluctuations in lead magnesium niobate relaxor, J. Appl. Phys. 68 (1990) 2916–2921.
[49] Y. Wu, M.J. Forbess, S. Seraji, S.J. Limmer, T.P. Chou, C. Nguyen, et al., Doping effect in layer structured $SrBi_2Nb_2O_9$ ferroelectrics, J. Appl. Phys. 90 (2001) 5296–6002.
[50] S. Huang, Ch Feng, L. Chen, Q. Wang, Relaxor behavior of $Sr_{1-x}Ba_xBi_2Nb_2O_9$ ceramics, J. Am. Cer. Soc. 89 (2006) 328–331.
[51] J.G. Wu, D.Q. Xiao, J.G. Zhu, Potassium–sodium niobate lead-free piezoelectric materials: past, present, and future of phase boundaries, Chem. Rev. 115 (2015) 2559–2595.
[52] S.E. Cummins, L.E. Cross, Electrical and optical properties of ferroelectric $Bi_4Ti_3O_{12}$ single crystals, J. Appl. Phys. 39 (1968) 2268–2274.
[53] P.C. Joshi, S.B. Krupanidhi, A. Mansingh, Rapid thermally processed ferroelectric $Bi_4Ti_3O_{12}$ thin films, J. Appl. Phys. 72 (1992) 5517–5519.
[54] H.S. Shulman, M. Testorf, D. Damjanovic, N. Setter, Microstructure, electrical conductivity and piezoelectric properties of bismuth titanate, J. Am. Ceram. Soc. 79 (1996) 3124–3128.
[55] J.K. Kim, H.S. Lee, J.H. Kim, Fabrication of Nb-doped bismuth titanate thin films using nonionic block copolymer: enhanced ferroelectric properties, Mat. Lett. 60 (2006) 2090–2093.
[56] Z. Peng, Q. Chen, Y. Chen, D. Xiao, J. Zhu, Microstructure and electrical properties in W/Nb co-doped Aurivillius phase $Bi_4Ti_3O_{12}$ piezoelectric ceramics, Mat. Res. Bull. 59 (2014) 125–130.
[57] Y. Zhao, H. Fan, G. Dong, Z. Liu, Enhanced electromechanical properties and conduction behaviors of Aurivillius $Bi_4Ti_{2.95}(B_{1/3}Nb_{2/3})_{0.05}O_{12}$ (B = Mg, Zn,Cu) ceramics, Mat. Lett. 174 (2016) 242–245.
[58] U. Chon, J.S. Shim, H.M. Jang, Compositional dependence of ferroelectric properties of highly c-axis oriented $Bi_{4-x}Nd_xTi_3O_{12}$ film capacitors, Sol. Sta. Comm. 129 (2004) 465–468.
[59] U. Chon, H.M. Jang, M.G. Kim, C.H. Chang, Layered perovskites with giant spontaneous polarizations for nonvolatile memories, Phys. Rev. Lett. 89 (2002) 087601.
[60] R.E. Newnham, R.W. Wolfe, J.F. Dorrian, Structural basis of ferroelectricity in the bismuth titanate family, Mat. Res. Bull. 6 (1971) 1029–1033.
[61] H. Zhang, H. Yan, H. Ning, M.J. Reece, M. Eriksson, Z. Shen, et al., The grain size effect on the properties of Aurivillius phase $Bi_{3.15}Nd_{0.85}Ti_3O_{12}$ ferroelectric ceramics, Nanotechnology 20 (2009) 385708.

[62] F. Yang, Y. Guo, L. Liand, M. Tang, Anisotropic polarization fatigue in $Bi_{3.15}Nd_{0.85}Ti_3O_{12}$ thin films, J. Phys. D: Appl. Phys. 49 (2016) 195305.
[63] S. Kumar, K.B.R. Varma, Structural and dielectric properties of $Bi_4Ti_2Nb_{0.5}Fe_{0.5}O_{12}$ ceramics, Sol. Sta. Comm. 146 (2008) 137–142.
[64] E. Masiukaitė, J. Banys, R. Sobiestianskas, T. Ramoska, V.A. Khomchenko, D.A. Kiselev, Conductivity investigations of Aurivillius-type $Bi_{2.5}Gd_{1.5}Ti_3O_{12}$ ceramics, Sol. Sta. Ion. 188 (2011) 50–52.
[65] H. Zhang, H. Yan, M.J. Reece, High temperature lead-free relaxor ferroelectric: intergrowth Aurivillius phase $BaBi_2Nb_2O_9$–$Bi_4Ti_3O_{12}$ ceramic, J. Appl. Phys. 107 (2010) 104111.
[66] D. Ben Jennet, P. Marchet, M. El Maaoui, J.P. Mercurio, From ferroelectric to relaxor behaviour in the Aurivillius-type $Bi_{4-x}Ba_xTi_{3-x}Nb_xO_{12}$ $(0 \leq x \leq 1.4)$ solid solutions, Mat. Lett. 59 (2005) 376–382.
[67] J. Hou, R. Vaish, Y. Qu, D. Krsmanovic, K.B.R. Varma, R.V. Kumar, Dielectric, pyroelectric and ferroelectric properties of $Bi_4Ti_{2.98}Nb_{0.01}Ta_{0.01}O_{12}$ ceramics, Mater. Chem. Phys. 121 (2010) 32–36.
[68] L.B. Kong, J. Ma, W. Zhu, O.K. Tan, Preparation of $Bi_4Ti_3O_{12}$ ceramics via a high-energy ball milling process, Mater. Lett. 51 (2001) 108–114.
[69] C.P. Shaw, S. Gupta, S.B. Stringfellow, A. Navarro, J.R. Alcock, R.W. Whatmore, Pyroelectric properties of Mn doped lead zirconate–lead titanate–lead magnesium niobate ceramics, J. Eur. Ceram. Soc. 22 (2002) 2123–2132.
[70] J.G. Lisoni, P. Millan, E. Vila, J.L. Martín de Vidales, T. Hoffmann, A. Castro, Synthesis of ferroelectric $Bi_4Ti_3O_{12}$ by alternative routes: wet no-coprecipitation chemistry and mechanochemical activation, Chem. Mater. 13 (2001) 2084–2091.
[71] B.D. Stojanovic, C.O. Paiva-Santos, C. Jovalekic, A.Z. Simoes, F.M. Filho, Z. Lazarevic, Mechanically activating formation of layered structured bismuth titanate, Mater. Chem. Phys. 96 (2006) 471–476.
[72] M. Zdujic, D. Poleti, C. Jovalekic, L. Karanovic, The evolution of structure induced by intensive milling in the system $2Bi_2O_3$-$3TiO_2$, J. Non-Cryst. Solids 352 (2006) 3058–3068.
[73] S.-T. Zhang, W.-J. Ji, L. Wang, L.-Y. Ding, Y.-F. Chen, Z.-G. Liu, Improved ferroelectric polarization of $KSrBi_2Ta_3O_{12}$ thin films, Sol. Sta. Comm. 149 (2009) 434–437.
[74] C.A. Paz de Araujo, J.D. Cuchiaro, L.D. McMillan, M.C. Scott, J.F. Scott, Fatigue-free ferroelectric capacitors with platinum electrodes, Nature 374 (1995) 627–629.
[75] H.N. Lee, A. Visinoiu, S. Senz, C. Harnagea, A. Pignolet, D. Hesse, et al., Structural and electrical anisotropy of (001)-, (116)-, and (103)-oriented epitaxial $SrBi_2Ta_2O_9$ thin films on $SrTiO_3$ substrates grown by pulsed laser deposition, J. Appl. Phys. 88 (2000) 6658–6664.
[76] S. Kumar, K.B.R. Varma, Influence of lanthanum doping on the dielectric, ferroelectric and relaxor behavior of barium bismuth titanate ceramics, J. Phys. D: Appl. Phys. 42 (2009) 075405–075414.
[77] S.K. Rout, E. Sinha, A. Hussian, J.S. Lee, C.W. Ahn, I.W. Kim, et al., Phase transition in $ABi_4Ti_4O_{15}$ (A = Ca, Sr, Ba) Aurivillius oxides prepared through a soft chemical route, J. Appl. Phys. 105 (2009) 024105.
[78] A. Chakrabarti, J. Bera, Effect of La-substitution on the structure and dielectric properties of $BaBi_4Ti_4O_{15}$ ceramics, J. All. Comp. 505 (2010) 668–674.
[79] M. Reddyprakash, S.K. Rout, A. Satapathy, T.P. Sinha, S. Md Sariful, Dielectric and ferroelectric properties of samarium substituted $BaBi_4Ti_4O_{15}$ Aurivillius oxides, Cer. Inter. 42 (2016) 8798–8803.

[80] C.-M. Wang, J.-F. Wang, Z.-G. Gai, Enhancement of dielectric and piezoelectric properties of $M_{0.5}Bi_{4.5}Ti_4O_{15}$ (M = Na, K, Li) ceramics by Ce doping, Scr. Mat. 57 (2007) 789–792.

[81] C.-M. Wang, J.-F. Wang, S. Zhang, T.R. Shrout, Electromechanical properties of A-site (LiCe)-modified sodium bismuth titanate ($Na_{0.5}Bi_{4.5}Ti_4O_{15}$) piezoelectric ceramics at elevated temperature, J. Appl. Phys. 105 (2009) 094110.

[82] Z.X. Cheng, X.L. Wang, S.X. Dou, Ferroelectric properties of $Bi_{3.25}Sm_{0.75}V_{0.02}T_{2.98}O_{12}$ thin film at elevated temperature, Appl. Phys. Lett. 90 (2007) 222902.

[83] C.M. Raghavan, J.W. Kim, T.K. Song, S.S. Kim, Microstructural and ferroelectric properties of rare earth (Ce, Pr, and Tb)-doped $Na_{0.5}Bi_{4.5}Ti_3O_{15}$ thin films, App. Sur. Sci. 355 (2015) 1007–1012.

[84] A. Tanwar, K. Sreenivas, V. Gupta, Effect of orthorhombic distortion on dielectric and piezoelectric properties of $CaBi_4Ti_4O_{15}$ ceramics, J. Appl. Phys. 105 (2009) 084105.

[85] H. Yan, Z. Zhang, W. Zhu, L. He, Y. Yu, Ch Li, et al., The effect of (Li,Ce) and (K, Ce) doping in Aurivillius phase material $CaBi_4Ti_4O_{15}$, Mater. Res. Bull. 39 (2004) 1237–1241.

[86] T.-L. Zhao, C.-M. Wang, C.-L. Wang, Y.-M. Wang, S. Dong, Enhanced piezoelectric properties and excellent thermal stabilities of cobalt-modified Aurivillius-type calcium bismuth titanate ($CaBi_4Ti_4O_{15}$), Mat. Sci. Eng. B 201 (2015) 51–56.

[87] Ch-M. Wang, L. Zhao, J.-F. Wang, M.-L. Zhao, Z.-G. Gai, W.-B. Su, et al., Piezoelectric and dielectric properties of cerium-modified Aurivillius type $K_{0.5}La_{0.5}Bi_4Ti_4O_{15}$ ceramics, Mater. Chem. Phys. 114 (2009) 1004–1007.

[88] Z.-G. Gai, J.-F. Wang, M.-L. Zhao, W.-B. Sun, S.-Q. Sun, B.-Q. Ming, et al., The effect of (Li,Ce) doping in aurivillius phase material $Na_{0.25}K_{0.25}Bi_{4.5}Ti_4O_{15}$, Scr. Mat. 59 (2008) 115–118.

[89] Ch-M. Wang, L. Zhao, J.-F. Wang, L.-M. Zheng, J. Du, M.-L. Zhao, et al., Cerium-modified Aurivillius-type sodium lanthanum bismuth titanate with enhanced piezoactivities, Mater. Sci. Eng. B 163 (2009) 179–183.

[90] R.J. Goff, P. Lightfoot, Structural phase transitions in the relaxor ferroelectric $Pb_2Bi_4Ti_5O_{18}$, J. Sol. Sta. Chem. 182 (2009) 2626–2631.

[91] T. Ismunandar, A. Kamiyama, Q. Hoshikawa, B.J. Zhou, Y. Kennedy, K. Kubota, Kato, Structural studies of five layer Aurivillius oxides: A2Bi4Ti5O18 (A = Ca, Sr, Ba and Pb), J. Sol. Sta. Chem. 177 (2004) 4188–4196.

[92] J.F. Fernandez, A.C. Caballero, M. Villegas, J. De Frutos, L. Lascano, Relaxor behavior of $Pb_xBi_4Ti_{3+x}O_{12+3x}$ (x = 2, 3) Aurivillius ceramics, Appl. Phys. Lett. 81 (2002) 4811–4813.

[93] P. Ferrer, M. Algueró, J.E. Iglesias, A. Castro, Processing and dielectric properties of $Bi_4Sr_{n-3}Ti_nO_{3n+3}$ (n = 3, 4 and 5) ceramics obtained from mechanochemically activated precursors, J. Eur. Ceram. Soc. 27 (2007) 3641–3645.

[94] C. Moure, V. Gil, J. Tartaj, P. Durán, Crystalline structure, dielectric and piezoelectric properties of bismuth-layer $Ca_xBi_4Ti_{3+x}O_{12+3x}$ (x = 1, 2) compounds, J. Eur. Ceram. Soc. 25 (2005) 2447–2451.

[95] Z. Xu, R. Chu, J. Hao, Y. Zhang, Q. Chen, L. Zhao, et al., Study on high temperature performances for bismuth layer-structured $(Sr_{1-x}Ca_x)_2Bi_4Ti_5O_{18}$ ($0 \leq x \leq 1$) ceramics, J. All. Comp. 487 (2009) 585–590.

[96] S. Dubey, R. Kurchania, Study of dielectric and ferroelectric properties of five-layer Aurivillius oxides: $A_2Bi_4Ti_5O_{18}$ (A = Ba, Pb and Sr) synthesized by solution combustion route, Bull. Mater. Sci. 38 (2015) 1881–1889.

[97] Z.G. Yi, Y.X. Li, Y. Liu, Ferroelectric and piezoelectric properties of Aurivillius phase intergrowth ferroelectrics and the underlying materials design, Phys. Sta. Sol. A 208 (2011) 1035–1040.
[98] M.A. Zurbuchen, V.O. Sherman, A.K. Tagantsev, J. Schubert, M.E. Hawley, D.D. Fong, et al., Synthesis, structure, and electrical behavior of $Sr_4Bi_4Ti_7O_{24}$, J. Appl. Phys. 107 (2010) 024106.
[99] Y. Noguchi, M. Miyayama, T. Kudo, Ferroelectric properties of intergrowth $Bi_4Ti_3O_{12}$– $SrBi_4Ti_4O_{15}$ ceramics, Appl. Phys. Lett. 77 (2000) 3639–3641.
[100] J.S. Zhu, D. Su, X.M. Lu, H.X. Qin, Y.N. Wang, D.Y. Wang, et al., La-doped effect on the ferroelectric properties of $Bi_4Ti_3O_{12}$–$SrBi_4Ti_4O_{15}$ thin film fabricated by pulsed laser deposition, J. Appl. Phys. 92 (2002) 5420–5424.
[101] W. Wang, S.P. Gu, X.Y. Mao, X.B. Chen, Effect of Nd modification on electrical properties of mixed-layer Aurivillius phase $Bi_4Ti_3O_{12}$-$SrBi_4Ti_4O_{15}$, J. Appl. Phys. 102 (2007) 024102.

Section II

Ferroelectric Metal Oxides: Synthesis and Deposition

Review of methods for powder-based processing

5

Barbara Malič[1], Danjela Kuščer[1], Marko Vrabelj[1] and Jurij Koruza[2]
[1]Jožef Stefan Institute, Ljubljana, Slovenia, [2]Technische Universität Darmstadt, Darmstadt, Germany

5.1 Introduction

Ferroelectric ceramics are at the heart of numerous electronic devices that surround us; indeed, it would be difficult to imagine our lives without these versatile materials. Their outstanding properties are the result of scientific breakthroughs in ceramics research, which makes possible the production of multilayer capacitors, piezoelectric transducers, actuators, and so on. In this review we describe the powder-based processing of ferroelectric perovskite ceramics.

Prototype ferroelectric barium titanate (BT), the basic material for ceramic multilayer capacitors, and lead zirconate titanate (PZT), with its excellent piezoelectric properties, share more than half a century of research [1]. However, since early 2000 environmental concerns supported by legislation have initiated research on environment-friendly, lead-free piezoelectrics that can replace lead-based perovskites, such as PZT and lead magnesium niobate titanate (PMN-PT). Within this group, materials based on sodium potassium niobate (KNN) and sodium bismuth titanate (NBT) have been rediscovered, and found to be worthy of further study (see [2,3]). More details about this can be found in Chapter 3, Lead-free perovskite ferroelectrics.

Ferroelectric materials contain at least two metal ions, and typically more; therefore, the first step in ceramic processing is the synthesis of the target material from a mixture of oxides or carbonates. We briefly discuss the basics of solid-state synthesis and explain the reactions taking place in individual material systems to yield the perovskite phase. It should be noted that the mechanochemical synthesis of ferroelectric oxide powders that enables the synthesis of the target materials at nominally room temperature is not discussed in this chapter.

The general principles of the sintering of bulk ferroelectric ceramics are presented, followed by the specific features of selected materials groups. In the last part of the chapter thick-film technology is addressed. With thick films, the ceramic powder is suspended in a liquid vehicle, deposited on a substrate, and sintered.

The processing of ferroelectric bulk ceramics involves a number of steps, which are schematically shown in Fig. 5.1. The reactant powders, usually oxides or carbonates, are weighed in a stoichiometric ratio, mixed and/or milled. Both processes usually take place in different mills, such as ball, planetary, or attrition, in a liquid medium, which does not react with the added powder. Typical choices are water or

Magnetic, Ferroelectric, and Multiferroic Metal Oxides. DOI: http://dx.doi.org/10.1016/B978-0-12-811180-2.00005-0

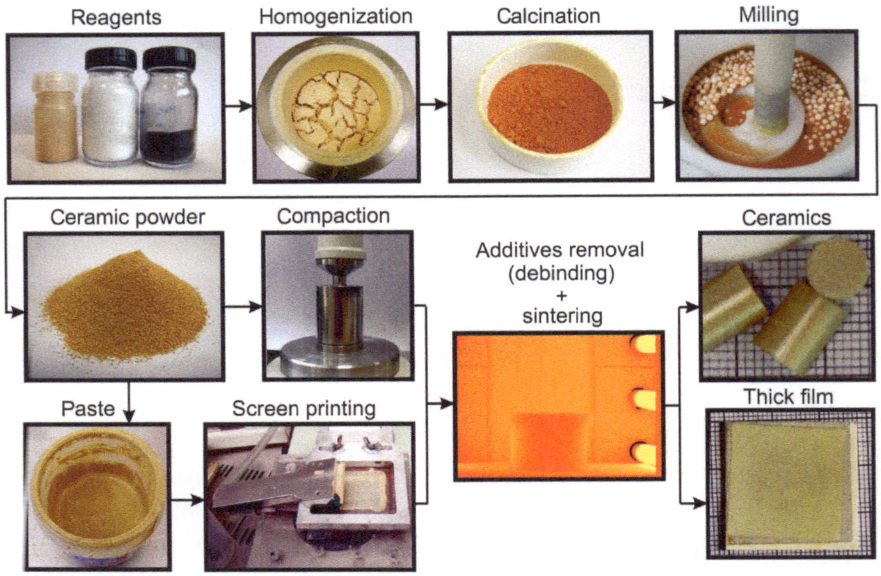

Figure 5.1 Schematic representation of the processing of ferroelectric ceramics in bulk and thick-film forms.

organic solvents, such as alcohols or acetone. Milling contributes to a reduction in the average particle size, narrowing of the particle size distribution, breaking of the agglomerates, and/or a change in the shape of the particles, but on the other hand it is also a source of contamination of the powder by the milling medium [4,5]. The powder mixture is dried and calcined for the solid-state reaction to take place. Calcination temperatures are usually a few hundreds of degrees lower than the sintering temperatures of individual materials. After calcination the powder is coarse as a consequence of partial sintering, so it is milled again. The calcination and milling steps may be repeated to bring the solid-state reaction to completion. In the next step the powder is compacted or shaped by dry pressing, which can be performed uniaxially or isostatically, or by wet-shaping methods. In the latter case, the ceramic powder is mixed with solvent(s) and other additives to obtain the required rheological behavior of the suspension for a selected shaping or patterning technology. The green body then undergoes a heat treatment that involves drying and additive-removal steps, if needed, and sintering, whereby densification and grain growth take place and the final microstructure is formed [4,6,7].

5.2 Solid-state synthesis of ferroelectric perovskites

The solid-state synthesis of complex oxides, such as perovskite ferroelectrics, occurs through a reaction between solid reactants in contact. Besides the metal

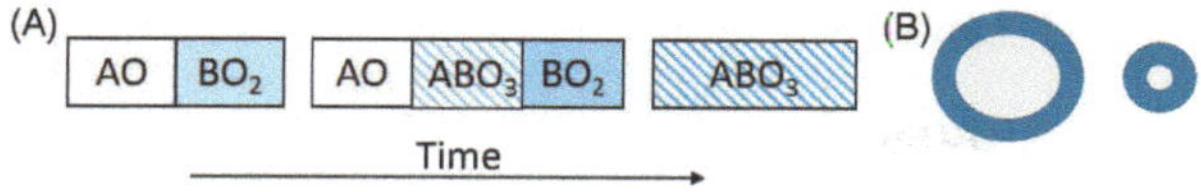

Figure 5.2 (A) Schematic representation of a reaction between two oxides, AO and BO_2, in a planar contact, yielding a perovskite ABO_3. In this general case it is assumed that both oxides diffuse through the reaction layer and that only one reaction product, ABO_3, is formed at the interface. (B) Reaction-product layer of equal thickness formed on a coarse and fine particle. The unreacted fraction is evidently much smaller in the latter case.

oxides, carbonates are common reactants in solid-state reactions, as they are often more stable and easily available. The product is formed at the interface between the reactants and it separates them spatially. The rate of the reaction-product formation is thus mainly controlled by the diffusion of the constituent species through the reaction layer. The interested reader is referred to the textbook by Schmalzried [8]. A schematic representation of the reaction between two metal oxides, AO and BO_2, in a planar contact yielding a perovskite ABO_3 is shown in Fig. 5.2A. Note that we assume that only one reaction product is formed with the A/B = 1/1 molar ratio and that both oxides diffuse through the reaction layer, which is often not the case in real oxide systems. The growth rate of the reaction layer follows the parabolic rate law (Eq. 5.1), meaning that the diffusion distance (y) increases with the square root of time (t) under isothermal conditions.

$$y^2 = KDt \text{ or } y = \sqrt{KDt}, \tag{5.1}$$

where D is the diffusion coefficient and K is a constant.

In the case of powder reactions, the reactant particles should be intimately mixed, which is performed in the homogenization/milling step (see Fig. 5.1). Under isothermal conditions the growth rate of the reaction layer, which forms on the surface of the reactant particles, is approximated by the parabolic law. For spherical particles the rate of the reaction, expressed as the volume fraction of the reacted material (x), is inversely proportional to the particle size (r) (Eq. 5.2). The Jander model [9], although containing many simplifications, indicates that by shortening the diffusion path—that is, by using fine particles—the reaction can proceed faster to completion (see Fig. 5.2B). Furthermore, as the diffusion coefficient is temperature-dependent according to the Arrhenius law ($D = D_0 e^{-Q/RT}$, where Q is activation energy, R is gas constant, and T temperature) the reaction is enhanced by increasing the temperature [8,10].

$$\left(1-\sqrt{(1-x)}\right)^2 = \left(\frac{KD}{r^2}\right)t \tag{5.2}$$

5.2.1 Barium titanate

A typical example of a carbonate-oxide reaction is the solid-state synthesis of a prototype ferroelectric BT. The synthesis takes place between 800°C and 1300°C for micron-sized powders [11]. The reaction between barium carbonate $BaCO_3$ and titanium oxide TiO_2 in a 1/1 molar ratio typically yields not only the perovskite $BaTiO_3$ phase, but also secondary phases [1,12–17]. The first reaction at the interface between the reactants occurs via the diffusion of Ba^{2+} and O^{2-} ions into the TiO_2 phase, resulting in the perovskite $BaTiO_3$ product (Eq. 5.3). The latter reaction-product layer hinders the further diffusion of Ba^{2+} ions toward the TiO_2 core and tends to react with the $BaCO_3$, yielding the Ba_2TiO_4 phase (Eq. 5.4). The perovskite forms in the reaction between Ba_2TiO_4 and TiO_2 (Eq. 5.5), with $Ba_6Ti_{17}O_{40}$ and $Ba_4Ti_{13}O_{30}$ as the side products.

$$BaCO_3 + TiO_2 \rightarrow BaTiO_3 + CO_2 \quad (5.3)$$

$$BaCO_3 + BaTiO_3 \rightarrow Ba_2TiO_4 + CO_2 \quad (5.4)$$

$$Ba_2TiO_4 + TiO_2 \rightarrow 2BaTiO_3 \quad (5.5)$$

The synthesis of BT has been extensively studied because of its importance in commercial applications, especially in multilayer ceramic capacitors. Besides the conventional solid-state route, wet-chemistry routes have been introduced (see also [11,17,18]) and are discussed in Chapter 6, Chemical synthesis and epitaxial growth methods for the preparation of ferroelectric ceramics and thin films. Studies of the solid-state synthesis of $BaTiO_3$ from the carbonate-oxide mixture have revealed that by reducing the particle size of both reactants to the nanoscale and efficient mixing the temperature of the reaction was greatly reduced, to about 700°C [19].

5.2.2 Lead-based perovskites

The solid-state synthesis of two representative lead-based perovskite ferroelectrics, $Pb(Zr,Ti)O_3$ (PZT) with compositions close to the morphotropic phase boundary (MPB, Zr/Ti molar ratio ≈53/47) and $(1-x)Pb(Mg_{1/3}Nb_{2/3})O_3-xPbTiO_3$ (PMN-PT) solid solutions, is briefly discussed in this section.

The solid-state synthesis of PZT solid solution from a $PbO-ZrO_2-TiO_2$ powder mixture has been investigated by many groups [20–24]. There is a general agreement that $PbTiO_3$ is formed first, but there is a discrepancy between respective studies regarding the subsequent reaction sequence, which is most probably related to the different characteristics of the reagents, such as the particle size, processing conditions, and PZT composition.

Hiremath et al. [24] proposed that $Pb(Zr_{0.6}Ti_{0.4})O_3$ forms via a sequence of reactions between $PbTiO_3$ and Zr- or Ti-rich PZT solid solutions. As the reaction progresses with temperature and/or time, intermediate reaction products react to form a stoichiometric PZT solid solution, as described by the following reactions.

$$PbO + TiO_2 \rightarrow PbTiO_3 \quad (5.6)$$

$$PbO + ZrO_2 + PbTiO_3\ (\text{traces}) \rightarrow PZT\ (\text{Zr rich}) \quad (5.7)$$

$$PbO + PbTiO_3 + ZrO_2 \rightarrow PZT\ (\text{Ti rich}) \quad (5.8)$$

$$PbTiO_3 + PZT\ (\text{Ti rich}) + PZT\ (\text{Zr rich}) \rightarrow PZT\ (\text{stoichiometric}) \quad (5.9)$$

The conventional solid-state synthesis of $Pb(Mg_{1/3}Nb_{2/3})O_3$ (PMN) by mixing constituent oxides yields, in addition to the perovskite phase, an undesired, thermodynamically stable, pyrochlore lead niobate phase with a low dielectric permittivity. The reaction mechanism of PMN from a stoichiometric oxide mixture was proposed by Inada [25]. The reaction between PbO and Nb_2O_5 yields the pyrochlore phase at temperatures between 530°C and 600°C (Eq. 5.10). Even though the pyrochlore subsequently reacts with MgO to form the perovskite PMN (Eq. 5.11), some pyrochlore remains in the mixture (Eq. 5.12). The low reactivity of MgO and its low weight fraction in the oxide mixture are additional impediments to obtaining a single-phase perovskite.

$$3PbO + 2Nb_2O_5 \xrightarrow{530-600^\circ C} Pb_3Nb_4O_{13}\ (\text{cubic pyrochlore}), \quad (5.10)$$

$$Pb_3Nb_4O_{13} + PbO \xrightarrow{600-700^\circ C} 2Pb_2Nb_2O_7\ (\text{rhombohedral pyrochlore}), \quad (5.11)$$

$$3Pb_2Nb_2O_7 + MgO \xrightarrow{700-800^\circ C} 3Pb(Mg_{1/3}Nb_{2/3})O_3\ (\text{perovskite}) + Pb_3Nb_4O_{13}\ (\text{cubic pyrochlore}). \quad (5.12)$$

In order to avoid, or at least to suppress, the amount of pyrochlore phase in PMN, Swartz and Shrout [26] proposed a two-step columbite route, where the undesired reaction between PbO and Nb_2O_5 is avoided. In the first step a columbite precursor $MgNb_2O_6$ is synthesized:

$$MgO + Nb_2O_5 \xrightarrow{1000^\circ C} MgNb_2O_6\ (\text{columbite}). \quad (5.13)$$

In the second step, the PbO reacts with $MgNb_2O_6$ to form the perovskite phase:

$$MgNb_2O_6 + 3PbO \xrightarrow{900^\circ C} 3Pb(Mg_{1/3}Nb_{2/3})O_3\ (\text{perovskite}). \quad (5.14)$$

The columbite synthesis route has also been successfully applied to PMN-PT ceramics [27]. In the second step, both PbO and TiO_2 are added to the $MgNb_2O_6$.

Even though the columbite method is the most frequently used for the synthesis of PMN or PMN-PT materials, other approaches have been introduced, usually with the aim of avoiding an additional calcination. For example, polyethylene-glycol-modified solid-state synthesis [28], auto-combustion synthesis [29], or synthesis with the controlled agglomeration of precursor particles [30,31] were proposed.

5.2.3 Alkaline niobates

The summary reaction for the solid-state synthesis of lead-free piezoelectric sodium potassium niobate $K_{0.5}Na_{0.5}NbO_3$ (KNN) is as follows:

$$K_2CO_3 + Na_2CO_3 + 2Nb_2O_5 \rightarrow 4K_{0.5}Na_{0.5}NbO_3 + 2CO_2. \quad (5.15)$$

According to a thermogravimetric analysis of the powder mixture, the reaction proceeds with the evolution of CO_2 between 400°C and 700°C [32,33]. The usually reported temperatures for the solid-state synthesis of KNN are well above this temperature interval, typically between 750°C and 950°C. Besides the perovskite phase, other polyniobate phases have also been frequently reported as the side products of the reaction between alkali carbonates and niobia. A diffusion-couples study revealed that the first phase to form at the interface between the mixture of carbonates and niobium oxide was a polyniobate $(K,Na)_2Nb_4O_{11}$ phase and not the perovskite. The reaction sequence in the reactant powder mixture is schematically shown in Fig. 5.3. The rate of the diffusion-controlled reaction in the ternary system was determined by the diffusion of the slower potassium ions [34]. The different diffusion rates of the sodium and potassium species could lead to local inhomogeneities in the solid solution [35,36]. Such a problem could, for example, be overcome by synthesizing the $K_{0.5}Na_{0.5}NbO_3$ from the binaries, using the so-called perovskite route [37,38].

The deficiency of alkalines has been a frequently reported problem related to the processing of KNN, stemming mainly from the strong hygroscopicity of alkali reagents, especially potassium carbonate. This could be avoided by careful drying, the manipulation of reagents in a dry atmosphere [37], using hydrated carbonates [39], or adding a slight excess of alkalines [40].

Figure 5.3 Schematic presentation of the reaction sequence in the solid-state synthesis of KNN from the carbonate–oxide powder mixture.

5.2.4 $(Na_{1/2}Bi_{1/2})TiO_3$-based materials

Pure $(Na_{1/2}Bi_{1/2})TiO_3$ (NBT) is a ternary compound [41] formed at 425–600°C by the following summary reaction [42,43]:

$$Bi_2O_3 + Na_2CO_3 + 4TiO_2 \rightarrow 4Na_{1/2}Bi_{1/2}TiO_3 + CO_2 \tag{5.16}$$

The Bi_2O_3 particles act as structural hosts for the formation of the NBT perovskite phase. Note that this particle-conversion mechanism is different from the two-step mechanism of $BaTiO_3$ formation. Secondary transient phases were observed in the range 500–650°C. Kainz et al. extended the analysis to the $(Na_{1/2}Bi_{1/2})TiO_3$–$(K_{1/2}Bi_{1/2})TiO_3$ system and identified the transient phases as pyrochlore $Bi_2Ti_2O_7$ and polytitanates of the type $(Na/K)_2Ti_6O_{13}$ [43]. The formation reaction is endothermic and diffusion controlled, while the exact sequence depends on the molar ratios of NBT and KBT.

Excellent electromechanical properties triggered extensive research of the solid solutions between NBT and other perovskite titanates, such as $BaTiO_3$ and $SrTiO_3$. The perovskite phase in the $0.94(Na_{1/2}Bi_{1/2})TiO_3$–$0.06BaTiO_3$ system starts to form at 600°C by a similar mechanism to that reported for pure NBT [44]. In addition, transient phases $BaTiO_3$, $Bi_4Ti_3O_{12}$, and Ba_2TiO_4 were observed throughout the heat treatment. The formation of $(1-x)(Na_{1/2}Bi_{1/2})TiO_3$–$x SrTiO_3$ (NBT-ST) was found to take place in two stages [45]. The NBT phase was formed between 410°C and 610°C, while the $SrCO_3$ decomposed and reacted only at higher temperatures (610–810°C). This resulted in the characteristic core-shell microstructure of NBT-ST, which gradually disappeared during prolonged sintering.

5.3 Sintering of ferroelectric bulk ceramics

Sintering is the procedure of consolidating powder compacts by thermal energy to obtain dense ceramics [7,10,46,47]. Solid-state sintering is driven by a reduction in the surface free energy and can be subdivided into several sintering stages with different densification kinetics. Most material-transport mechanisms are related to the diffusion along different paths—for example, surfaces, grain boundaries or grain interior—and have different activation energies. The diffusing species in ceramics are ions with different charges, diffusion rates, and diffusion paths, which considerably complicates the analysis. In general, the diffusion rate is governed by the slowest-diffusing species along its fastest path. During the initial and intermediate sintering stages most ceramic materials predominantly exhibit densification, while grain growth appears in the final stage, starting at about 92% of relative density. The basic approach to increase the sintering driving force is to increase the curvature of the particle surfaces, that is, to reduce the powder particle size. Alternatively, the driving force can also be increased by externally applied pressure or chemical reactions. The material transport, densification, and grain growth are enhanced by the formation of a liquid phase, referred to as liquid-phase sintering [48], or by the formation of A-site lattice vacancies [1]. The sintering of ceramics at temperatures below 200°C, referred to as cold sintering, was recently introduced by Guo et al. [49]. The densification is facilitated by a mediated dissolution-precipitation process using a transient aqueous environment.

In the following sections we briefly summarize what is known about the sintering behavior of the most important perovskite ferroelectric materials.

5.3.1 Barium titanate

The high melting point of 1618°C and the absence of volatile oxides result in a relatively high sintering temperature for $BaTiO_3$ (BT), typically between 1300°C and 1500°C [1,50]. A typical shrinkage curve is presented in Fig. 5.4. The sintering of a powder with an average particle size of 1.5 μm for 1 h at 1460°C in air results in a relative density of 96% and a grain size of 20 μm [51]. Under similar sintering conditions Ba excess reduces the grain size to 5−10 μm, while Ti excess promotes grain growth and results in 50−100 μm grains. Sintering temperatures above 1460°C, a reducing atmosphere, and impurities should be avoided to prevent the appearance of the undesired hexagonal polymorph [1,52]. The sintering mechanisms and the corresponding activation energies were studied by Genuist and Haussonne [53].

Many studies concentrated on obtaining dense BT ceramics with small grain sizes, mostly due to the increased need for the miniaturization of electronic devices. Dense, fine-grained ceramics were obtained with the use of growth inhibitors [54],

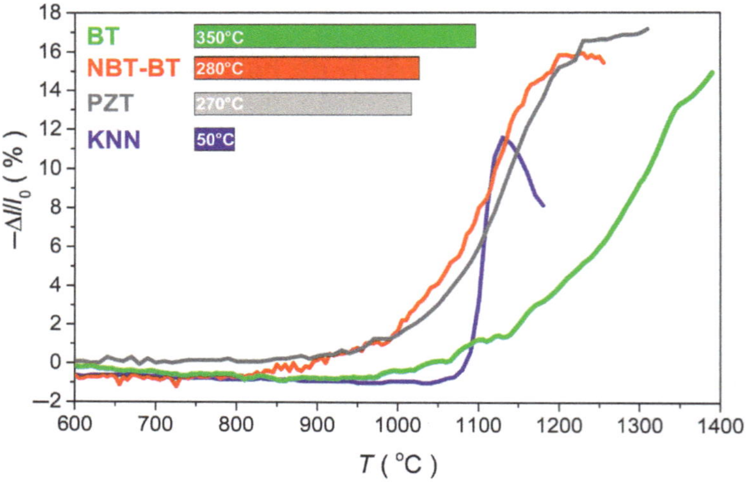

Figure 5.4 Shrinkage/sintering curves for common perovskite ferroelectrics: $Pb(Zr_{0.53}Ti_{0.47})_{0.98}Nb_{0.02}O_3$ (PZT), $BaTiO_3$ (BT), $(K_{0.5}Na_{0.5})NbO_3$ (KNN), and $0.94(Na_{1/2}Bi_{1/2})TiO_3{-}0.06BaTiO_3$ (NBT-BT). The samples were heated at 5 K/min in air. The inset bars and the corresponding temperatures represent the width of the shrinkage interval.

pressure-assisted sintering [55], fast-firing [56,57], spark-plasma sintering [58,59], microwave sintering [60], or a combined sintering method [61].

The recently reported excellent piezoelectric properties of (Ba,Ca)(Zr,Ti)O_3 [62] triggered extensive research on these BT-based compositions. The best properties were reported for sintering temperatures of 1450–1550°C and sintering times of 2–5 h [63,64], which was related to the coarse-grained microstructure with grain sizes of 10–50 μm [65]. The sintering temperature can be effectively lowered by using additives/dopants, for example, CuO [66], CeO_2 [67], Y_2O_3 [68], or Bi_2O_3 [69]. Moreover, the sintering of (Ba,Ca)(Zr,Ti)O_3 was carried out with microwave sintering [70], pressure-assisted sintering [71], and spark-plasma sintering [72]. Although low sintering temperatures, high densities, and small grain sizes were typically reported, the functional properties were not always improved, presumably due to the previously mentioned grain-size dependence.

5.3.2 Lead-based perovskites

The sintering interval for PZT-based ceramics with compositions close to the MPB extends between 1100°C and 1300°C; see Fig. 5.4.

The parameter that strongly influences the sintering of lead-based perovskites, with PZT as their representative, is the high vapor pressure of PbO at the sintering temperatures [73–75]. This may result in PbO loss and a change in the stoichiometry of the ceramic [1,76]. The approach to avoid PbO losses is to sinter PZT in the presence of the packing or atmosphere powder, such as $PbZrO_3 + ZrO_2$ or PZT + PbO, which should provide a high enough vapor pressure of PbO upon sintering, as explained by Kingon [74,75].

PbO loss from the material can also be avoided by adding an excess of PbO to the powder formulation. Its low melting point, 888°C, ensures the presence of a liquid phase at the typical sintering temperatures for PZT, and contributes to promoted densification in the early stages of sintering, but to lower final densities [74,77]. If the excess PbO does not evaporate during sintering it can remain at the grain boundaries [77,78]. Stoichiometric PZT densifies by solid-state sintering [77,79].

The multifunctionality of PZT and consequently the wide spectrum of properties suitable for numerous applications is achieved by donor or acceptor doping [1,18,80]. Alio-valent dopants modify the defect chemistry of the material, sintering, and consequently the microstructure: the grain size is reduced and the density is increased, as compared to the undoped samples [77,79]. In their study of the mechanism of sintering for undoped and La-doped PZT ceramics, both stoichiometric and with 3 mol% PbO excess, Hammer and Hoffmann found that doping with 0.5–4 mol% La contributed to an increased density and a reduction of the grain size as compared to the undoped PZT. This was related to the chemical inhomogeneities observed in the materials: Ti- and La-enriched grain cores and a PbO-rich secondary phase at the grain boundaries and triple junctions, and not to the earlier proposed dopant segregation at the grain boundaries. After sintering for 2 h at 1200°C in oxygen the undoped stoichiometric ceramics had ≈96% relative density and a grain size of 7.2 μm, while the 2 mol% La-doped sample was almost fully dense and with a grain size of 1.4 μm Note, however, that doping with very small amounts of La (≤0.2 mol%) resulted in a lower density and a larger grain size [77].

5.3.3 (K,Na)NbO_3-based ferroelectrics

Difficulties in obtaining dense, fine-grained, and homogeneous (K,Na)NbO_3 (KNN) ceramics have been reported as one of the main issues preventing the widespread application of KNN-based materials [35,81,82]. The sintering is characterized by a very narrow sintering range (Fig. 5.4), the proximity of the liquidus temperature [1], the appearance of secondary phases, and the difficulty in controlling the microstructure. The sintering of a stoichiometric KNN powder with a particle size of 0.6 μm at 1105°C for 2 h resulted in a relative density of 95.3% and a grain size of about 10 μm [39,83]. An excess of alkalis resulted in the formation of an intergranular liquid phase, a reduced density (86.5%), a lower sintering temperature, and considerable grain growth. On the other hand, an excess of Nb produced a fine-grained microstructure with a relative density of 90.9%. The stoichiometry of the powder can be controlled by drying the starting powders, while the reduction of the volatilization during sintering by the use of packing powders seems to be successful only for certain compositions [82,84]. It should be noted that the problem of volatilization does not seem to be as large as initially predicted. Recent experimental evidence shows that the vapor pressures of alkalis over their respective niobates are several orders of magnitude lower than the vapor pressures of PbO or Bi over respective PZT or alkali bismuth titanate materials [85,86].

A mechanistic description of sintering is difficult due to the presence of two volatile oxides with different vapor pressures [85], ambipolar diffusion [34], anisotropic surface energy [87,88], and the simultaneous activation of several material-transport mechanisms. A study of the model system $NaNbO_3$ revealed considerable grain growth during the initial sintering stage, prior to the activation of the main densification process, which resulted in abnormal sintering trajectories [82,89]. Haugen et al. reported a reduction in the sintering driving force in KNN due to the formation of a low-temperature liquid phase, which promotes particle coarsening and hinders the densification. The use of low pO_2 atmospheres during sintering was suggested as a way to avoid this problem [90].

The densification of KNN-based materials can be effectively enhanced by the formation of a liquid phase, through the addition of compounds containing Cu [91–93], ZnO [94], and borate- or germanate-based glassy phases [95,96]. An optimal amount of the liquid phase and its homogeneous distribution in the solid matrix are crucial for the prevention of abnormal grain growth. An alternative and inexpensive approach to improving densification is the introduction of A-site vacancies, which enhance the diffusion. The sintering of KNN doped with 0.5 at% Sr^{2+} for 2 h at 1115°C resulted in a relative density of 96.0%, as compared to 94.4% for pure KNN [97]. A modest improvement was also reported for Mg^{2+} [98] and Ba^{2+} [99] doping. Alternatively, Zuo and Rödel demonstrated that a high relative density of 98.4% can be obtained by reducing the powder particle size [100].

Beyond conventional sintering, KNN-based ceramics were also prepared using alternative sintering techniques. A relative density of 99%, good electromechanical properties, and reduced sintering temperatures were obtained by pressure-assisted sintering [101,102], spark-plasma sintering [103,104], and microwave sintering [105]. However, these techniques are often limited by other issues, such as high costs or reduction of the oxide ceramic by the commonly used carbon dies.

A study of different sintering atmospheres revealed an increase in the relative density of KNN ceramics sintered in $75N_2-25H_2$ or H_2, as compared to air or O_2, which was related to an increase in the oxygen vacancy concentration and changes in the structure of the grain boundary [106,107]. Moreover, the absence of abnormal grain growth and changes in the phase-transition behavior were reported. Sintering in reducing atmospheres is of particular importance for the production of multilayer actuators with base-metal electrodes. Such structures were successfully cofired with Ni [108–111] or Cu electrodes [112].

5.3.4 $(Na_{1/2}Bi_{1/2})TiO_3$-based ferroelectrics

Pure $(Na_{1/2}Bi_{1/2})TiO_3$ (NBT) requires a relatively high sintering temperature of over 1200°C, resulting in enhanced Bi evaporation, the formation of oxygen vacancies, and large grain sizes of 20–50 μm [113,114]. This effect can be significantly reduced by adding excess Bi_2O_3. The melting point of NBT is at 1290°C [115]. A thermogravimetric analysis revealed a measurable weight loss above 1130°C [114]. Sintering at 1100°C for 2 h resulted in low relative densities (87%), while prolonging the sintering time to 100 h improved the density to 96%. The highest density in pure NBT (98%) was obtained by pressure-assisted sintering at 1100°C for 0.5 h and $\approx$20 MPa. A reduction of the sintering temperature was achieved by doping with Co, which resulted in the generation of oxygen vacancies and the formation of a small amount of liquid phase [116]. Extensive studies were carried out to understand the influence of off-stoichiometry on the microstructure [114,117–120]. The isothermal grain growth was reported to be interface controlled, following the parabolic grain-growth law [121].

NBT-based solid solutions were first investigated to decrease the sintering temperature and thus also Bi volatilization and to reduce the coercive/poling fields. In this way many new systems with excellent electromechanical properties were discovered, such as NBT-BT, NBT-BT-KNN, $(Na_{1/2}Bi_{1/2})TiO_3-SrTiO_3$ (NBT-ST), and $(Na_{1/2}Bi_{1/2})TiO_3-(K_{1/2}Bi_{1/2})TiO_3$ [122,123]. In general, these materials are relatively easy to densify above 95% of the relative density by using sintering temperatures between 1050°C and 1200°C [43,124–126]. Fig. 5.4 shows a typical sintering curve for $0.94(Na_{1/2}Bi_{1/2})TiO_3-0.06BaTiO_3$, which is very similar to the one for PZT. A strong dependence of the electromechanical properties on the sintering parameters was reported for the NBT-ST system [45,127], which exhibits the aforementioned core-shell microstructure due to a two-stage solid-state reaction.

5.4 Thick films

The manufacturing of thick films is an additive process that involves the deposition of a precursor onto a substrate and subsequent postdeposition curing. The structure is considered a thick film when the film thickness is at least 1 μm, typically between 5 and 100 μm. However, the film thickness is not a determining factor for classifying the structure as a thick or thin film, rather it is the technology. The thick-film precursor is a powder, suspended in a certain solvent, while in the case of thin films, which can also reach micrometer thicknesses, a solution is deposited on a substrate [128].

The processing of thick films is to some extent similar to that of bulk ceramics. Since both processes are powder-based, they involve shaping of the powder or a suspension to a required geometry, followed by drying, debinding, and sintering (see Fig. 5.1). The shaping of bulk ceramics from powder suspensions includes technologies such as extrusion, injection-molding, and casting, which result in self-standing structures [4]. In contrast, thick films are deposited on various substrates, such as platinized or golded alumina, using technologies such as screen printing, inkjet printing, electrophoretic deposition, or tape casting followed by lamination. A thick-film sintering temperature of about 850°C has been adopted from thick-film technology, see [129]; however, higher sintering temperatures are also implemented to develop a dense film microstructure and the required properties.

During sintering a free-standing powder compact shrinks isotropically. In contrast, the thick film is clamped to the rigid substrate and therefore cannot shrink in-plane, but only in the direction perpendicular to the substrate (see Fig. 5.5). Consequently, a tensile stress develops in the plane of the film, which reduces the driving force for sintering and promotes the formation of defects [130–134].

The presence of a liquid phase during constrained sintering, to some extent, can relax the shear stress through particle rearrangement and/or enhanced

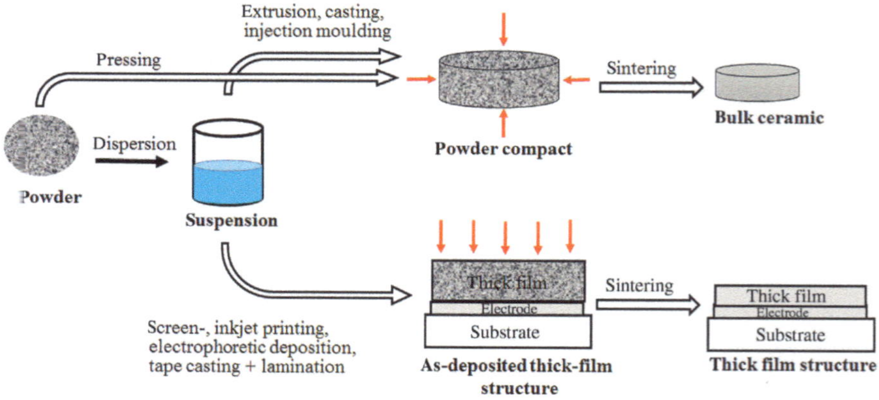

Figure 5.5 Schematic presentation of the processing of thick films as compared to bulk ceramics.

transport via a stress-supported dissolution-precipitation mechanism [134,135]. For example, lead-based thick films such as PZT are regularly sintered in the presence of a liquid phase provided by an excess of PbO or additives such as $Pb_5Ge_3O_{11}$, Cu_2O-PbO, Li_2CO_3-Bi_2O_3, Bi_2O_3-ZnO, and Li_2CO_3-Bi_2O_3-CuO, as summarized in [136]. Typical additives for lead-free materials include CuO [93], $K_{5.4}Cu_{1.3}Ta_{10}O_{29}$ [137], (Na,K)-germanate [95,138], and $BiFeO_3$ [139]. The presence of a liquid phase during the sintering usually lowers the processing temperature. It was shown for PMN-PT thick films that with the addition of 2 mol% PbO to the powder formulation the sintering temperature was reduced from 1100°C to 950°C [140]. Lowering the sintering temperature contributes to reducing the chemical reactions between the film and the substrate and reducing the sublimation of volatile oxides.

Screen printing is the most frequently used method for the processing of thick films. It is an in-contact, additive process that is based on pressing a suspension through a mask onto a substrate (see Fig. 5.1). The mask consists of a rigid frame and a polyester mesh with the defined size and position of the openings acting as a design carrier for the printed structure. The suspension is placed onto the mask, which is in direct contact with the substrate, and squeezed through the openings using a squeegee. The process is repeated until a certain thickness is reached, after which the mask is removed, resulting in a structure with a predefined geometry on the substrate.

Screen printing enables the processing of structures with thicknesses from a few to a few tens of μm with a reliable lateral resolution of about 100 μm. The composition of the suspension, consisting of the powder and organic components, should be carefully designed to obtain a pseudoplastic behavior of the suspension and to minimize the evaporation of solvents during the printing process and also reduce the shrinkage upon sintering [141]. Typically, the suspension contains about 60 wt% of the powder and 40 wt% of organic components, often α-terpineol and butylcarbitole as the solvents and ethyl-cellulose as the binder [129]. The organic components are removed from the as-deposited film by heating between 300°C and 600°C, while sintering is generally performed at temperatures at/above 850°C. Details about screen-printing technology can be found in [142–144].

Piezoelectric inkjet printing has recently attracted a lot of attention since it enables the patterning of structures with a micrometer-size lateral resolution. It is a noncontact, computer-controlled deposition technique that is based on the ejection of picoliter amounts of a fluid through nozzles onto a substrate [145]. The piezoelectric actuator, located in a reservoir of ink, generates a pressure pulse under an applied voltage, which ejects the fluid from the nozzle. The drops land on the substrate and, collectively, form a structure with a predefined architecture. In some cases, the ejected drop breaks into a main drop and a smaller drop (i.e., a satellite). The satellite lands on the substrate in the vicinity of the main drop, thus contributing to a reduced drop-position accuracy. In order to create well-defined structures with a high resolution, we need to control and optimize the jetting conditions, such as the waveform, the drop spacing, the distance between the print head and the substrate; the properties of the suspension, such as its stability, particle size, density, surface tension, viscosity, temperature; and the properties of the substrate, such as roughness and wettability [145–149].

Reports on the functional response of inkjet-printed ferroelectric thick films are rare. More often the papers address the processing of the suspension and/or the capability of jetting the suspension with a selected inkjet printer. For example, the processing of organic- and aqueous-based inkjet-printing inks with the required properties has been reported [150–152]. Some reports focus on the realization of ceramic structures with a predefined geometry [153–155]. A local piezoelectric response measured by piezoresponse microscopy was confirmed for PMN [156] and PZT thick films [157]. To obtain a thick film with an adequate functional response it is essential to prepare an ink with well-dispersed particles and rheological properties, prescribed for a selected inkjet printer. In addition, the sintering of a printed pattern under constrained conditions is a challenge that needs to be overcome.

Electrophoretic deposition (EPD) has been considered an effective technique for the processing of thick films on complex-shape substrates. It is based on the mobility of charged particles, dispersed in a solvent, toward the oppositely charged electrode, induced by a DC electric field, and the deposition of the particles onto the electrode. The mass and the properties of the deposit depend on the properties of the suspension, that is, the zeta-potential of the particles, the viscosity, the solid load, as well as on the deposition conditions, that is, the local electric field strength and the deposition time [157–160].

The possibility of preparing lead-based ferroelectric thick films on flat ceramic [161,162], nonflat metalized alumina substrates [163] or the preparation of patterned structures [164,165] by EPD has been reported in the literature. The reports on the processing of lead-free ferroelectric thick films by EPD are rare. Sodium potassium niobate thick films were prepared on platinum foil from acetone- and water-based suspensions [166,167].

Tape casting enables the processing of a few to a few-tens-of-micrometers-thick layers of green ceramic that can be used to produce self-standing, single-layer ceramics, or laminated into thicker, tridimensional structures [168]. In a suspension, or slurry, the powder is dispersed in an organic solvent or water with several additives, such as binders and softener [169,170]. The suspension is cast from a storage container onto a plastic foil, forming a deposit, the thickness of which is controlled by a doctor blade. The foil is continuously moved with a controlled velocity to a drying chamber [170].

Single-layer ceramics have commonly been used as ceramic substrates; for example, alumina. Multilayered structures have been used for the manufacturing of multilayered ceramic capacitors (MLCC). MLCCs, produced by the metallization of individual layers, interconnection, lamination, and sintering, contain a few tens or a few hundreds of individual layers. Tape casting is also suitable for the processing of functionally graded materials and textured materials, in particular to enhance the piezoelectric response of polycrystalline materials [171].

Acknowledgements

This work was supported by the Slovenian Research Agency under the program P2-0105. J.K. acknowledges the support of the Deutsche Forschungsgemeinschaft under Grant No. DFG KO5100/1-1. The authors acknowledge Jitka Hreščak and Ringailė Katiliūtė for help in editing the manuscript.

References

[1] B. Jaffe, W.R.J. Cook, H.L. Jaffe, Piezoelectric Ceramics, Academic Press Inc, London, United Kingdom, 1971.

[2] J. Rödel, W. Jo, K.T.P. Seifert, E.M. Anton, T. Granzow, D. Damjanović, Perspective on the development of lead-free piezoceramics, J. Am. Ceram. Soc. 92 (2009) 1153–1177. Available from: http://dx.doi.org/10.1111/j.1551-2916.2009.03061.x.

[3] J. Rödel, K.G. Webber, R. Dittmer, W. Jo, M. Kimura, D. Damjanović, Transferring lead-free piezoelectric ceramics into application, J. Eur. Ceram. Soc. 35 (2015) 1659–1681. Available from: http://dx.doi.org/10.1016/j.jeurceramsoc.2014.12.013.

[4] J.S. Reed, Principles of Ceramics Processing, second ed., John Wiley & Sons, New York, USA, 1995.

[5] J. Koruza, T. Rojac, B. Malič, Polar oxide nanopowders prepared by mechanical treatments, in: M. Aliofkhazraei (Ed.), Handb. Mech. Nanostructuring, Wiley-VCH Verlag GmbH & Co. KGaA, Weinheim, Germany, 2015, pp. 641–661. Available from: http://dx.doi.org/10.1002/9783527674947.ch27.

[6] A.J. Moulson, J.M. Herbert, Electroceramics: Materials, Properties, Applications, second ed., John Wiley & Sons, Chichester, United Kingdom, 2003.

[7] M.N. Rahaman, Ceramic Processing and Sintering, second ed. CRC Press, New York, USA, 2003.

[8] H. Schmalzried, Solid State Reactions, Verlag Chemie, Weinheim, Germany, 1974.

[9] W. Jander, Reaktionen im festen zustande bei höheren temperaturen, Zeitsch. Anorg. Allg. Chem. 163 (1927) 1–30. Available from: http://dx.doi.org/10.1002/zaac.19271630102.

[10] W.D. Kingery, H.K. Bowen, D.R. Uhlmann, Introduction to Ceramics, second ed. John Wiley & Sons, New York, USA, 1976.

[11] M.M. Vijatović, J.D. Bobić, B.D. Stojanović, History and challenges of barium titanate: Part II, Sci. Sinter 40 (2008) 235–244. Available from: http://dx.doi.org/10.2298/SOS0803235V.

[12] L.K. Templeton, J.A. Pask, Formation of $BaTiO_3$ from $BaCO_3$ and TiO_2 in Air and in CO_2, J. Am. Ceram. Soc. 42 (1959) 212–216. Available from: http://dx.doi.org/10.1111/j.1151-2916.1959.tb15455.x.

[13] A. Beauger, J.C. Mutin, J.C. Niepce, Synthesis reaction of metatitanate $BaTiO_3$, Part 1, J. Mater. Sci. 18 (1983) 3041–3046. Available from: http://dx.doi.org/10.1007/BF00700786.

[14] A. Beauger, J.C. Mutin, J.C. Niepce, Synthesis reaction of metatitanate $BaTiO_3$, Part 2, J. Mater. Sci. 18 (1983) 3543–3550. Available from: http://dx.doi.org/10.1007/BF00540726.

[15] J.C. Mutin, J.C. Niepce, About stoichiometry of polycrystalline $BaTiO_3$ synthesized by solid-solid reaction, J. Mater. Sci. Lett. 3 (1984) 591–592. Available from: http://dx.doi.org/10.1007/BF00719620.
[16] P.P. Phule, S.H. Risbud, Low-temperature synthesis and processing of electronic materials in the BaO-TiO_2 system, J. Mater. Sci. 25 (1990) 1169–1183. Available from: http://dx.doi.org/10.1007/BF00585422.
[17] C. Pithan, D. Hennings, R. Waser, Progress in the synthesis of nanocrystalline $BaTiO_3$ powders for MLCC, Int. J. Appl. Ceram. Technol. 2 (2005) 1–14. Available from: http://dx.doi.org/10.1111/j.1744-7402.2005.02008.x.
[18] A.J. Bell, Ferroelectrics: the role of ceramic science and engineering, J. Eur. Ceram. Soc. 28 (2008) 1307–1317. Available from: http://dx.doi.org/10.1016/j.jeurceramsoc.2007.12.014.
[19] M.T. Buscaglia, M. Bassoli, V. Buscaglia, R. Vormberg, Solid-state synthesis of nanocrystalline $BaTiO_3$: reaction kinetics and powder properties, J. Am. Ceram. Soc. 91 (2008) 2862–2869. Available from: http://dx.doi.org/10.1111/j.1551-2916.2008.02576.x.
[20] Y. Matsuo, H. Sasaki, Formation of lead zirconate-lead titanate solid solutions, J. Am. Ceram. Soc. 48 (1965) 289–291. Available from: http://dx.doi.org/10.1111/j.1151-2916.1965.tb14743.x.
[21] T. Ohno, M. Takahashi, N. Tsubouchi, Perovskite formation process in $Pb(Ti{\cdot}Zr)O_3$ ceramics containing SiO_2, J. Jpn. Soc. Powder Powder Met. 20 (1973) 154–160. Available from: http://dx.doi.org/10.2497/jjspm.20.154.
[22] D.L. Hankey, J.V. Diggers, Solid-state reactions in the system PbO-TiO_2-ZrO_2, J. Am. Ceram. Soc. 64 (1981), C-172–C-173. http://dx.doi.org/10.1111/j.1151-2916.1981.tb15903.x
[23] S.S. Chandratreya, R.M. Fulrath, J.A. Pask, Reaction mechanisms in the formation of PZT solid solutions, J. Am. Ceram. Soc. 64 (1981) 422–425. Available from: http://dx.doi.org/10.1111/j.1151-2916.1981.tb09883.x.
[24] B.V. Hiremath, A.I. Kingon, J.V. Biggers, Reaction sequence in the formation of lead zirconate-lead titanate solid solution: role of raw materials, J. Am. Ceram. Soc. 66 (1983) 790–793. Available from: http://dx.doi.org/10.1111/j.1151-2916.1983.tb10564.x.
[25] M. Inada, Analysis of the formation process of the piezoelectric PCM ceramics, Jpn. Natl. Tech. Rep. 27 (1977) 95–102.
[26] S.L. Swartz, T.R. Shrout, Fabrication of perovskite lead magnesium niobate, Mater. Res. Bull. 17 (1982) 1245–1250, http://dx.doi.org/10.1016/0025-5408(82)90159-3.
[27] S.L. Swartz, T.R. Shrout, W.A. Schulze, L.E. Cross, Dielectric properties of lead-magnesium niobate ceramics, J. Am. Ceram. Soc. 67 (1984) 311–314. Available from: http://dx.doi.org/10.1111/j.1151-2916.1984.tb19528.x.
[28] S. Yu, H. Huang, L. Zhou, Y. Ye, A polyethylene glycol-modified solid-state reaction route to synthesize relaxor ferroelectric $Pb(Mg_{1/3}\ Nb_{2/3})O_3$ $-PbTiO_3$ (PMN−PT), J. Am. Ceram. Soc. 91 (2008) 1057–1064. Available from: http://dx.doi.org/10.1111/j.1551-2916.2007.02241.x.
[29] M. Ghasemifard, S.M. Hosseini, G.H. Khorrami, Synthesis and structure of PMN-PT ceramic nanopowder free from pyrochlore phase, Ceram. Int. 35 (2009) 2899–2905. Available from: http://dx.doi.org/10.1016/j.ceramint.2009.03.036.
[30] G. Trefalt, B. Malič, D. Kuščer, J. Holc, M. Kosec, Synthesis of $Pb(Mg_{1/3}Nb_{2/3})O_3$ by self-assembled colloidal aggregates, J. Am. Ceram. Soc. 94 (2011) 2846–2856. Available from: http://dx.doi.org/10.1111/j.1551-2916.2011.04443.x.

[31] G. Trefalt, B. Malič, J. Holc, H. Uršič, M. Kosec, Synthesis of $0.65Pb(Mg_{1/3}Nb_{2/3})O_3$-$0.35PbTiO_3$ by controlled agglomeration of precursor particles, J. Am. Ceram. Soc. 95 (2012) 1858–1865. Available from: http://dx.doi.org/10.1111/j.1551-2916.2012.05142.x.

[32] B. Malič, D. Jenko, J. Bernard, J. Cilenšek, M. Kosec, Synthesis and sintering of $(K,Na)NbO_3$ based ceramics, MRS Proc. 755 (DD4) (2003) 4. Available from: http://dx.doi.org/10.1557/PROC-755-DD4.4.

[33] J. Hreščak, B. Malič, J. Cilenšek, A. Benčan, Solid-state synthesis of undoped and Sr-doped $K_{0.5}Na_{0.5}NbO_3$, J. Therm. Anal. Calorim. 127 (2017) 129–136. Available from: http://dx.doi.org/10.1007/s10973-016-5615-3.

[34] B. Malič, D. Jenko, J. Holc, M. Hrovat, M. Kosec, Synthesis of sodium potassium niobate: a diffusion couples study, J. Am. Ceram. Soc. 91 (2008) 1916–1922. Available from: http://dx.doi.org/10.1111/j.1551-2916.2008.02376.x.

[35] Y. Wang, D. Damjanović, N. Klein, E. Hollenstein, N. Setter, Compositional inhomogeneity in Li- and Ta-modified $(K, Na)NbO_3$ ceramics, J. Am. Ceram. Soc. 90 (2007) 3485–3489. Available from: http://dx.doi.org/10.1111/j.1551-2916.2007.01962.x.

[36] A. Benčan, E. Tchernychova, S. Šturm, Z. Samardžija, B. Malič, M. Kosec, Approaches for a reliable compositional analysis of alkaline-based lead-free perovskite ceramics using microanalytical methods, J. Adv. Dielectr. 1 (2011) 41–52. Available from: http://dx.doi.org/10.1142/S2010135X11000057.

[37] N.M. Hagh, B. Jadidian, A. Safari, Property-processing relationship in lead-free (K, Na, Li) NbO_3-solid solution system, J. Electroceramics 18 (2007) 339–346. Available from: http://dx.doi.org/10.1007/s10832-007-9171-x.

[38] B. Malič, J. Bernard, A. Benčan, M. Kosec, Influence of zirconia addition on the microstructure of $K_{0.5}Na_{0.5}NbO_3$ ceramics, J. Eur. Ceram. Soc. 28 (2008) 1191–1196. Available from: http://dx.doi.org/10.1016/j.jeurceramsoc.2007.11.004.

[39] J. Acker, H. Kungl, M.J. Hoffmann, Influence of alkaline and niobium excess on sintering and microstructure of sodium-potassium niobate $(K_{0.5}Na_{0.5})NbO_3$, J. Am. Ceram. Soc. 93 (2010) 1270–1281. Available from: http://dx.doi.org/10.1111/j.1551-2916.2010.03578.x.

[40] L. Egerton, D.M. Dillon, Piezoelectric and dielectric properties of ceramics in the system potassium-sodium niobate, J. Am. Ceram. Soc. 42 (1959) 438–442. Available from: http://dx.doi.org/10.1111/j.1151-2916.1959.tb12971.x.

[41] K. Uchida, T. Kikuchi, Subsolidus phase-equilibria in system Na_2O-Bi_2O_3-TiO_2 at 1000 °C, J. Am. Ceram. Soc. 61 (1978) 5–8. Available from: http://dx.doi.org/10.1111/j.1151-2916.1978.tb09217.x.

[42] E. Aksel, J.L. Jones, Phase formation of sodium bismuth titanate perovskite during solid-state processing, J. Am. Ceram. Soc. 93 (2010) 3012–3016. Available from: http://dx.doi.org/10.1111/j.1551-2916.2010.03977.x.

[43] T. Kainz, M. Naderer, D. Schuetz, O. Fruhwirth, F.A. Mautner, K. Reichmann, Solid state synthesis and sintering of solid solutions of BNT-xBKT, J. Eur. Ceram. Soc. 34 (2014) 3685–3697. Available from: http://dx.doi.org/10.1016/j.jeurceramsoc.2014.04.040.

[44] D. Hou, E. Aksel, C.M. Fancher, T.-M. Usher, T. Hoshina, H. Takeda, et al., Formation of sodium bismuth titanate-barium titanate during solid-state synthesis, J. Am. Ceram. Soc. (2017). Available from: http://dx.doi.org/10.1111/jace.14631.

[45] J. Koruza, V. Rojas, L. Molina-Luna, U. Kunz, M. Duerrschnabel, H.J. Kleebe, et al., Formation of the core-shell microstructure in lead-free $Bi_{1/2}Na_{1/2}TiO_3$-$SrTiO_3$ piezoceramics and its influence on the electromechanical properties, J. Eur. Ceram. Soc. 36

(2016) 1009–1016. Available from: http://dx.doi.org/10.1016/j.jeurceramsoc.2015.11.046.

[46] S.-J.L. Kang, Sintering: Densification, Grain Growth, and Microstructure, Elsevier Butterworth-Heinemann, Oxford, United Kingdom, 2005.

[47] R.M. German, Sintering: From Empirical Observations to Scientific Principles, Butterworth-Heinemann, Waltham, MA, USA, 2014.

[48] R. German, P. Suri, S. Park, Review: liquid phase sintering, J. Mater. Sci. 44 (2009) 1–39. Available from: http://dx.doi.org/10.1007/s10853-008-3008-0.

[49] J. Guo, H.Z. Guo, A.L. Baker, M.T. Lanagan, E.R. Kupp, G.L. Messing, et al., Cold sintering: a paradigm shift for processing and integration of ceramics, Ang. Chem.-Int. Ed., 55, 2016, pp. 11457–11461. Available from: http://dx.doi.org/10.1002/anie.201605443.

[50] L. Egerton, S.E. Koonce, Effect of firing cycle on structure and some dielectric and piezoelectric properties of barium titanate ceramics, J. Am. Ceram. Soc. 38 (1955) 412–418. Available from: http://dx.doi.org/10.1111/j.1151-2916.1955.tb14566.x.

[51] F. Kulcsar, A microstructure study of barium titanate ceramics, J. Am. Ceram. Soc. 39 (1956) 13–17. Available from: http://dx.doi.org/10.1111/j.1151-2916.1956.tb15591.x.

[52] D.E. Rase, R. Roy, Phase equilibria in the system $BaO-TiO_2$, J. Am. Ceram. Soc. 38 (1955) 102–113. Available from: http://dx.doi.org/10.1111/j.1151-2916.1955.tb14585.x.

[53] C. Genuist, F.J.M. Haussonne, Sintering of $BaTiO_3$ - dilatometric analysis of diffusion-models and microstructure control, Ceram. Int. 14 (1988) 169–179, http://dx.doi.org/10.1016/0272-8842(88)90006-5.

[54] K.S. Mazdiyasni, L.M. Brown, Microstructure and electrical properties of Sc_2O_3-doped, rare-earth-oxide-doped, and undoped $BaTiO_3$, J. Am. Ceram. Soc. 54 (1971) 539–543. Available from: http://dx.doi.org/10.1111/j.1151-2916.1971.tb12202.x.

[55] M.H. Frey, Z. Xu, P. Han, D.A. Payne, The role of interfaces on an apparent grain size effect on the dielectric properties for ferroelectric barium titanate ceramics, Ferroelectrics 206 (1998) 337–353. Available from: http://dx.doi.org/10.1080/00150199808009168.

[56] H. Mostaghaci, R.J. Brook, Fast-firing of non stoichiometric $BaTiO_3$, Trans. J. Br. Ceram. Soc. 80 (1981) 148–149.

[57] G. Arlt, D. Hennings, G. de With, Dielectric properties of fine-grained barium titanate ceramics, J. Appl. Phys. 58 (1985) 1619. Available from: http://dx.doi.org/10.1063/1.336051.

[58] T. Takeuchi, M. Tabuchi, H. Kageyama, Y. Suyama, Preparation of dense $BaTiO_3$ ceramics with submicrometer grains by spark plasma sintering, J. Am. Ceram. Soc. 82 (1999) 939–943. Available from: http://dx.doi.org/10.1111/j.1151-2916.1999.tb01857.x.

[59] Z. Zhao, V. Buscaglia, M. Viviani, M.T. Buscaglia, L. Mitoseriu, A. Testino, et al., Grain-size effects on the ferroelectric behavior of dense nanocrystalline $BaTiO_3$ ceramics, Phys. Rev. B (2004) 70. Available from: http://dx.doi.org/10.1103/PhysRevB.70.024107.

[60] H. Takahashi, Y. Numamoto, J. Tani, K. Matsuta, J.H. Qiu, S. Tsurekawa, Lead-free barium titanate ceramics with large piezoelectric constant fabricated by microwave sintering, Jpn. J. Appl. Phys. 45 (2006) L30–L32. Available from: http://dx.doi.org/10.1143/Jjap.45.L30.

[61] A. Polotai, K. Breece, E. Dickey, C. Randall, A. Ragulya, A novel approach to sintering nanocrystalline barium titanate ceramics, J. Am. Ceram. Soc. 88 (2005) 3008–3012. Available from: http://dx.doi.org/10.1111/j.1551-2916.2005.00552.x.

[62] W.F. Liu, X.B. Ren, Large piezoelectric effect in Pb-free ceramics, Phys. Rev. Lett. (2009) 103. Available from: http://dx.doi.org/10.1103/PhysRevLett.103.257602.

[63] P. Wang, Y.X. Li, Y.Q. Lu, Enhanced piezoelectric properties of $(Ba_{0.85}Ca_{0.15})(Ti_{0.9}Zr_{0.1})O_3$ lead-free ceramics by optimizing calcination and sintering temperature, J. Eur. Ceram. Soc. 31 (2011) 2005–2012. Available from: http://dx.doi.org/10.1016/j.jeurceramsoc.2011.04.023.

[64] J.G. Wu, D.Q. Xiao, W.J. Wu, J.G. Zhu, J. Wang, Effect of dwell time during sintering on piezoelectric properties of $(Ba_{0.85}Ca_{0.15})(Ti_{0.90}Zr_{0.10})O_3$ lead-free ceramics, J. Alloys Compd. 509 (2011) L359–L361. Available from: http://dx.doi.org/10.1016/j.jallcom.2011.08.024.

[65] J.G. Hao, W.F. Bai, W. Li, J.W. Zhai, Correlation between the microstructure and electrical properties in high-performance $(Ba_{0.85}Ca_{0.15})(Zr_{0.1}Ti_{0.9})O_3$ lead-free piezoelectric ceramics, J. Am. Ceram. Soc. 95 (2012) 1998–2006. Available from: http://dx.doi.org/10.1111/j.1551-2916.2012.05146.x.

[66] T. Chen, T. Zhang, G.C. Wang, J.F. Zhou, J.W. Zhang, Y.H. Liu, Effect of CuO on the microstructure and electrical properties of $Ba_{0.85}Ca_{0.15}Ti_{0.90}Zr_{0.10}O_3$ piezoceramics, J. Mater. Sci. 47 (2012) 4612–4619. Available from: http://dx.doi.org/10.1007/s10853-012-6326-1.

[67] Y.R. Cui, X.Y. Liu, M.H. Jiang, X.Y. Zhao, X. Shan, W.H. Li, et al., Lead-free $(Ba_{0.85}Ca_{0.15})(Ti_{0.9}Zr_{0.1})O_3$-$CeO_2$ ceramics with high piezoelectric coefficient obtained by low-temperature sintering, Ceram. Int. 38 (2012) 4761–4764. Available from: http://dx.doi.org/10.1016/j.ceramint.2012.02.063.

[68] Y.R. Cui, C.L. Yuan, X.Y. Liu, X.Y. Zhao, X. Shan, Lead-free $(Ba_{0.85}Ca_{0.15})(Ti_{0.9}Zr_{0.1})O_3$-$Y_2O_3$ ceramics with large piezoelectric coefficient obtained by low-temperature sintering, J. Mater. Sci. Electron. 24 (2013) 654–657. Available from: http://dx.doi.org/10.1007/s10854-012-0785-7.

[69] R. Hayati, M.A. Bahrevar, T. Ebadzadeh, V. Rojas, N. Novak, J. Koruza, Effects of Bi_2O_3 additive on sintering process and dielectric, ferroelectric, and piezoelectric properties of $(Ba_{0.85}Ca_{0.15})(Zr_{0.1}Ti_{0.9})O_3$ lead-free piezoceramics, J. Eur. Ceram. Soc. 36 (2016) 3391–3400. Available from: http://dx.doi.org/10.1016/j.jeurceramsoc.2016.05.033.

[70] S. Mahajan, O.P. Thakur, D.K. Bhattacharya, K. Sreenivas, A comparative study of $Ba_{0.95}Ca_{0.05}Zr_{0.25}Ti_{0.75}O_3$ relaxor ceramics prepared by conventional and microwave sintering techniques, Mater. Chem. Phys. 112 (2008) 858–862. Available from: http://dx.doi.org/10.1016/j.matchemphys.2008.06.054.

[71] S.K. Ye, J. Fuh, L. Lu, Y.L. Chang, J.R. Yang, Structure and properties of hot-pressed lead-free $(Ba_{0.85}Ca_{0.15})(Zr_{0.1}Ti_{0.9})O_3$ piezoelectric ceramics, R. Soc. Chem. Adv. 3 (2013) 20693–20698. Available from: http://dx.doi.org/10.1039/c3ra43429j.

[72] F. Benabdallah, C. Elissalde, U.C.C. Seu, D. Michau, A. Poulon-Quintin, M. Gayot, et al., Structure-microstructure-property relationships in lead-free BCTZ piezoceramics processed by conventional sintering and spark plasma sintering, J. Eur. Ceram. Soc. 35 (2015) 4153–4161. Available from: http://dx.doi.org/10.1016/j.jeurceramsoc.2015.06.030.

[73] K.H. Hardtl, H. Rau, PbO vapour pressure in $Pb(Ti_{1-x}Zr_x)O_3$ system, Solid State Commun. 7 (1969) 41–45, http://dx.doi.org/10.1016/0038-1098(69)90688-7.

[74] A.I. Kingon, J.B. Clark, Sintering of PZT ceramics: II, effect of PbO content on densification kinetics, J. Am. Ceram. Soc. 66 (1983) 256–260. Available from: http://dx.doi.org/10.1111/j.1151-2916.1983.tb15709.x.

[75] A.I. Kingon, J.B. Clark, Sintering of PZT ceramics: I, atmosphere control, J. Am. Ceram. Soc. 66 (1983) 253–256. Available from: http://dx.doi.org/10.1111/j.1151-2916.1983.tb15708.x.

[76] S. Fushimi, T. Ikeda, Phase equilibrium in the system PbO-TiO_2-ZrO_2, J. Am. Ceram. Soc. 50 (1967) 129–132. Available from: http://dx.doi.org/10.1111/j.1151-2916.1967.tb15061.x.

[77] M. Hammer, M. Hoffmann, Sintering model for mixed-oxide-derived lead zirconate titanate ceramics, J. Am. Ceram. Soc. 84 (1998) 3277–3284. Available from: http://dx.doi.org/10.1111/j.1151-2916.1998.tb02768.x.

[78] E.K.W. Goo, R.K. Mishra, G. Thomas, Transmission electron microscopy of $Pb(Zr_{0.52}Ti_{0.48})O_3$, J. Am. Ceram. Soc. 64 (1981) 517–519. Available from: http://dx.doi.org/10.1111/j.1151-2916.1981.tb10316.x.

[79] R.B. Atkin, R.M. Fulrath, Point defects and sintering of lead zirconate-titanate, J. Am. Ceram. Soc. 54 (1971) 265–270. Available from: http://dx.doi.org/10.1111/j.1151-2916.1971.tb12286.x.

[80] D. Damjanović, Ferroelectric, dielectric and piezoelectric properties of ferroelectric thin films and ceramics, Rep. Prog. Phys. 61 (1998) 1267–1324.

[81] S. Priya, S. Nahm (Eds.), Lead-Free Piezoelectrics, Springer-Verlag, New York, USA, 2012.

[82] B. Malič, J. Koruza, J. Hreščak, J. Bernard, K. Wang, J. Fisher, et al., Sintering of lead-free piezoelectric sodium potassium niobate ceramics, Materials 8 (2015) 8117–8146. Available from: http://dx.doi.org/10.3390/ma8125449.

[83] M.J. Hoffmann, H. Kungl, J. Acker, C. Elsässer, S. Körbel, P. Marton, et al., Influence of the A/B stoichiometry on defect structure, sintering, and microstructure in undoped and Cu-doped KNN, in: S. Priya, S. Nahm (Eds.), Lead-Free Piezoelectrics, Springer, New York, USA, 2012, pp. 209–251.

[84] J. Acker, H. Kungl, R. Schierholz, S. Wagner, R.A. Eichel, M.J. Hoffmann, Microstructure of sodium-potassium niobate ceramics sintered under high alkaline vapor pressure atmosphere, J. Eur. Ceram. Soc. 34 (2014) 4213–4221. Available from: http://dx.doi.org/10.1016/j.jeurceramsoc.2014.06.021.

[85] A. Popovič, L. Bencze, J. Koruza, B. Malič, A. Popovič, L. Bencze, et al., Vapour pressure and mixing thermodynamic properties of the $KNbO_3$-$NaNbO_3$ system, R. Soc. Chem. Adv. 5 (2015) 76249–76256. Available from: http://dx.doi.org/10.1039/c5ra11874c.

[86] A. Popovič, L. Bencze, J. Koruza, B. Malič, M. Kosec, Knudsen effusion mass spectrometric approach to the thermodynamics of Na_2O-Nb_2O_5 system, Int. J. Mass Spectrom. 309 (2012) 70–78. Available from: http://dx.doi.org/10.1016/j.ijms.2011.08.028.

[87] C.R. Cho, A. Grishin, Background oxygen effects on pulsed laser deposited $Na_{0.5}K_{0.5}NbO_3$ films: from superparaelectric state to ferroelectricity, J. Appl. Phys. 87 (2000) 4439–4448. Available from: http://dx.doi.org/10.1063/1.373089.

[88] A. Kupec, B. Malič, J. Tellier, E. Tchernychova, S. Glinšek, M. Kosec, Lead-free ferroelectric potassium sodium niobate thin films from solution: composition and structure, J. Am. Ceram. Soc. 95 (2012) 515–523. Available from: http://dx.doi.org/10.1111/j.1551-2916.2011.04892.x.

[89] J. Koruza, B. Malič, Initial stage sintering mechanism of $NaNbO_3$ and implications regarding the densification of alkaline niobates, J. Eur. Ceram. Soc. 34 (2014) 1971–1979. Available from: http://dx.doi.org/10.1016/j.jeurceramsoc.2014.01.035.

[90] A.B. Haugen, F. Madaro, L.P. Bjorkeng, T. Grande, M.A. Einarsrud, Sintering of sub-micron $K_{0.5}Na_{0.5}NbO_3$ powders fabricated by spray pyrolysis, J. Eur. Ceram. Soc. 35 (2015) 1449–1457. Available from: http://dx.doi.org/10.1016/j.jeurceramsoc.2014.11.011.

[91] M. Matsubara, T. Yamaguchi, K. Kikuta, S. Hirano, Sinterability and piezoelectric properties of $(K,Na)NbO_3$ ceramics with novel sintering aid, Jpn. J. Appl. Phys. 43 (2004) 7159–7163. Available from: http://dx.doi.org/10.1143/Jjap.43.7159.

[92] M. Matsubara, T. Yamaguchi, K. Kikuta, S. Hirano, Synthesis and characterization of $(K_{0.5}Na_{0.5})(Nb_{0.7}Ta_{0.3})O_3$ piezoelectric ceramics sintered with sintering aid $K_{5.4}Cu_{1.3}Ta_{10}O_{29}$, Jpn. J. Appl. Phys. 44 (2005) 6618–6623. Available from: http://dx.doi.org/10.1143/Jjap.44.6618.

[93] M. Matsubara, T. Yamaguchi, W. Sakamoto, K. Kikuta, T. Yogo, S. Hirano, Processing and piezoelectric properties of lead-free (K,Na) (Nb,Ta) O_3 ceramics, J. Am. Ceram. Soc. 88 (2005) 1190–1196. Available from: http://dx.doi.org/10.1111/j.1551-2916.2005.00229.x.

[94] S.H. Park, C.W. Ahn, S. Nahm, J.S. Song, Microstructure and piezoelectric properties of ZnO-added $(Na_{0.5}K_{0.5})NbO_3$ ceramics, Jpn. J. Appl. Phys. 43 (2004) L1072–L1074. Available from: http://dx.doi.org/10.1143/Jjap.453.L1072.

[95] J. Bernard, A. Benčan, T. Rojac, J. Holc, B. Malič, M. Kosec, Low-temperature sintering of $K_{0.5}Na_{0.5}NbO_3$ ceramics, J. Am. Ceram. Soc. 91 (2008) 2409–2411. Available from: http://dx.doi.org/10.1111/j.1551-2916.2008.02447.x.

[96] K.P. Chen, F.L. Zhang, J.Q. Zhou, X.W. Zhang, C.W. Li, L.N. An, Effect of borax addition on sintering and electrical properties of $(K_{0.5}Na_{0.5})NbO_3$ lead-free piezoceramics, Ceram. Int. 41 (2015) 10232–10236. Available from: http://dx.doi.org/10.1016/j.ceramint.2015.04.131.

[97] B. Malič, J. Bernard, J. Holc, D. Jenko, M. Kosec, Alkaline-earth doping in (K,Na) NbO_3 based piezoceramics, J. Eur. Ceram. Soc. 25 (2005) 2707–2711. Available from: http://dx.doi.org/10.1016/j.jeurceramsoc.2005.03.127.

[98] M. Kosec, D. Kolar, Activated sintering and electrical properties of $NaKNbO_3$, Mater. Res. Bull. 10 (1975) 335–339.

[99] J. Taub, L. Ramajo, M.S. Castro, Phase structure and piezoelectric properties of Ca- and Ba-doped $K_{1/2}Na_{1/2}NbO_3$ lead-free ceramics, Ceram. Int. 39 (2013) 3555–3561. Available from: http://dx.doi.org/10.1016/j.ceramint.2012.10.181.

[100] R.Z. Zuo, J. Rödel, R.Z. Chen, L.T. Li, Sintering and electrical properties of lead-free $Na_{0.5}K_{0.5}NbO_3$ piezoelectric ceramics, J. Am. Ceram. Soc. 89 (2006) 2010–2015. Available from: http://dx.doi.org/10.1111/j.1551-2916.2006.00991.x.

[101] R.E. Jaeger, L. Egerton, Hot pressing of potassium-sodium niobates, J. Am. Ceram. Soc. 45 (1962) 209–213. Available from: http://dx.doi.org/10.1111/j.1151-2916.1962.tb11127.x.

[102] G.H. Haertling, Properties of hot-pressed ferroelectric alkali niobate ceramics, J. Am. Ceram. Soc. 50 (1967) 329. Available from: http://dx.doi.org/10.1111/j.1151-2916.1967.tb15121.x.

[103] K. Wang, B.P. Zhang, J.F. Li, L.M. Zhang, Lead-free $Na_{0.5}K_{0.5}NbO_3$ piezoelectric ceramics fabricated by spark plasma sintering: annealing effect on electrical properties, J. Electroceram. 21 (2008) 251–254. Available from: http://dx.doi.org/10.1007/s10832-007-9137-z.

[104] T. Hungria, J. Galy, A. Castro, Spark plasma sintering as a useful technique to the nanostructuration of piezo-ferroelectric materials, Adv. Eng. Mater. 11 (2009) 615–631. Available from: http://dx.doi.org/10.1002/adem.200900052.

[105] M.R. Bafandeh, R. Gharahkhani, M.H. Abbasi, A. Saidi, J.S. Lee, H.S. Han, Improvement of piezoelectric and ferroelectric properties in (K,Na)NbO_3- based ceramics via microwave sintering, J. Electroceram. 33 (2014) 128−133. Available from: http://dx.doi.org/10.1007/s10832-014-9951-z.
[106] J.G. Fisher, S.J.L. Kang, Microstructural changes in $(K_{0.5}Na_{0.5})NbO_3$ ceramics sintered in various atmospheres, J. Eur. Ceram. Soc. 29 (2009) 2581−2588. Available from: http://dx.doi.org/10.1016/j.jeurceramsoc.2009.02.006.
[107] J.G. Fisher, D. Rout, K.S. Moon, S.J.L. Kang, High-temperature X-ray diffraction and Raman spectroscopy study of $(K_{0.5}Na_{0.5})NbO_3$ ceramics sintered in oxidizing and reducing atmospheres, Mater. Chem. Phys. 120 (2010) 263−271. Available from: http://dx.doi.org/10.1016/j.matchemphys.2009.11.001.
[108] M.S. Kim, S. Jeon, D.S. Lee, S.J. Jeong, J.S. Song, Lead-free NKN-5LT piezoelectric materials for multilayer ceramic actuator, J. Electroceram. 23 (2009) 372−375. Available from: http://dx.doi.org/10.1007/s10832-008-9470-x.
[109] S. Kawada, M. Kimura, Y. Higuchi, H. Takagi, (K,Na)NbO_3-based multilayer piezoelectric ceramics with nickel inner electrodes, Appl. Phys. Express 2 (2009), 111401-1111. http://dx.doi.org10.1143/Apex.2.111401.
[110] K. Kobayashi, Y. Doshida, Y. Mizuno, C.A. Randall, Possibility of cofiring a nickel inner electrode in a $(Na_{0.5}K_{0.5})NbO_3$-LiF piezoelectric actuator, Jpn. J. Appl. Phys. (2013) 52. Available from: http://dx.doi.org/10.7567/Jjap.52.09kd07.
[111] C. Liu, P. Liu, K. Kobayashi, C.A. Randall, Base metal co-fired (Na,K)NbO_3 structures with enhanced piezoelectric performance, J. Electroceram. 32 (2014) 301−306. Available from: http://dx.doi.org/10.1007/s10832-014-9899-z.
[112] L.S. Gao, S.W. Ko, H.Z. Guo, E. Hennig, C.A. Randall, Demonstration of copper co-fired (Na,K)NbO_3 multilayer structures for piezoelectric applications, J. Am. Ceram. Soc. 99 (2016) 2017−2023. Available from: http://dx.doi.org/10.1111/jace.14207.
[113] X.X. Wang, X.G. Tang, K.W. Kwok, H.L.W. Chan, C.L. Choy, Effect of excess Bi_2O_3 on the electrical properties and microstructure of $(Bi_{1/2}Na_{1/2})TiO_3$ ceramics, Appl. Phys. A-Mater. 80 (2005) 1071−1075. Available from: http://dx.doi.org/10.1007/s00339-003-2350-y.
[114] H. Nagata, T. Shinya, Y. Hiruma, T. Takenaka, I. Sakaguchi, H. Haneda, Piezoelectric properties of bismuth sodium titanate ceramics, in: K.M. Nair, R. Guo, A.S. Bhalla, S.-I. Hirano, D. Suvorov (Eds.), Dev. Dielectr. Mater. Electron. Devices, vol. 167, John Wiley & Sons, Inc, Hoboken, NJ, USA, 2006, pp. 213−221.
[115] S.E. Park, S.J. Chung, I.T. Kim, K.S. Hong, Nonstoichiometry and the long-range cation ordering in crystals of $(Na_{1/2}Bi_{1/2})TiO_3$, J. Am. Ceram. Soc. 77 (1994) 2641−2647. Available from: http://dx.doi.org/10.1111/j.1151-2916.1994.tb04655.x.
[116] V. Schmitt, F. Raether, Effect of cobalt doping on the sintering mechanisms of the lead-free piezoceramic $(Bi_{0.5}Na_{0.5})TiO_3$, J. Eur. Ceram. Soc. 34 (2014) 21−27. Available from: http://dx.doi.org/10.1016/j.jeurceramsoc.2013.07.021.
[117] M. Li, M.J. Pietrowski, R.A. De Souza, H.R. Zhang, I.M. Reaney, S.N. Cook, et al., A family of oxide ion conductors based on the ferroelectric perovskite $Na_{0.5}Bi_{0.5}TiO_3$, Nat. Mater. 13 (2014) 31−35. Available from: http://dx.doi.org/10.1038/Nmat3782.
[118] Y. Hiruma, H. Nagata, T. Takenaka, Thermal depoling process and piezoelectric properties of bismuth sodium titanate ceramics, J. Appl. Phys. 105 (2009) 84112. Available from: http://dx.doi.org/10.1063/1.3115409.
[119] Y.S. Sung, J.M. Kim, J.H. Cho, T.K. Song, M.H. Kim, H.H. Chong, et al., Effects of Na nonstoichiometry in $(Bi_{0.5}Na_{0.5+x})TiO_3$ ceramics, Appl. Phys. Lett. 96 (2010) 22901. Available from: http://dx.doi.org/10.1063/1.3275704.

[120] R.Z. Zuo, S. Su, Y. Wu, J. Fu, M. Wang, L.T. Li, Influence of A-site nonstoichiometry on sintering, microstructure and electrical properties of $(Bi_{0.5}Na_{0.5})TiO_3$ ceramics, Mater. Chem. Phys. 110 (2008) 311–315. Available from: http://dx.doi.org/10.1016/j.matchemphys.2008.02.007.
[121] M. Muthuramalingam, D.E.J. Ruth, M.V.G. Babu, N. Ponpandian, D. Mangalaraj, B. Sundarakannan, Isothermal grain growth and effect of grain size on piezoelectric constant of $Na_{0.5}Bi_{0.5}TiO_3$ ceramics, Scr. Mater. 112 (2016) 58–61. Available from: http://dx.doi.org/10.1016/j.scriptamat.2015.09.011.
[122] W. Jo, R. Dittmer, M. Acosta, J.D. Zang, C. Groh, E. Sapper, et al., Giant electric-field-induced strains in lead-free ceramics for actuator applications - status and perspective, J. Electroceram. 29 (2012) 71–93. Available from: http://dx.doi.org/10.1007/s10832-012-9742-3.
[123] K. Reichmann, A. Feteira, M. Li, Bismuth sodium titanate based materials for piezoelectric actuators, Materials 8 (2015) 8467–8495. Available from: http://dx.doi.org/10.3390/ma8125469.
[124] T. Takenaka, K. Maruyama, K. Sakata, $(Bi_{1/2}Na_{1/2})TiO_3$-$BaTiO_3$ system for lead-free piezoelectric ceramics, Jpn. J. Appl. Phys. Part 30 (1991) 2236–2239. Available from: http://dx.doi.org/10.1143/Jjap.30.2236.
[125] H.D. Li, C.D. Feng, W.L. Yao, Some effects of different additives on dielectric and piezoelectric properties of $(Bi_{1/2}Na_{1/2})TiO_3$-$BaTiO_3$ morphotropic-phase-boundary composition, Mater. Lett. 58 (2004) 1194–1198. Available from: http://dx.doi.org/10.1016/j.matlet.2003.08.034.
[126] Y. Hiruma, Y. Imai, Y. Watanabe, H. Nagata, T. Takenaka, Large electrostrain near the phase transition temperature of $(Bi_{0.5}Na_{0.5})TiO_3$-$SrTiO_3$ ferroelectric ceramics, Appl. Phys. Lett. 92 (2008), Artn 262904, http://dx.doi.orf/10.1063/1.2955533.
[127] M. Acosta, L.A. Schmitt, L. Molina-Luna, M.C. Scherrer, M. Brilz, K.G. Webber, et al., Core-shell lead-free piezoelectric ceramics: current status and advanced characterization of the $Bi_{1/2}Na_{1/2}TiO_3$-$SrTiO_3$ system, J. Am. Ceram. Soc. 98 (2015) 3405–3422. Available from: http://dx.doi.org/10.1111/jace.13853.
[128] R.A. Dorey, Ceramic Thick Films for MEMS and Microdevices., Elsevier, Waltham, USA, 2012. Available from: http://dx.doi.org/10.1016/B978-1-4377-7817-5.10001-X.
[129] M. Prudenziati (Ed.), Thick Film Sensors. Handb. Sensors Actuators, Vol. 1, Elsevier Science, London, United Kingdom, 1994.
[130] R.K. Bordia, R. Raj, Sintering behavior of ceramic films constrained by a rigid substrate, J. Am. Ceram. Soc. 68 (1985) 287–292. Available from: http://dx.doi.org/10.1111/j.1151-2916.1985.tb15227.x.
[131] G.W. Scherer, T. Garino, Viscous sintering on a rigid substrate, J. Am. Ceram. Soc. 68 (1985) 216–220. Available from: http://dx.doi.org/10.1111/j.1151-2916.1985.tb15300.x.
[132] D.J. Green, O. Guillon, J. Rödel, Constrained sintering: a delicate balance of scales, J. Eur. Ceram. Soc. 28 (2008) 1451–1466. Available from: http://dx.doi.org/10.1016/j.jeurceramsoc.2007.12.012.
[133] R.K. Bordia, A. Jagota, Crack growth and damage in constrained sintering films, J. Am. Ceram. Soc. 76 (1993) 2475–2485. Available from: http://dx.doi.org/10.1111/j.1151-2916.1993.tb03969.x.
[134] O. Guillon, R.K. Bordia, C.L. Martin, Sintering of thin films/constrained sintering, in: Z.Z. Fang (Ed.), Sinter. Adv. Mater., Elsevier, 2010, pp. 415–434. Available from: http://dx.doi.org/10.1533/9781845699949.3.415.

[135] C.L. Martin, H. Camacho-Montes, L. Olmos, D. Bouvard, R.K. Bordia, Evolution of defects during sintering: discrete element simulations, J. Am. Ceram. Soc. 92 (2009) 1435–1441. Available from: http://dx.doi.org/10.1111/j.1551-2916.2009.03014.x.

[136] M. Kosec, D. Kuščer, J. Holc, Processing of ferroelectric ceramic thick films, in: L. Pardo, J. Ricote (Eds.), Multifunct. Polycryst. Ferroelectr. Mater. Process. Prop., Dordrecht, Netherland, Springer, Netherlands, 2011, pp. 39–61. Available from: http://dx.doi.org/10.1007/978-90-481-2875-4_2.

[137] M. Matsubara, T. Yamaguchi, K. Kikuta, S.I. Hirano, Effect of Li substitution on the piezoelectric properties of potassium sodium niobate ceramics, Jpn. J. Appl. Phys. 1 44 (2005) 6136–6142. Available from: http://dx.doi.org/10.1143/JJAP.44.6136.

[138] J. Pavlič, B. Malič, T. Rojac, Small reduction of the piezoelectric d_{33} response in potassium sodium niobate thick films, J. Am. Ceram. Soc. 97 (2014) 1497–1503. Available from: http://dx.doi.org/10.1111/jace.12797.

[139] R. Zuo, C. Ye, X. Fang, $Na_{0.5}K_{0.5}NbO_3$–$BiFeO_3$ lead-free piezoelectric ceramics, J. Phys. Chem. Solids 69 (2008) 230–235. Available from: http://dx.doi.org/10.1016/j.jpcs.2007.08.066.

[140] D. Kuščer, M. Kosec, $0.65Pb(Mg_{1/3}Nb_{2/3})O_3$–$0.35PbTiO_3$ thick films prepared by electrophoretic deposition from an ethanol-based suspension, J. Eur. Ceram. Soc. 30 (2010) 1437–1444. Available from: http://dx.doi.org/10.1016/j.jeurceramsoc.2009.11.002.

[141] R. Rudež, J. Pavlič, S. Bernik, Preparation and influence of highly concentrated screen-printing inks on the development and characteristics of thick-film varistors, J. Eur. Ceram. Soc. 35 (2015) 3013–3023. Available from: http://dx.doi.org/10.1016/j.jeurceramsoc.2015.04.035.

[142] C.A. Harper, Handbook of Thick Film Hybrid Microelectronics: A Practical Sourcebook for Designers, Fabricators, and Users, McGraw-Hill, New Jersey, USA, 1974.

[143] R.A. Dorey, R.W. Whatmore, Electroceramic thick film fabrication for MEMS, J. Electroceram. 12 (2004) 19–32. Available from: http://dx.doi.org/10.1023/B:JECR.0000033999.74149.a3.

[144] R.N. Torah, S.P. Beeby, M.J. Tudor, N.M. White, Thick-film piezoceramics and devices, J. Electroceram. 19 (2007) 95–110. Available from: http://dx.doi.org/10.1007/s10832-007-9040-7.

[145] S. Magdassi, The Chemistry of Inkjet Inks, Word Scientific Publishing Company, Singapore, Singapore, 2010.

[146] D. Jang, D. Kim, J. Moon, D. Jang, D. Kim, J. Moon, Influence of fluid physical properties on ink-jet printability influence of fluid physical properties on ink-jet printability, Langmuir 25 (2009) 2629–2635. Available from: http://dx.doi.org/10.1021/la900059m.

[147] J. Kettle, T. Lamminmäki, P. Gane, A review of modified surfaces for high speed ink-jet coating, Surf. Coat. Tech. 204 (2010) 2103–2109. Available from: http://dx.doi.org/10.1016/j.surfcoat.2009.10.035.

[148] H. Wijshoff, The dynamics of the piezo inkjet printhead operation, Phys. Rep. 491 (2010) 77–177. Available from: http://dx.doi.org/10.1016/j.physrep.2010.03.003.

[149] Z. Pan, Y. Wang, H. Huang, Z. Ling, Y. Dai, S. Ke, Recent development on preparation of ceramic inks in ink-jet printing, Ceram. Int. 41 (2015) 12515–12528. Available from: http://dx.doi.org/10.1016/j.ceramint.2015.06.124.

[150] A.R. Bhatti, M. Mott, J.R.G. Evans, M.J. Edirisinghe, PZT pillars for 1-3 composites prepared by ink-jet printing, J. Mater. Sci. Lett. 20 (2001) 1245–1248. Available from: http://dx.doi.org/10.1023/A:1010987209703.

[151] J.H. Song, H.M. Nur, Defects and prevention in ceramic components fabricated by inkjet printing, J. Mater. Process. Tech. 155–156 (2004) 1286–1292. Available from: http://dx.doi.org/10.1016/j.jmatprotec.2004.04.292.

[152] O. Noshchenko, D. Kuščer, O.C. Mocioiu, M. Zaharescu, M. Bele, B. Malič, Effect of milling time and pH on the dispersibility of lead zirconate titanate in aqueous media for inkjet printing, J. Eur. Ceram. Soc. 34 (2014) 297–305. Available from: http://dx.doi.org/10.1016/j.jeurceramsoc.2013.08.002.

[153] W.J. Tseng, S.Y. Lin, S.R. Wang, Particulate dispersion and freeform fabrication of $BaTiO_3$ thick films via direct inkjet printing, J. Electroceram. 16 (2006) 537–540. Available from: http://dx.doi.org/10.1007/s10832-006-9913-1.

[154] A. Friederich, J.R. Binder, W. Bauer, Rheological control of the coffee stain effect for inkjet printing of ceramics, J. Am. Ceram. Soc. 96 (2013) 2093–2099. Available from: http://dx.doi.org/10.1111/jace.12385.

[155] T. Bakarič, B. Malič, D. Kuščer, Lead-zirconate-titanate-based thick-film structures prepared by piezoelectric inkjet printing of aqueous suspensions, J. Eur. Ceram. Soc. 36 (2016) 4031–4037. Available from: http://dx.doi.org/10.1016/j.jeurceramsoc.2016.06.038.

[156] Kruger R. IMAP/ACer, 9th International Conference and Exhibition on Ceramic Interconnect and Ceramic Microsystems Technologies (CICMT 2013). Inkjet Print. piezoelectric lead Magnes. niobate-lead titanate Thick Film, Orlando, Florida: 2013.

[157] D. Kuščer, T. Bakarič, B. Kozlevčar, M. Kosec, Interactions between lead – zirconate titanate, polyacrylic acid, and polyvinyl butyral in ethanol and their influence on electrophoretic deposition behavior, J. Phys. Chem. B 117 (2013) 1651–1659. Available from: http://dx.doi.org/10.1021/jp305289u.

[158] P. Sarkar, P.S. Nicholson, Electrophoretic deposition (EPD): mechanisms, kinetics and application to ceramics, J. Am. Ceram. Soc. 79 (1996) 1987–2002. Available from: http://dx.doi.org/10.1111/j.1151-2916.1996.tb08929.x.

[159] L. Besra, M. Liu, A review on fundamentals and applications of electrophoretic deposition (EPD), Prog. Mater. Sci. 52 (2007) 1–61. Available from: http://dx.doi.org/10.1016/j.pmatsci.2006.07.001.

[160] B. Ferrari, R. Moreno, EPD kinetics: a review, J. Eur. Ceram. Soc. 30 (2010) 1069–1078. Available from: http://dx.doi.org/10.1016/j.jeurceramsoc.2009.08.022.

[161] C. Kaya, F. Kaya, B. Su, B. Thomas, A.R. Boccaccini, Structural and functional thick ceramic coatings by electrophoretic deposition, Surf. Coat. Tech. 191 (2005) 303–310. Available from: http://dx.doi.org/10.1016/j.surfcoat.2004.03.042.

[162] D. Kuščer, F. Levassort, M. Lethiecq, A.P. Abellard, M. Kosec, Lead-zirconate-titanate thick films by electrophoretic deposition for high-frequency ultrasound transducers, J. Am. Ceram. Soc. 95 (2012) 892–900. Available from: http://dx.doi.org/10.1111/j.1551-2916.2011.04911.x.

[163] A.P. Abellard, D. Kuščer, J.M. Gregoire, M. Lethiecq, B. Malič, F. Levassort, Lead zirconate titanate-based thick films for high-frequency focused ultrasound transducers prepared by electrophoretic deposition, IEEE T. Syst. Man. Cy C 61 (2014) 547–556. Available from: http://dx.doi.org/10.1109/TUFFC.2014.2938.

[164] A.J. Pascall, K.T. Sullivan, J.D. Kuntz, Morphology of electrophoretically deposited films on electrode strips, J. Phys. Chem. B 117 (2013) 1702–1707. Available from: http://dx.doi.org/10.1021/jp306447n.

[165] D. Kuščer, M.S. Bernardo, M.S. Zarnik, B. Malič, Patterning of lead-zirconate-titanate thick-film structures by electrophoretic deposition from ethanol-based dispersions, J. Eur. Ceram. Soc. 36 (2016) 291–297. Available from: http://dx.doi.org/10.1016/j.jeurceramsoc.2015.08.033.

[166] M. Dolhen, A. Mahajan, R. Pinho, M.E. Costa, G. Trolliard, P.M. Vilarinho, Sodium potassium niobate ($K_{0.5}Na_{0.5}NbO_3$, KNN) thick films by electrophoretic deposition, RSC Adv. 5 (2015) 4698–4706. Available from: http://dx.doi.org/10.1039/c4ra11058g.

[167] A. Mahajan, R. Pinho, M. Dolhen, M.E. Costa, P.M. Vilarinho, Unleashing the full sustainable potential of thick films of lead-free potassium sodium niobate ($K_{0.5}Na_{0.5}NbO_3$) by aqueous electrophoretic deposition, Langmuir 32 (2016) 5241–5249. Available from: http://dx.doi.org/10.1021/acs.langmuir.6b00669.

[168] M. Jabbari, R. Bulatova, A.I.Y. Tok, C.R.H. Bahl, E. Mitsoulis, J.H. Hattel, Ceramic tape casting: a review of current methods and trends with emphasis on rheological behaviour and flow analysis, Mater. Sci. Eng. B Solid 212 (2016) 39–61. Available from: http://dx.doi.org/10.1016/j.mseb.2016.07.011.

[169] D. Hotza, P. Greil, Aqueous tape casting of ceramic powders, Mater. Sci. Eng. A-Struct. 202 (1995) 206–217, http://dx.doi.org10.1016/0921-5093(95)09785-6.

[170] E. Richard, E.R.T. Mistler, Tape Casting: Theory and Practice, Wiley-The American Ceramic Society, Ohio, USA, 2000.

[171] G.L. Messing, S. Trolier-McKinstry, E.M. Sabolsky, C. Duran, S. Kwon, B. Brahmaroutu, et al., Templated grain growth of textured piezoelectric ceramics, Crit. Rev. Solid State Mater. Sci. 29 (2004) 45–96. Available from: http://dx.doi.org/10.1080/10408430490490905.

Chemical synthesis and epitaxial growth methods for the preparation of ferroelectric ceramics and thin films

6

Guilhermina F. Teixeira, Glauco M.M.M. Lustosa, Sônia M. Zanetti and Maria A. Zaghete
Institute of Chemistry, São Paulo State University (Unesp), Araraquara, Brazil

6.1 Introduction

The advances in modern technology have essentially created the need for the improvement of our knowledge regarding new routes for the processing of ferroelectric materials. Ferroelectrics have been present in our society since ferroelectricity was discovered in Rochelle salt in 1921 [1]. Nonetheless, the discovery of ferroelectricity in perovskite barium titanate ($BaTiO_3$) for military use in World War II was, undoubtedly, a great development for that period [1–3]. Nowadays, the existence of various ferroelectric-based multifunctional materials is feasible thanks to the possibility of controlling particle size and morphology besides the ability to grow epitaxial films.

The first step to be considered when obtaining ferroelectric ceramics has to do with the powder preparation, which may be done via the chemical method known as the bottom-up method. Through this method, it is possible to control the shape, stoichiometry, size, and surface area of the particles, among other relevant features [4]. The epitaxial thin film growth favors the use of ferroelectrics in miniaturized devices, which can be achieved by growth methods via the chemical route. In the present review, the readers will find important information about the chemical routes for obtaining ferroelectric powders and epitaxial thin films with the best characteristics. We shall present the most relevant concepts of the methods as well as the importance of processing in the materials properties.

6.2 Chemical synthesis of ferroelectric ceramic powders

6.2.1 Polymeric precursor method (Pechini method)

The polymeric precursor method (derived from the method developed by Pechini in 1967) is basically the obtaining of polyester from citrates [5]. In this route, metallic cations are complexed or chelated by hydroxycarboxylic acid (for instance,

Magnetic, Ferroelectric, and Multiferroic Metal Oxides. DOI: http://dx.doi.org/10.1016/B978-0-12-811180-2.00006-2

citric acid (CA)) in an aqueous solution at 90−100°C. The solution is mixed with ethylene glycol and heated between 130°C and 140°C to promote polymerization and water evaporation [6]. Thermal treatment at higher temperatures, as can be noted, helps to eliminate the excess of organic components in addition to promoting the cation oxidation and producing the ceramic stoichiometric phase of interest. This route homogeneously disperses the metallic ions and immobilizes them in a rigid polymeric network at atomic scale avoiding precipitation and phase segregation [7]. The Pechini method has been extensively used to produce several piezoelectric and ferroelectric materials passing through some modifications to make it fit the needed purposes. In most cases, CA is used as a chelating agent though it can be substituted by organic acid with a smaller molecule such as the DL-malic acid (MA) [8]. Based on this method, it is possible to synthesize several ferroelectric powders, such as lead-based and lead-free perovskites.

$Pb(Zr_xTi_{1-x})O_3$ (PZT) powders are obtained after thermal treatment at 400°C for the elimination of organic substances, and the resulting material is heated at varying temperatures. Higher temperatures promote the better crystallization of PZT with particle size around 1.1−1.75 μm and hysteresis loop showing remnant polarization (Pr) of 7.2 μC/cm^2 [9]. PZT particle size can be reduced by controlling the pH. Abreu Jr. et al. [10] studied the influence of NH_4OH and urea on PZT particle size, and demonstrated that when the pH is adjusted using NH_4OH, particles of 0.25 μm are obtained and urea is able to reduce the PZT size to 0.1 μm. The molecular weight of polyesters influences the crystallite size of PZT. Qiu et al. [7] also showed that using the same CA/cation molar ratio and different CA/ethylene glycol molar ratio polyesters with different molecular weight are produced. The crystallite size, morphology, and PZT phase formation are also influenced. Small crystallite size and hard agglomerated particles are formed when the molecular weight is low, while soft agglomerated particles are obtained from the same molar ratio of CA and ethylene glycol.

The Pechini method has been clearly shown to be more advantageous when it comes to the production of $BaTiO_3$ (BTO) particles in comparison with the products obtained from other methods. BTO particles produced from this route present a more uniform grain size distribution with particles size around 44 nm. However, the particles are seen to be agglomerated owing to the high synthesis temperature [11]. The electrical properties of BTO-based ceramics are improved by doping or substituting elements. Tian et al. [12] investigated the influence of Ir^{4+} doping $0.5Ba_{0.9}Ca_{0.1}TiO_3$-$0.5BaTi_{0.88}Zr_{0.12}O_3$−0.12%La ceramics obtained from a modified Pechini method. The presence of $Ir^{4+}$is found to modify the ceramics structure, which consequently leads to the modification of the ferroelectric and the dielectric properties [12].

Ultrafine pure phase $NaNbO_3$ nanopowders can be obtained using DL-MA as chelating agent at considerably low calcination temperature [8]. The use of CA as chelating agent produces pure phase $NaNbO_3$ at 350°C and average particle size of about 10−80 nm when thermally treated at 750°C [13]. Bismuth ferrite-potassium sodium niobate-based multiferroic ceramic composites ($(1-x)BiFeO_3$-$xNa_{0.5}K_{0.5}NbO_3$) are obtained in a single step of the Pechini method, dissolving the raw materials in CA adjusting the pH to 4–5 with ammonia. $BaTiO_3$ is also considered a good ferroelectric material, yet its application is restricted to low temperatures. To circumvent this shortcoming related to the application temperature of $BaTiO_3$, a KN-BTO solid solution can be synthesized based on a modified Pechini method according to a study conducted by Pribosic et al. [14] where they showed that a solid solution between KN and BTO perovskite phase is formed at temperatures below 900°C.

As mentioned earlier, the Pechini method is an easily performed route that is adaptable to each case. Notwithstanding this outstanding feature, the method cannot be applied when a specific shape is required. Growing particles in a specific shape can improve the ferroelectricity, and it is achieved by employing the routes explored next.

6.2.2 Sol-gel method

The sol-gel method promotes the atomic level mixing of reagents leading to chemically homogeneous complex ceramic materials at low temperature and short synthesis time. According to Kakihana [15], there are three important kinds of sol-gel processes: (1) colloidal sol-gel, (2) inorganic polymeric gel derived from organometallic compounds, and (3) gel routes involving the formation of organic polymer glass. Recently, Danks et al. [16] summarized the sol-gel process in vital key steps: the sol synthesis, the gel formation, the aging time to remove the solvent, the formation of xerogel, and the removal of surface hydroxyl groups whenever necessary.

Dense bismuth sodium titanate ($Na_{0.5}Bi_{0.5}TiO_3$) ceramics have been obtained from nanometric powder precursor synthesized by the sol-gel route. $Na_{0.5}Bi_{0.5}TiO_3$ ceramics showed higher Pr (47 $\mu C/cm^2$) [17] compared to $Na_{0.5}Bi_{0.5}TiO_3$ ceramics obtained from the mixed oxide method (26 $\mu C/cm^2$) [18] and the citrate method (33.3 $\mu C/cm^2$) [19].

The modified sol-gel process using ethylene diaminetetraacetic (EDTA)/CA as a stabilizing agent has been largely efficient in the production of $KNbO_3$ nanopowders. The stability of CA, EDTA, and ethylene glycol affects the stability of the sol. The energy released during the decomposition of the organic substances and EDTA/CA complexes contribute toward $KNbO_3$ crystallization. The K/Nb ratio influences the formation of orthorhombic pure-phase $KNbO_3$ powders besides exerting an influence on the agglomeration degree, size, and particle morphology [20]. Ultralong $KNbO_3$ ferroelectric nanofibers can be synthesized using a sol-gel-based far-field electrospinning process. For the nanofibers formation, the sol-gel precursor is extruded using a syringe under application of an electric field equal to 1.5 kV/cm. In sequence, the product is dried and calcined in atmosphere [21].

Nanocrystalline PZT has been prepared by the sol-gel method. The single tetragonal phase of PZT is formed at 800°C since at 750°C there is still the coexistence of PZT in the rhombohedral and tetragonal phases [22]. Wu et al. [23] studied the effect of alkoxide stabilizers and lead precursors on PZT synthesis aiming at determining the better conditions for preparing monophasic submicrometric PZT 52/48 powders. The products consist of fine nanoparticles, which are used to produce PZT ceramics with P_r equal to 20 μC/cm^2. PZT nanotubes are produced using multiwalled carbon nanotubes (MWCNTs) as a sacrificial template. To this proposal, a MWCNT obtained by chemical vapor deposition is coated with PZT sol with weight ratio of MWCNT:PZT = 1:5. The crystallization of PZT nanotubes occurs at 410°C while the microscopy techniques confirmed the tubular nanostructure of the material [24].

The first report about ferroelectric perovskite nanotubes synthesized using a sol-gel template was conducted by Hernandez et al. [25], who produced $PbTiO_3$ and $BaTiO_3$ nanotubes Anodic alumina oxide membranes are also employed to obtain nanotubes containing $PbTiO_3$ in their composition [26]. $PbTiO_3$ with a granular morphology is obtained by template-free sol-gel synthesis [27] while $Pb_2Ti_2O_6$ nanoparticles with cubic crystalline structure are converted to tetragonal $PbTiO_3$ after heating at 500°C [28].

$BiFeO_3$ nanoparticles have been successfully obtained using MA as chelating agent. Based on differential thermoanalysis, the researchers noted the transition from ferroelectric to paraelectric phase occurring at 821°C and 932°C, respectively [29]. When the aim is to produce one-dimensional $BiFeO_3$, the sol-gel-based electrospinning process [30] can be employed, which is quite similar to the method used by Ganeshkumar et al. [21] in the production of $KNbO_3$ nanofibers.

6.2.3 Wet-chemical precipitation method (coprecipitation method)

The coprecipitation method has been widely used to synthesize ultrafine ferroelectric ceramic powders with narrow particle size distribution. By this method, the precipitation of materials of interest from alkaline solution occurs. Barium calcium titanate-barium zirconate titanate solid solution ($x[(Ba_{0.7}Ca_{0.3})TiO_3]$-$(1-x)[Ba(Zr_{0.2}Ti_{0.8})O_3]$) has been obtained from $(Ba_{0.7}Ca_{0.3})TiO_3$ and $Ba(Zr_{0.2}Ti_{0.8})O_3$ powders. The pure phase was crystallized into a tetragonal structure with space group (*P*4mm). In bulk ceramic, the average grain size is around 100 nm, which is smaller than the same product obtained from solid state reaction. In addition, the $(Ba_{0.7}Ca_{0.3})TiO_3$ content exerts influence over the grain size and ferroelectric behavior [31].

To prepare $NaNbO_3$ and $KNbO_3$, niobium oxide (Nb_2O_5) should be dissolved in fluoridric acid (HF-40%) forming niobium fluoride (NbF_5) solution to precipitate niobium hydroxide ($Nb(OH)_5 \cdot xH_2O$) in alkaline medium of NH_4OH. The $Nb(OH)_5 \cdot xH_2O$ precipitate acts as niobium source in the reaction with NaOH or KOH to form the respective niobates. The temperature to form orthorhombic $NaNbO_3$ is lower than that to form the orthorhombic $KNbO_3$ pure phase [32]. Hexagonal crystal structure of nanometric $LiNbO_3$ has been obtained through the precipitation of lithium carbonate and niobium hydroxide inaqueous medium of NH_4OH and $(NH_4)2CO_3$ [33].

The coprecipitation method is additionally applied in the synthesis of nanocrystalline ferroelectric $SrBi_2Ta_2O_9$ powders [34]. The replacement of Sr^{2+} cation by Ba^{2+}produces nanometric irregular-shaped particles under the same temperature as $SrBi_2Ta_2O_9$ powders synthesis [35].

PZT with a different ratio of Zr/Ti is synthesized based on coprecipitation of metal hydroxides (Pb(OH), Zr(OH), and Ti(OH)). The starting solution contains metal precursors; ammonia solution is used as precipitant agent for respective hydroxides in addition to washing the precipitate in order to avoid Pb effusing [36,37]. PZT formation can be processed in one or two steps: (1) coprecipitation of zirconium, titanium, and lead hydroxides ($PbZr_{0.52}Ti_{0.48}O(OH)_4$) in a single step, and (2) $(Zr_{0.52}Ti_{0.48})O(OH)_2$ precipitate is prepared and then Pb(OH) is precipitated in ammonia solution containing $(Zr_{0.52}Ti_{0.48})O(OH)_2$ producing ($PbZr_{0.52}Ti_{0.48}O(OH)_4$). As observed by the TG-DTA curves, the PZT crystallization temperature via the two-step synthesis is higher than that prepared through the single-step precipitation route. During the process, the PbO formation and weight loss are avoided through the encapsulation of $Pb(OH)_2$ particle by $(Zr_{0.52}Ti_{0.48})O(OH)_2$ and through the great combination of Zr^{4+} and Ti^{4+} ions [37]. By coprecipitation in oxalic acid medium, strontium and barium are easily incorporated into PZT structure by partial replacement of Pb^{2+} ions. In this case, pure or modified PZT exhibit single-phase PZT at 900°C or above, and a decline is observed in the Curie temperature [38].

Rod-shaped pure and Eu-doped ZnO particles are obtained by the wet-chemical precipitation method. The ferroelectric behavior of the samples is attributed to the difference in the ionic size between Zn^{2+} and Eu^{3+} cations, giving rise to the permanently located electric dipoles [39]. This route is also found to be effective when it comes to the production of Cu-doped ZnO nanoparticles [40], Sn-doped ZnO [41], and Al-doped ZnO [42] in alkaline medium of NaOH. According to Giovannelli et al. [42], the formation of different intermediary phase as well as the particles size and morphology are related to the way the precursors are prepared. Pure ZnO exhibits micrometric sand rose morphology while Al-doped ZnO presents nanometric isotropic particles. In a study conducted by Farahmandjou and Jurablu [43], nanometric ZnO was produced from $Zn(NO_3)_2.6H_2O$ and K_2CO_3 as precipitant agent. In this case, ZnO, as it was prepared, presents spherical shape and particle size in the range of 20–80 nm. After thermal treatment, the particles show pyramidal shape and higher sizes.

In spite of this method being versatile and easily performed, its use may bear certain demerits in some cases. $Pb_{0.98}Bi_{0.02}(Zr_{0.52}Ti_{0.48})O_3$ ceramics can be prepared by the co-precipitation and sol-gel methods. Both synthesis methods produce compounds with mixed phases in the morphotropic phase boundary. Thermal analysis and X-ray diffraction indicated the merits that go with the use of the sol-gel method. At temperatures as low as 250°C it is sufficient to achieve the desired phase whereas a higher temperature is required to obtain the pure perovskite phase in the case of samples derived by the coprecipitation method [44]. Yet when the comparison is made between the coprecipitation and conventional solid-state method, the first one is seen to be quite more advantageous since the temperature required for the perovskite phase formation is around 850°C in the solid-state method [45].

6.2.4 Molten salt synthesis

Molten salt synthesis (MSS) is another method used for obtaining ferroelectric powders with defined morphology. The high diffusion of the components in salt enables the execution of this method at low temperature and in a short synthesis time [46].

MSS involves the use of molten salts as the medium to obtain complex oxides from their constituents such as carbonates or oxides. Unlike the other synthesis methods described in the present chapter that use liquids like water as solvent, what makes the MSS different is the solvent applied in the process—molten salt. The melting process occurs through the addition of salt with low melting point to other reactants followed by thermal treatment at temperatures above the melting point of the salt added. The powder characteristics are associated with the amount of salts used as solvent. In this route, the salts that are mostly employed are sulfates and chlorides. In many cases, the temperature of reaction can be decreased by the formation of eutectic mixture of salts. The particles are formed in two stages: during reaction and during particle growth. In the reaction stage, solid particles are dissolved in the molten salt to form the product. The second stage involves particle growth following the dissolution of all reactants. During the growth stage, the particle size increases through Ostwald ripening process. In a review published by Kimura [47], ample details and good descriptions about the MSS steps are given.

Li et al. [48] obtained wire-like KNb_3O_8 particles that were used as a template in the synthesis of one-dimensional $ANbO_3$ (A = K, Na, and (Na,K)) by self-sacrificing template synthesis. To carry out the synthesis, Nb_2O_5 and KCl were mixed in ethanol, dried, and heated. The prepared KNb_3O_8 was then added to the HNO_3 solution, promoting the change of K^+ ions by H_3O^+ ions forming $H_3ONb_3O_8$. Through thermal treatment $H_3ONb_3O_8$ is transformed into rod-like Nb_2O_5 powder. Finally, the one-dimensional $ANbO_3$ is obtained using rod-like Nb_2O_3 as a template. In addition, $(Na,K)NbO_3$ ceramics obtained from the KNb_3O_8 template have a high piezoelectric constant (d_{33}).

$(Na,K)NbO_3$ nanoplates were obtained in a single-step MSS in a study conducted by Ge et al. [49]. Although the content of Na^+ and K^+ ions is the same, the better diffusion rate of Na^+ ions favors the formation of $Na_{0.9}K_{0.1}NbO_3$ rather than $Na_{0.5}K_{0.5}NbO_3$. Li et al. [46] obtained polyhedral-shaped particles grown according to the Ostwald ripening process [46,47]. The increasing of sintering temperature of $0.96(Na,K)(Nb_{0.9}Ta_{0.1})O_3$-$0.04LiSbO_3$ ceramics causes the volatilization of cations, leading to the rise in the quantity of the secondary phases, thus resulting in a decrease of d_{33} and P_r values.

Anisotropic multiferroic $Bi_4NdTi_3Fe_{0.7}Ni_{0.3}O_{15}$ with plate-like morphology was also prepared via the MSS. The plates presented preferential *c*-axis orientation having a large size distribution. The grain growth is related to the increase in the sintering temperature; the texture, however, decreases at higher temperatures [50]. MSS processing parameters have considerable influence over $BiFeO_3$ powders and their derived ceramics. Rhombohedral crystalline pure phase is achieved with the increase in synthesis temperature, holding time at a given temperature, and the molten-salt ratios with respect to the starting oxides. The particle growth is also based on Ostwald ripening. Furthermore, the salt ratios and holding time do not bear any significant effect on the particle size and morphology [51]. $BiFeO_3$ product can be achieved at only 500 °C when the MSS method is active by high-energy milling [52].

PZT-PZN-PMS ceramics prepared by the MSS method present better homogeneity, density, and piezoelectric properties than those prepared by the conventional mixed-oxide method [53]. Bortolani and Doray [52] reported obtaining spherical PZT particles that are suitable for higher resolution ink jet printing. Irregular-shaped particles are obtained in shorter synthesis time while spherical particles are products of longer synthesis time. The reaction involving the formation of PZT can be explained in two stages: (1) intermediate $PbTiO_3$ is formed through the reaction between dissolved Pb with insoluble TiO_2, and (2) upon the formation of $PbTiO_3$, Zr ions diffuse and react with $PbTiO_3$, forming the PZT product.

6.2.5 Hydrothermal synthesis

One of the most interesting synthesis methods for obtaining ferroelectric materials is the hydrothermal synthesis. Byrappa and Yoshimura [54] proposed the following for the definition of hydrothermal reaction: "any heterogeneous chemical reaction in the presence of a solvent (whether aqueous or nonaqueous) above room temperature and at pressure greater than 1 atm in a closed system." However, some researchers prefer the use of the term *solvothermal* for chemical reactions that occur in a nonaqueous solvent medium.

Ferroelectric films or ceramic powders with very well-controlled morphology can be obtained by hydrothermal synthesis. Because the crystallization occurs directly in solution, the calcination step is deemed unnecessary and this contributes toward decreasing both the temperature and the synthesis time for obtaining crystalline materials. This process can be considered self-purifying owing to the fact that the environment tends to reject impurities during the material growth [55]. An extension of hydrothermal synthesis is the microwave-assisted hydrothermal (MAH) synthesis, which has the same advantages of the conventional hydrothermal (CH) method besides requiring an additional reduction in synthesis time and temperature. The fast-heating in MAH involves dipolar polarization and ionic conduction [56], which enables the products to be attained in few minutes instead of several hours as in CH synthesis. In the CH method, chemical synthesis is performed through thermal conduction, which is slow and inefficient for transferring energy into the systems, thus resulting in long synthesis time. In MAH, contrary to the CH method, the heating is faster because the microwaves interact directly with molecules present in the reaction mixture, leading to a rapid increase in temperature [57]. Several parameters such as temperature, mineralizer, precursor concentration, pH, or templates have great influence on nucleation, morphology, and particle growth.

Problems related to high synthesis temperature, such as poor stoichiometry control due to the volatilization of components, can be avoided by low reaction temperature. The temperature used for obtaining PZT by CH or MAH is much lower compared to the methods mentioned earlier. Huang et al. [58] reported controlling PZT particle size, distribution, and morphology by CH synthesis via ramping and cooling rates. Particles with cubic morphology were obtained at 200°C under a processing time of 3 h. KOH concentration controls the degree of directional growth for PZT. In this case, the synthesis was carried out for 12 h due to the long period necessary for the outside of the reactor to heat up to the set temperature [59]. Through the MAH method, it is possible to obtain microcube-shaped PZT at 180°C in 2 h [9,60]. Cubic-particle PZT architecture is composed of well-aligned nanoplates that self-assemble to form the cubic morphology [60]. The influence of the polymer on PZT crystallization was studied by Teixeira et al. [61], Glass et al. [59], and Wang et al. [62], who obtained different morphologies for PZT introducing polyvinyl alcohol and/or polyacrylic acid in the reaction medium.

The ultrasonically assisted CH method has been efficient in the production of $BiFeO_3$ powders. This procedure consists of modifying the hydrothermal method by ultrasonic treatment of precursor solution prior to the synthesis. In this sense, $BiFeO_3$ particles were obtained at 200°C during 6 h without exhibiting a defined morphology [63]. Under microwave heating, the time required for obtaining $BiFeO_3$ particles decreases to 20 min at 200°C [64].The pH value has great influence on $BiFeO_3$ particle shape [65]. The $BiFeO_3$ morphologic characteristics can be controlled using polyvinyl pyrrolidone [66], polyethylene glycol [64,67], or EDTA [68] as surfactants. The morphology may vary from ball-like particles without surfactant to spindle or flower-like particles in surfactant medium.

Aiming at studying the morphologic evolution of $NaNbO_3$, the synthesis temperature was increased from 100°C to 250°C for 12 h. At 100°C, nanorod-like $NaNbO_3$ was obtained while cubic shaped particles were obtained at 200°C (microcubes) and at 250°C (nanocubes), indicating that the heating process plays a key role in the control of $NaNbO_3$ morphology [69]. Wire-like $Na_2Nb_2O_6 \cdot H_2O$ metastable phase was obtained at 150°C for 10 h and 15 M NaOH [70]. After heating, $Na_2Nb_2O_6 \cdot H_2O$ wires were converted to one-dimensional $NaNbO_3$ [70,71]. Electrical properties of flexible composites based on $NaNbO_3$ particles obtained by MAH and PVDF polymer are greatly influenced by $NaNbO_3$ anisotropy. Composites containing $NaNbO_3$ fiber-like particles in their composition show better d_{33} and P_r values than those containing cubic-like $NaNbO_3$ particles. This can be explained by the fact that the fiber morphology allows the polarization to reach a long distance over the length of the fiber [72].

KNb_3O_8 produced by the MSS method [48] can be used as sacrificial template in hydrothermal synthesis of single-crystalline $KNbO_3$ nanowires. In this case, the hydrothermal synthesis has no surfactants, and the driving force for particle growth is derived from the structure of the materials. The anisotropic Nb_2O_5 is converted to soluble $Nb_6O_{19}{}^{8-}$ while the atomic arrangement in the template is unaffected during the reaction. Hence $KNbO_3$ particles grow with the same anisotropic shape of the Nb_2O_5 precursor [73]. According to Cao et al. [74], the anisotropic growth of

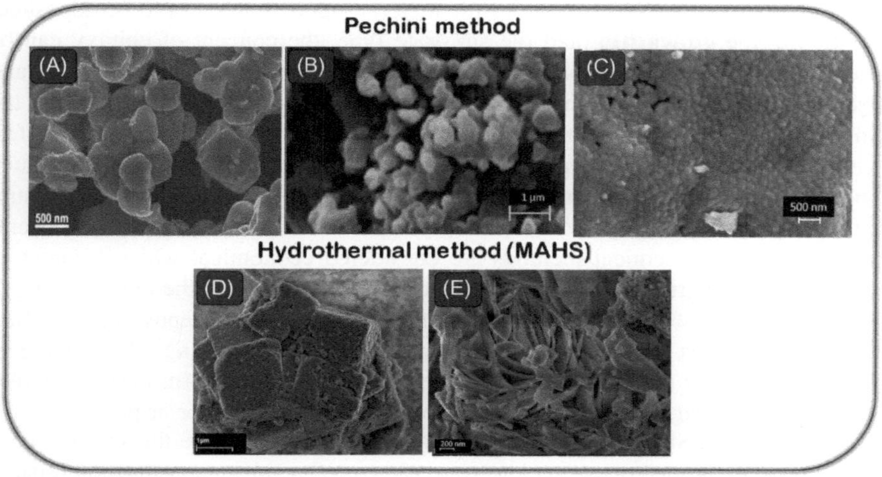

Figure 6.1 PZT particles: (A) without pH control (particle size ~1.1–1.5 μm) [9]; (B) pH controlled by NH_4OH (size ~0.25 μm) [10]; (C) pH controlled by urea (size ~0.10 μm) [10]; (D) template-free microcube-shaped [60]; and (E) polyvinyl alcohol as template [61].

$KNbO_3$ is related to the amount of KCl used as raw material. A smaller amount of KCl produces $KNbO_3$ as cubic-like particles, while the increase of KCl concentration promotes the crystallization of rod-like particles. In another point of view, the rod's growth can be prevented by stirring during the synthesis. In this way, it is possible to grow nanoplate-shaped $KNbO_3$ [75] regardless of the reactants being used. In MAH, the one-dimensional crystallization of $KNbO_3$ is related to the synthesis time, since long synthesis time causes the increase of particles size and the loss of anisotropy [76].

As it was shown here, there are several chemical routes for obtaining ferroelectric powders. Clearly, the method chosen along with its parameters tend to exert great influence over the material's characteristics. Fig. 6.1 shows the influence of synthesis parameters on the morphology and size of PZT particles obtained by Pechini and MAH methods. The synthesis method discussed here gives rise to particles with multifunctional applications, such as in energy harvesting devices, solar cells, photoluminescent and photocatalytic materials.

Ferroelectric development also entails the fabrication of epitaxial thin films to be applied in several technological systems. The film's efficiency depends on aspects that will be discussed next.

6.3 Epitaxial ferroelectric films: growth methods

In general, epitaxial growth involves growing a single crystalline thin film organized atomically on the surface of a crystalline substrate that has the same or similar crystallographic orientation of the material upon which it is to be deposited. Considering the grown film and the substrate type, the concept of epitaxy can be divided into homoepitaxy and heteroepitaxy. The former is used when the film and substrate are found to have the same composition. The latter is applied when the film and substrate are different in composition. In heteroepitaxial growth, the difference between the layer film and substrate can lead to a significant layer mismatch of both, causing a strain as a result of the larger interface energies between the substrate and the film [77]. Epitaxial films can be obtained by both physical and chemical methods. The chemical methods are considered economical when it comes to obtaining ferroelectric films with desired properties [78]. When the chemical deposition method is a wet route, it is known as chemical solution deposition (CSD), a process regarded as an essentially advantageous way of obtaining films owing to the fact that it does not require sophisticated equipment to perform besides involving smaller free energy compared to other processes such as vapor or plasma.

Basically, the CSD consists of the following steps: (1) prepare the solution containing the suitable precursors and solvent; (2) coat the substrate with the prepared solution; (3) dry the film and remove any residual organic matter by thermal treatment; and (4) perform the final thermal treatment to promote film densification and crystallization. It is worth pointing out that these steps ought to be carefully executed in order to obtain films with homogeneous surface, being crack-free and dense. The nucleation process starts during step 3 around 300−400°C. Generally, after nucleation the film crystallization is enforced by the increase in temperature in the range of 600−850°C [79].

The most common precursors used in CSD belong to the metal-organic class (metal alkoxides, metal carboxylates, and metal β-diketonates). A number of solutions from sol-gel and Pechini methods can be used as precursors for the CSD process. The Pechini and sol-gel processes have already been discussed in the previous section about obtaining ferroelectric powders. Different from powder synthesis, in the case of film deposition the solution is dried after the substrate coating. The viscosity of the solution is clearly an important parameter to be controlled at a value neither too high nor too low and in many cases it must not exhibit Newtonian behavior [80].

Among the well-known coating techniques, we can highlight the spin coating and dip coating processes.

Spin coating: Basically, it consists of dripping a viscous solution at the center of the substrate and subsequently subjecting the substrate to a rotation at high speed for few seconds. The solution is spread over the surface of the substrate by centrifugal force, where the speed rotation and deposition time are crucial to produce film with a desired thickness [81].

Dip coating: This is an easy deposition technique that consists of substrate immersion in precursor solution during a time period while subsequently pulling the substrate upward. The film thickness is influenced, among other factors, by the speed at which the substrate is pulled upward, the viscosity of the solution, the solid content, and the angle formed between the substrate and the solution surface during the process [82].

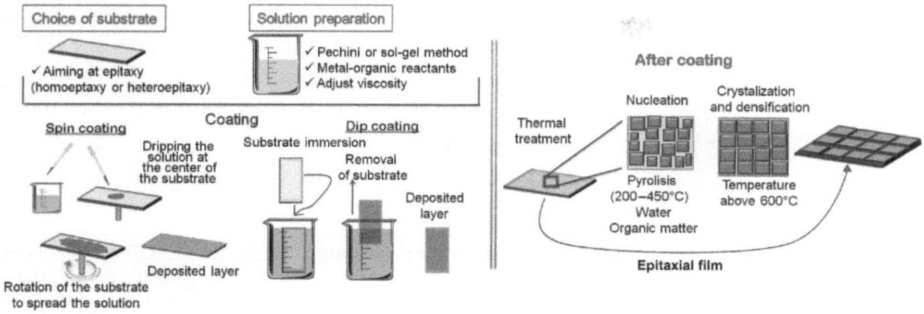

Figure 6.2 Scheme of epitaxial film processing by chemical solution deposition.

Following deposition, the films are subjected to two stages of thermal treatment. The first stage is the pyrolysis, which involves the elimination of water and residual organic matter [81,82]. This process can be accompanied by volume shrinkage leading to the cracking of the film, thus compromising the quality of the material. To prevent this outcome, the control of temperature, heating time, and heating rate ought to be crucially regarded as vital stages of the film processing. Besides that, during pyrolysis the nucleation process can be started [79] while the second stage of the thermal treatment takes place at temperatures above 600°C. The second thermal treatment promotes the film crystallization and densification. The grain size and density increase proportionately with the temperature. The crystallization process involves Gibbs free energy besides being related to nucleation and grain growth [80]. Surface energy of the interface between the substrate and the layer deposited is an important factor to be considered once the growth of an epitaxial film can lead to a lower interfacial energy [80]. According to this concept, it is clear that the final structure of the films will be influenced by these factors. In previous work reported by Attia et al. [82], Brinker et al. [83], Tyona [81], Wang et al. [78], Bassiri-Gharb [79], Elshof [80], Palisaits and Vasiliauskas [77], and Brune [84], readers will find well-detailed descriptions about epitaxial growth and the CSD process. A scheme of CSD epitaxial film preparation is also shown in Fig. 6.2. Here, some examples of epitaxial ferroelectric thin films can be found in [85–90].

6.4 Conclusion

In this review, we have provided clear and concise information about chemical methods employed in obtaining ferroelectric ceramic powders and epitaxial growth of ferroelectric films. The methods for obtaining powders addressed here were Pechini, sol-gel, wet-chemical, molten salt, and hydrothermal. It has been clear that the properties of the materials are related to the synthesis parameters. In spite of the fact that a huge number of ferroelectric materials may be obtained from the Pechini method, this method is regarded unsuitable for the synthesis of powders with a wide range of morphologies. Epitaxial ferroelectric films can be grown via the chemical solution methods. Lastly, we saw that the choice of substrate is one of the most important steps for attaining the best film characteristics.

Acknowledgments

The authors would like to thank the Brazilian research agencies CAPES (Process: 88881.068060/2014-01), CNPq, FAPESP-CEPID/CDMF 2013/07296-2, and FAPESP 2014/11314-9 for granting the financial support to the research group.

References

[1] G.H. Haertling, Ferroelectric ceramics: history and technology, J. Am. Ceram. Soc. 82 (4) (1999) 797−818.

[2] A. Lüker, A short history of ferroelectricity, Ther. Ggw. 115 (10) (1976) 1796−1802.

[3] J. Varghese, R.W. Whatmore, J.D. Holmes, Ferroelectric nanoparticles, wires and tubes: synthesis, characterisation and applications, J. Mater. Chem. C 1 (5) (2013) 2618−2638.

[4] R. Safi, H. Shokrollahi, Physics, chemistry and synthesis methods of nanostructured bismuth ferrite ($BiFeO_3$) as a ferroelectro-magnetic material, Prog. Solid State Chem. 40 (1−2) (2012) 6−15.

[5] M.P. Pechini, Method of preparing lead and alkaline earth titanates and niobates and coating methods using the same to form a capacitor. U.S. Patent 3,330,697, July 11, 1967.

[6] E.R. Leite, C.M.G. Sousa, E. Longo, J.A. Varela, Influence of polymerization on the synthesis of $SrTiO_3$: Part I. Characteristics of the polymeric precursors and their thermal decomposition, Ceram. Int. 21 (3) (1995) 143−152.

[7] S. Qiu, H. Fan, X. Zheng, $Pb(Zr_{0.95}Ti_{0.05})O_3$ powders synthesized by Pechini method: effect of molecular weight of polyester on the phase and morphology, J. Sol-Gel Sci. Technol. 42 (1) (2007) 21−26.

[8] J.M. Calderon-Moreno, E.R. Camargo, Electron microscopy studies on the formation and evolution of sodium niobate nanoparticles from a polymeric precursor, Catal. Today 78 (1−4) (2003) 539−542.

[9] C.A. Oliveira, E. Longo, J.A. Varela, M.A. Zaghete, Synthesis and characterization of lead zirconate titanate (PZT) obtained by two chemical methods, Ceram. Int. 40 (1B) (2014) 1717−1722.

[10] A. Abreu, S.M. Zanetti, M.A.S. Oliveira, G.P. Thim, Effect of urea on lead zirconate titanate-$Pb(Zr_{0.52}Ti_{0.48})O_3$-nanopowders synthesized by the Pechini method, J. Eur. Ceram. Soc. 25 (5) (2005) 743−748.

[11] L.R. Prado, N.S. de Resende, R.S. Silva, S.M.S. Egues, G.R. Salazar-Banda, Influence of the synthesis method on the preparation of barium titanate nanoparticles, Chem. Eng. Process. Process Intensif. 103 (2015) 12−20.

[12] Y. Tian, Y. Gong, D. Meng, S. Cao, Structure and electrical properties of Ir^{4+}-doped $0.5Ba_{0.9}Ca_{0.1}TiO_3$-$0.5BaTi_{0.88}Zr_{0.12}O_3-0.12\%La$ ceramics via a modified Pechini method, Mater. Lett. 153 (2015) 44−46.

[13] W. Liu, Dielectric and sintering properties of $NaNbO_3$ ceramic prepared by Pechini method, J. Electroceram. 31 (3–4) (2013) 376–381.
[14] I. Pribošič, D. Makovec, M. Drofenik, Chemical synthesis of $KNbO_3$ and $KNbO_3$-$BaTiO_3$ ceramics, J. Eur. Ceram. Soc. 25 (12) (2005) 2713–2717 (special issue).
[15] M. Kakihana, Invited review: sol-gel preparation of high temperature superconducting oxides, J. Sol-Gel Sci. Technol. 6 (1) (1996) 7–55.
[16] A.E. Danks, S.R. Hall, Z. Schnepp, The evolution of "sol–gel" chemistry as a technique for materials synthesis, Mater. Horiz. 3 (2) (2016) 91–112.
[17] N.A. Halim, T.S. Velayutham, W.H.A. Majid, Pyroelectric, ferroelectric, piezoelectric and dielectric properties of $Na_{0.5}Bi_{0.5}TiO_3$ ceramic prepared by sol-gel method, Ceram. Int. 42 (14) (2016) 15664–15670.
[18] T. Kainz, M. Naderer, D. Schütz, O. Fruhwirth, F.A. Mautner, K. Reichmann, Solid state synthesis and sintering of solid solutions of BNT-xBKT, J. Eur. Ceram. Soc. 34 (15) (2014) 3685–3697.
[19] Q. Xu, X. Chen, W. Chen, S. Chen, B. Kim, J. Lee, Synthesis, ferroelectric and piezoelectric properties of some $(Na_{0.5}Bi_{0.5})TiO_3$ system compositions, Mater. Lett. 59 (2005) 2437–2441.
[20] Y. Cao, K. Zhu, J. Qiu, X. Pang, H. Ji, Sol-gel processing and characterization of potassium niobate nano-powders by an EDTA/citrate complexing method, J. Solid State Sci. 14 (5) (2012) 655–660.
[21] R. Ganeshkumar, K.V. Sophia, Wu Ping, C.W. Cheah, R. Zhao, Ferroelectric $KNbO_3$ nanofibers: synthesis, characterization and its application as humidity nanosensor, Nanotechnology 27 (39) (2016) 395607 (7pp).
[22] A. Sachdeva, M. Arora, R.P. Tandon, Synthesis and characterization of sol-gel derived PZT nano powder, J. Nanosci. Nanotechnol. 9 (11) (2009) 6631–6636.
[23] A. Wu, P.M. Vilarinho, I.M.M. Salvado, J.L. Baptista, Sol-gel preparation of lead zirconate titanate powders and ceramics: effect of alkoxide stabilizers and lead precursors, J. Am. Ceram. Soc. 83 (6) (2000) 1379–1385.
[24] M.R. Mohammadi, S.A. Tabei, A. Nemati, D. Eder, T. Pradeep, Synthesis and crystallization of lead-zirconium-titanate (PZT) nanotubes at the low temperature using carbon nanotubes (CNTs) as sacrificial templates, Adv. Powder Technol. 23 (5) (2012) 647–654.
[25] B. Hernandez, K. Chang, E.R. Fisher, P.K. Dorhout, Sol-Gel template synthesis and characterization of $BaTiO_3$ and $PbTiO_3$ nanotubes, Chem. Mater. 14 (2) (2002) 480–482.
[26] C. Zhong, Z. Lu, X. Wang, L. Li, Template-based synthesis and piezoelectric properties of $BiScO_3$-$PbTiO_3$ nanotube arrays, J. Alloys Compd. 655 (2016) 28–31.
[27] J.H. Kwak, J.K. Han, S.W. Kang, T.A. Johnson, S.D. Bu, Dielectric relaxation properties of $PbTiO_3$-multiwalled carbon nanotube composites prepared by a sol-gel process, Ceram. Int. 42 (7) (2016) 8165–8169.
[28] C.Y. Lee, N.H. Tai, H.S. Sheu, H. Ten Chiu, S.H. Hsieh, The formation of perovskite $PbTiO_3$ powders by sol-gel process, Mater. Chem. Phys. 97 (2–3) (2006) 468–471.
[29] P.N. Francis, S. Dhanuskodi, M. Muneeswaran, A.R. Thomas, N.V. Giridharan, Optical nonlinearity in multiferroic bismuth ferrite, J. Alloys Compd. 688 (A) (2016) 796–802.
[30] L. Wu, W. Sui, C. Dong, C. Zhang, C. Jiang, One-dimensional $BiFeO_3$ nanotubes: Preparation, characterization, improved magnetic behaviors, and prospects, Appl. Surf. Sci. 384 (2016) 368–375.
[31] S.M. Mane, P.M. Tirmali, D.J. Salunkhe, P.B. Joshi, C.B. Kolekar, S.B. Kulkarni, Synthesis, structural and dielectric characterization of lead free $x[(Ba_{0.7}Ca_{0.3})TiO_3]$-$(1-x)[Ba(Zr_{0.2}Ti_{0.8})O_3]$ composite, J. Mater. Sci. Mater. Electronics 27 (7) (2016) 7204–7210.

[32] H. Muthurajan, H.H. Kumar, V. Samuel, U.N. Gupta, V. Ravi, Novel hydroxide precursors to prepare $NaNbO_3$ and $KNbO_3$, Ceram. Int. 34 (3) (2008) 671–673.
[33] S. Buddhudu, Synthesis and analysis of $LiNbO_3$ ceramic powders by co-precipitation method, Indian J. Pure Appl. Phys. 50 (5) (2012) 320–324.
[34] S.R. Dhage, Y.B. Khollam, S.B. Deshpande, V. Ravi, Co-precipitation technique for the preparation of nanocrystalline ferroelectric $SrBi_2Ta_2O_9$, Mater. Res. Bull. 38 (11–12) (2003) 1601–1605.
[35] S.R. Dhage, R. Pasricha, A. Vadivel Murugan, V. Ravi, A co-precipitation technique for the preparation of ferroelectric $BaBi_2Ta_2O_9$, Mater. Chem. Phys. 98 (2–3) (2006) 344–346.
[36] G. Xu, W. Weng, J. Yao, P. Du, G. Han, Low temperature synthesis of lead zirconate titanate powder by hydroxide co-precipitation, Microelectron. Eng. 66 (1–4) (2003) 568–573.
[37] G. Xu, G. Han, Synthesis and thermal stability of perovskite $PbZr_{0.52}Ti_{0.48}O_3$ (PZT) powders via precipitation routes, J. Electroceram 21 (1–4) (2008) 867–870 (special issue).
[38] K.R.M. Rao, A.V.P. Rao, S. Komameni, Reactive PZT precursor powder by coprecipitation, Mater. Lett. 28 (1996) 463–467.
[39] H. Yadav, N. Sinha, S. Goel, B. Kumar, Eu-doped ZnO nanoparticles for dielectric, ferroelectric and piezoelectric applications, J. Alloys Compd. 689 (2016) 333–341.
[40] M. Mukhtar, Co-Precipitation Synthesis, and Characterization of nanocrystalline zinc oxide particles doped with Cu^{2+} ions, Mater. Sci. Appl. 3 (8) (2012) 543–551.
[41] P. Junlabhut, W. Mekprasart, R. Noonuruk, K. Chongsri, W. Pecharapa, Characterization of ZnO:Sn nanopowders synthesized by co-precipitation method, Energy Procedia 56 (2014) 560–565.
[42] F. Giovannelli, A. Ngo Ndimba, P. Diaz-Chao, M. Motelica-Heino, P.I. Raynal, C. Autret, et al., Synthesis of Al doped ZnO nanoparticles by aqueous coprecipitation, Powder Technol. 262 (2014) 203–208.
[43] M. Farahmandjou, S. Jurablu, Co-precipitation synthesis of zinc oxide (ZnO) nanoparticles by zinc nitrate precursor, Int. J. Bio-Inorg. Hybr. Nanomater. 3 (3) (2016) 179–184.
[44] P. Goel, K.L. Yadav, A comparative analysis of PBZT synthesized by co-precipitation and sol-gel method, Indian J. Eng. Mater. Sci. 12 (6) (2005) 552–556.
[45] J.G. Chen, Z. Xu, X. Yao, Effect of bismuth doping on dielectric, piezoelectric and ferroelectric properties of PZT ceramics, Mater. Res. Innov. 14 (3) (2010) 234–237.
[46] L. Li, J. Deng, J. Chen, R. Yu, L. Qiao, X. Xing, Piezoelectric and ferroelectric properties of $0.96(Na,K)(Nb_{0.9}Ta_{0.1})O_3$-$0.04LiSbO_3$ ceramics synthesized by molten salt method, J. Alloys Compd. 471 (1–2) (2009) 428–431.
[47] T. Kimura, Molten salt synthesis of ceramic powders, in: C. Sikalidis (Ed.), Advances in Ceramics-synthesis and Characterization Processing and Specific Applications, InTech, Rijeka, 2011, pp. 75–100.
[48] L. Li, J. Deng, J. Chen, X. Sun, R. Yu, G. Liu, et al., Wire structure and morphology transformation of niobium oxide and niobates by molten salt synthesis, Chem. Mater. 21 (7) (2009) 1207–1213.
[49] H. Ge, Y. Hou, J. Yang, M. Zhu, H. Wang, H. Yan, Fabrication and properties of $Na_{0.9}K_{0.1}NbO_3$ nanostructures by molten salt synthesis, Powder Technol. 246 (2013) 144–147.

[50] X. Chen, Z. Lu, F. Huang, J. Min, J. Li, J. Xiao, et al., Molten salt synthesis and magnetic anisotropy of multiferroic $Bi_4NdTi_3Fe_{0.7}Ni_{0.3}O_{15}$ ceramics, J. Alloys Compd. 693 (2017) 448–453.

[51] Z. Liu, S. Liang, S. Li, Y. Zhu, X. Zhu, Synthesis, microstructural characterization, and dielectric properties of $BiFeO_3$ microcrystals derived from molten salt method, Ceram. Int. 41 (Supp. 1) (2014) S19–S25.

[52] F. Bortolani, R.A. Dorey, Molten salt synthesis of PZT powder for direct write inks, J. Eur. Ceram. Soc. 30 (10) (2010) 2073–2079.

[53] Z. Yang, Y. Chang, H. Li, Piezoelectric and dielectric properties of PZT-PZN-PMS ceramics prepared by molten salt synthesis method, Mater. Res. Bull. 40 (12) (2005) 2110–2119.

[54] K. Byrappa, M. Yoshimura (Eds.), Handbook of Hydrothermal Technology: A Technology for Crystal Growth and Material Processing, William Andrew Publishing, Norwich, 2001.

[55] R.E. Riman, W.L. Suchanek, Hydrothermal synthesis of advanced ceramic powders in: P. Vincenzini (Ed.), Proceedings of the 11th International Ceramics Congress, part of CIMTEC 2006—11th International Ceramics Congress and 4th Forum on New Materials, Acireale, Sicily, Italy, June 4–9, Adv. Sci. Technol. 45 (2006) 184–193.

[56] I. Bilecka, M. Niederberger, Microwave chemistry for inorganic nanomaterials synthesis, Nanoscale 2 (8) (2010) 1358–1374.

[57] B.L. Hayes, Microwave Synthesis-Chemistry at the Speed of Light, CEM Publishing, Matthews, 2002.

[58] H.-L. Huang, G.Z. Cao, I.Y. Shen, Hydrothermal synthesis of lead zirconate titanate (PZT or $Pb(Zr_{0.52}Ti_{0.48})O_3$) nano-particles using controlled ramping and cooling rates, Sensors Actuators A Phys. 214 (2014) 111–119.

[59] C. Glass, W. Ahmed, J. van Ruitenbeek, Tuning the morphology of lead zirconate titanate (PZT) nanostructures, Mater. Lett. 125 (2014) 71–74.

[60] G.F. Teixeira, M.A. Zaghete, G. Gasparotto, M.G.S. Costa, J.W.M. Espinosa, E. Longo, et al., Photoluminescence properties and synthesis of a PZT mesostructure obtained by the microwave-assisted hydrothermal method, J. Alloys Compd. 512 (1) (2012) 124–127.

[61] G.F. Teixeira, G. Gasparotto, E.C. Paris, M.A. Zaghete, E. Longo, J.A. Varela, Photoluminescence properties of PZT 52/48 synthesized by microwave hydrothermal method using PVA with template, J. Lumin. 132 (1) (2012) 46–50.

[62] J. Wang, A. Durussel, C.S. Sandu, M.G. Sahini, Z. He, N. Setter, Mechanism of hydrothermal growth of ferroelectric PZT nanowires, J. Cryst. Growth. 347 (1) (2012) 1–6.

[63] D.L. Golić, A. Radojković, J. Ćirković, A. Dapčević, D. Pajić, N. Tasić, et al., Structural, ferroelectric and magnetic properties of $BiFeO_3$ synthesized by sonochemically assisted hydrothermal and hydro-evaporation chemical methods, J. Eur. Ceram. Soc. 36 (7) (2015) 1623–1631.

[64] K. Chybczyńska, E. Markiewicz, M. Błaszyk, B. Hilczer, B. Andrzejewski, Dielectric response and electric conductivity of ceramics obtained from $BiFeO_3$ synthesized by microwave hydrothermal method, J. Alloys Compd. 671 (2016) 493–501.

[65] X. Gao, Y. Dai, F. Fu, X. Hua, 2D laminated cylinder-like $BiFeO_3$ composites: Hydrothermal preparation, formation mechanism, and photocatalytic properties, Solid State Sci. 62 (2016) 6–12.

[66] X. Wang, W. Mao, Q. Zhang, Q. Wang, Y. Zhu, J. Zhang, et al., PVP assisted hydrothermal fabrication and morphology-controllable fabrication of $BiFeO_3$ uniform nanostructures with enhanced photocatalytic activities, J. Alloys Compd. 677 (2016) 288–293.

[67] K. Chybczynska, M. Blaszyk, B. Hilczer, T. Lucinski, M. Matczak, B. Andrzejewski, PEG-controlled thickness of $BiFeO_3$ crystallites in microwave hydrothermal synthesis, Mater. Res. Bull. 86 (2017) 178−185.
[68] J. Huang, G. Tan, W. Yang, L. Zhang, H. Ren, A. Xia, Microwave hydrothermal synthesis of $BiFeO_3$: impact of different surfactants on the morphology and photocatalytic properties, Mater. Sci. Semicond. Process. 25 (3) (2014) 84−88.
[69] E.S. Baeissa, Photocatalytic degradation of malachite green dye using Au/$NaNbO_3$ nanoparticles, J. Alloys Compd. 672 (2016) 564−570.
[70] A. Yu, J. Qian, L. Liu, H. Pan, X. Zhou, Surface sprouting growth of $Na_2Nb_2O_6 \cdot H_2O$ nanowires and fabrication of $NaNbO_3$ nanostructures with controlled morphologies, Appl. Surf. Sci. 258 (8) (2012) 3490−3496.
[71] G.F. Teixeira, T.R. Wright, D.C. Manfroi, E. Longo, J.A. Varela, M.A. Zaghete, Photoluminescence in $NaNbO_3$ particles and films, Mater. Lett. 139 (2015) 443−446.
[72] G.F. Teixeira, R.A. Ciola, W.K. Sakamoto, M.A. Zaghete, Perovskite-based mesostructures and related composites-influence exerted by morphology and interface, in: A.P. Barranco (Ed.), Ferroelectric Materials-Synthesis and Characterization, InTech, Rijeka, 2015, pp. 59−83.
[73] C. Huiqun, Y. Cheng, Y. Jie, C. Bo, X. Hong, Y. Bin, Synthesis of single-crystalline $KNbO_3$ nanowires with a sacrificial template method and their second harmonic generation response, Rare Met. Mater. Eng. 45 (6) (2016) 1391−1395.
[74] H. Cao, C. You, J. Yu, B. Cao, B. Yu, H. Xin, et al., Effect of potassium chloride amount on structures of hydrothermally grown $KNbO_3$ nanostructure, Mater. Res. Innov. 18 (Suppl. 2) (2014), S2-696−S2-699.
[75] I. Krad, O. Bidault, S. Said, M. El Maaoui, Characterization of $KNbO_3$ nanoplates synthesized by a stirred hydrothermal process, Mater. Lett. 159 (2015) 237−240.
[76] A.J. Paula, R. Parra, M.A. Zaghete, J.A. Varela, Synthesis of $KNbO_3$ nanostructures by a microwave assisted hydrothermal method, Mater. Lett. 62 (17−18) (2008) 2581−2584.
[77] J. Palisaits, R. Vasiliauskas, Epitaxial growth of thin films, in: Physics of Advanced Materials Winter School, 2008.
[78] Y. Wang, W. Chen, B. Wang, Y. Zheng, N. Physics, Ultrathin ferroelectric films: growth, characterization, physics and applications, Materials 7 (9) (2014) 6377−6485.
[79] N. Bassiri-Gharb, Y. Bastani, A. Bernal, Chemical solution growth of ferroelectric oxide thin films and nanostructures, Chem. Soc. Rev. 43 (7) (2014) 2125−2140.
[80] J.E. Elshof, Chemical solution deposition techniques for epitaxial growth of complex oxides, in: G. Koster, Mark Huijben, G. Rijnders (Eds.), Epitaxial Growth of Complex Metal Oxides, Woodhead Publishing, Cambridge, UK, 2015, pp. 69−93.
[81] M.D. Tyona, A theorical study on spin coating technique, Adv. Mater. Res. 2 (4) (2013) 195−208.
[82] S.M. Attia, J. Wang, G. Wu, J. Shen, J. Ma, Review on sol-gel derived coatings: process, techniques and optical applications, J. Mater. Sci. Technol. 18 (3) (2002) 211−218.
[83] C.J. Brinker, G.C. Frye, A.J. Hurd, C.S. Ashley, Fundamentals of sol-gel dip coating, Thin Solid Films 201 (1) (1991) 97−108.
[84] H. Brune, Epitaxial growth of thin films, in: K. Wandelt (Ed.), Surface and Interface Science: Solid-SolidInterfaces and Thin Films, vol. 4, Wiley, Weinheim, 2014, pp. 421−491.

[85] K. Vojisavljević, G. Branković, T. Srećković, A. Rečnik, Z. Branković, Preparation of ultrathin PZT films by a chemical solution deposition method from a polymeric citrate precursor, J. Eur. Ceram. Soc. 30 (2) (2010) 485–488.
[86] V.V. Sidsky, A.V. Semchenko, A.G. Rybakov, V.V. Kolos, A.S. Turtsevich, A.N. Asadchyi, et al., La^{3+}-doped $SrBi_2Ta_2O_9$ thin films for FRAM synthesized by sol-gel method, J. Rare Earths 32 (3) (2014) 277–281.
[87] S.Y. Lee, C.W. Ahn, J.S. Kim, A. Ullah, H.J. Lee, H.I. Hwang, et al., Enhanced piezoelectric properties of Ta substituted-$(K_{0.5}Na_{0.5})NbO_3$ films: a candidate for lead-free piezoelectric thin films, J. Alloys Compd. 509 (20) (2011) L194–L198.
[88] C. Feng, C.H. Yang, Y.Y. Zhou, F.J. Geng, P.P. Lv, Q. Yao, Substrate-dependent ferroelectric and dielectric properties of Mn doped $Na_{0.5}Bi_{0.5}TiO_3$ thin films derived by chemical solution decomposition, J. Alloys Compd. 679 (2016) 133–137.
[89] A.Z. Simões, M.A. Zaghete, M. Cilense, J.A. Varela, B.D. Stojanivic, Preparation of 9/65/35 PLZT thin films deposited by a dip-coating process, J. Eur. Ceram. Soc. 21 (9) (2001) 1151–1157.
[90] Q. Zhang, N. Valanoor, O. Standard, Epitaxial (001) $BiFeO_3$ thin-films with excellent ferroelectric properties by chemical solution deposition-the role of gelation, J. Mater. Chem. C 3 (3) (2015) 582–595.

Nanosized ferroelectrics: Preparation, properties, and applications

7

Glauco M.M.M. Lustosa, Guilhermina F. Teixeira, Wagner B. Bastos, Sonia M. Zanetti, Leinig A. Perazolli and Maria A. Zaghete
Institute of Chemistry, São Paulo State University (Unesp), Araraquara, Brazil

7.1 Synthesis of nanostructured ferroelectrics

Top-down and bottom-up are two kinds of processes employed to obtain ferroelectric nanostructures. Top-down processes are physical methods involving the use of a bulk ferroelectric material to create coherently and continuously ordered nanosized structures (such as focused ion beam milling, nanoimprint lithography) [1–5]. On the other hand, bottom-up synthesis methods involve solution-based routes such as sol-gel, molten salt synthesis, solution-phase decomposition, and hydro/solvothermal synthesis, which lead to the production of different nanostructures (i.e., nanorings, nanowires, nanotubes) beginning at the atomic or molecular level [6–13].

The synthesis based on bottom-up routes has clearly become attractive as a result of the possibility of obtaining nanosized crystal or nanoparticles in aqueous/colloidal suspension. These particles can be deposited onto any substrate, forming ordered up three-dimensional crystal arrays with properties for specific applications following solvent evaporation.

The hydrothermal synthesis is a suitable technique for the preparation of nanomaterials involving many reactions in a closed system (autoclave), generating an internal pressure (0.1–15 MPa) that leads to an increase in the solubility of the reactants during the process with temperature that generally does not exceed the range of 100–400°C. In some cases, after the synthesis, a conventional heat treatment is required to complete the crystallization of the material [14,15]. Recently, Xie et al. [16] reported the hydrothermal synthesis of $BaTiO_3$-shaped nanowires for application in nanocomposites with poly(vinylidene fluoride-co-chlorotrifluoroethylene) (PVDF-CTFE) in their composition. It is worth noting that this material showed interesting energy storage behavior as it is improved under low electric field application, suggesting its promising suitability for application in the automobile and aerospace fields.

Niobate-based nanocrystals are a class of materials that present high piezoelectric performance and have been gaining wider attention in the scientific community.

Magnetic, Ferroelectric, and Multiferroic Metal Oxides. DOI: http://dx.doi.org/10.1016/B978-0-12-811180-2.00007-4

One such material in the limelight is Na,K(NbO_3) (KNN) a one-dimensional structure that can be synthesized by the molten salt method, among other methods previously discussed in Chapter 6 [17–19]. The synthesis of KNN by the molten salt route can be summarized in a few steps: the Nb_2O_5 powder is initially converted to a complex precursor that acts as a template for obtaining the nanorods, followed by thermal decomposition prior to obtaining the nanostructured powder [20].

Many research studies have been conducted on niobate-based materials owing to the fact that their properties are regarded as quite promising. In the future, these materials are expected to replace pollutant materials, especially the lead-based ones like $Pb(Zr_xTi_{1-x})O_3$ (PZT). The advantages of the use of lead-free material include low toxicity, low cost, biocompatibility, and the relative ease at which this material is processed [21].

KNN nanoparticles were produced by Lusiola et al. [22] from a modified molten salt technique using hydroxides instead of salts. This adaptation consists of replacing NaCl/KCl by NaOH/KOH due to the lower melting points of hydroxides, being an alternative for diminishing the molten phase temperature in addition to being an advantage for obtaining a fully reacted KNN product without the formation of unreacted niobium oxide.

A less common approach of nanoferroelectric particle synthesis is the use of biological synthesis methods, which are considered to be clean-type syntheses since the physical and/or chemical methods make use of toxic reagents, thus making it impossible to apply these nanoparticles in the biomedical field—a recently studied area for the application of nanostructured ferroelectrics. Although $BaTiO_3$ is one of the most studied materials produced by these routes, few reports have been published focusing on it [23–25]. In a study conducted by Ahmad et al. [26], $BaTiO_3$ nanocrystals with a size of around 50 nm were obtained through the use of peptides that induced crystal precipitation. As reported by Jha [27], microorganisms (such as *Lactobacillus*) have been employed in the synthesis of spherical $BaTiO_3$ nanoparticles with the use of microorganisms, leading to the production of nanoparticles with a size around 20–80 nm of diameter and tetragonal crystalline structure. Smaller $BaTiO_3$-sized nanoparticles (range of 8–21 nm) [28] were obtained as a substitute to *Lactobacillus* by *Saccharomyces cerevisiae*. According to Uddin et al. [29], during the synthesis, the microorganisms secrete extracellular proteins, which are responsible for the particles' size reduction. Indeed, although more in-depth studies are required to better understand the biological synthesis of nanostructured materials, there is no doubt that it is a very promising area mainly for the application of ferroelectrics in the health field as a result of the shorter reaction time involved, as well as the use of room temperature (or even at low temperatures).

Characterizing electromechanical responses in a variety of materials is crucial for understanding and improving technologies ranging from bioscience to energy production. Scanning microscopy has emerged as a universal tool to probe such structures and functionality at the nanometer scale. The ability of piezoforce microscopy allows the characterization of an infinite variety of materials and devices that previously could not be measured using conventional piezoresistive force microscopy. Research with this new tool allows advances in many fields, from biology to semiconductors, and allows ongoing work in a variety of areas, from data storage devices and molecular machines to improved renewable energy materials, to be enhanced in assembly and efficiency.

7.2 Piezoresponse force microscopy

7.2.1 Piezoresponse force microscopy technique: A brief overview

In recent decades, several techniques have been used for visualizing the ferroelectric nanostructured systems since these kinds of materials have multioriented domains, as shown in Fig. 7.1. Among these techniques, piezoresponse force microscopy (PFM) has become one of the most applied techniques in the study of the structures of ferroelectric domains in ferroelectric and multiferroic materials because their behavior is directly correlated with the knowledge and control of the domains (size and orientation).

Nowadays, the growing need to miniaturize electro-electronic devices has allowed great advances in the preparation of ferroelectric materials with different shapes and sizes [30–33]. Due to high spatial resolution, PFM has become an important tool for the investigation of ferroelectric materials since this technique promotes the observation and the localized modification of the structures from nanoscale ferroelectric domains [34]. In addition, the analysis of polarization orientation makes three-dimensional reconstruction of ferroelectric domains possible [35–39]. This approach is important because macroscopic measurements are not efficient for understanding the behavior presented by individual domains, since size domains are normally below 10 μm [40,41].

Several researchers have reported the observation of monodomains in thin-film ferroelectric nanograins. Durkan et al. have succeeded in localized quantitative measurements of domains with radii around 40 nm in sol-gel–produced PZT thin

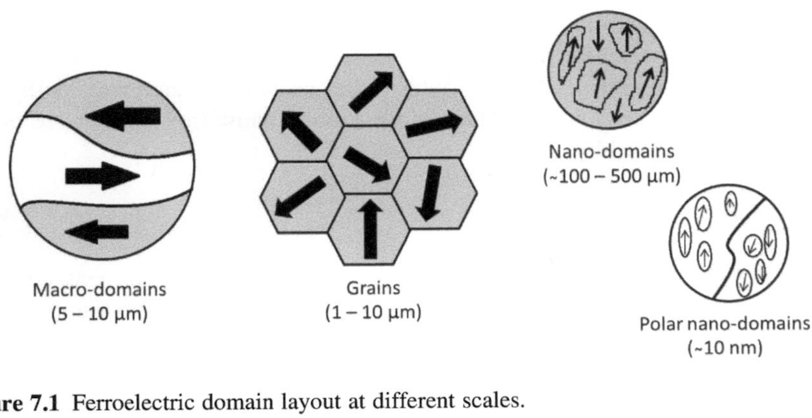

Figure 7.1 Ferroelectric domain layout at different scales.

films. They have shown evidence that the localized formation of domains is related to surface film composition [41]. In addition, based on PFM measurements, great progress toward one-dimensional and ferroelectric materials also have been made [42]. One-dimensional ceramic particle-based composites are widely exploited as piezoelectric nanogenerator energy. Even when one-dimensional particles are combined with polymeric material, it is possible to measure and quantify the piezoelectric response of individual ceramic components. In this sense, Yun et al. obtained the different piezoelectric coefficients d_{33} and d_{31} for $LiNbO_3$ nanowires combined with poly(dimethylsiloxane) (PDMS) polymer [43]. Jung et al. could easily control the piezoelectric domains by electric field of one-dimensional $NaNbO_3$ suggesting that this compound is a good candidate for use in a nanogenerator [44]. The piezoelectric property of ZnO nanostructure is the object of several studies. In spite of some difficulties in applying the PFM technique to nanorods, Xiao et al. have proved that one-dimensional ZnO presents ferroelectric response. In addition, it was verified that this behavior can be related to oxygen vacancies associated with domain switching [45].

Basically, PFM is a modification of a typical atomic force microscopy (AFM) based on the detection of very subtle vibrations produced by the inverse piezoelectric effect exhibited by ferroelectric materials. This technique allows observation of the ferroelectric domain structures at nanoscale through the local polarization of the domains on the sample surface. In 1992, Guthner and Dransfeld observed, for the first time, the domains in ferroelectric polymers in scans of about $15 \times 15\ \mu m^2$ and $30 \times 30\ \mu m^2$ through this technique [46]. Furthermore, PFM enables us to study the ferroelectric switching and domain nucleation, and to determine the direction of the local polarization. The analysis of the polarization direction allows the reconstruction of the three-dimensional structure of ferroelectric domains through the orientation of the polarization in the three directions (x-, y-, z-axes) [36,37,47]. These possibilities associated with the advantage of a nondestructive way of obtaining high resolution images made PFM an indispensable tool in the investigation of ferroelectric materials.

PFM operates in AFM contact mode using a conductive tip by applying a bias, $V_{TIP} = V_{dc} + V_{ac}\cos(\omega t)$, to locally excite the sample during scanning. In this condition, the bottom electrode of the ferroelectric sample is grounded. All ferroelectric materials are piezoelectric, thus the sample surface mechanically deforms when subjected to an external electric field. This causes a cantilever deflection, $h_{cantilever} = h_{dc} + h_{ac}\cos(\omega t + \phi)$. This deformation is caused by the reverse piezoelectric effect described by:

$$x_i = d_{ij}E_j \tag{7.1}$$

where x_i is the mechanical deformation, d_{ij} is the piezoelectric coefficient, and E_j is the external electric field. The deformation of the surface is proportional to the intensity of the applied bias.

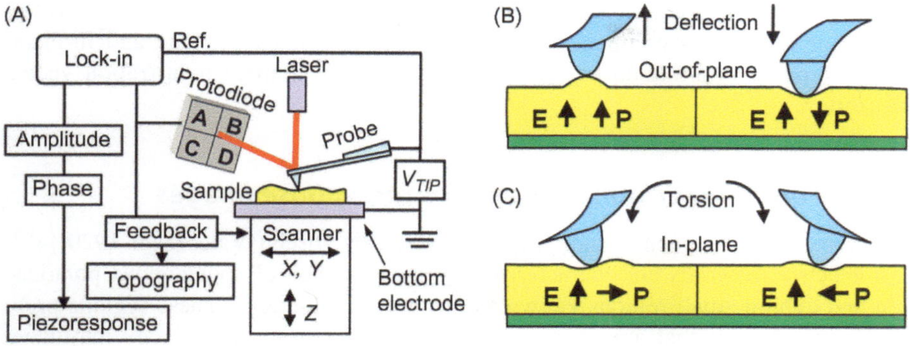

Figure 7.2 (A) Typical PFM setup, (B) and (C) movement of the cantilever according to the ferroelectric domain polarization orientation.

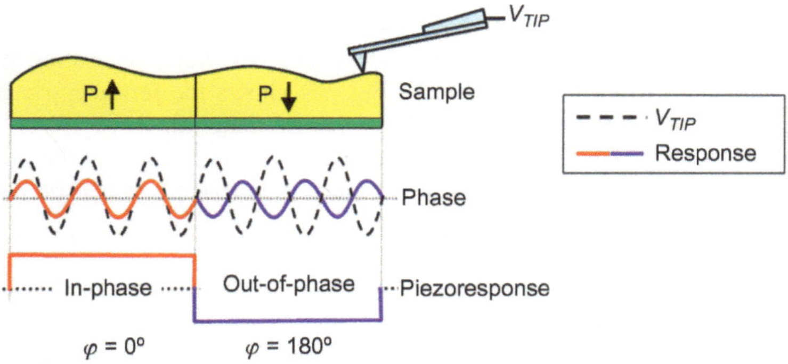

Figure 7.3 Basic scheme of phase detection between the bias and electromechanical response.

Fig. 7.2 shows a typical commercial AFM setup modified to work as a PFM, and the movement of the cantilever over several orientations of the ferroelectric domains polarization.

If a DC bias is applied, the deformation in the sample will be undetectable. In this case, an AC bias associated with lock-in techniques allows the monitoring of extremely small signals. Hence, the lock-in provides information about the phase difference φ as the response to AC bias V_{TIP}. If the local polarization direction of the ferroelectric domain is accompanied by the electric field direction, the phase difference is $\varphi = 0°$. If local polarization direction is opposite to the electric field, the phase difference becomes $\varphi = 180°$ (see Fig. 7.3).

PFM provides two types of images: out-of-plane and in-plane piezoresponse. The out-of-plane image corresponds to the vertical component of the ferroelectric domain polarization. If the electric field and the polarization are in the same direction, there is an expansion of the ferroelectric domain. Otherwise, a compression occurs if polarization is opposite to the electric field (see Fig. 7.2B). The in-plane image is related to a component parallel to the sample surface that is known as the lateral (or horizontal) component (see Fig. 7.2C). Both out-of-plane and in-plane images provide information regarding the ferroelectric domain polarization vector orientation.

7.2.2 Ferroelectric domains and piezoresponse images

The ferroelectricity phenomenon was discovered by Joseph Valasek in 1920 [48]. Ferroelectric materials are characterized by the presence of spontaneous polarization and exhibit small regions known as ferroelectric domains. These regions present electric dipoles that are oriented in the same direction, though they may have their direction modified when an external electric field is applied [49,50]. The boundary between two adjacent domains is called the domain wall [51]. These walls are classified according to the angles formed between the directions of the adjacent domains. The angle depends on the crystalline structure of the material. For example, tetragonal materials have two types of domain walls: 180 degrees walls separating opposing polarizations and 90 degrees walls separating polarizations perpendicular to each other [52,53]. Fig. 7.4 shows the ferroelectric domains and domain walls of a typical tetragonal ferroelectric material.

The relationship between the ferroelectric domain polarization and the piezoelectric effect used in PFM involves a thermodynamic approach and can be expressed by the tensor relationship:

$$d_{ij} = 2\varepsilon_{im} M_{jmk} P_k \tag{7.2}$$

where ε_{im} is the electrical permissivity, M_{jmk} is the electrostrictive tensor, and P_k is the polarization component. In the case of a tetragonal material, this relationship provides two important relations associated with the piezoresponse images through the nonzero piezoelectric coefficients d_{33} and d_{15}.

The first relation is expressed as $d_{33} = 2\varepsilon_{33} M_{33} P_3$ and is related to the out-of-plane image. This relationship describes the vertical movement of the cantilever that allows the PFM to detect the vertical polarization component of the ferroelectric domain (see Fig. 7.2B). The other relation is given by $d_{15} = 2\varepsilon_{11} M_{44} P_3$. This tensor relation can obtain the in-plane images and is associated with the shear

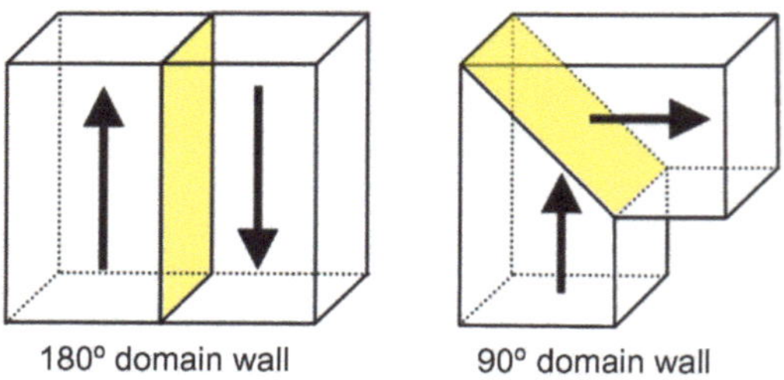

Figure 7.4 Domain wall of typical tetragonal ferroelectric material.

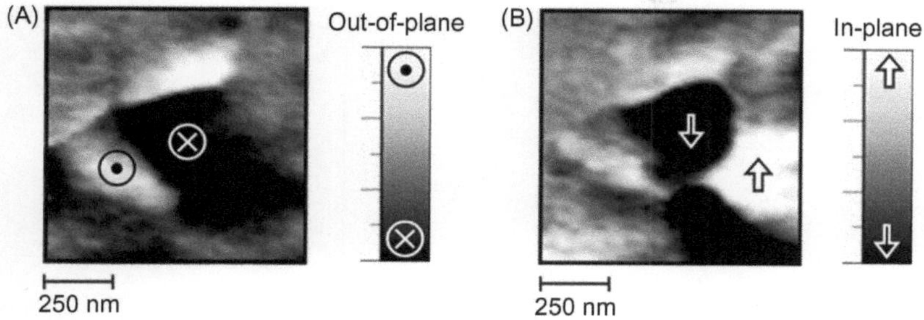

Figure 7.5 Piezoresponse images of the $NaNbO_3$ fiber: (A) vertical and (B) lateral polarization component.

motion (horizontal movement) of the ferroelectric domain that causes torsion in the cantilever (see Fig. 7.2C). In general, out-of-plane and in-plane piezoresponse images are found to be different owing to the different contributions of the polarization vector components in the monocrystalline materials and, particularly, in the polycrystalline materials. Fig. 7.5 shows piezoresponse images over the same area of a sodium niobate ($NaNbO_3$) fiber and the local polarization direction of the ferroelectric domain.

7.2.3 Polarity controlled

When a ferroelectric material is subjected to an external electric bias (V_{dc}) the reorientation of its polarization occurs. Ferroelectric lithography is a writing technique based on this principle in piezoresponse microscopy [54]. Ferroelectric lithography can construct several polarization patterns in the material that are determined by the trajectory of the tip on the surface and by the orientation of the applied electric field. Fig. 7.6 shows some patterns written in ferroelectric thin films. The recording was carried out by applying 15 V_{dc} on a $Pb_{0.9}Ca_{0.1}TiO_3$ thin film sample.

The ferroelectric feature of maintaining a remaining polarization gives the ferroelectrics the ability to store information (memory). This capability is well studied by PFM in the development of the so-called nonvolatile ferroelectric memories (NV-FeRAM) [55,56]. Fig. 7.6C shows a light/dark pattern corresponding to upward/downward polarization. This pattern was achieved after applying a ± 12 V_{dc} bias in PZT thin film to study the behavior of ferroelectric switching in a grain population of the material.

7.3 Potential applications of nanosized ferroelectrics

With the advances in nanoscale ferroelectrics, novel applications have been developed for these materials. This development is attributed to the reversible

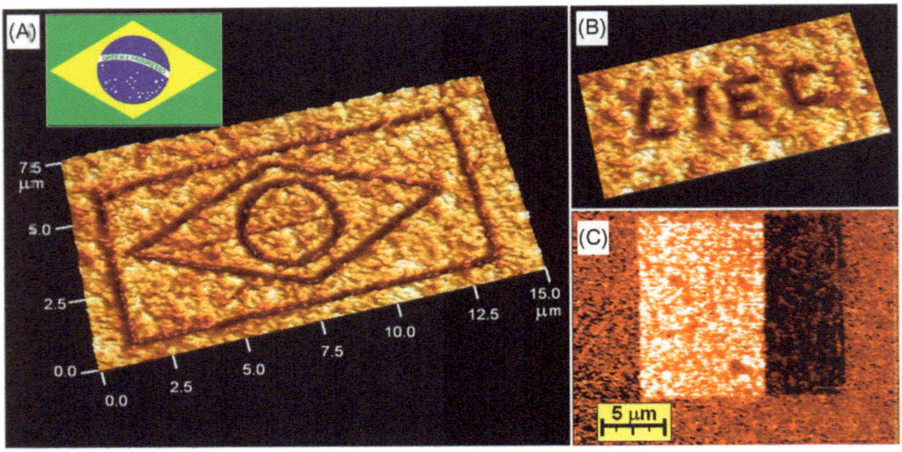

Figure 7.6 PFM images after ferroelectric lithography in thin film samples. (A) Writing of the Brazilian flag (15 × 7.5 μm²), (B) recording of the letters LIEC (3 × 10 μm²), (C) polarization pattern containing two opposed polarization states (15 × 15 μm²).

polarization of a ferroelectric material, which can be changed and subsequently provided with pyroelectric and piezoelectric characteristics [56–61].

Nanocomposites containing polymers and nanostructured ferroelectric particles have been receiving considerable attention because they can represent the answer to the processing of materials with high permittivity [62]. The use of the polymer provides flexibility to the material, while aiding in recovering nanoparticles with high dielectric permittivity. Notwithstanding these merits, there are certain difficulties in the film processing with homogeneous particle distribution due to the phase separation as well as aggregate formation. Among research that merits mentioning Kim et al. [62] cited the obtaining of a $BaTiO_3$/PVDF-HFP (poly(vinylidene fluoride-cohexafluoropropylene)) composite; they reported obtaining nanocomposites with low dielectric permittivity constituted by (PVDF-HFP) polymer and $BaTiO_3$ particles with surface modified by phosphoric acid. In the literature, there are several reports relating to the influence exerted by polymers over $BaTiO_3$ properties. Nunes-Pereira et al. reported obtaining $BaTiO_3$/PVDF-HFP nanocomposite for energy harvesting applications. The efficiency of the composite materials is related to the size and quantity of particles of $BaTiO_3$ immersed in the polymeric matrix. Moreover, the power output values allow the implementation of these composites in nanogenerator energy devices [63]. The use of nanoferroelectrics in energy harvesting materials is an interesting application as it enables the use of energy from the environment as a mechanical source to be converted into electrical energy. This makes it possible to reload the battery of mobile electronic devices with human body movement, for instance.

In recent years, ferroelectric nanostructures have attracted attention as a result of their photocatalytic abilities. Zang et al. [64] demonstrated the use of $KNbO_3$

nanowires for photocatalytic degradation of rhodamine B. The advances in the synthesis of ferroelectric nanocrystals improves the way toward the fine control of the size and shape of the nanoparticles, enhancing the photoreactivity through the affinity of the photogenerated electron-hole pairs (e^-/h^+), thereby avoiding recombination, leading to the increase in the efficiency levels. In the study published by Zang et al., the photocatalytic experiments enabled them to determine that orthorombic-$KNbO_3$ nanowires exhibited twice as much of the rate constant for photodegradation than monoclinic-$KNbO_3$ nanowires; this was attributed to the synergy between the structure at atomic level and the ferroelectric polarization.

Another important application for ferroelectric nanoparticles is in the biological/medical area for the detection of carcinogenic tumors in tissues. The particles have the ability to generate light due to their dielectric properties, thereby enabling them to be used as imaging/diagnostic agents, an area that has not been explored until recently, following the publication of few studies in that regard [65–67]. The light (second-harmony) is generated after converting two photons of light into one photon, where the produced photon has half of the incident wavelength [68,69]. This application is related to the increase of the dielectric constant of the material, and allows the development of noninvasive techniques for diagnosis in internal organs of the human body. Scientifically inspiring and socially benefit-accruing as it may appear, one of the intrinsic challenges posed by this kind of application is to confer biocompatibility to ferroelectric particles so as to avoid having a toxic effect on human cells [69–71]. Semenov et al. [72] have demonstrated the use of $CaTiO_3$/$BaTiO_3$ ferroelectric nanoparticles ($\sim$50 nm size) for detection (based on the contrast in dielectric properties between normal and malignant tissue) of cancer in microwave tomography. The study reported a difference in terms of visualization between two cancer areas: (1) one with 10 mm of diameter and another (2) with a diameter of 20 mm. Besides (1) being smaller than (2), the ferroelectrics nanoparticles in (1) helped to enlarge the volume of the tumor in the image raising it to about 50% higher than the image in (2),. This work, however, did not undertake studies related to biocompatibility nor biosafety issues, yet it is clear that it demonstrates the potential use of ferroelectric nanoparticles as an agent capable of improving the diagnosis of cancer.

7.4 Final considerations

The development and characterization of new nanosized ferroelectric materials have promoted wide application opportunities for their use as devices for energy storage and conversion, biosensors, nonvolatile memories, and many others. By virtue of that, many researchers have been devoting their efforts toward improving their knowledge and expertise regarding the synthesis, fundamental understanding, and applications of ferroelectric nanomaterials (mainly zero- and one-dimensional).

Ferroelectric nanoparticles have been found to remain ferroelectric to sizes of at least 50 nm, though the production costs and the properties ought to be improved so as to make the production of these devices essentially worthy of an endeavor. Nanosized ferroelectrics have a potential for application in many areas, especially in nanofabrication approaches and devices in the electronic field such as high-density nonvolatile memories due largely to the possibility of manipulating ferroelectric domains. The PFM technique provides a wide range of possibilities for investigating the physical properties and functionality of ferroelectric materials at the nanoscale level. The development of ferroelectric nanostructures and the ability to reduce their dimensions unfold a novel generation of devices that require thorough investigation regarding the effects of the particles size on the properties of the ferroelectric materials.

The study of ferroelectric nanoparticles is at the onset of its potentiality, where many problems need to be solved for the emerging applications.

Acknowledgments

The author is thankful to the coauthors for their invaluable dedication and collaboration in this project. We would like to express our deepest gratitude and indebtedness to the LMA-IQ for providing the FEG-SEM facilities and to the Brazilian research funding agencies CAPES (Process: 88881.068060/2014-01), CNPq, CEPID/CDMF 2013/07296-2, and FAPESP 2014/11314-9 for granting the financial support to the research group.

References

[1] R.G.P. McQuaid, M. McMillen, L.W. Chang, A. Gruverman, J.M. Gregg, Domain wall propagation in meso- and nanoscale ferroelectrics, J. Phys. Condens. Mat. 24 (2) (2012) 024204.

[2] R.G.P. McQuaid, L.J. McGilly, P. Sharma, A. Gruverman, J.M. Gregg, Mesoscale flux-closure domain formation in single-crystal $BaTiO_3$, Nat. Commun. 2 (2011) 404.

[3] A. Schilling, R.M. Bowman, G. Catalan, J.F. Scott, J.M. Greeg, Morphological control of polar orientation in single-crystal ferroelectric nanowires, Nano Lett. 7 (12) (2007) 3787–3791.

[4] Y. Liu, D.N. Weiss, J. Li, Rapid nanoimprinting and excellent piezoresponse of polymeric ferroelectric nanostructures, ACS Nano 4 (1) (2010) 83–90.

[5] C.C. Hong, S.Y. Huang, J. Shieh, S.H. Chen, Enhanced piezoelectricity of nanoimprinted sub-20 nm poly(vinylidene fluoride–trifluoroethylene) copolymer nanograss, Macromolecules 45 (3) (2012) 1580–1586.

[6] Y.Y. Chen, B.Y. Yu, J.H. Wang, R.E. Cochran, J.J. Shyue, Template-based fabrication of $SrTiO_3$ and $BaTiO_3$ nanotubes, Inorg. Chem. 48 (2) (2009) 681–686.

[7] H. Li, H. Wu, D. Lin, W. Pan, High Tc in electrospun $BaTiO_3$ nanofibers, J. Am. Ceram. Soc 92 (9) (2009) 2162–2164.

[8] T. Ould-Ely, M. Luger, L.K. Reinig, K. Niesz, M. Doherty, D.E. Morse, Large-scale engineered synthesis of $BaTiO_3$ nanoparticles using low-temperature bioinspired principles, Nat. Protoc. 6 (1) (2011) 97–104.

[9] J.J. Urban, J.E. Spanier, L. Ouyang, W.S. Yun, H. Park, Single-crystalline barium titanate nanowires, Adv. Mater. 15 (5) (2003) 423–426.
[10] J.J. Urban, W.S. Yun, Q. Gu, H. Park, Synthesis of single-crystalline perovskite nanorods composed of barium titanate and strontium titanate, J. Am. Chem. Soc. 124 (7) (2002) 1186–1187.
[11] Y.F. Zhu, L. Zhang, T. Natsuki, Y.Q. Fu, Q.Q. Ni, Facile synthesis of $BaTiO_3$ nanotubes and their microwave absorption properties, ACS Appl. Mater. Inter. 4 (4) (2012) 2101–2106.
[12] W. Sun, Y. Pang, J. Li, W. Ao, Particle coarsening II: growth kinetics of hydrothermal $BaTiO_3$, Chem. Mater. 19 (7) (2007) 1772–1779.
[13] U.A. Joshi, J.S. Lee, Template-free hydrothermal synthesis of single-crystalline barium titanate and strontium titanate nanowires, Small 1 (12) (2005) 1172–1176.
[14] J.O. Landesor, C.G. Yáñez, R.L. Juárez, I.D. Velasco, H. Pfeiffer, Synthesis of advanced ceramics by hydrothermal crystallization and modified related methods, J. Adv. Ceram. 1 (2012) 204–220.
[15] S.Z. Ji, H. Liu, Y.H. Sang, W. Liu, G.W. Yu, Y.H. Leng, Synthesis, structure, and piezoelectric properties of ferroelectric and antiferroelectric $NaNbO_3$ nanostructures, Cryst. Eng. Comm. 16 (2014) 7598–7604.
[16] X. Bing, Q. Zhang, H. Zhang, S. Jiang, Largely enhanced ferroelectric and energy storage performances of P(VDF-CTFE) nanocomposites at a lower electric field using $BaTiO_3$ nanowires by stirring hydrothermal method, Ceram. Int. 42 (16) (2016) 19012–19018.
[17] P.M. Rørvik, T. Grande, M.A. Einarsrud, One-dimensional nanostructures of ferroelectric perovskites, Adv. Mater. 23 (2011) 4007–4034.
[18] L. Li, J. Deng, J. Chen, X. Sun, R. Yu, G. Liu, et al., Wire structure and morphology transformation of niobium oxide and niobates by molten salt synthesis, Chem. Mater. 21 (2009) 1207–1213.
[19] F. Madaro, J.R. Tolchard, Y. Yu, M.A. Einarsrud, T. Grande, Synthesis of anisometric $KNbO_3$ and $K_{0.5}Na_{0.5}NbO_3$ single crystals by chemical conversion of non-perovskite templates, Cryst. Eng. Comm. 13 (2011) 1350–1359.
[20] A.C. Santulli, H. Zhou, S. Berweger, M.B. Raschke, E. Sutter, S.S. Wong, Synthesis of single-crystalline one-dimensional $LiNbO_3$ nanowires, Cryst. Eng. Comm. 12 (2010) 2675–2678.
[21] J. Rodel, W. Jo, K.T.P. Seifert, E.M. Anton, T. Granzow, D. Damjanovic, Perspectiveon the development of lead-free piezoceramics, J.Am. Ceram. Soc. 92 (2009) 1153–1177.
[22] T. Lusiola, F. Bortolani, Q. Zhang, R.A. Dorey, Molten hydroxide synthesis as an alternative to molten salt synthesis for producing $K_{0.5}Na_{0.5}NbO_3$ lead free ceramics, J. Mater. Sci. 47 (4) (2012) 1938–1942.
[23] X. Li, H. Xu, Z.S. Chen, G. Chen, Biosynthesis of nanoparticles by microorganisms and their applications, J. Nanomater. (2011) 1–16, 2011.
[24] B.D. Briggs, M.R. Knecht, Nanotechnology meets biology: peptide-based methods for the fabrication of functional materials, J. Phys. Chem. Lett. 3 (2012) 405–418.
[25] G.S. Dhillon, S.K. Brar, S. Kaur, M. Verma, Green approach for nanoparticle biosynthesis by fungi: current trends and applications, CRC Cr. Rev. Biotechn. 32 (2011) 49–73.
[26] G. Ahmad, M.B. Dickerson, Y. Cai, S.E. Jones, E.M. Ernst, J.P. Vernon, et al., Rapid bioenabled formation of ferroelectric $BaTiO_3$ at room temperature from an aqueous salt solution at near neutral pH, J. Am. Chem. Soc. 130 (1) (2007) 4–5.

[27] A.K. Jha, K. Prasad, Ferroelectric $BaTiO_3$ nanoparticles: biosynthesis and characterization, Colloid Surface B75 (1) (2010) 330–334.
[28] A.K. Jha, K. Prasad, Synthesis of $BaTiO_3$ nanoparticles: a new sustainable green approach, Integr. Ferroelectr. 117 (2010) 49–54.
[29] I. Uddin, A. Jaiswal, P. Poddar, Novel green bio-milling technique for the synthesis of $BaTiO_3$ nanoparticle using *Saccharomyces cerevisiae*, Int. J. Innov. Biol. Res. 2 (1) (2013) 1–5.
[30] B. Kumar, S.W. Kim, Energy harvesting based on semiconducting piezoelectric ZnO nanostructures, NanoEnergy 1 (2012) 342–355.
[31] Y. Wakasa, I. Kanno, R. Yokokawa, H. Kotera, K. Shibata, T. Mishima, Piezoelectric properties of microfabricated (K,Na)NbO_3 thin films, Sensor Actuat. A-Phys. 171 (2011) 223–227.
[32] X. Liu, M. Deng, X. Wang, Nanoscale domain imaging and local piezoelectric coefficient d_{33} studies of single piezoelectric polymeric nanofibers, Mater. Lett. 189 (2017) 66–69.
[33] J. Xiao, T.S. Herng, J. Ding, K. Zeng, Polarization rotation in copper doped zinc oxide (ZnO:Cu) thin films studied by piezoresponse force microscopy (PFM) techniques, Acta Mater. 123 (2017) 394–403.
[34] A. Rüdiger, T. Schneller, A. Roelofs, S. Tiedke, T. Schmitz, R. Waser, Nanosize ferroelectric oxides - tracking down the super paraelectric limit, Appl. Phys. A Mater. Sci. Process. 80 (2005) 1247–1255.
[35] M. Alexe, C. Harnagea, D. Hesse, U. Gösele, Polarization imprint and size effects in mesoscopic ferroelectric structures, Appl. Phys. Lett. 79 (2001) 242–244.
[36] A. Gruverman, A. Kholkin, A. Kingon, H. Tokumoto, Asymmetric nanoscale switching in ferroelectric thin films by scanning force microscopy, Appl. Phys. Lett. 78 (2001) 2751–2753.
[37] S.V. Kalinin, B.J. Rodriguez, S. Jesse, J. Shin, A.P. Baddorf, P. Gupta, et al., Vector piezoresponse force microscopy, Microsc. Microanal. (2006) 206–220.
[38] G. Suyal, E. Colla, R. Gysel, M. Cantoni, N. Setter, Piezoelectric response and polarization switching in small anisotropic perovskite particles, Nano Lett. 4 (2004) 1339–1342.
[39] X. Meng, W. Wang, H. Ke, J. Rao, Y. Zhou, Synthesis of potassium sodium niobate nanostructures by hydrothermal combining with the sol-gel method, Mater. Sci. Eng. B 212 (2016) 1–6.
[40] D. Damjanovic, H. Search, C. Journals, A. Contact, M. Iopscience, I.P. Address, Ferroelectric, dielectric and piezoelectric properties of ferroelectric thin films and ceramics, Rep. Prog. Phys. 61 (1998) 1267.
[41] C. Durkan, M.E. Welland, D.P. Chu, P. Migliorato, Probing domains at the nanometer scale in piezoelectric thin films, Phys. Rev. B 60 (1999) 16198.
[42] C. Harnagea, M. Azodi, R. Nechache, C.V. Cojocaru, V. Buscaglia, M.T. Buscaglia, et al., Characterization of individual multifunctional nanoobjects with restricted geometry, Phase Transit. 86 (2013) 635–650.
[43] B.K. Yun, Y.K. Park, M. Lee, N. Lee, W. Jo, S. Lee, et al., Lead-free $LiNbO_3$ nanowire-based nanocomposite for piezoelectric power generation, Nano Express 9 (2014) 4–10.
[44] J.H. Jung, M. Lee, J. Il Hong, Y. Ding, C.Y. Chen, L.J. Chou, et al., Lead-free $NaNbO_3$ nanowires for a high output piezoelectric nanogenerator, ACS Nano 5 (2011) 10041–10046.

[45] J. Xiao, W.L. Ong, Z. Guo, G.W. Ho, K. Zeng, Resistive switching and polarization reversal of hydrothermal-method-grown undoped zinc oxide nanorods by using scanning probe microscopy techniques, ACS Appl. Mater. Interfaces. 7 (2015) 11412–11422.

[46] P. Güthner, K. Dransfeld, Local poling of ferroelectric polymers by scanning force microscopy, Appl. Phys. Lett. 61 (1992) 1137.

[47] F.M. Pontes, A.J. Chiquito, W.B. Bastos, M.A. Pereira-da-Silva, E. Longo, Structural, optical, magnetic, ferroelectric, and piezoelectric properties of $(Pb,Ba)(Ti,Fe)O_3$ perovskites: a macroscopic and nanoscale properties approach, J. Mater. Chem. C 4 (2016) 9331–9342.

[48] J. Valasek, Piezoelectric and allied phonomena in Rochelle Salt, Phys. Rev. 15 (1920) 537–538.

[49] E. Fatuzzo, W.J. Merz, Ferroelectricity, North-Holland Pub. Co., Amsterdam, 1967.

[50] IEEE, Standards Committee of the IEEE standard definitions of terms associated with ferroelectric and related materials, IEEE Transactions on Ultrasonics Ferroelectrics and Frequency Control 50 (12) (2003) 1613–1646.

[51] Y. Xu, Ferroelectric Materials and Their Applications, North-Holand, Amsterdan, 1991.

[52] G. Arlt, Twinning in ferroelectric and ferroelastic ceramics – stress relief, J. Mater. Sci. 25 (6) (1990) 2655–2666.

[53] B. Jaffe, W.R. Cook, H. Jaffe, Piezoelectric Ceramics, Academic Press, London, 1971.

[54] D.B. Li, D.A. Bonnell, Ferroelectric lithography, Ceram. Int. 34 (1) (2008) 157–164.

[55] P. Paruch, T. Tybell, J.M. Triscone, Nanoscale control of ferroelectric polarization and domain size in epitaxial $Pb(Zr_{0.2}Ti_{0.8})O^{-3}$ thin films, Appl. Phys. Lett. 79 (4) (2001) 530–532.

[56] Y. Cho, S. Hashimoto, N. Odagawa, K. Tanaka, Y. Hiranaga, Nanodomain manipulation for ultrahigh density ferroelectric data storage, Nanotechnology 17 (7) (2006) S137–S141.

[57] J.F. Scott, Applications of modern ferroelectrics, Science 315 (2007) 954–959.

[58] M. Ionescu, Nanoelectronics: ferroelectric devices show potential, Nat. Nanotechnol. 7 (2012) 83–85.

[59] M. Bibes, Nanoferronics is a winning combination, Nat. Mater. 11 (2012) 354–357.

[60] R. Guduru, P. Liang, C. Runowicz, M. Nair, V. Atluri, S. Khizorev, Magneto-electric nanoparticles to enable field-controlled high-specificity drug delivery to eradicate ovarian cancer cells, Sci. Rep. 3 (2013) 2953.

[61] A. Rodzinski, R. Guduru, P. Liang, A. Hadjikhani, T. Stewart, E. Stimphil, et al., Targeted and controlled anticancer drug delivery and release with magnetoelectric nanoparticles, Sci. Rep. 6 (2016) 20867.

[62] P. Kim, N.M. Doss, J.P. Tillotson, P.J. Hotchkiss, M.J. Pan, S.R. Mardet, et al., High energy density nanocomposites based on surface-modified $BaTiO_3$ and a ferroelectric polymer, ACS Nano 3 (9) (2009) 2581–2592.

[63] J. Nunes-Pereira, V. Sencadas, V. Correia, V.F. Cardoso, W. han, J.G. Rocha, et al., Energy harvesting performance of $BaTiO_3$/poly(vinylidene fluoride–trifluoroethylene) spin coated nanocomposites, Compos. Part B-Eng. 72 (2015) 130–136.

[64] T. Zang, W. Lei, P. Liu, J.A. Rodrigues, J. Yu, Y. Qi, et al., Insights into the structure-photoreactivity relationships in well-defined perovskite ferroelectric $KNbO_3$ nanowires, Chem. Sci. 6 (2015) 4118–4123.

[65] P. Pantazis, J. Maloney, D. Wu, S.E. Frases, Second harmonic generating (SHG)nanoprobes for in vivo imaging, P. Nat. Acad. Sci. 107 (33) (2010) 14535–14540.

[66] N. Horiuchi, Imaging: second-harmonic nanoprobes, Nat. Photonics 5 (7) (2011), 7-7.
[67] C.C. Wang, F.C. Li, V.A. Hovhannisyan, W.C. Lin, S.J. Lin, P.T. So, et al., Differentiation of normal and cancerous lung tissues by multiphoton imaging, J. Biomed. Opt. 14 (4) (2009) 044034.
[68] E. Kim, M.T. Steinbrück, M.T. Buscaglia, V. Buscaglia, T. Pertsch, R. Grange, Second-harmonic generation of single $BaTiO_3$ nanoparticles down to 22 nm diameter, ACS Nano 7 (6) (2013) 5343−5349.
[69] W.P. Dempsey, S.E. Fraser, P. Pantazis, SHG nanoprobes: advancing harmonic imaging in biology, BioEssays 34 (5) (2012) 351−360.
[70] G. Yuriy, O. Zribi, A. Glushchenko, Emerging Applications of Ferroelectric Nanoparticles in Materials Technologies, Biology and Medicine, Advances in Ferroelectrics, EdInTech, 2012.
[71] G. Ciofani, L. Ricotti, V. Mattoli, Preparation, characterization and in vitro testing of poly(lactic-co-glycolic) acid/barium titanate nanoparticle composites for enhanced cellular proliferation, Biomed. Microdevices 13 (2) (2011) 255−266.
[72] S. Semenov, N. Pham, S. Egot-Lemaire, Ferroelectric nanoparticles for contrast enhancement microwave tomography: feasibility assessment for detection of lung cancer, IFMBE Proc. 25 (8) (2009) 311−313.

Nanosized $BaTiO_3$-based systems

8

Adelina-Carmen Ianculescu[1] *and Catalina-Andreea Stanciu*[1,2]
[1]"Politehnica" University of Bucharest, Bucharest, Romania
[2]National Institute for Lasers, Plasma and Radiation Physics, Magurele, Romania

8.1 Fundamentals of undoped $BaTiO_3$ systems

In the context of limiting or even replacing dangerous substances such as lead (a volatile, toxic, and nonrecyclable element) in electric and electronic devices, in the last decade many efforts were devoted to identify new lead-free oxides and to improve or adjust the electrical behavior of the already known ferroelectrics, in order to obtain functional properties of higher performance, close to those of piezoelectric $Pb(Zr,Ti)O_3$ (PZT) and relaxor $Pb(Mg_{1/3}Nb_{2/3})O_3$ (PMN) systems. A viable alternative, from this point of view, is represented by the $BaTiO_3$ (BT)-based systems, one of the most widely investigated classes of perovskites, due to their environmentally-friendly character along with unique and multiple useful properties, such as high permitivity, high pyro- and piezoelectric coefficients, switching properties, positive temperature coefficient of resistivity (PTCR effect), and high-voltage tunability. These properties make them suitable for many applications, in high performance ceramic capacitors, information storage devices, field effect devices (FeET), piezoelectric and ultrasonic actuators, transducers, sensors, pyroelectric detectors, and thermistors [1–4].

Despite the large number of exhaustive studies and reports on elaboration and characterization of materials derived from BT, they continue to represent a challenging and "hot" topic, due to their versatile electric behavior, and being very sensitive to many factors such as dopants/solutes, stoichiometry, processing methods, among others. In the last few years, the development of several innovative preparation and sintering techniques caused a revival of the interest in BT-based systems, with controlled and highly reproducible electrical behavior. This trend become even more compelling and explicable in the context of the miniaturization and of the high integrability degree imposed by nanotechnology, which involves either preservation of ferroelectricity or development of new functionalities in highly densified, nanosized BT products with restrictive geometries. Thus, in multilayer ceramic capacitors (MLCCs), for a high volumetric capacity C_v in a volume as small as possible, since $C_v \sim \varepsilon_r \cdot n/d^2$, (where n is the number of layers, d is the thickness of a ceramic layer, and ε_r is the relative permittivity), the thickness of the dielectric layer should be as small as possible and the number of layers as high as possible [5–8]. Nowadays, the thickness of a dielectric layer is in the submicron range and the number of dielectric layers can exceed 2000 in MLCCs.

It is well known that BT exhibits a perovskite structure in which Ti ion occupies the central position in its paraelectric (cubic, nonpolar) state. Cooling down from the nonpolar state, the coarse-grained polycrystalline BT undergoes a succession

Magnetic, Ferroelectric, and Multiferroic Metal Oxides. DOI: http://dx.doi.org/10.1016/B978-0-12-811180-2.00008-6

of crystalline modifications at $\approx 130°C$ (cubic-tetragonal: C-T), $\approx 5°C$ (tetragonal-orthorhombic: T-O), and $\approx -90°C$ (orthorhombic-rhombohedral: O-R). All the low-temperature phases are ferroelectric, with the spontaneous polarization oriented along the <001>, <011>, and <111> crystallographic directions, respectively. The paraelectric−ferroelectric phase transition is accompanied by a sharp increase of the dielectric constant, which for polycrystalline ceramics rises from ~1500, at room temperature (RT), up to ~10,000 at Curie temperature. In the paraelectric state the relative permittivity decreases with temperature, according to the Curie−Weiss law.

For nanoscaled systems such as nanopowders, thin films, nanowires, nanotubes, and nanocrystalline ceramics the ferroelectric, dielectric, and piezoelectric characteristics change significantly, relative to those specific to bulk microstructured ceramics. This means that effects determined by the small size of structural units, involving a high density of interface states in the case of nanostructured ceramics and thick films, and additional features related to restrictive geometries and other extrinsic contributions in the case of 2D and 1D nanostructures such as thin films, nanowires, nanotubes, and nanoshell tubes, induce a particular electrical behavior and, sometimes, even peculiar and unexpected functional characteristics in such nanostructured BT-based products. Since ferroelectricity represents a long-range cooperational phenomenon, based on the interaction with the neighboring permanent dipoles in the crystalline lattice, an increase in surface area by nanostructuring could induce drastic changes in the long- and short-range ordering of dipoles such as depolarization field inside a material. Consequently, size effects occurring in nanoscaled BT systems contribute in suppressing ferroelectricity at RT. The value of the minimum size of a structural feature (grain in ceramics and thin films or particle in powders) required to still preserve ferroelectricity in BT nanosized systems currently represents a highly debated and controversial topic of the fundamental physics of the solid state. This critical dimension strongly depends on the intrinsic factors, as well as on extrinsic contributions, especially related to morphological/microstructural features induced by the preparation and sintering strategies used. The Landau−Ginsburg−Devonshire theoretical models, in which surface effects were taken into consideration, foresaw a critical size value for ferroelectricity vanishing (spontaneous polarization suppression), which varies from a few nanometers to several tens of nanometers [9,10]. It was reported that when particle/grain size decreases, the domain structure is strongly affected and the tetragonality of the BT unit cell is gradually reduced, so that below the mentioned critical value, a stress-induced change from the tetragonal to the so-called pseudocubic symmetry (i.e., a ferroelectric-superparaelectric (FE-SPE) phase) transition takes place [11].

8.2 State of the art of nanosized $BaTiO_3$-based systems

8.2.1 $BaTiO_3$ nanopowders: Synthesis methods, parameters controlling particle size and morphology, size effects

The conventional mixed oxide method for preparing BT powders by solid state reaction between $BaCO_3$ and TiO_2 produces relatively coarse and agglomerated particles with broad particle size distributions [12,13]. These starting powders are less suitable for processing dense, single-phase ceramics, with uniform and controlled microstructures, required for advanced applications. For this reason and in order to decrease the sintering temperature, various innovative procedures such as hydrothermal/solvothermal methods [14–17], different variants of chemical coprecipitation (from inorganic aqueous solution, oxalate/citrate route) [18–21], self-propagating high-temperature synthesis [22], combustion [23], microwave synthesis [24], polymeric precursor methods [25–29], sol-gel [30–33] and sol-precipitation techniques [34,35], microemulsion route [36,37], molten salt synthesis [38], mechanically activated processes [39,40], flame-assisted spray pyrolysis [41,42], sonochemical synthesis [43], freeze-drying [44] and even biosynthesis routes [45] have been developed as alternative approaches for preparing BT nanopowders of high purity, with controlled stoichiometry, narrow particle size distribution, and low tendency of agglomeration. In this sense, many efforts have been focused on choosing the appropriate synthesis route and optimizing the processing parameters in order to produce powders of high quality and, consequently, to control microstructure and to tailor or to improve the functional properties of the related ceramics [46]. It was found that synthesis and thermal treatment factors strongly influence the particle size and particle size distribution, morphology, porosity, specific surface area, agglomeration tendency, and surface state of particles in barium titanate powders [47].

In the case of the hydrothermal/solvothermal method, involving the progress of chemical reaction under the conjugated effect of temperature and pressure by heating the reactants for several hours in a bomb at temperatures up to 300°C and pressures up to 100 MPa, parameters such as type and characteristics of reagents, initial Ba/Ti ratio in the precursor solutions, concentration of the starting solutions, type of solvent, presence and type of surfactants, and synthesis parameters (reaction time and temperature, pH value) seem to play a key role, not only in adjusting particle morphology, phase composition, and stoichiometry of the final product, but also in determining the kinetics and type of mechanisms that lead to barium titanate formation [48a–c]. It is well known that for hydrothermal synthesis, a high pH value (basic environment) is required in order to obtain BT [48a]. However, a slight increase of pH value can strongly affect the size and morphology of BT particles resulting from hydrothermal synthesis at 150°C for 7 h [49]. Thus, the isolated, single-crystal BT particles (average size of 88.6 nm) obtained using a pH value of 14 (Fig. 8.1A) change into raspberry-like mesocrystals (average size of 231.1 nm) resulting from self-assembly phenomena of similarly-oriented small crystallites, when the pH value decreased to 13.5 (Fig. 8.1B).

The density of hydrothermally processed BT powders determined by means of helium pycnometer exhibits lower values of $\sim 5.65-5.75\ g/cm^3$ in comparison with polycrystalline BT ceramics ($\rho \sim 6.0\ g/cm^3$). These low density values were explained by Hennings et al. [50,51] in terms of the formation of highly defective BT particles, with an equal number of double-ionized barium vacancies and tetra-ionized titanium vacancies with effective negative charge, which electrically compensate the protons trapped into the oxygen sublattice, forming $OH^{\bullet}$ groups with effective positive charge. This defective BT can be described by the formula:

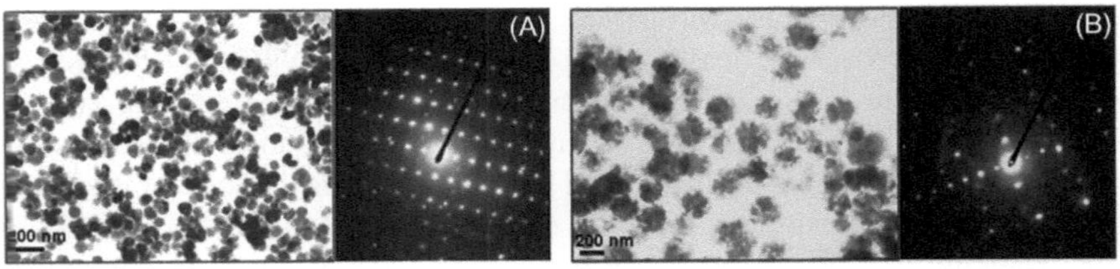

Figure 8.1 Transmission electron microscopy (TEM) images and selected area electron diffraction (SAED) patterns of $BaTiO_3$ powders hydrothermally synthesized at 150°C for 7 h at: (A) pH = 14; (B) pH = 13.5 (bar in TEM images = 200 nm).

$\left[Ba_{1-x/6}(V''_{(Ba)})_{x/6}\right]\left[Ti_{1-x/6}(V''''_{(Ti)})_{x/6}\right]O_{3-x}(OH^{\bullet})_x$. During heating at temperatures up to 500°C the water release leads to the disappearance of lattice defects, according to Eq. (8.1), so that stoichiometric BT grains with intragranular porosity are formed. It is worthy to mention that because of the internal structural/microstructural strains generated by this process, BT powders prepared by the hydrothermal method are not suitable for the elaboration of products with confined geometries such as thick films. These exhibit "bloating" effects caused by the migration of pores toward the inner electrodes in the final sintering stage, thus inducing cracks and delamination in multilayer ceramic capacitors [51].

$$6\left[Ba_{1-x/6}(V''_{(Ba)})_{x/6}\right]\left[Ti_{1-x/6}(V''''_{(Ti)})_{x/6}\right]O_{3-x}(OH^{\bullet})_x \rightarrow (6-x)BaTiO_3 + 3xH_2O\uparrow \tag{8.1}$$

In the case of the coprecipitation technique via the oxalate route, BT is formed by thermal decomposition during the heating of the molecular precursor, barium titanyl oxalate, resulting in the final step of synthesis, according to Eq. (8.2).

$$BaTiO(C_2O_4)_2 \cdot xH_2O \rightarrow BaTiO_3 + 2CO_2\uparrow + 2CO\uparrow + xH_2O \tag{8.2}$$

The characteristics of the molecular precursor particles govern the size and morphology of BT particles resulting from annealing. The crystallinity degree, structure, and morphology of the oxalate precursor particles are strongly influenced by several synthesis parameters such as reagent type, solvent nature, drying temperature, aging time, pH, and mixing order of the reagents [52–54]. For example, the polarity (dielectric permittivity) of the solvent of the precipitation agent (oxalic acid) solution strongly influences the characteristics of the precipitate particles. The lower is the dielectric constant ε_r of the solvent; the smaller will be the size of BT particles [49]. A polar solvent such as water leads to the formation of a few nuclei, which grow rapidly, resulting in larger mixed oxalate particles, whose thermally-induced crystallization determines the formation, by self-assembly phenomena, of rose-like BT mesocrystals (Fig. 8.2A). Unlike the mentioned case, nonpolar (organic) solvents such as ethanol or isopropanol induce supersaturation (i.e., the

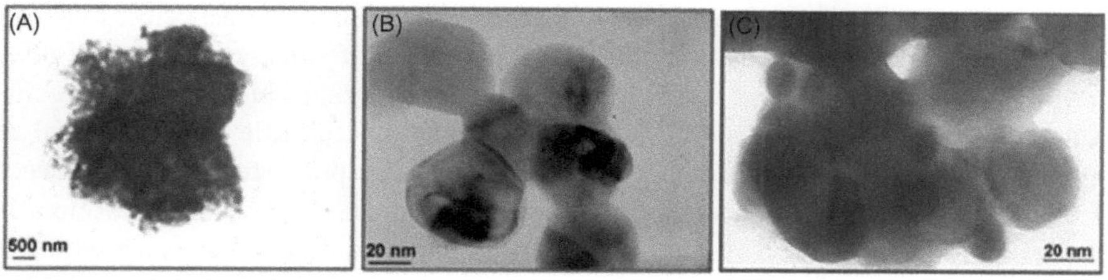

Figure 8.2 Transmission electron microscopy (TEM) images of $BaTiO_3$ particles resulted after annealing at 850°C for 2 h from molecular precursors prepared by coprecipitation via oxalate route, using as solvent: (A) water (bar = 500 nm); (B) ethanol (bar = 20 nm) and (C) isopropanol (bar = 20 nm).

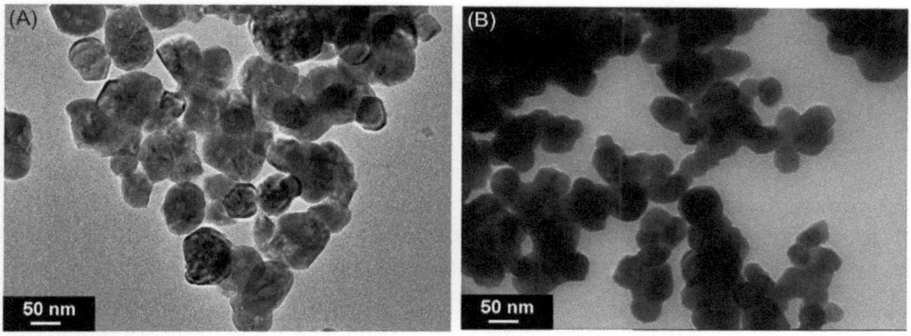

Figure 8.3 Transmission electron microscopy (TEM) images of $BaTiO_3$ particles resulted after annealing at 850°C for 2 h, from precursors prepared by: (A) the alkoxide variant of the sol-gel method (bar = 50 nm) and (B) the modified sol-gel (Pechini) method (bar = 50 nm).

formation of a large number of small nuclei whose growth is strongly inhibited), leading to small precursor particles, and, consequently, to small, single-crystal BT particles (Fig. 8.2B and C).

Concerning the powders prepared by the sol-gel method, the size and morphology of their particles are also very sensitive to the type of starting reagents and synthesis/thermal treatment parameters [34]. The polymeric variant of the sol-gel route, involving the use of barium and titanium alkoxides used as starting reagents, leads to nanosized, stoichiometric particles with low agglomeration tendency (Fig. 8.3A). However, because of their sensitivity to hydrolysis, an undesirable competition between hydrolysis and condensation may occur, according to reaction in Eq. (8.3).

$$Ti(OR)_4 + Ba(OR')_2 + 3H_2O \rightarrow BaTiO_3 + 4ROH + 2R'OH \tag{8.3}$$

In order to avoid this disadvantage in the case of bimetallic oxides, the semialkoxide route, which consists of using as starting materials the barium acetate or nitrate and an alkoxide of titanium, was proposed [34].

Later, the so-called modified sol-gel method, known also as the Pechini method, was developed. These methods consist of the formation of some complexes of alkaline, earth-alkaline, and even transitional metals with organic bi- or tridentate chelating agents such as citric acid (CA). In the reaction mixture a polyalcohol as ethylene glycol (EG) is added in order to form bonds by polyesterification between the chelate compounds, thus causing the mixture gelation. After a heat treatment of polymerization (120–135°C/11 h), the gel is converted in a glassy resin, which is lightly ground and calcined in several steps at temperatures ranging between 300°C and 400°C, in order to burn off the organic material. Annealing at 850°C for 2 h is carried out in order to obtain a single phase, stoichiometric BT powder. Unlike the traditional sol-gel techniques, when the metal belongs to the gel network, in the Pechini method the gel network is formed exclusively by esterification reaction between the chelating agent and the polyalcohol, the metallic ions being retained in this network only by means of some weak bonds [55]. Therefore, by Pechini synthesis structural elements such as chains or ribbons, specific to sol-gel process, are not formed. The particles obtained after precursor annealing are often spherical or almost spherical, exhibiting a pronounced tendency to form agglomerates or even partially sintered aggregates (Fig. 8.3B).

Recent studies showed that BT nanoparticles can be prepared by the ordinary as well as a modified variant of the conventional solid state reaction method [56,57]. In the first case, single-phase, barium titanate particles with an average size of 100 nm were obtained by calcining a mixture of nanosized $BaCO_3$ and TiO_2 particles (of 50 and 70 nm, respectively), for 8 h at temperature of 800°C [56]. In the second case, tetragonal BT nanoparticles were produced by an innovative and versatile approach, which combines colloidal chemistry with solid state reaction, with the formation of core-shell $BaCO_3$@TiO_2 particles as the intermediate phase. In the first stage of this second procedure a uniform layer of amorphous titania has been formed on the surface of fine, rod-shaped barium carbonate crystals suspended in a peroxotitanium(IV) solution by a precipitation process. Formation of single phase BT from the $BaCO_3$@TiO_2 particles occurs by single step calcination at 600°C–650°C [57]. The lowering of the reaction temperature was associated with the maximization of the contact surface between the reactants and by the minimization of diffusion distances. In the second step, the $BaCO_3$@TiO_2 particles undergo spontaneous fragmentation and spheroidization during the reaction, and consequently, the resulting BT nanoparticles have no memory of the initial shape of the barium carbonate crystals. A schematic illustration of this process is presented in Fig. 8.4 [57].

Results of research regarding preparation by solid state reaction of BT nanoshell hollow particles were also reported in the literature [58,59]. The same procedure discussed earlier was adopted, but in this case a coarser powdered $BaCO_3$ reagent was used. Formation of the cavity occurs because the out-diffusion of the core phase ($BaCO_3$) is much faster than in-diffusion of the shell material. Initially, the strong imbalance of the diffusion fluxes determines the formation of Kirkendall porosity close to the original $BaCO_3$/TiO_2 interface. Later, surface diffusion of the carbonate leads to growth of pores and coalescence with the formation of a single

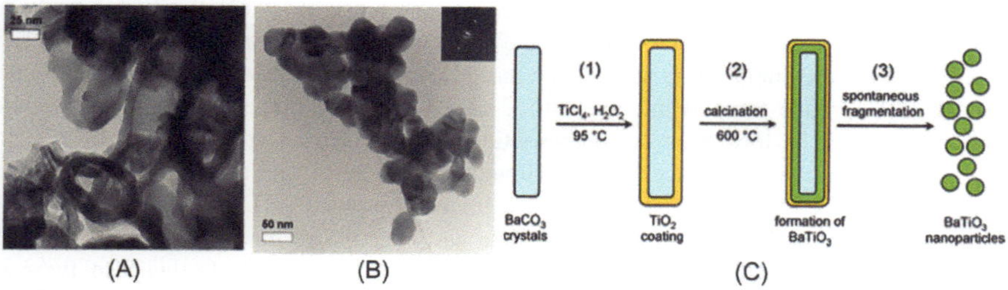

Figure 8.4 Transmission electron microscopy (TEM) images of: (A) core-shell $BaCO_3@TiO_2$ particles (bar = 25 nm); (B) $BaTiO_3$ nanoparticles—inset: selected area electron diffraction pattern (bar = 50 nm); and (C) schematic illustration of the combined colloidal-solid state process for preparation of dense, nanosized $BaTiO_3$ nanoparticles. Reprinted with permission from M.T. Buscaglia, V. Buscaglia, R. Alessio, Coating of $BaCO_3$ crystals with TiO_2: versatile approach to the synthesis of $BaTiO_3$ tetragonal nanoparticles, Chem. Mater. 19 (2007) 711–718.

cavity. Elongated nanoshell BT hollow particles, which retain the morphology of the initial $BaCO_3$ particles, were formed in this case after thermal treatment at 700°C for 24 h [58]. The size of the starting $BaCO_3$ crystals is a critical factor in the fabrication of BT hollow particles. If the diameter of the particles is much lower than 1 μm, the shell would undergo a rapid collapse during the solid-state reaction with the formation of dense nanoparticles.

A free-template method, consisting of the use of a molten hydrated salt ($Ba(OH)_2 \cdot 8H_2O$) as the solvent and rutile TiO_2 as the reactant was developed by Tian et al. to produce at 180°C core-shell and hollow BT particles under sealed conditions, similar to those of solvothermal/hydrothermal processes [59]. They found that, in the mentioned conditions, the mechanism that governs the morphology and structuring of the final product is strongly dependent on the TiO_2 dosage. Thus, a high TiO_2 dosage leads to the formation of larger (of ~400–800 nm) composite core-shell particles, while smaller (<250 nm) hollow particles resulted when a low rutile dosage was used.

The ferroelectric nature of BT nanoparticles represents a hot topic, still under debate. Thus, even if earlier calculation of Zhong et al. [9] revealed that ferroelectricity should persist in particles of ~40 nm, the experimental results of Uchino et al. [11] indicated a critical value of the particle size of 0.12 μm, below which ferroelectricity is suddenly suppressed in BT nanopowders. On the other hand, based on their structural data obtained by high intensity synchrotron X-ray diffraction (XRD) technique, Wada et al. reported a critical size for BT nanoparticles between 17 and 30 nm, when a gradual stress-induced transition from a typical ferroelectric (FE) phase toward a superparaelectric (SPE) state occurs [60].

Two different approaches were proposed in order to explain the ferroelectricity lack in barium titanate nanoparticles:

- Pure phase model. Explains this phenomenon in terms of (1) the particle size effect or the so-called intrinsic-size effect, as a driving force of the drastic decrease of the cooperative long-range interactions of dipoles (i.e., phonon softening) [61]; (2) the surface role by depolarization effects, which means that the surface tension associated with the fine particles stabilizes the cubic phase at RT [11]; and (3) internal strains imposed by the presence of the $OH^{•}$ groups trapped into the lattice of BT particles prepared in aqueous media, which involves the formation of compensating defects (i.e., cation vacancies) in order to maintain the electrical balance [50,62]. However, the role of the hydroxyl groups formed by protons in the oxygen sublattice still remains controversial [63].

- Surface layer model. Based on the XRD data and Rietveld refined method, this assumes the formation of structurally nonuniform composite particles [60,64]. In this case, two variants were proposed: (1) the two-phase model, which stipulates the presence of a tetragonal core and a thin (of ~10 nm) surface shell of cubic unit cell ("dead" layer); and (2) the three-phase model, which takes into account an additional well-marked intermediate layer of low and variable tetragonality [60].

By means of their powder dielectric measurement using the slurry method, Wada et al. [60] reported a dielectric constant maximum of 15,000 for BT nanoparticles of 70 nm prepared via the oxalate route, and they also predicted that in the absence of the surface dead layers, an ideal dielectric behavior with a permittivity maximum of around 35,000 at a critical particle size between 17 and 30 nm should be obtained [65]. The evolution of the structure of BT nanoparticles with temperature seems to be even more complicated, taking into account that, concurrently, a shift of orthorhombic-tetragonal phase transition toward higher temperature values could also take place. Therefore, the so-called pseudocubic form, related to low temperature–chemically prepared BT powders, differs from the prototypic cubic structure that characterizes a large single crystal in its paraelectric state (above T_C). It is well known that this pseudocubic modification precedes the formation of the typical tetragonal phase, when higher processing temperatures (above 200°C in the case of hydrothermal route and above 750°C in the case of the citrate/oxalate and sol-gel methods) are used and, consequently, when coarser particles are formed by the Ostwald ripening process.

It seems more likely that, at RT, nanosized barium titanate particles consist of a mixture of slightly distorted orthorhombic and tetragonal forms, as Raman and transmission electron microscopy (TEM) investigations revealed [66a,b]. This more realistic view, also sustained by various theoretical approaches, shows that with decreasing particle size, the ferroelectric order is gradually reduced through a disordering process, which involves small and complex local distortions, whose higher-length scale averaging reveals an atomic ordering, similar to a cubic-like structure [66–70]. Calculations in barium titanate quantum dots have predicted the existence of a vortex domain configuration, with a toroidal ordering of dipoles, with zero net polarization [70,71]. Contrary to these predictions, the recent experimental results of Polking et al., consisting of maps of ferroelectric structural distortions obtained from aberration-corrected TEM combined with holographic polarization imaging indicated the persistence of an overall, linearly ordered, and monodomain polarization state, which is still stable at nanometer scale. Thus, direct electrical measurements of individual, cube-shaped BT nanocrystals demonstrated room-temperature polarization switching and revealed the disappearance of the polar state below a critical size of 5–10 nm [72].

Another interesting feature observed in nanosized BT particles is that, even if ferroelectricity at RT is significantly lower than in coarse-grained ceramics, some distortions corresponding to a polar order are still present at temperatures far above the Curie temperature specific to the microstructured bulk ceramics. Thermo-Raman investigations revealed diffuse paraelectric (PE)-FE and FE-FE phase transitions, so that modifications with dissimilar symmetries coexist over relatively large temperature ranges. It was found that the smaller the particle size, the higher the temperature, until the residual polar order persists [66a,b]. Li et al. studied the phase structures of very small BT nanoparticles with sizes ranging from 2.5 to 10 nm at various temperatures by using aberration-corrected TEM equipped with an in situ heating holder [66b]. Their experimental results indicated that each BT nanoparticle was composed of different phases and, unlike the previous view, the polar ones were observed even in the shells due to the atomic shifts induced by the surface relaxation. The thickness of the relaxed shell was estimated to be about 1 nm, accounting for a sharp increasing share of the particle, taking into account that the particle size decreases to only several nanometers. The ferroelectric phases in such small BT nanoparticles are maintained at an unexpectedly high temperature of 600°C, suggesting the significant increase of T_C. Based on their results and the data reported by others, Li et al. showed that for particles with sizes below 10 nm a complex mixture of several distortions like tetragonal, rhombohedral, and orthorhombic, coexist with the nonpolar cubic phase in a temperature range of 120–600°C.

Concerning new functionalities of ferroelectric nanopowders, recent studies have shown that ferromagnetism occurs in small-sized nanoparticles of otherwise non-magnetic oxides such as BT [73]. The ferromagnetic behavior of barium titanate nanoparticles at RT could be induced by the presence, at their surface, of oxygen vacancies, whose electrical compensation involves the appearance of reduced titanium species such as Ti^{2+} and Ti^{3+} [74–76]. It was reported that while the electrical polarization decreases with decreasing particle size, the magnetization increases below a critical particle size, which is of the order of a few nanometers [74]. A recent finding of residual ferroelectricity in small nanoparticles of only 12 nm [77] stimulated the exploration of the simultaneous occurrence of ferromagnetism and ferroelectricity in barium titanate nanopowders, where the former is expected to arise from the surface and the latter from the particle core [74]. It is worthy to mention that, because of the opposite tendency of the two phenomena, that is, increasingly higher ferromagnetism, and increasingly lower ferroelectricity with the decreasing particle size, we can conclude that nanosized BT particles exhibit multiferroic properties in a narrow range of particle size.

8.2.2 $BaTiO_3$ nanostructured ceramics: Elaboration, correlations structure–microstructure–electrical behavior, size effects

From the as-prepared oxide nanopowders, dense, single-phase, and microstructurally controlled ceramics can be prepared using alternative sintering strategies. In the case of undoped BT ceramics, a large spread of values of the dielectric permittivity can be found in the literature. It was reported that the dielectric constant at RT and the switching properties are significantly influenced by the domain structure, domain wall mobility, and internal strains induced by grain size, porosity, and lattice defects [78–80]. Previous works showed that coarse-grained pure BT ceramics (average grain size higher than 20 μm) exhibit, at RT, a permittivity value of ~1500, almost independent of grain size, and behave like typical ferroelectrics [79]. As the average grain size of BT ceramics is reduced below 10 μm, the RT permittivity strongly increases, reaching a maximum of ~5500 μm, for a grain size of ~1 μm [79,81]. This anomaly was explained by Arlt et al. in terms of the action of internal stresses originated in the transition from a state characterized by multidomain grains (twinned domain structure) specific to coarse-grained ceramics, toward a state involving single-domain grains, which occurs at this grain size value [79]. A further decrease of the grain down to the submicron range induces a gradual decrease of dielectric constant at room temperature, an obvious flattening of the permittivity maximum at the ferroelectric-paraelectric (F-P)FE-PE phase transition, indicating a FE-relaxor crossover, as well as a decrease of the Curie temperature, T_C [78,82].

The critical grain size for the stability of the ferroelectric phase in BT ceramics is strongly dependent on the preparation method of the starting oxide powders and varies between 0.2 and 0.12 μm. It was found that the T_C value is always lower in fine-grained BT ceramics than that specific to powders with particle size similar to the ceramic grain size [11,83,84]. In the case of nanostructured BT ceramics, an almost constant and significantly lower permittivity (~1000–1500) over a wide temperature range, as a result of the strong reduction of ferroelectricity, was reported [85]. For this reason, nanocrystalline barium titanate bulk materials are considered potential candidates for new, potential applications in tunable high-frequency devices, such as phase shifters, filters, and antennas [86]. Frey and Payne showed that, in BT ceramics with an average grain size of 70 nm, the FE-PE phase transition takes place at 120°C, although the $\varepsilon_r = f(T)$ variation is quite flattened. Subsequently, they determined a Curie temperature of 110°C for 35 nm-grained ceramics. The ferroelectric properties (especially the remanent polarization and the area of the hysteresis loop) are also strongly affected by the grain size decrease [78]. Based on literature data regarding the size effects in undoped BT ceramics, it was concluded that the great diversity of the reported results (Fig. 8.5), which sometimes seem contradictory, might be determined by dissimilar microstructures and local stoichiometry deviations that compel the system to evolve in different electrical and mechanical boundary conditions [87a,b,88]. In the case of BT nanoceramics, beside the intrinsic size effects that lead to the decrease of the unit cell tetragonality and spontaneous polarization and thus, to the suppression of the ferroelectricity toward the SPE limit, other extrinsic effects, such as mechanical strains induced by particular microstructural and geometrical features, electrode-ceramic interfaces, porosity, density of grain boundaries, defect chemistry, compositional gradients in the core or in the interface regions of the grains also have to be taken into account. Polotai et al. showed that a ceramic sample with an average grain size of 30 nm exhibits tetragonal structure in the temperature range of 50–105°C, while

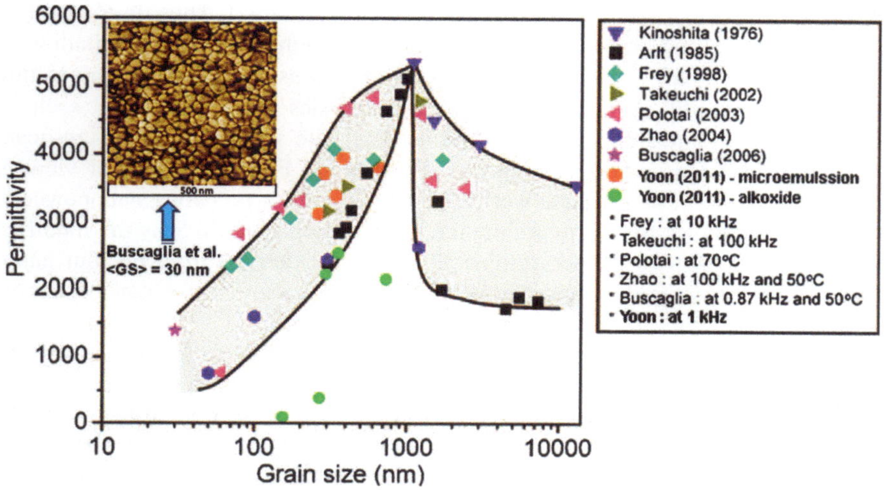

Figure 8.5 Dielectric constant at room temperature for $BaTiO_3$ products of different grain size. Inset: microstructure of a nanostructured ceramic with grain size $<GS> = 30$ nm. Reprinted with permission from (a) S. Yoon, J. Dornseiffer, Y. Xiong, D. Grüner, Z. Shen, S. Iwaya, et al., Spark plasma sintering of nanocrystalline $BaTiO_3$-powders: consolidation behavior and dielectric characteristics, J. Eur. Ceram. Soc. 31 (2011) 1723–1731; (b) M.T. Buscaglia, M. Viviani, V. Buscaglia, L. Mitoseriu, A. Testino, P. Nanni, et al., High dielectric constant and frozen macroscopic polarization in dense nanocrystalline $BaTiO_3$ ceramics, Phys. Rev. B 73(6) (2006) 064114.

at temperatures above 110°C the nonpolar cubic phase becomes stable [89]. Consequently, we can conclude that unlike the BT nanopowders with a similar particle size, where a certain polar order is maintained up to 320°C, in nanostructured ceramics the Curie temperature decreases, relative to the T_C value specific to coarse-grained ceramics.

Fine-grained ceramics with average grain size below 100 nm have been obtained only in recent years, particularly for pure BT, by using special densification methods, like two-step pressureless sintering [90] (grain size 70 nm, 97% relative density), ultrahigh pressure sintering in a multianvil press (grain size down to 70 nm, 98% relative density) [91], combined sintering method (grain size down to 90 nm, 99% relative density) [89], and most recently by using the spark plasma sintering (SPS) technique [92]. In the last years, SPS was quite often used as an alternative method to the classical sintering procedure in order to obtain highly densified, nanostructured ceramics from powders prepared by various wet chemical methods [87,92–101]. BT ceramic bodies with a minimum grain growth and, therefore, exhibiting grain size down to 30 nm and relative density of 95%–99% were produced by this procedure [87,99–101].

Previous papers on SP sintered barium titanate ceramics pointed out that the grain growth rate during this type of sintering is very sensitive to the type of wet-chemical method used for the powder preparation [93,94]. Thus, in the case of certain routes such as the sol-gel acetate-methanol method or sol-precipitation, in spite of relative small starting particles, their growth is very fast as temperature increases during SPS. Moreover, other characteristics of the particles such as morphology, agglomeration tendency, and surface state seem to have a significant influence on the densification during the SPS process. From this point of view, it was found that the most adequate wet chemical method for preparing nanopowders, consisting of particles that show a minimal growth during the SPS, is the modified Pechini method [93]. In nanocrystalline BT ceramics derived from Pechini nanopowders and SP-sintered in appropriate conditions, the grain size can be almost similar to the particle size in the starting oxide powder.

Another advantage of the SPS procedure is the possibility to produce easily, by changing sintering parameters (temperature and plateau), barium titanate ceramics with a large spectrum of controlled sizes of grains, ranging from micrometer toward nanometer scale, in order to carefully investigate the influence of microstructure, especially of size effects as extrinsic contribution, on the dielectric and ferroelectric properties.

At the end of 2006, Deng et al. [102] reported experimental data on the dielectric/ferroelectric properties of the most fine-grained BT ceramic obtained by SPS. Their ceramic sample with an average grain size of 20 nm exhibits diffuse phase transition, a maximum value of permittivity of 930 at 1 kHz and a T_C value of $\sim 110°C$ (Fig. 8.6A). In this nanoceramic the strongly reduced grain size affects the switching phenomena, so that the $P(E)$ dependence is almost linear and the hysteresis loop becomes extremely thin and sloping, with a remanent polarization of only 0.05 $\mu C/cm^2$ (Fig. 8.6B). However, as it was reported by Ma et al. [103], the lowest grain size ever obtained in a nanostructured ceramic was of 8 nm and it was produced by a pressureless two-step sintering (TSS) process. This process consists

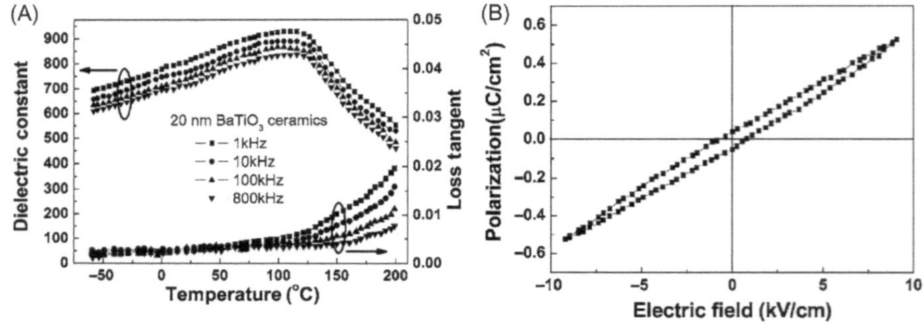

Figure 8.6 (A) Evolution of relative permittivity and dielectric losses versus temperature at some fixed frequency values and (B) $P(E)$ dependence for a ceramic with a grain size of 20 nm prepared by spark plasma sintering.
Reprinted with permission from X. Deng, X. Wang, H. Wen, L. Chen, L. Chen, L. Li, Ferroelectric properties of nanocrystalline barium titanate ceramics, Appl. Phys. Lett. 88 (2006) 252905.

of heating the samples that resulted from pressing a BT powder prepared via an oxalate route with a rate of 10°C/min to a sintering temperature of (950°C–1250°C), followed by a quick cooling with a rate of 30°C/min to a lower temperature of (850°C–1150°C), at which the sample is maintained for 2–20 h. It was assumed that in the first sintering stage grain growth occurs, while in the second stage, the grains have the required mobility to diffuse for densification, but grain boundary migration is inhibited. The advantage of the TSS strategy is that the grain size and density of the sample can be properly controlled by adjusting the temperature and the soaking time of each thermal treatment stage.

8.2.3 $BaTiO_3$ thin films: Preparation techniques, the influence of intrinsic and extrinsic contributions on the functional properties

Particularly, thin films have received much attention in recent years, for their numerous applications in microelectronics and integrated optics technologies, ranging from buffer layers for the integration of high temperature superconductors and voltage-tunable microwave filters, to pyroelectric imaging arrays and multifunctional highly integrated CMOS-based devices, such as ferroelectric nonvolatile random access memories and dynamic random access memories (DRAMs). Thin film ferroelectric devices offer several advantages over their bulk counterparts, such as lower driving voltages, higher speed, and the potential for monolithic integration with Si.

Various methods, each of them with particular advantages and inconveniences, like molecular beam epitaxy [104–106], pulsed laser deposition [107–109], RF-magnetron sputtering [110–114], chemical vapor deposition (CVD) and its variant metal organic chemical vapor deposition [115,116], chemical solution deposition (CSD) by sol-gel process [117–119] or polymeric precursor method [120] (using different deposition techniques such as spin-coating [121], dip-coating [122], or electrophoretic deposition [123]), spray pyrolysis [124], hydrothermal-electrochemical route [125–127], and atomic layer deposition [128] were proposed for the fabrication of BT thin films.

In the case of BT thin films, besides the large scatter in the reported values of the dielectric permittivity caused by the use of different deposition techniques and processing parameters, other issues related to drastic deviations of structural and electrical properties relative to those of bulk ceramics have been noticed. As well as in the case of BT nanograined ceramics, size effects, inducing diffuse FE-PE phase transition, act in fine-grained BT films, irrespective of the processing route.

Unlike other nanoscaled systems, in the case of thin films some particular features related to the high surface/bulk ratio, growth mechanisms, texturing, surface topography, effects determined by substrate and proximity of electrodes induce additional extrinsic contributions to the dielectric and ferroelectric behavior. Because of the confined geometry, even the influence of the structural defects (vacancies, dislocations) and chemical gradients, which also act in nanostructured barium titanate ceramics, is much more complex in the case of thin films. Thus, the point defect chemistry in thin films is responsible for some undesired phenomena like (DC) resistance degradation, ferroelectric imprints, and fatigue [129,130]. In addition, another feature specific mainly to polycrystalline films is the "double" size effect induced by two different dimensional parameters—grain size and thickness [131]. At first sight, we can estimate that the thickness actually represents an apparent size effect, because a reduced thickness is expected to induce grain size decrease, so that the role of size effects on the functional properties might be explained in similar terms as in nanocrystalline ceramics. However, it must be mentioned that a small thickness also involves other phenomena related to enhanced microstructurally induced stresses and various constraints such as electrode interactions, space charge layers, and lattice mismatches between film and substrate, which will exert a stronger influence on the functional properties in ferroelectric thin films in comparison with ceramics with similar grain size [132]. In order to quantify the size effect associated with the thickness contribution on the electrical behavior, microstructurally uniform and chemically homogeneous thin films are required. It was found that, at a fixed temperature, the dielectric constant of ferroelectric thin films decreases continuously with film thickness [133].

The grain size and cross-section microstructure of thin films is strongly influenced by particular growth mechanisms induced by the processing method. Low-temperature techniques that imply the limitation of surface diffusion (e.g., CVD) usually lead to columnar grains by island growth mechanism (Volmer–Weber growth), while high-temperature processes, implying an enhanced surface diffusion, stimulate nucleation and consequently, the formation of a fine-grained polycrystalline microstructure with uniformly distributed grain boundaries by means of a layer-by-layer growth mode (Frank van der Merve growth; e.g., temperature-induced crystallization through amorphous films as in the case of the CSD method). If there is no good compatibility between the substrate and film, a combined growth mechanism (Stranski–Krastanov), which first involves a two-dimensional growth in layer-by-layer mode, and then a three-dimensional growth in island mode, can occur [134]. The competition between nucleation and growth processes in thin films is controlled by the quality of the substrate-film interface. As a result not only the size of the grains, but also their morphology induced by the type of growth mechanism plays a key role in influencing the dielectric and ferroelectric behavior. As well as for nanostructured ceramics, in polycrystalline BT thin films with thickness below 1 μm and with grain size in submicronic or nanometric range, the dielectric permittivity decreases with the grain size decrease and the FE-PE phase transition becomes increasingly more diffuse, with broad temperature-dependent permittivity maxima, most likely due to the presence of the dead layers in the grain boundary regions. However, even if the grain size dependence is still present, because of the additional contributions, BT thin films exhibit lower permittivity values, relative to ceramics with similar grain size.

In the case of BT thin films with columnar structure, the preferentially oriented polarization along the [001] direction and the lower effect exerted by microstructural stresses associated with a significantly lower density of interface states between larger grains oriented parallel with respect to each other and perpendicularly with respect to electrodes, induces less diffuse FE-PE phase transitions and higher permittivity values [135]. In these films, both the domain configuration and the domain wall mobility are affected by size effects and mechanical boundary conditions imposed by the substrate characteristics (as lattice parameters, roughness, texturing, defects) and therefore, by the quality of the substrate-film interface. Local and global stresses resulted from small grain size, reduced thickness, growth mode, density of surface steps in epitaxial films, or associated to thermal expansion mismatch during the cooling from the growth temperature, as well as in-plane strains (of the order of MPa−GPa) originated in film-substrate misfits are responsible for the domain walls clamping, which strongly affect the switching phenomena associated with the ferroelectric state. For energetical reasons, at grain sizes near a tenth of a micron, progressively finer domain structures induce the disappearance of 90 degrees domain walls. Therefore, BT thin films often exhibit high coercivity and low remanent polarization, leading to thin, tilted, and unsaturated hysteresis loops at room temperature.

In epitaxial BT thin films, if the stress due to the mismatch between the unit cell parameters of the film and substrate is not relaxed by the formation of misfit dislocations, the influence of in-plane strains are even more significant, leading to the appearance of additional or more complicated local distortions. A diagram representing the strain-induced evolution of structural distortions in epitaxial BT thin films was proposed based on *ab-initio* calculation of Pertsev et al. [136] (Fig. 8.7).

It seems clear that size effects are controlled especially by the type and quality of substrate and by the parameters related to the processing method, rather than by the intrinsic limits for the stability of the ferroelectric state. Consequently, over time, the minimum thickness at which effects are observed is continuously dropping. Contrary to current opinion, some studies involving *ab-initio* simulation in the case of BT thin films between two metallic $SrRuO_3$ electrodes in short circuit revealed that ferroelectric order persists at a very low scale, an order of 6−7 unit cells (~24 Å). Below this critical thickness, ferroelectric instability vanishes, most likely because of the depolarizing electrostatic field caused by dipoles at the ferroelectric−metal interfaces [137,138].

8.2.4 $BaTiO_3$ one-dimensional nanostructures: Preparation and properties

One-dimensional (1D) ferroelectric nanostructures exhibiting high dielectric permittivity and electromechanical coupling behavior have demonstrated unique transducer applications in nanoelectromechanical systems (NEMSs) as energy harvesters and advanced sensors [139−143]. NEMSs are generally used to convert forms of mechanical/vibrational energy from the environment into electric power, resulting in reliable power sources for ultralow power wireless electronics.

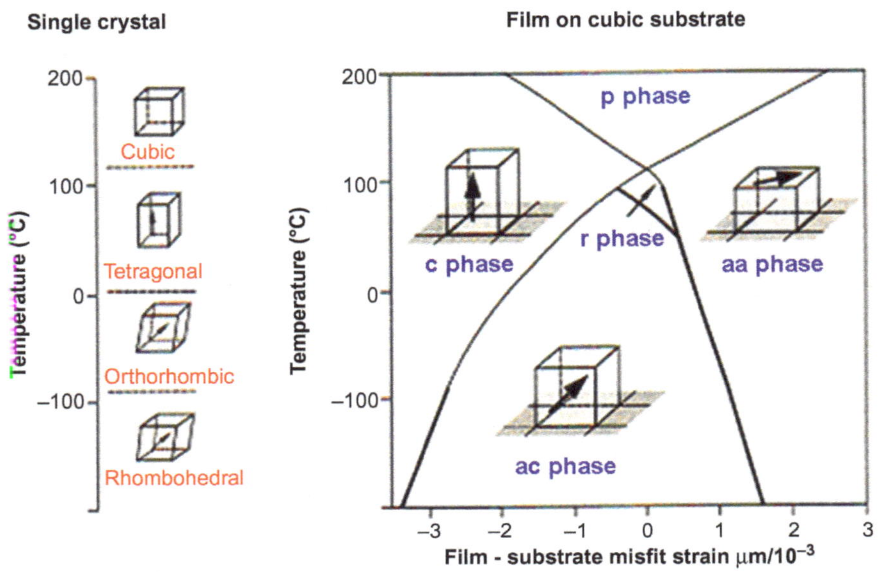

Figure 8.7 Diagram showing a comparative representation of structure evolution versus temperature in $BaTiO_3$ single crystal and in (001) single-domain epitaxial $BaTiO_3$ film, where additional distortions are induced by in-plane strains caused by film-substrate misfits; "r" phase exhibits components of polarization in all three directions.
Reprinted with permission from N.A. Pertsev, A.G. Zembilgotov, A.K. Tagantsev, Effect of mechanical boundary conditions on phase diagrams of epitaxial ferroelectric thin films, Phys. Rev. Lett. 80 (1998) 1988–1991.

Extensive research was focused in developing innovative synthesis routes and characterization techniques in order to exploit the full potential of nanoscaled ferroelectrics with particular structuring, such as nanowires, nanorods, nanofibers, nanotubes, and self-assembled hierarchical architectures [139,144–148]. There are two main approaches to prepare nanoscaled 1D structures: bottom-up methods, based on self-assembly phenomena [149], and template-based methods [150]. The most used wet chemical techniques for preparing ferroelectric 1D nanostructures are (1) different variants of the hydrothermal method [151–159] and (2) combined sol-gel processing/template-mediated routes, in conjunction with simple infiltration [144,148,150,160], electrophoretic deposition [161], and spin-coating techniques [162]. For this purpose, nanoporous or mesoporous membranes of silicon [144,148,160,163], anodic aluminum oxide (AAO) [164–167], or polycarbonate [168] were used as negative templates.

Sometimes more specific methods, involving electrospinning, for elaboration of BT-based nanofibers [169,170], diffusion-controlled topochemical solid state reactions to produce nanowires [171], and the use of positive templates (such as silicon and ZnO nanowires) for fabrication of ferroelectric nanotubes [172], were also reported in the literature.

In the case of the hydrothermal method, the growth mechanisms that generate hierarchical BT structures by particular self-assembly phenomena are strongly influenced by different factors, such as type of 1D nanoproduct used as titanium precursor, concentration of barium hydroxide suspension, and synthesis temperature [146] (Fig. 8.8A–D).

The structural properties and morphology of BT 1D nanostructures were intensively investigated by XRD, Raman spectroscopy, microscope techniques such as FE-scanning electron microscopy (FE-SEM), as well as by atomic force microscopy (AFM), while ferroelectric/piezoelectric properties were investigated

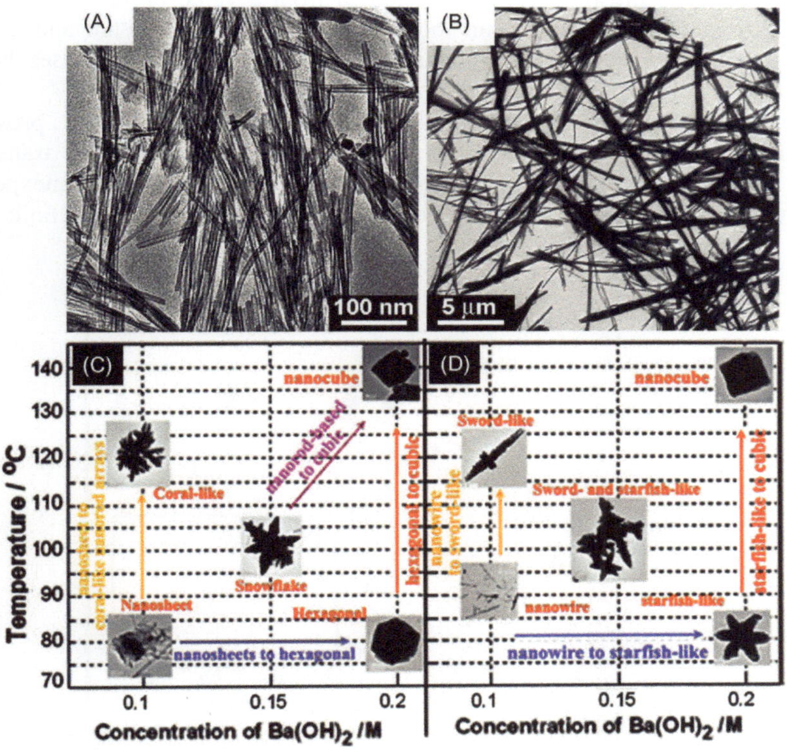

Figure 8.8 Transmission electron microscopy (TEM) images of the titanium precursors, (A) $Na_2Ti_3O_7$ nanotubes and (B) $Na_2Ti_3O_7$ nanowires; (C) the influence of the concentration of $Ba(OH)_2$ solution; and (D) the influence of the temperature of hydrothermal synthesis on the morphology and structuring of $BaTiO_3$ products.
Reprinted with permission from N. Bao, L. Shen, G. Srinivasan, K. Yanagisawa, A. Gupta, Shape-controlled monocrystalline ferroelectric barium titanate nanostructures: from nanotubes and nanowires to ordered nanostructures, J. Phys. Chem. C 112 (2008) 8634–8642.

by piezoresponse force microscopy (PFM) and electrostatic force microscopy (EFM).

Compared to nanoparticles, thin films, and nanostructured ceramics, influence of size effects on ferroelectricity in 1D BT nanostructures has been much less studied experimentally due to the difficulties in elaborating this kind of product. Based on first-principle calculations (FPCs) carried out to study size dependence of ferroelectric properties in BT nanowires, Geneste et al. found that stable polarization along the nanowire axis exists above a critical diameter of 1.2 nm [173]. Other FPC results reported by Fu et al. predicted the existence of a configuration of multivortex polarization in BT nanorods and nanowires, with one-direction ordering of dipoles [70]. EFM investigations performed by Spanier et al. on the polarization perpendicular to the nanowire axis show that the Curie temperature progressively decreases from a value of 116°C for nanowires with a diameter of ~50 nm to 68°C for a diameter of 5 nm and then it drops at room temperature as the diameter of the nanowire decreases below 3 nm (Fig. 8.9) [174].

From a structural point of view, recent data reported by Louis et al. provided evidence for the suppression of the orthorhombic-rhombohedral phase transition in BT nanowires. They predicted the existence of a low-temperature unexpected and previously unreported ferroelectric ground state of monoclinic *Cm*

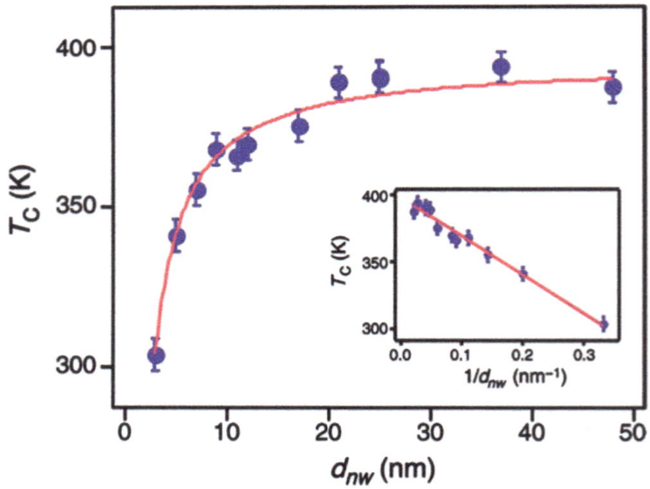

Figure 8.9 Evolution of the Curie temperature (T_C) versus $BaTiO_3$ nanowire diameter (d_{nw}). The solid circles are the experimental data, while the solid red line represents the result of the fitting process; inset shows the plot of T_C versus the inverse diameter (d_{nw}). Reprinted with permission from J.E. Spanier, A.M. Kolpak, J.J. Urban, I. Grinberg, L. Ouyang, W.S. Yun, et al., Ferroelectric phase transition in individual single-crystalline $BaTiO_3$ nanowires, Nano Lett. 6(4) (2006) 735–739.

symmetry, instead of the typical rhombohedral ground state that is stable in bulk ceramics [175]. This monoclinic modification seems to be interesting due to its piezoelectric response.

Regarding the patterning of ferroelectric domains in 1D BT nanoproducts, Schilling et al. have used scanning transmission electron microscopy investigations to observe the domain structure in nanocolumns with square cross-sections that were cut from BT single crystals using a focused ion beam [176]. In this length-constrained particular geometry, elastic strains associated with the cubic/tetragonal transition during cooling through the Curie temperature are released by the formation of 90 degrees domains (Fig. 8.10A). These alternate domains possess polarization directions that are perpendicular to the axis of the nanocolumn. Decreasing the dimension along the *y*-axis, while keeping constant the dimension along the *x*-axis resulted in a reorientation of the nonaxial polarization perpendicular to the shortest dimension in the column, in order to decrease the depolarization energy (Fig. 8.10B). This means that by changing morphology of such an individual 1D nanostructure, we can control and manipulate the local nonaxial polarization direction in a manner that involves the minimization of the depolarizing field.

Some authors reported that in the ferroelectric state, the spontaneous polarization develops parallel to the nanowire elongation axis, which becomes the [001] direction specific to tetragonal distortion [149,177–181]. Thus, by comparing

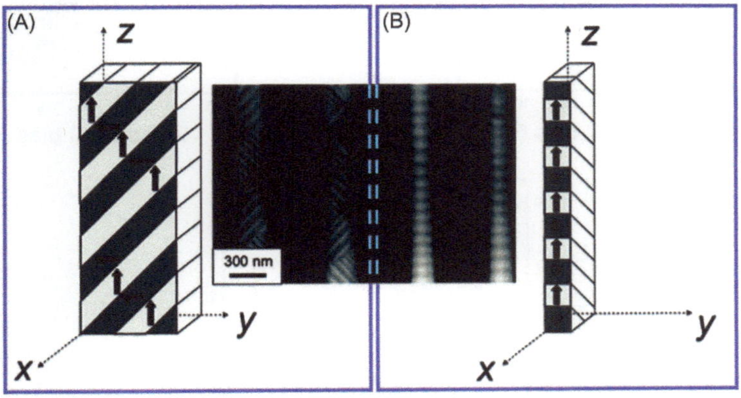

Figure 8.10 Schematic illustration and scanning transmission electron microscopy (STEM) images of the domain patterns observed in the $BaTiO_3$ nanocolumns under STEM investigation with the electron beam parallel to the *x*-axis (A); reorientation of the nonaxial polarization and the associated change in the domain contrast when the column dimension along the *y*-axis was intentionally decreased, while the dimension along the *x*-axis was kept constant (B).

Reprinted with permission from A. Schilling, R.M. Bowman, G. Catalan, J.F. Scott, J.M. Gregg, Morphological control of polar orientation in single-crystal ferroelectric nanowires, Nano Lett. 7(12) (2007) 3787–3791.

amplitude and phase loops obtained by lateral and vertical PFM measurements, Wang et al. found that the polarization is oriented along the major axis and therefore, they demonstrated that they have obtained single-domain BT nanowires [177] (Fig. 8.11).

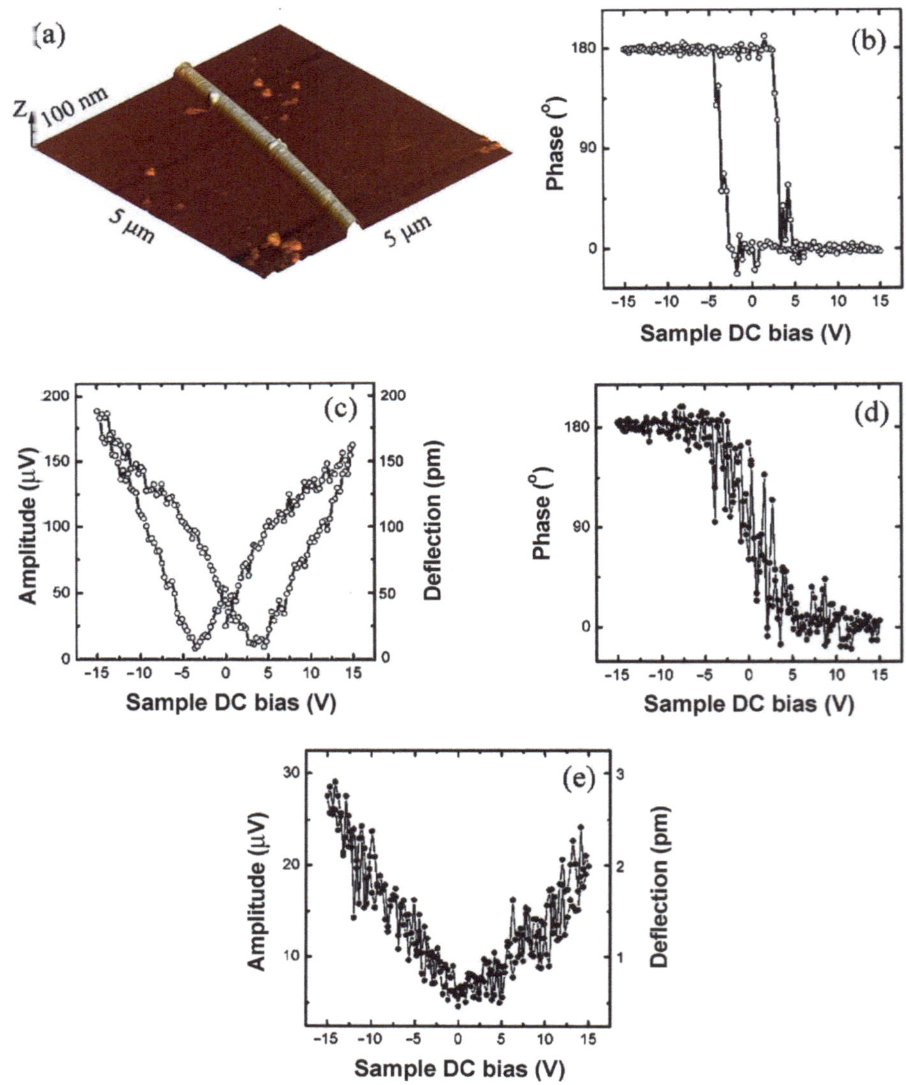

Figure 8.11 (A) 3D atomic force microscopy image of a $BaTiO_3$ nanowire; (B), (C) lateral piezoresponse force microscopy in-field measurements: (B) phase loops and (C) amplitude loop; (D), (E) vertical PFM measurements: (D) phase loop and (E) and amplitude loop. Reprinted with permission from Z.Y. Wang, J. Hu, M.F. Yu, One-dimensional ferroelectric monodomain formation in single crystalline $BaTiO_3$ nanowire, Appl. Phys. Lett. 89 (2006) 263119.

8.3 Recent approach to nanosized $BaTiO_3$-based systems

8.3.1 Introduction

BT is known as a chameleonic inorganic compound because its structural flexibility induces an unusual capability of the A- and B-sites of the perovskite lattice to accommodate a wide range of chemical elements. Adopting an adequate compositional design by choosing carefully the doping strategy represents the main approach to tailor and control functional behavior of BT-based products. It was found that electrical behavior of coarse-grained doped BT ceramics are strongly dependent by composition (stoichiometry, type, and content of dopants and solutes) as well as by microstructural features (density, grain size, porosity) induced both by the preparation route and factors and sintering strategy.

Size effects—driven functional properties in fine-grained compositionally modified barium titanate ceramics are much less studied relative to undoped, nanostructured products. Therefore, the electrical behavior and its dependence on the conjugated effect of the doping strategy and microstructural design, particularly at submicron and nanometric scale, are topics of high interest and novelty.

8.3.2 *Undoped and doped $BaTiO_3$ nanopowders prepared by wet-chemical methods*

Single phase $(Ba_{1-x}Sr_x)TiO_3$ (BST) ($x = 0$–0.35) powders were synthesized by the polymeric precursor (Pechini) method and by the acetate variant of the sol-gel method, using similar annealing conditions (850°C for 2 h) [28,29,182]. In both cases, the evolution of the lattice parameters was similar to that reported by Suasmoro et al. [183] for their $(Ba,Sr)TiO_3$ powders synthesized via oxalate route. The variation of the tetragonality degree (c/a) of the unit cell versus the solute content shows that with the increase of strontium proportion, the unit cell symmetry (which is tetragonal for the BT powder) changes gradually toward a cubic symmetry for the solid solutions with $x = 0.35$ (Fig. 8.12). This evolution suggests that, above a certain concentration of Sr^{2+} ions replacing the Ba^{2+} ions, the unit cell shrinks on the c-axis, due to the ionic radius of Sr^{2+} ions smaller than that one specific to Ba^{2+} ions. The monotonically decreasing variation is steeper for the particles synthesized by the sol-gel route, which means that, in this case, the tetragonality of the unit cell is higher at lower strontium concentration (Fig. 8.12).

This result can be explained in terms of the particle size. Thus, taking into account that a reduced size affects the tetragonality, it is normal that the slightly larger $Ba_{0.8}Sr_{0.2}TiO_3$ particles with an average size of 56.8 nm obtained by the sol-gel route are more tetragonal relative to those with an average particle size of 44.7 nm resulting from the modified Pechini method (Fig. 8.13A and B). Both types of particles present high crystallinity, as the corresponding selected area electron diffraction (SAED) patterns demonstrated (insets of Fig. 8.13A and B) but, the morphology is slightly different. More rounded particles, with a higher agglomeration tendency, were observed in the case of the Pechini method (Fig. 8.13A), in

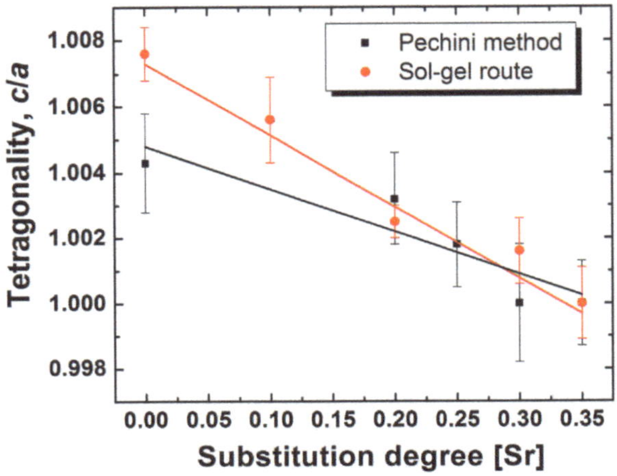

Figure 8.12 Comparative representation of the evolution of unit cell tetragonality versus strontium content for $(Ba_{1-x}Sr_x)TiO_3$ nanopowders prepared by the modified Pechini method and acetate variant of the sol-gel route, respectively [unpublished work].

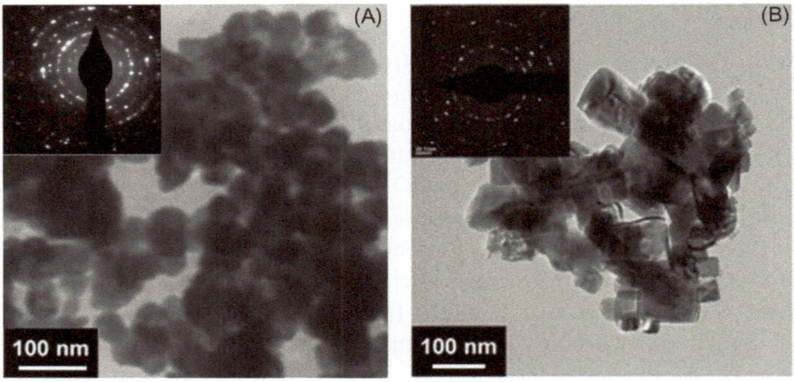

Figure 8.13 Transmission electron microscopy (TEM) images showing the size and morphology of $B_{0.8}Sr_{0.2}TiO_3$ particles prepared by (A) modified Pechini method and (B) acetate variant of the sol-gel route [182; Suppl. Information].

comparison with the tabular-square shape, with well-defined edges and right angles, specific to the thinner (and consequently more transparent) BST particles obtained by the sol-gel route (Fig. 8.13B). For both procedures, the single crystal nature of the BST particles was demonstrated by the values of the average crystallite size ($<D>_{Pechini} = 462(66)$ Å and $<D>_{sol\text{-}gel} = 530(15)$ Å), very close to those corresponding to the average particle size values. Besides, in both cases the particle size is gradually reduced with the increase of the strontium content [28,182].

Because of its chemical stability, a valuable B-site homovalent solute for barium titanate is zirconium. $BaTi_{1-x}Zr_xO_3$ (BTZ) particles, with Zr content in the compositional range specific to the ferroelectric state at room temperature ($x = 0.10$, 0.15, and 0.20), were also prepared by the modified Pechini method, in the same conditions as the BST nanopowders [27,184]. In this case, at RT, the evolution toward a cubic structure (the decrease of the tetragonality) with the increase of Zr content is even more pronounced than one observed for the strontium solute. Thus, all the $BaTi_{1-x}Zr_xO_3$ nanoparticles exhibit a cubic symmetry of the unit cell, with increasingly higher lattice parameter values with the increase of Zr proportion, due to the higher ionic radius of Zr^{4+} relative to that of Ti^{4+} ions.

Aliovalently doped BT powders were also synthesized by different wet chemical methods [33,185]. Thus, pure and lanthanum-doped barium titanate nanopowders (BLT) described by two different formulae: $Ba_{1-x}La_xTiO_3$ for lower La concentrations ($0 \leq x \leq 0.005$) and $Ba_{1-x}La_xTi_{1-x/4}O_3$ for higher donor dopant proportions ($x = 0.025$) were prepared by the alkoxide variant of the sol-gel method, using barium and titanium isopropoxides as starting reagents [33]. For the undoped composition, crystallization at RT of BT as major phase occurs. In the case of La-doped powders, the presence of lanthanum prevents crystallization. The increase of the solute content causes significant structural and morphological changes in the BLT powders, like evolution of the unit cell from tetragonal toward a cubic symmetry and particle size decrease, as in the case of the already discussed BST particles (Fig. 8.14). As a result of the decrease of particle size with the increase of solute content and, therefore, of a higher surface free energy of the smaller particles, in lanthanum-doped BT powders, the formation mechanism of the perovskite during the thermal decomposition of the precursors is more complex, involving, beside the higher temperature solid state reaction between the $BaCO_3$ and TiO_2 decomposition products, a lower temperature process, which leads to the formation, as intermediate, of a metastable (Ba,Ti,La) oxycarbonate phase, by continuous CO_2 release in a

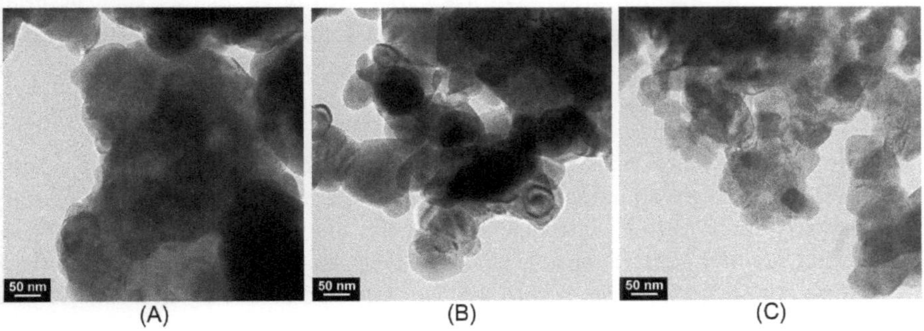

Figure 8.14 Transmission electron microscopy (TEM) images of barium titanate powders prepared by the alkoxide variant of the sol-gel method and annealed at 900°C/2 h (A) $x = 0.0025$ (stoichiometric); (B) $x = 0.005$ (stoichiometric); and (C) $x = 0.025$ (with built-in Ti vacancies) [33].

large temperature range. Single phase powders were obtained after annealing in air, at temperatures above 800°C, with a plateau of two hours.

8.3.3 Undoped and doped nanostructured $BaTiO_3$ ceramics consolidated by spark plasma sintering

From the nanopowder-synthesized soft chemistry methods, bulk ceramics were produced by SPS [186,187]. Dense, undoped BT ceramics with different grain sizes were obtained by adjusting SPS parameters (temperature and plateau) [82]. It was found that by reducing grain size, the dielectric constant, dielectric loss, and Curie temperature decreased. The lowering of dielectric constant is determined by the dilution effect of the low permittivity, nonferroelectric grain boundaries. For grain sizes below 100 nm, no Debye-relaxation phenomena (ascribed to the movement of domain walls pinned by oxygen vacancies) below T_C were observed, which suggests that nanostructured ceramics with grain size of the order of 100 nm are in a single-grain domain state. A Curie temperature of 106°C, a low frequency-dispersion, and only a slightly increasing trend of the dielectric permittivity (from ~1040 to 1080 at 20 kHz) over the temperature range specific to the ferroelectric state (30–106°C) was found in the case of a nanostructured ceramic with an average grain size of 99.1 nm, derived from ultrafine powders synthesized by a low-temperature precipitation route, involving the use of $BaCl_2$ and $TiCl_4$ as starting reagents (Fig. 8.15A and B). On the other hand, a nanostructured ceramic sample with an almost similar grain size of 99.5 nm (Fig. 8.15E), derived from a barium titanate powder synthesized by the already mentioned modified Pechini method (Fig. 8.15D), exhibits a quite similar T_C of 108°C, but at the same frequency of 20 kHz, it shows a significantly higher (more than double) value of the dielectric permittivity at Curie temperature and even at room temperature (~1750) and a better defined FE-PE phase transition (Fig. 8.15C and F). This means that in ceramics with similar average grain size, but derived from powders prepared by dissimilar routes, not only the average grain size itself, but also other factors such as grain size distribution, or particular grain boundary characteristics, induced by different grain growth kinetics during SPS, most likely because of some specific features as morphology and surface state of the starting BT nanoparticles, might play a key role in influencing dielectric behavior.

It is worthy to mention that even if the size of the starting BT particles prepared by low-temperature precipitation is smaller (average particle size $<d>$ of only 16 nm), their growth during sintering seems to be much faster, relative to that of the coarser ($<d> = 44$ nm) particles prepared by the modified Pechini method. This finding is in good agreement with the data reported earlier by Li et al. [93] (as already shown in Section 8.2).

Nanostructured ceramics were also produced by SPS from doped BT powders. Data regarding dense $Ba_{1-x}Sr_xTiO_3$ ceramics were reported by several authors, especially for the $Ba_{0.4}Sr_{0.4}TiO_3$ composition [188–192]. As in the case of undoped BT, in fine-grained $Ba_{0.4}Sr_{0.4}TiO_3$ ceramics dielectric permittivity is strongly

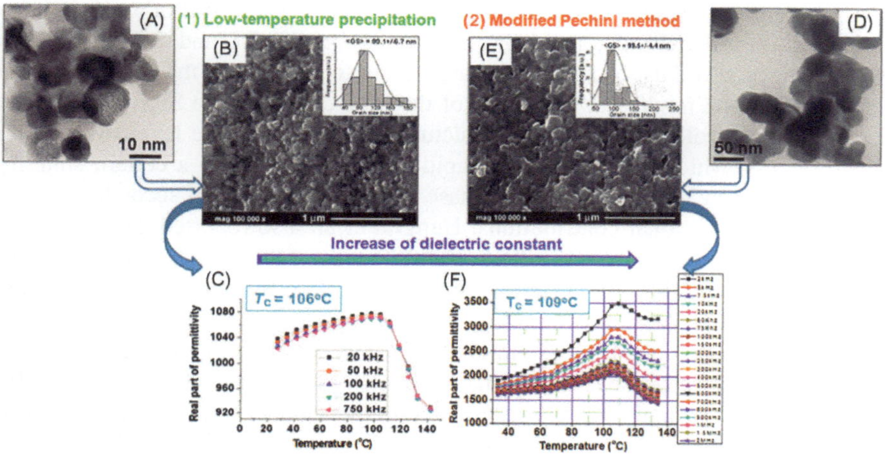

Figure 8.15 Relation between powder morphology–ceramic microstructure–dielectric permittivity in nanoscaled $BaTiO_3$ ceramics with almost similar grain size (<*GS*> = 99 nm), but derived from nanopowders synthesized by different routes: (1) low-temperature precipitation (<*d*> = 16 nm) and (2) modified Pechini method (<*d*> = 44 nm): (A), (D) transmission electron microscopy images images showing the size and morphology of the starting particles (bar = 10 nm in (A) and 50 nm in (D); (B), (E) scanning electron microscopy images showing the microstructure of the related ceramics resulted after spark plasma sintering; insets—grain size distribution histograms and (C), (F) temperature dependence of the relative permittivity; <*d*> represents average particle size; <*GS*> represents average grain size.
(B) and (C) reprinted with permission from L. Curecheriu, S.B. Balmus, V. Nica, M.T. Buscaglia, V. Buscaglia, A. Ianculescu, et al., Grain-size dependent properties of dense nanocrystalline barium titanate ceramics, J. Am. Ceram. Soc. 95 (2012) 3912–3921; (D, E and F) from (unpublished work).

dependent on grain size. Only Gao et al. reported data referring to the electrical properties in a BST ceramic with grain size below 100 nm [188]. As well as for BT ceramics with grain sizes below 50 nm, a broadened temperature-dependent permittivity maximum was obtained, together with small dielectric losses (tan $\delta < 5\%$). This maximum corresponds to the diffuse ferroelectric-paraelectric phase transition, which, for a ceramic sample with an average grain size of 38 nm, takes place at a temperature value of 232K, significantly lower than that specific to a coarse-grained ceramic with the same strontium content ($T_C = 278$K). No anomalies associated with the other two phase transitions (T-O and O-R), specific to the coarse-grained BST ceramics, were detected.

For the same composition, an almost constant value of the real part of permittivity over a large temperature range was reported by Su et al. [189]. However, taking into account that their ceramics consist of grains with sizes in the submicronic, or even in the micrometric range, it is more likely that the temperature-independent evolution of the permittivity is determined only in a lesser extent by grain size effects, but it is especially due to a typical duplex microstructure, with large BST grains surrounded by submicron BT and $SrTiO_3$ grains, as a result of an incomplete reaction during the SPS process between the commercial BT and $SrTiO_3$ powders used as reagents. It must be specified that in order to understand the role of intrinsic size effects on the functional properties of the BST ceramics, the SPS process must be carefully controlled to ensure (1) structuring at the nanoscale level by avoiding kinetic factors, which could lead to a rapid grain growth above a certain sintering temperature; and (2) obtaining single-phase compositions, free of secondary phases able to induce additional contributions. Hungría et al. also reported the lack of the anomaly associated with the FE-PE transition between 40°C and 200°C, for BST ceramics, with variable Ba/Sr ratio produced by SPS from mechanosynthesized nanopowders [193].

Regarding B-site doped BT ceramics, only Maiwa et al. [194,195] reported some results on the dielectric behavior of the $BaTi_{1-x}Zr_xO_3$ ceramics fabricated by SPS. They also found strongly flattened, temperature-dependent permittivity maxima in their samples structured in submicronic range. Experimental data regarding nanostructured BST and BTZ ceramics, particularly those with solute additions into the compositional range specific to the ferroelectric state at room temperature ($x < 0.35$ for Sr and $x < 0.20$) are almost missing in the literature. However, it was concluded that the main characteristics of the homovalently modified nanostructured BT ceramics, like dilution effect exerted by the nonferroelectric grain boundaries on lowering the dielectric permittivity, low dielectric losses, higher temperature stability of the dielectric constant and high tunability, make them promising candidates for microwave applications, even for compositions with lower solute addition, otherwise specific to the ferroelectric state in coarse-grained ceramics.

The electrical behavior in donor doped-BT ceramics produced by SPS was even less investigated. Data regarding properties in a nanostructured BT ceramic (average grain size of 50 nm) doped with a lanthanum amount of 5 at% (far above the critical concentration of 0.5 at% specific to the semiconductor-insulator transition) were reported by Valdez-Nava et al. [196]. The lack of the FE-PE phase transition, the colossal values of the effective dielectric permittivity ($\varepsilon_r \sim 10^6$), as well as the high dielectric losses at ambient temperature in the low frequency range in their nanostructured $Ba_{0.95}La_{0.05}TiO_{3-x}$ ceramics were explained in terms of the conduction mechanisms involving small polarons associated to the presence of Ti^{3+} ions in the grain boundary regions. No data referring to the nanoscale switching phenomena, preservation, or change of the semiconducting behavior and PTCR effect in nanostructured donor doped-BT ceramics with dopant content below the critical concentration were reported.

8.3.4 Undoped and homovalently doped multilayer $BaTiO_3$ thin films

8.3.4.1 Multilayer $BaTiO_3$ thin films prepared by RF-magnetron sputtering

As already mentioned, the main issues in BT thin films are related to the dependence of the electrical properties on thickness and on the characteristics of the bottom electrode, respectively. The thinner the film, the more significant the influence of the substrate. Electrodes could strongly affect ferroelectric switching phenomena by pinning the domain walls. In order to improve ferroelectric and dielectric behavior in BT thin films deposited by RF-magnetron sputtering, an innovative procedure, implying a multistep deposition–annealing cycle, was proposed. In the cross-section microstructurally graded films as-prepared, the gradual decrease of grain size from the bottom region, adjacent to the substrate, toward the top surface, has many beneficial effects: (1) facilitates ferroelectric switching by minimizing the influence of the bottom electrode; this is due to the presence of some larger grains, in the proximity of the film-electrode interface region, whose increased sizes, suitable for domain walls movement, have been induced by the successive thermal treatments suffered by the internal layer; (2) ensures higher compactness and a corresponding decrease of the overall porosity because of the grain size gradient, which means that the increasingly smaller grains of each newly deposited layer, relative to the previous one, fill the pores in the interface region of the two adjacent layers; this induces lower dielectric losses and higher permittivity values; and (3) reduces the roughness of the surface, due to the low size of grains composing the last deposited layer [197].

Two single-deposit polycrystalline films, with thickness of 0.6 and 1 μm, respectively, were fabricated by RF-magnetron sputtering. In order to investigate the influence of the number of deposits on the dielectric response, the thickness of 0.6 μm was kept as the constant parameter and a four-deposit film with this thickness was also elaborated by the new procedure already mentioned. A schematic illustration of this approach is presented in Fig. 8.16. In this way, we can have a comparative picture of the separate influence of the two parameters, the thickness and number of layers on the microstructure, surface topography, and dielectric behavior of the as-prepared BT thin films. As expected, in the case of single-deposit BT thin films, the lowering of thickness from 1 to 0.6 μm, induces the

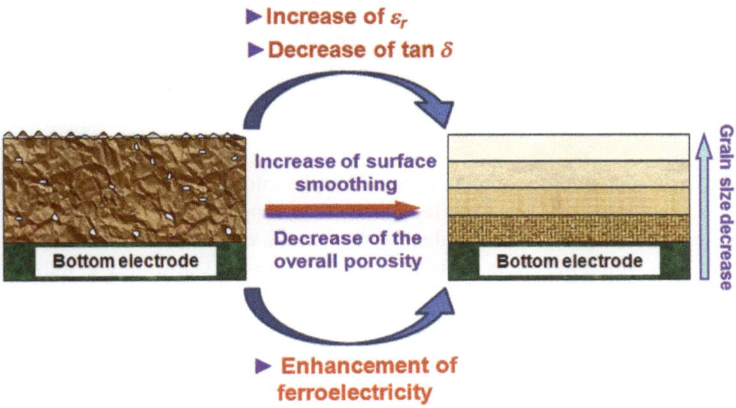

Figure 8.16 Schematic illustration of the new approach, consisting in preparation of multilayer microstructurally graded $BaTiO_3$ thin films.

diminishing of the grain size and consequently, the decrease of the dielectric permittivity over the whole temperature range investigated (Fig. 8.17A–D). For the multilayer (four-deposit) film of similar thickness (0.6 μm), the dielectric response is improved, even if, apparently, a smaller grain size was obtained (Fig. 8.17E and F). It has to be specified that the 2D AFM images show only the microstructure of the surface layer. Therefore, the rise of the dielectric permittivity (which approaches the permittivity value of the thicker single-deposit film) and the decrease of the dielectric loss (not shown here) [197] can be explained in terms of a higher microstructural compactness and also a higher overall crystallinity induced by the cross-section gradient of grain size. For the single-deposit thin films, irrespective of their thickness, an almost temperature-independent permittivity, with no typical anomaly related to the FE-PE phase transition, was noticed in the temperature range of 30–90°C (Fig. 8.17B and D). In the case of the four-deposit film, the cross-section microstructual gradient induces a more complicated temperature dependence of permittivity, with two anomalies, a significant one at ~50°C and the second, much less marked, at ~70°C.

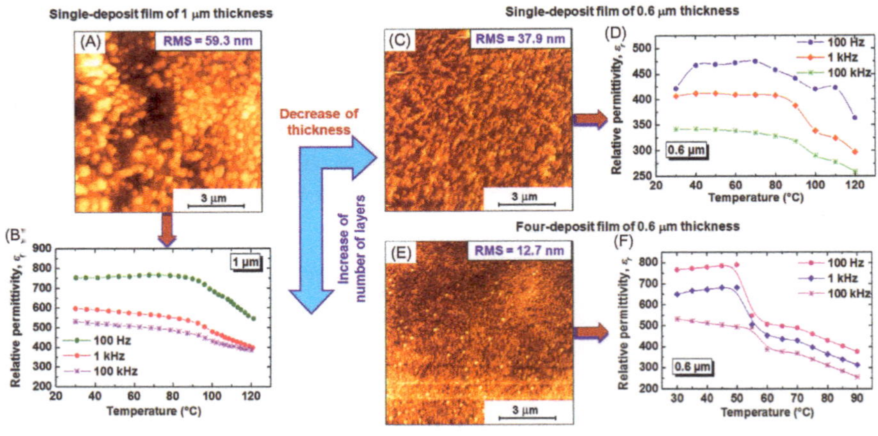

Figure 8.17 Representation of complex correlations thickness–number of deposits–surface topography–dielectric response in multilayered $BaTiO_3$ thin films prepared by a multistep rf-sputtering procedure: (A), (B) single-deposit film with a thickness of 1 μm; (C), (D) single-deposit film with a thickness of 0.6 μm; (E), (F) four-deposit film with a thickness of 0.6 μm. (A), (C), (E) 2D atomic force microscopy images (bar = 3 μm) and (B), (D), (F) temperature dependence of relative permittivity. In the top-right corner, values of the root mean square roughness (RMS) are specified on the 2D-AFM images.
Adapted from A. Ianculescu, B. Despax, V. Bley, Th. Lebey, R. Gavrilă, N. Drăgan, Structure–properties correlations for barium titanate thin films obtained by RF-sputtering, J. Eur. Ceram. Soc. 27(2–3) (2007) 1129–1135.

8.3.4.2 Multilayer $Ba(Ti,Zr)O_3$ thin films prepared by the sol-gel method

Numerous literature data refer to homogeneous and compositionally graded (Ba,Sr) TiO_3 films due to their use in DRAMs, tunable microwave applications, as well as sensing applications [198–203]. Fewer papers describing correlation between the structure, microstructure, and functional properties in Zr-doped BT (BTZ) films were found. Most of these refer either to BTZ thin films with low Zr content ($x = 0.05$) [204], or to those of higher Zr concentration ($0.20 \leq x \leq 0.50$) [205–208]. A study about the preparation and characteristics of multilayered $BaTi_{0.85}Zr_{0.15}O_3$ (BTZ15) thin films was carried out by Vasilescu et al. [209]. They found that, as in the case of the undoped BT thin films already discussed, the increase of the deposit number in BTZ films also induces a smoothing of the surface, so that a lower roughness value was found for the six-deposit BTZ15 film, relative to that specific to a four-deposit film processed in similar conditions. The higher the layers number, the more obvious the trend toward a broad, but unimodal grain size distribution. Thus, while the four-deposit BTZ15 film showed a bimodal grain size distribution consisting of larger grains of 93 nm, which coexist with smaller grains of ~36 nm, the six-deposit BTZ film exhibits an average grain size of 66 nm and a more uniform surface topography (Fig. 8.18A and C).

The frequency response in the BTZ15 films analyzed here shows a power law dependence specific to highly disordered structures (Fig. 8.18B and D) and obviously different from the pure Debye resonances displayed by the bulk BT. This

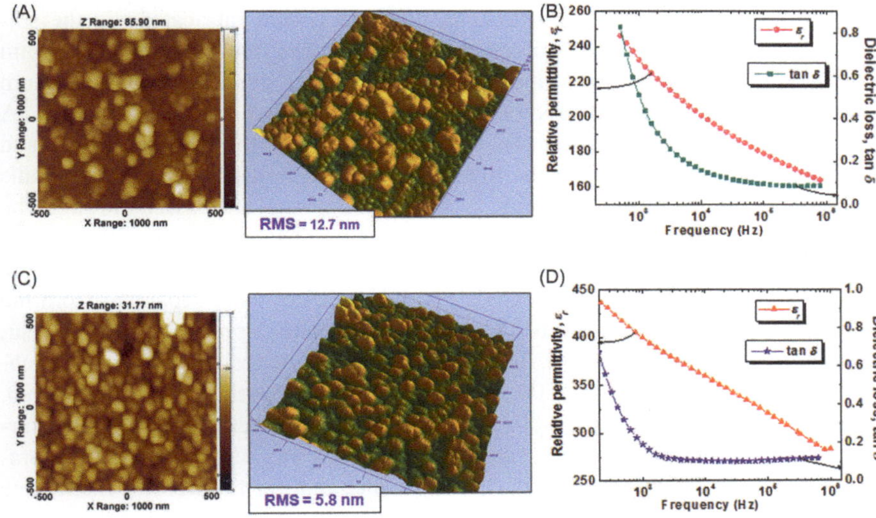

Figure 8.18 (A), (C) 2D- and 3D-atomic force microscopy (AFM) images and (B), (D) dielectric properties of the sol-gel BTZ15 thin films: (A), (B) a four-deposit film and (C), (D) six-deposit film [209]. In the bottom-left corner values of root mean square roughness are specified on the 3D-AFM images.

behavior has been attributed to the inhomogeneities resulting from either the presence of boundaries or point defects, which modify the local charge density. For both thin films, the high dielectric losses recorded especially in the low frequency range were attributed to the contribution of the grain boundary phenomena (Fig. 8.18B and C). The values of the dielectric losses are stabilized and remain almost constant ($\tan\delta = 0.1$) at a frequency of $\sim 10^5$ Hz for the four-deposit film, and at a significantly lower frequency of 2×10^3 Hz for the six-deposit film. We can assume that the higher permittivity values over all the investigated frequency range exhibited by the six-deposit BTZ15 film, relative to the four-deposit one, originate in its higher thickness and increased microstructural uniformity.

Regarding donor-doped systems, papers reporting data on the influence of La^{3+} and Ce^{3+} on Ba sites and Nb^{5+} on Ti-sites on the characteristics of BT-based thin films were also found in the literature [210–214]. It was shown that the structure, microstructure, and electrical properties strongly depended on the processing route and parameters, substrate type, and dopant content.

8.3.5 Donor-doped $BaTiO_3$ one-dimensional nanostructures prepared by template-mediated colloidal chemistry

Although the structure, morphology, and functional properties of pure BT nanowires and nanotubes are well described in some of the papers already mentioned as references in Section 8.2.4, however the reports on such of donor-doped 1D BT structures are almost missing.

Five mol% Ce^{3+}-doped barium titanate nanowires were fabricated via the sol-gel method using as a template polycarbonate membranes with channels of 100 nm diameter, thickness of ~ 17 μm, and pore density of 10^8 pores/cm^2 [168]. Barium acetate, titanium isopropoxide, and cerium acetate were used as starting reagents. A stable sol of 0.3 M concentration was prepared by mixing the starting Ba, Ti, and Ce solutions. The sol poured on the membrane surface moistened the pores' walls and filled the channels. It is worthy to mention that the sol passed through the membrane pores only by infiltration, without spinning. The membrane with sol was kept for three days in air in order to transform the sol to gel. Then, the membrane was dissolved in dichloromethane (CH_2Cl_2), the gel wires were separated from the solution through evaporation at 100°C, and washed with 2-propanol until a clear, transparent solution was obtained. The dried wires were deposited on the Si-Pt substrate and calcined in air, at 700°C for 1 h; with a heating rate of 5°C/min. Crystallized wires of $Ba_{0.95}Ce_{0.05}Ti_{0.9875}O_3$ composition were obtained after furnace cooling at room temperature. The same procedure was adopted for the elaboration of Ce-doped BT nanotubes with the difference that, in this case, a membrane with channels of 200 nm diameter was used and the storage time of the sol into membrane pores was only 2 h [215].

FE-SEM analyses showed that continuous, uniform, and slightly bent gel wires of length up to 17 μm and an average diameter of 81.1 (±3.2) nm, as well as amorphous nanoshell tubes of similar length, exhibiting an average outer diameter of

188.6 (±6.5) nm and a wall thickness of 15.1 (±1.1) nm, were obtained after the removal of polycarbonate membrane used as template (Fig. 8.19A–F). High aspect ratios, *L*/*D*, of ~210, in the case of nanowires, and of ~90, for the nanoshell tubes were calculated taking into account the values of length (*L*), average diameter $<D>$ for nanowires, and average outer diameter $<D_{outer}>$ for nanoshell tubes.

These amorphous 1D structures were converted into polycrystalline, single-phase products by calcination in the mentioned conditions. We can observe that after calcination the initial 1D morphology is preserved. The thermal treatment determines a stiffening process and an obvious contraction, so that rigid nanowires with an average diameter of 59.7 nm (Fig. 8.20A–C) as well as nanoshell tubes with an average outer diameter of 157.4 (±7.0) were noticed (Fig. 8.20D–F). The low magnification FE-SEM images of Fig. 8.20A and D) show bundles of Ce^{3+}-doped BT nanowires and nanoshell tubes which maintained the vertically aligned ordering even after thermal treatment. The detail views presented in the FE-SEM images of the highest magnification (×200,000) show that, actually, due to crystallization, the wire surface as well as the outer and the inner surfaces of the nanoshell tubes are not smooth anymore, but rough, suggesting that the calcined 1D nanostructures possess a granular microstructure (Fig. 8.20C and F). In the case of nanowires, taking into account the estimated value of the average diameter, the wire thickness in cross-section seems to consist of rows of 2–3 tightly-packed, equiaxial (especially spherical) grains, with sizes of 20–30 nm, which sometimes coalesce into single grains with sizes almost similar with the diameter of the wires, due to a partial pre-sintering process (Fig. 8.20B and C). For the nanoshell tubes, FE-SEM images

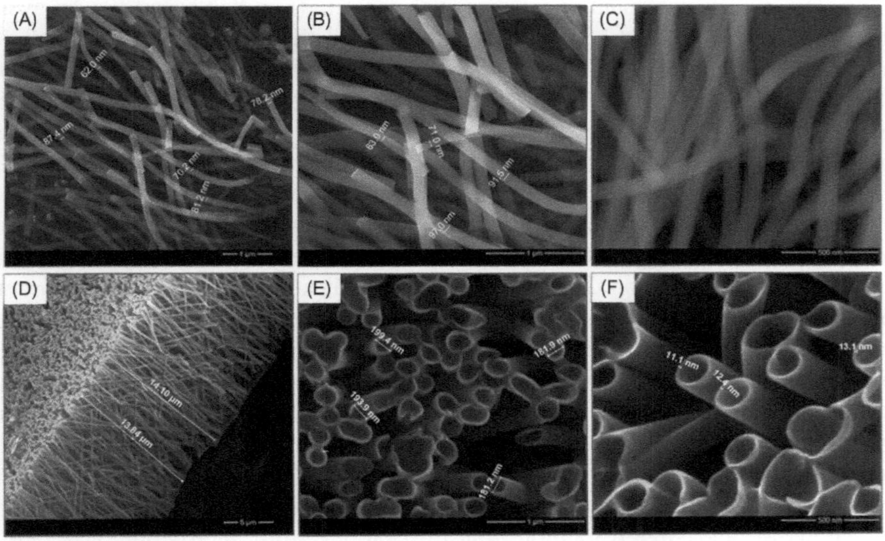

Figure 8.19 Field emission-scanning electron microscopy (FE-SEM) images of different magnifications of amorphous 1D nanostructures: (A), (B), (C) nanowires; and (D), (E), (F) nanoshell tubes [168,215].

Figure 8.20 Ferroelectric-scanning electron microscopy images of different magnifications of polycrystalline Ce^{3+}-doped $BaTiO_3$ 1D nanostructures resulted after calcination at 700°C for 1 h: (A), (B), (C) nanowires and (D), (E), (F) nanoshell tubes; inset of (F) is a detailed view of the extremity of a tube, showing the hollow nature of this 1D structure [168,215].

clearly indicate a fine-grained (in the nanometer range) microstructure of the tube walls, with a certain amount of intergranular porosity (Fig. 8.20E and F). An average grain size of 43.4 ($\pm$1.9) nm was estimated in this case. The polycrystalline nature of these tubes made an accurate estimation of the wall thickness difficult. The FE-SEM image of Fig. 8.20E and the inset of Fig. 8.20F seem to suggest that the wall thickness of the polycrystalline tubes consists of a single grain layer and, therefore, it can be roughly considered equivalent to the average grain size. Consequently, an average inner diameter of 114 ($\pm$5.2) nm was estimated.

The XRD investigations indicated for both 1D products a single-phase composition, with the perovskite structure of BT involving a low tetragonality of the unit cells. Raman investigation completed the view, indicating at room temperature an intricate structure, which most likely consists of tetragonal and orthorhombic modifications. For the nanoshell tubes, the evolution of Raman spectra versus temperature emphasized diffuse phase transitions, and, as a specific feature, a stable polar order, which persists even at temperatures far above 100°C. This polar state induced by the confined geometry, generating microstrains and subsequent structural distortions, hinders the effect of cerium content in inducing RT FE-PE phase transition, which normally occurs in microstructured ceramics with similar composition.

The TEM and high-resolution transmission electron microscopy (HRTEM) images, as well as the SAED pattern of the Ce^{3+}-doped BT 1D structures, indicate more clearly their nanoscale granular structure and high crystallinity (Figs. 8.21A–F and 8.22A–D). In the case of nanoshell tubes, lower and higher

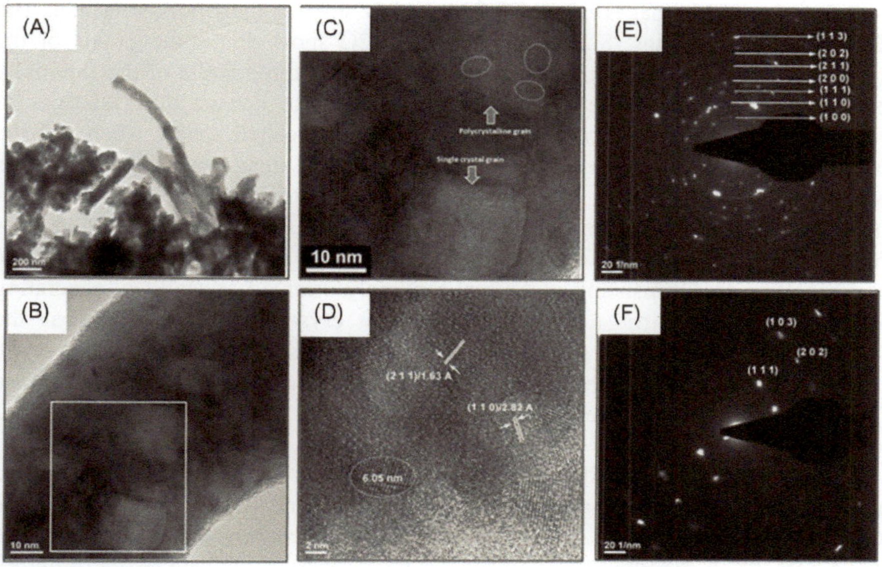

Figure 8.21 Transmission electron microscopy (TEM)/high-resolution TEM (HRTEM) images and selected area electron diffraction (SAED) patterns of 5 mol% Ce^{3+}-doped barium titanate nanowires calcined at 700°C for 1 h: (A) low magnification TEM image; (B) and (C) higher magnification TEM images ((C) represents the area marked in (B) by the white-outlined rectangle; the white ellipses in (C) mark areas corresponding to several crystallites with dissimilar orientation); (D) HRTEM image inside a polycrystalline grain; (E) SAED pattern of a polycrystalline area; and (F) SAED pattern of an area with local ordering in the same direction of some adjacent crystalline grains [168].

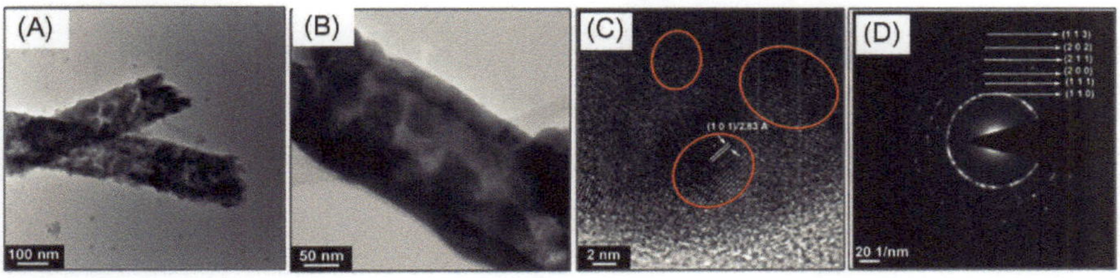

Figure 8.22 Transmission electron microscopy (TEM)/high-resolution TEM (HRTEM) images and selected area electron diffraction (SAED) patterns of 5 mol% Ce^{3+}-doped barium titanate nanoshell tubes calcined at 700°C for 1 h: (A) lower magnification TEM image; (B) higher magnification TEM image, showing the fine-grained microstructure of the tube wall; (C) HRTEM image inside a polycrystalline grain consisting of randomly oriented crystallites marked by red ellipses; and (D) SAED pattern of a polycrystalline area [215].

magnification TEM images of Fig. 8.22A and B suggest their hollow aspect. The HRTEM images of Figs. 8.21C and D and 8.22C show that some grains of the nanowires, as well as those that build up the nanostructured walls of the thermally treated tubes, are not single crystals, but they seem to consist of crystallites of ~4–7 nm with different orientations. Unlike the nanoshell tubes, where all the grains investigated were polycrystalline, as the HRTEM images and SAED patterns revealed (Fig. 8.22C and D), in the nanowires the structuring is even more complicated, showing the coexistence of single crystal grains with polycrystalline ones (Fig. 8.21C). Thus, for nanowires, the SAED pattern of Fig. 8.21E corresponds to a polycrystalline area consisting of several randomly oriented crystallites, generating concentric diffraction rings specific to different crystalline planes, whereas the SAED pattern of Fig. 8.21F shows well-aligned bright spots, each of them specific to a crystalline plane, as in the case of a single crystal or, most likely, as in the case of a local ordering in the same direction of some crystalline grains specific to 1D mesocrystals [216].

The piezoelectric and ferroelectric character of the Ce^{3+}-doped BT nanowires and nanoshell tubes was investigated by PFM (Figs. 8.23 and 8.24). The butterfly-type strain loops of Figs. 8.23A and 8.24A and the ferroelectric hysteresis cycles of Figs. 8.23B and 8.24B confirm the piezoelectric properties, as well as the presence of switchable domains in Ce^{3+}-doped BT nanowires and nanoshell tubes. In the case of the nanoshell tube investigated here (insets of Fig. 8.24A and B), the imprint of both loops along the positive side of the voltage axis indicates the presence of a strong internal bias field due to the mechanical tensile stress or charge trapping (Fig. 8.24A and B). Stress-induced polarization along the positive direction of bias might be explained in terms of the flexoelectric effect, where the stress gradient, rather than the stress itself, affected the domain structure [217,218]. For the nanowire of similar composition (insets of Fig. 8.23A and B) only the imprint of the

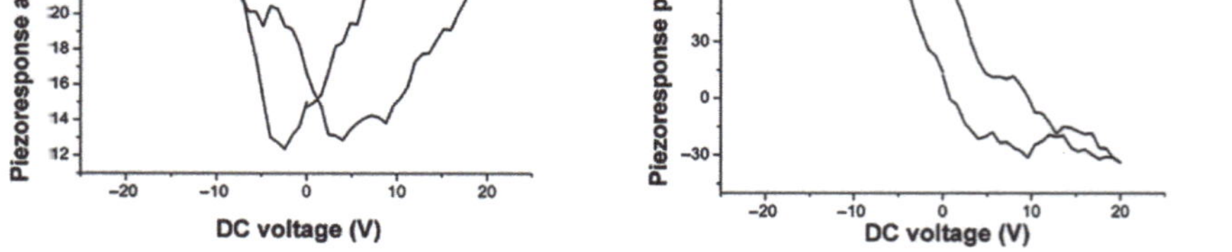

Figure 8.23 Local piezoresponse force microscopy hysteresis loops measured as a function of the applied DC field on 5 mol% Ce^{3+}-doped barium titanate nanowires: (A) amplitude signal— inset, amplitude image; and (B) phase signal—inset, phase image [168].

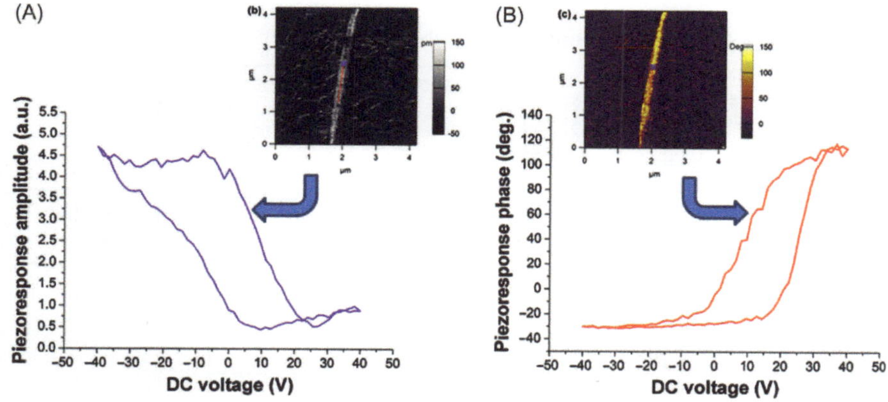

Figure 8.24 Local piezoresponse force microscopy hysteresis loops measured as a function of the applied DC field on 5 mol% Ce^{3+}-doped barium titanate nanoshell tubes: (A) amplitude signal—inset, amplitude image; and (B) phase signal—inset, phase image [215].

ferroelectric cycle was observed, while imprint in the butterfly loop of the piezoresponse amplitude signal is almost missing [168] (Fig. 8.23A and B), indicating that the restrictive tubular geometry might play a key role in generating the flexoelectric effect.

8.4 Conclusions and trends

Tailoring and controlling functional properties of BT-based products by an adequate microstructural design and by a careful choice of doping strategy represents a challenging and hot topic in the domain of chemistry, physics, and engineering of electroceramics. The change of structuring from microscale toward nanoscale in BT-based bulk ceramics leads to new functionalities, while elaboration of products with confined geometry such as nanowires, nanotubes, and thin films ensures high integrability degree, providing new alternatives for the manufacture of nanoactuators, transducers, ultrasonic devices, and memories. The control of microstructure and the effectiveness of the doping process can be achieved by adopting suitable synthesis routes and by optimizing processing parameters. Innovative preparation methods, alternative sintering procedures, and sophisticated investigation techniques were developed for the elaboration and complex characterization of nanosized and/or nanostructured BT products.

In order to establish correlation between structure, microstructure, and electrical behavior, it is absolutely necessary to understand the effects exerted by the nanoscale phenomena and by many other extrinsic factors that could affect, or modify the dielectric and ferroelectric response in BT nanostructured materials.

In the case of the compositionally modified BT systems, to quantify the influence of intrinsic versus extrinsic contributions on the functional properties, a more

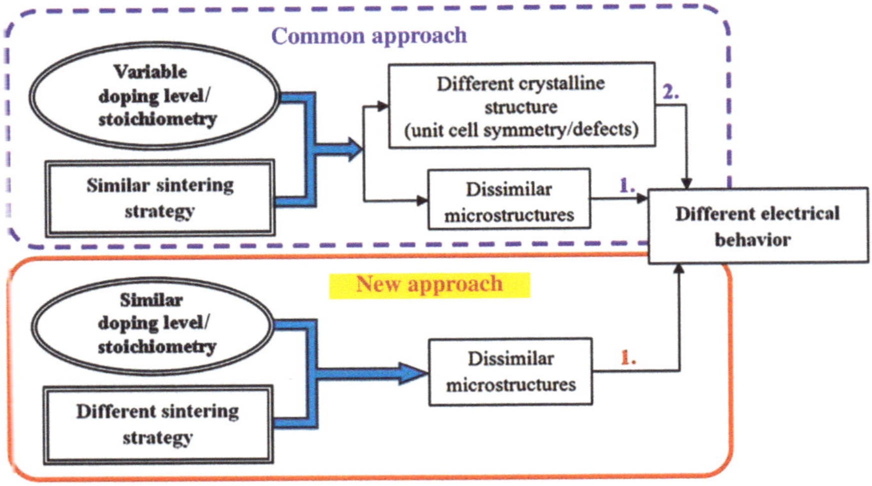

Figure 8.25 Schematic illustration of new approach versus common approach in analyzing functional properties in multiscale compositionally-modified $BaTiO_3$ systems.

logical approach consisting of a comparative analysis of the effect of grain size as a single variable induced by different sintering strategies in products with a fixed solute concentration have to be taken into account against the common approach. The common approach generally proposes to analyze and describe the influence of various solute contents via structural changes in ceramics sintered in similar conditions, but in which dissimilar microstructures, arising from additional effects induced by different doping levels, act as a secondary variable. A comparative and conclusive view representing our own approach in analyzing the functional properties of compositionally modified BT systems, which can be also extended in the analysis of other multiscale ferroelectrics, is schematically illustrated in Fig. 8.25.

Acknowledgements

This work was supported by a grant from the Romanian National Authority for Scientific Research, CNCS-UEFISCDI, Project No. PN-III-P4-ID-PCE-2016-0072 (contract no. **154/ 2017**).

References

[1] Y. Saito, H. Takao, T. Tani, T. Nonoyama, K. Takatori, T. Homma, et al., Lead-free piezoceramics, Nature 432 (2004) 84–87.

[2] J. Rödel, W. Jo, K.T.P. Seifert, E.-M. Anton, T. Granzow, D. Damjanovic, Perspective on the development of lead-free piezoceramics, J. Am. Ceram. Soc. 92 (6) (2009) 1153–1177.
[3] T. Tsurumi, T. Hoshina, H. Takeda, Y. Mizuno, H. Chazono, Size effect of barium titanate and computer-aided design of multilayered ceramic capacitors, IEEE Trans. Ultrason. Ferroelectrics Freq. Contr. 56 (8) (2009) 1513–1522.
[4] Z. Tian, X. Wang, L. Shu, T. Wang, T.H. Song, Z. Gui, et al., Preparation of nano $BaTiO_3$-based ceramics for multilayer ceramic capacitor application by chemical coating method, J. Am. Ceram. Soc. 92 (4) (2009) 830–833.
[5] T.G. Reynod III, Application space influences electronic *ceramic* materials, Am. Ceram. Soc. Bull. 80 (10) (2001) 29–33.
[6] C.A. Randall, Scientific and engineering issues of the state-of-the-art and future multilayer capacitors, J. Ceram. Soc. Jpn. 109 (2001) S2–S6.
[7] A.J. Moulson, J.M. Herbert, Electroceramics: Materials, Properties, Applications, second ed., Wiley & Sons Ltd, West Sussex, England, 2003.
[8] K. Uchino, Ferroelectric Devices, Marcel Dekker, Inc, New York, USA, 2000.
[9] L. Zhong, Y.G. Wang, P.L. Zhang, B.D. Qu, Phenomenological study of the size effect on phase transitions in ferroelectric particles, Phys. Rev. B 50 (2) (1994) 698–703.
[10] S. Li, J.A. Eastman, Z. Li, C.M. Foster, R.E. Newnham, L.E. Cross, Size effects in nanostructured ferroelectrics, Phys. Lett. A 212 (6) (1996) 341–346.
[11] K. Uchino, E. Sadanaga, T. Hirose, Dependence of the crystal structure on particle size in barium titanate, J. Am. Ceram. Soc. 72 (1989) 1555–1558.
[12] L.K. Templeton, J.A. Pask, Formation of $BaTiO_3$ from $BaCO_3$ and TiO_2 in air and in CO_2, J. Am. Cerm. Soc. 42 (5) (1959) 212–216.
[13] D.F.K. Hennings, B.S. Schreinemacher, H. Schreinemacher, Solid-state preparation of $BaTiO_3$-based dielectrics, using ultrafine raw materials, J. Am. Cerm. Soc. 84 (12) (2001) 2777–2782.
[14] M. Boulos, S. Guillemet-Fritsch, F. Mathieu, B. Durand, Th Lebey, V. Bley, Hydrothermal synthesis of nanosized $BaTiO_3$ powders and dielectric properties of corresponding ceramics, Solid State Ionics 176 (2005) 1301–1309.
[15] L. Qi, B.I. Lee, P. Badheka, D.-H. Yoon, W.D. Samuels, G.J. Exarhos, Short-range dissolution–precipitation crystallization of hydrothermal barium titanate, J. Eur. Ceram. Soc. 24 (2004) 3553–3557.
[16] X. Wei, G. Xu, Z. Ren, Y. Wang, G. Shen, G. Han, Synthesis of highly dispersed barium titanate nanoparticles by a novel solvothermal method, J. Am. Ceram. Soc. 91 (1) (2008) 315–318.
[17] Y.-I. Kim, K.-S. Ryu, S.-H. Nahm, J.-S. Park, Structure and electron charge-density analysis of nano-sized $BaTiO_3$ powder prepared by solvothermal method, Curr. Appl. Phys. 6S1 (2006) 266–270.
[18] M. Viviani, M.T. Buscaglia, A. Testino, V. Buscaglia, P. Bowen, P. Nanni, The influence of concentration on the formation of $BaTiO_3$ by direct reaction of $TiCl_4$ with $Ba(OH)_2$ in aqueous solution, J. Eur. Ceram. Soc. 23 (2003) 1383–1390.
[19] S. Gablenz, H.-P. Abicht, E. Pippel, O. Lichtenberger, J. Woltersdorf, New evidence for an oxycarbonate phase as an intermediate step in $BaTiO_3$ preparation, J. Eur. Ceram. Soc. 20 (2000) 1053–1060.
[20] W.-S. Jung, B.-K. Min, J. Park, D.-H. Yoon, Formation mechanism of barium titanate by thermal decomposition of barium titanyl oxalate, Ceram. Int. 37 (2011) 669–672.
[21] L.A. Pérez-Maqueda, M.J. Diánez, F.J. Gotor, M.J. Sayagués, C. Real, J.M. Criado, Synthesis of needle-like $BaTiO_3$ particles from the thermal decomposition of a citrate

precursor under sample controlled reaction temperature conditions, J. Mater. Chem. 13 (2003) 2234–2241.
[22] K.S. Martirosyan, D. Nawarathna, J.R. Claycomb, J.H. Miller Jr., D. Luss, Complex dielectric behavior during formation of $BaTiO_3$ by combustion synthesis, J. Phys. D: Appl. Phys. 39 (2006) 1–6.
[23] A.V. Komarov, I.P. Parkin, New routes in the self-propagating high-temperature synthesis of barium titanium oxide, Polyhedron 15 (8) (1996) 1349–1353.
[24] Y. Ma, E. Vileno, S.L. Suib, P.K. Dutta, Synthesis of tetragonal $BaTiO_3$ by microwave heating and conventional heating, Chem. Mater. 9 (1997) 3023–3031.
[25] W.-S. Cho, Structural evolution and characterization of $BaTiO_3$ nanoparticles synthesized from polymeric precursor, J. Phys. Chem. Solids 59 (5) (1998) 659–666.
[26] V. Vinothini, P. Singh, M. Balasubramanian, Synthesis of barium titanate nanopowder using polymeric precursor method, Ceram. Int. 32 (2006) 99–103.
[27] M. Deluca, C.A. Vasilescu, A.C. Ianculescu, D.C. Berger, C.E. Ciomaga, L.P. Curecheriu, et al., Investigation of the composition-dependent properties of $BaTi_{1-x}Zr_xO_3$ ceramics prepared by the modified Pechini method, J. Eur. Ceram. Soc. 32 (13) (2012) 3551–3566.
[28] A. Ianculescu, D. Berger, L. Mitoşeriu, L.P. Curecheriu, N. Drăgan, D. Crişan, et al., Properties of $Ba_{1-x}Sr_xTiO_3$ ceramics prepared by the modified-Pechini method, Ferroelectrics 369 (1) (2008) 22–34.
[29] A. Ianculescu, D. Berger, M. Viviani, C. Ciomaga, L. Mitoseriu, E. Vasile, et al., Investigation of $Ba_{1-x}Sr_xTiO_3$ ceramics prepared from powders synthesized by the modified Pechini route, J. Eur. Ceram. Soc. 27 (13-15) (2007) 3655–3658.
[30] R. Kavian, A. Saidi, Sol-gel derived $BaTiO_3$ nanopowders, J. Alloy Compd. 468 (2009) 528–532.
[31] X. Xing, J. Deng, J. Chen, G. Liu, Phase evolution of barium titanate from alkoxide gel-derived precursor, J. Alloy Compd. 384 (2004) 312–317.
[32] H.P. Beck, W. Eiser, R. Haberkorn, Pitfalls in the synthesis of nanoscaled perovskite type compounds. Part I: Influence of different sol-gel preparation methods and characterization of nanoscaled $BaTiO_3$, J. Eur. Ceram. Soc. 21 (2001) 687–693.
[33] A.C. Ianculescu, C.A. Vasilescu, M. Crisan, M. Raileanu, B.S. Vasile, M. Calugaru, et al., Formation mechanism and characteristics of lanthanum-doped $BaTiO_3$ powders and ceramics prepared by the sol-gel process, Mater. Charact. 106 (2015) 195–207.
[34] A. Ianculescu, A. Brăileanu, M. Crişan, P. Budrugeac, N. Drăgan, G. Voicu, et al., Influence of barium source on the characteristics of sol-precipitated $BaTiO_3$ powders and related ceramics, J. Therm. Anal. Calorim. 88 (1) (2007) 251–260.
[35] S. Yoon, S. Baik, M.G. Kim, N. Shin, Formation mechanisms of tetragonal barium titanate nanoparticles in alkoxide–hydroxide sol-precipitation synthesis, J. Am. Ceram. Soc. 89 (6) (2006) 1816–1821.
[36] J. Wang, J. Fang, S.-C. Ng, L.-M. Gan, C.-Hr Chew, X. Wang, et al., Ultrafine barium titanate powders via microemulsion processing routes, J. Am. Ceram. Soc. 82 (4) (1999) 873–881.
[37] Y. Sakabe, Y. Yamashita, H. Yamamoto, Dielectric properties of nano-crystalline $BaTiO_3$ synthesized by micro-emulsion method, J. Eur. Ceram. Soc. 25 (2005) 2739–2742.
[38] Y. Zhang, L. Wang, D. Xue, Molten salt route of well dispersive barium titanate nanoparticles, Powder Technol. 217 (2012) 629–633.
[39] B.D. Stojanović, A.Z. Simoes, C.O. Paiva-Santos, C. Jovalekić, V.V. Mitić, J.A. Varela, Mechanochemical synthesis of barium titanate, J. Eur. Ceram. Soc. 25 (2005) 1985–1989.

[40] C. Gomez-Yañez, C. Benitez, H. Balmori-Ramirez, Mechanical activation of the synthesis reaction of $BaTiO_3$ from a mixture of $BaCO_3$ and TiO_2 powders, Ceram. Int. 26 (2000) 271–277.
[41] A. Purwanto, W.-N. Wang., I.W. Lenggoro, K. Okuyama, Formation of $BaTiO_3$ nanoparticles from an aqueous precursor by flame-assisted spray pyrolysis, J. Eur. Ceram. Soc. 27 (2007) 4489–4497.
[42] D.S. Jung, S.K. Hong, J.S. Cho, Y.C. Kang, Nano-sized barium titanate powders with tetragonal crystal structure prepared by flame spray pyrolysis, J. Eur. Ceram. Soc. 28 (2008) 109–115.
[43] M. Xu, Y.-N. Lu, Y.-F. Liu, S.-Z. Shi, T.-S. Qian, D.-Y. Lu, Sonochemical synthesis of monosized spherical $BaTiO_3$ particles, Powder Technol. 161 (2006) 185–189.
[44] J.M. McHale, P.C. McIntyre, K.E. Sickafus, N.V. Coppa, Nanocrystalline $BaTiO_3$ from freeze-dried nitrate solutions, J. Mater. Res. 11 (5) (1996) 1199–1209.
[45] V. Bansal, P. Poddar, A. Ahmad, M. Sastry, Room-temperature biosynthesis of ferroelectric barium titanate nanoparticles, J. Am. Chem. Soc. 128 (2006) 11958–11963.
[46] A. Polotai, K. Breece, E. Dickey, C. Randall, A. Novel, Approach to sintering nanocrystalline barium titanate ceramics, J. Am. Ceram. Soc. 88 (11) (2005) 3008–3012.
[47] A. Ianculescu, D. Berger, C. Matei, P. Budrugeac, L. Mitoseriu, E. Vasile, Synthesis of $BaTiO_3$ by soft chemistry routes, J. Electroceram. 24 (2010) 46–50.
[48] (a) J.O. Eckert Jr., C.C. Hung-Houston, B.L. Gersten, M.M. Lencka, R.E. Riman, Kinetics and mechanisms of hydrothermal synthesis of barium titanate, J. Am. Ceram. Soc. 79 (11) (1996) 2929–2939.
(b) Y. Mao, S. Mao, et al., Solvothermal synthesis and Curie temperature of monodispersed barium titanate nanoparticles, Mater. Chem. Phys. 124 (2010) 1232–1238.
(c) D. Caruntu, T. Rostamzadeh, T. Costanzo, S.S. Parizi, G. Caruntu, Solvothermal synthesis and controlled self-assembly of monodisperse titanium-based perovskite colloidal nanocrystals, Nanoscale 7 (2015) 12955–12969.
[49] Adelina Ianculescu, Sophie Guillemet-Fritsch, Bernard Durand, Ana Brăileanu, Maria Crişan and Daniela Berger, Chap. IV. $BaTiO_3$ nanopowders and nanocrystalline ceramics: I. Nanopowders, in: "New Developments in Advanced Functional Ceramics", Ed. Liliana Mitoşeriu, Transworld Research Network India, 2007, p. 89–118.
[50] D. Hennings, S. Schreinemacher, Characterization of hydrothermal harium hitanate, J. Eur. Ceram. Soc. 9 (1992) 41–46.
[51] D.F.K. Hennings, Ch Metzmacher, B.S. Schreinemacher, Defect chemistry and microstructure of hydrothermal barium titanate, J. Am. Ceram. Soc. 84 (1) (2001) 179–182.
[52] H. Yamamura, A. Watanabe, S. Shirasaki, Y. Moriyoshi, M. Tanada, Preparation of barium titanate by oxalate method in ethanol solution, Ceram. Int. 11 (1985) 17–22.
[53] T.-T. Fang, H.-B. Lin, J.-B. Hwang, Thermal analysis of precursors of barium titanate prepared by coprecipitation, J. Am. Ceram. Soc. 73 (11) (1990) 3363–3367.
[54] T.-T. Fang, H.-B. Lin, Factors affecting the preparation of barium titanyl oxalate tetrahydrate, J. Am. Ceram. Soc. 72 (10) (1989) 1899–1906.
[55] F.V. Motta, A. Paula, A. Marques, C.A. Paskocimas, M.R.D. Bomio, A.S.F. Santos, et al., Elson Longo, Chap. 12. synthesis of $Ba_{1-x}Ca_xTiO_3$ by complex polymerization method (CPM), Polymerization, Ailton De Souza Gomes, InTech, 2012, pp. 261–278.
[56] M.T. Buscaglia, M. Bassoli, V. Buscaglia, R. Alessio, Solid-state synthesis of ultrafine $BaTiO_3$ powders from nanocrystalline $BaCO_3$ and TiO_2, J. Am. Ceram. Soc. 88 (9) (2005) 2374–2379.
[57] M.T. Buscaglia, V. Buscaglia, R. Alessio, Coating of $BaCO_3$ crystals with TiO_2: versatile approach to the synthesis of $BaTiO_3$ tetragonal nanoparticles, Chem. Mater. 19 (2007) 711–718.

[58] M.T. Buscaglia, V. Buscaglia, M. Viviani, G. Dondero, S. Röhrig, Andreas Rüdiger, et al., Ferroelectric hollow particles obtained by solid-state reaction, Nanotechnology 19 (2008) 225602.
[59] X. Tian, J. Li, K. Chen, J. Han, S. Pan, Template-free and scalable synthesis of core-shell and hollow $BaTiO_3$ particles: using molten hydrated salt as a solvent, Cryst. Growth Des. 9 (11) (2009) 4927−4932.
[60] Satoshi Wada, Takuya Hoshina, Hiroaki Yasuno, Masanori Ohishi, Hirofumi Kakemoto, Takaaki Tsurumi, et al., Chap. 9. Dielectric properties of nm-sized barium titanate fine particles and their size dependence, in: Sheng Yao, Bruce Tuttle, Clive Randall, Dwight Viehland (Eds.), Advances in Electronic Ceramic Materials: Ceramic Engineering and Science Proceedings, vol. 26 (5), Wiley, 2008, pp. 89−100
[61] R.E. Newnham, K.R. Udayakumar, S. Trolier-McKinstry, Chemical Processing of Advanced Materials, in: L. L. Hench and J.K. West (Eds.), Wiley, New York, 1992, p. 379−393.
[62] R. Waser, Solubility of hydrogen defects in doped and undoped $BaTiO_3$, J. Am. Ceram. Soc. 71 (1) (1988) 58−63.
[63] D. Begg, E.R. Vance, J. Nowotn, Effect of particle size on room temperature crystal structure of barium titanate, J. Am. Ceram. Soc. 77 (12) (1994) 3186−3192.
[64] T. Takeuchi, K. Ado, T. Asai, H. Kageyama, Y. Saito, C. Masquelier, et al., Thickness of cubic surface phase on barium titanate single-crystalline grains, J. Am. Ceram. Soc. 77 (6) (1994) 1665−1668.
[65] S. Wada, H. Yasuno, T. Hoshina, S.-M. Nam, H. Kakemoto, T. Tsurumi, Preparation of nm-sized barium titanate fine particles and their powder dielectric properties, Jpn. J. Appl. Phys. 42 Part 1 (9B) (2003) 6188−6195.
[66] (a) M.B. Smith, K. Page, Th Siegrist, P.L. Redmond, E.C. Walter, R. Seshadri, et al., Crystal structure and the paraelectric-to-ferroelectric phase transition of nanoscale $BaTiO_3$, J. Am. Chem. Soc. 130 (2008) 6955−6963.
(b) Y. Li, Z. Liao, F. Fang, X. Wang, L. Li, J. Zhu, Significant increase of Curie temperature in nano-scale $BaTiO_3$, Appl. Phys. Lett. 105 (2014) 182901.
[67] V. Petkov, M. Gateshki, M. Niederberger, Atomic-scale structure of nanocrystalline $Ba_xSr_{1-x}TiO_3$ ($x = 1$, 0.5, 0) by X-ray diffraction and the atomic pair distribution function technique, Y. Ren, Chem. Mater. 18 (3) (2006) 814−821.
[68] V. Petkov, V. Buscaglia, M.T. Buscaglia, Z. Zhao, Y. Ren, Structural coherence and ferroelectricity decay in submicron- and nano-sized perovskites, Phys. Rev. B 78 (2008) 054107.
[69] K. Page, T. Proffen, M. Niederberger, R. Seshadri, Probing local dipoles and ligand structure in $BaTiO_3$ nanoparticles, Chem. Mater. 22 (15) (2010) 4386−4391.
[70] H. Fu, L. Bellaiche, Ferroelectricity in barium titanate quantum dots and wires, Phys. Rev. Lett. 91 (2003) 257601.
[71] I.I. Naumov, L. Bellaiche, H. Fu, Unusual phase transitions in ferroelectric nanodisks and nanorods, Nature 432 (2004) 737−740.
[72] M.J. Polking, M.-G. Han, A. Yourdkhani, V. Petkov, Ch.F. Kisielowski, V.V. Volkov, et al., Ferroelectric order in individual nanometrescale crystals, Nat. Mater. 11 (2012) 700−709.
[73] A. Sundaresan, R. Bhargavi, N. Rangarajan, U. Siddesh, C.N.R. Rao, Ferromagnetism as a universal feature of nanoparticles of the otherwise nonmagnetic oxides, Phys. Rev. B 74 (16) (2006) 161306.
[74] R.V.K. Mangalam, N. Rayb, U.V. Waghmare, A. Sundaresan, C.N.R. Rao, Multiferroic properties of nanocrystalline $BaTiO_3$, Solid State Commun. 149 (2009) 1−5.

[75] S.G. Bahoosh, S. Trimper, J.M. Wesselinowa, Origin of ferromagnetism in $BaTiO_3$ nanoparticles, Phys. Status Solidi RRL 5 (10-11) (2011) 382–384.
[76] D. Ishii, M. Ohta, Uehara, Y. Kimishima, Vacancy induced ferromagnetism in nano-$BaTiO_3$, Procedia Eng. 36 (2012) 578–582.
[77] N. Nuraje, K. Su, A. Haboosheh, J. Samson, E.P. Manning, N.L. Yang, et al., Room temperature synthesis of ferroelectric barium titanate nanoparticles using peptide nanorings as templates, Adv. Mater. 18 (6) (2006) 807–811.
[78] M.H. Frey, D.A. Payne, Grain size effect on structure and phase transformations for barium titanate, Phys. Rev. B 54 (1996) 3158–3168.
[79] G. Arlt, D. Hennings, G. de With, Dielectric properties of fine-grained barium titanate ceramics, J. Appl. Phys. 58 (1985) 1619–1625.
[80] T. Hiramatsu, T. Tamura, N. Wada, H. Tamura, Y. Sakabe, Effects of grain boundary on dielectric properties in fine-grained $BaTiO_3$ ceramics, Mater. Sci. Eng. B 120 (2005) 55–58.
[81] K. Kinoshita, A. Yamaji, Grain size effects on dielectric properties in barium titanate ceramics, J. Appl. Phys. 47 (1976) 371–373.
[82] L. Curecheriu, S.B. Balmus, V. Nica, M.T. Buscaglia, V. Buscaglia, A. Ianculescu, et al., Grain-size dependent properties of dense nanocrystalline barium titanate ceramics, J. Am. Ceram. Soc. 95 (2012) 3912–3921.
[83] S. Wada, T. Suzuki, T. Noma, Preparation of barium titanate ceramics from amorphous fine particles of the Ba–Ti–O system and its dielectric properties, J. Mater. Res. 10 (2) (1995) 306–311.
[84] R.P. Lobo, N.D.S. Mohallem, R.L. Moreira, Grain-size effects on diffuse phase transitions of sol-gel prepared barium titanate ceramics, J. Am. Ceram. Soc. 78 (5) (1995) 1343–1346.
[85] Z. Zhao, V. Buscaglia, M. Viviani, M.T. Buscaglia, L. Mitoseriu, A. Testino, et al., Grain-size effects on the ferroelectric behavior of dense nanocrystalline $BaTiO_3$ ceramics, Phys. Rev. B 70 (2) (2004) 024107.
[86] A. Feteira, D.C. Sinclair, I.M. Reaney, Y. Somiya, M.T. Lanagan, $BaTiO_3$-based ceramics for tunable microwave applications, J. Am. Ceram. Soc. 87 (2004) 1082–1087.
[87] (a) S. Yoon, J. Dornseiffer, Y. Xiong, D. Grüner, Z. Shen, S. Iwaya, et al., Spark plasma sintering of nanocrystalline $BaTiO_3$-powders: consolidation behavior and dielectric characteristics, J. Eur. Ceram. Soc. 31 (2011) 1723–1731.
(b) M.T. Buscaglia, M. Viviani, V. Buscaglia, L. Mitoseriu, A. Testino, P. Nanni, et al., High dielectric constant and frozen macroscopic polarization in dense nanocrystalline $BaTiO_3$ ceramics, Phys. Rev. B 73 (6) (2006) 064114.
[88] E.K. Akdogan, M.R. Leonard, A. Safari, Size effects in ferroelectric ceramics, in: H.S. Nalwa (Ed.), Handbook of Low and High Dielectric Constant Materials for Applications, vol. 2, Acad. Press, 1999, pp. 61–112.
[89] A.V. Polotai, A.V. Ragulya, C.A. Randall, Preparation and size effect in pure nanocrystalline barium titanate ceramics, Ferroelectrics 288 (1) (2003) 93–102.
[90] X.-H. Wang, X.-Y. Deng, H. Zhou, L.-T. Li, I.-W. Chen, Bulk dense nanocrystalline $BaTiO_3$ ceramics prepared by novel pressureless two-step sintering method, J. Electroceram. 21 (2008) 230–233.
[91] M.H. Frey, Z. Xu, P. Han, D.A. Payne, The role of interfaces on an apparent grain size effect on the dielectric properties for *ferroelectric* barium titanate ceramics, Ferroelectrics 206 (1) (1998) 337–353.
[92] T. Takeuchi, M. Tabuchi, H. Kageyama, Preparation of dense $BaTiO_3$ ceramics with submicrometer grains by spark plasma sintering, J. Am. Ceram. Soc. 82 (4) (1999) 939–943.

[93] B. Li, X. Wang, M. Cai, L. Hao, L. Li, Densification of uniformly small-grained $BaTiO_3$ using spark-plasma-sintering, Mater. Chem. Phys. 82 (1) (2003) 173–180.
[94] B. Li, X. Wang, L. Li, H. Zhou, X. Liu, X. Han, et al., Dielectric properties of fine-grained $BaTiO_3$ prepared by spark-plasma-sintering, Mater. Chem. Phys. 83 (1) (2004) 23–28.
[95] R. Licheri, S. Fadda, Roberto Orrù, G. Cao, V. Buscaglia, Self-propagating high-temperature synthesis of barium titanate and subsequent densification by spark plasma sintering (SPS), J. Eur. Ceram. Soc. 27 (5) (2007) 2245–2253.
[96] W. Luan, L. Gao, H. Kawaoka, T. Sekino, K. Niihara, Fabrication and characteristics of fine-grained $BaTiO_3$ ceramics by spark plasma sintering, Ceram. Int. 30 (3) (2004) 405–410.
[97] T. Takeuchi, C. Capiglia, N. Balakrishnan, Y. Takeda, H. Kageyama, Preparation of fine grained $BaTiO_3$ ceramics by spark plasma sintering, J. Mater. Res. 17 (3) (2002) 575–581.
[98] X. Deng, X. Wang, H. Wen, A. Kang, Z. Gui, L. Li, Phase transitions in nanocrystalline barium titanate ceramics prepared by spark plasma sintering, J. Am. Ceram. Soc. 89 (3) (2006) 1059–1064.
[99] L. Mitoşeriu, C. Harnagea, P. Nanni, A. Testino, M.T. Buscaglia, V. Buscaglia, et al., Local switching properties of dense nanocrystalline $BaTiO_3$ ceramics, Appl. Phys. Lett. 84 (13) (2004) 2418–2420.
[100] X. Deng, X. Wang, D. Li, L. Li, Electronic structure of nanograin barium titanate ceramics, Front. Mater. Sci. China 1 (3) (2007) 316–318.
[101] M.T. Buscaglia, V. Buscaglia, M. Viviani, J. Petzelt, M. Savinov, L. Mitoşeriu, et al., Ferroelectric properties of dense nanocrystalline $BaTiO_3$ ceramics, Nanotechnology 15 (9) (2004) 1113–1117.
[102] X. Deng, X. Wang, H. Wen, L. Chen, L. Chen, L. Li, Ferroelectric properties of nanocrystalline barium titanate ceramics, Appl. Phys. Lett. 88 (2006) 252905.
[103] S. Ma, X. Wang, J. Zhou, L. Li, Q. Sun, Thermal stability of the nanostructured $BaTiO_3$ determined by long and short range interactions: a dual-shell model, J. Appl. Phys. 107 (2010) 064102.
[104] K.J. Choi, M. Biegalski, Y.L. Li, A. Sharan, J. Schubert, R. Uecker, et al., Enhancement of ferroelectricity in strained $BaTiO_3$ thin films, Science 306 (2004) 1005–1009.
[105] R.A. McKee, F.J. Walker, J.R. Conner, E.D. Specht, D.E. Zelmon, Molecular beam epitaxy growth of epitaxial barium silicide, barium oxide, and barium titanate on silicon, Appl. Phys. Lett. 59 (1991) 782–784.
[106] Y. Yoneda, T. Okabe, K. Sakaue, H. Terauchi, H. Kasatani, K. Deguchi, Structural characterization of $BaTiO_3$ thin films grown by molecular beam epitaxy, J. Appl. Phys. 83 (1998) 2458–2461.
[107] H. Tabata, H. Tanaka, T. Kawai, Formation of artificial $BaTiO_3/SrTiO_3$ superlattices using pulsed laser deposition and their dielectric properties, Appl. Phys. Lett. 65 (1994) 1970–1972.
[108] Y. Yang, Z. Wang, J.-F. Li, D. Viehland, Pulsed laser deposition of $BaTiO_3$ thin films on different substrates, J. Nanomater. 2010 (2010) 756319.
[109] H.-F. Cheng, M.-H. Yeh., K.-S. Liu, I.-N. Lin, Characteristics of $BaTiO_3$ films prepared by pulsed laser deposition, Jpn. J. Appl. Phys. 32 Part 1 (12A) (1993) 5656–5660.
[110] X.M. Lu, J.S. Zhu, W.Y. Zhang, G.Q. Ma, Y.N. Wang, The energy gap of r.f.-sputtered $BaTiO_3$ thin films with different grain size, Thin Solid Films 274 (1-2) (1996) 165–168.

[111] W.-C. Shih, M.-H. Chiang, Structural and optical characterization of c-axis textured $BaTiO_3$ thin films on MgO fabricated by RF magnetron sputtering, J. Mater. Sci.: Mater. Electron. 21 (2010) 844–848.
[112] L. Preda, B. Despax, L. Courselle, J. Bandet, A. Ianculescu, Structural characteristics of RF–sputtered $BaTiO_3$ thin films, Thin Solid Films 389 (1-2) (2001) 43–50.
[113] A. Ianculescu, M. Gartner, B. Despax, V. Bley, Th Lebey, R. Gavrilă, et al., Optical characterization and microstructure of $BaTiO_3$ thin films obtained by RF-magnetron sputtering, Appl. Surf. Sci. 253 (1) (2006) 344–348.
[114] S. Kim, S. Hishita, Y.M. Kang, S. Baik, Structural characterization of epitaxial $BaTiO_3$ thin films grown by sputter deposition on MgO(100), J. Appl. Phys. 78 (9) (1995) 5604–5608.
[115] D.L. Kaiser, M.D. Vaudin, L.D. Rotter, Z.L. Wang, J.P. Cline, C.S. Hwang, et al., Epitaxial growth of $BaTiO_3$ thin films at 600°C by metalorganic chemical vapor deposition, Appl. Phys. Lett. 66 (21) (1995) 2801–2803.
[116] T. Tohma, H. Masumoto, T. Goto, Microstructure and dielectric properties of barium titanate film prepared by MOCVD, Mater. Trans. 43 (11) (2002) 2880–2884.
[117] M. Nagasaka, D. Iwasaki, N. Sakamoto, N. Wakiya, H. Suzuki, $BaTiO_3$ thin film by CSD from molecular-designed precursor solution, Funct. Mater. Lett. 05 (02) (2012) 1260007.
[118] T.M. Stawski, W.J.C. Vijselaar, O.F. Göbel, S.A. Veldhuis, B.F. Smith, D.H.A. Blank, et al., Influence of high temperature processing of sol-gel derived barium titanate thin films deposited on platinum and strontium ruthenate coated silicon wafers, Thin Solid Films 520 (2012) 4394–4401.
[119] R. Thomas, D.C. Dube, M.N. Kamalasanan, S. Chandra, Optical and electrical properties of $BaTiO_3$ thin films prepared by chemical solution deposition, Thin Solid Films 346 (1-2) (1999) 212–225.
[120] E.J.H. Lee, F.M. Pontes, E.R. Leite, E. Longo, J.A. Varela, E.B. Araujo, et al., Preparation and properties of ferroelectric $BaTiO_3$ thin films produced by the polymeric precursor method, J. Mater. Sci. Lett. 19 (2000) 1457–1459.
[121] M. Manso-Silván, L. Fuentes-Cobas, R.J. Martín-Palma, M. Hernández-Vélez, J.M. Martínez-Duart, $BaTiO_3$ thin films obtained by sol-gel spin coating, Surf. Coat. Technol. 151-152, 2002, 118–121.
[122] R. Ashiri, A. Nemati, M. Sasani Ghamsari, Crack-free nanostructured $BaTiO_3$ thin films prepared by sol-gel dip-coating technique, Ceram. Int. 40 (6) (2014) 8613–8619.
[123] Y.J. Wu, J. Li, H. Tanaka, M. Kuwabara, Preparation of nano-structured $BaTiO_3$ thin film by electrophoretic deposition and its characterization, J. Eur. Ceram. Soc. 25 (12) (2005) 2041–2044.
[124] R. Ayouchi, F. Martín, J.R. Ramos-Barrado, D. Leinen, Compositional, structural and electrical characterization of barium titanate thin films prepared on fused silica and Si(111) by spray pyrolysis, Surf. Interface Anal. 30 (2000) 565–569.
[125] K.W. Seo, H.G. Kong, Hydrothermal preparation of $BaTiO_3$ thin films, Korean J. Chem. Eng. 17 (4) (2000) 428–432.
[126] W. Suchanek, T. Watanabe, M. Yoshimura, Preparation of $BaTiO_3$ thin films by the hydrothermal–electrochemical method in the flowing solution, Solid State Ionics 109 (1-2) (1998) 65–72.
[127] M. Yoshimura, S. Yoo, M. Hayashi, N. Ishizawa, Preparation of $BaTiO_3$ thin film by hydrothermal electrochemical method, Jpn. J. Appl. Phys. 28 (1989) L2007–L2009.

[128] M. Vehkamäki, T. Hatanpää, M. Ritala, M. Leskelä, S. Väyrynen, E. Rauhala, Atomic layer deposition of $BaTiO_3$ thin films – effect of barium hydroxide formation, Chem. Vapor Depos. 13 (5) (2007) 239–246.
[129] M. Alexe, C. Harnagea, D. Hesse, U. Gosele, Polarization imprint and size effects in mesoscopic ferroelectric structures, Appl. Phys. Lett. 79 (2001) 242–244.
[130] T.M. Shaw, S. Trolier-McKinstry, The properties of ferroelectric films at small dimensions, Annu. Rev. Mater. Sci. 30 (2000) 263–298.
[131] B.A. Strukov, S.T. Davitadze, S.G. Shulman, B.M. Goltzman, V.V. Lemanov, Clarification of size effects in polycrystalline $BaTiO_3$ thin films by means of the specific heat measurements: grain size or film thickness?, Ferroelectrics 301 (1) (2004) 157–162.
[132] T.M. Shaw, S. Trolier-McKinstry, P.C. McIntyre, The properties of ferroelectric films at small dimensions, Annu. Rev. Mater. Sci. 30 (2000) 263–298.
[133] S. Hwang, B.T. Lee, C.S. Kang, K.H. Lee, H.J. Cho, Depletion layer thickness and Schottky type carrier injection at the interface between Pt electrodes and $(Ba, Sr)TiO_3$ thin films, J. Appl. Phys. 85 (1) (1999) 287–295.
[134] Y. Wang, W. Chen, B. Wang, Y. Zheng, Ultrathin ferroelectric films: growth, characterization, physics and applications, Materials 7 (2014) 6377–6485.
[135] U. Böttger, Dielectric properties of polar oxides, in: R. Waser, U. Böttger, S. Tiedke (Eds.), Polar Oxides: Properties, Characterization, and Imaging, Wiley-VCH Verlag GmbH & Co.KGaA, Weinheim, Germany, 2005, pp. 12–38 (Chapter 1).
[136] N.A. Pertsev, A.G. Zembilgotov, A.K. Tagantsev, Effect of mechanical boundary conditions on phase diagrams of epitaxial ferroelectric thin films, Phys. Rev. Lett. 80 (1998) 1988–1991.
[137] D.D. Fong, G.B. Stephenson, S.K. Streiffer, J.A. Eastman, O. Auciello, P.H. Fuoss, et al., Ferroelectricity in ultrathin perovskite films, Science 304 (2004) 1650–1653.
[138] J. Junquera, P. Ghosez, Critical thickness for ferroelectricityin perovskite ultrathin films, Nature 422 (2003) 506–509.
[139] Y.N. Xia, P.D. Yang, Y.G. Sun, Y.Y. Wu, B. Mayers, B. Gates, et al., One-dimensional nanostructures: synthesis, characterization, and applications, Adv. Mater. 15 (2003) 353–389.
[140] P.X. Gao, J. Song, J. Liu, Z. Wang, Nanowire piezoelectric nanogenerators on plastic substrates as flexible power sources for nanodevices, Adv. Mater. 19 (2007) 67–72.
[141] Z.L. Wang, Self-powered nanosensors and nanosystems, Adv. Mater. 24 (2012) 280–285.
[142] J. Zhou, P. Fei, Y. Gao, Y. Gu, J. Liu, G. Bao, et al., Mechanical-electrical triggers and sensors using piezoelectric micowires/nanowires, Nano Lett. 8 (2008) 2725–2730.
[143] A. Koka, Z. Zhou, H. Tang, H.A. Sodano, Controlled synthesis of ultra-long vertically aligned $BaTiO_3$ nanowire arrays for sensing and energy harvesting applications, Nanotechnology 25 (37) (2014) 375603.
[144] F.D. Morisson, Y. Luo, I. Szafraniak, V. Nagarajan, R.B. Wehrspohn, M. Steinhart, et al., Ferroelectric nanotubes, Rev. Adv. Mater. 4 (2003) 114–122.
[145] J.J. Urban, W.S. Yun, Q. Gu, H. Park, Synthesis of single crystalline perovskite nanorods composed of barium titanate and strontium titanate, J. Am. Chem. Soc. 124 (2002) 1186–1187.
[146] N. Bao, L. Shen, G. Srinivasan, K. Yanagisawa, A. Gupta, Shape-controlled monocrystalline ferroelectric barium titanate nanostructures: from nanotubes and nanowires to ordered nanostructures, J. Phys. Chem. C 112 (2008) 8634–8642.

[147] N. Bao, L. Shen, A. Gupta, A. Tatarenko, G. Srinivasan, K. Yanagisawa, Size-controlled one-dimensional monocrystalline $BaTiO_3$ nanostructures, Appl. Phys. Lett. 94 (2009) 253109.

[148] Y. Luo, I. Szafraniak, V. Nagarajan, R.B. Wehrspohn, M. Steinhart, J.H. Wendorff, et al., Ferroelectric lead zirconate titanate and barium titanate nanotubes, Integr. Ferroelectr. 59 (2003) 1513−1520.

[149] J.J. Urban, J.E. Spanier, L. Ouyang, W.S. Yun, H. Park, Single-crystalline barium titanate nanowires, Adv. Mater. 15 (2003) 423−426.

[150] G. Cao, D. Liu, Template-based synthesis of nanorod, nanowire, and nanotube arrays, Adv. Colloid. Interf. Sci. 136 (2008) 45−64.

[151] A. Lamberti, N. Garino, K. Bejtka, S. Bianco, S. Stassi, A. Chiodoni, et al., Synthesis of ferroelectric $BaTiO_3$ tube-like arrays by hydrothermal conversion of a vertically aligned TiO_2 nanotube carpet, New. J. Chem. 38 (2014) 2024−2030.

[152] F. Zhang, S.G. Chen, Y.S. Yin, C.R. Xue, C. Lin, Hydrothermal synthesis of $BaTiO_3$ nanotubes on Ti substrates, Adv. Mater. Res. 79-82 (2009) 617−620.

[153] Y. Yang, X. Wang, C. Sun, L. Li, Structure study of single crystal $BaTiO_3$ nanotube arrays produced by the hydrothermal method, Nanotechnology 20 (2009) 055709.

[154] R. Poyato, B.D. Huey, Local piezoelectric and ferroelectric responses in nanotube-patterned thin films of $BaTiO_3$ synthesized hydrothermally at 200°C, J. Mater. Res. 21 (2006) 547−551.

[155] N.P. Padture, X. Wei, Hydrothermal synthesis of thin films of barium titanate ceramic nano-tubes at 200°C, J. Am. Ceram. Soc. 86 (2003) 2215−2217.

[156] Y. Wang, L. Yang, Y. Wang, X. Wang, Synthesis of $PbTiO_3$ nanowires via a simple hydrothermal method, J. Ceram. Proc. Res. 14 (2013) 114−116.

[157] Y.F. Zhu, L. Zhang, T. Natsuki, Y.Q. Fu, Q.Q. Ni, Facile synthesis of $BaTiO_3$ nanotubes and their microwave absorption properties, ACS Appl. Mater. Interf. 4 (2012) 2101−2106.

[158] Y. Yang, X. Wang, C. Sun, L. Li, Phase transition in $BaTiO_3$ nanotube arrays, J. Appl. Phys. 109 (2011) 014109.

[159] A. Koka, Z. Zhou, H.A. Sodano, Vertically aligned $BaTiO_3$ nanowire arrays for energy harvesting, Energy. Environ. Sci. 7 (2014) 288−296.

[160] M. Alexe, D. Hesse, One-dimensional ferroelectrics: nanowires and nanotubes, Ferroelectrics 433 (2012) 53−64.

[161] S.J. Limmer, G. Cao, Sol-gel electrophoretic deposition for the growth of oxide nanorods, Adv. Mater. 15 (2013) 427−431.

[162] J. Kim, S.A. Yang, Y.C. Choi, J.K. Han, K.O. Jeong, Y.J. Yun, et al., Ferroelectricity in highly ordered arrays of ultra-thin-walled $Pb(Zr,Ti)O_3$ nanotubes composed of nanometer-sized perovskite crystallites, Nano. Lett. 8 (2008) 1813−1818.

[163] Y. Luo, I. Szafraniak, N.D. Zakharov, V. Nagarajan, M. Steinhart, R.B. Wehrspohn, et al., Nanoshell tubes of ferroelectric lead zirconate titanate and barium titanate, Appl. Phys. Lett. 83 (2003) 440−442.

[164] L. Liu, T. Ning, Y. Ren, Z. Sun, F. Wang, W. Zhou, et al., Synthesis, characterization, photoluminescence and ferroelectric properties of $PbTiO_3$ nanotube arrays, Mater. Sci. Eng. B 149 (2008) 41−46.

[165] Y.Y. Chen, B.Y. Yu, J.H. Wang, R.E. Cochran, J.J. Shyue, Template-based fabrication of $SrTiO_3$ and $BaTiO_3$ nanotubes, Inorg. Chem. 48 (2009) 681−686.

[166] B.A. Hernandez-Sanchez, K.S. Chang, M.T. Scancella, J.L. Burris, S. Kohli, E.R. Fisher, et al., Examination of size-induced ferroelectric phase transitions in template synthesized $PbTiO_3$ nanotubes and nanofibers, Chem. Mater. 17 (2005) 5909−5919.

[167] B.A. Hernandez-Sanchez, K.S. Chang, E.R. Fisher, P.K. Dorhout, Sol-gel template synthesis and characterization of $BaTiO_3$ and $PbTiO_3$ nanotubes, Chem. Mater. 14 (2002) 480−482.
[168] C.A. Vasilescu, L. Trupina, B.S. Vasile, R. Trusca, M. Cernea, A.C. Ianculescu, Characteristics of 5 mol% Ce^{3+}-doped barium titanate nanowires prepared by a combined route involving sol-gel chemistry and polycarbonate membrane-templated process, J. Nanopart. Res. 17 (2015) 434.
[169] S. Maensiri, W. Nuansing, J. Klinkaewnarong, P. Laokul, J. Khemprasit, Nanofibers of barium strontium titanate (BST) by sol-gel processing and electrospinning, J. Colloid. Interf. Sci. 29 (2006) 578−583.
[170] Y. Zhuang, Z. Xu, F. Li, Z. Liao, W. Liu, Improve piezoelectricity and elasticity of Ce-doped $BaTiO_3$ nanofibers - towards energy harvesting application, RSC Adv. 5 (2015) 55269.
[171] M.T. Buscaglia, C. Harnagea, M. Dapiaggi, V. Buscaglia, A. Pignolet, P. Nanni, Ferroelectric $BaTiO_3$ nanowires by a topochemical solid-state reaction, Chem. Mater. 21 (2009) 5058−5065.
[172] M. Alexe, D. Hesse, V. Schmidt, S. Senz, H.J. Fan, M. Zacharias, et al., Ferroelectric nanotubes fabricated using nanowires as positive templates, Appl. Phys. Lett. 89 (2006) 172907.
[173] G. Geneste, E. Bousquet, J. Junquera, P. Ghosez, Finite-size effects in $BaTiO_3$ nanowires, Appl. Phys. Lett. 88 (2006) 112906.
[174] J.E. Spanier, A.M. Kolpak, J.J. Urban, I. Grinberg, L. Ouyang, W.S. Yun, et al., Ferroelectric phase transition in individual single-crystalline $BaTiO_3$ nanowires, Nano Lett. 6 (4) (2006) 735−739.
[175] L. Louis, P. Gemeiner, I. Ponomareva, L. Bellaiche, G. Geneste, W. Ma, et al., Low-symmetry phases in ferroelectric nanowires, Nano Lett. 10 (2010) 1177−1183.
[176] A. Schilling, R.M. Bowman, G. Catalan, J.F. Scott, J.M. Gregg, Morphological control of polar orientation in single-crystal ferroelectric nanowires, Nano Lett. 7 (12) (2007) 3787−3791.
[177] Z.Y. Wang, J. Hu, M.F. Yu, One-dimensional ferroelectric monodomain formation in single crystalline $BaTiO_3$ nanowire, Appl. Phys. Lett. 89 (2006) 263119.
[178] J. Yang, J. Zhang, C. Liang, M. Wang, P. Zhao, M. Liu, et al., Ultrathin $BaTiO_3$ nanowires with high aspect ratio: a simple one-step hydrothermal synthesis and their strong microwave absorption, ACS Appl. Mater. Interfaces 5 (15) (2013) 7146−7151.
[179] P.M. Rørvik, T. Grande, M.-A. Einarsrud, One-dimensional nanostructures of ferroelectric perovskites, Adv. Mater. 23 (2011) 4007−4034.
[180] W.S. Yun, J.J. Urban, Q. Gu, H. Park, Ferroelectric properties of individual barium titanate nanowires investigated by scanned probe microscopy, Nano Lett. 2 (2002) 447−450.
[181] Z. Zhou, H.X. Tang, H.A. Sodano, Vertically aligned arrays of $BaTiO_3$ nanowires, ACS Appl. Mater. Inter. 5 (22) (2013) 11894−11899.
[182] A. Ianculescu, I. Pintilie, C.A. Vasilescu, M. Botea, A. Iuga, A. Melinescu, et al., Intrinsic pyroelectric properties of thick, coarse grained $Ba_{1-x}Sr_xTiO_3$ ceramics, Ceram. Int. 42 (8) (2016) 10338−10348.
[183] S. Suasmoro, S. Pratapa, D. Hartanto, D. Setyoko, U.M. Dani, The characterization of mixed titanate Ba_{1-x} Sr_xTiO_3 phase formation from oxalate coprecipitated precursor, J. Eur. Ceram Soc. 20 (2000) 309−314.
[184] Adelina Ianculescu, Liliana Mitoşeriu, $Ba(Ti,Zr)O_3$ - Functional Materials: From Nanopowders to Bulk Ceramics, Nova Science Publishers Inc, Hauppauge New York, USA, 2010.

[185] A.-C. Ianculescu, D.C. Berger, C.A. Vasilescu, B.S. Vasile, M. Olariu, L.P. Curecheriu, et al., Chap. 1. Incorporation mechanism and functional properties of Ce-doped $BaTiO_3$ ceramics derived from nanopowders prepared by the modified-Pechini method, in: M. Algueró, L. Marty Gregg (Eds.), Nanoscale Ferroelectrics and Multiferroics: Key Processing and Characterization Issues, and Nanoscale Effects, Mitoseriu, John Wiley & Sons Ltd, UK, 2016, pp. 13–43.
[186] A. Testino, M.T. Buscaglia, M. Viviani, V. Buscaglia, P. Nanni, Synthesis of $BaTiO_3$ particles with tailored size by precipitation from aqueous solutions, J. Am. Ceram. Soc. 87 (2004) 79–83.
[187] A. Testino, M.T. Buscaglia, V. Buscaglia, M. Viviani, C. Bottino, P. Nanni, Kinetics and mechanism of aqueous chemical synthesis of $BaTiO_3$ particles, Chem. Mater. 16 (2004) 1536–1543.
[188] Y. Gao, V.V. Shvartsman, D. Gautam, M. Winterer, D.C. Lupascu, Nanocrystalline barium strontium titanate ceramics synthesized via the "organosol", route and spark plasma sintering, J. Am. Ceram. Soc. 97 (7) (2014) 2139–2146.
[189] B. Su, J.Y. He, B.L. Cheng, T.W. Button, J. Liu, Z. Shen, et al., Dielectric properties of spark plasma sintered (SPS) barium strontium titanate (BST) ceramics, Integrated Ferroelectrics 61 (2004) 117–122.
[190] J. Liu, Z. Shen, M. Nygren, Spark plasma sintering behavior of nanosized $(Ba,Sr)TiO_3$ powders: determination of sintering parameters yielding nanostructured ceramic, J. Am. Ceram. Soc. 89 (2006) 2689–2694.
[191] H. Xue, Z. Xiong, The structure and dielectric tunable properties of fine-grained $Ba_{0.6}Sr_{0.4}TiO_3$ ceramics prepared by spark plasma sintering, J. Am. Ceram. Soc. 90 (2007) 2653–2656.
[192] C. Elissalde, U.C. Chung, A. Artemenko, C. Estournes, R. Costes, M. Pat_e, et al., Stoichiometry and grain boundaries control by spark plasma sintering in $Ba_{0.6}Sr_{0.4}TiO_3$:Mn/MgO composites, J. Am. Ceram. Soc. 95 (2012) 3239–3245.
[193] T. Hungría, M. Algueró, A.B. Hungría, A. Castro, Dense, fine-grained $Ba_{1-x}Sr_xTiO_3$ ceramics prepared by the combination of mechanosynthesized nanopowders and spark plasma sintering, Chem. Mater. 17 (2005) 6205–6212.
[194] H. Maiwa, Structure and properties of $Ba(Zr_{0.2}Ti_{0.8})O_3$ ceramics prepared by spark plasma sintering, J. Mater. Sci. 43 (2008) 6385–6390.
[195] H. Maiwa, Dielectric and electromechanical properties of $Ba(Zr_xTi_{1-x})O_3$ (x = 0.1 and 0.2) ceramics prepared by spark plasma sintering, Jpn. J. Appl. Phys. 46 (2007) 7013–7017.
[196] Z. Valdez-Nava, S. Guillemet-Fritsch, Ch Tenailleau, T. Lebey, B. Durand, J.Y. Chane-Ching, Colossal dielectric permittivity of $BaTiO_3$-based nanocrystalline ceramics sintered by spark plasma sintering, J. Electroceram. 22 (2009) 238–244.
[197] A. Ianculescu, B. Despax, V. Bley, Th Lebey, R. Gavrilă, N. Drăgan, Structure – properties correlations for barium titanate thin films obtained by RF-sputtering, J. Eur. Ceram. Soc. 27 (2-3) (2007) 1129–1135.
[198] S. Ezhilvalavan, T.-Y. Tseng, Progress in the developments of $(Ba,Sr)TiO_3$ (BST) thin films for Gigabit era DRAMs, Mater. Chem. Phys. 65 (2000) 227–248.
[199] K. Nadaud, C. Borderon, R. Gillard, E. Fourn, R. Renoud, H.W. Gundel, Temperature stable $BaSrTiO_3$ thin films suitable for microwave applications, Thin Solid Films, 591 Part A 30 (2015) 90–96.
[200] J. Zhai, H. Chen, Orientation control and dielectric properties of sol-gel deposited $(Ba,Sr)TiO_3$ thin films for room-temperature tunable element applications, J. Korean Ceram. Soc. 40 (4) (2003) 380–384.
[201] S. Agarwal, G.L. Sharma, Humidity sensing properties of $(Ba,Sr)TiO_3$ thin films grown by hydrothermal-electrochemical method, Sensor Actuat. B 85 (2002) 205–211.

[202] Z. Jiwei, Y. Xi, Z. Liangying, S. Bo, H. Chen, Dielectric properties of the compositionally graded (Ba,Sr)TiO_3 thin film, Ferroelectrics 329 (1) (2005) 43–48.

[203] X. Zhu, J. Zhu, S. Zhou, Z. Liu, N. Ming, H.L.-W. Chan, et al., Microstructure and dielectric properties of compositionally-graded $(Ba_{1-x}Sr_x)TiO_3$ thin films, Mat. Sci. Eng. B 118 (2005) 219–224.

[204] W.X. Cheng, A.L. Ding, X.Y. He, X.S. Zheng, P.S. Qiu, Characterization of Ba$(Zr_{0.05}Ti_{0.95})O_3$ thin film prepared by sol-gel process, J. Electroceram. 16 (2006) 523–526.

[205] L.S. Cavalcante, J.C. Sczancoski, F.S. De Vicente, M.T. Frabbro, M. Siu Li, J.A. Varela, et al., Microstructure, dielectric properties and optical band gap controlon the photoluminescence behavior of $Ba[Zr_{0.25}Ti_{0.75}]O_3$ thin films, J. Sol-Gel Sci. Technol. 49 (2009) 35–46.

[206] L.S. Cavalcante, M.F.C. Gurgel, E.C. Paris, A.Z. Simões, M.R. Joya, J.A. Varela, et al., Combined experimental and theoretical investigations of the photoluminescent behavior of Ba(Ti,Zr)O_3 thin films, Acta Mater. 55 (2007) 6416–6426.

[207] A. Liu, J. Xue, X. Meng, J. Sun, Z. Huang, J. Chu, Infrared optical properties of $Ba(Zr_{0.20}Ti_{0.80})O_3$ and $Ba(Zr_{0.30}Ti_{0.70})O_3$ thin films prepared by sol-gel method, Appl. Surf. Sci. 254 (2008) 5660–5663.

[208] M. Anicete-Santos, L.S. Cavalcante, E. Orhan, E.C. Paris, L.G.P. Simões, M.R. Joya, et al., The role of structural order-disorder for visible intense photoluminescence in the $BaZr_{0.5}Ti_{0.5}O_3$ thin films, Chem. Phys. 316 (2005) 260–266.

[209] C.A. Vasilescu, M. Crişan, A.C. Ianculescu, M. Răileanu, M. Gartner, M. Anastasescu, et al., Structure, morphology and optical properties of multilayered sol-gel $BaTi_{0.85}Zr_{0.15}O_3$ thin films, Appl. Surf. Sci. 265 (2013) 510–518.

[210] A.I. Ali, K. Park, A. Ullah, R. Huh, Y.S. Kim, Ferroelectric enhancement of La-doped $BaTiO_3$ thin films using $SrTiO_3$ buffer layer, Thin Solid Films 551 (2014) 127–130.

[211] M. Cernea, A. Ianculescu, O. Monnereau, L. Argème, V. Bley, C. Logofătu, Preparation and characterization of Ce-doped $BaTiO_3$ thin films by pulsed laser deposition, J. Mater. Sci. 39 (8) (2004) 2755–2759.

[212] M. Cernea, I. Matei, A. Iuga, C. Logofatu, Preparation and characterization of Ce-doped $BaTiO_3$ thin films by RF sputtering, J. Mater. Sci. 36 (20) (2001) 5027–5030.

[213] Y. Liu, Z. Chen, C. Li, D. Cui, Y. Zhou, G. Yang, et al., Electrical and electro-optical properties of Ce-doped barium titanate thin films prepared by pulsed laser deposition, J. Appl. Phys. 81 (9) (1997) 6328–6331.

[214] L. Liu, H. Guo, H. Lü, S. Dai, B. Cheng, Z. Chen, Effects of donor concentration on the electrical properties of Nb-doped $BaTiO_3$ thin films, J. Appl. Phys. 97 (2005) 054102.

[215] A.-C. Ianculescu, C.-A. Vasilescu, L. Trupină, B.S. Vasile, R. Truşcă, M. Cernea, et al., Characteristics of Ce^{3+}-doped barium titanate nanoshell tubes prepared by template-mediated colloidal chemistry, J. Eur. Ceram. Soc. 36 (7) (2016) 1633–1642.

[216] V. Kalyani, B.S. Vasile, A. Ianculescu, M.T. Buscaglia, V. Buscaglia, P. Nanni, Hydrothermal synthesis of $SrTiO_3$ mesocrystals: single crystal to mesocrystal transformation induced by topochemical reactions, Cryst. Growth Des. 12 (2012) 4450–4456.

[217] A. Gruverman, A. Kholkin, Nanoscale ferroelectrics: processing, characterization and future trends, Rep. Prog. Phys. 69 (2006) 2443–2474.

[218] B. Yin, S. Qu, An *ab initio* investigation of flexoelectric effect in ultrathin $BaTiO_3$ nanotubes, J. Appl. Phys. 115 (2014) 074102.

Ecological, lead-free ferroelectrics

9

Lorena Pardo[1], María E. Villafuerte-Castrejón[2], Emilio Morán[3], Armando Reyes-Montero[2], Rodrigo Vivar-Ocampo[2] and Amador M. Gonzalez[4]

[1]Instituto de Ciencia de Materiales de Madrid (ICMM). CSIC. c/ Sor Juana Inés de la Cruz, 3. Cantoblanco. 28049-Madrid, Spain,

[2]Instituto de Investigaciones en Materiales, Universidad Nacional Autónoma de México, Circuito Exterior S/N, A.P. 70-360, México, CDMX, México,

[3]Departamento de Química Inorgánica. Facultad de Ciencias Químicas. Universidad Complutense de Madrid. 28040-Madrid, Spain,

[4]CEMDATIC, ETS Ingeniería y Sistemas de Telecomunicación, Campus Sur UPM, c/ Nikola Tesla s/n. 28031-Madrid, Spain

Legislation to replace lead-based ferroelectric ceramics for electromechanical transducers (commonly made of lead titanate-zirconate (PZT)) started in Europe shortly before the year 2000 in the context of elimination of toxic substances from electric and electronic equipment. From that date there has been a large, worldwide research effort to develop new ecological, lead-free compositions that can substitute for PZT. Some issues involved in this effort will be reviewed here, from the synthesis procedures to functional properties and trends. New properties on lead-free systems that surpass those of PZT (bismuth sodium titanates (BNTs) for actuators) and enhanced properties of other compositions in which lead-free materials are already at work in devices will be analyzed in this chapter. Present limits of well-known lead-free materials (barium titanate (BT), Aurivillius-type structure compounds, sodium potassium niobate (KNN), BNT, and barium calcium titanate-zirconate (BCZT)) will be summarized, and new and recently studied compositions based on materials with high spontaneous polarization (Ps) will be considered as future lead-free alternatives.

9.1 Lead-free ferroelectrics

The origin of ferroelectricity and the foundation for the numerous applications of these multifunctional materials is Ps. Ps arises from the crystal structure distortions that create an unbalance of ionic charges and it is switchable by application of a sufficiently high electric field. Ferroelectricity of the perovskite-type structure lead-titanate ($PbTiO_3$) was discovered independently by Smolenskii and Shirane et al. in 1950 [1]. It remained the ferroelectric with the highest Ps (81 $\mu C/cm^2$), associated with the off-centering of ions and strong tetragonal distortion ($c/a = 1.064$) of its crystal lattice, until bismuth ferrite ($BiFeO_3$) was studied and researchers measured a remanent polarization of $Pr \sim 100$ $\mu C/cm^2$ in 2003 [2] in thin films of this composition. Solid solutions based on $PbTiO_3$—among others, the one with lead zirconate ($PbZrO_3$), known as PZT—dominate the industrial applications of ferroelectrics in

Magnetic, Ferroelectric, and Multiferroic Metal Oxides. DOI: http://dx.doi.org/10.1016/B978-0-12-811180-2.00009-8

polycrystal form, also called ferropiezoelectric ceramics. The use of PbO in the fabrication of PZT and the possible decomposition at the end of the life cycle of this universally present material involves hazards for human health and the environment [3].

European Union regulations arising in 2003 [4,5] were the first to request PZT piezoceramics replacement due to the hazards of PbO in the processing and its end of the life cycle. Exemption 7(c) of such legislation concerns "lead in electronic ceramic parts (e.g., piezoelectronic devices)" and expired July 31, 2014. This exemption could have been applicable only if there were a lead-free alternative available. Recommendation of the European industry was to extend the exemption beyond that date, since it was foreseen that a feasible alternative to lead-based piezoelectric ceramics will not be ready for commercialization until 2018. Nevertheless, these regulations triggered the study of lead-free piezoelectric compositions [6]. This research field grew rapidly from that time [7]. It has today reached a level of maturity in some scientific themes, industrial transference [8], and spreading of the topic from the scientific community to society and business schools [9].

Among the lead-free ferropiezoelectric materials studied in the last years, here we consider those based on $BaTiO_3$ (BT), Aurivillius type-structure compounds, also known as bismuth layer-structured ferroelectrics (BLSF), $(K_{0.5}Na_{0.5})NbO_3$ (KNN), $(Bi_{0.5}Na_{0.5})TiO_3$ (BNT), and $Ba_{1-x}Ca_xTi_{1-y}Zr_yO_3$ (BCZT). These are five groups of compounds whose properties are competitive with those of PZT for similar applications. In addition to other chapters in this book, there are several recent reviews on lead-free ferroelectrics in the literature [10–13] and a book [14] that can also be recommended to the reader.

9.2 Preparation of lead-free piezoelectric ceramics with perovskite structure

Some challenges in the synthesis and processing of lead-free ferropiezoelectric ceramics are yet to be solved. Stoichiometric control, when volatile elements are involved, search for morphotropic phase boundary (MPB) compositions, softer synthesis conditions, adequate grain size, adequate ferroelectric domain size and distribution, sintering conditions, together with scalability and cost are among the main challenges for their industrial transfer. Properties strongly depend on the processing issues that ultimately determine the crystal structure and ceramic microstructure of these materials. Some processing issues exemplified in BT will be reviewed in this section.

$BaTiO_3$ is the first ferropiezoelectric oxide with perovskite structure and has been continuously studied and used since the 1940s [15], when it was discovered in the context of World War II [16]. It is an archetypical electroceramic widely used in the capacitor industry for its exceptional properties: very high dielectric constant, high quality factors, and low temperature coefficient of the dielectric constant. Worth noting, several structural phase transitions occur when changing the temperature: cubic (above 120°C), tetragonal (between 120°C and 5°C), orthorhombic (between 5°C and −90°C), and rhombohedral (below −90°C). Because of the high symmetry, the cubic phase of BT exhibits paraelectricity and an isotropic dielectricity albeit with a high dielectric constant. Below the Curie temperature (120°C), the crystal structure transforms to the distorted tetragonal, which causes a spontaneous polarization and a ferroelectric and piezoelectric behavior [17,18]. However, the low Curie temperature reduces the range of applications. Regarding its piezoelectric properties, it has a high electromechanical coupling factor, $k_{33} \approx 0.50$, and piezoelectric strain constant, $d_{33} \approx 190$ pC/N and, among other applications, it has been used in sonars. Nevertheless, its performance as piezoelectric material is well below that of PZT, and this has been the main obstacle for wider application as actuators and sensors. Nowadays, because BT is the simplest lead-free piezoelectric ceramic, because many data are available, and because some new piezoelectric materials present a similar crystal structure, much work is being carried out. New methods of synthesis are tested to improve BT piezoelectric performance. Much of this effort is made looking for nanocrystalline BT powders, which will be further sintered and the results so far obtained indicate that BT (or related) ceramics possess a high potential for being used as lead-free piezoelectric materials [7,8].

The most commonly used method for BT synthesis is the conventional ceramic route using $BaCO_3$ and TiO_2 as raw materials and high temperature conditions that provoke grain growth, a resulting microstructure with a wide grain-size distribution and some degree of porosity and secondary phases, like Ba_2TiO_4, among others. However, as an example of careful work, high piezoelectric properties and adequate domain configuration in $BaTiO_3$ ceramics obtained through the solid-state reaction route have been reported [19]. The study as a function of the sintering temperature and grain size showed the best results ($d_{33} = 419$ pC/N; $k_p = 0.453$ and $\tan\delta = 1.36\%$ at room temperature) for a sintering temperature of 1210°C for 2 h with average domain width in the poled BT of about 240 nm and a decrease of the grain size down to 2 μm.

Owing to the previously mentioned drawbacks of the ceramic method, a great effort is being devoted to new production methods [20,21]. Noteworthy alternative, lower temperature synthetic methods such as coprecipitation, sol-gel, Pechini, hydrothermal, microwave-assisted are being explored, aiming at the preparation of high purity nanopowders with controlled stoichiometry and better sinterability and, consequently, with better and reproducible quality. Once the BT submicron or nanopowders are obtained, sintering conditions are an important issue that strongly affects the properties as they correlate with the final grain and ferroelectric domain sizes [22,23]. Starting with the sol-gel processes, wet synthesis routes, using different precursors such as acetate, stearate [24], nitrates [25], and oxide [26] for barium and titanium (IV) alkoxides (i.e., Ti-isopropoxide) as the source for titanium are employed; organic alcohols are the reaction media [27]. It is recognized that the sol-gel method is complex and involves several steps—hydrolysis, condensation, drying, and calcination—and the final properties will be affected by changes made during the entire process (i.e., concentrations, pH, temperature, time, etc.). A common challenge for this kind of sol-gel methodology is the agglomeration of nanoparticles, and some authors claim that, by using surfactants such as oleic acid, this problem can be overcome and tetragonal BT can be obtained with room temperature permittivity higher than 3000 and 98.5% density [28]. Precipitation methods from aqueous solutions of inorganic compounds such as precursors $Ba(OH)_2$ and $TiCl_4$ can also be used and, interestingly, the concentration of Ba^{2+} ions has a strong influence on the particles homogeneity and crystallite size [29,30].

Some authors report on the kinetics and mechanism of this direct reaction [31]. In this connection, an ambient condition sol process, which produces a nanocrystalline cubic phase, has been reported [32]. With a different approach but the same goal, an electrochemical route has been explored, using Ti metal plate in KOH and $Ba(OH)_2$ aqueous electrolyte with absolute ethanol, nanoparticles of the cubic phase were obtained [33]. For $BaTiO_3$ synthesized from various salts (NO_3^-, Cl^-, and acetates), the permittivity and the dielectric loss values, measured as a function of the sintering temperature, has been reported, with the highest permittivity value found for samples prepared from Cl^- salts, with the smallest particle size and 1250°C sintering temperature. Nevertheless, the role of the anions on this effect is not well understood. Since the search for lead-free ceramics is driven by environmental concerns, a comment on the various levels of toxicity of the precursors in these methods must be added here and, consequently, all mentioned routes cannot be considered in the same manner.

Another successful wet method is the hydrothermal synthesis due to its ability to provide highly pure and crystalline powders. Actually, hydrothermally-made BT is already of commercial use and some companies such as Sakai Chemical Industry Co., Ltd. (Osaka, Japan) and Toda Industry (Ho Chi Minh, Vietnam) provide high quality powders of 0.1 μm grain size. Hydrothermal synthesis is carried out in closed vessels (usually made in stainless steel and lined with Teflon) at temperatures between 80°C and 240°C and with autogenous pressure [34]. Therefore, the method is a low-temperature, low-energy one, and it is environmentally benign since the synthesis is accomplished in a closed system from which different chemicals can be recovered and recycled. Nonetheless, working in sealed conditions provides a CO_2-free environment, which avoids $BaCO_3$ formation, a common impurity.

For the synthesis of BT, barium hydroxide or soluble barium salts are used as the source of barium and anatase or amorphous TiO_2^- hydrated gels are used as the source of titanium in the presence of concentrated hydroxide ion; mixed precursors such as Ba-Ti acetate can also be used. These procedures usually result in submicron cubic BT—the paraelectric form—at room temperature, which needs to be heated afterward at higher temperatures of 800°C−1100°C to obtain the ferroelectric tetragonal form. Hydrothermal synthesis, when performed at low temperatures (95°C) using $BaCl_2$ and anatase in basic NaOH solution, always yields the cubic form. Working at higher temperatures (240°C) with $BaCl_2$ and hydrous titanium oxide (prepared by hydrolyzing titanium isopropoxide with hydrochloric acid) produces the tetragonal form. In this regard, some authors conclude that the powder obtained from $BaCl_2$ and $TiCl_4$, with 80 nm particle size and 40% tetragonal content has the highest dielectric permittivity (6200) and the lowest dielectric loss [33].

Interestingly enough, it seems that hydrothermal BT may contain large amounts of protons in the oxygen sublattice and these defects are compensated by vacancies on metal sites; by annealing the powder at $T < 600°C$ water is released and the point defects disappear [35].

A variation of this method, the so-called microwave-assisted hydrothermal synthesis, is becoming increasingly used since it has several advantages such as significantly lowering reaction times (minutes or hours rather than days), uniform nucleation of the powders in suspension, energy and cost efficiency, among others [36]. Worth mentioning is the pioneering work that uses this methodology for a plethora of inorganic compounds, among them BT [37]: the microwave-assisted hydrothermal synthesis of BT was done at 138°C with reaction time as low as 15 min, using barium nitrate, titanium chloride, and KOH as starting chemicals [38,39]. Several works using laboratory-made microwave heating autoclaves (or domestic ovens) working at the usual frequency of 2.45 GHz were reported [40,41]. Nowadays, microwave furnaces with pressure, temperature, and power control can be used. As starting materials $Ba(OH)_2$ $Ba(NO_3)_2$, $BaCl_2$, and so on can be used as source for barium and $TiOCl_2$, $TiCl_4$, and such, for titanium; different equimolar or slightly higher concentrated Ba^{2+} solutions are employed. The most common mineralizer used is KOH in different concentrations. Depending of the conditions, particles ranging from 0.1–0.5 and submicron-sized equiaxed $BaTiO_3$ particles with monosized distribution are obtained [42–45]. Last but not least, microwaves can also be employed for sintering the ceramic powders previously obtained.

The sintering process plays an important role in the BT properties and many works have reported the influence of the conventional sintering process and the microwave sintering on the BT properties. As it is well known, microwaves interact readily with polar species producing heat and in this case, dielectric materials such as BT absorb the radiation in a very efficient way. As many authors point out, it is difficult to prepare $BaTiO_3$ dense ceramics via the conventional sintering process even if nanopowders are used [45]. The nanosized particles have the tendency to form agglomerates that cause defects on the microstructure. In samples prepared by the Pechini route, a two-step sintering method, first at 1300°C and then cooled down and held at 1100°C for 0–20 h, results in much greater dielectric permittivity than that of the normal sintering method [45]. The reported piezoelectric properties changed with the synthesis method from 190 pC/N when obtained by the conventional solid-state route, to 350 pC/N when synthesized by the hydrothermal method and later microwave sintered at 1300°C [46]. The authors attributed the high d_{33} value to the sintering density (98.3%), the appropriate grain size: 3.1–3.4 μm, and the high dielectric constant over 4000. Worth noting, the same authors correlate the excellent piezoelectric performance of the microwave-sintered BT with the microstructure (grain boundaries and domains size) as observed by different microscopic techniques: the size of the ferroelectric domains is less than 50 nm and the fraction of random boundaries is approximately 10% higher than that of conventionally sintered samples. This work has been followed by more recent papers using microwaves both for the synthesis and the sinterization (at 1150°C/30 min) with similar piezoelectric properties [47,48].

To conclude this section we can state that BT, whose piezoelectric properties are not competitive with those of PZT, is now coming into focus since it is a lead-free ferropiezoelectric whose properties are being improved using alternative methods of synthesis, many of which require lower temperatures and yield nanomaterials. The adequate sintering of these nanomaterials results in improved piezoelectric properties. Worth noting is that the use of microwave-assisted methods (both for synthesis and sintering) and, in general, all the work performed in BT could in principle be extended to BT-derived materials, such as the solid solutions with $Bi_{0.5}Na_{0.5}TiO_3$, among others.

9.3 Properties of lead-free piezoelectric ceramics

Most of the reviews on lead-free piezoelectrics focus on their properties as ultrasonic generator, actuator, and sensor within the linear range. In the following, a review of lead-free ferroelectric properties with emphasis on their potential use as high-power transducers, a field that has not acquired the same maturity level with lead-free ceramics as with lead-based ones, is presented.

There are many works in the literature that analyze the optimal properties of commercial, lead-based, piezoelectric ceramics for use as high-power transducers [49,50]. For that purpose the so-called hard PZT-doped compositions are used [7]. Materials to be used, for example, in ultrasonic motors, transformers, and high intensity focused ultrasound requires, in addition to sufficiently high piezoelectric coefficients, low mechanical and dielectric losses (i.e., high mechanical quality factor Q and low tanδ), which prevent excessive heat generation and hysteresis at use. The hardening effect of PZT materials is obtained by modifying the composition with acceptor dopants such as $Mn^{3+,4+}$ and Fe^{3+}, due to the induced internal bias field, which pins the domain wall and stabilizes the domains, thus increasing the coercive field and decreasing the losses [7].

To date, there is limited work [51] focusing on the hard BT-based piezoelectric ceramics with potential use for high-power applications. In the following the work on development of hard materials and characterization of the ceramics of four of the most explored families of lead-free ceramics will be revised. Production of ceramics by alternative synthesis routes will also be mentioned when available.

9.3.1 *Aurivillius-type structure ceramics*

Development of BLSFs or Aurivillius-type structure ceramics for piezoelectric use is relatively recent [52]. Ferroelectrics with this type of perovskite layered structure are characterized by the highest Curie temperatures, which can reach up to 1525°C [53]. However, it has been considered that for applications requiring good sensitivity or high power, they cannot be used due to their commonly very low sensitivity, for example, $d_{33}\sim 19$ pC/N for a typical bismuth titanate–based commercial composition for continuous use at 500°C (Pz46) [54]. Such piezoelectric coefficients, although small, are stable until very high temperatures, such as 800°C, when the Curie point is at 943°C [55]. BLSFs has been synthesized by several routes as molten salts [56] and by microwave-assisted hydrothermal, initially, the $Bi_4Ti_3O_{12}$ powder was synthesized via solid-state reaction [57]. However there is scarce characterization as piezoelectric of such materials.

BLSF compositions with higher piezoelectric properties are available, for example with $K_t \sim 25\%$, stable up to 600°C [58]. Some efforts to obtain compositions with hard characteristics [59] were done in the last years. In particular, it was found that with an increase in the vibration velocity, the rise in temperature for some ceramics was lower than that for PZT ceramics.

9.3.2 Alkaline niobates

This is the most promising and widely studied family of hard lead-free ceramics based on substitutions of the $(K,Na)NbO_3$ (KNN) solid-solution system [60, 61] that, similarly to PZT, presents a composition separating two polymorphs of the perovskite-type structure (polymorphic phase boundary), but this is temperature dependent one. Lithium and tantalum substituted KNN ceramics have been prepared by texturing ceramics [62] but also by simple solid state and pressureless sintering [63], without sintering aids or special powder handling, to reach high densities and high piezoelectric response. The latter method was expected to be easily scaled up making possible their industrial production. The piezoelectric properties are comparable to those in hard PZT ceramics for the converse piezoelectric effect (strain-field) ($k_t = 52\%$ and d_{33} over 300 pm/V).

Pure KNN ceramics present some disadvantages such as a poor densification and off-stoichiometry induced by volatilization during sintering. Actually, soft-chemistry routes succeeded in preparing phase-pure KNN or modified. One of them was the spray-drying method, as an alternative to the ceramic method, yielding nanosized powders that can be sintered to prepare ceramics of high density (96%) with improved properties [64]. With the same strategy, trying to avoid potassium or sodium losses by sublimation, these ceramics have been prepared by fast chemistry, by means of the microwave-hydrothermal method, at 160°C–220°C from 5 min to 7 h, with alkalinity of 6 M [65-68]. Also, KNN was successfully sintered by spark-plasma sintering (SPS) [68]. The specimen sintered by the SPS technique at 900°C for 15 min showed a higher density and better piezoelectric properties than the sample sintered by the conventional method.

One of the major drawbacks of this composition is the thermal stability, because of the onset of a phase transition (~200°C) between an orthorhombic and tetragonal polymorphs before its Curie temperature (~420°C), above which the piezoelectric properties are found to be greatly deteriorated. Appropriated doping can, however, extend the thermal stability range up to 250°C [69].

Many researchers have been focusing also on hard characteristics of KNN-based materials, modified by $K_{5.4}Cu_{1.3}Ta_{10}O_{29}$, $K_4CuNb_8O_{23}$ (KCN), or CuO, with piezoelectric coefficients $d_{33} < 00$ pC/N but high mechanical quality factors. The full set of material constants in the linear range, with limited treatment of losses, was determined for the KNN-KCN materials [70]. Matrix characterization is needed when numerical methods such as finite element analysis are used for 3D modeling of the material [71] in the design of devices. Even in the linear range, this characterization must include all losses to give place to accurate modeling. Compared with conventional hard PZT ceramics, the lead-free possessed lower dielectric and piezoelectric properties; however, comparable values of Q (~ 1500), dielectric loss (0.6%), and coercive fields (~ 12 kV/cm) were obtained [70]. Later works have further improved these properties in the linear range [72], but much work is still pending to improve those for high power use. Meanwhile, these ceramics have already been implemented in actual devices [73].

9.3.3 Bismuth-sodium titanates

The lead-free composition $Bi_{0.5}Na_{0.5}TiO_3$ (BNT) is known as ferroelectric since 1961 [74]. BNT presents high conductivity associated with loss of volatiles (Bi or $(Bi_{0.5}Na_{0.5})$) at the sintering temperature [75], and coercive fields that cause incomplete poling, thus preventing its use as piezoceramic. For these reasons, a large number of solid-solution systems of increasing complexity (BNT-$ATiO_3$ (A = Ca, Sr, Ba and Pb), BNT−$KNbO_3$(KN), BNT-$Bi_{0.5}Li_{0.5}TiO_3$(BNLT), BNT-$Bi_{0.5}K_{0.5}TiO_3$(BNKT), BNT-$K_{0.5}Na_{0.5}NbO_3$(BNT-KNN), BNT-BKT-KNN, BNT-BT-KNN, BNT-BKT-$BiFeO_3$, BNT-BKT-$BaTiO_3$-$SrTiO_3$, etc.) have been tested and the improvement of their piezoelectric properties with respect to BNT demonstrated [76].

In particular, the study of the $(1-x)(Bi_{0.5}Na_{0.5})TiO_3$-$xBaTiO_3$ (BNBT100x) system of solid solutions showed that there is a MPB, between rhombohedral and tetragonal ferroelectric phases at room temperature, for poled ceramics of compositions close to $x = 0.06$ (BNBT6). Even if there are still a number of open questions about this composition due to structural complexity and relaxor properties arising from a large coexistence of polymorphs [77,78] BNBT near MPB is considered a good candidate for lead-free piezoelectric ceramics due to the properties reported in the early 1990s [79]. Namely, an electromechanical coupling factor for thickness resonance of thin disks of $k_t = 52\%$ and a d_{33} piezoelectric coefficient of 125 pC/N were measured for BNBT6 ceramics sintered at 1200°C for 2 h in air. BNBT ceramics have been also fully characterized in matrix form [80], though not including all losses and their possibilities as high-power ceramic analyzed. To date the main limitation of the use of pure BNBT ceramics at the MPB for high power is the low depolarization temperature linked to a phase transition to a weakly polar phase at ~ 100°C [81], above which piezoelectricity quickly vanishes.

A $Q_m \sim 500$, Curie temperature of 260°C, and a dielectric constant of 520 was reported for $0.82(Bi_{0.5}Na_{0.5})TiO_3-0.15BaTiO_3-0.03(Bi_{0.5}Na_{0.5})(Mn_{1/3}Nb_{2/3})O_3$ (BNT-BT_BMNM) [82]. Recently, it was reported that Mn or Fe doping resulted in a considerable enhancement of Q_m in both planar and thickness resonance modes [83]. In 1.5 mol% Mn-doped ceramics sintered at 1100°C, a planar Q_m of about 970 and $\tan\delta$ of 0.88% were obtained. In Fe-doped ceramics, a planar Q_m as high as 900 was achieved. Acceptor dopants also resulted in decreasing the coupling coefficients, the piezoelectric charge coefficient, and the dielectric constant.

However, the application in which BNT-based led-free piezoelectric ceramics really defeats PZT for applications is in actuator, where the major drawback was that the hysteretic behavior had recently been minimized [84].

9.3.4 Barium-calcium titanate-zirconate

The newest family of high-performance lead-free ceramics ($d_{33}\sim 620$ pC/N), was found to arise from the pseudobinary solid-solution system $(1-x)Ba(Zr_yTi_{1-y})O_3-x(Ba_{1-z}Ca_z)TiO_3$ [85], and it is being extensively studied nowadays. Their matrix characterization, not taking into account all losses, was carried out for a representative composition in this system [86]. The drawbacks of these compositions for being amenable for commercial processing and practical use are the low Curie temperature $T_C\sim 93°C$ and the high synthesis (1300°C for 2 h) and ceramic sintering (1450°C for 3 h) temperatures required for their production. Some works have been focused on increasing T_C (up to 114°C) at the expense of the piezoelectric performance (down to 450 pC/N) by changing the composition, but still sintering at 1450°C for 3 h [87]. Sintering these ceramics at lower temperatures by conventional methods, thus reducing large grain size and enhancing the mechanical properties, is a difficult task [88]. For this purpose, alternative processing routes by using reactive precursors were successful to produce high sensitivity, dense, fine-grained materials by sintering below 1300°C [89]. Hardening of these ceramics and their high power applications is a relatively unexplored field [90].

9.3.5 Comparative data on properties for lead-free compositions

In this section a nonexhaustive summary of properties of the lead-free ferropiezoelectric under consideration is given with the purpose of identifying their differences and comparing them with lead-based ceramics. Tables 9.1–9.3 list the literature reference and type of composition, the mechanical quality factor (Q), and the electromechanical coupling coefficient (k_p) for the planar mode, together with the corresponding piezoelectric d_{33} coefficient and the relative dielectric permittivity (ε_r) and losses ($\tan\delta$) of selected compositions of the various families under consideration (excluding those of $Q < 100$).

BNT-, KNN-, and BCZT-based ceramics, for the different reasons already mentioned, do not have a working temperature range as high as that of lead-based ceramics. Only BLSFs present higher transition temperatures than those of lead-based materials, but they have the drawback of much lower piezoelectric activity. On the one hand, and notwithstanding the fact that KNN-based ceramics possess the highest Q values reported in the literature for ferropiezoelectric ceramics (Table 9.1), lead-free are not competitive for high-temperature, high-power applications [50] to date. On the other hand, these applications at the moment are led by single crystals instead of ceramics, some of which are also lead-free [115].

Lead-free ceramics can be compared with PZT with advantages from other points of view. For example, is has been shown that the change in mechanical quality factor with increasing vibration velocity is lower for some KNN- and BNT-based ceramics than for PZT [50].

Table 9.1 KNN-based ceramics

Material	ε_r	tan δ	d_{33} (pC/N)	K_p	Q	References
KNN	290	0.040	80	0.36	130	[91]
KNN(HP)	420	–	160	0.45	240	[92]
Sb-KNN	850	0.020	204	0.47	150	[93]
Ta-KNN	917	0.005	270	0.44	455	[94]
KNNLT	540	0.020	206	0.38	140	[95]
KNN-KCN	292	0.006	90	0.36	1500	[70]
KNN-$K_4CuNb_8O_{23}$	–	–	–	0.39	1200	[96]
0.75 mol% CuO-KNN	280	0.015		0.41	2100	[97]
0.02Li-KNN + 0.45 mol% CuO	–	–	93	0.25	1400	[98]
0.04Li-NN + 0.45 mol% CuO	–	–	100	–	414	[99]
0.02Li-KNN + 2 mol% CuO	–	–	70	–	742	[99]
0.5% CuO–KNN + KCT	280	0.02	–	0.14	2665	[100]

Table 9.2 BNT-based ceramics

Material	ε_r	tan δ	d_{33} (pC/N)	K_p	Q	References
BNT-BT-BMNM	520	0.007	110	0.41	500	[82]
Fe-doped-(BNT-BKT-BLT)	300	0.009	81	0.20	900	[83]
Mn-doped (BNT-BKT-BLT)	310	0.009	85	0.22	970	[83]
BNKLT88	380	0.011	95	0.26	400	[101]
BNT (Sr,Ca) + MnO 4%	518	–	–	0.12	365	[102]
BNLKT4-8	≈500	≈0.050	94	0.24	400	[103]
BNLKT4-8 + Mn 6%	≈500	≈0.040	85	0.22	740	[103]

Table 9.3 $(Ba_{1-y}Ca_y)$ $(Ti_{1-x}Zr_x)TiO_3$-based ceramics

Material	ε_r	tan δ	d_{33} (pC/N)	k_p	Q	References
$x = 0.07$ $y = 0.05$	1950	–	387	0.44	140	[104]
$x = 0.15$ $y = 0.10$ Ba excess	2700	0.015	310	0.39	122	[105]
$x = 0.15$ $y = 0.10$	4821	0.015	572	0.57	125	[106,107]
$x = 0.05$ $y = 0.04$	2070	0.015	338	0.36	–	[108]
$x = 0.08$ $y = 0.05$	2200	–	365	0.48	140	[109]
$x = 0.16$ $y = 0.10$	4800	–	328	0.38	230	[110]
$x = 0.15$ $y = 0.10$ 0.7% Co_2O_3	–	0.062	392	0.46	435	[111]
$x = 0.15$ $y = 0.085$ CuO 4%	3657	–	454	0.51	239	[112]
$x = 0.15$ $y = 0.10$	4000	5	540	0.46	160	[113]
$x = 0.15$ $y = 0.10$	2892	1.53	423	0.51	165	[114]

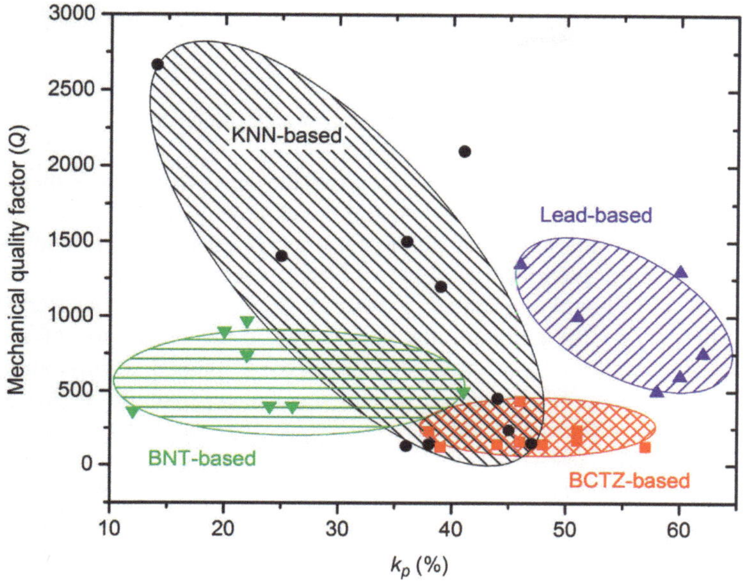

Figure 9.1 Mechanical quality factor (Q) versus electromechanical quality factor (k_p) for the planar mode of thin disks for lead-free and lead-based ferropiezoelectric ceramics.

As another example, the values of Q versus the electromechanical coupling factor (k_p) from Tables 9.1–9.3 are plotted in Fig. 9.1. and compared with some lead-based materials, as reported elsewhere [50].

For applications in which the ceramic is driven at a resonance mode, like flow meters, medical probes, ultrasonic machining tools, and so on, the product of the electromechanical coupling factor and the square of Q is a figure of merit. This product informs about a high performance together with low-heat generation during the material operation. For a given electromechanical coupling factor needed for the application, the higher their Q values the better the materials. In general, lead-free compositions do not have as high coupling combined with high Q as lead-based compositions. Efforts in hardening BCZT-based ceramics need to be done in order to compete with lead-based compositions in applications requesting high coupling coefficients. Fig. 9.1 shows that for a medium range of k_p (35%–45%), KNN compositions could be superior to lead-based ones in some applications [11,73] due to the higher values of Q. In this range, BNT-based ceramics are already of commercial use in ultrasonic cleaning baths [82].

Finally, considering nonresonant applications of ferropiezoelectrics like buzzers, ink jet printers, loudspeakers, camera lens modules, and so on, the considered figure of merit is the achievable ratio strain/electric field (S_{max}/E_{max}). After years of research on BNT-based compositions it has been proven that these ceramics are superior to PZTs as actuators due to the gigantic strain, up to 0.7% with low hysteresis [84], with respect to the typical 0.2% of remanent strain in PZT.

9.4 Future trends in the development of lead-free ferropiezoelectric ceramics

Research and development of lead-free piezoelectric materials remain the hottest topics in the field of piezoelectric ceramics [13,116].

The development of lead-free materials and the need to increase their properties for practical applications has led to the development of new processing techniques aiming to be alternatives to the expensive crystal growth techniques in the search of maximum polarization alignment with the electric field in ceramics. These techniques generate oriented growth of the ceramic grains (crystallites) that, nevertheless, do not eliminate the poling step in the fabrication of the piezoelectric ceramic. Textured ceramics also have zero remanent polarization, due to domain formation needed for electrical and mechanical stability of the ceramic. One of the most effective techniques is the reactive templated grain growth, which makes use of crystallites with a plate-like growth habit and compatible composition with the material to be obtained. These are mixed with the ceramic powder to be molded and sintered to obtain the ceramic promoting an oriented crystal growth during the formation of the ceramic grains. Another technique to obtain textured ceramics is the growth of a seed crystal inside a ceramic precursor, the so-called solid-state single crystal growth, using the abnormal grain growth of ceramic grains at the expense of the neighbor grains. Both techniques were applied to materials with high melting point and volatile components, as the alkaline niobates [117] and bismuth sodium titanates [118]. The composition control of the textured ceramics and single crystals by these methods is good, which ensures excellent properties.

It seems clear, nevertheless, that there is not a lead-free system that can directly replace commercial PZT, which can be properly doped for enhancement of the properties required in each application. Instead, the present approach is the development of the most appropriate lead-free composition for each application.

The lead-free perovskites with higher spontaneous polarization than lead-titanate, $BiFeO_3$ [119], among others are very promising for use as piezoelectrics as was established in previous chapters of this book. A number of problems remain to be solved for industrial applications (e.g., the control of the electrical resistivity or special conditions of stability). Other interesting materials as potential candidates to replace PZT, but not so widely explored yet, are the lead-free tungsten-bronze ferroelectric compounds [120,121].

Finally, it should not be forgotten that in the search of good piezoelectric materials, ferroelectric and antiferroelectric lead-free polar perovskites play an important role as materials for energy applications. Worth mentioning are energy harvesting [122], energy storage [123,124], and electrocaloric effects [125], applications on which there is also a strong research effort nowadays.

References

[1] L. Pardo, J. Ricote, M. Algueró, M.L. Calzada. Ferroelectric materials based on lead titanate. in: Handbook of Low and High Dielectric Constant Materials and Their Applications, vol. 1: Materials and Processing, pp. 457–499. Ed. by H.S. Nalwa. Academic Press, San Diego (USA), 1999 (Chapter 10).

[2] J. Wang, J.B. Neaton, H. Zheng, V. Nagarajan, S.B. Ogale, B. Liu, et al., Epitaxial $BiFeO_3$ multiferroic thin film heterostructures, Science 299 (5613) (2003) 1719–1722.

[3] M. Kosec, B. Malic, W. Wolny, A. James, C. Alemany, L. Pardo, Effect of a chemically aggressive environment on the electromechanical behaviour of modified lead titanate ceramics, J Korean Phys. Soc. 32 (1998) S1163–S1166.

[4] EU-Directive 2002/96/EC: Waste Electrical and Electronic Equipment (WEEE), Off. J. Eur. Union, 46 [L37] 24–38 (2003).

[5] EU-Directive 2002/95/EC: Restriction of the Use of Certain Hazardous Substances in Electrical and Electronic Equipment (RoHS) Off. J. Eur. Union, 46 [L37] 19–23 (2003).

[6] E. Ringgaard, T. Wurlitzer, Lead-free piezoceramics based on alkali niobates, J. Eur. Ceram. Soc. 25 (12) (2005) 2701–2706.

[7] J. Rödel, W. Jo, K.T.P. Seifert, E.M. Anton, T. Granzow, D. Damjanovic, Perspective on the development of lead-free piezoceramics, J. Am. Ceram. Soc. 92 (2009) 1153.

[8] J. Rödel, K.G. Webber, R. Dittmer, W. Jo, M. Kimura, D. Damjanovic, Transferring lead-free piezoelectric ceramics into application, J. Eur. Ceram. Soc. 35 (2015) 1659–1681.

[9] T. Ibn-Mohammed, S.C.L. Koh, I.M. Reaney, D.C. Sinclair, K.B. Mustapha, A. Acquaye, et al., Are lead-free piezoelectrics more environmentally friendly?, MRS Commun. (2017).

[10] S.O. Leonstev, R.E. Eitel, Progress in engineering high strain lead-free piezoelectric ceramics, Sci. Technol. Adv. Mater. 11 (4) (2010) 044302.

[11] E. Taghaddos, H. Mehdi, A. Safari, Lead-free piezoelectric materials and ultrasonic transducers for medical imaging, J. Adv. Dielect. 5 (2) (2015) 1530002.

[12] J. Wu, D. Xiao, J. Zhu, Potassium–sodium niobate lead-free piezoelectric materials: past, present, and future of phase boundaries, Chem. Rev. 115 (7) (2015) 2559–2595.

[13] C.-H. Hong, H.-P. Kim, B.-Y. Choi, H.-S. Han, J.S. Son, C.W. Ahn, et al., Lead-free piezoceramics, where to move on? J. Materiomics 2 (2016) 1e24.

[14] S. Priya, S. Nahm, Lead-Free Piezoelectrics, Springer, New York, 2011.

[15] H. Thurnauer, J. Deaderick. "Insulating Materials". American Lava Co. Patent 2.429.588, October 2, 1941.

[16] A. Von Hippel, Ferroelectricity, domain structure, and phase transitions of barium titanate, Rev. Mod. Phys. 22 (1950) 221.

[17] M.B. Smith, K. Page, T. Siegrist, P.L. Redmond, E.C. Walter, R. Seshadri, et al., Crystal structure and the paraelectric-to-ferroelectric phase transition of nanoscale $BaTiO_3$, J. Am. Chem. Soc. 130 (2008) 6955–6963.

[18] R. Bechmann, Elastic, piezoelectric, and dielectric constants of polarized barium titanate ceramics and some applications of the piezoelectric equations, J. Acoust. Soc. Am. 28 (1956) 347–350.

[19] S. Shao, J. Zhang, Z. Zhang, P. Zheng, M. Zhao, J. Li, et al., High piezoelectric properties and domain configuration in $BaTiO_3$ ceramics obtained through the solid-state reaction route, J. Phys. D Appl. Phys. 41 (2008) 125408.

[20] P.P. Phule, S.H. Risbud, Low-temperature synthesis and processing of electronic materials in the BaO-TiO_2 system, J. Mat. Sci. 25 (1990) 1169–1183.

[21] K. Kaur, Synthesis of various barium titanates, Int. J. IT Eng. Appl. Sci. Res. 4 (2015) 61–66.
[22] W.R. Buessem, L.E. Cross, A.K. Goswami, Phenomenological theory of high permittivity in fine-grained barium titanate, J. Am. Ceram. Soc. 49 (1966) 33–36.
[23] S.E. Park, S. Wada, L.E. Cross, T.R. Shrout, Crystallographically engineered $BaTiO_3$ single crystals for high-performance piezoelectrics, J. Appl. Phys. 86 (1999) 2746.
[24] L. Wang, L. Liu, D. Xue, H. Kang, C. Liu, Wet routes of high purity $BaTiO_3$ nanopowders, JALCOM 440 (2007) 78–83.
[25] W. Lu, M. Quilitz, H. Schmidt, Nanoscaled $BaTiO_3$ powders with a large surface area synthesized by precipitation from aqueous solutions: preparation, characterization and sintering, J. Eur. Cer. Soc. 24 (2007) 3149–3159.
[26] J. Moreno, J.M. Dominguez, A. Montoya, L. Vicente, T. Viveros, Synthesis and characterization of $MTiO_3$ (M = Mg, Ca, Sr, Ba) sol-gel, J. Mat. Chem. 5 (1995) 509–512.
[27] G. Pfaff, Sol–gel synthesis of barium titanate powders of various compositions, J. Mat. Chem. 2 (1992) 591–594.
[28] P. Yu, B. Cui, Q. Shi, Preparation and characterization of $BaTiO_3$ powders and ceramics by sol–gel process using oleic acid as surfactant, Mat. Sci. Eng. A 473 (2008) 34–41.
[29] M. Viviani, M.T. Buscaglia, A. Testino, V. Buscaglia, P. Bowen, P. Nanni, The influence of concentration on the formation of $BaTiO_3$ by direct reaction of $TiCl_4$ with $Ba(OH)_2$ in aqueous solution, J. Eur. Ceram. Soc. 23 (2003) 1383–1390.
[30] A. Testino, M.T. Buscaglia, V. Buscaglia, M. Viviani, C. Bottino, P. Nanni, Kinetics and mechanism of aqueous chemical synthesis of $BaTiO_3$ particles, Chem. Mater. 16 (2004) 1536–1543.
[31] X. Wang, B.I. Lee, M. Hu, E.A. Payzant, D.A. Blom, Nanocrystalline $BaTiO_3$ powder via a sol process ambient conditions, J. Eur. Ceram. Soc. 26 (2006) 2319–2326.
[32] J. Tao, J. Ma, Y. Wang, X. Zhu, J. Liu, X. Jiang, Synthesis of barium titanate nanoparticles via a novel electrochemical route, Mater. Res. Bull. 43 (2008) 639–644.
[33] H. Xu, L. Gao, Hydrothermal synthesis of high-purity $BaTiO_3$ powders: control of powder phase and size, sintering density, and dielectric properties, Mater. Lett. 58 (2004) 1582–1586.
[34] P.K. Dutta, R. Asiaie, S.A. Akbar, W. Zhy, Hydrothermal synthesis and dielectric properties of tetragonal $BaTiO_3$, Chem. Mater. 6 (1994) 1542–1548.
[35] D.F.K. Hennings, C. Metzmacher, B.S. Schreinemacher, Defect chemistry and microstructure of hydrothermal barium titanate, J. Am. Ceram. Soc. 84 (2001) 179–182.
[36] J. Prado-Gonjal, R. Schmidt, Morán, Microwave-assisted routes for the synthesis of complex functional oxides, Inorganics, 3, (2015) 101–117.
[37] S. Komarneni, Q. Li, K.M. Stefansson, R. Roy, Microwave-hydrothermal processing for synthesis of electroceramic powders, J. Mater. Res. 8 (1993) 3176–3183.
[38] B.L. Newalkar, S. Komarneni, H. Katsuki, Microwave-hydrothermal synthesis and characterization of barium titanate powders, Mater. Res. Bull. 36 (2001) 2347–2355.
[39] S.F. Liu, I.R. Abothu, S. Komarneni, Barium titanate ceramics prepared from conventional and microwave hydrothermal powders, Mater. Lett. 38 (1999) 344–350.
[40] S. Komarneni, H. Katsuki, Microwave-hydrothermal synthesis of barium titanate under stirring condition, Ceram. Int. 36 (2010) 1165–1169.
[41] W. Sun, J. Li, Microwave-hydrothermal synthesis of tetragonal barium titanate, Mater. Lett. 60 (2006) 1599–1602.
[42] A.Z. Simões, F. Moura, T.B. Onofre, M.A. Ramirez, J.A. Varela, E. Longo, Microwave-hydrothermal synthesis of barium strontium titanate nanoparticles, J. Alloys Compd. 508 (2010) 620–624.

[43] Y.B. Khollam, A.S. Deshpande, A.J. Patil, H.S. Potdar, S.B. Deshpande, S. Date, Microwave-hydrothermal synthesis of equi-axed and submicron-sized $BaTiO_3$ powders, Mater. Chem. Phys. 71 (2001) 304−308.
[44] W. Sun, W. Liu, J. Li, Effects of chloride ions on hydrothermal synthesis of tetragonal $BaTiO_3$ by microwave heating and conventional heating, Powder Technol. 166 (2006) 55−59.
[45] H.T. Kim, Y.H. Han, Sintering of nanocrystalline $BaTiO_3$, Ceram. Int. 30 (2004) 1719−1723.
[46] K. Sadhana, T. Krishnaveni, K. Praveena, S. Bharadwaj, S.R. Murthy, Microwave sintering of nanobarium titanate, Scr. Mater. 59 (2008) 495−498.
[47] H. Takahashi, Y. Numamoto, J. Tani, K. Matsuta, J. Qiu, S. Tsurekawa, Lead-free barium titanate ceramics with large piezoelectric constant fabricated by microwave sintering, Jpn. J. Appl. Phys. Part 2 45 (2006) 43−46.
[48] H. Takahashi, Y. Numamoto, J. Tani, S. Tsurekawa, Piezoelectric properties of $BaTiO_3$ ceramics with high performance fabricated by microwave sintering, Jpn. J. Appl. Phys. Part 1 45 (2006) 7405−7408.
[49] H.N. Shekhani, E.A. Gurdal, S.O. Ural, K. Uchino, Analysis of high power behavior in piezoelectric ceramics from a mechanical energy density perspective, arXiv preprint arXiv 1605 (2016) 06685.
[50] S. Zhang, R. Xia, L. Lebrun, D. Anderson, T.R. Shrout, Piezoelectric materials for high power, high temperature applications, Mat. Lett. 59 (2005) 3471−3475.
[51] L. Chen, H. Fan, Q. Li, Characterization of acceptor-doped (Ba, Ca)TiO_3 "hard" piezoelectric ceramics for high-power applications, Ceram. Int. 43 (2017) 5579−5584.
[52] L. Pardo, A. Castro, P. Millán, C. Alemany, R. Jiménez, B. Jiménez, $(Bi_3TiNbO_9)_x$-$(SrBi_2Nb_2O_9)_{1-x}$ Aurivillius type structure piezoelectric ceramics obtained from mechanochemically activated oxides, Acta Mater. 48 (9) (2000) 2421−2428.
[53] H. Yan, H. Ning, Y. Kan, P. Wang, M.J. Reece, Piezoelectric strontium niobate and calcium niobate ceramics with super-high curie points, J. Am. Ceram. Soc. 93 (5) (2009) 1409−1413.
[54] http://ferroperm-piezo.com/.
[55] H. Yan, H. Zhang, R. Ubic, M. J. Reece, J. Liu, Z. Shen, et al., A lead-free high-curie -point ferroelectric ceramic, $CaBi_2Nb_2O_9$, Adv. Mat. 17 (10) (2005) 1261−1265.
[56] M. García-Guadarrama, L. Fuentes, M.E. Montero-Cabrera, A. Márquez-Lucero, M.E. Villafuerte-Castrejón, Molten salts synthesis and crystal structure of $Bi_5Ti_3FeO_{15}$, Int. Ferroelectrics 71 (2005) 233−239.
[57] D. Thomazini, M.V. Gelfuso, J.A. Eiras, Microwave assisted hydrothermal synthesis of $Bi_4Ti_3O_{12}$ nanopowders from oxide as raw materials, Powder Tech. 222 (2012) 139−142.
[58] Z.G. Gai, J.F. Wang, M.L. Zhao, C.M. Wang, G.Z. Zang, B.Q. Ming, et al., High temperature $(NaBi)_{0.48}\square_{0.04}Bi_2Nb_2O_9$-based piezoelectric ceramics, Appl. Phys. Lett. 89 (2006) 012907.
[59] Y. Noumura, Y. Hiruma, H. Nagata, T. Takenaka, High-power piezoelectric characteristics at continuous driving of $Bi_4Ti_3O_{12}$−$SrBi_4Ti_4O_{15}$-based ferroelectric ceramics, Jpn. J. Appl. Phys. 50 (2011), 07HB06.
[60] G. Shirane, R.E. Newnham, R. Pepinsky, Dielectric properties and phase transitions of $NaNbO_3$ and (Na, K)NbO_3, Phys. Rev. 96 (1) (1954) 581.
[61] J.F. Li, K. Wang, F.Y. Zhu, L.Q. Cheng, F.Z. Yao, K,Na)NbO_3-based lead-free piezoceramics: fundamental aspects, proc technologies, and remaining challenges, J. Am. Ceram. Soc. 96 (12) (2013) 3677−3696.

[62] Y. Saito, H. Takao, T. Tani, T. Nonoyama, K. Takatori, T. Homma, et al., Lead-free piezoceramics, Nature 432 (2004) 84.

[63] E. Hollenstein, M. Davis, D. Damjanovic, N. Setter, Piezoelectric properties of Li- and Ta-modified $K_{0.5}Na_{0.5}NbO_3$ ceramics, Appl. Phys. Lett. 87 (2005) 182905.

[64] R. López, F. González, M.E. Villafuerte-Castrejón, Structural and electrical characterization of $(K_{0.48}Na_{0.52})_{0.96}Li_{0.04}Nb_{0.85}Ta_{0.15}O_3$ synthesized by spray drying, J. Eur. Ceram. Soc. 30 (2010) 1549−1553.

[65] R. López-Juárez, R. Castañeda-Guzmán, M.E. Villafuerte-Castrejón, Fast synthesis of $NaNbO_3$ and $K_{0.5}Na_{0.5}NbO_3$ by microwave hydrothermal method, Ceram. Int. 40 (2014) 14757−14764.

[66] Y. Zhou, J. Yu, M. Guo, M. Zhang, Microwave hydrothermal synthesis and piezoelectric properties investigation of $K_{0.5}Na_{0.5}NbO_3$ lead-free ceramics, Ferroelectrics 404 (1) (2010) 69−75.

[67] M. Zhang, M. Guo, Y. Zhou, Low-temperature preparation of $K_xNa_{1-x}NbO_3$ lead-free piezoelectric powders by microwave-hydrothermal synthesis, Int. J. Appl. Ceram. Technol. 8 (3) (2011) 591−596.

[68] R. López-Juárez, F. González-García, J. Zárate-Medina, R. Escalona-González, S.D. de la Torre, M.E. Villafuerte-Castrejón, Piezoelectric properties of Li−Ta co-doped potassium−sodium niobate ceramics prepared by spark plasma and conventional sintering, J. Alloys Compd. 509 (2011) 3837−3842.

[69] J. Fu, R. Zuo, X. Wang, L. Li, Polymorphic phase transition and enhanced piezoelectric properties of $LiTaO_3$-modified $(Na_{0.52}\ K_{0.48})(Nb_{0.93}\ Sb_{0.07})O_3$ lead-free ceramics, J. Phys. D: Appl. Phys. 42 (2009) 012006.

[70] S. Zhang, J.B. Lim, H.J. Lee, T.R. Shrout, Characterization of hard piezoelectric lead-free ceramics, IEEE Trans. Ultrasonics Ferroelectrics Freq. Control 56 (8) (2009) 1523.

[71] L. Pardo, K. Brebøl. Properties of ferro-piezoelectric ceramic materials in the linear range: determination from impedance measurements at resonance, in: L. Pardo and J. Ricote (Eds). Multifunctional Polycrystalline Ferroelectric Materials: Processing and Properties Springer Series in Materials Science, vol. 140. Springer, Londres (UK), 2011 (Chapter 13).

[72] K. Xu, J. Li, X. Lv, J. Wu, X. Zhang, D. Xiao, et al., Superior piezoelectric properties in potassium−sodium niobate lead-free ceramics, Adv. Mater. 28 (2016) 8519−8523.

[73] A. Mathieson, D. DeAngelis, Analysis of lead-free piezoceramic-based power ultrasonic transducers for wire bonding, IEEE Trans. UFFC 63 (1) (2016) 156−164.

[74] G.A. Smolenskii, V.A. Isupov, A.I. Agranovskaya, N.N. Krainik, New ferroelectrics of complex composition IV, Soviet Physics Solid State 2 (11) (1961) 2651−2654.

[75] Y. Hiruma, H. Nagata, T. Takenaka, Thermal depoling process and piezoelectric properties of bismuth sodium titanate ceramics, J. Appl. Phys. 105 (2009), 084112- 084112-8.

[76] T. Takenaka, H. Nagata, Y. Hiruma, Phase transition temperatures and piezoelectric properties of $(Bi_{1/2}Na_{1/2})TiO_3$- and $(Bi_{1/2}K_{1/2})TiO_3$-based bismuth perovskite lead-free ferroelectric ceramics, IEEE Trans. Ultrasonics, Ferroelectrics Freq. Control 56 (8) (2009) 1595−1612.

[77] L. Pardo, E. Mercadelli, K. Brebøl, A. García, C. Galassi, Piezoelectric properties of lead-free submicron structured $(Bi_{0.5}Na_{0.5})_{0.94}Ba_{0.06}TiO_3$ ceramics, Smart Mater. Struct. 19 (11) (2010) 11507.

[78] L. Pardo, E. Mercadelli, A. García, K. Brebøl, C. Galassi, Field-induced phase transition and relaxor character in submicron structured lead-free $(Bi_{0.5}Na_{0.5})_{0.94}Ba_{0.06}TiO_3$ piezoceramics at the morphotropic phase boundary, IEEE Trans. UFFC 58 (9) (2011) 1893−1904.

[79] T. Takenaka, K. Maruyama, K. Sakata, $(Bi_{1/2}Na_{1/2})TiO_3$-$BaTiO_3$ system for lead-free piezoelectric ceramics, Jpn. J. Appl. Phys. 30 (9B) (1991) 2236–2239.
[80] H.J. Lee, S.O. Ural, L. Chen, K. Uchino, S. Zhang, High power characteristics of lead-free piezoelectric ceramics, J. Am. Ceram. Soc. 95 (11) (2012) 3383–3386.
[81] L. Pardo, A. García, K. Brebøl, E. Mercadeli, C. Galassi. Anisotropy and dynamic thermal depolarization of $0.96(Bi_{0.5}Na_{0.5})TiO_3 - 0.04\ BaTiO_3$ lead-free piezoceramics, Book Series: IEEE International Symposium on Applications of Ferroelectrics (2012).
[82] T. Tou, Y. Hamaguti, Y. Maida, H. Yamamori, K. Takahashi, Y. Terashima, Properties of $(Bi_{0.5}Na_{0.5})TiO_3$-$BaTiO_3$-$(Bi_{0.5}Na_{0.5})(Mn_{1/3}Nb_{2/3})O_3$ lead-free piezoelectric ceramics and its application to ultrasonic cleaner, Jpn. J. Appl. Phys. (2009), 48 07GM03, 4pp.
[83] E. Taghaddos, M. Hejazi, A. Safari, Electromechanical properties of acceptor-doped lead-free piezoelectric ceramics, J. Am. Ceram. Soc. 97 (6) (2014) 1756–1762.
[84] T. Li, X. Lou, X. Ke, S. Cheng, S. Mi, X. Wang, et al., Giant strain with low hysteresis in A-site-deficient $(Bi_{0.5}\ Na_{0.5})TiO_3$-based lead-free piezoceramics, Acta Mater. 128 (2017) 337–344.
[85] W. Liu, X. Ren, Large piezoelectric effect in Pb-free ceramics, Phys. Rev. Lett. 103 (2009) 257602.
[86] D. Xue, Y. Zhou, H. Bao, C. Zhou, J. Gao, X. Ren, Elastic, piezoelectric, and dielectric properties of $Ba(Zr_{0.2}Ti_{0.8})O_3$-$50(Ba_{0.7}Ca_{0.3})TiO_3$ Pb-free ceramic at the morphotropic phase boundary, J. Appl. Phys. 109 (2011) 054110.
[87] A modified lead-free piezoelectric BZT–xBCT system with higher T_C, in: H. Bao, C. Zhou, D. Xue, J. Gao, X. Ren (Eds.), J. Phys. D: Appl. Phys. 43, 2010.
[88] M. Jiang, Q. Lin, D.M. Lin, Q.J. Zheng, X.M. Fan, X.C. Wu, et al., Effects of MnO_2 and sintering temperature on microstructure, ferroelectric, and piezoelectric properties of $Ba_{0.85}Ca_{0.15}Ti_{0.90}Zr_{0.10}O_3$ lead-free ceramics, J. Mater. Sci. 48 (2013) 1035–1041.
[89] A. Reyes, L. Pardo, R. López, A.M. Gonzalez, M.P. Cruz, M.E. Villafuerte-Castrejón, Lead-free $Ba_{0.9}Ca_{0.1}Ti_{0.9}Zr_{0.1}O_3$ piezoelectric ceramics processed below 1300°C, J. Alloys Compd. 584 (2014) 28–33.
[90] Y. Zhang, H. Sun, W. Chen, Improved electrical properties of low-temperature sintered Cu doped $Ba_{0.99}Ca_{0.01}Zr_{0.02}Ti_{0.98}O_3$ ceramics, J. Electron. Mat. 45 (10) (2016) 5006–5016.
[91] L. Egerton, D.M. Dillon, Piezoelectric and dielectric properties of ceramics in the system potassium sodium niobate, J. Am. Ceram. Soc. 42 (9) (1959) 438–442.
[92] R.E. Jaeger, L. Egerton, Hot pressing of potassium-sodium niobates, J. Am. Ceram. Soc. 45 (5) (1962) 209–213.
[93] D.M. Lin, K.W. Kwok, H.Y. Tian, H.W.L.W. Chan, Phase transitions and electrical properties of $(Na_{1-x}K_x)(Nb_{1-y}Sb_y)O_3$ lead-free piezoelectric ceramics with a MnO_2 sintering aid, J. Am. Ceram. Soc. 90 (5) (2007) 1458–1462.
[94] D.M. Lin, K.W. Kwok, H.L.W. Chan, Phase transition and electrical properties of $(K_{0.5}Na_{0.5})(Nb_{1-x}Ta_x)O_3$ lead-free piezoelectric ceramics, Appl. Phys. A 91 (1) (2008) 167–171.
[95] Y.H. Zhen, J.F. Li, Normal sintering of (K, Na)NbO_3-based ceramics: influence of sintering temperature on densification, microstructure, and electrical properties, J. Am. Ceram. Soc. 89 (12) (2006) 3669–3675.
[96] M. Matsubara, T. Yamaguchi, K. Kikuta, S.I. Hirano, Sinterability and piezoelectric properties of (K,Na)NbO_3 ceramics with novel sintering aid, Jpn. J. Appl. Phys. 43 (10) (2004) 7159 (Part 1).
[97] S.L. Yang, C.C. Tsai, Y.C. Liou, C.S. Hong, B.J. Li, S.Y. Chu, Investigation of CuO-doped NKN ceramics with high mechanical quality factor synthesized by a B-site oxide precursor method, J. Am. Ceram. Soc. 95 (3) (2012) 1011–1017.

[98] E. Li, H. Kakemoto, T. Hoshina, T. Tsurumi, A shear-mode ultrasonic motor using potassium sodium niobate-based ceramics with high mechanical quality factor, Jpn. J. Appl. Phys. 47 (9) (2008) 7702–7706.

[99] E. Li, H. Kakemoto, S. Wada, T. Tsurumi, Enhancement of Q_m by co-doping of Li and Cu to potassium sodium niobate lead-free ceramics, IEEE Trans. Ultrasonics Ferroelectrics Freq. Control 55 (5) (2008) 980–987.

[100] H.S. Han, J. Koruza, E.A. Patterson, J. Schultheiß, E. Erdem, W. Jo, et al., Hardening behavior and highly enhanced mechanical quality factor in $(K_{0.5}Na_{0.5})NbO_3$–based ceramics, J. Eur. Cer. Soc. 37 (5) (2017) 2083–2089.

[101] M. Hejazi, E. Taghaddos, E. Gurdal, K. Uchino, A. Safari, High power performance of manganese-doped BNT-based Pb-free piezoelectric ceramics, J. Am. Ceram. Soc. 97 (10) (2014) 3192–3196.

[102] N. Ichinose, K. Udagawa, Piezoelectric properties of $(Bi_{1/2}Na_{1/2})TiO_3$ based ceramics, Ferroelectrics 169 (1) (1995) 317–325.

[103] Y. Hiruma, T. Watanabe, H. Nagata, T. Takenaka, Piezoelectric properties of $(Bi_{1/2}Na_{1/2})TiO_3$ -based solid solution for lead-free high-power applications, Jpn. J. Appl. Phys. 47 (9S) (2008) 7659.

[104] W. Li, Z. Xu, R. Chu, P. Fu, G. Zang, High piezoelectric d_{33} coefficient of lead-free $(Ba_{0.93}Ca_{0.07})(Ti_{0.95}Zr_{0.05})O_3$ ceramics sintered at optimal temperature, Mat. Sci. Eng. B 176 (2011) 65–67.

[105] P. Wang, Y. Wang, Y. Lu, Y. Li, Microstructure and electrical properties of lead-free piezoelectric Ba-excess $(Ba_{0.85(1+0.7\%)}Ca_{0.15})(Ti_{0.9}Zr_{0.1})O_3$ ceramics by microwave sintering, Ferroelectrics 458 (2014) 163–173.

[106] Y. Tian, X. Chao, L. Wei, P. Liang, Z. Yang, Phase transition behavior and electrical properties of lead-free $(Ba_{1-x}Ca_x)(Zr_{0.1}Ti_{0.9})O_3$ piezoelectric ceramics, J. Appl. Phys. 113 (2013) 184107.

[107] Y. Tian, L. Wei, X. Chao, Z. Liu, Z. Yang, Phase transition behavior and large piezo-electricity near the morphotropic phase boundary of lead-free $(Ba_{0.85}Ca_{0.15})(Zr_{0.1}Ti_{0.9})$ O_3 ceramics, J. Am. Ceram. Soc. 96 (2) (2013) 496–502.

[108] S.W. Zhang, H. Zhang, B.P. Zhang, S. Yang, Phase-transition behavior and piezoelectric properties of lead-free $(Ba_{0.95}Ca_{0.05})(Ti_{1-x}Zr_x)O_3$ ceramics, JALCOM 506 (2010) 131–135.

[109] W. Li, Z. Xu, R. Chu, P. Fu, G. Zang, Piezoelectric and dielectric properties of $(Ba_{1-x}Ca_x)$ $(Ti_{0.95}Zr_{0.05})O_3$ lead-free ceramics, J. Am. Ceram. Soc. 93 (10) (2010) 2942–2944.

[110] W. Li, Z. Xu, R. Chu, P. Fu, G. Zang, Polymorphic phase transition and piezoelectric properties of $(Ba_{1-x}Ca_x)(Ti_{0.9}Zr_{0.1})O_3$ lead-free ceramics, Phys. B 405 (2010) 4513–4516.

[111] Y. Kun, J. Xiangping, C. Chao, L. Lu, L. Xiaohong, Effects of Co_2O_3 doping on properties of BCTZ lead-free piezoelectric ceramics, Piezoelectris Acoustooptics 36 (3) (2014) 441–444.

[112] S.H. Shin, J.H. Yoo, Electrical properties of $(Ba,Ca)(Ti,Zr)O_3$ ceramics for bimorph-type piezoelectric actuator, Trans. Elec. Elect. Mat. 15 (4) (2014) 226–229.

[113] L. Pei-Qing, L. Xi-Tao, Y. Zhi-Guo, Effect of sintering process on microstructure and properties of $(Ba_{0.85}Ca_{0.15})(Ti_{0.9}Zr_{0.1})O_3$ lead-free ceramics, J. Inorg. Mat. 32 (3) (2017) 300–304.

[114] J. Wu, D. Xiao, W. Wu, Q. Chen, J. Zhu, Z. Yang, et al., Composition and poling condition-induced electrical behavior of $(Ba_{0.85}Ca_{0.15})(Ti_{1-x}Zr_x)$, J. Eur. Ceram. Soc. 32 (2012) 891–898.

[115] Y. Yokota, T. Kudo, Y. Ohashi, A. Medvedev, S. Kurosawa, K. Kamada, et al., Crystal growth of $Ca_3Nb(Ga_{1-x}Al_x)_3Si_2O_{14}$ piezoelectric single crystals with various Al concentrations, Materials 8 (9) (2015) 5597−5605.

[116] M.E. Villafuerte-Castrejón, E. Morán, A. Reyes-Montero, R. Vivar-Ocampo, J.A. Peña-Jiménez, S.O. Rea-López, et al., Towards lead-free piezoceramics: facing a synthesis challenge, Materials 9 (1) (2016) 21 (27pp).

[117] J.G. Fisher, A. Benčan, M. Kosec, S. Vernay, D. Rytz, Growth of dense single crystals of potassium sodium niobate by a combination of solid-state crystal growth and hot pressing, J. Am. Ceram. Soc. 91 (5) (2008) 1503.

[118] H. Yilmaz, S. Trolier-Mckinstry, G.L. Messing, (Reactive) templated grain growth of textured sodium bismuth titanate ($Na_{1/2}$ $Bi_{1/2}$ TiO_3-$BaTiO_3$) ceramics—II dielectric and piezoelectric properties, J. Electr. 11 (2003) 217−226.

[119] T. Rojac, A. Bencan, B. Malic, G. Tutuncu, J.L. Jones, J.E. Daniels, et al., $BiFeO_3$ ceramics: processing, electrical, and electromechanical properties, J. Am. Ceram. Soc. 97 (7) (2014) 1993−2011.

[120] Y. Doshida, H. Shimizu, Y. Mizuno, K. Itoh, S. Hirose, H. Tamura, Non-linear behaviour and high-power properties of $(Bi,Na,Ba)TiO_3$ and $(Sr,Ca)_2$ $NaNb_5O_{15}$ piezoelectric ceramics, Jpn. J. Appl. Phys. 50 (2011), 09ND06.

[121] J.R. Oliver, R.R. Neurgaonkar, L.E. Cross, Ferroelectric properties of tungsten bronze morphotropic phase bounary systems, J. Am. Ceram. Soc. 72 (1989) 202−211.

[122] X. Yan, M. Zheng, Y. Hou, M. Zhu, Composition-driven phase boundary and its energy harvesting performance of BCZT lead−free piezoelectric ceramic, J. Eur. Ceram. Soc. 37 (7) (2017) 2583−2589.

[123] A. Mishra, B. Majumdar, R. Ranjan, A complex lead-free (Na, Bi, Ba)(Ti, Fe)O_3 single phase perovskite ceramic with a high energy-density and high discharge-effiischarge-efty and high dceramic with a high e, J. Eur. Ceram. Soc. 37 (6) (2017) 2379−2384.

[124] H. Yang, F. Yan, Y. Lin, T. Wang, F. Wang, Y. Wang, et al., Lead-free $BaTiO_3$-$Bi_{0.5}$ $Na_{0.5}$ $TiO_3Na_{0.73}$ $Bi_{0.09}$ NbO_3 relaxor ferroelectric ceramics for high energy storage, J. Eur. Ceram. Soc. 37 (2017) 3303.

[125] A.K. Axelsson, F. Le Goupil, M. Valant, N. Mc.N. Alford, Electrocaloric effect in lead-free Aurivillius relaxor ferroelectric ceramics, Acta Mater. 124 (2017) 120−126.

Section III

Ferroelectric Metal Oxides Application

Compositionally-graded ferroelectric ceramics and multilayers for electronic and sensing applications

10

Adelina-Carmen Ianculescu[1], Ioana Pintilie[2] and Lucian Pintilie[2]
[1]"Politehnica" University of Bucharest, Bucharest, Romania
[2]National Institute of Materials Physics, Magurele, Romania

10.1 Review of the current situation

The concept of graded ferroelectrics became of significant interest for researchers working in the field of ferroelectric materials and related applications during the 1990s, when a number of published papers reported a series of intriguing phenomena observed in this type of structure [1–5]. Among these are the shift of the hysteresis loop along the polarization axis, the presence of a giant effective pyroelectric coefficient, and the possibility of using the graded structures as ferroelectrics similar to semiconductor p–n junctions. Some pioneering experimental work was performed on simple structures composed of two layers with different values of polarization, called also ferroelectric bimorph [6,7]. During the years, the interest for graded structures remained significant both from the point of view of the basic research as well as considering the potential applications in electronics, sensing, and energy. The concept of graded ferroelectric structures (GFSs) includes all structures in which a polarization gradient is present. This gradient can be obtained in several ways, such as (1) producing a concentration gradient along the direction normal to the electrodes [8]; (2) producing a temperature gradient between the opposite electrodes of the capacitor [9]; and (3) producing a strain gradient along the direction normal to the electrodes [10].

The interest in GFSs stemmed primarily from the experimentally observed shift of the hysteresis loop along the polarization axis (see Fig. 10.1). This shift was named the polarization offset and was considered an intrinsic property of GFS. Therefore, polarization offset was observed on GFS samples manufactured from various materials by different techniques. Almost all the studies related to GFS were performed on thin films and only a few were performed on thick films or bulk ceramics. Among the investigated materials, the most studied are ones of the types $Pb(Zr,Ti)O_3$ (PZT) and $(Ba,Sr)TiO_3$ (BST). The polarization gradient was induced at the beginning by changing the composition, namely by changing the Zr/Ti ratio or the Ba/Sr ratio, respectively [11,12]. Variants of these materials, obtained by replacing the component

Magnetic, Ferroelectric, and Multiferroic Metal Oxides. DOI: http://dx.doi.org/10.1016/B978-0-12-811180-2.00010-4

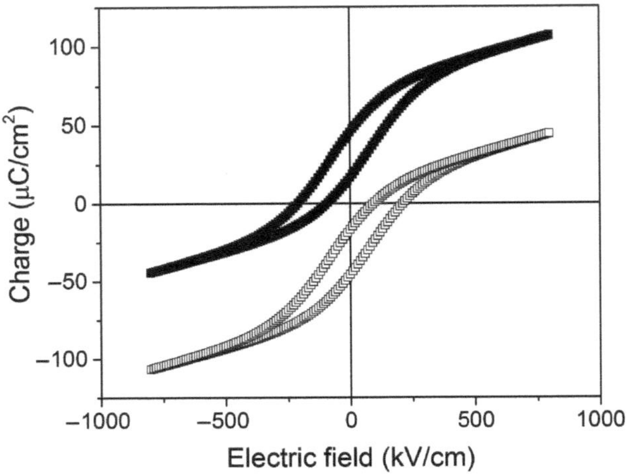

Figure 10.1 Graphical representation of the shift of the hysteresis loop along the polarization axis. The direction of the shift depends on the direction of the polarization gradient inside the GFS sample.

elements with other homo- or heterovalent substitutions, were also studied (e.g., substitution of Pb in PZT with La or Ca, or substitution of Ti in BST with Zr) [13–15]. Graded structures were obtained by a variety of deposition techniques, such as spin-coating (sol-gel method), tape casting, metal-organic deposition, metal-organic vapor deposition, or pulse laser deposition (PLD) [16–18].

Depending on the orientation of the concentration gradient the GFS were considered upgraded if the concentration increases from the substrate toward the surface, or downgraded if the concentration increased from the surface toward the substrate. It was observed that the polarization offset is dependent on the orientation of the concentration gradient being positive for upgraded structures and negative for the downgraded ones [19]. Further, it was found that the polarization offset has temperature dependence, leading to giant values for the so-called effective pyroelectric coefficient, reaching values of 0.05 C/m^2K [20]. Therefore, GFSs were expected to generate a new type of pyroelectric detectors with enhanced characteristics. There were also other reports claiming that pyroelectric properties in GFS are enhanced compared to the component single-phase materials, but not because of the polarization offset [21]. Other properties of interest for applications are the dielectric ones, since it was found that GFS can have enhanced tenability compared to single-phase samples [22], making them attractive for microwave devices.

Many theoretical models were developed to explain the properties of GFS, based mainly on Landau–Ginzburg–Devonshire thermodynamic theory. A comprehensive review of thermodynamic-based models can be found in Mantese and Alpay [23]. Other models developed to explain the polarization offset have considered electrostatic conditions inside GFS, time-dependent space charge currents, or asymmetric leakage currents, due to the presence of asymmetric Schottky-like contacts at the electrode interfaces [24–28]. However, the polarization offset was not observed in all the studied GFSs, as there are reports presenting hysteresis loops without any shift along the polarization axis [29,30]. Moreover, it was shown that shifts can be obtained in nominally homogeneous samples if heated at elevated temperatures and if simple Sawyer-Tower circuits are used to record the hysteresis [31]. It appears that the polarization offset is more likely an artifact of the method used to record the hysteresis or may be related to some asymmetries in the sample not related to polarization gradient (e.g., asymmetric electrodes, or asymmetric distribution of defects across the sample thickness).

In any case most of the theoretical models predict that important built-in fields should exist in samples with polarization gradient, either electric fields shifting the hysteresis loop along the voltage axis or stress fields generating anomalous distributions of ferroelectric domains [32,33]. Various models were developed in recent years to explain the properties of GFS, including Ising-type models or others based on first principle [34,35]. It was even suggested that ferroelectrics with a polarization gradient can be included in the class of materials presenting geometric frustration [36].

10.2 Recent results

10.2.1 Graded bulk BST (Ba,Sr)TiO_3 ceramics

As previously mentioned, most of the studies on GFS were performed on multilayers formed by successive deposition of films with different compositions, so that a concentration gradient occurs on the direction normal to the substrate. However, there are situations when thicker samples are required, such as for monitoring high-power, high-intensity laser beams. Bulk BST ceramics were recently obtained from powders prepared by sol-gel and it was shown that these bulk samples possess very good pyroelectric properties [37]. Same homogeneous ceramics were then prepared by spark plasma sintering (SPS). Graded samples were also prepared by SPS, varying the Sr content across the thickness from 0% to 30%. The microstructural and compositional details of the graded structure are presented in Fig. 10.2.

The structural analysis has revealed that the homogeneous ceramics obtained by SPS present a good density of about 98% from the crystal density, and that the average size of the crystalline grains are below 1 μm, decreasing from about 0.65 μm for 0% Sr to about 0.17 μm for 30% Sr. In the graded structures, the average size of the crystalline grains varies from about 3 μm in the region with low Sr content to about 2 μm in the region with high Sr content (regions 1 and 3 in Fig. 10.2A).

The dielectric properties of the homogeneous and graded ceramics were investigated by measuring the capacitance as function of temperature. The extracted relative permittivity and loss tangent are presented in Fig. 10.3 for a frequency of 1 kHz. We can see that the three phase transitions specific for the materials derived from $BaTiO_3$ are clearly visible, suggesting a good crystalline structure of the

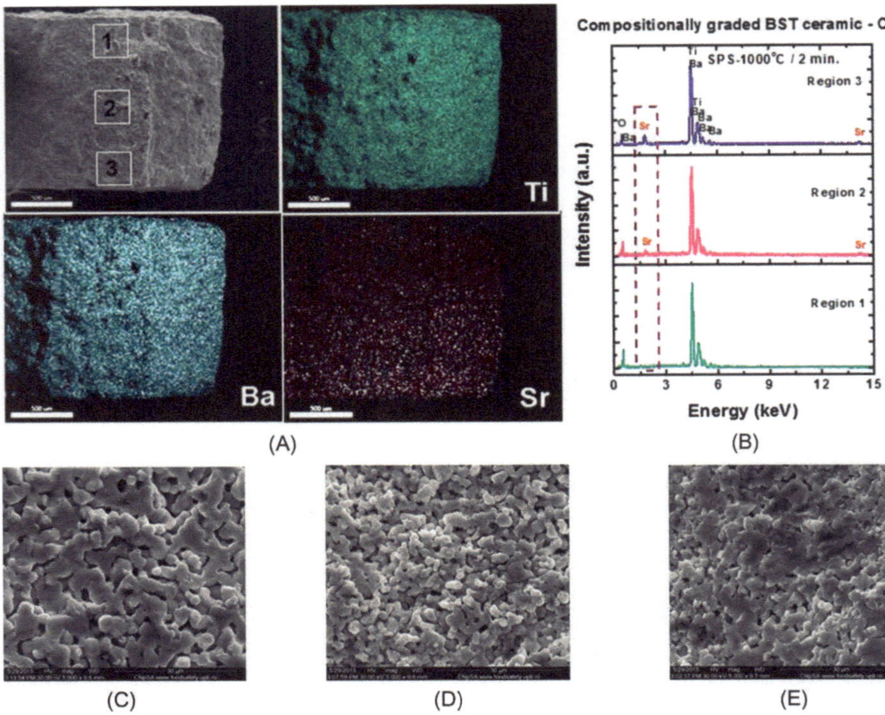

Figure 10.2 (A) Cross-section SEM images for the graded structure sintered by spark plasma sintering at 1000°C for 2 min, together with the EDX mapping showing the compositional gradient in Sr content; (B) EDX spectra corresponding to regions numbered as 1, 2, and 3 in (A); (C)–(E) SEM images evidencing the microstructures of region as 1, 2, and 3.

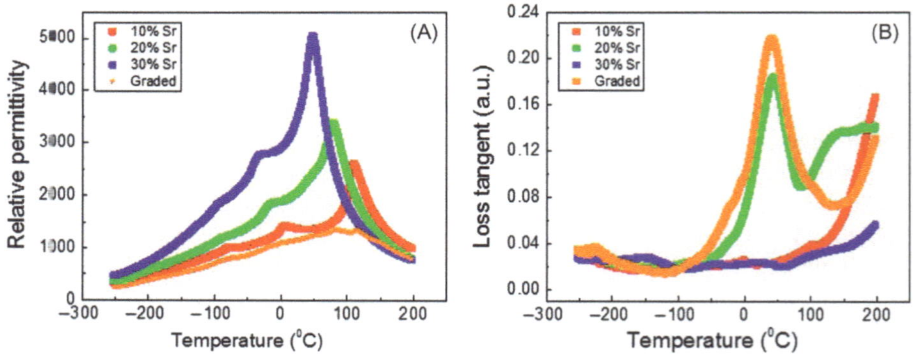

Figure 10.3 The temperature dependence of the relative permittivity (A) and loss tangent (B) for the homogeneous BST ceramics and for the graded structure [37].

ceramic. The transition temperature may vary, depending on whether the measurement was performed in air or in a vacuum. In the last case the transition temperatures can be up to 10°C higher than in air. Regarding the losses around room temperature, these are low for compositions with 10% and 30% Sr content but are higher for 20% Sr content and for the graded structure. The losses drop below 5% at frequencies above 500 kHz, making these materials attractive for devices working at high frequencies. The relative permittivity is lower and relatively constant on a broad temperature range for the graded structure.

Pyroelectric properties were also investigated by measuring the pyroelectric current function of temperature and extracting the pyroelectric coefficient as in the Byer-Roundy method [38]. The results obtained for the graded BST ceramics mentioned earlier are presented in Fig. 10.4A. For a relatively large temperature domain (from 0°C to about 55°C) the pyroelectric coefficient of the graded structure is larger than the values obtained for the homogeneous, single-phase BST ceramics.

The pyroelectric signal measured in the voltage mode shows a clear increase with the Sr content. However, the signal measured for the graded structure, whichever face was exposed to the IR beam, is lower. This can be explained by the high losses at low frequencies, considering that the pyroelectric signal is measured at frequencies well below 1 kHz. Another observation is that the value for the pyroelectric coefficient is lower than for coarse-grained ceramics obtained by classical sintering, a result that it is in agreement with a previous report showing that the pyroelectric properties are enhanced as the size of the crystalline grains increases [39,40]. Nevertheless, this preliminary results shows that it is possible to obtain enhanced pyroelectric properties in bulk-graded ceramics. Similar results, namely higher and relatively uniform pyroelectric properties on a broader range of temperatures, were obtained for PZT-type ceramics processed by SPS [41].

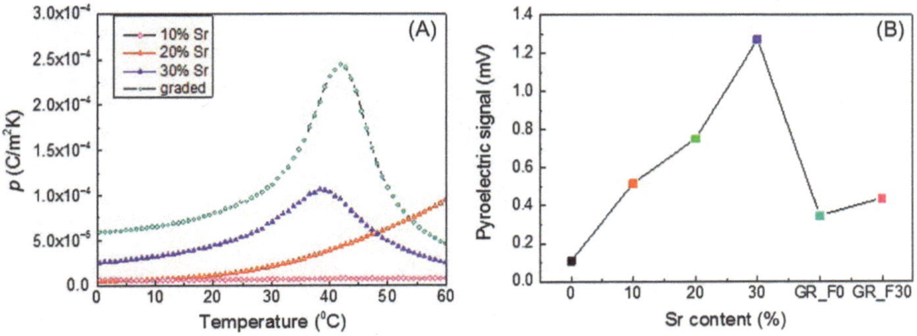

Figure 10.4 (A) The temperature dependence of the pyroelectric coefficient for homogeneous BST ceramics obtained by spark plasma sintering, and for the graded structure; (B) the pyroelectric signal measured at fixed frequency of 25 Hz for the BST ceramics with different Sr content and for the graded structure (GR_FO is for the case when the structure was exposed to the IR beam with the face containing 0% Sr; GR_F30 is for the case when the exposed face was with 30% Sr content).

10.2.2 Graded epitaxial multilayers

Epitaxial multilayers of PZT type were grown by PLD (see the schematic of the structures in Fig. 10.5A). The thickness of the component layers was selected in such a way that the films are relaxed, knowing that in the case of the PZT materials the relaxation occurs at thicknesses of about 50 nm [42]. This is to remove the strain as a possible source of changes in the graded structure and to observe only the effect of the concentration gradient.

Lattice parameters were determined by X-ray diffraction (see Fig. 10.5C), based on the correlation of information from 2θ-ω diffraction with those obtained from the map of intensity in reciprocal space. In the case of the downgraded sample the last deposited layers (Zr/Ti ratio of 52/48 in the middle and 20/80 on top) increased as *a* domain, meaning that the direction of the polarization is in the plane of the sample. Similar results were obtained from transmission electron microscopy investigations (see Fig. 10.5B). This could be the main reason for the significant difference between the properties of the two types of multilayer structures, as observed from the electrical measurements presented in Fig. 10.5C.

We can see from Fig. 10.6 that the ferroelectric and dielectric properties are much better for the upgraded structure. A significant shift along the voltage axis can be observed in this case, suggesting that the layers are not fully relaxed and that a strain field still exists in the structure. Similar results were reported in [43]. The hysteresis loop for the downgraded structure is symmetric along the voltage axis, suggesting that the films are fully relaxed in this case, in good agreement with the results of the structural investigations showing a small out-of-plane component of the polarization. The shape of the hysteresis loop is affected by leakage current in this case. Also, the C-V behavior is different, with higher capacitance for the

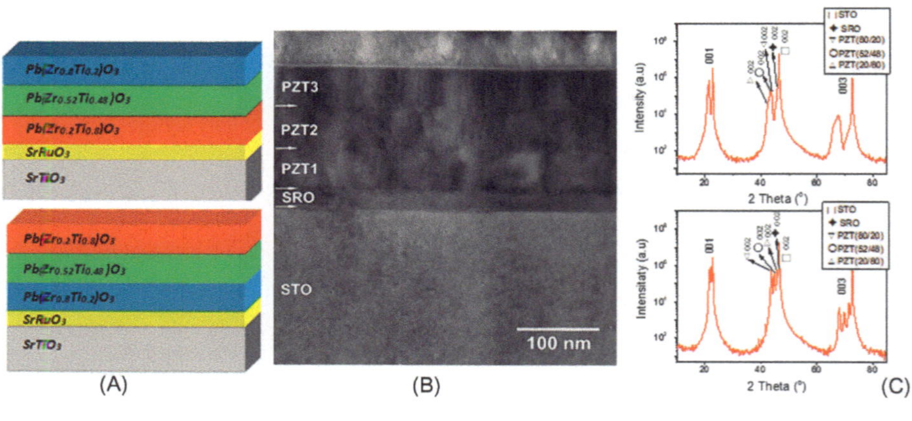

Figure 10.5 (A) The schematic of the epitaxial graded structures; (B) low magnification TEM image revealing the films in the upgraded structures; (C) X-ray diffractograms for the upgraded and downgraded structures presented in (A). Upgraded or downgraded refers to the variation of the Zr/Ti ratio starting from the substrate (increasing for upgraded and decreasing).

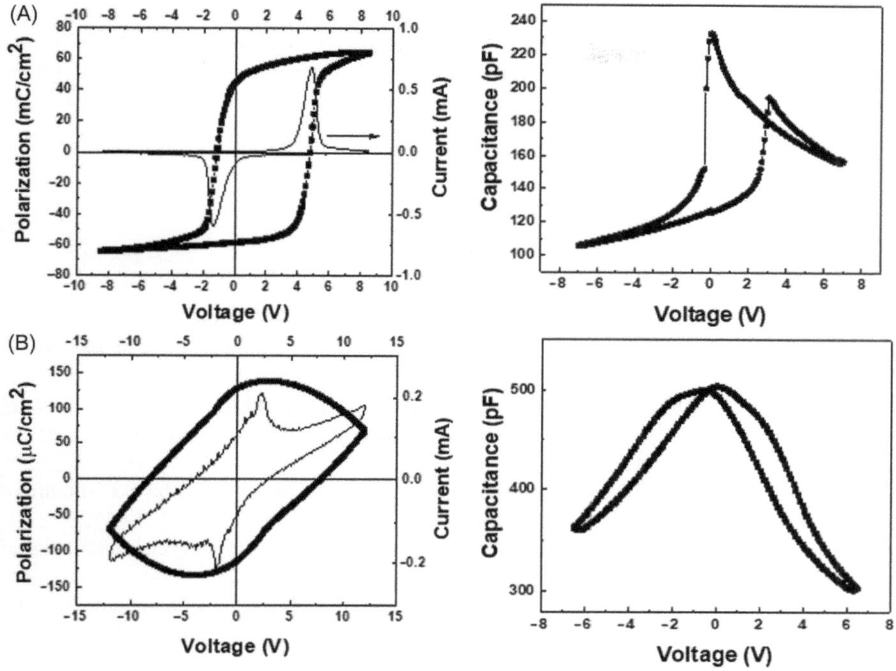

Figure 10.6 (A) Hysteresis loop and C-V characteristic for the upgraded structure; (B) hysteresis loop and C-V characteristic for the downgraded structure, respectively [42].

downgraded structure (total thickness and electrode area was the same in both cases). This can be explained by a larger amount of charged structural defects in the downgraded structure, as suggested by structural investigations.

I-V measurements have revealed the fact that the leakage current is about two order of magnitude larger in the downgraded structure compared to the upgraded one. This is a good explanation for the fact that no pyroelectric signal was obtained for the downgraded structure. As for the upgraded structures, it presents good pyroelectric properties, with a value of the pyroelectric coefficient of about 1.47×10^{-4} C/m^{2}K as determined from the frequency dependence of the pyroelectric signal measured in the voltage mode [44].

10.3 Conclusions and trends

GFSs proved to be of significant interest for research in the field of ferroelectric materials, due to their potential applications in areas such as microwave devices (enhanced tenability and dielectric properties), pyroelectric detection and energy generation through pyroelectric effect, or for energy storage [45]. Photovoltaic applications are also considered for GFS [46]. The concept of graded structures can be extended to other ferroic structures, possible leading to completely new functionalities [47]. Also, the gradient concept can be extended to defect or impurity gradients, as well as to graded nanocomposites [48,49]. However, there is still a lot of work to do in order to optimize the preparation conditions for the graded structures, in close relation with their functional properties.

References

[1] J.V. Mantese, N.W. Schubring, A.L. Micheli, A.B. Catalan, Ferroelectric thin films with polarization gradients normal to the growth surface, Appl. Phys. Lett. 67 (5) (1995) 721–723.

[2] J.V. Mantese, N.W. Schubring, A.L. Micheli, A.B. Catalan, M.S. Mohammed, R. Naik, et al., Slater model applied to polarization graded ferroelectrics, Appl. Phys. Lett. 71 (14) (1997) 2047–2049.

[3] M. Brazier, M. McElfresh, S. Mansour, Unconventional hysteresis behavior in compositionally graded $Pb(Zr,Ti)O_3$ thin films, Appl. Phys. Lett. 72 (1998) 1121–1123.

[4] Z. Chen, K. Arita, M. Lim, C.A. Paz De Araujo, Asymmetric C-V characteristics of graded PZT thin film capacitors, Integr. Ferroelectr. 24 (1999) 189–194.

[5] N.W. Schubring, J.V. Mantese, A.L. Micheli, A.B. Catalan, M.S. Mohammed, R. Naik, et al., Graded ferroelectrics: a new class of steady-state thermal/electrical/mechanical energy interchange devices, Integr. Ferroelectr. 24 (1999) 155–168.

[6] L. Pintilie, C. Constantin, Dielectric and pyroelectric properties of a ceramic bimorph structure, Ferroelectics 173 (1995) 111–124.

[7] L. Pintilie, I. Pintilie, Temperature dependence of the pyroelectric voltage in a 2-2 connectivity pyroelectric bimorph, Ferroelectrics 200 (1997) 219–235.

[8] D. Bao, X. Yao, L. Zhang, Dielectric enhancement and ferroelectric anomaly of compositionally graded $(Pb,Ca)TiO_3$ thin films derived by a modified sol–gel technique, Appl. Phys. Lett. 76 (2000) 2779–2781.

[9] W. Fellberg, J. Mantese, N. Schubring, A. Micheli, Origin of the "up" and "down" hysteresis offsets observed from polarization graded ferroelectric materials, Appl. Phys. Lett. 78 (2001) 524–526.

[10] J.V. Mantese, N.W. Schubring, A.L. Micheli, M.P. Thompson, R. Naik, G.W. Auner, et al., Stress-induced polarization-graded ferroelectrics, Appl. Phys. Lett. 81 (2002) 1068–1070.

[11] A. Khodorov, S.A.S. Rodrigues, M. Pereira, M.J.M. Gomes, Impedance spectroscopy study of a compositionally graded lead zirconate titanate structure, J. Appl. Phys. 102 (2007) 114109.

[12] M.S. Mohammed, G.W. Auner, R. Naik, J.V. Mantese, N.W. Schubring, A.L. Micheli, et al., Temperature dependence of conventional and effective pyroelectric coefficients for compositionally graded $Ba_xSr_{1-x}TiO_3$ films, J. Appl. Phys. 84 (1998) 3322–3325.

[13] I. Boerasu, L. Pintilie, M. Kosec, Ferroelectric properties of $Pb_{1-3y/2}La_y(Zr_{0.4}Ti_{0.6})O_3$ structures with La concentration gradients, Appl. Phys. Lett. 77 (2000) 2231–2233.

[14] X.G. Tang, J. Wang, Y.W. Zhang, H.L.W. Chan, A.L. Ding, C.L. Choy, Leakage current and pyroelectric properties of compositionally graded $(Pb,Ca)TiO_3$ films, Appl. Phys. A 78 (2004) 1205–1209.

[15] C. Wang, B.L. Cheng, S.Y. Wang, H.B. Lu, Y.L. Zhou, Z.H. Chen, G.Z. Yang, Improved dielectric properties and tunability of multilayered thin films of $(Ba_{0.80}Sr_{0.20})(Ti_{1-x}Zr_x)O_3$ with compositionally graded layer, Appl. Phys. Lett. 84 (2004) 765–767.

[16] J. Zhai, X. Yao, Z. Xu, H. Chen, Enhancement of ferroelectricity in the compositionally graded $(Pb,Sr)TiO_3$ thin films derived by a sol–gel process, J. Cryst. Growth 286 (2006) 37–41.

[17] N.M. Sbrockey, M.W. Cole, T.S. Kalkur, M. Luong, J.E. Spanier, G.S. Tompa, MOCVD growth of compositionally graded $Ba_xSr_{1-x}TiO_3$ thin films, Integr. Ferroelectr. 126 (2011) 21–27.

[18] X.H. Zhu, H.L.W. Chan, C.L. Choy, K. Wong, D. Hesse, A comparative microstructural study of compositionally up- and down-graded $(Ba,Sr)TiO_3$ thin films epitaxially grown on (La, Sr)CoO_3-covered MgO(100) substrates by pulsed laser deposition, Appl. Phys. A 82 (2006) 709–713.

[19] D. Wu, H. Wu, Z. Fu, C. Zhao, A. Li, Polarization offsets of compositionally graded Nd-substituted $Bi_4Ti_3O_{12}$ ferroelectric thin films, Appl. Phys. Lett. 23 (2008) 062904.

[20] F. Jin, G.W. Auner, R. Naik, N.W. Schubring, J.V. Mantese, A.B. Catalan, et al., Giant effective pyroelectric coefficients from graded ferroelectric devices, Appl. Phys. Lett. 73 (1998) 2838–2840.

[21] R.V.K. Mangalam, J.C. Agar, A.R. Damodaran, J. Karthik, L.W. Martin, Improved pyroelectric figures of merit in compositionally graded $PbZr_{1-x}Ti_xO_3$ thin films, ACS Appl. Mater. Inter. 5 (2013) 13235–13241.

[22] C.V. Weiss, M.B. Okatan, S.P. Alpay, M.W. Cole, E. Ngo, R.C. Toonen, Compositionally graded ferroelectric multilayers for frequency agile tunable devices, J. Mater. Sci. 44 (2009) 5364–5374.

[23] J.V. Mantese, S.P. Alpay, Graded ferroelectrics, transpacitors and transponents, in: O. Auciello, R. Ramesh (Eds.), Multifunctional Thin Films Series, Springer Science + Business Media Inc, New Yotk, USA, 2005.

[24] L. Pintilie, I. Boerasu, M.J.M. Gomes, Simple model of polarization offset of graded ferroelectric structures, J. Appl. Phys. 93 (12) (2003) 9961–9967.

[25] C.K. Wong, F.G. Shin, Modeling of anomalous hysteresis behavior of compositionally graded ferroelectric films at low fields, J. Appl. Phys. 98 (2005) 024104.

[26] H.K. Chan, C.H. Lam, F.G. Shin, Time-dependent space-charge-limited conduction as a possible origin of the polarization offsets observed in compositionally graded ferroelectric films, J. Appl. Phys. 95 (5) (2004) 2665–2671.

[27] R. Bouregba, G. Poullain, B. Vilquin, G. Le Rhun, Asymmetrical leakage currents as a possible origin of the polarization offsets observed in compositionally graded ferroelectric films, J. Appl. Phys. 93 (2003) 5583–5591.

[28] J. Barrel, E. Stytsenko, M. Viviani, K. MacKenzie, Origin of PE hysteresis offsets in compositionally graded $Ba_{(1-x)}Sr_xTiO_3$ thick films, Adv. Mat. Res. 55-57 (2008) 15–22.

[29] J. Zhai, X. Yao, Z. Xu, H. Chen, Ferroelectric properties of $Pb_xSr_{1-x}TiO_3$ and its compositionally graded thin films grown on the highly oriented $LaNiO_3$ buffered $Pt/Ti/SiO_2/Si$ substrates, J. Appl. Phys. 100 (2006) 034108.

[30] L. Jiankang, Y. Xi, Dependence of the properties of compositionally graded $Pb(Zr,Ti)O_3$ ferroelectric films of the bottom electrode, Ceram. Int. 34 (2008) 1031–1034.

[31] H. Wu, D. Wu, A. Li, Polarization offset of homogeneous $Bi_{3.15}Nd_{0.85}Ti_3O_{12}$ ferroelectric thin films, J. Appl. Phys. 104 (2008) 074117.

[32] J.C. Agar, A.R. Damodaran, G.A. Velarde, S. Pandya, R.V.K. Mangalam, L.W. Martin, Complex evolution of built-in potential in compositionally-graded $PbZr_{1-x}Ti_xO_3$ thin films, ACS Nano 9 (2015) 7332−7342.
[33] J.C. Agar, A.R. Damodaran, M.B. Okatan, J. Kacher, C. Gammer, R.K. Vasudevan, et al., Highly mobile ferroelastic domain walls in compositionally graded ferroelectric thin films, Nat. Mater. 15 (2016) 549−556.
[34] Q. Zhang, I. Ponomareva, Microscopic insight into temperature-graded ferroelectrics, Phys. Rev. Lett. 105 (2010) 147602.
[35] H.-X. Cao, V.C. Lo, Z.Y. Li, Dielectric response of temperature-graded ferroelectric films, J. Appl. Phys. 98 (2005) 114105.
[36] N. Choudhury, L. Walizer, S. Lisenkov, L. Bellaiche, Geometric frustration in compositionally modulated ferroelectrics, Nature 470 (2011) 513−517.
[37] A. Ianculescu, I. Pintilie, C.A. Vasilescu, M. Botea, A. Iuga, A. Melinescu, et al., Intrinsic pyroelectric properties of thick, coarse grained $Ba_{1-x}Sr_xTiO_3$ ceramics, Ceram. Int. 42 (2016) 10338−10348.
[38] R.L. Byer, C.B. Roundy, Pyroelectric coefficient direct measurement technique and application to a nsec response time detector, Ferroelectrics 3 (1972) 333−338.
[39] T.M. Kamel, G. de With, Grain size effect on the poling of soft $Pb(Zr,Ti)O_3$ ferroelectric ceramics, J. Eur. Ceram. Soc. 28 (2008) 851−861.
[40] C. Mao, S. Yan, S. Cao, C. Yao, F. Cao, G. Wang, et al., Effect of grain size on phase transition, dielectric and pyroelectric properties of BST ceramics, J. Eur. Ceram. Soc. 34 (2014) 2933−2939.
[41] K. Kakegawa, S.S. Suzuki, N. Uekawa, Design of pyroelectric properties by controlling compositional distribution, J. Eur. Ceram. Soc. 26 (2006) 613−617.
[42] L. Pintilie, C. Ghica, C. Teodorescu, I. Pintilie, C. Chirila, I. Pasuk, et al., Polarization induced self-doping in epitaxial $Pb(Zr_{0.20}Ti_{0.80})O_3$ thin films, Sci. Rep. 5 (2015) 14974.
[43] J. Karthik, R.V.K. Mangalam, J.C. Agar, L.W. Martin, Large built-in electric fields due to flexoelectricity in compositionally graded ferroelectric thin films, Phys. Rev. B 87 (2013) 024111.
[44] M. Botea, A. Iuga, L. Pintilie, Giant pyroelectric coefficient determined from the frequency dependence of thepyroelectric signal generated by epitaxial $Pb(Zr_{0.2}Ti_{0.8})O_3$ layers grown on single crystal $SrTiO_3$ substrates, Appl. Phys. Lett. 103 (2013) 232902.
[45] Y. Zhao, X. Hao, Q. Zhang, Enhanced energy-storage performance and electrocaloric effect in compositionally graded $Pb_{(1\_3x/2)}La_xZr_{0.85}Ti_{0.15}O_3$ antiferroelectric thick films, Ceram. Int. 42 (2016) 1679−1687.
[46] Y. Zhou, Enhancement of photovoltaic effect in nanoscale polarization graded ferroelectrics, Sol. Energy 86 (2012) 811−815.
[47] X. Wang, E. Pan, J.D. Albrecht, Exact closed-form electromagnetic Green's functions for graded uniaxial multiferroic materials, J. Appl. Phys. 103 (2008) 113511.
[48] P. Chakraborty, S.B. Krupanidhi, Polarization enhancement in compositionally graded vanadium doped bismuth titanate, thin films, J. Appl. Phys. 107 (2010) 124105.
[49] N. Wongdamnern, D. Maurya, Y. Zhou, M. Sanghadasa, R. Yimnirun, S. Priya, Complex permittivity scaling of functionally graded composites, Mater. Res. Express. 1 (2014) 016305.

Review of the most common relaxor ferroelectrics and their applications

11

Jelena D. Bobic, Mirjana M. Vijatovic Petrovic and Biljana D. Stojanovic
Institute for Multidisciplinary Research, University of Belgrade, Belgrade, Serbia

11.1 Introduction

A relaxor ferroelectric material was first synthesized by Smolenskiiin 1961 [1]. Since then, ferroelectric materials with diffuse phase transition (DPT) characteristics and/or relaxor properties have been extensively studied mainly due to their very interesting and still not completely explained physical properties. Understanding the origin of these properties and physical background is a key for useful applications. In despite of their controversial physical behavior, relaxor ferroelectrics with DPTs had found applications in different devices such as piezoelectric actuators, multilayer capacitors, medical imaging devices, nonvolatile memories, pyroelectric detectors, and microwave tunable applications [2,3].

A generic feature of relaxors is a broad maximum in temperature dependence of the dielectric permittivity, whose position, T_m, is shifted to lower temperatures as the frequency of the probing field decreases. Different models have been proposed by the various authors to understand the diffuse phase transition of the relaxor materials: order-disorder, compositional fluctuation, dipole glass, and random field theory [4]. On the atomic scale all relaxors are compositionally (chemically) disordered; that is, different cations are distributed more or less randomly among the equivalent lattice sites, forming quenched, randomly oriented dipoles and creating the associated local random field. These dipoles and field exist even in the paraelectric phase, which is macroscopically and mesoscopically nonpolar. Upon cooling paralectric phase transform into the ergodic relaxor state (state with the lowest free energy, which is a similar state regardless of the initial conditions) in which polar regions of nanometer scale (polar nanoregions; PNRs) with randomly distributed directions of dipole moments appear. These regions of locally correlated polarizations appear at the Burns temperature, T_B, which is typically far above the maximum point of permittivity, T_m [5]. In spite of the fact that the term phase transition is still used for relaxors, the maximum of the permittivity does not correspond to the real phase transition to the ferroelectric state, since the long-range order does not arise here. Instead, in relaxors, the polarization is correlated on a local scale resulting in the appearance of polar nanometer-size regions.

Magnetic, Ferroelectric, and Multiferroic Metal Oxides. DOI: http://dx.doi.org/10.1016/B978-0-12-811180-2.00011-6

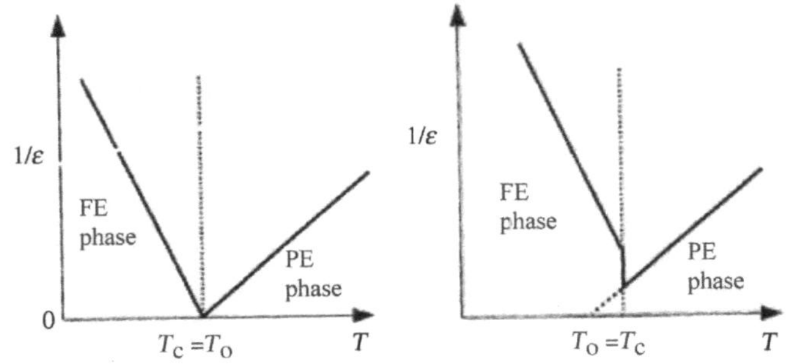

Figure 11.1 Temperature dependence of inverse permittivity in normal ferroelectrics with second order (left) and first order (right) phase transition at Curie point.

It is a well-known fact that the dielectric constant of a normal ferroelectric follows the Curie–Weiss law, which is described by the relation:

$$1/\varepsilon = (T - T_c)/C \qquad (11.1)$$

where T_c is the Curie temperature and C is the Curie constant. The Curie–Weiss plot between $1/\varepsilon$ and temperature gives a straight line with a slope of $(1/C)$ and the x-axis intercept at T_c (Fig. 11.1). In normal ferroelectrics a phase transition can be of the first and second order. In the case of continuous (second order) FE transitions T_o coincide with the phase transition (Curie) point T_c and relation (Eq. 11.1) is applicable also at $T < Tc$ but with different C. The first order transition occurred at $Tc > T_0$, and T_0 (which is theoretically the stability limit of the paraelectric phase) can be determined experimentally only by extrapolation.

The broadness in the ε_r versus temperature curve is one of the most important characteristics of the disordered perovskite structure showing relaxor behavior. A modified Curie–Weiss relationship is used to study the diffuseness behavior of a ferroelectric phase transition, given as: [6]

$$1/\varepsilon - 1/\varepsilon_{\max} = (T - T_{\max})^{\gamma}/C, \text{ for } (T > T_{\max}) \qquad (11.2)$$

where C is the modified Curie–Weiss constant and the value of γ $(1 \le \gamma \le 2)$ indicates the degree of diffuseness of the material. In the case of normal ferroelectrics $\gamma = 1$, and $\gamma = 2$ for relaxors that correspond to polar domain fluctuations in the neighborhood of the maximum permittivity.

The parameter ΔT_m, which describes the degree of the deviation from Curie–Weiss law, is defined as:

$$\Delta T_m = T_{cw} - T_m \qquad (11.3)$$

where T_{cw} denotes the temperature from which the permittivity starts to deviate from Curie–Weiss law and T_m represents the temperature of the dielectric

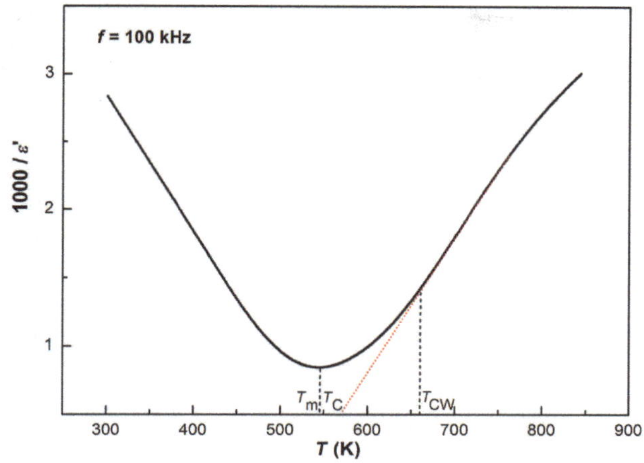

Figure 11.2 Temperature dependence of $1/\varepsilon'$ at 1 kHz, as an example of the general Curie–Weiss-like behavior.

maximum. Fig. 11.2 presents the temperature dependence of $1/\varepsilon'$ at 1 kHz as an example of the general Curie–Weiss-like behavior.

Recently, several relations have been attempted to describe the dielectric permittivity $\varepsilon(T)$ of relaxors at $T > T_m$, one of which is the Lorentz-type empirical relation: [7]

$$\frac{\varepsilon_A}{\varepsilon} = 1 + \frac{(T - T_A)^2}{2\delta_A^2} \tag{11.4}$$

where the temperature (T_A) and the magnitude (ε_A) of the Lorenz height generally differ from the T_m and ε_m of the experiments. The parameter δ_A, being practically independent of frequency, can be considered as a convenient measure for the degree of diffuseness of the permittivity peaks in relaxors. Eq. (11.4) describes the high temperature ($T > T_m$) dielectric permittivity quite well in a number of relaxor ferroelectrics [8].

Frequency dispersion in relaxor ferroelectrics can be characterized by the Vogel–Fulcher relation. The relation characterizes normal ferroelectrics, if the $\ln\nu$ versus $1/T$ plot is linear. However, in relaxors, $\ln\tau$ versus $1/T$ is nonlinear and the Vogel–Fulcher relation [9],

$$\nu = \nu_o \exp\left\{\frac{-E_a}{k_B(T_m - T_f)}\right\} \tag{11.5}$$

is used to describe the experimental results. ν_0 is the Debye frequency—the attempt frequency of dipole reorientation, E_a is the activation energy (i.e., the energy barrier between two equivalent polarization states), k_B is the Boltzmann's constant, and T_f is the static freezing temperature of the polarization fluctuations (i.e., the temperature below which dynamic reorientation of the dipolar cluster polarization can no longer be thermally activated). The degree of relaxation in the frequency range of relaxor ferroelectrics could be described by an empirical parameter ΔT_{relax}:

$$\Delta T_{relax} = T_{m\ (high\ frequency)} - T_{m\ (low\ frequency)} \tag{11.6}$$

The objective of the present chapter is to show the studies carried out on several of the most investigated relaxor ferroelectrics, Pb-based perovskites. Lead magnesium niobate ($Pb(Mg_{1/3}Nb_{2/3})O_3$ (PMN), lanthanum-doped lead zirconate titanate $Pb_{1-y}La_y(Zr_{1-x}Ti_x)_{1-y/4}O_3$ (PLZT $100_{(y/1-x/x)}$), and lead zinc titanate Pb $(Zn_{1/3}Nb_{2/3})O_3$ (PZN) are discussed. Also, a general overview of the structure, dielectric and piezoelectric studies of the Aurivillius relaxor materials has been carried out in his chapter. The results are mostly given from the literature data.

11.2 Lead-based perovskite relaxors

11.2.1 Lead magnesium niobate (PMN)

$Pb(Mg_{1/3}Nb_{2/3})O_3$ (PMN) is an important ferroelectric with an extremely high dielectric permittivity, ~30,000, broad dielectric transition near room temperature, and a large electrostrictive effect. It serves as a model object in studies of relaxor behavior. The disorder that is responsible for the relaxor character of PMN is brought about by differences in valence (5+ vs 2+), ionic radii (0.64 vs 0.72 Å), and electronegativities (1.6 vs 1.2 on the Pauling scale) between the Nb^{5+} and Mg^{2+} ions, respectively, that introduce charge fluctuations and local disorder [10]. As shown in Fig. 11.3 [11], the dielectric permittivity ε of the PMN crystal exhibits a typical relaxor characteristic at temperatures slower than T_m, where strong frequency dispersion takes place. T_m increases from 270 at 1 kHz to 290 K at 1 MHz, so the degree of relaxation is $\Delta T_m = 20$ K.

Generally, as already mentioned, a relaxor state is characterized by the presence of polar clusters of nanometric size. Bokov and Ye [7,11,12] discovered universal relaxor dispersion in PMN and related materials, and showed that it is an important common property of relaxor ferroelectrics. The universal relaxor polarization is described by a microscopic model of soft PNRs with unit cells that can freely choose several different directions, while the direction of the total moment of the nanoregion remains the same [13]. Such an approach makes it possible to apply a standard spherical model to relaxor ferroelectrics, which predicts the experimentally observed quadratic divergence of the universal part of the susceptibility above the critical temperature.

Complementary to this model, Pirc and Blinc recently proposed the spherical random bond random field model. They assumed that PMRs are spherical and interact randomly and they proposed a frozen dipole glass state for relaxor [14]. In contrast, the random field model of Fisch [13], which assumes nonrandom two-spin exchange, predicts a ferroelectric or ferroelectric domain state in relaxor. Fu et al. [15] proposed a physics model for PMN relaxors. They observed by

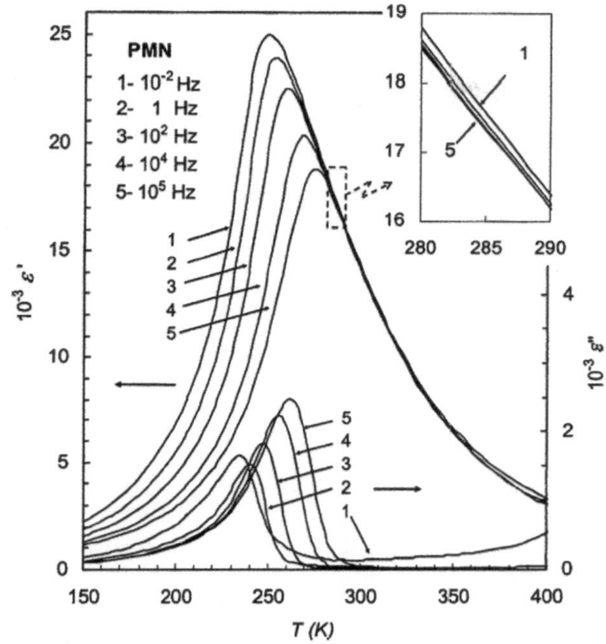

Figure 11.3 Temperature dependencies of the real and imaginary parts of the relative dielectric permittivity measured at different frequencies in a crystal of the prototypical relaxor $Pb(Mg_{1/3}Nb_{2/3})O_3$. Enlarged view in the insert shows the universal relaxor dispersion. Reprinted from A.A. Bokov, Z.-G. Ye, Recent progress in relaxor ferroelectrics with perovskite structure, J. Mater. Sci. 41 (2006) 31–52, Copyright (2006) with permission from Springer.

transmission electron microscopy that, in addition to PNR mentioned earlier, chemically ordering region (COR) with a size of several nm has been found in PMN crystal. It should be mentioned that PNR and COR belong to different symemetry groups and are considered to have noncentrosymmetric symmetry R3*m* and centrosymmetric symmetry F*m*3*m*, respectively. Therefore, there is spontaneous polarization Ps along the $<111>_c$ direction of pseudocubic structure in PNRs. COR was found to be a spherical shape and smaller than 5 nm. It is very stable and remains unchanged within the temperature range of 130–675 K. In contrast to COR, PNRs exhibit remarkable change with temperature, and show continuous growth as the temperature decreases. At approximately room temperature, PNRs grow to a size of about 10 nm. For $T < T_c$ (~225 K) neighboring PNRs merge together and form elliptical-shaped regions. These elliptical PNRs tend to team up together to form large, micrometric domains in ferroelectric state. Such multiscale inhomogeneity of domain structure provides a key point to understand the huge electromechanical coupling effects in relaxor and piezoelectrics with morphotropic phase boundary.

It is well known that a stable pyrochlore phase often forms when producing PMN ceramics. The presence of this lead-based pyrochlore phase causes the dielectric constant to decrease [16,17]. Similar to the case of PZN, relaxor PMN-based solid solutions such as (1-x)$PbMg_{1/3}Nb_{2/3}O_3$-$x PbTiO_3$ (PMN-PT) and $x Pb(Mg_{1/3}Nb_{2/3})O_3$-$(1-x)Pb(Zr_{0.4}7Ti_{0.53})O_3$ (PMN-PZT) attracted much more attention. They exhibit great ferroelectric/relaxor and piezoelectric/electrostrictive properties, especially important for the application in the field of energy harvesting. It has been proven that the piezoelectric charge constant d_{33} of PMN-PT could reach the range from 2000 to 2500 pC/N [18,19], which is almost 4 times higher than that of PZT, and 90 times higher than that of ZnO. Also, it was reported that the recoverable energy-storage density (W) of the PMN-PT thin films could be as high as 11.8 J/cm^3 [20], which indicated that PMN-PT possessed good energy storage ability. Therefore, the superior properties of solid solution PMN-PT make them useful for capacitors and energy electric generators.

11.2.2 PLZT

Lead zirconate titanate ($Pb(Zr_{1-x}Ti_x)O_3$ or PZT) is a typical ferroelectric perovskite showing normal ferroelectric-paraelectric (FE-PE) phase transition [21]. The typical dielectric constant versus temperature behavior is shown in Fig. 11.4 with a Curie temperature around 730 K. PZT materials are widely used in many applications. This is because the PZT compositions (1) possess high piezoelectric and electromechanical coupling coefficients (d_{33} = 152 pC/N), (2) exhibit a high Curie

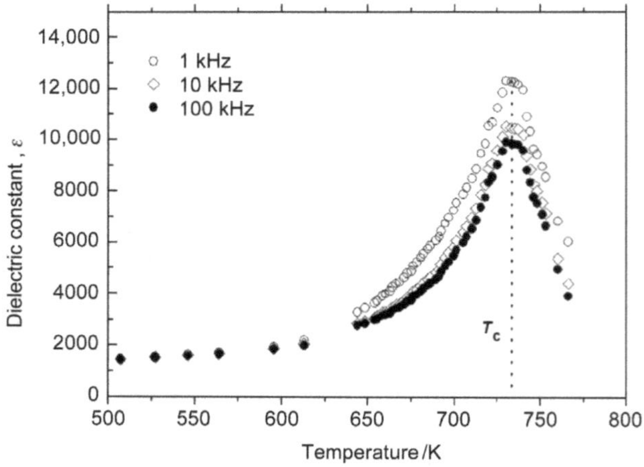

Figure 11.4 Temperature dependences of the real and imaginary parts of the relative dielectric permittivity measured at different frequencies of $Pb(Zr_{0.52}Ti_{0.48})O_3$ samples prepared by a chemical method.

Reprinted from M.D. Gonçalves, F.L. Souza, E. Longo, E.R. Leite, E.R. Camargo, Dielectric characterization of microwave sintered lead zirconate titanate ceramics, Ceram. Int. 42 (2016) 14423–14430, Copyright (2016) with permission from Elsevier.

point, T_C, which permits a high temperature of operation (above 200°C), (3) can be easily poled, (4) possess a wide range of dielectric constants, (5) are relatively easy to sinter at lower temperatures, and (6) form solid solution compositions with many different constituents, thus allowing a wide range of achievable properties [22].

PZT-based solid solutions exhibit excellent electrochemical properties and are widely used as actuators, transducers, ceramic filters, resonators, sensors, and other electronic devices due to their excellent piezoelectric properties. Although there has been a concerted effort to develop lead-free piezoelectric ceramics, no effective alternative to PZT has yet been found. Most other materials that possess high dielectric and piezoelectric coefficients still contain lead ions. Depending on the requirement of the applications, commercially used PZT ceramics are always modified by different dopants. Therefore, the relaxor properties can be varied considerably by introducing impurities that affect the charge state of the compounds.

Those doped PZT ceramics are mainly divided into soft and hard PZTs, which are normally prepared by substituting the A-site or B-site ions with donor and acceptor dopants, respectively. Nevertheless, the partial substitution by different elements, such as lanthanum, contributes to enhance relaxor characteristics of these compounds. PLZT ($Pb_{1-y}La_y(Zr_{1-x}Ti_x)_{1-y/4}O_3$), where the substitution of La^{3+} for Pb^{2+} ions necessarily leads to vacancies on the A-sites that are related to dielectric relaxation phenomenon, is one of the typical ferroelectric perovskites showing relaxor FE-PE phase transition. Temperature dependences of the real part of the permittivity $\varepsilon'(T)$ for different lanthanum concentrations y in PLZT(y/65/35) and different frequencies were investigated by Otonicar et al. [23]. It was shown that with increasing La concentration from 5–10 at% in the x/65/35 PLZT the values of permittivity maximum decrease toward lower temperatures. The results show the emerging relaxor character at 7/65/35 PLZT, becoming more pronounced toward the 10/65/35 composition.

Concerning a wide range of applications of PLZT, Lee et al. [24] obtained large energy and power densities on PLZT samples by performing the Olsen cycle undergoing an ergodic relaxor–ferroelectric phase transition. According to these results, PLZT (x/65/35) compositions with x ranging from 5 to 9 mol% can be used in multistage devices operating between 70°C and 240°C. A maximum energy density of 1014 ± 16 J/L per cycle was produced by 7/65/35 PLZT under quasiequilibrium conditions. This is the largest energy density generated by any pyroelectric materials to date. A maximum power density of 48 ± 1 WJ/L was obtained from 6/65/35 PLZT at 0.060 Hz.

The La-modified lead zirconate titanate ceramics possess lower phonon energy and a higher refractive index, thereby reducing nonradiative relaxation rates and raising spontaneous emission probabilities, and from this point of view, the PLZT compositions represent a model system to develop active electrooptic devices [25]. PLZT ceramics containing trivalent rare-earth ions can also become a potential and alternative material in the development of active solid-state lasers in the near-infrared spectral region, with increased functionality in the future thanks to its electrooptical characteristics [26,27].

11.2.3 Lead zinc niobate (PZN)

Lead zinc niobate ($Pb(Zn_{1/3}Nb_{2/3})O_3$; PZN), a zinc analog of the typical relaxor lead magnesium niobate PMN, is also a relaxor ferroelectric compound with a rhombohedral disordered perovskite structure at room temperature. It has a high dielectric constant (~22,000) around the Curie temperature of 140°C and shows DPT behaviors presented by Bing and coauthors [28]. PZN ceramics with pure perovskite are relatively difficult to prepare by conventional ceramic techniques. Although many attempts have been made to prepare pyrochlore-free PZN ceramics including synthesis under high pressure conditions, sol-gel processing, and forming solid solution with more stable perovskite additions, the last method seems to be the most promising and popular [29]. So far, however, attempts to prepare PZN with perovskite structure by a conventional calcining-mixed-oxide process under atmospheric pressure were unsuccessful. Instead, synthesis of perovskite phase PZN was realized under high pressure (2.5 GPa) and temperature of 800°C–1000°C [30], while single crystalline forms of PZN can be synthesized by using flux method [31,32] with excellent dielectric, optical, and electrostrictive properties.

The preparation of a PZN system usually needs the addition of $PbTiO_3$ (PT) to obtain pure phases. Therefore, the binary and ternary systems (PZN-PT [33,34], PZN-PZT [35,36], PZN-PT-BT [37], PMN-PZN-PZT [38–40]) have received wider attention showing higher values of the dielectric permittivity and the pyroelectric coefficient. Materials from this group in single crystal form have significantly better piezoelectric properties than the ceramics and because of that there is much interest in fabrication and investigation of large single crystals.

The solid solution PZN–PT single crystals of ferroelectric compounds PT and relaxor ferroelectric compounds PZN exhibit more advantages due to their special compositions and symmetry structures. It is believed that the PT compound changes the short-range ordering of PZN, and reduces its phase transition diffuseness and dielectric relaxation. In other words, upon the addition of ferroelectric PT to relaxor PZN there is a transition from short-range ordered relaxor behavior at low PT content to long-range ordered normal ferroelectric behavior at higher PT content. Besides the characteristic frequency dispersion of their dielectric constants, millimeter size PZN–PT single crystals show remarkable piezoelectric and electromechanical properties as reported in 1982 [41].

Piezoelectric-based sensing and measuring systems have been the subject of increasing interest in recent years for their sensitivity and reliability. New generation piezo-materials such as PZN-PT and PMN-PT have a better performance on the regarding sensitivity and reliability in comparison to traditional PZT sensing. Zhang and coauthors [33] fabricated smart sensing probes with both PZN-PT and PMN-PT single crystals for high temperature viscosity measurements. An investigation confirms that the proposed smart probes have very high sensitivity to the viscosity change of viscous fluids at high temperature (up to 90°C). Also, results confirmed that PZN-PT shows an even more stable and reliable performance at high temperatures than PMN-PT due to the higher depolarization temperature, which is more applicable for high-temperature measurements.

One of the many possible applications of relaxor PZN-PT single crystal is energy harvesting. It is well known that vibration energy harvesting has great potential for achieving self-powered operations for wireless sensors, wearable devices, and medical electronics, and thus has attracted much attention in academia and industry. Yang and coauthors [34] worked on comparing the PMN-PT and PZN-PT-single crystal energy harvesters against PZT-ceramic harvesters. The experimental results indicate that energy harvesters using the PMN-PT and PZN-PT single crystals can significantly outperform those using the PZT ceramics.

11.3 Bismuth-layered perovskite relaxors

Bismuth layer-structured ferroelectric materials (BLSFs) have received great attention due to their large remanent polarization, lead-free nature, relatively low processing temperatures, high Curie temperatures, and excellent piezoelectric properties, which make them good candidates for high-temperature piezoelectric applications and memory storage [42].

The Aurivillius family compounds are layered bismuth with the general formula $(Bi_2O_2)(A_{n-1}B_nO_{3n+1})$, where n is an integer between 1 and 5 indicating the number of perovskite-like blocks stacking between two adjacent $(Bi_2O_2)^{2+}$ slabs. Twelve-coordinated A-sites are usually occupied by Pb^{2+}, Bi^{3+}, alkaline, or alkaline earth cations, whereas in 6-coordinated B-sites cations of transition metals are occupied by cations such as Ti^{4+}, Nb^{5+}, or Ta^{5+}. Various A and B cations are allowed, so this family could have numerous representatives. Examples include Bi_2WO_6 ($n=1$), $SrBi_2Ta_2O_9$ ($n=2$), $Bi_4Ti_3O_{12}$ ($n=3$), $SrBi_4Ti_4O_{15}$ ($n=4$), $Ba_2Bi_4Ti_5O_{18}$ ($n=5$), and so on. Additional members can be generated by allowing the stacking of perovskite-like blocks of different sizes.

Many Aurivillius compounds present ferroelectric properties up to temperatures much higher than the most commonly used ferroelectrics. For example, $Bi_4Ti_3O_{12}$ is ferroelectric up to 675°C, $SrBi_4Ti_4O_{15}$ up to 530°C, $SrBi_2Nb_2O_9$ up to 440°C, and so on, whereas the transition temperature of PZT ($Pb(Zr_{1-x}Ti_x)O_3$) in the morphotropic phase boundary does not exceed 365°C [43–45]. However, the low electric conductivity of these compounds hinders the poling process and makes difficult the development of piezoelectric properties in ceramic materials. The layered structure is described by a high value of c-parameter in comparison with parameters a and b of the orthorhombic cell and leads to a high anisotropy of the crystallographic structure, which is directly connected to the high anisotropy of the electric properties. In the ceramics in which the microstructure of such materials consists of plate-like shaped grains, the piezoelectric effect is the highest in the a–b plane [46]. For these crystallites, the smallest dimension of the grain corresponds to the crystallographic c-axis so that the polarization lies in the plane of the grains.

Among BLSFs, most members have a normal FE-PE phase transition, while only a few of them (Ba-based ferroelectrics with a bilayered structure), such as $BaBi_4Ti_4O_{15}$, $BaBi_2Ta_2O_9$, and $BaBi_2Nb_2O_9$, are interesting from the point of view of the relaxor behavior. Unlike perovskite $A(B_1B_2)O_3$ relaxors whose B-site is occupied by heterogeneous cations (which causes structural disorder responsible for relaxation), Aurivillius compounds with a relaxor character would have no cation disorder at first glance at the formula $(Bi_2O_2)^{2+}(A_{m-1}B_mO_{3m+1})^{2-}$. It seems that each site is occupied by only one kind of cation. But a characteristic behavior of materials from the Aurivillius family is that some of them allow cation site mixing among atom positions, especially between the bismuth and the A-site of the perovskite block. So, similar to materials with the perovskite structure, the relaxor behavior in Ba-based bilayered compounds is attributed to the structural disorder, which is either inherent or introduced by doping. The major properties of relaxor Ba-based ferroelectrics discussed in this chapter are presented in Table 11.1.

11.3.1 $BaBi_2Ta_2O_9$ and $BaBi_2Nb_2O_9$

Ferroelectric materials with bilayered structure such as $SrBi_2Nb_2O_9$ (SBN) and $SrBi_2Ta_2O_9$ (SBT) belong to the group of normal ferroelectric materials with a Curie temperature around 450°C and 280°C, respectively (Figs. 11.5 [47,50] and 11.6 [47,51]). With incorporation of barium to the structure into the site of the perovskite block and/or the incorporation to the bismuth site into the $(Bi_2O_2)^{2+}$ layers relaxor behavior is observed.

Huang et al. confirmed that the transition from a normal ferroelectric to a relaxor-like ferroelectric is induced by the addition of Ba according to the formula $Sr_{1-x}Ba_xBi_2Nb_2O_9$ [50]. For the pure SBN, ΔT_m equals zero, which shows the character of normal ferroelectrics. For specimens with $x = 0.2$, 0.4, 0.6, 0.8, and 1, ΔT_m equals 12°C, 36°C, 68°C, 941°C, and 110°C, respectively. The result indicates that the degree of frequency dispersion has been enhanced by increasing Ba content. The ionic radius of Ba^{2+} (1.61 Å) in 12th coordination (A-site) is higher than that of Sr^{2+} (1.44 Å), which should provide an increase of the lattice parameters. Its ionic radius in 8th coordination (bismuth site) is higher (1.42 Å) than that of Bi^{3+} (1.17 Å), but is smaller than that of its radius of the A-site. Substitution of Ba^{2+} on the Bi^{3+} site, therefore, will reduce the interlayer strain. An X-ray and neutron

Table 11.1 Dielectric and relaxor characteristics of some Ba-based Bilayered compounds

Composition	Phase transition	ε_m (100 kHz)	T_m, °C (100 kHz)	ΔT_{relax}, K (1 kHz-1 MHz)	T_f (K)	E_a (eV)	ν_o (Hz)
$BaBi_2Nb_2O_9$, $n = 3$ [47]	Relaxor	600	~170	100	97	0.633	9.77×10^{13}
$BaBi_2Ta_2O_9$, $n = 3$ [48]	Relaxor	525	~80	60	113	0.250	8.00×10^{12}
$BaBi_4Ti_4O_{15}$, $n = 4$ [49]	Relaxor	2429	415	16	654	0.034	1.85×10^{10}

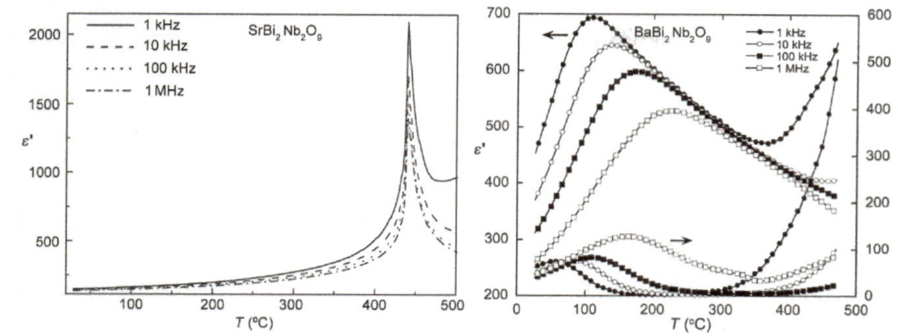

Figure 11.5 Dielectric constant as a function of temperature for $SrBi_2Nb_2O_9$ and $BaBi_2Nb_2O_9$ ceramics [47,50].
Reprinted from S. Huang, C. Feng, L. Chen, Q. Wang, Relaxor behavior of Sr1−xBaxBi2Nb2O9 ceramics, J. Am. Ceram. Soc. 89 (1) (2006) 328−331, Copyright (2005) with permission from John Wiley and Sons. Reprinted from C. Miranda, M.E.V. Costa, M. Avdeev, A.L. Kholkin, J.L. Baptist, Relaxor properties of Ba-based layered perovskites, J. Eur. Ceram. Soc. 21 (2001) 1303−1306, Copyright (2001) with permission from Elsevier.

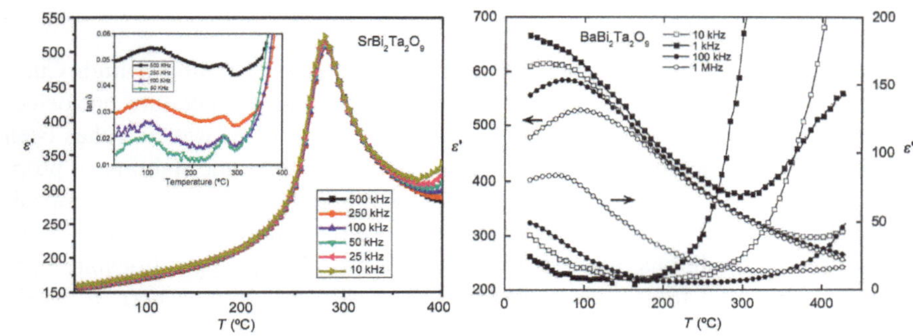

Figure 11.6 Temperature dependence of the real (ε') and imaginary parts (ε'') of the dielectric permittivity in $SrBi_2Ta_2O_9$ and $BaBi_2Ta_2O_9$ ceramics at different frequency [47,51].
Reprinted from C. Miranda, M.E.V. Costa, M. Avdeev, A.L. Kholkin, J.L. Baptist, Relaxor properties of Ba-based layered perovskites, J. Eur. Ceram. Soc. 21 (2001) 1303−1306; V. Senthil, T. Badapanda, A. Chandrabose, S. Panigrahi, Dielectric and ferroelectric behavior of cerium modified SrBi2Ta2O9 ceramic, Mater. Lett. 159 (2015) 138−141, Copyright (2015, 2001) with permission from Elsevier.

powder diffraction study of $BaBi_2Nb_2O_9$ suggested that about 13.4% of A-sites are occupied by bismuth atoms [52]. A-site bismuth has proven to contribute much to the ferroelectricity in bismuth layer-structured compounds, while barium atoms in the A-site are too large to allow orthorhombic distortion and retain tetragonal symmetry around them [53]. Thus, the displacement of bismuth along the *a*-axis from the equilibrium position is the source of electric dipoles. It is hard for such a low concentration of electron dipoles to induce a long-range electric ordering. In addition, point charges (positive point charges in perovskite block and negative point charges in $(Bi_2O_2)^{2+}$ layers) result in an inner electric field along the *c*-axis, which would destroy the long-range polar clusters in a $BaBi_2Nb_2O_9$. So the low dielectric constant maximum of these compounds (around 700) may be the result of the low concentration and the two-dimensional configuration of electric dipoles in the layered structure.

Shvartsman et al. [54] compared the obtained parameters of relaxation spectrum results of $BaBi_2Nb_2O_9$ and $BaBi_2Ta_2O_9$ with those reported for PLZT (9.5/65/35). In PLZT the local polarization may rotate over eight possible orientations along (111) axis. This allows for relatively easy switching and coalescence of clusters. On the other hand, in bilayered perovskites the polarization switching occurs only in the (*a*,*b*) plane of orthorhombic cell. Moreover, due to high anisotropy of layered structure and large *c*-parameter, the distance between neighboring clusters is large and hence the electrostatic interaction between them can be weak. Therefore, the cluster growth is interfered with and the formation of a glass-like state is delayed. The same mechanism might explain the inability to induce ferroelectricity by an external electric field, both in BBT and BBN at temperatures as low as $\sim$90 K.

11.3.2 $BaBi_4Ti_4O_{15}$

As previously noted, many compounds of Aurivillus phases exhibit a high Curie temperature (up to 980°C) and have been proposed for high temperature piezoelectric applications and nonvolatile memory applications [55,56]. However, it has been noted that different members of the family do not exhibit the same stability of properties with respect to amplitude and frequency of the driving field. Some compositions, such as $SrBi_4Ti_4O_{15}$, exhibit a remarkably constant response over a wide range of driving conditions, whereas the properties of others compositions are nonlinear, hysteretic, and frequency dependent [57]. $SrBi_4Ti_4O_{15}$ has a normal FE-PE phase transition at elevated temperature (around 530°C; Fig. 11.7), linear piezoelectric properties under low fields, and relatively low room-temperature conductivity. All these properties make it very attractive for precision sensors.

The relaxor behavior observed for BLSF ceramic material has been analyzed, considering the incorporation of barium to the structure into the A-site of the perovskite block and/or the incorporation to the bismuth site into the $(Bi_2O_2)^{2+}$ layers. It is noticed that A-site modification of a BLSF ceramic is more effective than that of B-site modification in enhancing the structural as well as electrical properties. Nayak et al. [59] investigated influence of Ba^{2+} substitution on a Sr^{2+} site in $Sr_{1-x}Ba_xBi_4Ti_4O_{15}$ ceramics ($x = 0.05$, 0.1, 0.15, 0.2, and 0.25). With this low-doped concentration of Ba^{2+} they found that the sharp dielectric peaks corresponding to the Curie temperature gradually decrease with an increase in Ba^{2+} content, but there is no proof for increase in relaxor behavior. Higher concentration of Ba^{2+} substitution may give the diffuse phase transition in this system while the

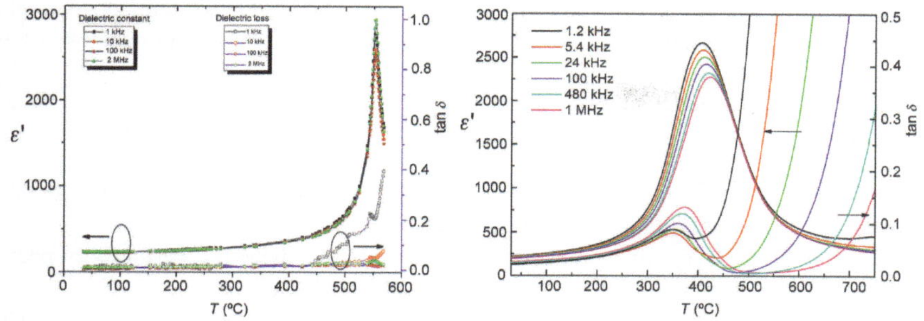

Figure 11.7 Temperature dependence of the real (ε') and imaginary parts (ε'') of the dielectric permittivity in $SrBi_4Ti_4O_{15}$ and $BaBi_4Ti_4O_{15}$ ceramics at different frequencies [49,58].

Reprinted from J.D. Bobic, M.M. Vijatovic, J. Banys, B.D. Stojanovic, Effect of La substitution on the structural and electrical properties of $BaBi_{4-x}La_xTi_4O_{15}$, Ceram. Int. 39 (2013) 8049–8057; A.R. James, Effect of oxygen assisted sintering on piezoelectric properties of $SrBi_4Ti_4O_{15}$ ceramics prepared via high energy mechanochemical processing, Ceram. Int. 41 (2015) 5100–5106, Copyright (2015, 2013) with permission from Elsevier.

highest concentration ($x = 1$) turns this compound in $BaBi_4Ti_4O_{15}$ where the phase transition typically corresponds to relaxors. The degree of relaxation behavior of $BaBi_4Ti_4O_{15}$ is found to be around 16 K [49,60,61]. This value is comparable with ΔT_{relax} of perovskite relaxors where the degree of relaxation is ~20 K and ~14 K in PMN and PLZT (8/65/35) ceramics [62], while much more pronounced relaxation is observed in Aurivillius $BaBi_2Nb_2O_9$ compounds ($\Delta T_{relax} = 100$ K).

Kennedy et al. demonstrated the presence of the A-type cation (Sr, Pb, Ba) within the Bi_2O_2 layers through structural refinements [63]. The amount of the alkaline earth cation within the Bi_2O_2 layers is apparently dependent on the size of the cation increasing from zero in $CaBi_4Ti_4O_{15}$ to 28% in $BaBi_4Ti_4O_{15}$. This increase is similar to that observed in the two layer oxides. In the case of $BaBi_4Ti_4O_{15}$, but in contrast to familiar compounds $ABi_4Ti_4O_{15}$ (A = Ca, Sr, and Pb), which are not relaxors, relaxation is a consequence of the disorder between Bi from $[Bi_2O_2]^{2+}$ layers and the perovskite-like A-type cations.

References

[1] G.A. Smolenskii, V.A. Isupov, A.I. Agranovskaya, S.N. Popov, Soviet, Phys. Solid State 2 (1961) 2584.

[2] A.A. Bokov, Z.G. Ye, Dielectric relaxation in relaxor ferroelectrics, J. Adv. Dielectr. 2 (2) (2012) 1241010-1-1241010-24.

[3] M.D. Gonçalves, F.L. Souza, E. Longo, E.R. Leite, E.R. Camargo, Dielectric characterization of microwave sintered lead zirconate titanate ceramics, Ceram. Int. 42 (2016) 14423–14430.

[4] I. Rivera, A. Kumar, N. Ortega, R.S. Katizar, S. Lushnikov, Divide line between relaxor, diffused ferroelectric, ferroelectric and dielectric, Solid State Commun. 149 (2009) 172−176.
[5] J.D. Bobic, J. Macutkevic, R. Grigalaiti, M. Ivanov, M.M. Vijatovic Petrovic, J. Banys, et al., Correlation Between Microstructure and Electrical Properties of Ferroelectric Relaxors, Wiley Publisher, 2016, pp. 554−587.
[6] K. Uchino, S. Namura, Critical exponents of the dielectric constant in diffuse-phase transition crystal, Ferroelectrics Lett. 44 (1982) 55−61.
[7] A.A. Bokov, Z.G. Ye, Phenomenological description of dielectric permittivity peak in relaxor ferroelectrics, Solid State Commun. 116 (2000) 105−108.
[8] S. Ke, H. Fan, H. Huang, H.L.W. Chan, Lorentz-type relationship of the temperature dependent dielectric permittivity in ferroelectrics with diffuse phase transition, Appl. Phys. Lett. 93 (2008) 112906.
[9] R. Pric, R. Blinc, Vogel-Fulcher freezing in relaxor ferroelectrics, Phys. Rev. B 76 (2007) 020101 (R).
[10] G.A. Samara, E.L. Venturini, Ferroelectric/relaxor crossover in compositionally disordered perovskites, Phase Transit. 79 (1−2) (2006) 21−40.
[11] A.A. Bokov, Z.-G. Ye, Recent progress in relaxor ferroelectrics with perovskite structure, J. Mater. Sci. 41 (2006) 31−52.
[12] Y.H. Bing, A.A. Bokov, Z.G. Ye, Diffuse and sharp ferroelectric phase transitions in relaxors, Curr. Appl. Phys. 11 (2011) S14−S21.
[13] C. Lei, A.A. Bokov, Z.-G. Ye, Relaxor behavior in $Ba(Ti_{0.72}Sn_{0.28})O_3$ solid solution, Ferroelectrics 339 (2006) 129−136.
[14] R. Fisch, Random-field models for relaxor ferroelectric behavior, Phys. Rev. B 67 (17) (2003) 094110.
[15] D. Fu, H. Taniguchi, M. Itoh, S. Mori, $Pb(Mg_{1/3}Nb_{2/3})O_3$ (PMN) relaxor: dipole glass or nano-domain ferroelectric?, Advances in Ferroelectrics, (InTech, Rijeka, Croatia) (2012) 51−67.
[16] L. Wu, Y.-C. Liou, The effect of heating rate on the properties of PMN relaxor ceramics, Ceram. Int. 21 (1995) 335−338.
[17] K. Singh, T. Quazi, Preparation and characterization of perovskite PMN [$Pb(Mg_{1/3}Nb_{2/3})O_3$], Indian J. Eng. Mater. Sci. 15 (2008) 176−180.
[18] W.B. Luo, K. He, D. Xu, C. Li, X.Y. Bai, Y. Shuai, et al., Reactive temperature and growth time effects on the morphology of PMN-PT nanorods by hydrothermal synthesis, J. Alloy. Compd. 671 (2016) 122−126.
[19] X. Wang, L. Zhang, X. Hao, S. An, High energy-storage performance of $0.9Pb(Mg_{1/3}Nb_{2/3})O_3$-$0.1PbTiO_3$ relaxor ferroelectric thin films prepared by RF magnetron sputtering, Mater. Res. Bull. 65 (2015) 73−79.
[20] E. Sun, W. Cao, Relaxor-based ferroelectric single crystals: growth, domain engineering, characterization and applications, Prog. Mater. Sci. 65 (2014) 124−210.
[21] T.R. Shrout, S.J. Zhang, Lead-free piezoelectric ceramics: alternatives for PZT? J. Electroceram. 19 (2007) 111−124.
[22] G.H. Haertling, Ferroelectric ceramics: history and technology, J. Am. Ceram. Soc. 82 (4) (1999) 797−818.
[23] M. Otonicar, A. Reichmann, K. Reichmann, Electric field-induced changes of domain structure and properties inLa-doped PZT—from ferroelectrics towards relaxors, J. Eur. Ceram. Soc. 36 (2016) 2495−2504.

[24] F.Y. Lee, H.R. Jo, C.S. Lynch, L. Pilon, Pyroelectric energy conversion using PLZT ceramics and the ferroelectric–ergodic relaxer phase transition, Smart Mater. Struct. 22 (2013), 025038 (16pp).
[25] M. Plonska, W.A. Pisarski, Excitation and emission of Pr^{3+}: PLZT ceramics, Ceram. Int. 142 (2016) 17822–17826.
[26] Z. Zheng, X. Li, J. Liu, Z. Feng, B. Li, J. Yang, et al., Optical properties of Er^{3+}/Yb^{3+}-codoped transparent PLZT ceramic, Physica B 403 (2008) 44–49.
[27] A.S.S. De Camargo, L.A.O. Nunes, I.A. Santos, D. Garcia, J.A. Eiras, Structural and spectroscopic properties of rare-earth (Nd^{3+}, Er^{3+}, and Yb^{3+}) doped trans- parent lead lanthanum zirconate titanate ceramics, J. Appl. Phys. 95 (2004) 2135–2140.
[28] Y.-H. Bing, A.A. Bokov, Z.-G. Ye, B. Noheda, G. Shirane, Structural phase transition and dielectric relaxation in $Pb(Zn_{1/3}Nb_{2/3})O_3$ single crystals, J. Phys.: Condens. Mat. 17 (2005) 2493–2507.
[29] H. Fan, The influence of fabrication processing on domain structure and polarization switching in PZN-based ferroelectric ceramics, Mat. Sci. Eng. B 120 (2005) 114–118.
[30] D.H. Lee, N.K. Kim, Crystallographic, dielectric, and diffuseness characteristics of PZN–PT ceramics, Mat. Lett. 34 (1998) 299–304.
[31] S.E. Park, T.R. Shrout, Ultrahigh strain and piezoelectric behavior in relaxor based ferroelectric single crystals, J. Appl. Phys. 82 (4) (1997) 1804–1811.
[32] W.Z. Zhu, P.Q. Mantas, J.I. Baptista, Polarization response of PZN-based relaxor ferroelectrics, J. Mat. Sci. Lett. 19 (2000) 491–493.
[33] C. Zhang, H. Zhang, S. Kaluvas, PZN-PT based smart probe for high temperature fluid viscosity measurements, Measurements 94 (2016) 753–758.
[34] Z. Yang, J. Zu, Comparison of PZN-PT, PMN-PTsingle crystals and PZT ceramic forvibration energy harvesting, Energ. Convers. Manage. 122 (2016) 321–329.
[35] S.H. Lee, S.E. Lee, H.C. Lee, Application of CIP processing and sol-gel surface coating to PZN-PZT piezoelectric ceramics for use as energy harvesters, Curr. Appl. Phys. 11 (2011) S6–S10.
[36] N. Vittayakorn, G. Rujijanagul, D.P. Cann, The improvement in dielectric and ferroelectric performance of PZT–PZN ceramics by thermal treatment, Curr. Appl. Phys. 7 (2007) 582–585.
[37] A. Pelaiz-Barranco, I. Gonzalez-Carmenate, F. Calderon-Pinar, Relaxor behaviour in PZN–PT–BT ferroelectric ceramics, Solid State Commun. 134 (2005) 519–522.
[38] C.H. Wang, The piezoelectric and dielectric properties of PZT–PMN–PZN, Ceram. Int. 30 (2004) 605–611.
[39] Z. Yang, X. Chaoa, R. Zhang, Y. Changa, Y. Chen, Fabrication and electrical characteristics of piezoelectric PMN–PZN–PZT ceramic transformers, Mater. Sci. Eng. B 138 (2007) 277–283.
[40] X. Chaoa, Z. Yang, G. Li, Y. Cheng, Fabrication and characterization of low temperature sintering PMN–PZN–PZT step-down multilayer piezoelectric transformer, Sensors Actuat. A 144 (2008) 117–123.
[41] H. Cao, V. Hugo Schmidt, R. Zhang, W. Cao, H. Luo, Elastic, piezoelectric, and dielectric properties of $0.58Pb(Mg_{1/3}Nb_{2/3})O_3$-$0.42PbTiO_3$ single crystal, J. Appl. Phys. 96 (1) (2004) 549–554.
[42] A. Pelaiz-Barranco, Y. Gonzalez-Abreu, Ferroelectric ceramic materials of the Aurivillius family, J. Adv. Dielectr. 3 (4) (2013) 1330003 (11 pp).
[43] K.S. Rao, D.M. Prasad, P.M. Krishna, B.H. Bindu, K. Suneetha, Frequency and temperature dependence of electrical properties of barium and gadolinium substituted $SrBi_2Nb_2O_9$ ceramics, J. Mater. Sci. 42 (2007) 7363–7374.

[44] C. Voisard, D. Damjanovic, N. Setter, Electrical conductivity of strontium bismuth titanate under controlled oxygen partial pressure, J. Eur. Ceram. Soc. 19 (1999) 1251–1254.

[45] T. Stevenson, D.G. Martin, P.I. Cowin, A. Blumfield, A.J. Bell, T.P. Comyn, et al., Piezoelectric materials for high temperature transducers and actuators, J. Mater. Sci.: Mater. Electronics 26 (2015) 9256–9267.

[46] T. Jardiel, A.C. Caballero, M. Villegas, Aurivillius ceramics: $Bi_4Ti_3O_{12}$-based piezoelectrics, J. Ceram. Soc. Jpn. 116 (4) (2008) 511–518.

[47] C. Miranda, M.E.V. Costa, M. Avdeev, A.L. Kholkin, J.L. Baptist, Relaxor properties of Ba-based layered perovskites, J. Eur. Ceram. Soc. 21 (2001) 1303–1306.

[48] V.V. Shvartsman, M.E.V. Costa, M. Avdeev, A.L. Kholkin, Relaxor behavior of $BaBi_2Ta_2O_9$and $BaBi_2Nb_2O_9$ ceramics, Ferroelectrics 296 (2003) 187–197.

[49] J.D. Bobic, M.M. Vijatovic, J. Banys, B.D. Stojanovic, Effect of La substitution on the structural and electrical properties of $BaBi_{4-x}La_xTi_4O_{15}$, Ceram. Int. 39 (2013) 8049–8057.

[50] S. Huang, C. Feng, L. Chen, Q. Wang, Relaxor behavior of $Sr_{1-x}Ba_xBi_2Nb_2O_9$ ceramics, J. Am. Ceram. Soc. 89 (1) (2006) 328–331.

[51] V. Senthil, T. Badapanda, A. Chandrabose, S. Panigrahi, Dielectric and ferroelectric behavior of cerium modified $SrBi_2Ta_2O_9$ ceramic, Mater. Lett. 159 (2015) 138–141.

[52] S.M. Blake, M.J. Falconer, M. McCreedy, P. Lightfoot, Cation disorder in ferroelectric Aurivillius phases of the type $Bi_2ANb_2O_9$ (A = Ba, Sr, Ca), J. Mater. Chem. 7 (8) (1997) 1609–1613.

[53] R.Z. Hou, X.M. Chen, Y.W. Zeng, Diffuse ferroelectric phase transition and relaxor behavior in Ba-based bismuth layer-structured compounds and La-substituted $SrBi_4Ti_4O_{15}$, J. Am. Ceram. Soc. 89 (9) (2006) 2839–2844.

[54] V.V. Shvartsman, D.C. Lupascu, Lead-free relaxor ferroelectrics, J. Am. Ceram. Soc. 95 (1) (2012) 1–26.

[55] D. Dhak, P. Dhak, T. Ghorai, S.K. Biswas, P. Pramanik, Preparation of nano-sized $ABi_2Nb_2O_9$ ($A = Ca^{2+}$, Sr^{2+}, Ba^{2+}) ferroelectric ceramics by soluble Nb(V) tartarate precursor and their dielectric characteristics after sintering, J. Mater. Sci.: Mater. El. 19 (2008) 448–456.

[56] D. Damjanovic, Materials for high temperature piezeoelctric tranducers, Curr. Opin. Solid St. Mater. Sci. 3 (1998) 469.

[57] M.D. Maeder, D. Damjanovic, N. Setter, Lead free piezoelectric materials, J. Electroceramics 13 (2004) 385–392.

[58] A.R. James, Effect of oxygen assisted sintering on piezoelectric properties of SrBi4Ti4O15 ceramics prepared via high energy mechanochemical processing, Ceram. Int. 41 (2015) 5100–5106.

[59] P. Nayak, S.R. Mohapatra, P. Kumar, S. Panigrahi, Effect of Ba^{2+} substitution on the structural and electrical properties of $SrBi_4Ti_4O_{15}$ceramic, Ceram. Int. 41 (2015) 9361–9372.

[60] A. Khokhar, M.L.V. Mahesh, A.R. James, et al., Sintering characteristics and electrical properties of $BaBi_4Ti_4O_{15}$ ferroelectric ceramics, J. Alloy. Compds. 581 (2013) 150–159.

[61] A. Chakrabarti, J. Bera, Structure and relaxor behavior of $BaBi_4Ti_{4-x}Zr_xO_{15}$ ceramics, Curr. Appl. Phys. 10 (2) (2010) 574–579.

[62] A.L. Kholkin, M. Avdeev, M.E.V. Costa, J.L. Baptista, Dielectric relaxation in Ba-based layered perovskites, Appl. Phys. Lett. 79 (2001) 662–664.

[63] B.J. Kennedy, Q. Zhou, Ismunandar, Y. Kubota, K. Kato, Cation disorder and phase transition in the four-layer ferroelectric Aurivillius phase $ABi_4Ti_4O_{15}$ (A = Ca, Sr, Ba, Pb), J. Solid State Chem. 181 (2008) 1377–1386.

Further reading

A. Dabkowski, H.A. Dabkowska, J.E. Greedan, W. Ren, B.K. Mukherjee, Growth and properties of single crystals of relaxor PZN–PT materials obtained from high-temperature solution, J. Cryst. Growth 265 (2004) 204–213.

Tunable ferroelectrics for frequency agile microwave and THz devices

12

Félix Lorenzo Martínez-Viviente and Juan Hinojosa
Technical University of Cartagena, Murcia, Spain

12.1 Introduction

Ferroelectrics are interesting materials for the design of new microwave devices. The potential of ferroelectrics and other high-permittivity (high-*k*) dielectrics for such high-frequency devices is due to the field-dependent permittivity of these materials, which can be of application to frequency agile devices whose operation frequency can be tuned by a control signal. In this way, the same circuit can be reconfigured or adapted to operate at different frequencies. This offers considerable advantages for the design of reconfigurable passive and active devices, especially in terms of reduction of size, complexity, and cost of systems with multiband operation capabilities.

In the past few years, the range of frequencies of interest for communication applications has been extended from microwave and millimeter waves to the domain of sub-mm or THz waves. This has been possible thanks to the development of new and efficient ways of generating THz waves, which open up new applications not only in communications but also in other fields of science and technology such as security, quality control, materials and semiconductor research, and so on. Ferroelectrics can play an important role in the expanding field of THz applications, and for this purpose an accurate knowledge of their properties in this frequency range is crucial. Therefore, the last section will be dedicated to the new techniques for the determination of the permittivity at THz frequencies and their application to ferroelectric materials.

12.2 Techniques for measuring permittivity at microwave frequencies

The first step in the design of microwave circuits and components with tunable capabilities is to have a good knowledge of the dielectric properties of the materials that are going to be used and their tunable characteristics. Since the dielectric permittivity and its tunability in the frequency range of interest is not always available in the published literature, design engineers and materials scientists need to have a

Magnetic, Ferroelectric, and Multiferroic Metal Oxides. DOI: http://dx.doi.org/10.1016/B978-0-12-811180-2.00012-8

basic understanding of the measurement techniques and instrumentation available to determine such properties at microwave frequencies. Thus, this section describes the general properties of ferroelectric materials and several nonresonant and resonant characterization techniques for different ferroelectric material states (bulk, thick-film, and thin-film).

12.2.1 General properties of ferroelectric materials and figure of merit for microwave applications

As mentioned in the introduction, ferroelectrics are an important class of dielectric materials in the field of microwave engineering. Numerous electronics components and circuits such as capacitors, resonators, tunable filters and oscillators, phase shifters, and others have been developed and continue to be developed thanks to the physical properties of these materials at microwave frequencies. Their most remarkable property for these applications is that their dielectric characteristics have a dependence on temperature and electric field (tunability).

Fig. 12.1 shows an illustrative example of temperature dependence of the permittivity curve for ferroelectric materials. As it can be seen, the permittivity increases as temperature is incremented and it reaches its maximum value at the transition temperature T_c, called Curie temperature. Above the Curie temperature T_c, the ferroelectric material becomes paraelectric and the permittivity decreases. The variation of the permittivity is due to the microstructure of the ferroelectric material, which undergoes various phase transformations at different temperatures. Above the Curie temperature T_c, its structure is of cubic type, while below T_c its structure transforms in different types as the temperature decreases: tetragonal, rhombohedral, and so on.

Ferroelectric materials can have one or more polar axes along which a spontaneous polarization controllable by an applied external electric field can appear below the Curie temperature. The response of the polarization P versus electric field E of these materials is similar to the magnetic induction B versus magnetic field H for ferromagnetic materials, and hence the name ferroelectric. Thus, a ferroelectric material exhibits a hysteresis loop of the polarization P versus electric field E as can be seen on the left of Fig. 12.2. At zero E field strength, the polarization P is not equal to zero, since it is offset by an amount called remanent polarization, which can have two stable values $+P_r$ and $-P_r$. To override the remanent polarization $+P_r$ ($-P_r$), a reverse electric field $-E_c$ (E_c), called coercive field, must be

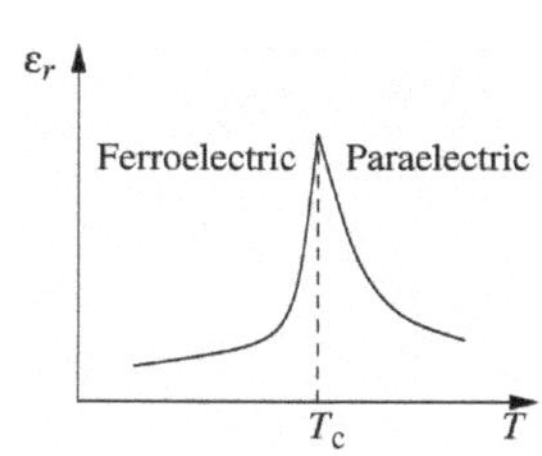

Figure 12.1 Curve of permittivity ε_r versus temperature T for ferroelectric materials.

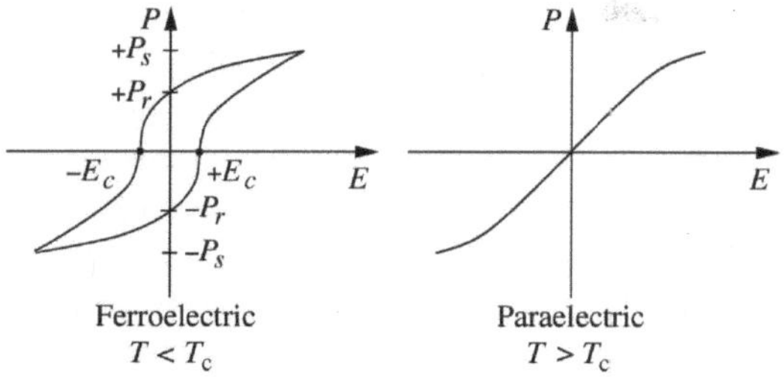

Figure 12.2 Polarization P versus electric field E for ferroelectric and paraelectric materials.

applied to the ferroelectric material. When the field E is increased, the polarization increases toward a saturation value $+P_s$ and a single dielectric response is reached. This behavior is also observed when the field E is reversed and the polarization decreases to reach a saturation value $-P_s$. As the temperature increases above the Curie temperature, the structural phase vanishes and the material becomes paraelectric. The response of the polarization P versus electric field E of a paraelectric material is similar to that shown on the right of Fig. 12.2. Its response depends on the electric field and is nonlinear. It should be noted that for tunable microwave devices a paraelectric phase is more suitable. Even though the material is in the paraelectric phase, with no remanent polarization, in microwave applications it is still referred to as a ferroelectric.

The main characteristic of ferroelectric materials for the development of tunable microwave devices is the dependence of their dielectric permittivity ε_r with respect to an external applied DC bias electric field E. In order to evaluate this property, a tunability parameter n, which allows describing the degree in which the permittivity of the material can be controlled by means an external electric field, is commonly used:

$$n = \frac{\varepsilon_r(0)}{\varepsilon_r(E)} \tag{12.1}$$

where $\varepsilon_r(0)$ and $\varepsilon_r(E)$ correspond, respectively, to the permittivity of the ferroelectric material at zero electric field and at some electric field. This characteristic is also described by the relative dielectric tunability:

$$n_r = \frac{\varepsilon_r(0) - \varepsilon_r(E)}{\varepsilon_r(0)} = \left(1 - \frac{1}{n}\right) \times 100\% \tag{12.2}$$

Generally, the permittivity without biasing field ($E = 0$) is higher than with biasing field ($E \neq 0$). For microwave ferroelectric devices, large tunability and low losses are usually required. To take into account both parameters, a figure of merit K, named quality factor of a tunable component, has been defined [1]:

$$K = \frac{(n-1)^2}{n \tan \delta_0 \tan \delta_E} \tag{12.3}$$

where tan δ_0 and tan δ_E are, respectively, the loss tangent of the ferroelectric material without the external applied electric field ($E=0$) and with the external applied electric field ($E \neq 0$). Note that the figure of merit K depends on the characteristic of the ferroelectric material and not of the tunable component. The values of the loss tangent for $E=0$ and maximum E are used to determine K.

The ferroelectric materials are used in different states depending on the microwave device from which they are part: bulk, thick-film, and thin-film. The thicknesses can be larger than several hundred microns for bulk samples, in the order of several microns for thick films, and lower than 1 μm for thin films, respectively. Due to many factors that influence their internal behavior, thin ferroelectric films in comparison with bulk materials present different electric characteristics such as lower real permittivity and higher loss tangent, dielectric properties less dependent on frequency, ferroelectric-paraelectric transition at lower temperature, and lower external applied electric field needed to change its permittivity. Thus, to better understand the characteristics of ferroelectric materials in different states at microwave frequencies, a good knowledge of the measurement techniques of dielectric properties is required by both materials scientists and design engineers. The following section is dedicated to such techniques.

12.2.2 Microwave characterization techniques of ferroelectric materials

The first electromagnetic characterization techniques to extract the intrinsic properties of materials in the radiofrequency (RF) and microwave frequency ranges date back to the 1950s [2]. However, it was when network analyzers gained in popularity (scalar at the beginning, and later with vector magnitude and phase), with higher performance and affordable prices, that a lot of RF/microwave electromagnetic characterization techniques of materials were developed and applied. The intense work on these techniques started and still continues because of the great interest from various scientific and technical sectors such as electronics, information technology and communications, quality control of many materials and products, and so on. The methods of characterization in the microwave range contribute to providing information about microscopic and macroscopic properties of a large variety of solid, semisolid (powder, gel), and liquid materials.

Microwave techniques for electromagnetic characterization of materials in RF/microwave frequencies may be classified into two categories, resonant and nonresonant methods, and each category may be divided into two subcategories. Generally, the resonant methods with respect to nonresonant have higher accuracy and are more suitable for measurements of materials with low losses. Nonresonant methods are usually used to obtain electromagnetic properties of a material in a wide frequency range, while resonant methods can only extract electromagnetic characteristics of a material at a fixed frequency or various discrete frequencies.

Ferroelectric materials usually present a high permittivity and their dielectric properties are temperature and electric field dependent. Additionally, depending upon its state (bulk, thick-film, or thin-film), ferroelectrics may present different dielectric properties at the same frequency. However, there is not a multipurpose characterization method using a single measuring cell capable of measuring different material states with a high accuracy. Thus, various characterization techniques must be applied to take into account all these aspects. In this section, we present some of the techniques most widely used for the study of ferroelectric materials according to their state at microwave frequencies.

12.2.2.1 Nonresonant methods

The nonresonant methods for the electromagnetic characterization of materials at microwave frequencies include measurement techniques in reflection and in transmission/reflection.

Reflection method

Fig. 12.3 shows the experimental setup for the electromagnetic characterization of ferroelectric materials with an open-ended coaxial probe. This probe is a cut-off section of a coaxial transmission line and it is used as a dipole. A flat surface of the sample to be characterized is placed in contact at the open end of the coaxial probe. Thus, this technique is also considered a nondestructive method. The vector network analyzer (VNA) realizes the measurements of the S_{11} reflection parameter as a function of frequency at the interface P_1 between the open-ended coaxial probe and the surface of the sample under test. The reference plane P_1 is set with an initial calibration at the end of the coaxial probe before making any measurement. The calibration standards usually applied are open and short terminations, and a standard material with known dielectric permittivity [3]. The dependence of the

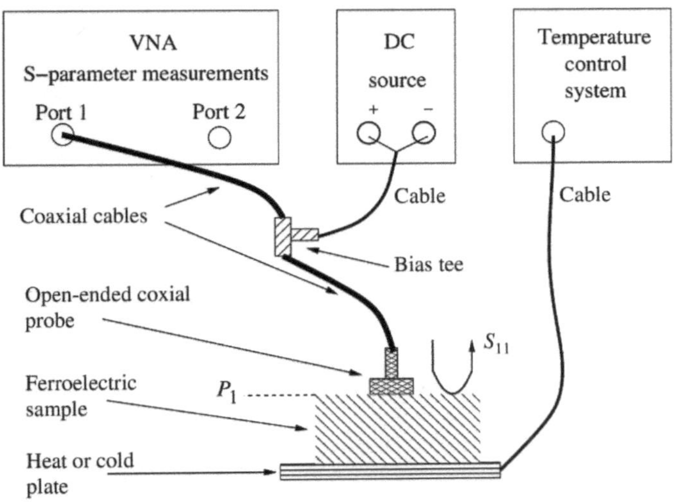

Figure 12.3 Measurement setup for the electromagnetic characterization of ferroelectric materials with an open-ended coaxial probe.

dielectric properties with an external electric field is determined by applying a radial DC bias to the ferroelectric sample under test by means of a voltage source and a bias tee component. Dielectric properties can also be measured as a function of the temperature by placing the ferroelectric sample on a heat or cold plate, which is connected to a temperature control system.

The permittivity is extracted from the measurements of the S_{11} coefficient performed at the interface P_1 between the coaxial probe and the sample under test, together with the analysis of the open-ended coaxial probe modeled by means of an equivalent circuit [4]. This technique and similar configurations have been used to study ferroelectrics in different states: bulk and thick films [5,6]. The main advantages of the presented configuration are its wide working frequency range, its simplicity, and the ability of making fast nondestructive measurements. However, large errors in the measurements of the dielectric properties can occur because of air gaps caused by a nonplanar surface of the sample to be characterized.

Transmission/reflection method

Fig. 12.4 shows a general experimental setup of a transmission/reflection method. The VNA realizes the measurements of the complex reflection (S_{11}, S_{22}) and transmission (S_{21}, S_{12}) parameters as a function of frequency at the reference planes P_1 and P_2 of the measuring cell, which is usually a transmission line. The reference planes P_1 and P_2 are set with an initial calibration of the measuring cell. A line-reflect-line (LRL) calibration is usually used and the applied standards with known properties are open or short terminations and different line lengths. The dependence of the dielectric properties with the applied electric field or with temperature can be obtained in the same way as with the previous technique. Any type of transmission line can be used in a transmission/reflection method. The choice of transmission

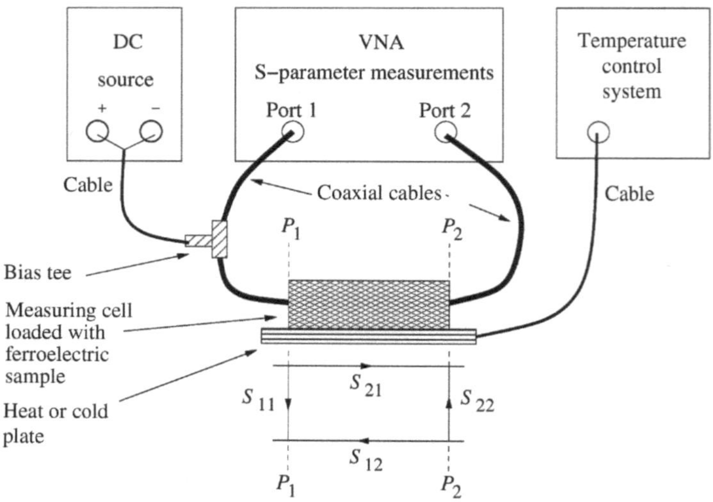

Figure 12.4 Measurement setup of a transmission/reflection method.

line depends on the state of the material to be characterized. Depending on the used cell, the material to be measured (solid, powder, film) will be inserted into a shielded line structure, placed on the top of a semiopen line or used as propagation media into a semiopen line. In this section, we will focus on the measurements of dielectric properties of bulk ferroelectric samples and thin ferroelectric films.

The dielectric properties of the ferroelectric sample are extracted from S_{11} and S_{21} parameter measurements performed at the reference planes P_1 and P_2 of the measuring cell and the Nicolson–Ross–Weir (NRW) procedure [7,8]. An additional refinement is proposed in [9] to resolve the instability of the NRW procedure for a dielectric material ($\mu_r = 1$). This transmission/reflection method has been extended to measuring cells based on a rectangular or stripline section, and a coplanar sensor [7,10–13].

The transmission/reflection technique based on a coaxial line section can be used to study the dielectric relaxation of ferroelectric composites and ferroelectric polycrystalline ceramics. The main advantage of this method is its wide working frequency range. Moreover, the accuracy of the complex permittivity is usually greater than with the reflection methods, since two complex coefficients instead of one are used to extract it. However, it is a destructive method, since it requires machining the samples to be characterized and, therefore, this increases the cost and the time to carry out the measurements. Moreover, gaps between measuring cell walls and the sample under test are unavoidable. These gaps produce large errors in the measurements of the dielectric properties when the sample under test presents high dielectric permittivity. These errors are reduced by painting with a gold- or silver-based ink the faces of the sample under test in contact with the measuring cell walls.

At microwave frequencies, the tunable ferroelectric-based devices in conventional microstrip and coplanar circuits are made from thin films. The dielectric properties of thin ferroelectric films are different with respect to the same bulk ferroelectric. To develop these devices, the dielectric properties of thin ferroelectric films are needed. Thus, a coplanar waveguide (CPW) or microstrip line can be used as measuring cell in the previous transmission/reflection method (Fig. 12.4) [2,14–17]. The CPW measuring cell presents several advantages with respect to the microstrip measuring cell for the electromagnetic characterization of materials. The CPW cell requires only one layer of metallization and it has a very low dispersion of the fundamental mode as the frequency increases. Therefore, the broadband electromagnetic characterization of thin ferroelectric films is mainly focused on the CPW measuring cell. Fig. 12.5 shows a measuring cell based on a two-layered CPW allowing the electromagnetic characterization of thin ferroelectric film samples [18]. The thin ferroelectric film of permittivity ε_{r1} and thickness h_1 is deposited onto a dielectric substrate of permittivity ε_{r2} and thickness h_2. The dielectric properties of the substrate ε_{r2} are known as well as the thicknesses h_1 and h_2. The S-parameters are measured by means of a VNA (Fig. 12.4). The coaxial to coplanar transitions between the coaxial cables and the measuring cell are made by an on-wafer measurement fixture [19]. The reference planes P_1 and P_2 are set on the outputs of the two coplanar probes of the on-wafer measurement fixture with an

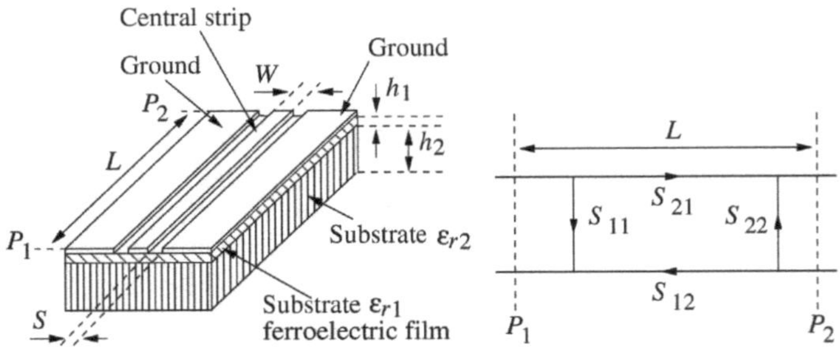

Figure 12.5 Measuring cell based on a two-layered coplanar waveguide for the electromagnetic characterization of thin ferroelectric film samples.

initial LRL calibration. Both coplanar probes, and therefore the reference planes P_1 and P_2, are approximately located at both ends of the two-layered CPW measuring cell (Fig. 12.5).

The dielectric properties of the thin ferroelectric films (ε_{r1}) are derived from a procedure similar to the previous coaxial measuring cell. Due to the inhomogeneous configuration of the CPW, a complex effective permittivity (ε_{ref}) is extracted from the measured S_{11} and S_{21} parameters. Then, the complex permittivity ε_{r1} of the thin ferroelectric film can be derived from the complex effective permittivity by solving the relationships given in [20,21].

The transmission/reflection methods based on planar circuits can be used to study the dielectric relaxation of thin and thick ferroelectric films [18,22]. Coplanar and microstrip lines used as measuring cells allow very broadband measurements and do not present air gaps between the sample and the conductors, since they are produced onto the sample to be characterized [19,23]. However, the cost and the time to carry out the measurements are increased, since they require depositing a thin ferroelectric film onto a known substrate and also metallization layers to make the structures of the planar circuit. Moreover, the measurements of low-loss thin ferroelectric samples are difficult with these techniques. To achieve reasonable accuracy, the dielectric losses of the thin ferroelectric sample must be greater than the substrate and conductor losses.

12.2.2.2 Resonant methods

The resonant methods for the electromagnetic characterization of materials are more accurate and sensitive than nonresonant methods. They are usually used to characterize low-loss materials. However, the material properties can only be measured at a fixed frequency or at several discrete frequencies. Thus, the resonant methods are not suitable to study the dielectric relaxation of ferroelectrics. Resonant methods can be classified into two categories: resonator method (or dielectric resonator) and resonant-perturbation method. For the resonator method, the dielectric sample under test is used as resonator in the measurement structure, and electric properties are acquired from its resonant frequency and quality factor.

In the case of the resonant-perturbation method, the electromagnetic properties are determined from the changes of the resonant frequency and quality factor between the empty resonator and the resonator filled with the sample to be characterized. In this section, we will focus on the measurements of dielectric properties of bulk ferroelectric samples by means of a resonator method [24] and a resonant-perturbation method [5], as well as thin ferroelectric films through a coplanar resonator [25].

Resonator method

Fig. 12.6 shows a typical configuration of a dielectric resonator. This resonator was initially proposed by Hakki and Coleman [24], and then developed by Courtney [26]. The structure consists of two parallel conducting plates separated by a thickness h. The sample to be characterized, with a cylindrical shape, is sandwiched between both conducting plates. The VNA realizes the measurements of the S_{21} transmission parameter as a function of the frequency by means of two coupling antennas symmetrically placed with respect to the sample. The dependence of the dielectric properties of the ferroelectric sample according to an external electric field is carried out by applying a DC bias electric field between the two parallel conducting plates.

The dielectric properties of the cylindrical sample under test are obtained from the measured parameter S_{21} by means of the relationships given in [27]. This resonator technique presents high accuracy in the measurements of dielectric properties and, therefore, it is often used to characterize ferroelectric composites and ceramics, and also bulk ferroelectric crystals [5]. However, it is not suitable for the electromagnetic characterization of lossy samples with a low quality factor Q, since it is based on the identification of the resonant frequency. This dielectric resonator requires machining the samples to be characterized in a cylindrical shape, and only a single frequency point for each sample can be provided. The cost and the time to measure several frequency points are increased, since several sample sizes are required.

Resonant-perturbation method

This method is based on a cylindrical cavity resonating in the TM010 mode [5]. The sample to be tested has a shape of a long thin rod. It is inserted in the cavity by means of a removable top plate. The VNA realizes the measurements of the S_{21}

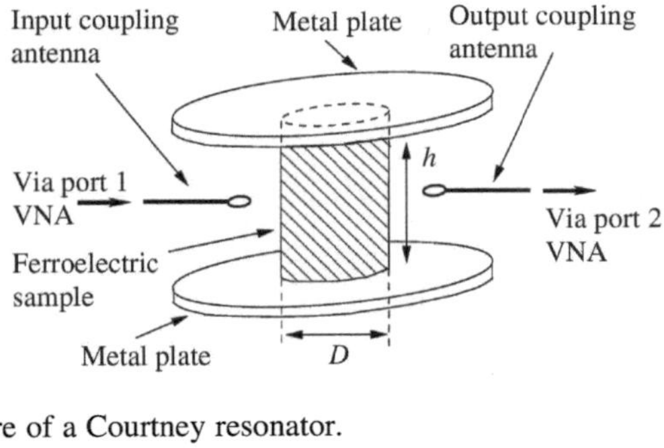

Figure 12.6 Structure of a Courtney resonator.

transmission parameters as a function of frequency by means of two coupling loops. In this method it is difficult to obtain the dependence of the dielectric properties on an external electric field. Therefore, it is not so useful for the investigation of the tunability of the material.

The resonant-perturbation technique presents good accuracy in the measurements of dielectric properties of ferroelectric composites and ceramics, and also bulk ferroelectric crystals [5]. Contrary to the earlier dielectric resonator, the resonant-perturbation technique allows measuring several frequency points. In addition, the computation of the dielectric properties is simpler. However, the resonant-perturbation technique is not suitable for the measurements of low loss samples. As for the previous dielectric resonator technique, it also requires machining the samples to be characterized, although the sample size is smaller (thin rod). This influences the cost and the time to carry out the measurements.

Coplanar resonator method

As for the transmission/reflection methods based on planar circuits, different types of resonators based on microstrip or CPW structures can be used for the electromagnetic characterization of thin ferroelectric films [25]. The CPW resonator is easier to manufacture than the microstrip resonator, since it needs only one layer of metallization. Therefore, the resonant method of thin ferroelectric films will be focused on the CPW resonator. Fig. 12.7 shows a typical coplanar resonator for the electromagnetic characterization of ferroelectric film samples [25]. The resonator is achieved from the open circuit of the CPW at both ends. The thin ferroelectric film of permittivity ε_{r1} and thickness h_1 is deposited onto a dielectric substrate of permittivity ε_{r2} and thickness h_2. The dielectric properties of the substrate ε_{r2} are known as well as the thicknesses h_1 and h_2. The VNA measures the S_{21} parameter as a function of frequency by means of two coaxial coupling loops that are located above the CPW resonator. The dependence of the dielectric properties according to the electric field is obtained by applying a dc bias voltage between the central conductor and the ground planes of the CPW by means of bias tee component. The bias voltage can also be applied at the halfway point of the central conductor of the CPW resonator, since this is not physically connected to the coupling loops.

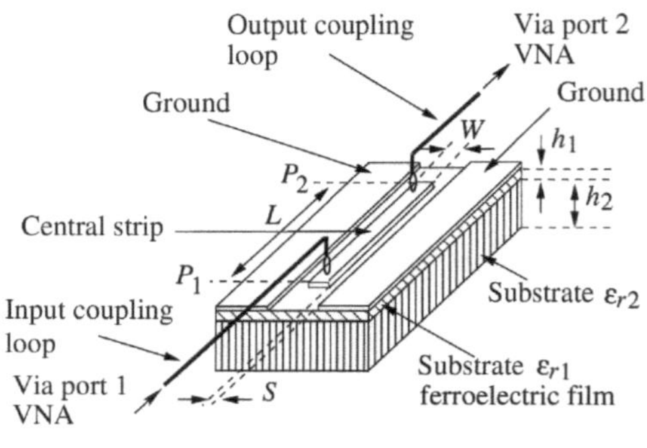

Figure 12.7 Coplanar resonator loaded with a thin ferroelectric film sample.

Due to the inhomogeneous configuration of the CPW resonator, a complex effective permittivity (ε_{ref}) is extracted from the measured S_{21} parameter and the resonant frequency (f_0). Once determined the effective permittivity, the complex permittivity ε_{r1} of the thin ferroelectric film under test can be derived by solving the relationships in [20,21]. A good accuracy is obtained for the dielectric loss if it is higher than conductor and radiation losses.

This coplanar resonator method allows measuring several frequency points of thin ferroelectric films. The computations of the dielectric properties are simple. However, to obtain accurate values of the dielectric properties of the thin ferroelectric film under test, it is necessary to manufacture several CPW resonator structures to take into account undesirable effects (conductor and radiation losses, parasitic capacitances, etc.). This increases the cost and the time to carry out the measurements of thin ferroelectric films.

12.3 Ferroelectrics at THz frequencies

The THz region of the electromagnetic spectrum has emerged as a subject of intense research during the few years before this book's publication. This is the so-called THz gap between microwaves and the optical (infrared) region of the spectrum, approximately from 0.1 THz up to a few tens of THz (in the wider, most general definition). The reason why this region was not really explored until recently has been the lack of convenient intensive, directional THz sources of electromagnetic energy in this frequency range. If the THz range is approached with electronic sources from the microwave side of the spectrum, the oscillators and frequency multipliers needed to generate such signals become inefficient and limited in terms of tunability as the THz frequencies are reached. On the other hand, if the approach to the THz range is done from the optical (far infrared) side of the spectrum, then different types of lasers can be used, each of them having their particular limitations. For example, THz gas lasers with significant power (in the order of watts) have been available for many years, but they can only provide a continuous wave single frequency beam, so they are inconvenient for many applications (including spectroscopy), apart from their excessive cost and size. Quantum cascade lasers are another option, although they must operate at cryogenic temperatures and they suffer from some optical limitations in terms of quality of beam profile and spectral purity.

However, the amount of THz power needed clearly depends on the application, and in the case of broadband THz transmission or reflection spectroscopy this power doesn't need to be particularly high. Because of this reason, spectroscopic applications have been dramatically simplified with the advent of cost effective femtosecond fiber lasers whose ultrashort optical pulses at telecommunication wavelengths (1550 nm) can be conveniently transformed into THz pulses by means of photoconductive switches.

The principle of operation is relatively simple. The optical pulse generates charge carriers at the emitter antenna, which consists of a multilayer InAlAs/InGaAs heterostructure (in case the laser emits at 1.55 μm, although other wavelengths and semiconductor structures are also possible). These carriers are accelerated by an externally applied electric field and they recombine very fast, producing a pulse of THz radiation whose intensity is proportional to the time derivative of the photocurrent.

On the detector side, photocarriers are also generated by the same femtosecond optical pulse (for this purpose lasers with two simultaneous outputs or a beam splitter are used). These carriers at the detector are accelerated by the electric field of the incoming THz pulse, producing a very weak photocurrent that can be amplified and measured. Since the laser pulses have a repetition rate of about 100 MHz (the exact frequency is not critical), an average value of this photocurrent is registered.

It should be noted that only a small fraction of the THz pulse is probed in this way, since this pulse has duration in the order of picoseconds (compared to femtoseconds for the laser pulse). Hence, in order to reconstruct the whole pulse, a variable time delay is introduced between the emitter and detector pulses. This is done by an optical delay line, which introduces a controllable time shift between the emission of the pulse and the detection. Then, a Fourier transform allows us to change from the time domain to the frequency domain and a spectrum with frequency content of a few THz can be registered in this way.

Ferroelectrics will certainly be needed in the future development of devices and systems for emerging applications in the THz field, especially because of their tunable properties. Although still in its infancy, the technique of broadband THz spectroscopy has already been applied to the study of some ferroelectrics [28,29] and even to ferroelectric thin films [30,31]. Data of dielectric constants of ferroelectric materials in this frequency range is still scarce, although some studies can be found for some particular ferroelectrics such as PZT [32] or some titanates and tantalates ($BaTiO_3$, $Ba_{1-x}Sr_xTiO_3$, $SrTiO_3$, and $LiTaO_3$) [33]. There are also some publications on phonon-polariton dispersion at THz frequencies in various ferroelectrics such as lithium niobate [34]. If we shift to practical applications, the strong electrooptical and nonlinear optical effects present in many ferroelectrics can be applied to polarization rotation and frequency conversion of light in the THz range [35]. Finally, advancements in ferroelectric-gated memories and transistors have been applied to graphene devices with ferroelectric contacts, where strong coupling can occur at THz frequencies between two-dimensional plasmons in graphene and phonon-polaritons in the ferroelectrics [36], leading to compact on-chip plasmonic waveguides with negligible crosstalk at only 20 nm of separation. Since applications of THz radiation become more and more common every time and ferroelectrics are expected to play an important role in many of them, much more work will certainly be done in this field in the near future.

References

[1] I.B. Vendik, O.G. Vendik, E.L. Kollberg, Commutation quality factor of two-state switchable devices, IEEE Trans. Microw. Theory Tech. 45 (5) (2000) 802–808.

[2] A.R. Von Hippel (Ed.), Dielectric Materials and Applications, Wiley, New York, 1954.

[3] S.S. Stuchly, C.E. Bassey, Microwave coplanar sensors for dielectric measurements, Meas. Sci. Technol. 9 (1998) 1324–1329.
[4] M.A. Stuchly, S. Stuchly, Coaxial line reflection methods for measuring dielectric properties of biological substances at radio and microwave frequencies – a review, IEEE Trans. Microw. Theory Tech. 29 (3) (1980) 176–183.
[5] M.P. McNeal, S.J. Jang, R.E. Newnham, The effect of grain and particule size on the microwave properties of barium titanate ($BaTiO_3$), J. Appl. Phys. 83 (1998) 3288–3297.
[6] F.L. Martínez, J. Hinojosa, G. Doménech, F.J. Fernández-Luque, J. Zapata, R. Ruiz, et al., Dielectric constant tunability at microwave frequencies and pyroelectric behaviour of lead-free submicrometer-structured $(Bi_{0.5}Na_{0.5})_{1-x}Ba_xTiO_3$ ferroelectric ceramics, IEEE Trans. Ultrason. Ferroelectr. Freq. Control 60 (2013) 1595–1602.
[7] A.M. Nicolson, G.F. Ross, Measurement of the intrinsic properties of materials by time-domain techniques, IEEE Trans. Instrum. Meas. 19 (1970) 377–382.
[8] W.B. Weir, Automatic measurement of complex dielectric constant and permeability at microwave frequencies, Proc. IEEE 62 (1974) 33–66.
[9] A.H. Boughriet, C. Legrand, A. Chapoton, Noniterative stable transmission/reflection method for low-loss material complex permittivity determination, IEEE Trans. Microw. Theory Tech. 45 (1997) 52–57.
[10] J. Baker-Jarvis, M.D. Janezic, J.H. Grosvenor, R.G. Geyer, Transmission/Reflection and Short-Circuit Line Methods for Measuring Permittivity and Permeability, NIST Technical note 1355, National Institute of Standards and Technology, 1993.
[11] W. Barry, A broad-band, automated, stripline technique for simultaneously measurement of complex permittivity and permeability, IEEE Trans. Microw. Theory Tech. 34 (1986) 80–84.
[12] P. Queffelec, P. Gelin, J. Gieraltowski, J. Loäec, A microstrip device for the broadband simultaneous measurement of complex permeability and permittivity, IEEE Trans. Magn. 30 (1994) 224–231.
[13] M. Shete, M. Shaji, M. Jaleel-Akhtar, Design of a coplanar sensor for RF characterization of thin dielectric samples, IEEE Sens. J. 13 (2013) 4706–4715.
[14] N.K. Das, S.M. Voda, D.M. Pozar, Two methods for the measurement of substrate dielectric constant, IEEE Trans. Microw. Theory Tech. 35 (1987) 636–641.
[15] M.D. Janezic, D.F. Williams, Permittivity characterization from transmission line measurements, IEEE Microwave Symposium Digest, vol. 3, June 8–13, 1997, Denver, USA, pp. 1343–1345.
[16] J. Hinojosa, L. Faucon, P. Queffelec, F. Huret, S-parameter broadband measurements of microstrip lines and extraction of the substrate intrinsic properties, Microw. Opt. Technol. Lett. 30 (2001) 65–69.
[17] J. Hinojosa, S-parameter broadband measurements on-coplanar and fast extraction of the substrate intrinsic properties, IEEE Microw. Wirel. Compon. Lett. 11 (2001) 80–82.
[18] R.A. Chakalov, Z.G. Ivanov, Y.A. Boikov, P. Larsson, E. Carlsson, S. Gevorgian, et al., Fabrication and investigation of YBCO/BSTO thin film structures for voltage tunable devices, Physica C 308 (1998) 279–288.
[19] J. Hinojosa, K. Lmimouni, S. Lepilliet, G. Dambrine, Very high broadband electromagnetic characterization method of film-shaped materials using coplanar, Microw. Opt. Technol. Lett. 33 (2002) 352–355.
[20] S. Gevorgian, L.J.P. Linnér, E. Kollberg, CAD models for shielded multi-layered CPW, IEEE Trans. Microw. Theory Tech. 43 (1995) 772–779.

[21] W. Hilberg, From approximation of exact relations for characteristic impedances, IEEE Trans. Microw. Theory Tech. 13 (1965) 29–38.
[22] K.B. Carroll, J.M. Pond, D.B. Chrisey, J.S. Horwitz, R.E. Leuchtner, Microwave measurement of the dielectric constant of $Sr_{0.5}Ba_{0.5}TiO_3$ ferroelectric thin films, Appl. Phys. Lett. 62 (1993) 1845–1847.
[23] J. Hinojosa, J.F. Kruck, G. Dambrine, Ridged waveguide to microstrip transition for electromagnetic characterisation of materials in V-band, Electron. Lett. 36 (2000) 1468–1470.
[24] B.W. Hakki, P.D. Coleman, A dielectric resonator method of measuring inductive capacities in the millimetre range, IRE Trans. Microw. Theory Tech. 8 (1960) 402–410.
[25] M.J. Lancaster, J. Powell, A. Porch, Thin film ferroelectric microwave devices, Supercond. Sci. Technol. 11 (1998) 1323–1334.
[26] W.E. Courtney, Analysis and evaluation of a method of measuring the complex permittivity and permeability of microwave insulators, IEEE Trans. Microw. Theory Tech. 18 (1970) 476–485.
[27] Y. Kobayashi, M. Katoh, Microwave measurement of dielectric properties of low-loss materials by the dielectric rod resonator method, IEEE Trans. Microw. Theory Tech. 33 (1995) 586–592.
[28] V. Skoromets, S. Glinsek, V. Bovtun, M. Kempa, J. Petzelt, S. Kamba, et al., Ferroelectric phase transition in polycrystalline $KTaO_3$ thin film revealed by terahertz spectroscopy, Appl. Phys. Lett. 99 (5) (2011) 2908.
[29] A. Pashkin, P. Samoukhina, T. Ostapchuk, P. Kuzel, P. Vilarinho, J. Petzelt, Submillimeter and far infrared dielectric response of Bi-doped SrTiO3 ceramics, Ferroelectrics 294 (2003) 133–139.
[30] S.B. Kang, M.H. Kwak, M. Choi, S. Kim, T. Kim, E.J. Cha, et al., Terahertz dielectric response of ferroelectric $Ba_xSr_{1-x}TiO_3$ thin films, IEEE Trans. Ultrason. Ferroelectr. Freq. Control 58 (11) (2011) 2276–2280.
[31] P. Mounaix, M. Tondusson, L. Sarger, D. Michau, V. Reymond, M. Maglione, High-frequency response in ferroelectric $BaSrTiO_3$ thin films studied by terahertz time-domain spectroscopy, Jpn. J. Appl. Phys. 44 (7A) (2005) 5058–5061.
[32] M. Naftaly, M.G. Cain, S. Lepadatu, T. Buchacher, J. Allam, Dielectric constants of bulk ferroelectric PZT measured by terahertz time-domain spectroscopy, Adv. Appl. Ceram. 115 (5) (2016) 260–263.
[33] T. Fujii, A. Ando, Y. Sakabe, Dielectric characteristics of ferroelectric materials in submillimeter-wave regions, Jpn. J. Appl. Phys. 43 (2004) 6765–6768.
[34] S. Kojima, N. Tsumura, H. Kitahara, M.W. Takeda, S. Nishizawa, Terahertz time domain spectroscopy of phonon-polaritons in ferroelectric lithium niobate crystals, Jpn. J. Appl. Phys. 41 (2002) 7033–7037.
[35] T. Miyamoto, H. Yada, H. Yamakawa, H. Okamoto, Ultrafast modulation of polarization amplitude by terahertz fields in electronic-typo organic ferroelectrics, Nat. Commun. 4 (2013) 2586. Available from: http://dx.doi.org/10.1038/ncomms3586.
[36] D. Jin, A. Kumar, K.H. Fung, J. Xu, N.X. Fang, Terahertz plasmonics in ferroelectric-gated graphene, Appl. Phys. Lett. 102 (2013) 201118.

Piezoelectric energy harvesting device based on quartz as a power generator

13

Luis A.O. Araujo[1], *Cesar R. Foschini*[2] *and Carlos A. Fortulan*[3]
[1]Federal University of São Carlos (UFSCar), São Carlos, Brazil, [2]São Paulo State University - UNESP, School of Engineering, Bauru, Brazil, [3]University of São Paulo (USP), São Carlos, Brazil

13.1 Introduction

The idea of harvesting ambient vibration energy to generate electricity has become a promising research field over the past decade due to the reduced power requirements of small electronic components. The motivation of vibration-based energy harvesting (EH) is to power such devices (e.g., sensor networks used in monitoring applications) by using the vibration energy available in their environment such as electromagnetic, electrostatic, piezoelectric, and magnetostrictive transduction mechanisms. Due to its large power densities, ease of application, and feasibility of fabrication in microscale, piezoelectric EH has received the attention.

EH has great potential when applied to a range of areas, for example, automotive, medical, military, and industrial purposes (among others). One of the reasons for this trend is the increasing performance of energy consumption. EH can be applied in embedded systems, autonomous or not, complementing or even replacing batteries. Usually, this can be referred to as an EH power generator [1].

Although there are many kinds of electric power generators, many fields remain to be explored, mainly relating to mechanical stress or vibration using a piezoelectric material as a power generating element. The advantages of an EH piezo mechanism are direct conversion of mechanical stress into electric power, simplicity in the final form of the generator, proportionality between the dimension of generator and the quantity of electricity obtained, and the fact there is no need for extra power activation.

The main disadvantage is device fatigue. Stress can reduce life with high frequency or amplitude. This can impact the choice of material, geometry, and permissible stress. Different materials can be applied, for example, ceramics, polymers, composites, or single crystals. Each one has its own relation: stress versus power, called electromechanical coupling [2,3].

In the group of piezo materials, single crystals have attractive mechanical characteristics, including dimensional stability in high temperatures, low weight, high hardness. However, one of the biggest challenges relating to crystals is process manufacturing. Due to their hardness, tenacity/fragility, and anisotropy, process manufacturing requires very specific parameters.

Magnetic, Ferroelectric, and Multiferroic Metal Oxides. DOI: http://dx.doi.org/10.1016/B978-0-12-811180-2.00013-X

Life resistance under stress is another challenge. Although monocrystalline quartz is the most representative piezoelectric crystal, the application in piezo EH is reduced. Currently it has applications as a resonator for electronics and load/inertial sensors. It can be machined by a range of processes, including dicing, grinding, polishing, lapping, and ultrasound, according to the crystal plan of interest and final design of the device [4–7].

The most usual materials in piezo power generation are $Pb(Zr,Ti)O_3$ (PZT) and $BaTiO_3$ ceramics [8]. Of these, quartz has one of the highest electromechanical quality factors ($Q_m > 10^5$), and this characteristic can boost the efficiency of the piezoelectric harvesting power generation, making it acceptable to low stress and, consequently, a lower amount of power generation.

This chapter addresses a renewable electric low-power piezoelectric EH generator based on a quartz single crystal, and the impact of the abrasive cutting process manufacturing on the functionality of this kind of generator. It is a contribution to EH power generation studies using quartz. Additionally, it also compares the results of cuts in alumina polycrystalline 99.8% and silicon (111). The abrasive cutting process has important potential applications in the manufacturing of microelectromechanical systems based on hard and brittle materials.

13.2 Low-power piezoelectric EH generator

The main part of the low-power piezoelectric EH generator in the presenting case is a quartz crystal that needs to be cut by a special procedure to be employed in the EH generator.

13.2.1 The quartz

The quartz has a molecular tetrahedron SiO_4 structure, with a trigonal crystalline structure (or rhombohedral) unit cell composed of three atoms of silicon and six of oxygen. The structure is polymorphic, with three basic compositions in standard conditions (temperature and pressure): the α quartz, the cristobalite, and the tridymite [8,9].

Those three basic compositions are allotropic. Differences came from crystal growth under different conditions of the environment. The α-quartz prevails in standard pressure and temperature conditions. Up to 573°C, the α-quartz has a tetrahedron structure. Between 573°C and 870°C, the quartz is named β-quartz and assumes a hexagonal structure, with dimensional expansion [9].

There are two phase transformation mechanisms: displacive and reconstructive transformation. The former is a transformation without chemical bonding breakage. In the case of reconstructive transformation, there is a chemical bonding breakage. It is a process that requires a lot of energy [9]. Therefore, in the manufacturing of products based on quartz crystal, much attention must be given to pressure and temperature applied, or developed, during the process. It is possible to cause damage to

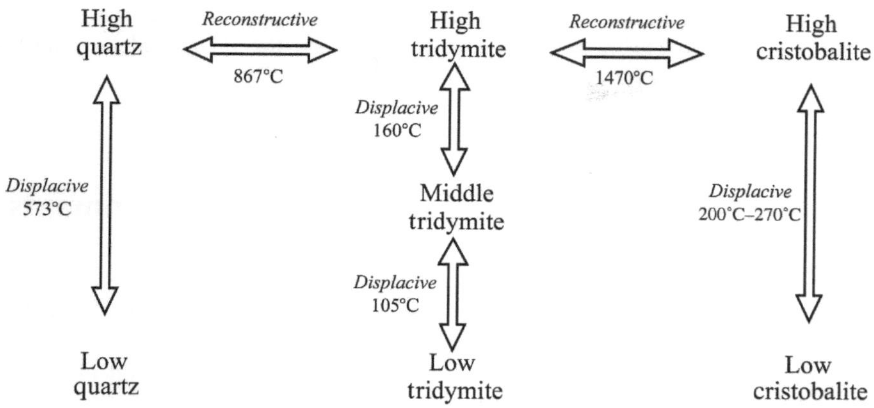

Figure 13.1 Classical representation of the silica polymorphism.

the material by the phase transformation mechanisms [10–13]. Fig. 13.1 presents the polymorphic variations of quartz under different temperature conditions.

From α-quartz to β-quartz (and from tridimite to cristobalite), it is possible to observe displacive transformation, a rearrangement of molecules resulting in dimensional expansion [14].

13.2.2 Cutting hard and brittle materials

The mechanism of material removal in the abrasive dicing process of hard and brittle materials is very similar to the mechanism observed in the longitudinal grinding process of those materials [15–18]. Efrat [19] observed three important factors in this mechanism: the interaction between the abrasive blade and the workpiece, the blade specification, and the workpiece material.

The interaction between abrasive blade and substrate was detailed by Luo and Wang [17] in an observation of scratching procedures along the surface of silicon wafers. The material had three stages in the interaction with an indenter: first, an elastic regime, second, a plastic deformation and, third, a cracking/chipping.

The study indicates that the three stages will be different according to the cutting direction and the wafer's cutting plan. Similar observations were made by Malkin and Hwang [15], Brinksmeier et al. [16], and Jahanmir et al. [20].

Abrasive blades are composed of abrasive grains and a resin bond. The most common grain applied is diamond (with a hardness of around 10 Ω) of different sizes (1–100 μm). The resin bond can be nickel, polymers, copper, or tin [21]. The space between grains is named the chip pocket or pore, and it is directly involved in the removal material mechanism.

According to Zhou et al. [22], there is a strong relationship between the size of the grain, the resin bond, and the pore size. This relationship can impact the quality of the kerf edges generated in the cut. The most common kerf edge damage is chipping (on the top material surface).

The workpiece material characteristics also have a great impact on the results of the cutting process. The removal mechanism between the blade and the material workpiece must be as elastic as possible to avoid damage.

13.3 Process manufacturing and functional experiments of quartz EH

To compare the properties of quartz with several other hard and brittle materials being used in the EH piezoelectric devices, three different specimens of hard and brittle materials were analyzed: alumina polycrystalline 99.8%, silicon (111), and quartz plan X (Fig. 13.2).

Alumina (Al_2O_3) specimens were prepared from commercial powders (Almatis Inc, A1000-SG, 99.8%, with an average particle diameter of 0.4 μm, a surface area of 7.7 m^2/g, and a theoretical density (ρ_{th}) of 3.99 g/cm^3). The water-based slurry was obtained by jar milling for 2 h before being spray dried. The dried powder was uniaxially pressed at 80 MPa followed by cold-isostatic pressing at 200 MPa during 0.5 min. The shape and dimensions of the as-pressed green specimens were tiles of $40 \times 40 \times 10$ mm. The dimensions after sintering in air at 1600°C for 2 h were approximately $34 \times 34 \times 8.5$ mm. The specimens were ground to similar flatness, dimensions, and roughness.

Silicon (111) was a 100 mm diameter wafer, 0.5 mm in thickness. The average surface roughness (Ra) was 10 a 20 Å. The orientation plan was 111. The quartz single crystal X was also a 100 and 0.5 mm wafer of similar roughness. The orientation plan was X.

All specimens were diced in a precision dicing saw from DISCO, model DAD 3350, using three blades from different suppliers: THERMOCARBON (model 2.25M-10C-54R7-3), DISCO (model R07-SDC280-B200-75), ADT (model 00777-8053-010-KUP), all of them of 0.254 mm thickness, with a 2″ outer diameter.

There were three cutting parameters: cut depth (p), feed speed (v), and rotation speed or cutting speed (w). See Fig. 13.3.

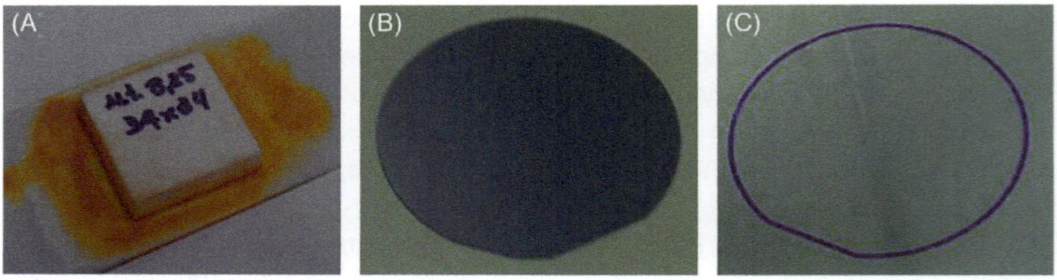

Figure 13.2 (A) Alumina polycrystalline 99.8%, (B) silicon (111) wafer, and (C) quartz X wafer

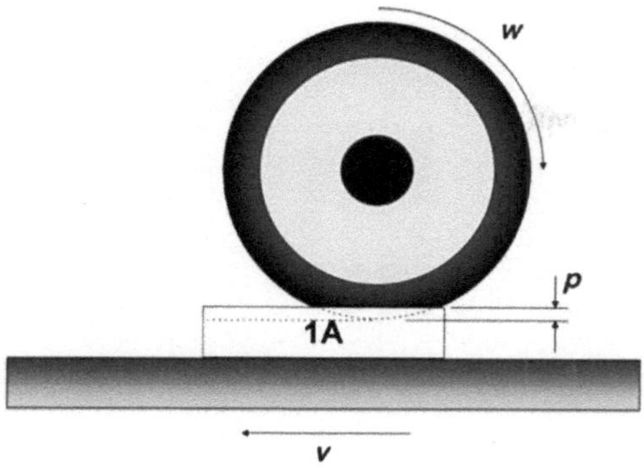

Figure 13.3 Scheme of the cutting process.

Each cutting parameter combination received a Group name according to a number and the workpiece material under cut. Table 13.1 describes the group of cuttings made from the quartz X specimen, combining the model of blade and cutting process parameters. Similar tables were applied to alumina and silicon.

The failures (usually chipping, although failures also included long cracks, voids, etc.) were measured by an average of 10 largest measurements along the kerf. Fig. 13.4 explains the procedure.

The failure measurements were plotted in two graphs: failure versus speed rotation and failure versus feed speed. After the machining experiments, a cantilever, made from quartz X, was manufactured with the lowest failures. The dimensions were $6.5 \times 32 \times 0.5$ mm, and both surfaces received a silver coating, as an electrode collector of voltage (Fig. 13.5).

To search the characteristic of used materials for EH generator, the specimen was analyzed as a power generator at two different frequencies of flexion, 5 and 900 Hz. The frequency of 5 Hz was tested in a hydraulic test machine MTS370.02. The frequency of 900 Hz was tested in a portable grinding RT650KA from Black & Decker (Fig. 13.6). The electric voltage obtained was measured by a National Instruments board USB-6009 and LabView program.

The best results were obtained with a DISCO blade for silicon and quartz. Graphs are presented in Figs. 13.7 and 13.8.

The cantilever that obtained the best cutting configuration is presented in Fig. 13.9.

The result of the flexion experiment at 5 and 900 Hz (both 0.01 mm flexion) are shown in Fig. 13.10. Although it reached a voltage of 500 mV with a 900 Hz flexion test, the cantilever did not resist very long and experienced a breakage after 2 s of work.

Table 13.1 Cutting groups/parameters of dicing quartz X

Cutting groups/cutting parameters for quartz X					
Blade supplier	**Rotation speed (rpm)**	**Cutting depth (mm)**	**Feed speed (mm/s)**	**Failures (μm)**	**Group**
Thermocarbon DISCO ADT	10,000	0.1	0.1	31/15/14	T.Q.G1*/D.Q.G1**/A.Q.G1***
			1	29/25/31	T.Q.G2/D.Q.G2/A.Q.G2
			5	64/26/51	T.Q.G3/D.Q.G3/A.Q.G3
		0.35	0.1	28/12/23	T.Q.G4/D.Q.G4/A.Q.G4
			1	49/24/32	T.Q.G5/D.Q.G5/A.Q.G5
			5	72/39/50	T.Q.G6/D.Q.G6/A.Q.G6
	30,000	0.1	0.1	34/11/19	T.Q.G7/D.Q.G7/A.Q.7
			1	43/21/38	T.Q.G8/D.Q.G8/A.Q.G8
			5	60/27/50	T.Q.G9/D.Q.G9/A.Q.G9
		0.35	0.1	34/7/13	T.Q.G10/D.Q.G10/A.Q.G10
			1	39/13/22	T.Q.G11/D.Q.G11/A.Q.G11
			5	56/35/32	T.Q.G12/D.Q.G.12/A.Q.G12
	50,000	0.1	0.1	–/13/13	T.Q.G13/D.Q.G13/A.Q.G13
			1	–/9/21	T.Q.G14/D.Q.G14/A.Q.G14
			5	–/22/33	T.Q.G15/D.Q.G15/A.Q.G15
		0.35	0.1	–/6/7	T.Q.G16/D.Q.G16/A.Q.G16
			1	–/20/24	T.Q.G17/D.Q.G17/A.Q.17
			5	–/25/47	T.Q.G18/D.Q.G18/A.Q.G18

T.Q.G1* T = thermocarbon; Q = quartz; 1 = group of cutting parameters 1.
D.Q.G1** D = DISCO; Q = quartz; G1 = group of cutting parameters 1.
A.Q.G1 *** A = ADT; Q = quartz; G1 = group of cutting parameters 1.

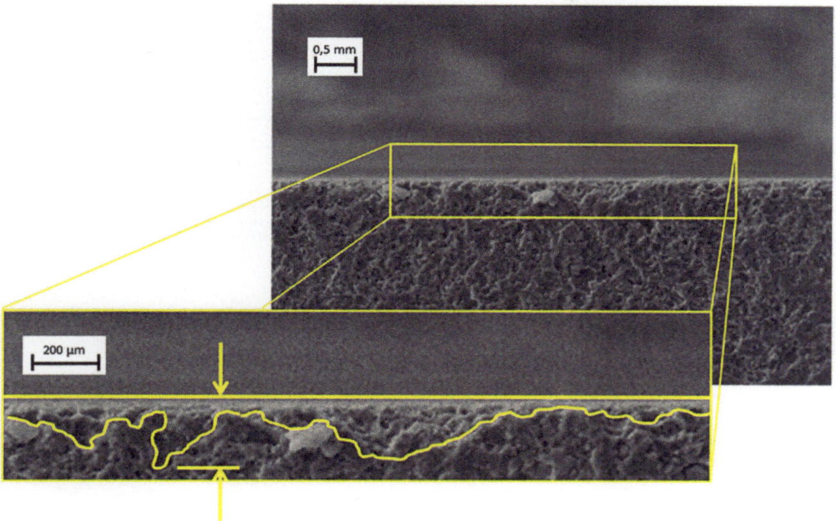

Figure 13.4 Detail of the procedure to measure failures.

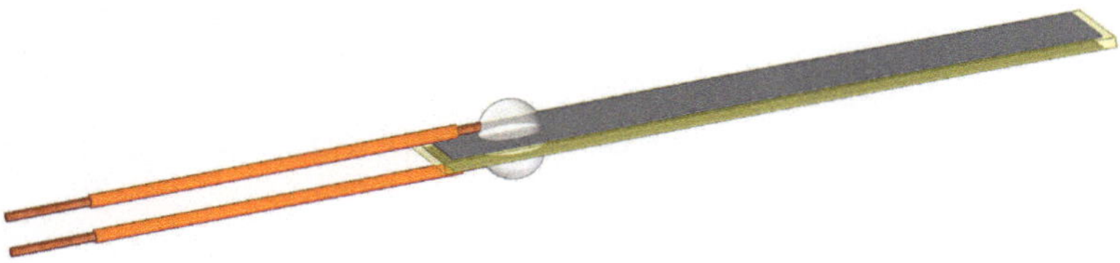

Figure 13.5 Energy harvesting piezo power generator cantilever.

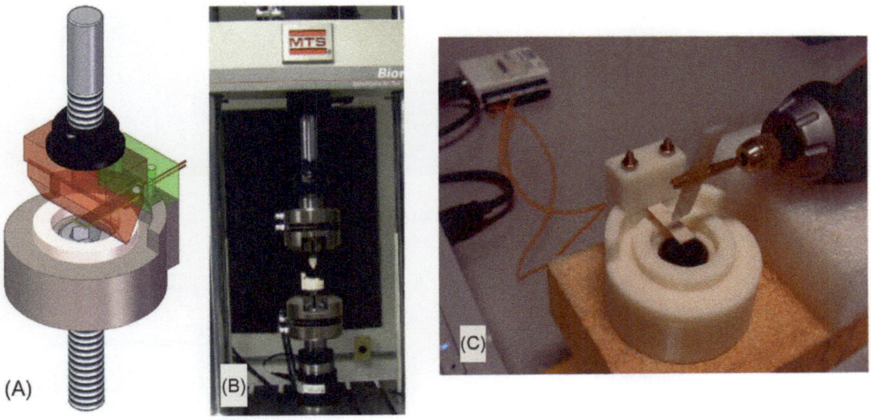

Figure 13.6 (A) Draft of the proposed support to develop the tests of the cantilever, (B) the experiment ready to be developed at MTS testing machine, and (C) the experiment developed for 900 Hz flexion test.

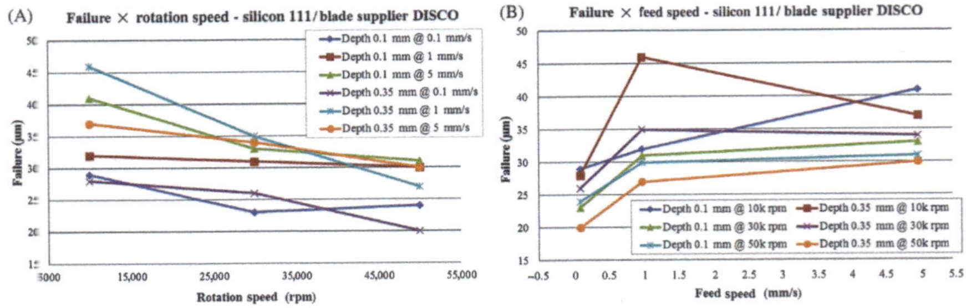

Figure 13.7 (A) Failure versus rotation speed—silicon (111)/DISCO abrasive blade and (B) failure versus feed speeds – Silicon (111)/DISCO abrasive blade.

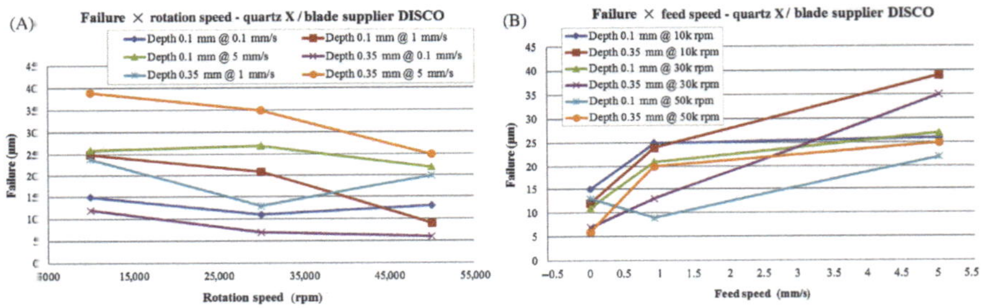

Figure 13.8 (A) Failure versus rotation speed—quartz X/DISCO abrasive blade and (B) failure versus feed speeds—Quartz X/DISCO abrasive blade.

Figure 13.9 Energy harvesting piezo power generator cantilever.

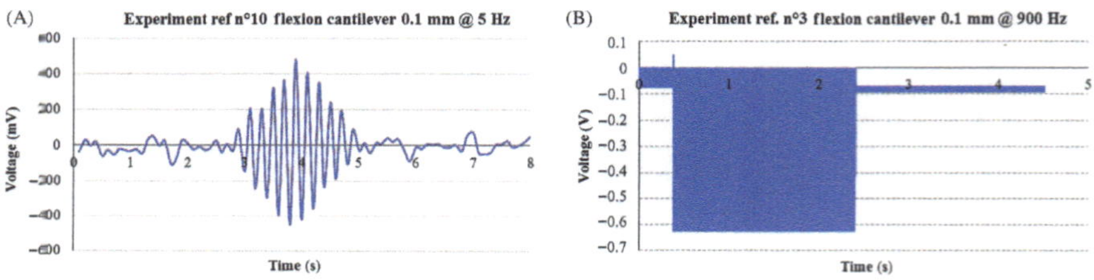

Figure 13.10 (A) Flexion experiment of cantilever at 0.01 mm flexion and 5 Hz and (B) flexion experiment of cantilever at 900 Hz and 0.01 mm flexion.

13.4 Conclusion

The cantilever power generator reached a maximum voltage of 500 mV. Considering that ordinary batteries can supply 1.5 V, this is a good amount. However, the maximum value of voltage was obtained with flexural amplitude at 1 mm and frequency at 900 Hz, where functional life was very short before total breakage, suggesting that flexion and mechanical impact may not be the best way to generate power in a piezoelectric single crystal. Even with much reduced chipping at the edges of the cantilever obtained in the dicing process manufacturing experiments, the mechanical stress was too high for a piezoelectric crystal.

The process of manufacturing using abrasive blades can be generalized for hard and brittle materials. Failure tends to reduce in proportion with the reduction in feed speed and the increase in rotation speed (Figs. 13.7 and 13.8). Small chipping and backside chipping failure occur in a process removal material working at the elastic regime of the material (or close to this condition).

The quartz X has the smallest number of failures in a blade rotation speed of 50,000 rpm (which represents 42.33 m/s in a 2″ blade), feed speed of 100 μm/s, and complete cutting depth of the wafer using a DISCO blade. This group of parameters, named D.Q.G16, had 100% failures classified as chipping with a size of 6 μm.

References

[1] E. Mounier, MEMS Markets & Applications: Focus on Wireless Sensor Networks & Energy Harvesting. In: 2nd Workshop on design, control and software implementation for distributed MEMS, Besançon, <http://dmems.univ-fcomte.fr/presentations/mounier.pdf>, 2012 (accessed 12.01.13).

[2] C.S. Mcgahey, Harnessing Nature's Timekeeper: A History of the Piezoelectric Quartz Crystal Technological Community (1880–1959). PhD, School of History, Technology and Society, Georgia Institute of Technology, 2009.

[3] H.S. Kim, J. Kim, J. Kim, A review of piezoelectric energy harvesting based on vibration, Int. J. Precis. Eng. Manuf. 6 (12) (2011) 1129–1141.

[4] T.G. Bifano, J.B. Hosler, Precision grinding of ultra-thin quartz wafers, J. Eng. Ind. 115 (3) (1993) 258–262.

[5] T. Watanabe, M. Kojima, K. Yamato, Study of quartz crystal slicing technology by using multi wire saw. In: 2001 IEEE International Frequency Control Symposium and PDA Exhibition, 2001.

[6] T. Hsu, MEMS and Microsystems: Design, Manufacture and Nanoscale Engineering, second ed., John Wiley & Sons, New Jersey, 2008.

[7] S. Chen, Z. Jiang, A force controlled grinding-milling technique for quartz-glass micromachining, J. Mater. Process. Technol. 216 (2015) 206–215.

[8] W.D. Callister, D.G. Rethwisch, Materials Science and Engineering: An Introduction, Wiley, New York, NY, 2007.

[9] P.L. Guzzo, et al., Characterization of growth sectors in synthetic quartz grown from cylindrical seeds parallel to [0001] direction, Mat. Res 7 (2004) 285–291.

[10] R.K. Cook, P.G. Weissler, Piezoelectric constants of alpha-and beta-quartz at various temperatures, Phys. Rev. 80 (4) (1950) 712.
[11] H. Grimm, B. Dorner, On the mechanism of the α-β phase transformation of quartz, J. Phys. Chem. Solids 36 (5) (1975) 407–413.
[12] M.T. Dove, M. Gambhir, V. Heine, Anatomy of a structural phase transition: theoretical analysis of the displacive phase transition in quartz and other silicates, Phys. Chem. Miner. 26 (4) (1999) 344–353.
[13] G.D. Quinn, K. Hoffman, J.B. Quinn, Strength and fracture origins of a feldspathic porcelain, Dent. Mater. 25 (5) (2012) 502–511.
[14] S. Luo, C.M. Swadener, O. Tschauner, Examining crystallographic orientation dependence of hardness of silica Stishovite, Physica B Condens. Matter 99 (2) (2007) 138–142.
[15] S. Malkin, T.W. Hwang, Grinding mechanisms for ceramics, CIRP Ann. Manuf. Technol. 45 (2) (1996) 569–580.
[16] E. Brinksmeier, et al., Ultra-precision grinding, CIRP Ann. Manuf. Technol. 59 (2) (2010) 652–671.
[17] S.Y. Luo, Z.W. Wang, Studies of chipping mechanisms for dicing silicon wafers, Int. J. Adv. Manuf. Technol. 35 (11–12) (2008) 1206–1218.
[18] R. Komanduri, S. Varghese, N. Chandrasekaran, On the mechanism of material removal at the nanoscale by cutting, Wear 269 (3) (2010) 224–228.
[19] U. Efrat, Optimizing the wafer dicing process. In: Electronic Manufacturing Technology Symposium, Fifteenth IEEE/CHMT International, 1993. 245–253.
[20] S. Jahanmir, M. Ramulu, P. Koshy, Machining of Ceramics and Composites, Marcel Dekker, USA, 1999.
[21] G. Levinson, Principles of dicing. ADT Tech Library, Kulicke & Soffa Industries Inc. <http://www.adt-co.com/ADT2014//userdata/SendFile.asp?DBID=1&LNGID=1&GID=709>, 2004 (accessed 10.11.16).
[22] H. Zhou, et al., High-speed dicing of silicon wafers conducted using ultrathin blades, Int. J. Adv. Manuf. Technol. 66 (5–8) (2013) 947–953.

Nonvolatile memories

14

Leandro S.R. Rocha[1], Alexandre Z. Simões[1], Johan A. Cortés[1], Miguel A. Ramirez[1], Francisco Moura[2], Cesar R. Foschini[3], Ronald Tararam[4], Mario Cilense[5], Elson Longo[6] and Carlos O. Paiva-Santos[5]

[1]São Paulo State University (Unesp), School of Engineering, Guaratinguetá, Brazil, [2]Federal University of Itajubá (Unifei), Itabira, Brazil, [3]São Paulo State University (Unesp), School of Engineering, Bauru, Brazil, [4]CEITEC S.A Semiconductors, Porto Alegre, Brazil, [5]São Paulo State University (Unesp), Institute of Chemistry, Araraquara, Brazil, [6]Federal University of São Carlos (UFSCar), São Carlos, Brazil

14.1 Introduction

The need for faster, smaller, cheaper and energy-efficient electronic devices has been growing continuously in the last decade, with the conventional data storage technologies (i.e., static random access memory (SRAM) and dynamic random access memory (DRAM) memory), which have been so far fulfilled by CMOS-charge storage-based circuits, approaching their fundamental limits, due to the lesser progress of technology in comparison to logic [1]. To overcome this challenge, increasingly high storage density memories has become one of the crucial approaches, aiming to improve storage capacity and reading/writing speed [2,3]. A semiconductor memory is an indispensable component of all modern electronic devices, with all recognizable computing platforms, from hand-held devices to large supercomputer storage systems being used for storing data, temporarily or permanently. The ability of a material to store information is defined as a solid-state memory effect, which requires at least two switchable memory states that can be addressed by an externally controlled parameter [4]. Based on storing data volatility, memories are basically classified into two categories, volatile and nonvolatile, with the former immediately losing the stored data after turning off the power, whereas the latter being capable of retaining the stored data for a longer period, even after the power is turned off [5,6]. To optimize the performance-cost trade-off, hierarchical systems made from devices with varying speed, density, and cost have been adopted and the novel nonvolatile memory (NVM) concepts, such as ferroelectric random access memory (FeRAM), phase-change RAM (PCRAM), magnetic RAM (MRAM), spin-transfer-torque RAM (STT-RAM), and resistive RAM (RRAM) are fulfilling the changing market trend requirements, from electronics to high performance computing, due to encouraging recent experimental demonstrations of high density, excellent scalability, low power consumption, endurance, and low cost [7–17]. An RRAM is normally referred as those NVM technologies built on the resistance changing mechanisms, which can be varied by applying a voltage pulse, other than phase-change memory and STT-RAM. Data is stored by changing the resistance

Magnetic, Ferroelectric, and Multiferroic Metal Oxides. DOI: http://dx.doi.org/10.1016/B978-0-12-811180-2.00014-1

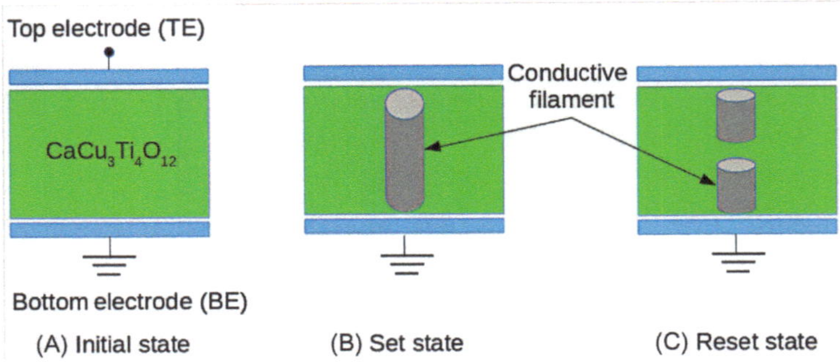

Figure 14.1 Illustrative representation of the nonvolatile memory device, consisting of a semiconductor oxide layer sandwiched between the top and bottom electrode in its initial state (A), and after SET (B), and RESET operations.

across a dielectric solid-state material in the RRAM cell, which presents two resistance states: a high-resistance state (HRS) and a low-resistance state (LRS), being respectively defined as R_{ON} and R_{OFF}. Thus, the device can be used as a Boolean logic switch returning (0) when the resistance is R_{OFF} and (1) when R_{ON} [18].

An RRAM device consists of an insulating layer, generally a semiconductor oxide, sandwiched between a top electrode (TE) and a bottom electrode (BE), both generally consisting of metallic layers, as seen in Fig. 14.1A. The device is initially subjected to the operation of electroforming, where a conductive filament (CF) is formed by dielectric breakdown (Fig. 14.1B). The current has to be limited by a series resistor/transistor during this process, controlling the size of the conductive filament and avoiding conductive (hard) breakdown of the switching layer. After forming, the device manifests improved conductance, thus resulting in LRS. The RESET operation can be done to disconnect the conductive filament (Fig. 14.1C), resulting in a HRS. Alternating SET/RESET operations results in repeatedly connected/disconnected conductive filaments, allowing multiple transition cycles between LRS and HRS. The two main methods of resistance switching differs by the polarity of the SET and RESET operations, in which both processes take place under positive voltage for unipolar switching, while alternated polarities in the SET/RESET processes are required on bipolar switching [19–22].

After decades of research, many exciting records regarding RRAM performance were reported, demonstrating scalability down to 10 nm in cell size in HfO_x and WO_x-based systems [23,24]. In addition, an ultrafast switching speed around 100 ps in TaO_x and Pt-dispersed SiO_2-based devices [25,26] along with a satisfactory retention of several months demonstrated experimentally, to be held for more than 10 years, when extrapolated [27], is far superior to that of flash memories and even comparable to that of DRAMs. Concerning their performance, these devices can be evaluated based on scalability, speed, power, and reliability, in which each category involves multiple factors, according to Fig. 14.2.

Among the varieties of materials used for RRAMs, the perovskite oxides received enormous attention for this application in the last decade. Lee et al. demonstrated that the oxides with dopants had higher device yield than the nonstoichiometric

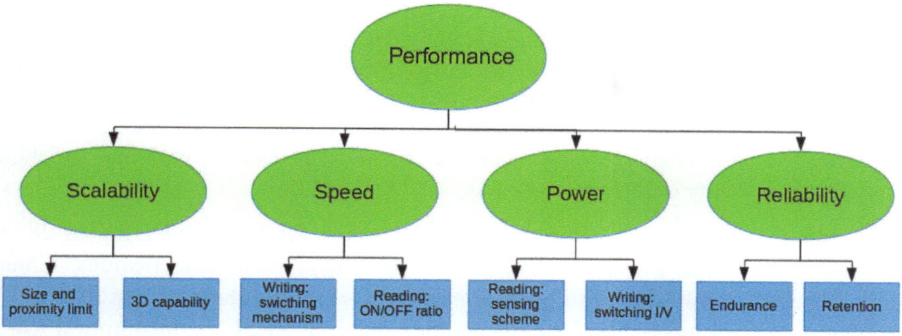

Figure 14.2 Multiple factors involving the evaluation of nonvolatile memory device performance and their main categories.

oxides, such as Nb_2O_{5-x}, ZrOx, and $SrTiO_x$ (STO) [28]. Reproducible resistive switching behaviors have been observed in doped perovskite oxide films, such as Cr-doped $SrZrO_3$ (Cr:SZO) [29–32], Cr-doped $SrTiO_3$ (Cr:STO) [33], Nb-doped $SrTiO_3$ (Nb:STO) [34,35], and V-doped $SrZrO_3$ (V:SZO) films [36,37].

14.2 Nonvolatile memory device operation

Taking into account the circuit requirements of today's high-density NVM, and the predictions about technology scaling for the next decade, a number of necessities are brought together [38]:

1. *Write operation.* Write voltages (V_{wr}) should be in the range of a few hundred mV to be compatible with scaled CMOS to a few V, to provide an advantage over flash, which suffers from the high programming voltages. The length of pulses are sought to be inferior to 100 ns in order to compete with DRAM specifications and to outperform flash (with time programming speed around 10 ms) or even less than 10 ns to compete with high-performance SRAM.
2. *Read operation.* Read voltages (V_{rd}) need to be significantly smaller than write voltages in order to prevent a change of the resistance during the read operation. Because of the constraints of circuit design, V_{rd} cannot be less than approximately $(1/10)V_{wr}$. An additional requirement originates from the minimum read current I_{rd}. In the ON-state, I_{rd} should not be less than approximately 1 mA to allow for a fast detection of the state by reasonably small sense amplifiers. The read time T_{rd} must be in the order or preferably shorter than write time T_{wr}.
3. *Resistance ratio.* Although an R_{OFF}/R_{ON} ratio of only 1.2–1.3 can be utilized in the circuit design as shown in MRAM, R_{OFF}/R_{ON} ratios above 10 are required to allow for small and highly efficient sense amplifiers and, hence, RRAM devices are cost competitive with flash.
4. *Endurance.* Contemporary flash shows a maximum number of write cycles between 10^3 and 10^7, depending on the type. RRAM should provide at least the same endurance, preferably a better one.
5. *Retention.* A data retention time T_{ret} of >10 years is required for the universal NVM. This retention time must be kept at thermal stress up to 85°C and small electrical stress

such as a constant stream of read voltage pulses. The writing speed depends on the switching mechanisms, while reading speed is determined by the sensing scheme and memory on/off ratio. Power is determined by operation voltage and current, and low power does not necessarily means low energy (e.g., flash memories have very small write current, due to tunneling mechanism, and therefore low power, but their long write time increases the write energy), while reliability (retention, endurance, etc.) and variability impact the feasibility of large memory array and products.

According to Canet et al. [39], the typical values for LRS and HRS in 28 nm technology are in the range of $R_{LRS} = 1\ k\Omega$ and $R_{HRS} = 100\ k\Omega$. Besides, Chang et al. [40] reported RRAM write operations with a supply voltage of less than 0.5 V on a 65 nm chip with a hybrid device scheme, enabling RESET consumption lower than 30 μA per cell at 1.5 V and a SET consumption lower than 20 μA at 2 V.

Nowadays the technological evolution of NVM devices with outstanding properties, including high density, excellent scalability, low power consumption for SET and RESET operations, and low cost are in progress. In this chapter we present the possibility of manufacturing NVM devices using perovskite $CaCu_3Ti_4O_{12}$ (CCTO) thin films obtained through chemical (spin-coated polymeric precursors) and physical (radio frequency (RF)-sputtering) routes. Those films show a small degradation of current ratio between the ON and OFF states along with a resistive switching behavior, allowing its application on RRAM devices.

14.3 Radio frequency-sputtered $CaCu_3Ti_4O_{12}$ thin film

CCTO thin film was RF-sputtered according to described parameters and procedures [41], and showed proper structural adjustment, as in accordance with many authors [42–48].

Fig. 14.3 shows a typical I–V hysteresis curve for the film with polycrystalline nature. The curve clearly shows the regions of accumulation, depletion, and inversion, according to Chung et al. [49], who demonstrated the presence of electrical potential boundaries at CCTO, attributing the nonlinear behavior to the grain boundary barrier layer, and also demonstrated that the transport of charge inside grains has a linear relation, and the nonlinear curve has resulted from grain boundaries serving as a barrier to the carriers. A clockwise rotation in the I–V characteristic of a dielectric film on a *p*-type substrate is expected when charge compensation on the surface is induced by the polarization present in the film, which is the desired switching mode for nonvolatile memory operation. An accumulation layer appeared at negative bias voltages since the dielectric film was deposited on a *p*-type substrate. As the applied voltage was increased and became positive, a depletion layer was formed inside platinum near the interface. A further increase in voltage caused inversion of the surface, and no significant difference was observed when the bias was reversed, which is reasonable because the system is completely symmetrical, and does not show evidence of ferroelectric behavior at room temperature. Thus, applying a voltage in one direction or the other would be equivalent, regardless of whether the conduction is limited by interfacial barriers or by bulk-like mechanisms.

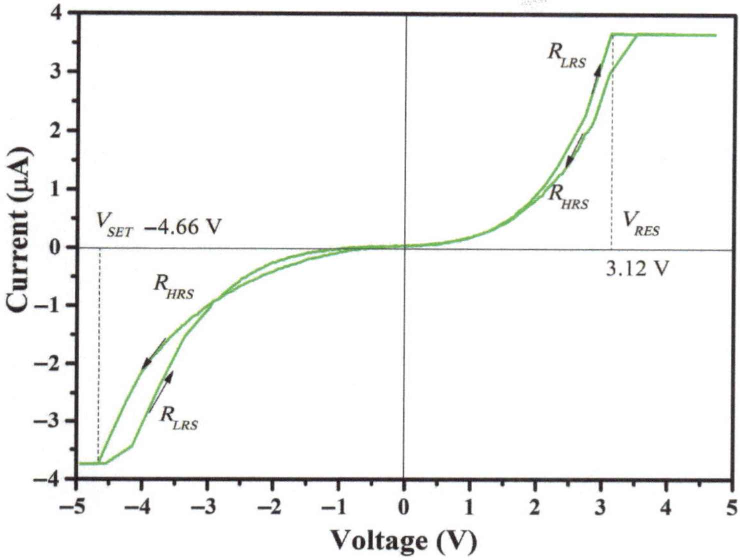

Figure 14.3 I−V curve for $CaCu_3Ti_4O_{12}$ film obtained by radio frequency-sputtering, depicting the SET and RESET required voltages, as well as the low and high-resistance states.

14.4 Spin-coated $CaCu_3Ti_4O_{12}$ thin films

Thin films of CCTO were synthesized and spin-coated on Pt/Ti/SiO_2/Si substrates according to [50], with no secondary phases observed, differing from most published results in the literature [44,51−53].

The respective I−V curves presented in Fig. 14.4 show a nonlinear resistive behavior. With higher temperature the breaking stress decreases, which has already been reported in the literature and is associated with the increase of grain size [52,54]. The film presents a nonvolatile memory behavior as already shown in the literature [55], indicating LRS corresponding to on-state. It can be observed that the erasing mechanism happens more slowly for the film when it is crystallized at a temperature of 500°C, different from that crystallized at 600°C, which occurred almost instantaneously, being explained by the relaxation mechanisms with an externally applied voltage [56−58].

Comparing the presented results for spin-coated polymeric precursors and RF-sputtered $CaCu_3Ti_4O_{12}$ thin films, LRS and HRS ranged from values below 500 kΩ to values above 1000 kΩ, with magnitudes 10 to 500× higher, while the consumption reached only 0.855 μA at 2 V for the RF-sputtered thin film, representing a consumption drop of more than 95%. On the other hand, Wang et al. [59] reported an RRAM device constituted of Au-doped TaO_x sputtered thin film, which allowed a SET operation to LRS at a negative bias of −0.7 V and a RESET to HRS at a positive bias of +1.2 V. Our RF-sputtered $CaCu_3Ti_4O_{12}$ thin film required

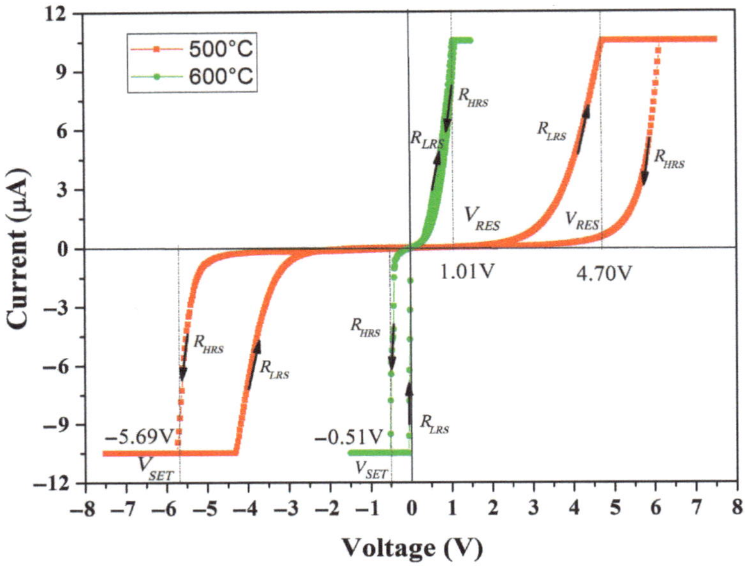

Figure 14.4 I–V curves for $CaCu_3Ti_4O_{12}$ films.

−4.66 V and +3.12 V and the spin-coated thin film crystallized at 600°C required −0.51 V and +1.01 V respectively, requesting less energy for its functioning due to lower voltages necessary for SET/RESET operations, along with lower power consumption due to the extremely reduced current, below 4 μA for the RF-sputtered and 10 μA for the spin-coated film. Concerning the RF-sputtered and spin-coated CCTO films, the obtained results are promising for their application on RRAM devices due to extreme low consumption and energy and their nanostructured dimensions, meeting the needs for future electronic devices.

References

[1] G. Sun, C. Zhang, P. Li, T. Wang, Y. Chen, IEEE Trans. Comput. 65 (2016) 3427–3440.
[2] A. Chen, Solid State Electron. 125 (2016) 25–38.
[3] M. Pešic, M. Hoffmann, C. Richter, T. Mikolajick, U. Schroeder, Adv. Funct. Mater. 26 (2016) 7486–7494.
[4] M.M. Vopson, G. Caruntu, X. Tan, Scr. Mater. 128 (2017) 61–64.
[5] F.M. Simanjuntak, D. Panda, K. Wei, T. Tseng, Nanoscale Res. Lett. 11 (2016) 368.
[6] B.H. Park, B.S. Kang, S.D. Bu, T.W. Noh, J. Lee, W. Jo, Nature 401 (1999) 682–684.
[7] Z. Sun, X. Bi, H. Li, Proc. Int. Symp. Low Power Electron. Des. (2012) 179–184.
[8] M.-J. Lee, C.B. Lee, D. Lee, S.R. Lee, M. Chang, J.H. Hur, et al., Nat. Mater. 10 (2011) 625–630.

[9] J. Li, S. Member, P. Ndai, S. Member, A. Goel, S. Salahuddin, et al., A. Spin-torque, S.T.T. Mram 18 (2010) 1710–1723.
[10] C.W. Smullen IV, V. Mohan, A. Nigam, S. Gurumurthi, M.R. Stan, Proc. Int. Symp. High-Performance Comput. Archit (2011) 50–61.
[11] International Technology Roadmap for Semiconductors (ITRS), Itrs 2013 (2013) 1–51.
[12] L. Wan, Q. Cao, F. Wang, S. Oral, J. Parallel Distrib. Comput. 100 (2017) 16–29.
[13] L. Zhu, J. Zhou, Z. Guo, Z. Sun, J. Mater. 1 (2015) 1–11.
[14] A. Jadidi, M. Arjomand, H. Sarbazi-azad, (2011) 79–84.
[15] D. Thuau, M. Abbas, G. Wantz, L. Hirsch, I. Dufour, C. Ayela, Org. Electron. Phys. Mater. Appl. 40 (2017) 30–35.
[16] U. Celano, L. Mirabelli, L. Goux, K. Opsomer, W. Devulder, F. Crupi, et al., Microelectron. Eng. 167 (2017) 47–51.
[17] Y. Hu, X. Zhu, H. Zou, J. Zhang, L. Yuan, J. Xue, et al., Appl. Phys. Lett. 108 (2016) 150–154.
[18] M. Congiu, L.G.S. Albano, O. Nunes-Neto, C.F.O. Graeff, Electron. Lett 52 (2016) 1871–1873.
[19] M.N. Kozicki, H.J. Barnaby, D. Ielmini, S. Balatti, S. Larentis, D. Ielmini, Semicond. Sci. Technol. 31 (n.d.) 1–25.
[20] F. Pan, S. Gao, C. Chen, C. Song, F. Zeng, Mater. Sci. Eng. R Reports 83 (2014) 1–59.
[21] M.N. Kozicki, M. Balakrishnan, C. Gopalan, C. Ratnakumar, M. Mitkova, Non-Volatile Mem. Technol. Symp. NVMTS05 (2005) 83–89.
[22] I.H. Inoue, S. Yasuda, H. Akinaga, H. Takagi, Phys. Rev. B - Condens. Matter Mater. Phys. 77 (2008).
[23] B. Govoreanu, G.S. Kar, Y.Y. Chen, V. Paraschiv, S. Kubicek, A. Fantini, et al., Tech. Dig. Int. Electron Devices Meet. IEDM (2011) 729–732.
[24] C. Ho, C. Hsu, C. Chen, J. Liu, C. Wu, Memory (2010), 19.1.1-19.1.4.
[25] A.C. Torrezan, J.P. Strachan, G. Medeiros-Ribeiro, R.S. Williams, Nanotechnology 22 (2011) 485203.
[26] B.J. Choi, A.C. Torrezan, K.J. Norris, F. Miao, J.P. Strachan, M.X. Zhang, et al., Nano Lett. 13 (2013) 3213–3217.
[27] S. Gao, F. Zeng, C. Chen, G. Tang, Y. Lin, Z. Zheng, et al., Nanotechnology 24 (2013) 335201.
[28] D. Lee, D.J. Seong, H.J. Choi, I. Jo, R. Dong, W. Xiang, et al., Tech. Dig. Int. Electron Devices Meet. IEDM (2006) 11–14.
[29] C.Y. Liu, P.H. Wu, A. Wang, W.Y. Jang, J.C. Young, K.Y. Chiu, et al., IEEE Electron Device Lett. 26 (2005) 351–353.
[30] J.W. Park, K. Jung, M.K. Yang, J.K. Lee, D.Y. Kim, J.W. Park, J. Appl. Phys. 99 (2006).
[31] C. Rossel, G.I. Meijer, D. Brémaud, D. Widmer, J. Appl. Phys. 90 (2001) 2892–2898.
[32] A. Beck, J.G. Bednorz, C. Gerber, C. Rossel, D. Widmer, Appl. Phys. Lett. 77 (2000) 139.
[33] Y. Watanabe, J.G. Bednorz, A. Bietsch, C. Gerber, D. Widmer, A. Beck, et al., Appl. Phys. Lett. 78 (2001) 3738–3740.
[34] A. Sawa, T. Fujii, M. Kawasaki, Y. Tokura, Strongly Correl. Electron Mater. Phys. Nanoeng. Ed. by Bozovic 5932 (2005) 366.
[35] T. Fujii, M. Kawasaki, A. Sawa, H. Akoh, Y. Kawazoe, Y. Tokura, Appl. Phys. Lett. 86 (2005) 2003–2006.
[36] S. Wang, T. Tseng, J. Adv. Dielectr. 1 (2011) 141–162.

[37] C.C. Lin, J.S. Yu, C.Y. Lin, C.H. Lin, T.Y. Tseng, Thin Solid Films 516 (2007) 402–406.

[38] J. Yang, M. Shen, L. Fang, Mater. Lett. 59 (2005) 3990–3993.

[39] P. Canet, J. Postel-Pellerin, H. Aziza, Microelectron. Reliab. 64 (2016) 36–41.

[40] M.-F. Chang, S.-J. Shen, Y.-L. Lu, Y.-S. Yang, J.-Y. Hung, C.-W. Wu, et al., IEEE Int. Conf. Electron Devices Solid-State Circuits (2016) 342–345.

[41] C.R. Foschini, R. Tararam, A.Z. Simões, L.S. Rocha, C.O.P. Santos, E. Longo, et al., J. Mater. Sci. Mater. Electron. 27 (2016) 2175–2182.

[42] A.K. Thomas, K. Abraham, J. Thomas, K.V. Saban, Ceram. Int. 41 (2015) 10250–10255.

[43] L. Ni, X.M. Chen, X.Q. Liu, R.Z. Hou, Solid State Commun. 139 (2006) 45–50.

[44] D. Capsoni, M. Bini, V. Massarotti, G. Chiodelli, M.C. Mozzatic, C.B. Azzoni, J. Solid State Chem. 177 (2004) 4494–4500.

[45] T. Li, R. Xue, J. Hao, Y. Xue, Z. Chen, J. Alloys Compd. 509 (2011) 1025–1028.

[46] C.M. Wang, S.Y. Lin, K.S. Kao, Y.C. Chen, S.C. Weng, J. Alloys Compd. 491 (2010) 423–430.

[47] S. Krohns, J. Lu, P. Lunkenheimer, V. Brizé, C. Autret-Lambert, M. Gervais, et al., Eur. Phys. J. B 72 (2009) 173–182.

[48] N. Barman, S. Tripathi, N. Ravishankar, K.B.R. Varma, Solid State Commun. 241 (2016) 7–13.

[49] S.-Y. Chung, I.-D. Kim, S.-J.L. Kang, Nat. Mater. 3 (2004) 774–778.

[50] M.A. Ramirez, A.Z. Simões, A.A. Felix, R. Tararam, E. Longo, J.A. Varela, J. Alloys Compd. 509 (2011) 9930–9933.

[51] D.P. Singh, Y.N. Mohapatra, D.C. Agrawal, Mater. Sci. Eng. B Solid-State Mater. Adv. Technol. 157 (2009) 58–65.

[52] R. Jia, X. Zhao, J. Li, X. Tang, Mater. Sci. Eng. B Solid-State Mater. Adv. Technol. 185 (2014) 79–85.

[53] X.H. Zheng, J. Xiao, X. Huang, D.P. Tang, X.L. Liu, J. Mater. Sci. Mater. Electron. 22 (2011) 1116–1119.

[54] K. Meeporn, T. Yamwong, S. Pinitsoontorn, V. Amornkitbamrung, P. Thongbai, Ceram. Int. 40 (2014) 15897–15906.

[55] Y. Shen, B. Chiou, C. Ho, Thin Solid Films 517 (2008) 1209–1213.

[56] C.R. Foschini, R. Tararam, A.Z. Simões, M. Cilense, E. Longo, J.A. Varela, J. Alloys Compd. 574 (2013) 604–608.

[57] J. Jumpatam, B. Putasaeng, T. Yamwong, P. Thongbai, S. Maensiri, Ceram. Int. 39 (2013) 1057–1064.

[58] D. Panda, T.-Y. Tseng, Ferroelectrics 471 (2014) 23–64.

[59] Z. Wang, J. Kang, Y. Fang, Z. Yu, X. Yang, Y. Cai, et al., 16th Non-Volatile Mem. Technol. Symp. (2016) 1–3.

Part II

Magnetic and Multiferroic Metal Oxides

Section IV

Magnetic Oxides: Ferromagnetics, Antiferromagnetics and Ferrimagnetics

Theory of ferrimagnetism and ferrimagnetic metal oxides

15

Ling Bing Kong[1,6], *Lie Liu*[2], *Zhihong Yang*[3], *Sean Li*[4], *Tianshu Zhang*[5] *and Chuanhu Wang*[6]
[1]Nanyang Technological University, Singapore, Nanyang, [2]General Test Systems Inc., Shenzhen, Guangdong, P.R. China, [3]Nanjing University of Aeronautics and Astronautics, Nanjing, P.R. China, [4]The University of New South Wales, Sydney, NSW, Australia, [5]Anhui Target Advanced Ceramics Technology Co. Ltd., Hefei, Anhui, P.R. China, [6]Bengbu University, Bengbu, Anhui, P. R. China

15.1 Introduction

Magnetism is a special physical phenomenon that is mediated by magnetic fields. There are five types of magnetisms: ferromagnetic, ferrimagnetic, antiferromagnetic, paramagnetic, and diamagnetic. The most familiar effects occur in ferromagnetic materials, which are strongly attracted by magnetic fields and can be magnetized to become permanent magnets, producing magnetic fields themselves. Only a few substances are ferromagnetic; the most common ones are iron, nickel, and cobalt and their alloys. The prefix *ferro-* refers to iron, because permanent magnetism was first observed in lodestone, a form of natural iron ore called magnetite, Fe_3O_4.

Although ferromagnetism is responsible for most of the effects of magnetism encountered in our daily life, all other materials are influenced to some extent by a magnetic field, by several other types of magnetism. Paramagnetic substances such as aluminum and oxygen are weakly attracted to an applied magnetic field; diamagnetic substances such as copper and carbon are weakly repelled; while antiferromagnetic materials such as chromium and spin glasses have a more complex relationship with a magnetic field. The force of a magnet on paramagnetic, diamagnetic, and antiferromagnetic materials is usually too weak to be detectable, so that it can only be detected by laboratory instruments. Therefore, these substances are often called nonmagnetic in everyday life.

15.2 Magnetic fields in materials

For a coil made of metallic wire, if an electrical current flows through it, the flowing electrical field will create a magnetic field, with magnitude of the field to be given by the following equation:

$$H = \frac{NI}{l}, \tag{15.1}$$

Magnetic, Ferroelectric, and Multiferroic Metal Oxides. DOI: http://dx.doi.org/10.1016/B978-0-12-811180-2.00015-3

where I is current, N and l are the number of turns and length of the coil, respectively.

In case of vacuum, the magnetic flux density going through the coil is given by:

$$B = \mu_0 H, \tag{15.2}$$

where μ_0 is called the permeability of vacuum, with a value of $\mu_0 = 4\pi \times 10^{-7}$ $(=1.257 \times 10^{-6})$ H/m.

If a material is inserted into the metallic wire coil, the magnetic flux density becomes:

$$B = \mu_0 H + \mu_0 M, \tag{15.3}$$

The term $\mu_0 M$ is called magnetic polarization or magnetization of the materials.

At relatively low magnetic field, the magnetization is approximately linear to magnetic field H; that is, there is:

$$B = \chi H, \tag{15.4}$$

where χ is a proportional constant, known as the magnetic susceptibility, which is measure of magnetic properties of a material. Therefore, the following equation is derived:

$$B = \mu_0 (1 + \chi) H. \tag{15.5}$$

Similar to Eq. (15.2), there is:

$$B = \mu H, \tag{15.6}$$

where μ is called permeability of the materials. As a result, there is the following relationship:

$$\mu = \mu_0 (1 + \chi). \tag{15.7}$$

Usually, it is more likely to use the relative permeability μ_r for convenience, which is defined as $\mu_r = \mu/\mu_0$. As a consequence, there is:

$$\mu_r = 1 + \chi. \tag{15.8}$$

In this case, for vacuum, $\mu_r = 1$. According to magnitude and sign of permeability and magnetic susceptibility, materials are classified into five groups of magnetisms—diamagnetism, paramagnetism, antiferromagnetism, ferromagnetism, and ferrimagnetism—which will be further elaborated in the following section.

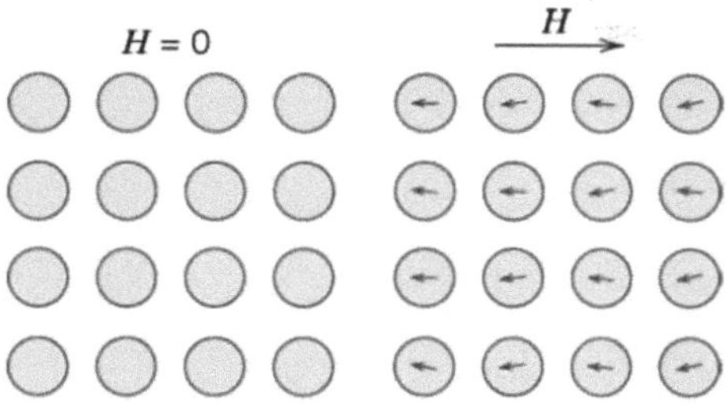

Figure 15.1 Schematic diagram of magnetization of diamagnetic materials, with a negative magnetic susceptibility and a permeability of slightly less than that of vacuum.

15.3 Magnetisms

15.3.1 Diamagnetism

Diamagnetic materials have no net magnetic moment without the presence of an external applied magnetic field (H). However, as an external magnetic field is applied to a diamagnetic material, the spinning electrons experience a motion, known as processing, producing electrical current and thus a magnetization (M), which is in the opposite direction to that of the external magnetic field, as schematically shown in Fig. 15.1. Therefore, diamagnetic materials have a small negative magnetic susceptibility (χ) and a permeability that is slightly lower than that of vacuum (μ_0). Although the diamagnetic effect occurs in all materials, it is too weak as compared with other magnetic effects in magnetic materials. The value of susceptibility is independent of temperature.

15.3.2 Paramagnetism

As an external magnetic field is applied to paramagnetic materials, an induced magnetic field is generated in the direction of the applied magnetic field. Paramagnetic materials have a relative magnetic permeability that is slightly larger than but very close to that of vacuum ($\mu_0 = 1$). Due to the presence of a nonnegative magnetic susceptibility, they are attracted to external magnetic fields. The magnetic moment induced by the external applied magnetic field is linearly proportional to in the field strength.

As the magnetic field is removed, the magnetic property is not retained. Fig. 15.2 shows a schematic diagram of paramagnetic phenomenon.

Paramagnetism has been well described by several theories for different types of materials. For example, the Langevin model is applicable to materials with noninteracting localized electrons, which indicates that every individual atom has a magnetic moment, while all the magnetic moments are randomly oriented due to the thermal agitation. As an external magnetic field is applied to the materials, the randomly oriented magnetic moments will be aligned to a certain degree, so that there is a low magnetization in the same direction as the magnetic applied field. As the

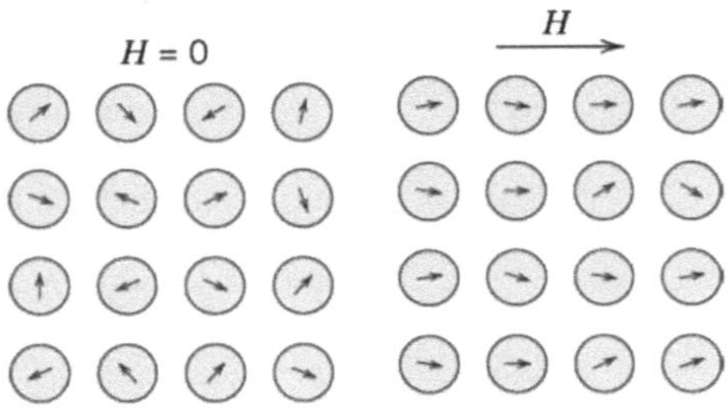

Figure 15.2 Schematic diagram of magnetization of paramagnetic materials, with low values of susceptibility and permeability.

materials are heated to a certain temperature, the thermal agitation will be enhanced. As a result, it is more difficult to align the atomic magnetic moments, thus leading to a decrease in the susceptibility. This phenomenon is known as the Curie law, given by:

$$\chi = \frac{C}{T}, \tag{15.9}$$

where C is a material constant that is called the Curie constant. Materials that obey this law have magnetic moments that are localized at the atomic or ionic sites, but there is no interaction between the neighboring magnetic moments.

More generally, the Curie law is a special case of the Curie–Weiss law, which is derived from the Weiss theory, by including a temperature constant (θ), expressed as

$$\chi = \frac{C}{T - \theta}. \tag{15.10}$$

In this case, θ has a value that can be either positive, zero, or even negative. Obviously, as $\theta = 0$, the Curie–Weiss law is simplified to the Curie law. If θ is nonzero, it implies that there is an interaction between the neighboring magnetic moments, so that the material is only in the paramagnetic state at temperatures above a critical level. If θ is positive, the material is ferromagnetic below the critical temperature, where the value of θ is actually the transition temperature—Curie temperature, T_C. If θ is negative, the material is antiferromagnetic below the critical temperature, which is known as Néel temperature, T_N. In this case, the value of θ is not related to the value of T_N. It is necessary to emphasize that this equation is only valid for the materials at the paramagnetic state.

15.3.3 Antiferromagnetism

Antiferromagnetic materials are very similar to ferromagnetic materials, in which the atomic magnetic moments are aligned in an antiparallel manner, due to the exchange interaction between the neighboring atoms. As a consequence, the magnetization is canceled out, so that the material seems to have a similar behavior in the paramagnetic material, that is, no net magnetization. As mentioned earlier, antiferromagnetic materials become paramagnetic above the Néel temperature. Fig. 15.3 shows a schematic diagram of antiferromagnetic material, MnO.

15.3.4 Ferromagnetism

In ferromagnetism, the prefix *ferro-* refers to iron (Fe), simply because permanent magnetism was first observed in lodestone, which is a form of natural iron ore called magnetite, Fe_3O_4. Schematically, Fig. 15.4 shows the atoms arranged in a lattice to have atomic magnetic moments that interact, so they are aligned parallel to each other. In classical theory, ferromagnetism is thought to be attributed to the presence of a molecular field in the material, which was first proposed by Weiss more than century ago. This field is so high that the material is magnetized to saturation. In quantum mechanics, the parallel alignment of magnetic moments can be well described by using the Heisenberg model, with the consideration of an exchange interaction between the neighboring moments.

Ferromagnetic materials contain magnetic domains, which are regions where the atomic magnetic moments are aligned in one direction. Between every two adjacent domains, the direction of the magnetic moment is continuously varied from one domain to the other. The response of the material to a magnetic field is highly

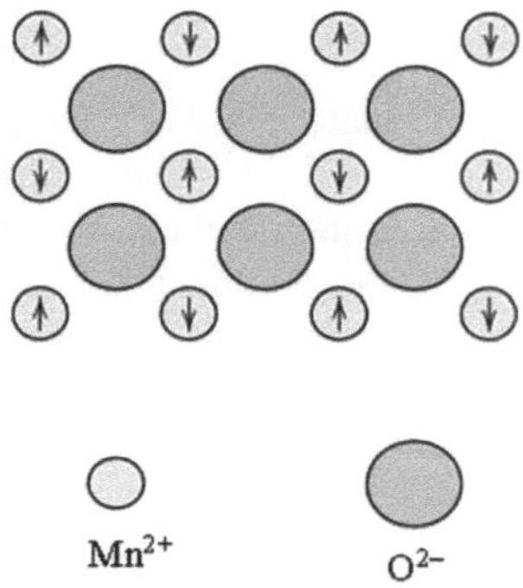

Figure 15.3 Schematic diagram of magnetization of paramagnetic materials (MnO), with zero net magnetization in the absence of an external magnetic field.

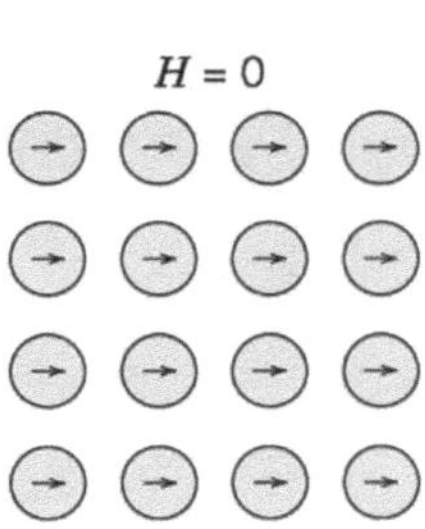

Figure 15.4 Schematic diagram of magnetization of ferromagnetic materials, with large values of susceptibility and permeability.

dependent on movement of the domains. One of the parameters that can be used to characterize the magnetic properties of ferromagnetic materials is saturation magnetization, that is, magnetization when all the domains are aligned. Elements that exhibit ferromagnetic behavior above room temperature include Fe, Co, and Ni. As ferromagnetic materials are heated to over a critical temperature, the thermal agitation of the atoms will disturb the alignment of the atomic magnetic moments, which is accompanied by the decrease in saturation magnetization. As the thermal agitation was sufficiently strong, the ferromagnetic material will become paramagnetic. Correspondingly, the critical temperature is called Curie temperature, T_C. For instance, Curie temperatures of Fe, Co, and Ni are 770°C, 1131°C, and 358°C, respectively. As mentioned before, above T_C, the variation of susceptibility follows the Curie–Weiss law.

15.3.5 Ferrimagnetism

Ferrimagnetism is only observed in compounds, which have more complex crystal structures than pure elements. In these materials, the exchange interactions lead to arrangement of magnetic moments in such a way that some of them have parallel alignment while others possess antiparallel alignment. Because of the difference in magnitude between the magnetic moments oriented in the two directions, there is net magnetization in ferrimagnetic materials. However, compared with that of ferromagnetic materials, this magnetization of ferrimagnetic materials is lower. Similar to ferromagnetic materials, ferrimagnetic materials will also break down into magnetic domains.

There are various ferrimagnetic materials. For example, magnetite (Fe_3O_4) has a unit cell, which consists of 8 Fe^{2+}, 16 Fe^{3+}, and 32 O^{2-}. The 16 Fe^{3+} ions are equally distributed at the tetrahedral and octahedral sites, while the 8 Fe^{2+} ions are at the octahedral sites. Therefore, the magnetic moments of the Fe^{3+} ions at the two sites are canceled out, whereas only the 8 Fe^{2+} ions contribute to the magnetic moment of the ferrite Fe_3O_4. Fig. 15.5 shows a schematic diagram demonstrating magnetic moment arrangement in Fe_3O_4. Similarly, the unit cell of barium ferrite

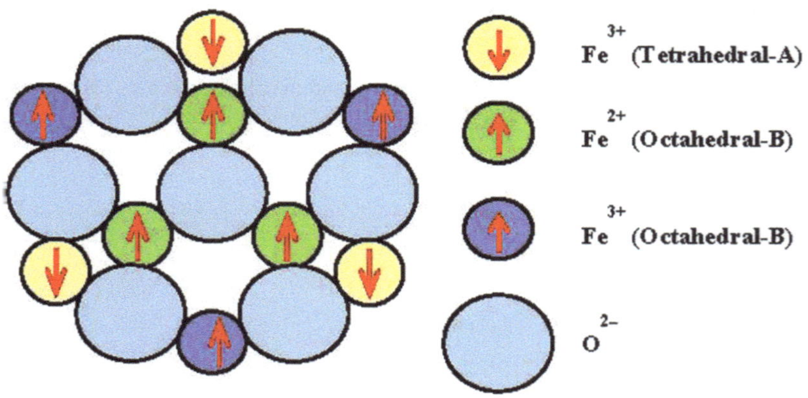

Figure 15.5 Schematic diagram of magnetization of ferrimagnetism materials (Fe_3O_4), with high values of susceptibility and permeability.

($BaFe_{12}O_{19}$) contains 64 ions, among which the Ba^{2+} and O^{2-} ions have no contribution to magnetic moment, while 16 Fe^{3+} ions have moments aligned parallel and 8 Fe^{3+} aligned antiparallel, thus leading to a net magnetization parallel to the applied field. In this case, the magnitude of the magnetic moment is relatively low, because only one-eighth of the ions contribute to the magnetization of the material.

15.4 Ferrites

15.4.1 Spinel ferrites

Spinel structured materials have a general formula, AB_2O_4, where A represents divalent cations and B trivalent cations, which are closely packed with oxygen ions to form a cubic crystal structure [1,2]. Because the exchange interaction between the A and B sites is negative and the strongest among the cations, net magnetization of spinel is determined by the difference in magnetic moment between the two sites. As a result, spinel ferrites are ferrimagnetic materials, with a general chemical composition of $MeO \cdot Fe_2O_3$ or $MeFe_2O_4$, where Me is divalent metal ions, such as Mn^{2+}, Fe^{2+}, Co^{2+}, Ni^{2+}, Cu^{2+}, Zn^{2+}, Mg^{2+}, Cd^{2+}, or their combinations, whereas Fe ions are the trivalent Fe^{3+}. Therefore, $MeFe_2O_4$ has a crystal structure that follows that of $MgAl_2O_4$, which has the mineral name "spinel." As a result, $MeFe_2O_4$ ferrites are called ferrospinel or spinel in simplicity.

Fig. 15.6 shows unit cell representing the smallest three-dimensional building block of spinel crystal structure, where the cubic symmetry can be observed [1]. The unit cell contains 32 oxygen anions (O^{2-}) with total negative charges of -64, 16 trivalent cations (Me^{3+}) with positive charges of 48 and 8 divalent cations (Me^{2+}) with positive charge of 16, so that charge balance is satisfied in the unit cell. The unit cell is constructed with the 32 O^{2-} ions, forming two crystallographic sites, (1) tetrahedral A-site that is surrounded by four O^{2-} ions (coordination number of 4) and (2) octahedral B-site surrounded by six O^{2-} ions (coordination number of 6). Totally, there are 64 tetrahedral and 32 octahedral sites, with one-eighth and one-half of them to be taken by the cations, respectively.

If all the 8 Me^{2+} ions occupy the tetrahedral A-sites, all the 16 Me^{3+} ions will take the octahedral B-sites. In this case, it is called normal spinel, with a chemical formula of $(Me^{2+})_{tet}[{Me_2}^{3+}]_{oct}O_4$. However, if all the 8 Me^{2+} ions occupy the octahedral B-sites, the 16 Me^{3+} ions will be equally distributed at the tetrahedral A- and octahedral B-sites, which results in a spinel that is called inverse spinel, with a chemical formula of $(Me^{3+})_{tet}[Me^{2+}Me^{3+}]_{oct}O_4$. There are also cases that are in between the normal and inverse spinels, which are called mixed spinels; that is, both the tetrahedral A- and octahedral B-sites are distributed with Me^{2+} and Me^{3+} ions, with a chemical formula of $({Me^{2+}}_{1-x}{Me^{3+}}_{x})_{tet}[{Me^{2+}}_{x}{Me^{3+}}_{2-x}]_{oct}O_4$. $MgAl_2O_4$ ($(Mg^{2+})_{tet}[{Al^{3+}}_2]_{oct}O_4$) and Fe_3O_4 ($(Fe^{2+})_{tet}[Fe^{2+}Me^{3+}]_{oct}O_4$) are representative normal and inverse spinels, respectively.

Usually, spinel ferrites can be well treated as the derivatives of magnetite (Fe_3O_4), which has been found in naturally occurring rocks. As stated earlier, one

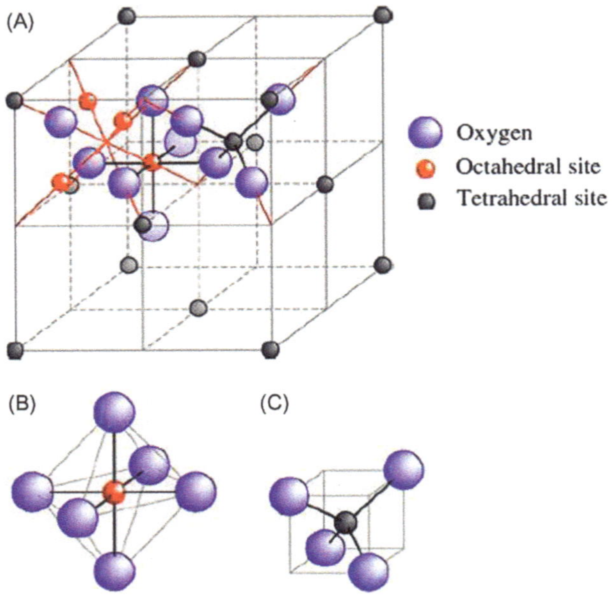

Figure 15.6 (A) Unit cell of spinel structure, (B) octahedral interstice (B-site: 32 per unit cell, 16 occupied), and (C) tetrahedral interstice (A-site: 64 per unit cell, 8 occupied). *Source*: Reproduced with permission from U. Ozgur, Y. Alivov, H. Morkoc, Microwave ferrites, part 1: Fundamental properties. J. Mater. Sci. Mater. Electron. 20 (9) (2009) 789–834, Copyright © 2009, Springer.

Fe atom is trivalent ferric (Fe^{3+}) and the other two are divalent ferrous (Fe^{2+}) ions. Magnetite is a cubic spinel, with a lattice constant, $a_0 = 8.39$ Å. One Fe atom has six *3d* electrons, four of which are unpaired. To form trivalent Fe^{3+} ions, two *4s* electrons and one *3d* electrons are used, thus leaving five unpaired electrons in the *d* shell. In comparison, to form divalent Fe^{2+} ions, only the two *4s* electrons are engaged in bonding, so that the six electrons in the *3d* shell are not involved. As a result, there are four unpaired electrons in the *3d* shell.

In Fe_3O_4, all the tetrahedral A-sites and half of the octahedral B-sites are occupied by Fe^{3+} ions, so that the other half of the octahedral B-sites are taken by Fe^{2+} ions. Due to the presence of five unpaired electrons, one Fe^{+3} ion contributes five Bohr magnetons, while one Fe^{2+} offers four Bohr magnetons because of the four unpaired electrons. Fe_3O_4 would have a net of four Bohr magnetons. This is because the spinel structure contains one Fe^{3+} in the A-site, and one Fe^{3+} and one Fe^{2+} in the B-sites. In terms of magnetic properties, the Fe^{3+} ion at the tetragonal B-site contributes $-5\ \mu_B$, while the Fe^{2+} ion at the octahedral B-site contributes 4 μ_B, and the Fe^{3+} ion at the octahedral B-site contribute 5 μ_B, thus leading to a net magnetization of 4 μ_B per molecular unit. Because one unit cell of Fe_3O_4 has eight molecular units, the total magnetization is 32 μ_B.

15.4.2 Garnet ferrites

Garnets exhibit orthorhombic crystal structure, with oxygen polyhedral to surround the cations, where the trivalent cations, including rare earth ions and Fe^{3+} ions, occupy tetrahedral (*d*), octahedral (*a*), or dodecahedral (a 12-sided distorted polyhedral) (*c*) sites. Similarly, the interaction between the tetrahedral and the octahedral sites is antiparallel, while the net magnetic moment is antiparallel to the rare earth ions on the *c* sites. The garnet structure is among the most complicated crystal structures, so that it is hard to draw a 2D diagram that can be used to clearly illustrate all the 160 ions in the unit cell. Fig. 15.7 shows a simplified diagram of the unit cell, with only one octant of a garnet structure to demonstrate just the cation positions [1].

Garnet structure consists of octahedral (trivalent cation surrounded by six O^{2-} ions), tetrahedral (trivalent cations surrounded by four O^{2-} ions), and 12-sided polyhedral dodecahedral (trivalent cations surrounded by eight O^{2-} atoms) sites. Garnets have a general chemical formula of $3Me_2O_3 \cdot 5Fe_2O_3$, corresponding to two $Me_3Fe_2(FeO_4)_3$, where Me stands for trivalent rare earth ions, such as nonmagnetic

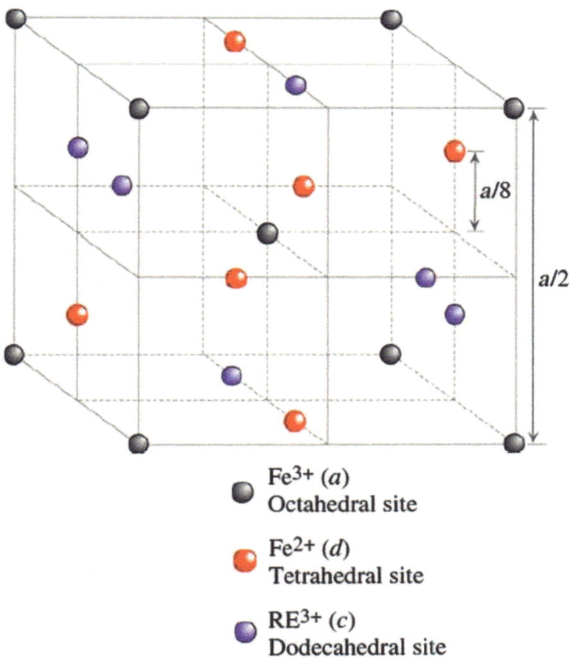

Figure 15.7 Schematic of an octant of the garnet crystal structure (lattice constant "a") showing only cation positions.
Source: Reproduced with permission from U. Ozgur, Y. Alivov, H. Morkoc, Microwave ferrites, part 1: Fundamental properties. J. Mater. Sci. Mater. Electron. 20 (9) (2009) 789–834, Copyright © 2009, Springer.

Y^{3+} or magnetic rare earth ions, such as those from La^{3+} to Yb^{3+}. The unit cell contains four sets of $3Me_2O_3 \cdot 5Fe_2O_3$. If the Me ion is nonmagnetic (e.g., Y^{3+} without *f* electrons), the materials will have net nonzero magnetization, due to the two nonidentical antiparallel sublattices of the Fe^{3+} ions that are in up and down states. Because of the antiferromagnetic superexchange interaction, there are three spin-up and two spin-down electrons, which leads to a net magnetic moment of 5 μ_B per molecule occurring, corresponding to a relatively low magnetization per unit volume.

It is well known that the characteristics of the superexchange interaction are closely related to the type of the magnetic ions and the bond length and bonding angle of the atoms. Therefore, the superexchange interaction (i.e., the magnetization and the Néel temperature of garnet ferrites) can be controlled by controlling the bonding strength through chemical substitutions with ions of different sizes and chemical properties. If magnetic rare earth ions are introduced to garnets, a third magnetic sublattice will be created, which makes the material have a compensation point that is below room temperature. This compensation point can be adjusted by substituting Fe^{3+} with Al^{3+} or Ga^{3+}. Since all rare earth cations in garnets are trivalent, there is electron hopping in between cations with different valences. As a consequence, garnet ferrites usually have relatively high resistivities.

15.4.3 Hexaferrites

Hexaferrites are those with a crystal structure similar to that of the mineral magnetoplumbite, $PbFe_{7.5}Mn_{3.5}Al_{0.5}Ti_{0.5}O_{19}$, where the Pb ion is divalent [3]. Without the doping of Mn, Al, and Ti, the chemical formula should be $PbO \cdot 6Fe_2O_3$. One example of hexaferrites is barium hexaferrite, $BaO \cdot 6Fe_2O_3$ or $BaFe_{12}O_{19}$. Due to the large size of Ba^{2+} ion, a slight perturbation in the lattice occurs around it, so that it could present large magnetic uniaxial anisotropy, with the crystallographic *c*-axis to be the magnetic easy axis. Furthermore, the combination of spinel ($MeO \cdot Fe_2O_3$) and ferroxdure ($BaO \cdot 6Fe_2O_3$) can deliver various other types of ferrites, denoted as M, W, Y, Z, U, and X, which are listed in Table 15.1 [4]. Almost all types of

Table 15.1 Chemical formula and interrelation of hexagonal ferrites [4]

Hexaferrite	Chemical formula		Stacking[a]	Relation
S (spinel)	$Me^{2+}{}_2Fe_4O_8$	$2\,MeO \cdot 2Fe_2O_3$	/	S
M	$Ba^{2+}Fe_{12}O_{19}$	$BaO \cdot 6Fe_2O_3$	RSR*S*	M
W	$Ba^{2+}\,Me^{2+}{}_2Fe_{16}O_{27}$	$BaO \cdot 2MeO \cdot 8Fe_2O_3$	RSSR*S*S*	M + S
Y	$Ba^{2+}{}_2\,Me^{2+}{}_2Fe_{12}O_{22}$	$2BaO \cdot 2MeO \cdot 6Fe_2O_3$	TSTSTS	Y
Z	$Ba^{2+}{}_3\,Me^{2+}{}_2Fe_{24}O_{41}$	$3BaO \cdot 2MeO \cdot 12Fe_2O_3$	RSTSR*S*T*S*	M + Y
U	$Ba^{2+}{}_4\,Me^{2+}{}_2Fe_{36}O_{60}$	$4BaO \cdot 2MeO \cdot 18Fe_2O_3$	RSR*S*T*S*	2M + Y
X	$Ba^{2+}{}_2\,Me^{2+}{}_2Fe_{28}O_{46}$	$2BaO \cdot 2MeO \cdot 14Fe_2O_3$	RSR*S*S*	2M + S

[a]Subunits for stacking order, with Me = bivalent element, S = Fe_6O_8 (spinel), R = $BaFe_6O_{11}$ (hexagonal), and T = $Ba_2Fe_8O_{14}$ (hexagonal). The asterisk (*) implies that the corresponding subunit is rotated by 180 degrees about the hexagonal axis.

barium ferrites have been extensively studied, especially for microwave absorbing applications [5–11].

As revealed in Table 15.1, all the different types of hexaferrites are nicely interrelated. For instance, with spinel structure $S = 2\,Me^{2+}Fe_2O_4$ or $2\,MeO \cdot 2Fe_2O_3$, where $Me = Fe^{2+}$, Co^{2+}, Mn^{2+}, Ni^{2+}, Cu^{2+}, Mg^{2+}, Zn^{2+}, and so on, and $M = BaFe_{12}O_{19}$ or $BaO \cdot 6Fe_2O_3$, W-type hexaferrite with a chemical formula of $BaMe_2Fe_{16}O_{27}$ can be formed; that is, $W = M + S$. Y-type hexaferrite, $Y = Ba_2Me_2Fe_{12}O_{22}$, is a planar hexaferrite. Another planar hexaferrite, Z-type ferrite, with a formula of $3BaO \cdot 2MeO \cdot 12Fe_2O_3$, is a combination of M and Y (i.e., $Z = M + Y$). The X-type hexaferrite, $X = 2M + S$, is formed by stacking R- and S-blocks along the hexagonal *c*-axis, with a sequence of RSR*S*S*, where R is an hexagonal three-oxygen-layer block with composition of $BaFe_6O_{11}$, whereas the asterisk denotes that the corresponding block is rotated by 180 degrees about the hexagonal axis. Therefore, the structure of X-type hexaferrite has a close relation to those of the M- and W-type ones. The U-type hexaferrite is formed through the combination of M- and Y-types; that is, $U = 2M + Y$.

The M-type hexagonal ferrite, with a formula of $BaO \cdot 6Fe_2O_3$, has lattice constants $a = 5.89$ Å and $c = 23.19$ Å, with an hexagonal lattice through the close packing of O^{2-} and Ba^{2+}, where Fe^{3+} ions occupy the octahedral ($12k$, $4f_2$, and $2a$), tetrahedral ($4f_1$), and trigonal bipyramidal/hexahedral ($2b$) sites. In this case, magnetic moment is only contributed by the Fe^{3+} cations. Fig. 15.8 shows a schematic diagram demonstrating the crystal structure of the M-type hexagonal ferrite, where the $12k$, $2a$, and $2b$ sites are spin up, and the $4f_1$ and $4f_2$ sites are spin down. M-type hexaferrite has a general formula of $AO \cdot 6Fe_2O_3$ or $AFe_{12}O_{19}$, where A is a divalent ion, such as Ba^{2+}, Sr^{2+}, Pb^{2+}, and so on.

Similar to M-type hexaferrite, W-type hexaferrite, or Me_2W, also has a magnetic uniaxial anisotropy, which is along the *c*-axis of the hexagonal structure. The Me_2W unit cell consists of one Ba^{2+} containing layer for every seventh O^{2-} layer of the spinel structure. Magnetic properties of the W-type hexaferrite, such as saturation magnetization and Néel temperature, can be modified by substituting various combinations of divalent metal ions. Similar strategies can be applied to both the Me_2Y (Y-type) and Me_2Z (Z-type) hexaferrites, which contain different combinations of barium layers and spinel blocks. The Me_2Y has a preferred plane perpendicular to the *c*-axis, in which the magnetization can move freely. As a result, Me_2Y is usually a soft magnetic material and cannot be used as a permanent magnet. MeZ ferrite is a combination of M and Y, in which the M portion tends to bring a preferential direction along the *c*-axis, whereas the Y portion prefers to offer a planar anisotropy in the plane perpendicular to the *c*-axis. Due to dominant uniaxial anisotropy of M, all Me_2Z ferrites, except for Co_2Z, have planar anisotropy.

15.5 Theoretical aspects of ferrimagnetism

As stated earlier, ferrimagnetism is somewhat similar to antiferromagnetism, so that the theory of ferrimagnetism developed by Néel is similar to that of

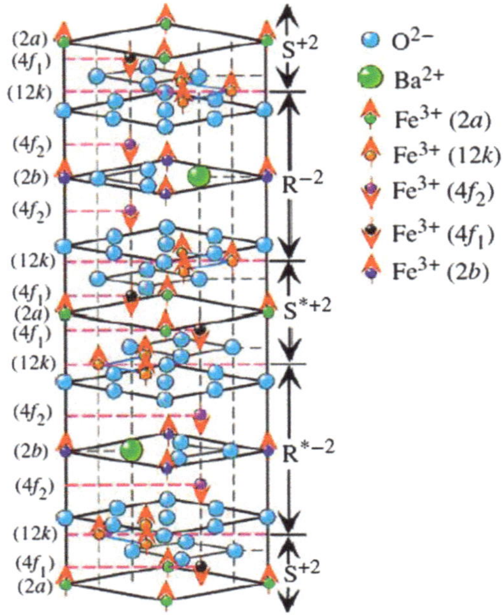

Figure 15.8 Schematic diagram illustrating structure of the M-type hexaferrite $BaFe_{12}O_{19}$. The arrows on Fe ions represent the direction of spin polarization. $2a$, $12k$, and $4f_2$ are octahedral; $4f_1$ are tetrahedral; and $2b$ are hexahedral (trigonal bipyramidal) sites. The unit cell contains a total of 38 O^{2-} ions, 2 Ba^{2+} ions and 24 Fe^{3+} ions. Fe^{3+} ions at the $12k$, $2a$, and $2b$ sites (totally 16 per unit cell) have spin-up moment, while the Fe^{3+} ions at the $4f_1$ and $4f_2$ sites (totally 8 per unit cell) have spin-down moment, thus leading to a net spin-up moment to be 8 totally, corresponding to a total moment of $8 \times 5\ \mu_B = 40\ \mu_B$ per unit cell. The R and S subunits have chemical formulae $R = (Ba_{2+}{Fe^{3+}}_6{O^{2-}}_{11})^{2-}$ and $S = ({Fe^{3+}}_8{O^{2-}}_8)^{2+}$. The asterisk (*) implies that the corresponding subunit is rotated by 180 degrees about the hexagonal axis.
Source: Reproduced with permission from U. Ozgur, Y. Alivov, H. Morkoc, Microwave ferrites, part 1: Fundamental properties. J. Mater. Sci. Mater. Electron. 20 (9) (2009) 789–834, Copyright © 2009, Springer.

antiferromagnetism [12,13]. Because the opposing magnetizations of two sublattices A and B in a ferrimagnetic material are different, in an antiferromagnetic material, they are equal to each other. As a consequence, there is a net magnetization in ferrimagnetic materials, but no net magnetization is present in antiferromagnetic materials. Theoretical aspects of ferrimagnetism are discussed by taking spinel ferrite as an example as follows.

15.5.1 Superexchange in spinel ferrites

As discussed earlier, in spinel ferrites, the cations occupy two different crystallographical sites, tetrahedral (A) and octahedral [B]. There are three types of possible magnetic interactions, between the cations, through the intermediate O^{2-} ions, and by a superexchange mechanism. For superexchange, there could be A–A interaction, B–B interaction, and A–B interaction. Experimental studies have confirmed that the energies of these interactions are negative, thus leading to an antiparallel orientation.

Generally, the magnitude of the interaction energy between the magnetic ions is dependent on two factors, (1) the distance between these ions and the O^{2-} ion for the interaction to occur and (2) the angle of cation-O-cation. Therefore, an angle of 180 degrees corresponds to the greatest exchange energy, whereas the energy is decreased quickly as the distance is increased. Fig. 15.9 shows a schematic diagram demonstrating various possible configurations of the cation pairs in spinel ferrites, with favorable distances and angles for an effective magnetic interaction [6]. The corresponding angles and the distances of cation–cation and cation–anion are listed in Table 15.2. Based on the values of the distance and the angles, it is readily expected that among the three interactions, the A–B interaction is the greatest one, followed by B–B and A–A interactions. The two configurations for A–B interaction have shorter distances and the values of the angle are relatively larger. For the

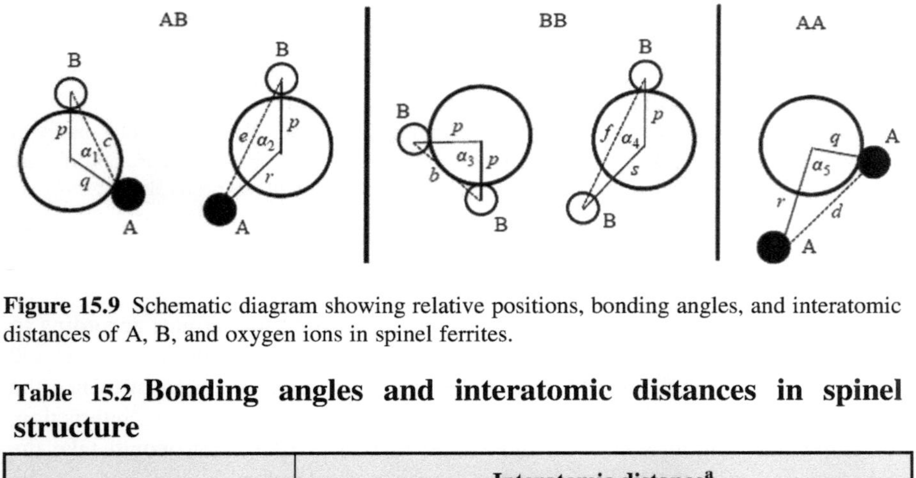

Figure 15.9 Schematic diagram showing relative positions, bonding angles, and interatomic distances of A, B, and oxygen ions in spinel ferrites.

Table 15.2 Bonding angles and interatomic distances in spinel structure

	Interatomic distance[a]	
Bonding angle	**Me–O**	**Me–Me**
$\alpha_1 = 125°9'$	$p = a\left(\frac{5}{8} - u\right)$	$b = \left(\frac{a}{4}\right)\sqrt{2}$
$\alpha_2 = 154°3'$	$q = a\left(u - \frac{1}{4}\right)\sqrt{3}$	$c = \left(\frac{a}{8}\right)\sqrt{11}$
$\alpha_3 = 90°$	$r = a\sqrt{\left(3u^2 - 2u + \frac{1}{2}\right)}$	$d = \left(\frac{3}{4}\right)\sqrt{3}$
$\alpha_4 = 125°2'$	$s = a\left(\frac{1}{3}u + \frac{1}{8}\right)\sqrt{3}$	$e = \frac{3a}{8}\sqrt{3}$
$\alpha_5 = 79°38'$		$f = \frac{a}{4}\sqrt{6}$

[a] a is lattice constant, and $u = 3/8 = 0.375$.

two configurations of B–B interaction, only the first one is effective, because the distance is too long for effective interaction in the second configuration. The A–A interaction is the weakest, since the distance is large and the angle is much smaller than 180 degrees.

In this case, because the A–B interaction is predominant in spinel ferrites, the spins of the A- and B-site ions are oppositely magnetized in the A and B sublattices, resulting in a magnetic moment that is equal to the difference between those of the A- and B-site ions, which is the origin of ferrimagnetism. Generally, the B lattice has higher magnitude of saturation magnetic moment (M_B) than that of the A lattice (M_A). Therefore, the net saturation magnetization (Ms) can be expressed as $Ms = |M_B - M_A|$. With this theory, experimentally measured magnetic susceptibility and magnetic saturation data of various ferrites have been satisfactorily explained.

In Néel's theory [3], there are four assumptions:

1. In spinel-structured ferrites, the ferrimagnetic crystal lattice is treated to have two sublattices, tetrahedral (A) and octahedral [B].
2. Among the three magnetic interactions, A–B, A–A, and B–B, the A–B interaction and B–A interaction are identical and they are predominant over the A–A interaction and B–B interaction. As a result, the net magnetic moment of a spinel ferrite is the difference between the magnetic moments of two sublattices.
3. If a magnetic ion in the material has a fraction λ to occupy the tetrahedral (A) site and remaining fraction μ will be at the octahedral [B] site, $\lambda + \mu = 1$.
4. Due to magnetic interactions within the materials, from the Weiss molecular point of view, the magnetic field applied to an ion is given by $H = H_0 + H_M$, where H_0 and H_M are externally applied magnetic field and internal or molecular field, respectively.

15.5.2 Ion distribution in spinel ferrites

As mentioned earlier, spinel ferrites can be classified into three groups according to the distribution of cations with divalent and trivalent states at the tetrahedral and octahedral sites, normal, inverse, and mixed spinels. Factors that can influence the cation distribution in spinel ferrites include ion radius, cation–anion interaction, and temperature. Intuitively, it is understood that the large cation would take the octahedral site and the smaller one would occupy the tetrahedral site, because the former has a larger space than the latter. Also, trivalent cations are usually larger than divalent cations. However, in fact, cation size is not the only factor to determine ion distribution of spinel ferrites. As a result, different types of spinels are present. For instance, Zn^{2+} has a radius of 0.82 Å, while that of Fe^{3+} is 0.67 Å. If size is the dominant factor, $ZnFe_2O_4$ should be $(Fe^{3+})_{tet}[Zn^{2+}Fe^{3+}]_{otc}O_4$, while it is preferably $(Zn^{2+})_{tet}[Fe^{3+}Fe^{3+}]_{otc}O_4$ or $(Zn^{2+})_{tet}[Fe^{3+}{}_2]_{otc}O_4$.

Ions with different radii at the lattice site make different contributions to the elastic energy of the crystals, together with the variations in Columbus energy and crystal field energy. From an energy point of view, cations are distributed in a crystal in such a way that the internal energy of the system is minimized. The internal energy of a crystal consists mainly of Columbus energy, crystal field energy, covalent energy, and repulsion energy, whereas other energies, such as exchange energy, spinning energy, orbital coupling energy, triangle crystal energy, Jahn-Teller effect, are relatively weak and thus make a very limited contribution to the internal energy.

15.5.2.1 Columbus energy

Columbus energy is also known as Madelung energy, which is static interaction energy between cation and anion, given by:

$$U_C = -\frac{N}{2}\sum\left(\frac{\pm\eta_l\eta_i e^2}{r_{li}}\right), \tag{15.11}$$

where N is the total number of ions, η is the valence of the ith ion, e is the electron charge, and the sign is dependent on the type of charges.

For spinel ferrites, the average Columbus energy of a molecule is:

$$E_C = -\frac{1}{2}(q_A M_A + 2q_B M_B + 8M_O)\frac{e^2}{a} = -M\frac{e^2}{a}, \tag{15.12}$$

where q_A and q_B represent average valences of the cations at A- and B-sites, respectively. M_A, M_B, and M_O are local Madelung constants of A-, B-site cations, and oxygen ion and a is lattice constant. M is a Madelung constant that can be expressed as the function of oxygen parameter, average valence of A-site cation q_A, given by:

$$M = \left[1522u - 430.8 + (172.2 - 488.2u)q_A + 2.61q_A^2\right]. \tag{15.13}$$

15.5.2.2 Crystal field effect

$3d$ transitional cations, with $n = 3$, $l = 2$, $m = 0, \pm 1, \pm 2$, have quintuple degenerate states, corresponding to wave functions of φ_0, $\varphi_{\pm 1}$, and $\varphi_{\pm 2}$, given by:

$$\varphi_n = R(r), \quad Y_i^{lm} = (\theta, \varphi). \tag{15.14}$$

Generally, a linear combination is applied for simple calculation, with widely used wave functions including:

$$d_{xy} = \frac{1}{i\sqrt{2}}(|2> - |-2>) = R(r)\sin^2\theta\sin\varphi = R(r)\left(\frac{xy}{r^2}\right), \tag{15.15}$$

$$d_{xz} = \frac{1}{i\sqrt{2}}(|1> + |-1>) = R(r)\sin\theta\cos\theta\cos\varphi = R(r)\left(\frac{xz}{r^2}\right), \tag{15.16}$$

$$d_{yz} = \frac{1}{i\sqrt{2}}(|1> - |-1>) = R(r)\sin\theta\cos\theta\sin\varphi = R(r)\left(\frac{yz}{r^2}\right), \tag{15.17}$$

$$d_{x^2-y^2} = \frac{1}{\sqrt{2}}(|2> + |-2>) = R(r)\sin^2\theta\cos 2\varphi = R(r)\left(\frac{x^2-y^2}{r^2}\right), \tag{15.18}$$

$$\begin{aligned} d_{z^2} &= |0> = R(r)(3\cos^2\theta - 1) = R(r)\left(\frac{3z^2-r^2}{r^2}\right) \\ &= R(r)\left(\frac{(z^2-x^2)+(z^2-y^2)}{r^2}\right) \end{aligned}. \tag{15.19}$$

For free cations, the five d orbitals are five-fold degenerate states. As the cations are within a crystal, a static electric field or crystal field will be present, caused by their surrounding ions. As a result, the degeneracies will be partially or totally broken, thus leading to energy splitting. For example, a cation is placed at the center of a square framed by four anions, where there are two sets of electron clouds, with one to be at the cross-direction and the other at the diagonal-direction. Due to the presence of the anions, electronic repelling effect forces the energy to be split into two levels, as shown in Fig. 15.10. As a result, the five-fold degeneracies are broken. With increasing strength of the crystal field, the degree of breaking of the degeneracy is gradually increased.

When a cation is at the center of the oxygen ion octahedral, the five d orbitals are split into two groups, triplet states (d_{xy}, d_{yz}, and d_{xz}, represented by t_{2g} or t_2, Γ_5, dε, γ_5) and doublet states ($d_{z^2}, d_{x^2-y^2}$, represented by e_{2g} or e, Γ_3, dγ, γ_3).

For d^1 and d^6 cations, t_{2g} is low energy state and e_{2g} is high energy state. For d^5 and d^{10} cations, because $L = 0$, the electron cloud has a spherical distribution, so that their energy states are affected by the crystal field. d^4 and d^9 cations can be treated as d^5 and d^{10} plus a positive hole. As a result, the effect of the crystal field is opposite to that on d^1 and d^6; that is, t_{2g} is high energy state and e_{2g} is low energy state.

According to crystal field theory, the decrease in energy of $3d^n$ electrons due to the energy splitting can be represented with D_q or Δ, which usually times a

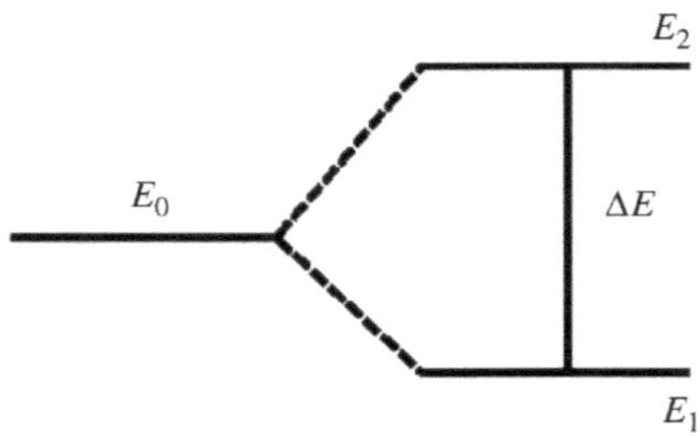

Figure 15.10 Schematic diagram showing the crystal field effect, with initial E_0 to be split into E_1 and E_2, leading to an energy difference of ΔE.

coefficient F. With D_q, it is possible to get the lowest energy of a crystal site, which can be used to identify the optimal energy of the octahedral configuration, thus leading to ion distribution. D_q can be derived from optical spectra.

15.5.2.3 Covalent effect

Covalent bond, also known as molecular bond, is a chemical bond that involves the sharing of electron pairs between two or more atoms. These electron pairs are also called shared pairs or bonding pairs. At stable balance of attractive and repulsive forces between the atoms they share electrons, which is called covalent bonding. In spinel ferrites, covalent bonding is usually present as coordination bonding.

It is not necessary that pure s, p, or d orbitals are used to form covalent bonding; a combination or hybridized orbitals can also be engaged in covalent bonding. In fact, hybridized orbitals have stronger bonding capability, so as to form stronger covalent bonding and thus more stable molecules. For instance, s orbital can be hybridized with p_x, p_y, and p_z orbitals for form sp^3 hybrid orbitals. In this case, the four sp^3 hybrid orbitals are in the respective four diagonal directions of a tetrahedron. Therefore, they are called tetrahedral hybridization. In spinel ferrites, there are also square hybridized orbitals dsp^2 with a coordination number of 4 and octahedral hybridized orbitals d^2sp^3 with a coordination number of 6. Because their orbitals are fully filled, cations with electronic configuration of d^{10} do not have an empty one to form for other hybridized orbitals, but only sp^3 ones for covalent bonding. As a result, they have tetrahedral hybridization. Examples of such cations include Zn^{2+} and Cd^{2+}, taking tetrahedral sites in spinel ferrites.

Cations with electronic configurations of $3d^9$ and $3d^4$ could have one electron to be excited to $4p$ orbital, thus leaving one empty d orbital, so that it is possible to form a stronger dsp^2 square hybridized orbital. For instance, Cu^{2+} and Mn^{3+} ions occupy the octahedral sites due to their dsp^2 hybridization behavior. They will cause distortion of the octahedral sites and even the crystals unite if they are orderly arranged. Cations with $3d^3$ electronic configuration, such as Cr^{3+}, are able to form a d^2sp^3 hybridized orbital, which is the strongest one, so that they usually tend to take the octahedral sites. Cation hybridization on crystal structure, ion distribution, and magnetic properties of spinel ferrites have been well documented in the open literature [14,15].

15.5.2.4 Short-range interaction energy

Short-range interaction can be approximately treated as the summation of elastic energy and polarization energy, which can be described with a formula similar to the Born's repulsion energy equation, given by:

$$E_B = \left(\frac{B_{II}}{a_0^n}\right)\left[4x\left(\frac{a_0}{r_t}\right)^n + 6(1-x)\left(\frac{a_0}{r_o}\right)^n\right] + \left(\frac{B_{III}}{a_0^n}\right)\left[4(1-x)\left(\frac{a_0}{r_t}\right)^n + 6x\left(\frac{a_0}{r_o}\right)^n\right], \tag{15.20}$$

where x is the molar fraction of the divalent cation at the tetrahedral (A) site in $(Me^{2+}{}_{1-x}Me^{3+}{}_{x})_{tet}[Me^{2+}{}_{x}Me^{3+}{}_{2-x}]_{oct}O_4$, B_{II}, B_{III} and n are experimentally obtained coefficients, while r_t and r_o are cation-anion distances of the cations at the tetrahedral and octahedral sites. Also, there are following interrelationship among them:

$$\frac{r_t}{a_0} = \sqrt{3}\left(u - \frac{1}{4}\right), \tag{15.21}$$

$$\frac{r_o}{a_0} = \left[2\left(u - \frac{3}{8}\right)^2 + \left(u - \frac{5}{8}\right)^2\right]^{1/2}. \tag{15.22}$$

According to Miller, with the consideration of Columbus energy (E_C), crystal field energy (E_D), and short-range interaction energy (E_B) (including repulsion and covalent energies), the derived ion distribution profile was in good agreement with the experimental results. For example, if three types of cations, C, D, and E, are distributed at both sites in a spinel ferrite, the chemical formula is expressed as $(C_{c/3}D_{d/3}E_{e/3})_{tet}[C_{2c/3}D_{2d/3}E_{2e/3}]_{oct}O_4$. If C and D cations with a fraction of x exchange their site occupation, the formula will become $(C_{c/3}D_{d/3}E_{e/3})_{tet}[C_{2c/3}D_{2d/3}E_{2e/3}]_{oct}O_4$. In this case, the C cations with x fraction are shifted to an A-site, while the D cations are moved to a B-site, which leads to a change in system free energy, given by:

$$\frac{dG}{dx} = \frac{\partial E_C}{\partial q_A} - \frac{\partial q_A}{\partial x} + \frac{\partial E_B}{\partial x} + \frac{\partial E_D}{\partial x} = P_D - P_C, \tag{15.23}$$

$$\begin{aligned} P_J = {} & \frac{e^2}{a_0}(172.2 - 4.88.2u + 5.22q_A)Z_J + O_J \\ & - \frac{1}{a^n}\left\{4\left[\sqrt{3}\left(u - \frac{1}{4}\right)\right]^{-n} - 6\left[2\left(u - \tfrac{3}{8}\right)^2 + \left(u - \tfrac{5}{8}\right)^2\right]^{-n/2}\right\}B_J, \end{aligned} \tag{15.24}$$

where Z_J is the valence of the J cation, O_J is the crystal field energy, B_J is the characteristic factor, and P_J is the optimization energy to occupy the B-site. Therefore, the tendencies of cations to take a B-site can be estimated. Fig. 15.11 shows the optimization energy of different cations tending to occupy the octahedral B-site in spinel ferrites. Generally, if $P_J \leq 3$, mixed spinel structures will be formed.

15.5.2.5 Ordering in spinel ferrites

It has been observed that long-range ordering could occur in spinel ferrites, totally with three types: (1) B-site 1:1 ordering, (2) B-site 1:3 ordering, and (3) A-site 1:1 ordering.

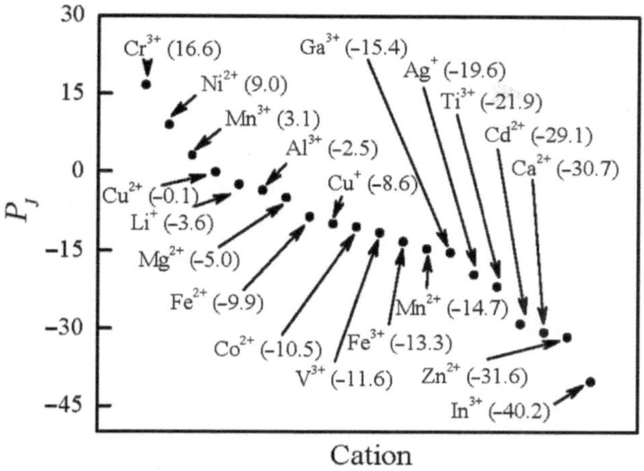

Figure 15.11 Optimization energy of different cations to take B-site in spinel ferrites.

In the case of B-site 1:1 ordering, the B-site Me^{2+} and Me^{3+} cation layers were arranged alternatively in the z-axis direction. For example, a Me^{2+} layer is at $z = 1/8$, then a Me^{3+} cation layer will be at $z = 3/8$, which are repeated alternatively. Fe_3O_4 adopts such an ordering structure at temperatures below 114K.

For B-site 1:3 ordering, on each plane perpendicular to the x-, y-, or z-axes, there are one Me^{2+} ion and three Me^{3+} ions at the four B-sites, with the following coordinations:

$$Me^{2+}: \quad \left(\frac{5}{8},\frac{5}{8},\frac{5}{8}\right), \left(\frac{1}{8},\frac{7}{8},\frac{3}{8}\right), \left(\frac{7}{8},\frac{3}{8},\frac{1}{8}\right), \left(\frac{3}{8},\frac{1}{8},\frac{7}{8}\right).$$

$$Me^{3+}: \quad \left(\frac{1}{8},\frac{5}{8},\frac{1}{8}\right), \left(\frac{5}{8},\frac{1}{8},\frac{1}{8}\right), \left(\frac{3}{8},\frac{7}{8},\frac{1}{8}\right), \left(\frac{7}{8},\frac{1}{8},\frac{3}{8}\right),$$
$$\left(\frac{3}{8},\frac{5}{8},\frac{3}{8}\right), \left(\frac{5}{8},\frac{3}{8},\frac{3}{8}\right), \left(\frac{1}{8},\frac{1}{8},\frac{5}{8}\right), \left(\frac{3}{8},\frac{3}{8},\frac{5}{8}\right),$$
$$\left(\frac{7}{8},\frac{7}{8},\frac{5}{8}\right), \left(\frac{1}{8},\frac{3}{8},\frac{7}{8}\right), \left(\frac{5}{8},\frac{7}{8},\frac{7}{8}\right), \left(\frac{7}{8},\frac{5}{8},\frac{7}{8}\right).$$

One example of ferrites with such an ordering structure is $(Fe)[Li_{0.5}Fe_{1.5}]O_4$ or $Li_{0.5}Fe_{2.5}O_4$.

$(Li^{+}{}_{0.5}Fe^{3+}{}_{0.5})[Cr^{3+}{}_2]O_4$ possesses A-site 1:1 ordering, with coordinations:

$$Me^{2+}: \quad (0,0,0), \left(0,\frac{1}{2},\frac{1}{2}\right), \left(\frac{1}{2},0,\frac{1}{2}\right), \left(\frac{1}{2},\frac{1}{2},0\right).$$

$$Me^{3+}: \quad \left(\frac{1}{4},\frac{1}{4},\frac{1}{4}\right), \left(\frac{1}{4},\frac{3}{4},\frac{3}{4}\right), \left(\frac{3}{4},\frac{1}{4},\frac{3}{4}\right), \left(\frac{3}{4},\frac{3}{4},\frac{1}{4}\right).$$

15.5.2.6 Superexchange in ferrimagnetic ferrites

It was observed in some compounds that magnetic ordering occurs at a certain temperature, called the Néel point, denoted as θ_N. Below θ_N, magnetic susceptibility becomes anomalous. At the same time, some parameters, such as specific heat and coefficient of thermal expansion, exhibit anomalous peaks. Above θ_N, the susceptibility as a function of temperature follows the Curie–Weiss law, as shown in Eq. (15.10).

A typical example of ferromagnetic materials is MnO, which has a NaCl crystal structure, with Mn^{2+} ions having an octahedral coordination with six O^{2-} ions. In this case, the adjacent Mn^{2+} ions have neither wave function overlapping nor *s-d* exchange. However, it has an ordered magnetic structure—that is, the ionic magnetic moments related to the bond with an angle of π are antiparallel—while those with angle of $\pi/2$ are randomly oriented. In the magnetically ordered state, the magnetic unit cell has a lattice constant that doubles that of the chemical unit cell. Above θ_N, it is in the magnetically disordered state; the magnetic and chemical units have the same parameters. This phenomenon cannot be explained by using the conventional exchange interaction theories. It was later proposed that the exchange interaction in ionic crystals could be realized indirectly through the excited state of anions, which led to the development of superexchange interaction theories [16,17].

In superexchange interaction theories, when O^{2-} ions approach their adjacent cations, the *p* electrons could be excited to *d* orbital, so as to couple with the d electrons of 3d transitional cations according to Hound's rule. Meanwhile, the remaining unpaired *p* electrons would have exchange interaction with the adjacent cations. Because the exchange is achieved via O^{2-} ions, it is an indirection exchange, the well-known superexchange. According to Hound's rule, since the number of 3*d* electrons of the cation is less than 5, the excited *p* electrons and the 3*d* electrons of cation Me_1 have parallel spinning. Otherwise, if the number is larger than or equal to 5, antiparallel spinning will be present. Because the paired 2*p* electrons have an antiparallel spinning, the other one *p* electron will have negative exchange with the *d* electrons of the cation. Therefore, as the number of electrons at the partially filled orbitals is equal to or larger than the half-filled value, superexchange will lead to antiparallel magnetic spinning of the cations. In most ferrites, superexchange is negative, so that the adjacent cations have antiparallel magnetic moment; that is, ferrimagnetism.

Following quantum mechanics, superexchange interaction can also be expressed in a similar form as that of direction exchange interaction, given by:

$$E_{ex} = -2AS_1S_2, \tag{15.25}$$

where S_1 and S_2 are magnetic moments of two adjacent cations; A is the exchange coefficient that is not determined by the direct overlapping of wave functions of the two cations, but is related to the interaction between the *p* electrons of O^{2-} ions and *d* electrons of the cations.

15.5.2.7 Néel linear model (molecular field theory)

Molecular field theory was applied to antiferromagnetism and ferrimagnetism by Néel in 1948 [12], where the magnetic cations in the crystals have two or more magnetic sublattices. In the simplest case of two magnetic sublattices, let the A sublattice and B sublattice intersperse each other. Assuming that A and B cations are closely neighboring, with a coordination number of z, exchange energy is given by:

$$E_{ex} = -2A\sum_{z} S_A \cdot S_B = -2AzS_A \cdot S_B. \tag{15.26}$$

Because $\mu = -gs\mu_B$, there is:

$$E_{ex} = -\frac{2Azu_A \cdot u_B}{(g\mu_B)^2} = -h_B \cdot \mu_B, \tag{15.27}$$

where there is:

$$h_B = \frac{2Azu_A \cdot u_B}{(g\mu_B)^2} = \frac{2AzN_A u_A}{N_A(g\mu_B)^2} = -\lambda M_A, \tag{15.28}$$

where N_A is the number of cation A per unit volume and μ_B is Bohr magneton, while u_A and u_B are magnetic moment vectors of A and B site cations and are equivalent to the molecular field of the A cation to apply to the B cation. Obviously, each given cation has interactions with not only the neighboring sublattice but also its own sublattice. Therefore, for the two magnetic sublattices, molecular fields can be expressed as:

$$h_A = -\lambda M_B - \varepsilon M_A, \tag{15.29}$$

$$h_B = -\lambda M_A - \varepsilon M_B. \tag{15.30}$$

In general, if there are more than two sublattices, the molecular fields applied to the cation of the ith sublattice is given by:

$$H_{\varpi i} = \sum_{j=1}^{n} W_{ij} M_j, \tag{15.31}$$

where M_j is saturation magnetization of jth sublattice, and W_{ij} is the Weiss coefficient of molecular field that characterizes exchange strength between the magnetic cations at the ith and jth sublattices.

For ferrites with two sublattices of A and B, there are three types of exchange interactions, A–A, B–B, and A–B, as stated earlier. Let $W_{AB} = -n$; n is positive because W_{AB} is negative. In this case, there are:

$$W_{AA} = -\alpha W_{AB} = \alpha n, \tag{15.32}$$

$$W_{BB} = -\beta W_{AB} = \beta n. \tag{15.33}$$

If A sublattice has magnetic cation fraction of X_A, $X_A M_A$ is its magnetization. Similarly, the magnetization of B sublattice with magnetic cation fraction of X_B is $X_B M_B$. Therefore, the total magnetization of the crystal is given by (vector):

$$M_S = X_A M_A + X_B M_B. \tag{15.34}$$

According to molecular field theory, the molecular fields applied to the A and B sublattices are:

$$\begin{aligned} H_{WA} &= W_{AB} X_B M_B + W_{AA} X_A M_A = n(-X_B M_B + \alpha X_A M_A) \\ &= n\left(-X_B \frac{M_B}{M_A} + \alpha X_A\right) M_A = W_A M_A, \end{aligned} \tag{15.35}$$

$$H_{WA} = n\left(-X_A \frac{M_A}{M_B} + \beta X_B\right) M_B = W_B M_B. \tag{15.36}$$

As a result, the spontaneous magnetizations in A and B sublattices are given by:

$$M_A = NgJ\mu_B B_J\left(\frac{gJ\mu_B}{kT} H_{WA}\right) = M_J B_J\left(\frac{mJ}{kT} H_{WA}\right), \tag{15.37}$$

$$M_B = M_J B_J\left(\frac{mJ}{kT} H_{WB}\right), \tag{15.38}$$

where B_J is Brillouin function, which can be expressed as:

$$B_J(x) = \frac{2J+1}{2J} \coth \frac{2J+1}{2J} x - \frac{1}{2J} \coth \frac{1}{2J} x, \tag{15.39}$$

where $x = gJ\mu_B H_W/kT$, with J to be the total quantum number. This theory can be used to predict and explain the magnetization of ferrites as a function of temperature.

Table 15.3 Magnetic moments of ions with unit of Bohr magneton (μ_B)

3d electrons	0	1	2	3	4	5	6	7	8	9	10
Cations	Sc^{3+} Ti^{4+} V^{5+} Cr^{6+}	Ti^{3+} V^{4+} Cr^{5+} Mn^{6+}	Ti^{2+} V^{3+} Cr^{4+} Mn^{5+}	V^{2+} Cr^{3+} Mn^{4+}	Cr^{2+} Mn^{3+} Fe^{4+}	Mn^{2+} Fe^{3+} Co^{4+}	Fe^{2+} Co^{3+} Ni^{4+}	Co^{2+} Ni^{3+}	Ni^{2+}	Cu^{2+}	Cu^{+} Zn^{2+}
μ_B	0	1	2	3	4	5	4	3	2	1	0

For ferrites, with a chemical formula of MFe_2O_4, if the cation distribution is expressed as $(M^{2+}{}_{1-x}Fe^{3+}{}_x)_{tet}[M^{2+}{}_xFe^{3+}{}_{2-x}]_{oct}O_4$, the molecular magnetic moment can be calculated by the following equation:

$$m = 10(1-x)\mu_B + n_M(2x-1)\mu_B, \tag{15.40}$$

where n_M is the magnetic moment of M^{2+} cation, with the unit of Bohr magneton.

Magnetic moment of a free ion is the coupling result of a spinning moment and orbital moment. However, as an ion is in a crystal, the orbital moment is quenched due to the effect of the crystal field. Therefore, the orbital moment has no contribution to the magnetic moment of the crystal, which can be neglected safely when calculating the magnetic moment of ferrites. According to Hound's rule, magnetic moment of different ions can be readily obtained from their electronic structures, as listed in Table 15.3.

According to Eq. (15.10) and Table 15.3, $ZnFe_2O_4$ as a normal spinel should have molecular magnetic moment of 10 μ_B. However, experimental results indicated that it has a zero magnetic moment. To explain this contradiction, it is necessary to use the superexchange interaction theory. Zn^{2+} has an electronic configuration of $3d^{10}$, a nonmagnetic ion. When it is at the A-site, Zn^{2+} ion cannot have superexchange interaction with the magnetic ions at the B-site. Therefore, with increasing content of Zn^{2+} ion, the strength of the A–B interaction is decreased. For $ZnFe_2O_4$, no A–B interaction is present at all. In this case, B–B interaction will become to be dominant. As a result, the parallel B sublattice is transferred to antiparallel, thus leading to an antiferromagnetic state and hence zero magnetization.

15.6 Summary

Ferrites have found various applications in different areas, such as magnetic sensors, actuators, electromagnetic shielding, and microwave absorption. Significant progress has been made in synthesis, fabrication, characterization, and application of ferrites and their composites. Due to their special ferrimagnetic characteristics, ferrites can be designed with desired properties and performances for specific applications. Theoretical studying and understanding of ferrimagnetic phenomenon have been well documented in the open literature. Classically, the currently available theories can be employed to explain almost all the experimental observations on ferrite materials, as mentioned in this chapter. However, with the development of nanotechnology and nanomaterials, ferrimagnetism or the magnetic properties of ferrites have also been found to show close relationships to the size, dimension, and morphology of the materials. It is still lacking information in this respect, which deserves more exploration as a research topic in future studies.

References

[1] U. Ozgur, Y. Alivov, H. Morkoc, Microwave ferrites, part 1: Fundamental properties, J. Mater. Sci.-Mater. Electron. 20 (9) (2009) 789–834.

[2] M. Sugimoto, The past, present, and future of ferrites, J. Am. Ceram. Soc. 82 (2) (1999) 269–280.

[3] R.C. Pullar, Hexagonal ferrites: a review of the synthesis, properties and applications of hexaferrite ceramics, Prog. Mater. Sci. 57 (7) (2012) 1191–1334.

[4] D. Lisjak, M. Drofenik, Thermal stability of (Co,Cu)Z-hexaferrite and its compatibility with Ag at 900°C, J. Am. Ceram. Soc. 90 (11) (2007) 3517–3521.

[5] Z.W. Li, Z.H. Yang, L.B. Kong, Ultrabroad bandwidth and matching characteristics for spinel ferrite composites with flaky fillers, J. Appl. Phys. 108 (6) (2010) 063927.

[6] Z.W. Li, Z.H. Yang, L.B. Kong, Ultrabroad bandwidth of single-layer electromagnetic attenuation composites with flaky fillers, Appl. Phys. Lett. 96 (9) (2010) 092507.

[7] Z.W. Li, Z.H. Yang, L.B. Kong, Enhanced microwave magnetic and attenuation properties for barium ferrite flaky-filler composites, IEEE Trans. Magn. 47 (10) (2011) 3712–3715.

[8] Z.W. Li, Z.H. Yang, L.B. Kong, Resonance like dispersion for W-type barium ferrite composites with c-axis anisotropy, J. Appl. Phys. 109 (3) (2011) 033916.

[9] Z.W. Li, et al., High-frequency magnetic properties at K and Ka bands for barium-ferrite/silicone composites, J. Magn. Magn. Mater. 325 (2013) 82–86.

[10] G.Q. Lin, et al., Influence of demagnetizing field on the permeability of soft magnetic composites, J. Magn. Magn. Mater. 305 (2) (2006) 291–295.

[11] G.Q. Lin, Y.P. Wu, Z.W. Li, Improvement of the electromagnetic properties in composites with flake-like Co_2Z powders by molten-salt synthesis, IEEE Trans. Magn. 42 (10) (2006) 3326–3328.

[12] L. Neel, Proprietes magnetiques des ferrites - ferrimagnetismd et antiferromagnetisme, Ann. De Phys. 3 (2) (1948) 137–198.

[13] L. Neel, Antiferromagnetism and ferrimagnetism, Proc. Phys. Soc. London Sec. A 65 (395) (1952) 869–885.

[14] J.B. Goodenough, A.L. Loeb, Theory of ionic ordering, crystal distortion and magneitc exchange due to covalent forces in spinels, Phys. Rev. 98 (2) (1955) 391–408.

[15] A.L. Loeb, J.B. Goodenough, The influence of covalence on ion ordering and distortion in spinels: 1, Phys. Rev. 95 (2) (1954), 616-616.
[16] P.W. Anderson, Antiferromagnetism—theory of superexchange interaction, Phys. Rev. 79 (2) (1950) 350–356.
[17] P.W. Anderson, New approach to the theory of superexchange interactions, Phys. Rev. 115 (1) (1959) 2–13.

Metal oxide structure, crystal chemistry, and magnetic properties

16

Vladimir V. Srdić[1], Zeljka Cvejić[2], Marija Milanović[1], Goran Stojanović[3] and Srdjan Rakić[2]
[1]Department of Materials Engineering, Faculty of Technology Novi Sad, University of Novi Sad, Novi Sad, Serbia, [2]Department of Physics, Faculty of Sciences, University of Novi Sad, Novi Sad, Serbia, [3]Department of Microelectronics, Faculty of Technical Sciences, University of Novi Sad, Novi Sad, Serbia

Magnetism is a force we experience every day without really being aware of it. The scientific interest in magnetism and magnetic oxides has been changed during years. The first magnetic material to be discovered was lodestone, which is better known today as magnetite (Fe_3O_4). The word *magnet* comes from the Greek word *magnes*, which itself may derive from the ancient colony of Magnesia (in Turkey) [1]. Magnetite was mined in Magnesia 2500 years ago. In its naturally occurring state lodestone is permanently magnetized and is the most magnetic mineral. The strange power of lodestone was well known in ancient times. Socrates (~400 BC) dangled iron rings beneath a piece of lodestone and found that the lodestone enabled the rings to attract other rings [1]. Even earlier (~2600 BC), a Chinese legend tells of the Emperor Hwang-ti being guided into battle through a dense fog by means of a small pivoting figure with a piece of lodestone embedded in its outstretched arm. The figure always pointed south and was probably the first compass.

Commercial interest in ceramic magnets started in the early 1930s with a Japanese patent describing application of copper and cobalt ferrites. The first Golden age of magnetic oxides was the 1950s and the 1960s, when the ferrites were explored and their properties optimized. Discovery of the first family of ferromagnetic (FM) oxides with a higher magnetization than ferrimagnets also dates from this period. A new age for magnetic oxides recently started, where multifunctionality, control of defects, interfaces, and thin-film device structures are the new challenges.

16.1 Magnetic elements/ions

The macroscopic magnetic properties of materials are a consequence of magnetic moments associated with individual electrons (i.e., its orbital motion around the nucleus and spinning around the spin axis). Some of these concepts are relatively

Magnetic, Ferroelectric, and Multiferroic Metal Oxides. DOI: http://dx.doi.org/10.1016/B978-0-12-811180-2.00016-5

complex and involve some quantum-mechanical principles, and will not be analyzed here. It is well known that in each individual atom, orbital moments of some electron pairs cancel each other; this also holds for the spin moments (i.e., an electron with spin-up will cancel the one with spin-down). The net magnetic moment, then, for an atom is just the sum of the magnetic moments of each of the constituent electrons, including both orbital and spin contributions, taking into account moment cancellation. The strength of spin–orbit coupling is a key factor in determining the extent to which the orbital moment contributes to the magnetic properties and conversely, to what extent the spins interact with the lattice [2]. For an atom having completely filled electron shells or subshells, when all electrons are considered, there is a total cancellation of both orbital and spin moments. Thus, materials composed of atoms (or ions) having completely filled electron shells are not capable of being permanently magnetized. From this point of view transitional elements with the unfilled inner shells (while electrons occupying outer shells participate in chemical bonding) are responsible for a variety of magnetic properties because of the magnetic moments carried by their unpaired spins.

The types of magnetism include diamagnetism, paramagnetism, and ferromagnetism (in addition, antiferromagnetism and ferrimagnetism are considered to be subclasses of ferromagnetism). All materials exhibit at least one of these types, and their behavior depends on the response of electron and atomic magnetic dipoles to the application of an externally applied magnetic field. Thus, most of the elements in the periodic table are diamagnetic and paramagnetic (Fig. 16.1), but exceptions are some *d*- and *f*-elements (transitional metals) from which most magnetic effects originate [3].

Characteristics of the *d*-elements are *d*-orbitals with up to five pairs of electrons and a relatively small difference in the energy of the different d-orbital electrons. This causes the number of electrons participating in chemical bonding to vary and results

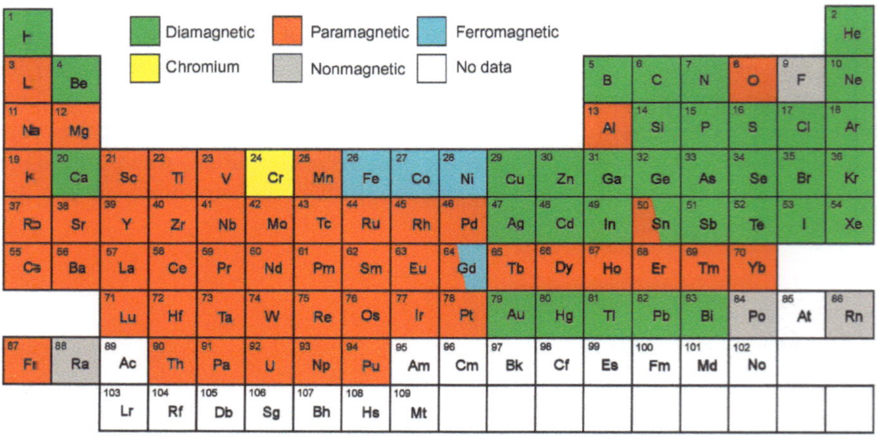

Figure 16.1 Magnetic behavior of elements in the periodic table.

in the same element exhibiting two or more oxidation states. The 3*d*-elements important for magnetism are iron, cobalt, and nickel, which are FM; and chromium, which is antiferromagnetic (AFM). Their ions, together with some ions of other *d*-elements (such as Fe^{2+}, Fe^{3+}, Co^{2+}, Ni^{2+}, Mn^{2+}, Mn^{3+}, Cr^{2+}, Cr^{3+}, V^{2+}, and V^{3+}) have high magnetic moments capable of producing strong spontaneous magnetism, when they are in sufficient densities, which allows exchange coupling to order magnetically as ferro-, ferri-, or antiferromagnets. However, the ions of the first two 3*d*-elements (Sc and Ti) are generally paramagnetic. The ions from 4*d*- and 5*d*-groups resemble the 3*d*-series in magnetic properties and can be used as alternatives for them in certain cases. However, in general these ions have not attracted much interest for their magnetic properties because their compounds do not exhibit strong spontaneous magnetic properties.

The second most important transition group from a magnetic standpoint has an unfilled 4*f*-shell. The f-orbitals can contain up to seven pairs of electrons that are less chemically active, causing high chemical similarity between f-elements. They are generally paramagnetic and only gadolinium shows FM behavior. An important distinction between the 4*f*-ions and those of the 3*d*-ion group is the shielding of the 4*f*-shell inside the 5*s* and 5*p* outer shells of the xenon core. In other words, the magnetically active electrons are buried inside the xenon core and are therefore shielded from electrostatic fields of the molecular environment. As a result, the ions act largely independent of one another, even in highly concentrated compounds, and are generally paramagnetic because the multiple lobes of the 4*f*-orbital wave functions do not extend far enough for covalent bonding and magnetic exchange to be significant [2].

16.2 Magnetic oxides

Oxides are ionic compounds where small, highly charged metal cations are embedded in a lattice of oxygen anions. The O^{2-} has its stable $2p^6$ closed-shell configuration with an ionic radius $r_O = 1.40$ Å. The oxide crystal structure is based on a close-pack array of oxygen anions, with metal cations occupying interstitial sites. The close-packed arrays are usually face-centered-cubic and hexagonal-close packed. They both accommodate smaller metal cations in octahedral and tetrahedral interstices. The smaller one is the tetrahedral site with four oxygen neighbors, which can accommodate, without distortion, a spherical cation of radius $r_{tet} = 0.32$ Å. The larger one is the octahedral site with six oxygen neighbors that can accommodate a spherical cation of radius $r_{oct} = 0.58$ Å. There are two tetrahedral interstices and one octahedral interstice for each oxygen in dense-packed lattices, so that only a fraction of them can ever be filled. The ionic radii of common 3*d*-cations and others that may be incorporated in oxide structures are listed in Table 16.1. Due to the small size the tetrahedral site can accommodate only smaller cations (Mg^{2+}, Zn^{2+}, Al^{2+}, Fe^{3+}, or Si^{4+}) with some distortion of the tetragonal site. The larger cations (0.55–0.70 Å) can occupy the octahedral sites (Fig. 16.2), while the cations such as Sr^{2+}or La^{3+}, having ionic radii comparable to that of

Table 16.1 **Ionic radii of cations in oxides**

4-fold coordination	*d* [Å]	6-fold coordination	*d* [Å]	6-fold coordination	*d* [Å]	12-fold coordination	*d* [Å]
Mg^{2+}	0.53	Cr^{4+}	0.55	Ti^{3+}	0.67	Ca^{2+}	1.34
Zn^{2+}	0.60	Mn^{4+}	0.53	V^{3+}	0.64	Sr^{2+}	1.44
Al^{3+}	0.42	Al^{3+}	0.54	Cr^{3+}	0.62	Ba^{2+}	1.61
Fe^{3+}	0.52	Mn^{2+}	0.83	Mn^{3+}	0.65	Pb^{2+}	1.49
Si^{4+}	0.40	Fe^{2+}	0.78 (0.61)	Fe^{3+}	0.64	Y^{3+}	1.19
		Co^{2+}	0.75 (0.65)	Co^{3+}	0.61 (0.56)	La^{3+}	1.36
		Ni^{2+}	0.69	Ni^{3+}	0.60	Gd^{3+}	1.22

Values in brackets indicate a low-spin state.

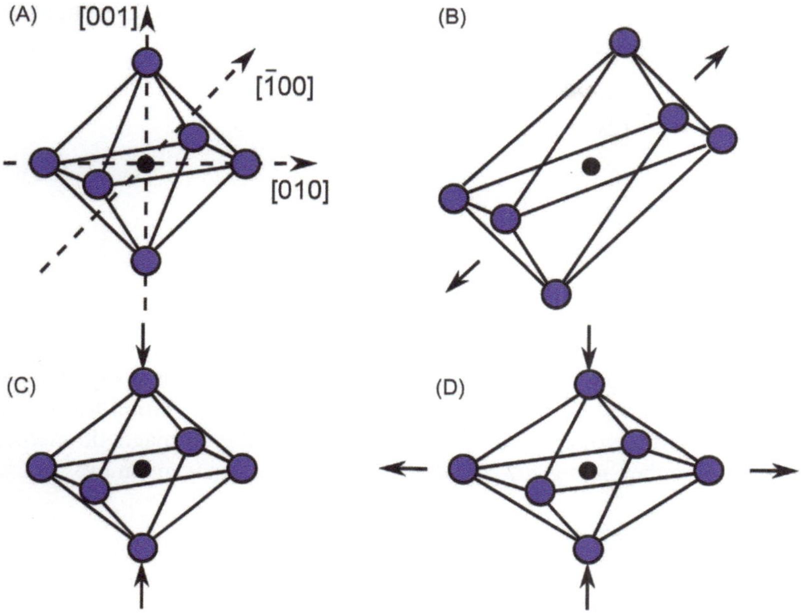

Figure 16.2 Octahedral site in a cubic symmetry (A) and its common distortions (B, C, D) [2].

oxygen, are incorporated within a 12-fold oxygen coordination [3]. Distortion of the lattice (Fig. 16.2) and presence of a variety of structural defects such as point defects (oxygen or metal vacancies and metal interstities) are usually following the incorporation of different cations in structure of metal oxides [2].

16.3 Magnetism of magnetic oxides

The magnetism of free atoms and ions can be associated with angular momentum. The electron has an intrinsic spin angular momentum s and intrinsic magnetic moment μ_s that can be expressed as $\mu_s = -2 \cdot s \cdot \mu_B/\hbar$, where $\mu_B = \frac{1}{2} \cdot e \cdot \hbar/m_e$ is the Bohr magneton. This quantity is the fundamental unit of atomic-scale magnetism. If a spinless electron is orbiting a nucleus in a state of angular momentum l, it behaves like a current loop with a magnetic moment $\mu_s = -l \cdot \mu_B/\hbar$.

When electron spin and orbital angular momentum are taken into account, the magnetic moment of each electron in an isolated atom becomes $\mu_e = -(2 \cdot s + l) \cdot \mu_B/\hbar$. The Zeeman Hamiltonian for the electron in magnetic field B is represented by:

$$\mathscr{H}_Z = -\mu \cdot B = \frac{\mu_B}{\hbar}(l_z + 2s_z) \tag{16.1}$$

with the B-field taken along with the z-direction. The total magnetic moment for an N-electron atom becomes:

$$\mu_e = \frac{\mu_B}{\hbar} \cdot \sum_{i=1}^{N} -(2s_i + l_i) \quad or \quad \mu = -(2S + L) \cdot \frac{\mu_B}{\hbar} \tag{16.2}$$

where

$$S = \sum_{i=1}^{N} s_i \quad \text{and} \quad L = \sum_{i=1}^{N} l_i \tag{16.3}$$

S and L are total spin and orbital angular momentum, respectively. If J represents total angular momentum (i.e., $J = S + L$), the previous equation for total magnetic moment can be written as $\mu = -(J + L) \cdot \mu_B/\hbar$. The total magnetic moment of the ion or atom is then $\mu = g \cdot \mu_B \cdot J$, where Lande's g-factor is given by:

$$g_L = 1 + \frac{J(J+1) + S(S+1) - L(L+1)}{2J(J+1)} \tag{16.4}$$

The Zeeman and spin–orbit interaction are represented, respectively, by the Hamiltonians:

$$\mathscr{H}_Z = -\mu \cdot B = \frac{\mu_B}{\hbar}(L_z + 2S_z) \cdot B \quad \text{and} \quad \mathscr{H}_{so} = -\lambda \cdot \vec{L} \cdot \vec{S} \tag{16.5}$$

Where λ is constant, $\vec{L}$ is the total orbital angular momentum, $\vec{S}$ is the spin angular momentum, and B-field is taken along with z-direction.

For a free atom or ion, the angular momentum and spin of each electron is arranged according to the semiempirical Hund's rules. Following these simple rules, it is then possible to calculate the magnetic moment without resorting to any particular knowledge of the individual atomic species for nearly all atoms. If we have a multielectron ion, the angular momentum of any filled (1*s*, 2*s*, 2*p*, 3*s*, 3*p*...) orbital is zero, because both the spin and orbital moments cancel each other. The net magnetic moment of atoms arises from unpaired electrons (Table 16.2).

When the ions are placed in a crystalline environment of oxygen neighbors there are some drastic changes. For 3*d*- and 4*d*-ions crystal–field interaction is stronger than spin–orbit interaction. For 4*f*-ions the crystal field is screened by the large filled 5*s*- and 5*p*-shells and it can influence on Hamiltonians only as a perturbation correction. For 3*d*-ions J is not a good quantum number, and from a magnetic viewpoint the 3*d*-ions may be represented as spin-only ions. The quenching of the orbital angular momentum in 3*d*-metal solids is due to the interaction of the *d*-orbitals with the crystal field formed by oxygen atoms. The crystal field leads to quenching by favoring the formation of superposition states with zero net magnetic moment.

Table 16.2 **The ground state and magnetic moments for 3*d* ions [4]**

Arrangements of electron in *d*-orbitals						T	S	L	J	g_L	$\mu_J/\mu_B = g_J[J(J+1)]^{1/2}$	$\mu_S/\mu_B = 2[S(S+1)]^{1/2}$
$3d^1$–Ti^{3+} V^{4+}	↑					$^2D_{3/2}$	1/2	2	3/2	4/5	1.55	1.73
$3d^2$–V^{3+} Cr^{4+}	↑	↑				$^3F_2$	1	3	2	2/3	1.63	2.83
$3d^3$–Cr^{3+} Mn^{4+}	↑	↑	↑			$^4F_{3/2}$	3/2	3	3/2	2/5	0.77	3.87
$3d^4$–Cr^{2+} Mn^{3+}	↑	↑	↑	↑		$^5D_0$	2	2	0	–	0	4.90
$3d^5$–Mn^{2+} Fe^{3+}	↑	↑	↑	↑	↑	$^6S_{5/2}$	5/2	0	5/2	2	5.92	5.92
$3d^6$–Fe^{2+} Co^{3+}	↓↑	↑	↑	↑	↑	$^5D_4$	2	2	4	3/2	6.70	4.90
$3d^7$–Co^{2+} Ni^{3+}	↓↑	↓↑	↑	↑	↑	$^4F_{9/2}$	3/2	3	9/2	4/3	6.54	3.87
$3d^8$–Ni^{2+} Co^{1+}	↓↑	↓↑	↓↑	↑	↑	$^3F_4$	1	3	4	5/4	5.59	2.83
$3d^9$–Cu^{2+} Ni^{1+}	↓↑	↓↑	↓↑	↓↑	↑	$^2D_{5/2}$	1/2	2	5/2	6/5	3.55	1.73

In order to explain how the crystal field influence magnetic properties of metal oxides we can start with the one-electron model [5,6], which is good approximation for d^1, d^4, d^6, and d^9-ions. Depending on the type of the crystalline coordination of the metal ions (cubic, tetrahedral, octahedral coordination), d-orbitals are splitting in a different way (Fig. 16.3). For example, the splitting of d-orbitals is reversed in octahedral and in tetrahedral coordination. At octahedral sites t_{2g} (d_{xy}, d_{yz}, and d_{xz})

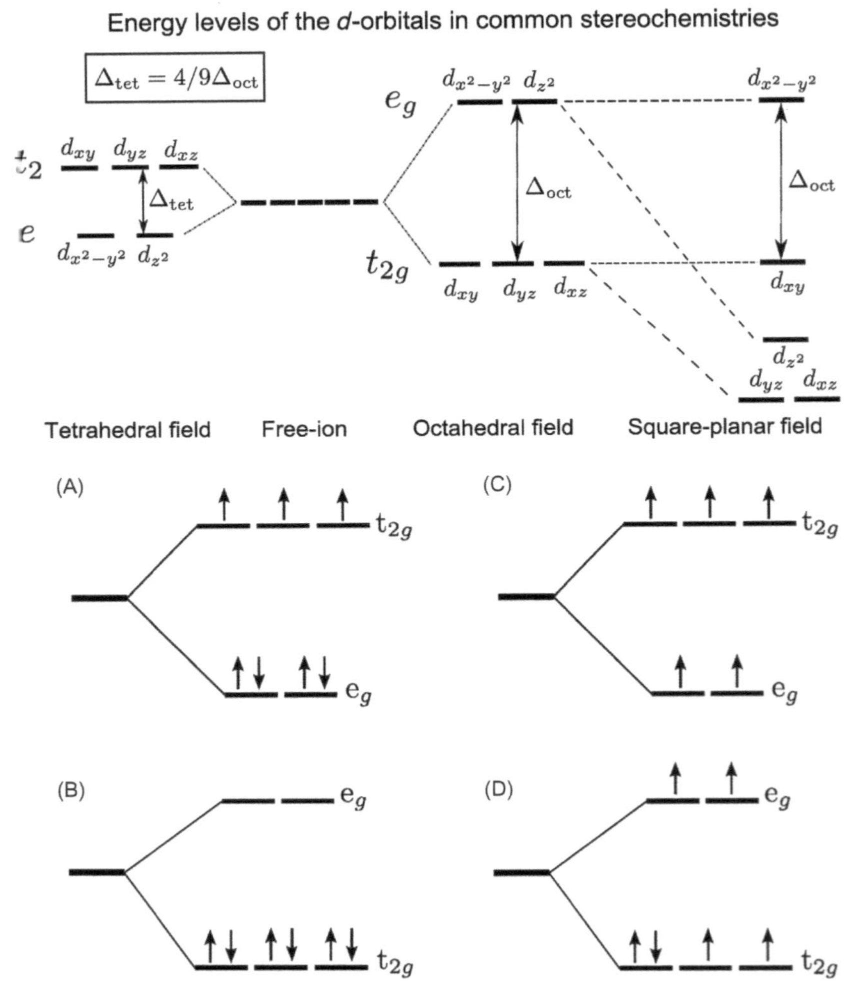

Figure 16.3 Schematic crystal field splitting and occupations of Co and Fe 3d-electrons in tetrahedral and octahedral coordination environments and energy levels of the d-orbitals common stereochemistries. (A) Co (II) tetrahedral d^7; (B) Co (III) octahedral d^9; (C) Fe (III) tetrahedral d^5; (D) Fe (II) octahedral d^9.

orbitals have lower energy compared to energy of e_g ($d_{x^2-y^2}$, $d_z{}^2$) orbitals. The magnitude of ligand field effect Δ_{oct} is of the order 1 eV. Contrary, the t_{2g} orbitals at tetrahedral sites are those that possess the higher energy in the ligand field. It should be noted that the Jahn-Teller effect is strong for d^4- and d^9-ions in octahedral coordination (Mn^{2+}, Cu^{2+}) and for d^1- and d^6-ions in tetrahedral coordination (V^{4+}, Co^{3+}). In the one-electron model, orbitals with the lowest energy are filled first, in a way that five ↓ (spin-down) lie above the five ↑ (spin-up) by amount J_H (the on-site exchange energy ~ 1.5 eV). The low-spin state is observed at ions for which the values of Δ_{oct} is higher than the values of J_H. In that case t_{2g} orbitals filled before e_g orbitals. The low-spin state, for example for the Co^{2+} ion, is $S = 1/2$ and the high-spin state is $S = 3/2$.

Exchange interaction is a symmetry-constrained Coulomb interaction that has the effect of coupling electronic spins. Intraionic exchange, also known as J_H or Hund's first rule exchange, couples electrons in partially filled orbitals in a way to maximize the total spin. This further means that the ↑ orbitals are filled first, before beginning to fill the ↓ orbitals. If the energy splitting caused by the crystal field is higher than the intraionic exchange, the ion enters in a low-spin state. The Heisenberg interaction (direct exchange) is described by Hamiltonian moments:

$$\mathscr{H}_Z = -\frac{1}{\hbar} J_{ij} \cdot \hat{S}i \cdot \hat{S}j \tag{16.6}$$

and is often used to describe the interaction between individual moments. If J_{ij} is positive, the parallel alignment is favored and material is FM. If J_{ij} is negative, the spin orientation is antiparallel and material is AFM. In this case the moments align when the overall moment is zero. There are many possible AFM alignments. If the magnitudes of the moments in an antiferromagnet are not equal (nonzero net magnetization), the material is known as a ferrimagnet. The mechanism that causes the order alignment is strong and temperature-dependent.

Direct exchange is a consequence of the Pauli exclusion principle. Consider two atoms, each with a single electron. When the atoms are close enough, their orbitals interact, electrons can move between the atoms, and according to the Pauli principle the total wave function must be antisymmetric. Having in mind that the symmetry of the total wave function is a product of the spatial wave function and spin wave function, one of them must be antisymmetric. An antisymmetric spatial wave function lowers the Coulomb energy of the system, which favors a FM (symmetric) spin wave function.

Indirect (super) exchange is responsible for the AFM order in metal oxides such as NiO, $CaMnO_3$, and Fe_2O_3. The 3d-orbitals of transition metal ion overlap with the p-orbitals of the oxygen. This overlap stabilizes an AFM coupling between the 3d-orbitals if the angle between metal and bridging oxygen is close to 180 degrees. The coupling becomes FM if the angle decreases toward 90 degrees. The occupancy and degeneracy of 3d-orbitals determine the strength and sign of superexchange, and possible cases are given in the Goodenough–Kanamari rules (later reformulated by Anderson) [7,8]. Superexchange interactions occur in insulators and make use of nonmagnetic ions to assist spin interactions between two magnetic ions. Also, this type of magnetic interaction is more commonly AFM than FM.

Ruderman–Kittel–Kasuya–Yosida (RKKY) exchange and superexchange both involve spin interactions accommodated by intermediate atoms. RKKY exchange occurs in metals, where the medium for the exchange interactions is the conduction electrons. The strength and sign of RKKY interaction varies with the distance, so it causes FM or AFM ordering.

Double exchange involves the transfer of an electron between two magnetic ions with different valences, and leads to metallic FM properties. For this type of magnetic interaction a mixed-valence configuration is required as in the case in mixed-valance manganite ($La_{1-x}A_xMnO_3$, A = Ba, Ca, Sr). Electrons from an oxygen orbital can hop to a Mn^{4+} ion, and its vacant orbital then can be filled by an electron from Mn^{3+}. The electron has moved between the neighboring metal ions, retaining its spin. The electron movement from one ion to another is "easier" when spin direction does not have to be changed (Hund's rules). Another well-known double-exchange pair is Fe^{3+} and Fe^{2+} ions. As Table 16.2 shows, Fe^{3+} ion has $\uparrow d^5$ configuration, and d-orbital is half-filled with spin parallel alignment. Fe^{2+} has six electrons and the sixth $\downarrow d$ electron is placed at the bottom of a t_{2g} band (if the ion is octahedrally coordinated by oxygen) from where it can jump directly from one d^5 ion to another.

There is clearly a difference between superexchange and double exchange. In a superexchange the occupancy of d orbital of two metal ions remains the same, so electrons are not actually moving between two metal ions. In double exchange electrons are hopping between ions of different valance states via the intermediate oxygen.

The antisymmetric part of exchange interaction between spins, characterized by a vector parameter D and called Dzyaloshinskii–Moriya interaction, is given by:

$$H_{DM} = 2D \cdot \sum_{i,j} (\hat{S}i \times \hat{S}j) \tag{16.7}$$

It gives rise to canted antiferromagnetismin iron oxides [9,10]. AFM iron oxides possessing weak ferromagnetism, such as iron rare-earth orthoferrites $RFeO_3$ (R represents rare-earth element) and hematite α-Fe_2O_3, are natural candidates for observing superexchange interactions. The presence of the Dzyaloshinskii–Moriya antisymmetric exchange interaction leads to a slight canting of the spins from the antiparallel orientation by an angle of 0.5–1 degrees.

16.3.1 Magnetism of metal-oxide nanoparticles

The magnetic properties of nanomaterials such as the Curie (T_C) or Néel (T_N) temperatures, and the coercivity field (H_C), can be significantly different than those of a bulk specimen [11]. It is well-known fact that a bulk FM material is composed of small regions called magnetic domains. The formation and orientation of these magnetic domains are the result of the balance of several energy terms: the exchange energy, magnetocrystalline anisotropy, and the magnetostatic (or dipolar) energy [12]. The first term, exchange energy, tries to align all magnetic moments in the same direction. The second one, magnetocrystalline anisotropy, tries to orient magnetic moments along specific directions, while the third term, magnetostatic energy, tries to eliminate the magnetization in the material. In each domain the magnetic moments of atoms are aligned in one direction giving a net magnetization of each domain. The directions of magnetizations of the domains are different. The net magnetization is obtained by adding the magnetization of individual domains. It was found that magnetic domains in FM material have a minimum critical size (~100 nm), below which the magnetic material cannot split up further into domains and are called single-domain particles [13,14]. Thus, super paramagnetism emerges when the system is composed of small particles called independent crystallites (Fig. 16.4). Since the particles are small, the direction of each magnetic moment is not rigidly confined. The crystalline anisotropy energy decreases as the size of the crystallite is reduced, and as a consequence of that each crystallite magnetic moment is relatively free to move between any of the easy axes in response to an applied field. When the applied field is removed, the thermal effects rapidly destabilize any overall moment and the material starts to act like a paramagnetic. The regime of paramagnetic behavior extends from the Curie (or Néel) temperature down to the so-called blocking temperature T_B. Thus in the superparamagnetic state, the magnetic nanoparticles behaves as a paramagnetic atom with a giant spin.

At temperatures below the blocking temperature, the thermal effect becomes small and cannot cause fluctuations in the orientations of the magnetic moments of the nanoparticles; in fact they freeze in random orientations. The blocking temperature can be estimated using dc magnetization measurements. These measurements involve the measurement of magnetization as a function of temperature in two different states. The first state is called the zero-field-cooled (ZFC) state, and the

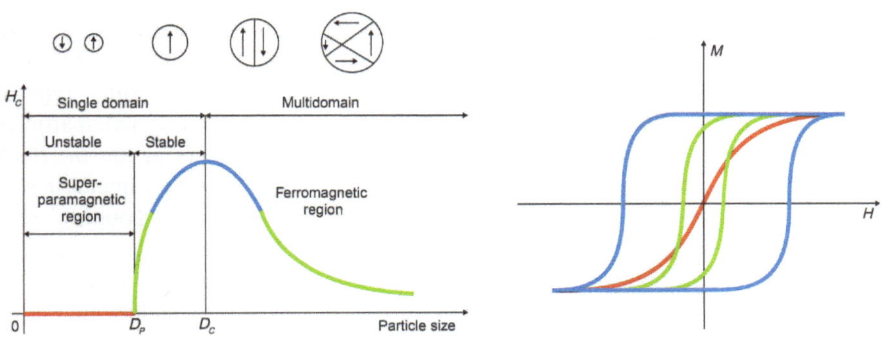

Figure 16.4 Superparamagnetic, single-domain, and multidomain magnetic regime as a function of particle size.

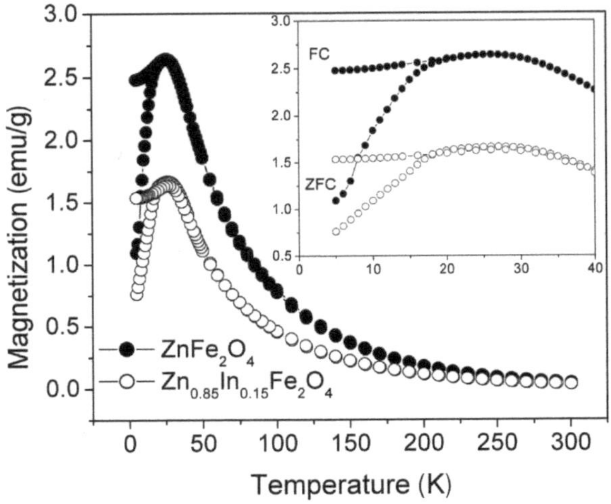

Figure 16.5 Temperature dependence of magnetization of In-doped $ZnFe_2O_4$ obtained in zero-field cooled (ZFC) and field cooled (FC) conditions [16].

second is called the field-cooled (FC) state. The value of the blocking temperature depends on the particle size, effective anizotropy constant, applied field, and experimental measurement time. The shape of the magnetization versus temperature plots in the ZFC and FC measurements can provide qualitative information about the size distribution and the strength of interaction among the particles in the sample (Fig. 16.5) [15,16].

It should be noted that due to the fine size effect in nanomaterials, the existence of randomly oriented uncompensated surface spins, the existence of canted spins and the existence of a spin-glass-like behavior of the surface spins can significantly affect magnetic behavior of metal oxides [17–19].

The total magnetization of the nanoparticle is thought to be composed of two components: a component due to the surface spins and a component due to the core of the particle. Thus, the magnetization of the magnetic nanoparticles can be modeled via a core–shell (or core–surface) magnetic model leading to another type of magnetic interaction at the interface between the core and the shell. In nanoparticles of AFM or ferrimagnetic (FIM) materials, this interaction occurs at the interface between the FM surface and the AFM (or FIM) core. The core–shell interaction is called the exchange bias or exchange coupling [20]. The exchange coupling provides an additional magnetic anisotropy to help align the ferromagnetic spins in certain directions. The exchange coupling is known to vanish above a critical temperature called the blocking temperature, but no satisfactory understanding of this interaction at the microscopic level exists. In order to obtain a satisfactory understanding of the theory of exchange bias, it is essential to understand the atomic interface structure [21].

16.4 Representative structures of magnetic oxides

The most valuable structures in magnetic oxides regarding their magnetic behavior are spinel, garnet, and magnetoplumbite structure. These structures will be explained in brief in following sections.

16.4.1 Spinel structure

Spinel structure most often occurs in ferrites and other binary oxides with a general chemical formula AB_2O_4, where usually Fe^{3+} (ferrite) is at the position B, but also some other transition metals Cr^{3+}, Mn^{3+}, and so on. The position A is filled with a divalent cation such as Mn^{2+}, Fe^{2+}, Co^{2+}, Ni^{2+}, Cu^{2+}, Zn^{2+}, or Mg^{2+}, or sometimes with the monovalent lithium cation (Li^{+}) or even vacancies, as long as these absences of positive charge are compensated for by additional trivalent iron cations (Fe^{3+}). The oxygen anions (O^{2-}) adopt a close-packed cubic crystal structure, while metal cations occupy two different interstitial sites: tetrahedral and octahedral. The spinel lattice cell is composed of 8 formula units that contain 32 oxygen anions, 8 tetrahedrally coordinated cations (A sublattice), and 16 cations in octahedral sites (B sublattice). The antiparallel alignment and incomplete cancellation of magnetic spins between the two sublattices A and B leads to a permanent magnetic moment (Fig. 16.6) [22,23].

The well-known ferrite that belongs to the spinel group is mineral magnetite, Fe_3O_4. Fe_3O_4 has a cubic inverse spinel structure that consists of a cubic close-packed array of oxide ions, where all the Fe^{2+} ions occupy half of the octahedral sites and the Fe^{3+} are split evenly across the remaining octahedral sites and the tetrahedral sites, leading to ferrimagnetism. In magnetite, the Fe^{2+} ions may be partly or fully replaced by other divalent ions (Co, Mn, Zn, etc.), thus creating a

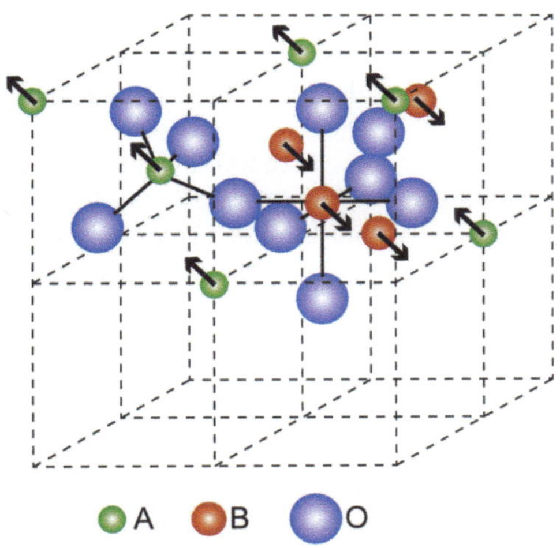

Figure 16.6 Unit cell and ferromagnetic ordering in spinel ferrites.

whole spectrum of magnetic interactions. The magnetic properties of spinels are highly sensitive to the cation distribution between two sublattices and these properties are a complex function of the total chemical composition, stoichiometry, the substitution degree of foreign ions, and inversion parameter. Weather a foreign cation substitutes for A-site or B-site iron depends upon the relative site-preference energies. Of the ions with partially filled d-shells, crystalline fields give a strong B-site preference to ions with d^3, low-spin d^5 or d^6, or high-spin d^8 configurations. Solute cations that can replace Fe^{3+} ions for the tetrahedral sites include Cu^+, Mg^{2+}, Zn^{2+}, Cd^{2+}, Mn^{2+}, Co^{2+}, Ga^{3+}, In^{3+}, Si^{4+}, Ge^{4+}, and so on [24]. Control of cation composition and inversion is essential for spinel properties [22,25].

Substitution on A-sites: The divalent cations such as Zn^{2+} and Cd^{2+}have a strong preference for tetrahedral A-site in spinel. Therefore, bulk $ZnFe_2O_4$ is a paramagnet that has a normal spinel structure and becomes antiferromagnetic at $T_N = 10$ K. However, when prepared in the form of nanoparticles, $ZnFe_2O_4$ shows some degree of inversion and different kinds of magnetic ordering connected with it [26,27]. Trivalent cations could reside at tetrahedral sites, but substitution in complete solid solution range is difficult to find. Substitution with Ga^{3+} has shown that it is possible to place Ga^{3+} ions only on A-sites for $x < 0.3$ [8]. Substitution of In^{3+} in $ZnFe_2O_4$ gives an even smaller compositional range ($x \leq 0.15$) for exclusive A-site substitution [26,28]. These substitutions have a profound impact on magnetic properties. For example, the saturation magnetization of spinel ferrites can be greatly enhanced by partial substitution of $ZnFe_2O_4$ for $NiFe_2O_4$ or $MnFe_2O_4$. The zinc cations prefer tetrahedral coordination and force additional Fe^{3+} onto the octahedral sites. This results in less cancellation of spins and greater saturation magnetization.

Substitution on B-sites: Substitution exclusively on B-sites can be achieved with ions having formal valence ranging from I to VI [8]. FIM $NiFe_2O_4$ is a good example of the complete inverse spinel with Ni^{2+} ions placed onto a B-site, and Fe ions distributed between tetrahedral and octahedral sites. Large cations, such as some of the rare-earth ions (e.g., Y^{3+}), almost exclusively reside on octahedral sites [23,29,30].

Substitution on both A- and B-sites: Magnesium and manganese ferrites are good examples of mixed spinel structures where substitution for iron may result in a distribution of the cations on both A- and B-sites. The physical properties of the mixed spinel structures obtained in such way vary strongly with temperature.

16.4.2 Garnet structure

The structure of garnets is a cubic structure, with a general chemical formula of $A_3B_2(XO_4)_3$ (A = Ca, Mg, Y, La, or rare earth; B = Al, Fe, Ga, Ge, Mn, Ni, or V; X = Si, Ge, Al). The A cations are placed on dodecahedral C-sites, B cations on octahedral A-sites, and X cations on tetrahedral D-sites [31] (Fig. 16.7). The most important and widely studied magnetic garnet is yttrium-iron garnet (YIG), which has the formula $Y_3Fe_5O_{12}$, where five Fe^{3+} ions occupy two octahedral and three

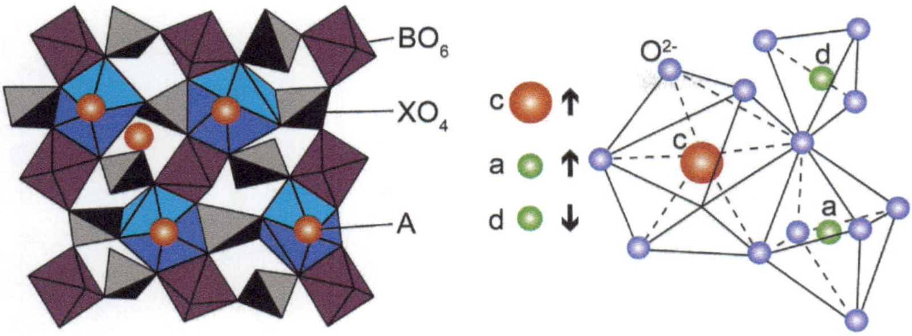

Figure 16.7 Structure of an ideal garnet with a general chemical formula of $A_3B_2(XO_4)_3$ and illustration of the cation coordination.

tetrahedral sites, and the Y^{3+} ions are surrounded by eight oxygen ions. The YIG unit cell is composed of 8 formula units with 160 atoms. The net magnetic moment in YIG arises from an uneven contribution from antiparallel spins; the magnetic moments of the a- and d-ions are aligned antiparallel as are those of the c- and d-ions (Fig. 16.7B).

By substituting specific sites with rare-earth elements, for example, interesting magnetic properties can be obtained. Other ions of the lanthanide group can be used generously in the *c*-sublattice to form what amounts to a distinct set of FIM compounds. Because the ionic spins that are introduced to the c-sublattice exchange couple to the iron in the *d*- and *a*-sublattices, a wide variety of thermomagnetic, magnetoelastic, microwave, and optical properties can be created in these rare-earth iron garnets.

16.4.3 Magnetoplumbite structure

Hexagonal ferrites adopt magnetoplumbite structure, which is similar to the inverse spinel but with hexagonal symmetry instead of cubic. The general formula of magnetoplumbite structure can be written as $AB_{12}O_{19}$, in which A is a divalent metal ion, such as Ba^{2+}, Pb^{2+}, or Sr^{2+}, and B is a trivalent metal ion, such as Al^{3+}, Ga^{3+}, Cr^{3+}, or Fe^{3+}. The hexagonal ferrites are all FIM materials, and their magnetic properties are intrinsically linked to their crystalline structures [32]. There are many structural forms of hexagonal ferrites such as W, X, Y, or Z type, but only M structure of hexaferrites are technically important.

Barium hexaferrite ($BaO \cdot 6Fe_2O_3$) is one of the most important M-type hexaferrite magnetic materials. The crystal structure of $BaFe_{12}O_{19}$ is shown in Fig. 16.8. It can be described by a periodically stacking sequence of two basic building blocks, S and R, along the *c*-axis, where an asterisk indicates rotation of the blocks by 180 degrees [33]. The Fe^{3+} ions occupy three different sites in this structure: octahedral and tetrahedral in S block and bipyramidal sites in R block [34]. Large divalent ions, such as Ba^{2+}, Pb^{2+}, or Sr^{2+} replace an oxygen atom in R block.

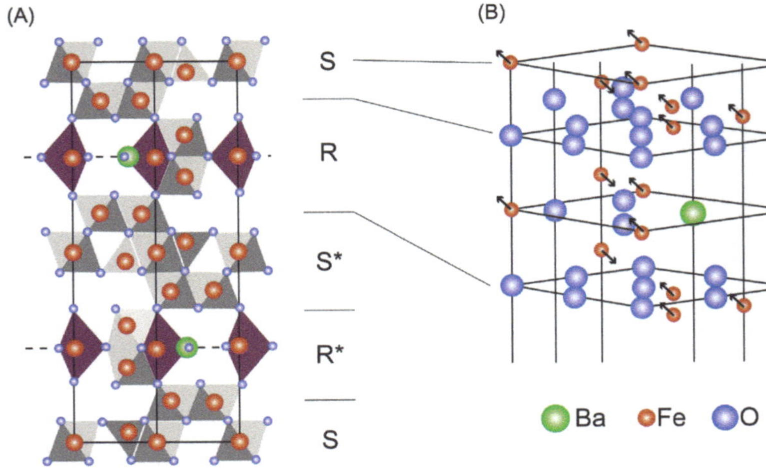

Figure 16.8 (A) Stacking sequence in M-type hexaferrite $BaFe_{12}O_{19}$, (B) crystal structure corresponding SR layers in barium hexaferrite $BaFe_{12}O_{19}$.

16.4.4 Other common structures in magnetic oxides

Besides the already-mentioned structures, magnetic oxides can adopt structures such as halite, corundum, rutile, perovskite, pyrochlore, and wurzite. These structures will be briefly explained.

Halite. The halite structure can be regarded as two interpenetrating face-centered cubic sublattices, with cations and anions alternating along [100] directions. Cations are octahedrally coordinated by anions and vice versa. Since there is only one type of cation, located in crystallographically equivalent positions, in these simple oxides, in accordance with the stoichiometric composition, ferrimagnetism cannot appear. However, fundamentally ferromagnetism is possible, but it has been found only in europium oxide, EuO. Typically magnetic moments are arranged in antiferromagnetic ordering, and in a set of planes (111) cations, situated in the same plane, have parallel magnetic moments and those in adjacent planes, antiparallel magnetic moments (Fig. 16.9A). The symmetry of these "magnetic structures" is no longer cubic, since the direction [111] is perpendicular to the set of planes with ordered magnetic moments and is not equivalent to the other three directions of the space diagonals of an elementary cube. Consequently, the occurrence of antiferromagnetism is accompanied by a distortion of the structure, which reduces the symmetry to rhombohedral (MnO, FeO, NiO) or to tetragonal (CoO). The real reason for these strains is exhibited in the fact that the ions are placed in such positions in which the exchange forces, as well as other interactions, are enhanced and accordingly, the free energy decreases. In the case of the rhombohedral distortion, the decrease of the free energy is achieved by reducing or increasing the distance between adjacent planes (111). If a portion of cations of a simple oxide, such as NiO, is substituted by other cations, such as Li^{+}cations, then the location of cations

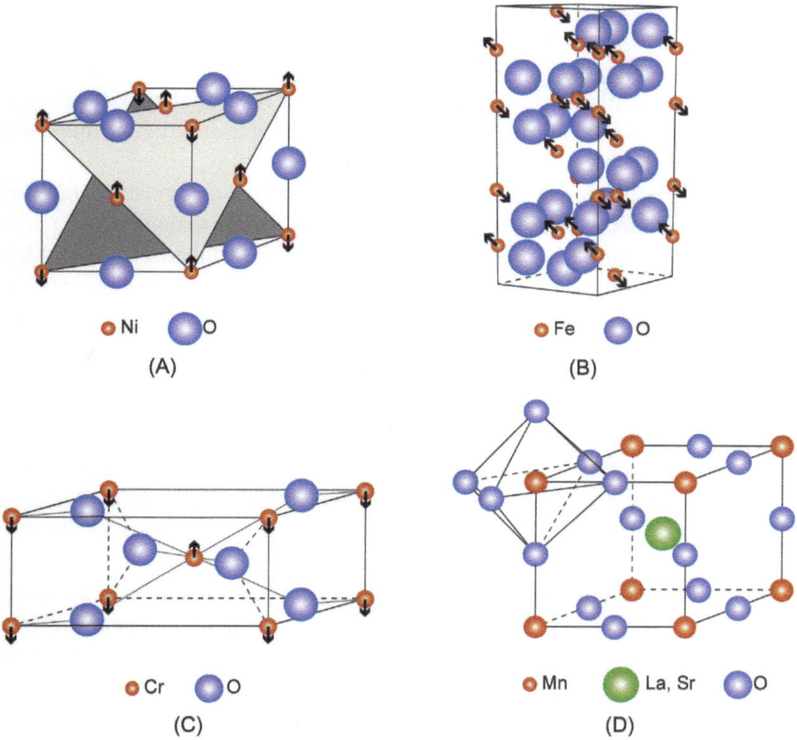

Figure 16.9 Structures of (A) NiO, (B) α-Fe_2O_3, (C) CrO_2, and (D) perovskite (La,Ca)MnO_3.

in adjacent planes (111) is often the same as the arrangement of the magnetic moments. The result is a type of layered sublattice (i.e., a structure), which in principle allows the emergence of ferrimagnetism. Just like the antiferromagnetic ordering of the magnetic moments, the ordering of cations often changes the symmetry of the crystal.

Corundum. In corundum structure, the oxygen atoms form a slightly distorted hexagonal close packing in which two-thirds of the octahedral interstices are occupied by metal ions. Hematite, α-Fe_2O_3, is a well-known magnetic oxide with corundum structure in its most common form (Fig. 16.9B) [35]. Since the cations in the corundum structure are mainly located in equivalent positions, in simple oxides of trivalent transition elements ferrimagnetism cannot occur. α-Fe_2O_3 and Cr_2O_3 compounds are indeed antiferromagnetic, whereby weak ferromagnetism is superimposed on the antiferromagnetism of the former. The structure of ilmenite, which is the most common in compositions of type ABO_3, is virtually identical to the structure of corundum and differs from it only by a regular arrangement of ions A and B (in ilmenite, $FeTiO_3$, half of the Fe atoms are replaced by Ti). For this reason it is enough to consider the simpler structure of corundum, although oxides with the structure of ilmenite are more important because of their magnetic properties.

Rutile. Rutile structure can be regarded as a distorted hexagonal close-packed oxygen array, where cations occupy one-half of the octahedral interstices. This structure has CrO_2 (Fig. 16.9C). The Cr^{4+} ions are located at the corners of the unit cell and there is a distorted CrO_6 octahedron in the center of the structure (Fig. 16.9C). Doped CrO_2 represents one of the most important ferromagnets used as magnetic media.

Perovskite. In the basic perovskite (ABO_3) unit cell, A designates a large cation site with 12-fold dodecahedral oxygen coordination, often occupied by a trivalent member of the lanthanide (rare-earth) series, and B is an octahedral site that harbors ions of the 3*d* transition series. Variations of the $A^{3+}B^{3+}O_3$ compounds, such as $A^{2+}B^{4+}O_3$, also occur (the original compound from which the term perovskite was coined is diamagnetic $Ca^{2+}Ti^{4+}O_3$), as well as solid solutions of multiple cations in each type of site that maintain the required 6+ total cation charge. The most frequently encountered example of a B^{4+} compound is the ferroelectric $Ba^{2+}Ti^{4+}O_3$ with its relatives $Sr^{2+}Ti^{4+}O_3$, $Ca^{2+}Ti^{4+}O_3$, and $Pb^{2+}Ti^{4+}O_3$, in which the Ti^{4+}ion is d^0 and therefore diamagnetic. The perovskite structure has two nonequivalent cationic sublattices, thus if both sublattices are formed by magnetic ions, FIM ordering of the magnetic moments is possible. Typical examples are $AFeO_3$ (A = Bi, Ho, Er, etc.). This situation is more complicated than with spinels because the exchange interaction within the sublattice B is in most cases much stronger than the exchange interactions A–B and A–A. Consequently, magnetic ordering appears in the sublattice B often, even at high temperatures. Depending on the type and valence of the ions, this ordering can be antiferromagnetic, ferromagnetic, or FIM. Ferrimagnetism occurs when planes are perpendicular to the directions [111] or [001], or are alternately occupied by ions of different elements, or by ions of the same element but with different valences. Another interesting magnetic oxide with perovskite structure is $LaMnO_3$, which has been the subject of detailed investigations regarding its crystallography and magnetic behavior, particularly when Sr^{2+} or Ca^{2+} ions are blended with La^{3+} in the A-sites to create a mixture of Mn^{3+} and Mn^{4+} ions in the B sublattice in (La,Ca)MnO_3 (Fig. 16.9D).

References

[1] C.C. Baryy, N.M. Grant, Ceramic Materials, Science and Engineering, Springer, 2007.

[2] G.F. Dionne, Magnetic Oxides, Springer, New York Dordrecht Heidelberg London, 2009.

[3] R. Skomski, J.M.D. Coey, Permanent Magnetism, Condensed Matter Physics, Institute of Physics Publishing, Bristol, 1999.

[4] S. Krupička, Elements of Theoretical Magnetism, Lincoln, UK, 1968.

[5] R.G. Burns, Intervalence transitions in mixed-valence minerals of iron and titanium, Annu. Rev. Earth Planet. Sci. 9 (1981) 345–383.

[6] I. Chamritski, G. Burns, Infrared- and Raman-active phonons of magnetite, maghemite, and hematite: a computer simulation and spectroscopic study, J. Phys. Chem. B 109 (2005) 4965–4968.

[7] P. Anderson, Magnetism, Academic Press, New York, 1963.
[8] J. Goodenough, Magnetism and the Chemical Bond, John Wiley And Sons, New York-London, 1963.
[9] I.E. Dzyaloshinskii, A thermodynamic theory of “weak” ferromagnetism of antiferromagnetics, J. Phys. Chem. Solids 4 (1958) 241–255.
[10] T. Moriya, Anisotropic superexchange interaction and weak ferromagnetism, Phys. Rev. 120 (1960) 91–98.
[11] A.P. Guimarães, Principles of Nanomagnetism, Springer, Berlin/Heidelberg, Germany, 2009.
[12] G. Bertotti, Hysterisis in Magnetism: For Physicists, Materials Scientists and Engineers, Academic Press-Elsevier, Waltham, MA, USA, 1998.
[13] S. Mørup, M.F. Hansen, C. Frandsen, Magnetic nanoparticles, in: G. Scholes, G. Wiederrecht, D. Andrews (Eds.), Comprehensive Nanoscience and Technology, Elsevier, Amsterdam, The Netherlands, 2011, pp. 437–491.
[14] A.G. Kolhatkar, A.C. Jamison, D. Litvinov, R.C. Willson, T.R. Lee, Tuning the magnetic properties of nanoparticles, Int. J. Mol. Sci. 14 (8) (2013) 15977–16009.
[15] Z. Cvejic, B. Antic, A. Kremenovic, S. Rakic, G.F. Goya, H.R. Rechenberg, et al., Influence of heavy rare earth ions substitution on microstructure and magnetism of nanocrystalline magnetite, J. Alloys Compd. 472 (1-2) (2009) 571–575.
[16] M. Maletin, E.G. Moshopoulou, V.V. Srdić, Magnetic properties of $ZnFe_2O_4$ and In-doped $ZnFe_2O_4$ nanoparticles, Phys. Status Solidi A 205 (8) (2008) 1831–1834.
[17] R. Kodama, A. Berkowitz, E. McNiff, S. Foner, Surface spin disorder in $NiFe_2O_4$ nanoparticles, Phys. Rev. Lett. 77 (1996) 394–397.
[18] R. Kodama, A. Berkowitz, Atomic-scale magnetic modeling of oxide nanoparticles, Phys. Rev. B. 59 (1999) 6321–6336.
[19] S. Mørup, E. Brok, C. Frandsen, Spin structures in magnetic nanoparticles, J. Nanomater. (2013) (2013) Article ID 720629, 8 pages.
[20] J. Noguésa, J. Sorta, V. Langlaisb, V. Skumryeva, S. Suriñachb, J.S. Muñozb, et al., Exchange bias in nanostructures, Phys. Rep. 422 (2005) 65–117.
[21] M. Kiwi, Exchange bias theory, J. Magn. Magn. Mater. 234 (2001) 584–595.
[22] Ž. Cvejić Ph.D., Thesis, Faculty of Sciences, University of Novi Sad, Serbia, 2008.
[23] M. Vucinic-Vasic, E.S. Bozin, L. Bessais, G. Stojanovic, U. Kozmidis-Luburic, M. Abeykoon, et al., Thermal evolution of cationdistribution/crystallite size and their correlation with the magnetic state of Yb-substituted zinc ferrite nanoparticles, J. Phys. Chem. C 117 (2013) 12358–12365.
[24] C. Gleitzer, J.B. Goodenough, Mixed-valence iron oxides, Cation Ordering and Electron Transfer Structure and Bonding, vol. 61, Springer, Berlin, Heidelberg, 1985, pp. 1–76.
[25] M. Milanović Ph.D., Thesis, Faculty of Technology, University of Novi Sad, Serbia, 2010.
[26] G. Stojanović, V.V. Srdić, M. Maletin, Electrical properties of yttrium-doped Zn and Ni–Zn ferrites, Phys, Status Solidi A 205 (10) (2008) 2464–2468.
[27] Ž. Cvejić, E. Đurđić, G. Ivković Ivandekić, B. Bajac, P. Postolache, L. Mitoseriu, et al., The effect of annealing on microstructure and cation distribution of $NiFe_2O_4$, J. Alloys Compd. 649 (2015) 1231–1238.
[28] A. Lakshman, K.H. Rao, R.G. Mendiratta, Magnetic properties of In^{3+} and Cr^{3+} substituted Mg–Mn ferrites, J. Magn. Magn. Mater. 250 (2002) 92–97.
[29] M. Milanović, K. Giannakopoulos, L. Nikolic, E. Devlin, D. Stamopoulos, E.G. Moshopoulou, et al., Structure and magnetic properties of $Zn_{1-x}In_xFe_2O_4$ and

$ZnY_xFe_{2-x}O_4$ nanoparticles prepared by coprecipitation, Ceram. Int. 37 (2011) 3235–3242.
[30] Z. Cvejić, S. Rapajić, S. Rakić, S. Jankov, S. Skuban, V.V. Srdić, Conductivity and dielectric behaviour of indium substituted zinc ferrites prepared by coprecipitation method, Phys.Scripta 90 (2015) 095802.
[31] V. Thangadurai, S. Narayanan, D. Pinzaru, Garnet-type solid-state fast Li ion conductors for Li batteries: critical review, Chem. Soc. Rev. 43 (2014) 4714–4727.
[32] R.C. Pullar, Hexagonal ferrites: a review of the synthesis, properties and applications of hexaferrite ceramics, Prog. Mater. Sci. 57 (2012) 1191–1334.
[33] A. Moitra, S. Kim, S.-G. Kim, S.C. Erwin, Y.-K. Hong, J. Park, Defect formation energy and magnetic properties of aluminum-substituted M-type barium hexaferrite, Comput. Condens. Matter 1 (2014) 45–50.
[34] S.-P. Shen, J.-C. Wu, J.-D. Song, X.-F. Sun, Y.-F. Yang, Y.-S. Chai, et al., Quantum electric-dipole liquid on a triangular lattice, Nat. Commun. 7 (2016) 10569.
[35] N.Y. Dzade, A. Roldan, N.H. de Leeuw, A density functional theory study of the adsorption of benzene on hematite (α-Fe_2O_3) surfaces, Minerals 4 (1) (2014) 89–115.

Review of methods for the preparation of magnetic metal oxides

17

Biljana D. Stojanovic, Adis S. Dzunuzovic and Nikola I. Ilic
Institute for Multidisciplinary Research, University of Belgrade, Belgrade, Serbia

17.1 Introduction

Metal oxides are a subject of interest because of their optical, electronic, and magnetic properties suitable for a variety of applications. Among the various metal oxides, magnetic metal oxides occupy an important place due to the numerous possibilities of their usage. Until 1740, the only useful permanent magnets known to man were lodestones largely composed of impure magnetite, the black spinel-structure oxide Fe_3O_4 with a ferrimagnetic arrangement of magnetic moments. The other common rock-forming iron oxide is hematite, the reddish corundum-structure sesquioxide α-Fe_2O_3, which is antiferromagnetic with magnetization about 200 times weaker than magnetite. The first more serious study of magnetism was made by Gilbert, who published his famous classic book, *On the Magnet*. But for a long time after that there were no discoveries of any fundamental importance, although many practical improvements in the manufacture of magnets appeared [1]. In 1820 Oersted made the great discovery that an electric current produces a magnetic property, opening the field of electromagnets.

Metallic alloys based on iron, or even magnetite itself (which contains both Fe^{2+} and Fe^{3+}), were unsuitable because of eddy-current losses associated with their conductivity. Ferromagnets have a higher magnetization than ferrimagnets, but they are conducting and not insulating. However, oxides from the cubic structural families of spinels and garnets, such as nickel–zinc and manganese–zinc ferrites, or yttrium–iron garnet, proved to have the greatest potential. Other winning combinations are hexagonal ferrites with the magnetoplumbite structure, which now account for 90% of permanent magnets produced worldwide [1].

Scientific interest in magnetic oxides has waxed and waned. The first golden age of magnetic materials was during the 1950s and the 1960s, when the ferrites were explored and their properties optimized. Discovery of the first family of ferromagnetic oxides, the mixed-valence manganites, also dates from this period [2]. The interest in them has decreased for some time, and then revived when the copper oxide high-temperature superconductors were found to have a similar, perovskite-related structure, and the magnetoresistance of thin film mixed-valence manganites was shown to be colossal. The manganites nicely illustrate the progression from

Magnetic, Ferroelectric, and Multiferroic Metal Oxides. DOI: http://dx.doi.org/10.1016/B978-0-12-811180-2.00017-7

studies on bulk ceramics to single crystals and to thin films and other nanostructures, which marks the historical evolution of research in magnetic oxides.

During several decades after 1970s, the interest in magnetic materials was quite reduced. Meanwhile, it should be noted that probably a new age for magnetic oxides research is happening now. Multifunctionality and thin-film device structures are the new challenges in the field. There is increasing interest in recent years in the structural chemistry and properties of metal magnetic oxide materials, especially layered structures. Many new oxide-based materials crystallize as AB_2O_4 type oxides with monoclinic and tetragonal structures with ferromagnetic or antiferromagnetic properties. The hexagonal ferrites were the first low-cost materials that could be manufactured into magnets of any desired shape. Around a million tons of these sintered and bonded ceramic magnets are being produced annually. The microwave application of ferrites or the drug and gene delivery will continue to impact significantly the practice of electronics and biomedicine. Magnetic drug-targeting will undoubtedly dramatically improve the therapeutic potential of many water-insoluble and unstable drugs. Further, the development of drugs able to target selected cells or organs within the body will also deeply improve the benefits of in vivo biomedical applications of metal-oxide nanomagnets. Recently, the developing of novel iron oxide magnetic nanoparticles as drug delivery systems allows the remarkable feature of ionically bounded pharmaceutical drugs, because the active low molecular weight substances can desorb from the carriers after a defined time period and hence diffuse from the vascular wall into the tissue.

Most magnetic metal oxides possess the structure where highly charged metal cations are embedded in a lattice of oxygen anions with stable $2p^6$ closed-shell configuration. There are many representative structures of magnetic metal oxides, probably 10 of them more pronounced, but only a few cover the most common materials. The most typical are perovskite, spinel, garnet, corundum, pyrochlore, and magnetoplumbite.

The perovskite solid solutions of $La^{3+}Mn^{3+}O_3$ and $Sr^{2+}Mn^{4+}O_3$ or $Ca^{2+}Mn^{4+}O_3$ are much studied due to fact that the mixed-valence compounds such as $La_{1-x}Sr_xMnO_3$ possess ferromagnetic properties and each of the end members possesses antiferromagnetic properties. The rare-earth orthoferrites $RFeO_3$ with an orthorhombic cell are antiferromagnets and exhibit weak ferromagnetism. Many ferrites crystallize in spinel structure. Fe_3O_4 magnetite, in addition to Fe_2O_3, is the most famous magnetic mineral and possesses the inverse spinel structure. The magnetic garnets are ferrites with formula $R_3Fe_5O_{12}$, where the antiparallel octahedral and tetrahedral sublattices are filled with Fe^{3+} ions. Strontium ($SrFe_{12}O_{19}$) and barium ($BaFe_{12}O_{19}$) ferrites crystallize in the hexagonal magnetoplumbite structure with characteristic large hexagonal unit cell.

Single or composite materials known as multiferroic (MF) materials, which possess two or more ferroic orders such as ferroelectric, ferromagnetic, or ferroelastic, can be widely considered as a special class of metal magnetic oxides. Due to the importance of MF materials, the significant attention has been drawn to their preparation. The simplest way for MF preparation is to mix pure ferroelectric and ferrite powders in a desired mass ratio. However, currently a lot of other synthesis methods are in use.

The review in this chapter will first focus on the methods for the synthesis of a wide class of various magnetic metal oxides with special attention to the materials with large or potential application, most of them possessing the spinel structure. Further, the synthesis methods for the preparation of multiferroic materials, which currently have the most important place in the field of single and composite MFs, will be particularly considered.

17.2 Synthesis of metal magnetic oxides

Generally, all the synthesis methods can be classified into physical and chemical ones. For nanomaterials, there are two fundamental approaches to preparation. The bottom-up approach represents the concept of constructing a material starting from atoms or molecules [3]. This approach illustrates the possibility of creating materials that are designed to have exactly the desired properties. The second approach, the top-down method, involves restructuring a bulk composition in order to create a new composition [4]. The top-down method usually does not allow preparing informally shaped particles, and very small sizes are difficult to realize. Bottom-up procedures are much better suited to generate uniform particles, often with outstanding control of the size, shape, and structure.

The physical methods are mostly based on thermally and mechanically induced chemical reactions (solid-state synthesis, mechanochemical reaction). For nanomaterials, the top-down physical methods imply subdivision of bulk components, including mechanical crushing, pulverization, grinding, milling, mechanical activation, and mechanochemical reaction. Thus, produced metal oxide particles are usually large in size and have wide size distribution. Several bottom-up physical methods have been reported for the synthesis of metal oxides. Most common are vapor condensation methods, spray pyrolysis, laser pyrolysis, mechanical deformation, thermochemical decomposition of metal-organic precursors in flame reactor, or aerosol processes based on external energy sources applied to provide the high temperature during gas-particle conversion [5].

The chemical methods are based on the precipitation or decomposition of precursors to form new compound usually with a narrow size distribution of particles. Increasing interest in chemical synthesis, in particular nanoparticles, is clearly indicated by the numerous published references on this subject in the recent years [5,6]. Various methods belong to chemical methods, such as coprecipitation, sol-gel, polymeric precursor synthesis (Pechini processing), hydrothermal, solvothermal, microemulsion, microbial, self propagation high temperature synthesis or autocombustion processing, chemical vapor reactions, among others. It is notable that some methods can be considered either chemical or physical depending on the media, precursors, and operating conditions such as milling. Mechanochemical synthesis, which involves solid-state reactions caused by milling of precursor powders, is investigated as an alternative chemical route.

Various methods of the synthesis can be distinguished based on the process media, including vapor, liquid, and solid-state processing routes; or combined methods such as the vapor-solid-liquid approach [7]. Solid-state reaction is a well-known and widely used method for the preparation of many materials. It is simple, less expensive than many other physical or chemical methods, and allows producing a larger amount of final compound. The disadvantage of this method is that the final product is often inhomogeneous in composition and microstructure, with wide particle and grain size distributions. This usually deteriorates the properties important for the application, such as electric, dielectric, and magnetic properties.

Besides the aforementioned processing, recently biological methods have been widely explored and have many advantages. However, the chart in Fig. 17.1 clearly illustrates that the chemical methods are still the most used for the preparation of magnetic metal oxides.

A review of most common methods conventionally applied for the synthesis of widely used materials including magnetic metal oxides will be presented in the following paragraphs. Also considered is how the synthesis procedures for preparation of magnetic metal oxides passed through innovations over time. The search for new synthetic routes able to produce reliable magnetic nanoparticles or magnetic metal oxide materials with improved characteristics, as well as the improvement of established ones, is in continuous progress. For example, the advance has taken place from simple precipitation to surfactant mediation, emulsion, microemulsion, hydrothermal, sol-gel, solvothermal, sonochemal, combustion techniques, among others.

The cost effectiveness of any synthesis procedure depends on the final product and its application. For high end use such as in drug delivery systems, comparatively higher cost of production is tolerable, but if the product has to be widely used at commercial market, the problem of cost can be essential, affecting the need to use low-cost chemicals and methods.

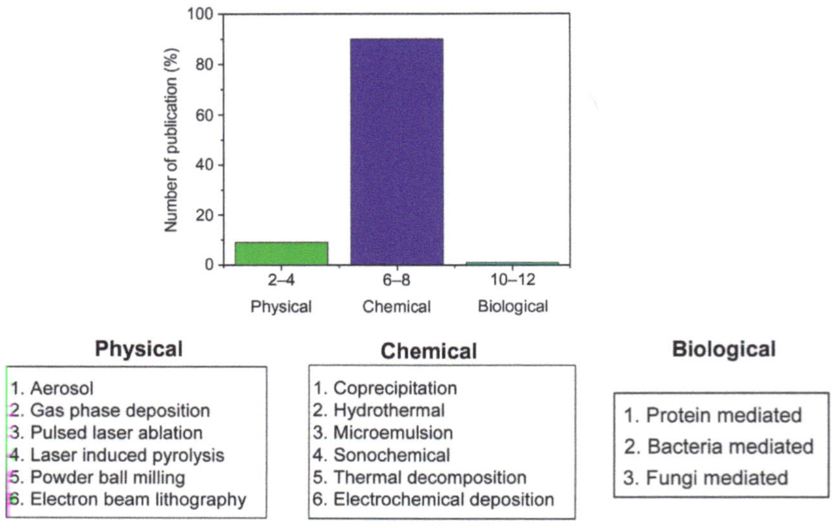

Figure 17.1 Three different approaches for the synthesis of metal magnetic oxides.

17.2.1 Chemical methods

Most chemical methods for the powder preparation are based on the liquid processing. Chemical routes for magnetic metal oxide preparation, mostly based on ferric compounds such as the hydrothermal reaction method, sol-gel processes, and chemical coprecipitation, usually involve synthesis of a precursor gel of iron followed by the decomposition of the gel or precursor into the designed crystalline iron oxide phase at an elevated temperature. The powders obtained by chemical processing are mostly nanosized, which is very convenient for the producing of magnetic metal oxide nanostructures. Precipitation, hydrothermal synthesis, sol-gel, polymeric precursor processing, and autocombustion reaction are probably most common methods for the iron-based oxides such as spinel and magnetoplumbite ferrites, magnetite, or some other magnetic metal oxides. However, not only the listed methods, but several others will be discussed more or less in detail, depending on the importance of the method itself and its suitability for the preparation of magnetic metal oxide materials.

17.2.1.1 Precipitation processing

Precipitation is a widely used wet-chemical technique that offers simple and rapid preparation of oxides with control of particle size and good overall homogeneity. The method avoids complex steps such as refluxing of alkoxides and is faster than many other techniques. The particle size of the precipitate is strongly dependent on pH and the molarity of the precursor solution. Precipitation or coprecipitation from aqueous salt solutions (nitrate, sulfate, chloride, perchlorate, etc.) by fine control of the pH yields a poorly crystallized hydroxide or an intimate mixture of the hydrated oxides, which is usually converted to oxides by heating.

The main advantage of the precipitation process is that a large number of nanoparticles can be synthesized [8]. However, the control of the particle size distribution is difficult because the growth of the crystal only depends on kinetic factors. Two stages are involved in the precipitation process. First, the concentration of the species reaches critical supersaturation, and a fast nucleation process takes place. The next process is characterized by a slow growth of the nuclei due to diffusion of the solutes to the surface of crystal, and the particles of compound are formed. Tartaj et al. [9] showed that to produce monodispersed iron oxide nanoparticles, these two stages should be separated; that is, nucleation should be avoided during the period of growth. Numerous studies have dealt with the influence of different factors in order to control particle size, magnetic characteristics, or surface properties. Some coprecipitation methods use aging of stoichiometric mixtures of components. A few examples of coprecipitation synthesis of metal magnetic oxides are presented in the following text.

The precipitation technique is probably the simplest and most efficient chemical method to obtain iron oxide particles such as magnetite (Fe_3O_4) or its oxidized form maghemite (γ-Fe_2O_3). Iron oxides are usually prepared by addition of alkali to iron salt solutions and letting the suspensions age. Tartaj et al. [9] noted that starting from ferrous and ferric hydroxides in aqueous media, spherical monomeric magnetite particles might be formed. The smallest particles can also be generated after adding poly(vinyl alcohol) to the iron salts. The size and shape of the nanoparticles can be tailored with relative success by adjusting pH, ionic strength, temperature, nature of the salts (perchlorates, chlorides, sulfates, and nitrates), or the Fe(II)/Fe(III) concentration ratio. Pure goethite (FeOOH) was synthesized using 1 M ferric nitrate and 10 M NaOH solutions vigorously stirred until the pH reached 12–12.5. The first controlled preparation of superparamagnetic iron oxide particles using alkaline precipitation of $FeCl_3$ and $FeCl_2$ was performed by Massart [8]. It is a rapid synthesis of homogeneous γ-Fe_2O_3 nanoparticles applicable for coating by a wide range of monomeric species, such as amino acids, R-hydroxyacids (citric, tartaric, and gluconic acids) hydroxamate, dimercaptosuccinic acid, or phosphoryl choline. Citrate ions allow a decrease in the diameter of citrate-coated nanoparticles.

Coprecipitation of iron and cobalt ions in alkaline medium (usually NaOH and NH_4OH) has proven to be a convenient and inexpensive synthesis route suitable to synthesize large batches of cobalt ferrite ($CoFe_2O_4$) with reasonable control of composition and particle size. However, local heterogeneous distribution of precipitant and metal ions during the reaction may lead to the precipitation with a relatively broad size distribution, low crystallinity, and highly reduced amount of product [10]. The homogeneous precipitation route was developed, starting with $FeCl_2 \cdot 4H_2O$, $Co(NO_3)_2 \cdot 6H_2O$, and using hexamethylenetetramine (HMT) as the precipitant. The $CoFe_2O_4$ powder with small particle size showed the tendency to aggregate owing to the magnetic attraction. Quasispherical instead of spherical particles were formed in case of weakly alkaline solution pH value (about 7.5), which stemmed the nucleation and growth rate of the spinel phase. Faster growth rate usually favors the formation of spherical particles as a result of less selective crystallographic growth direction. The saturation magnetization (M_S), coercivity (H_C), and remanence ratio (M_R/M_S) are partially changed compared to the properties in other published data. M_S changes from 13.6 to 61.5 emu/g, which is slightly lower than those reported for the bulk samples (>70 emu/g) [11].

In order to overcome difficulties with respect to precise control over the size and its distribution, Maaz et al. [12] used a nanometer-sized reactor to obtain homogeneous nanoparticles of nickel ferrites using 0.4 M $FeCl_3 \cdot 6H_2O$ and 0.2 M $NiCl_2.6H_2O$ solutions and 3 M NaOH as precipitating and pH controlling agent. Before dispersion, the particles were coated with surfactant sodium dodecyl sulfate or oleic acid and then dispersed in some carrier liquid like ethanol, methanol or ammonia. The advantage of this method over the others is that the production rate of $NiFe_2O_4$ nanoparticles, their size, and size distribution are admissible and there is no need for extra heat treatments.

Furthermore, applications in biology and medical diagnosis and therapy require the magnetic particles to be stable in water at neutral pH and physiological salinity. The colloidal stability of the fluid depends on the dimensions of the particles, which has to be sufficiently small to avoid the precipitation due to gravitation forces [12]. Additional restriction is whether these particles are going to be used for in vivo or in vitro applications. For in vivo application, the magnetic particles must be coated with a biocompatible polymer during or after the synthesis process to prevent the formation of large aggregates. In addition, they must be made from a nontoxic and nonimmunogenic material, with particle sizes small enough to remain in the circulation after injection and to pass through the capillary systems of organs and tissues, avoiding vessel embolism. They must also have a high magnetization so that their movement in the blood can be controlled with a magnetic field. For in vitro applications, the size restrictions are not as strict [13,14].

17.2.1.2 Microemulsion processing

The use of an inorganic phase in water-in-oil microemulsions has received extensive attention for the preparation of uniform and size-controlled magnet metal nanoparticles. A microemulsion is an isotropic and thermodynamically stable system formed by at least three components; two of them are nonmiscible, and the third one, called surfactant, is amphiphilic [15]. Several types of surfactants—cationic, anionic, or non-ionic—can be used, such as bis (2-ethylhexyl) sulfosuccinate, sodium dodecyl sulfate, cetyltrimethyl ammonium bromide, polyvinyl pyrrolidine, diethyl sulfosuccinate, pentadecaoxyethylene nonylphenyl ether, decaoxyethylene nonylphenyl ether, and poly(oxyethylene) (5) phenyl ether, among others. Organic solvents used for reverse micelle formation are usually alkanes or cycloalkanes with six to eight carbon atoms. The bulk properties of the surfactant, such as osmotic pressure, turbidity, solubility, surface tension, conductivity, and self-diffusion, depend on concentration of the solution.

Depending on the proportion of suitable components and hydrophilic–lipophilic balance value of the surfactant used, the formation of microdroplets can be in the form of oil-swollen micelles dispersed in the aqueous phase as the oil/water (O/W) microemulsion or water-swollen micelles dispersed in oil as the water/oil (W/O) microemulsion (reverse microemulsion). In the intermediate phase region between O/W and W/O microemulsions may exist bicontinuous microemulsions with aqueous and oil domains interconnected randomly in the form of sponge-like microstructures [16]. W/O microemulsion solutions are mostly transparent, isotropic liquid media with nanosized water droplets that are dispersed in the continuous oil phase and stabilized by surfactant molecules at the water/oil interface. A stabilizer (emulsifier) is a molecule that possesses both polar and nonpolar groups. The surfactant-stabilized nanocavities (typically in the range of 10 nm) provide a confinement effect that limits particle nucleation, growth, and agglomeration. The main advantage of the emulsion technology is the diversity of nanoparticles obtained by varying the nature and amount of surfactant and cosurfactant, the oil phase, or the reacting conditions. One of disadvantages of the microemulsion method lies in its high expense due to the large amounts of surfactant (as much as 20%–30%) added to the system. Another handicap is that the surfactant ensuring colloidal stability is adsorbed on the surface of nanoparticles, thereby decreasing their usability. The disadvantages may be prevented if the reaction takes place in the interior of micelles. Recent efforts are directed to decrease the amount of surfactant or even to avoid the use of surfactants at all. Another problem of using W/O microemulsions for nanoparticle synthesis is the separation and removal of some (high-boiling point) solvents from products and scale-up difficulties [3,17].

The preparation of magnetic metal oxides using microemulsion processing is based on inorganic salts ($FeCl_3$, $FeCl_2$, $CuCl_2$, etc.) in the oil phase mixed with an aqueous phase reducing agent (NH_4OH, $NaBH_4$, etc.) in oil phase, and followed by collision and coalescence of droplets, chemical reactions, and resulting precipitation of metal oxide [3]. The attractive interactions (percolation) between droplets play a dominant role in the particle nucleation and growth. Chemical reactions of metallic salt and reducing agent within the microdroplets may be very fast. However, they are often the rate-determining step in the overall reaction. The detailed mechanism of microemulsion processing is still not well understood. The preparation conditions such as pH, solution concentration, and mixing procedure influence the chemical composition, particle size, morphology, crystal structure, and, consequently, the magnetic properties. Salazar-Alvarez [18] has reported the synthesis of iron oxide nanoparticles using reverse emulsion. The number of aggregates is small and they are not sensitive to the surfactant concentration.

17.2.1.3 Sol-gel method

The sol-gel process is a suitable wet route for synthesis of nanostructure materials, including magnetic metal oxides [19]. The reaction usually involves the linkage of solution of metal alkoxides into a polymer matrix. This method is rather simple and based on the hydroxylation and condensation of molecular precursors in solution, that is, the formation of a sol, which gels after solvent removal or chemical reaction. Gel properties are dependent on the structure created during the sol stage. Base catalysis induces the formation of a colloidal gel, whereas acid catalysis yields a polymeric form of the gel. These reactions are performed at room temperature but heat treatments are needed to acquire the final crystalline state [20].

One of the advantages of the sol-gel method is the ability to coat surfaces of nanoparticles that allow the matrix to adhere. Since the sol-gel is based on the hydrolysis of metal alkoxides, it is possible to tailor various types of nanoparticles. There are, however, several disadvantages that prevent the use of sol-gel on the industrial scale. First, the hydrolysis rates of the precursors need to match well in order to produce the desired product and crystal phase. The final step of the reaction is calcination, which causes the agglomeration that results in polydispersity affected by temperature and time of heating. The formation of 3D oxide networks limits the formation of individual nanosize particles. In order to control and eliminate undesired reactions, protecting agents are frequently used but the systems later require additional purification. Finally, the control of particle morphology is difficult due to uneven solvent removal throughout the system.

Preparation of magnetic metal oxides mostly depends on the kinetic parameters that affect growth, hydrolysis, and condensation. Consequently, the structure and properties of the gel are influenced by solvent type, temperature, concentration of the precursors, pH, agitation, and so on [20,21]. The magnetic ordering is very sensitive to the size distribution, dispersion of the particles, and their volume fraction

[3]. Raileanu et al. obtained metal oxide monodisperse phase with good control of the particle size, microstructure, and homogeneity, maintaining stability and properties within the sol-gel matrix. Camponeschi et al. prepared iron oxides by the addition of common benzene sulfonate surfactant. The stable nanosized iron oxide particles were formed without formation of a 3D gel network [22].

Kumar et al. [23] prepared Ni-Zn ferrites by sol-gel reaction using Zn $(CH_3COO)_2 \cdot 2H_2O$, $Ni(NO_3)_2 \cdot 6H_2O$ dissolved by ethylene glycol and few drops of glycerol (to stabilize and increase the solubility of the solution) under stirring at 175°C. Isopropanol and triethylamine were added as catalysts. By heating at 150°C, the sol turns into gel, which becomes Ni-Zn ferrite during calcination at 350°C. Chen et al. [24] prepared $NiFe_2O_4$ nanoparticles using polyacrylic acid (PAA) as a chelating agent. Pure spinel $NiFe_2O_4$ nanoparticles with a diameter of 5–30 nm were produced using various molar ratios of PAA to total metal ions (0.5–2.0) by calcination of the gel precursors in air at 300°C for 2 h. $NiFe_2O_4$ particles exhibited saturation magnetization of 0.19 emu/mol at 298K.

Hu et al. [25] successfully synthesized Fe_3O_4 nanoparticles by the sol-gel method combined with annealing under vacuum. The starting material was ferric nitrate first dissolved in ethylene glycol with vigorous stirring. The sol was heated to 80°C and kept at the temperature to obtain brown gel, which was then aged and dried at 120°C. The xerogel was annealed at 200°C–400°C under vacuum. The as-synthesized Fe_3O_4 nanoparticles can be easily transformed into γ-Fe_2O_3 or α-Fe_2O_3 by treating in air or by annealing. The results indicate that the particle size, saturation magnetization, and coercivity of Fe_3O_4 increase with the increase of temperature.

17.2.1.4 Hydrothermal and solvothermal methods

Under hydrothermal conditions a broad range of nanostructure materials might be formed. This method exploits the solubility of almost all inorganic substances in water at elevated temperatures and pressures and subsequent crystallization of the dissolved material from the fluid. Hydrothermal methods rely on the ability of water to hydrolyze and dehydrate metal salts on elevated temperature. The hydrothermal reactions are performed in aqueous media in reactors or autoclaves where the pressure is higher than 1.4×10^4 kPa at temperatures above 200°C. The solubility of the resulting metal oxides in water may be very low to produce the supersaturation. The solvent is generally not limited to water and also includes other polar or nonpolar solvents (benzene). The process is in that case more appropriately called solvothermal synthesis. Recent trends in hydrothermal or solvothermal synthesis are directed toward more dispersed systems using solutions (wet systems) and/or gases (dry systems). During hydrothermal treatment metal cations initially precipitate in the form of polymeric hydroxides. Over time, these hydroxides undergo dehydration to form the metal oxide crystal structures. A detailed investigation of precursor concentration, temperature, time, and their effects on particle size and morphology was carried out. The time has a more significant influence on the average particle size than concentration; shorter time favors monodispersed particles.

Xu et al. investigated the influence of precursor (ferric nitrate) concentration on the size and shape of produced particles, indicating that at concentration of 0.03 M the small spherical particles of 11–20 nm were obtained. At higher ferric nitrate concentration (0.06 M) the particle size increased to 20–34 nm with mostly rhombic shape [26]. Lu et al. used the solid-liquid-solution hydrothermal synthesis for the preparation of variety of different nanocrystals [27]. The system of metal linoleate (solid), ethanol–linoleic acid (liquid), and water–ethanol (solution) phases was investigated at different reaction temperatures under hydrothermal conditions. Phase transfer and separation mechanism occurred at the interfaces of all phases and nanoparticles (9–12 nm) of Fe_3O_4 and $CoFe_2O_4$ were formed. Besides, monodisperse, hydrophilic, single-crystalline ferrite microspheres of 200–800 nm size were prepared starting from $FeCl_3$, ethylene glycol, sodium acetate, and polyethylene glycol. They were stirred vigorously to form a clear solution, then sealed in a teflon-lined stainless-steel autoclave, and heated at 200°C for 8–72 h [28]. The size and nature of the prepared α-Fe_2O_3 nanoparticles strongly depend on the pH, temperature, retention time, concentration of the complexing agent (oxalic acid), cationic surfactant, and alkali sources. Tetraethyl ammonium hydroxide was used to maintain the pH optimizing the crystallinity, size, and size distribution of the prepared nanoparticles [29].

Blanco-Gutierrez et al. [30] prepared $ZnFe_2O_4$ by hydrothermal synthesis, so-called microwave-assisted solvothermal method, starting from nitrates dissolved in ethylene glycol and KOH added until pH = 11. Later, a mixture was transferred into an autoclave to be treated under different solvothermal conditions (160°C for 2–288 h). Nanosize $ZnFe_2O_4$ with particles of 4–19 nm showed superparamagnetic and ferrimagnetic properties. Yu et al. [31] prepared $ZnFe_2O_4$ using metal Zn sheet and $FeCl_2$ in ammonia solutions at 180°C. As-prepared ferrite particles display high magnetization with maximum saturation magnetization of 61.2 emu/g at 80K and 54.6 emu/g at 300K, respectively. Puche et al. [32] prepared MFe_2O_4 (M = Ni and Zn) nanoparticles using iron and zinc or nickel nitrates dissolved in distilled water. By adding 2 M KOH solution until pH = 11, the iron, zinc, and nickel hydroxides were precipitated. The solution was heated in autoclave at 160°C (3–15 h) to obtain Zn-ferrite and at 180°C to obtain Ni-ferrite. $ZnFe_2O_4$ was superparamagnetic with maximum of susceptibility at 20K, while $NiFe_2O_4$ was superparamagnetic above room temperature. Below those temperatures, ferrites exhibited ferrimagnetic behavior arising from the total inversion of the ground state in $NiFe_2O_4$ to partially inverse character in $ZnFe_2O_4$. Hwang and Wang [33] prepared nanosized ferrites catalyst for CO_2 gas reduction. The $Mn(NO_3)_2 \cdot 6H_2O$, $Zn(NO_3) \cdot 6H_2O$, $Ni(NO_3)_2 \cdot 6H_2O$, and $Fe(NO_3)_3 \cdot 9H_2O$ salts were dissolved in distilled water and then neutralized with NH_4OH (or NaOH). The solution was hydrothermally treated in an autoclave at 150°C for 2 h. The MnZn ferrites and MnNi ferrites with particle size of less than 30 nm and high specific surface area were synthesized. The $(Mn_{0.33}Zn_{0.67})Fe_2O_4$ catalyst exhibited the best CO_2 decomposition efficiency.

Mazan et al. [34] prepared Mn-substituted cobalt ferrite nanoparticles by a fast-hydrothermal method with the advantage to significantly reduce the synthesis time, from 5 h (classical hydrothermal) to 30 min (fast hydrothermal). The method is based on a protective steel coating around an autoclave that permits better and faster heat transfer from the oil to the quartz autoclave. The $Co(NO_3)_2 \cdot 6H_2O$, $Fe(NO_3)_3 \cdot 9H_2O$, and $MnSO_4$ were dissolved in water and 1 M NaOH was used for precipitation. The temperature was adjusted to 200°C and after 30 min the autoclave was rapidly cooled. After filtering and washing in water and ethyl alcohol, the product was dried at 80°C for 4 h. The nanoparticles of 12–18 nm size showed a wide hysteresis loop, which is characteristic of hard magnetic materials. If the concentration of Mn in $CoFe_2O_4$ is increased, the saturation specific magnetization M_S increases and the coercive field H_C decreases. Liu et al. [35] prepared $CoFe_2O_4$ nanoplatelets and nanoparticles by a simple hydrothermal treatment of cobalt dodecyl sulfate, $FeCl_3$, and NaOH aqueous solution at 120°C. Compared with the H_C value of bulk $CoFe_2O_4$ reported in other studies (5400 Oe), the prepared $CoFe_2O_4$ nanocrystals exhibit much smaller coercivity (465 Oe), which could be attributed to their nanostructure.

17.2.1.5 Thermal decomposition processing

High-quality semiconductor nanocrystals and oxides were produced by thermal decomposition for years. Similar methods for the synthesis of magnetic particles with control over size and shape have been developed using thermal decomposition of organometallic compounds in high-boiling organic solvents containing stabilizing surfactants. The organometallic precursors include metal acetylacetonates, [M(acac)$_n$], (where M = Fe, Mn, Co, Ni, Cr; $n = 2$ or 3, and acac = acetylacetonate), metal cupferronates [M(Cup)$_x$] (where M = metal ion; Cup = N-nitroso phenylhydroxylamine, $C_6H_5N(NO)O$-), or carbonyls. Oleic acid and hexadecyl amine are often used as surfactants. Ratios of the starting reagents including organometallic compounds, surfactant, and solvent are the parameters that control the size and morphology of magnetic nanoparticles, but temperature, time of the reaction, and aging period may also be crucial. For example, addition of trimethylamine oxide at an elevated temperature results in formation of monodisperse γ-Fe_2O_3 nanocrystals, but further decomposition of precursors with cationic metal centers leads directly to Fe_3O_4 [27].

Peng et al. reported a decomposition approach for the synthesis of size- and shape-controlled magnetic oxide nanocrystals based on the pyrolysis of metal fatty acid salts in nonaqueous solution [27]. Hyeon et al. prepared monodisperse iron oxide nanoparticles using $FeCl_3$ and sodium oleate to generate an iron oleate complex in situ. Later, the complex was decomposed at temperatures of 240°C–320°C in different solvents, such as 1-hexadecene, octyl ether, 1-octadecene, 1-eicosene, or trioctylamine. Formation of nanoparticles (5–22 nm) depends on decomposition temperature and aging period [36]. The water-soluble magnetic nanoparticles are more desirable for applications in biotechnology. Recently, Li et al., using $FeCl_3 \cdot 6H_2O$ and 2-pyrrolidone as coordinating solvent, prepared water-soluble Fe_3O_4 nanocrystals in the range of 4–60 nm under reflux (245°C, 1–24 h) [37]. More recently, the same group developed a one-pot synthesis of water-soluble magnetite nanoparticles prepared under similar conditions by addition of α,β-dicarboxyl-terminated poly(ethylene glycol) as a surface-capping agent. Those nanoparticles might be potentially used as magnetic resonance imaging contrast agents for cancer diagnosis.

The thermal decomposing processing is also known as *surfactant mediated/template synthesis* because it may use neutral or charged template molecules. The hexagonal iron oxides were prepared using hexa-decylsulfonic acid mixed with an aqueous solution of $FeCl_2$. Lamellar iron oxide/surfactant composites were produced by the controlled precipitation and hydrolysis of aqueous iron cations into self-assembled iron/surfactant arrays [38]. They were yielded by adding ammonia to a $FeCl_3$ solution and mixing with sodium *n*-alkyl sulfate template. The α-Fe_2O_3 nanorods and nanotubes have been synthesized using a different surfactant, polyisobutylene bissuccinimide. High-quality one-dimensional crystalline products were obtained at 150°C using aqueous butanol solution as solvent and carbamide as base [39]. Besides long-chain surfactant templates, the inorganic anions like chloride, phosphate, or sulfate may be used.

17.2.1.6 Sonochemical methods

The sonochemical method has been extensively used to generate novel materials with unusual properties. The chemical effects of ultrasound have been found to be advantageous for the formation of new phases and to synthesize highly monodispersive nanoparticles. Consequently, sonochemistry is the processing in which molecules undergo chemical reactions due to the application of powerful ultrasound radiation (20 kHz to 10 MHz). The nanoparticle preparation is simple; the method operates at ambient conditions and it is easy to control the particle size. The physical phenomenon responsible for the sonochemical process is the acoustic cavitation. At very high temperatures, hot spots, generated by the rapid collapse of sonically generated cavities, allow the conversion of salts into nanoparticles. Since this collapse occurs in less than a nanosecond, a very high cooling rate of 10^{11} K/s hinders the organization and crystallization of the products and results in the amorphous phase. How the crystallization further occurred is not clear enough, and the theory of sonochemical processing is not completely understood. The advantage of sonochemistry is that the mixing of the constituent ions in the amorphous phase is on the atomic level, and the crystalline phase might be obtained by annealing at relatively low temperatures. Since the cavitation is a quenching process, the composition of particles formed is identical to the composition of the vapor in bubbles, which is very useful. Also, the mentioned method is favorable due to the absence of many reactants that usually remain as contaminants and require more chemicals or solvents to be removed. However, this method is still rarely used because the particles tend to agglomerate due to their high reactivity.

The sonochemical method is initially proposed for the synthesis of iron nanoparticles, but nowadays is used to synthesize various metal magnetic oxides. Sonochemical reactions of volatile organometallics have been exploited as a general approach by changing the reaction medium. Bang and Suslick [40] reported sonochemical synthesis of nanosized hollow hematite and Vijayakumar et al. [41]

considered preparation of pure superparamagnetic Fe_3O_4 powder with particle size of 10 nm and with low magnetization at room temperature. Superparamagnetic iron oxide nanoparticles with high magnetization have been synthesized from ferrofluids coated with oleic acid dispersed in chitosan.

17.2.1.7 Solution-combustion synthesis (autocombustion processing)

Combustion synthesis (CS), or self-propagating high-temperature synthesis (SHS), is an effective, low-cost method for production of various oxide materials. Recently, CS has become a very popular approach for preparation of nanomaterials [42]. The classification of CS processes is generally based on the physical nature of the initial reaction medium:

- Conventional SHS used as initial reactants in solid state (condensed phase combustion)
- SCS is processing where the initial reaction medium is aqueous solution
- Synthesis of nanoparticles in flame is known as gas-phase combustion

Our attention will be paid mostly to SCS, so-called autocombustion processing, as an important aim in the progress in SCS of magnetic metal oxides. In recent years many papers on SCS were published, but only the latest address the materials application. Generally, SCS is a versatile, simple, and rapid process, which involves a self-sustained reaction in homogeneous solution of different oxidizers (e.g., metal nitrates) and fuels (e.g., urea, glycine, hydrazides, citric acid). This process not only yields nanosize oxide materials but also allows uniform adding of small amounts of dopant ions. The reaction is affected by the nature of the organic complexing agent, stoichiometry of the solution, stability of the metal ion complex, and the flame temperature of the autocombustion reaction. Depending on the type of the precursors and on conditions used for the process organization, the SCS may occur as either volume-combustion synthesis (VCS) or layer-by-layer propagating combustion synthesis (LPL) modes.

The fuel is an important component for the preparation of oxides using SCS. Urea and glycine are the most popular and attractive fuels for producing highly uniform complex oxide powders with precisely controlled stoichiometry. The glycine nitrate process (GNP) is probably most used as environmentally compatible. However, a recent study by Pine et al. has shown that the amount of CO and NO_x gas products of incomplete combustion in the GNP process can appear as hazardous [43]. After the publication of the first paper on the concept of mixture of fuels [44], a large number of papers have been published on the use of the combination of fuels such as citric and succinic acids, citric acid and glycine, urea, monoethanolamine, alanine, and so on. Although complex fuels favor formation of nanosize particles, in many cases a further calcination is required to form organic free pure nanocrystalline powders.

The schematic presentation of the SCS in VCS mode is done in Fig. 17.2A. The process involves a reaction in an aqueous oxidizer–fuel solution uniformly heated by an external energy source (e.g., hot plate, electrical, or microwave oven). After

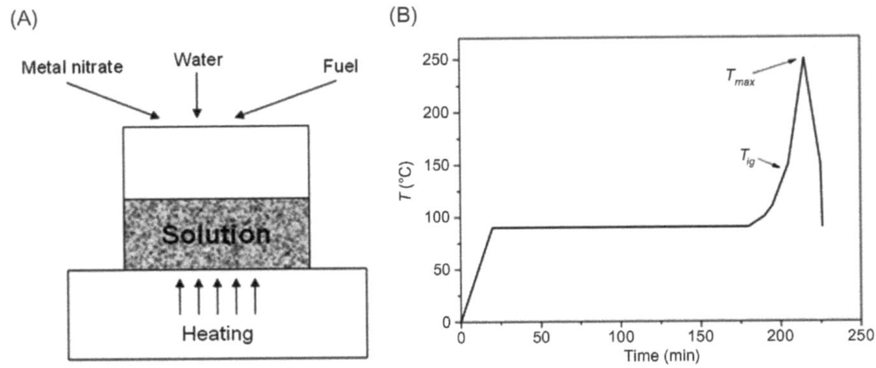

Figure 17.2 Schematic presentation of solution-combustion synthesis in volume-combustion synthesis mode (A) and temperature profile of the system during the synthesis process (B).

water evaporation, the temperature of formed viscous media rapidly increases, achieving ignition temperature (T_{ig} at Fig. 17.2B) at which reaction spontaneously initiates over the entire volume leading to the formation of solid product with the desired phase composition and process duration of 1–10 s. Characteristic phase transformations (melting or decomposition of oxidizer, fuel boiling) are typically responsible for the initiation of rapid chemical reactions. For example, the ignition temperature (T_{ig}) in the iron nitrite–glycine is equal to the decomposition temperature for $Fe(NO)_3$, while in systems where fuel is hydrazine, T_{ig} is the boiling point of the fuel. Ambient has also an important role because the interaction with oxygen in air occurs in a different stage of processing and different fuels do not act in the same way. Hydrazine starts to interact very fast in the first stage of processing, but glycine interacts with oxygen after ignition. By changing the ambient atmosphere, we can control the phase composition of the product. Magnetite (Fe_3O_4) is formed in argon, while maghemite (γ-Fe_2O_3) and hematite (α-Fe_2O_3) are formed in air. The subproduct gases of autocombustion can also influence the final product, which can be controlled by changing fuel/oxidizer ratio (F/O). When citric acid is used as a fuel, it is common to specify that as metal/citric acid ratio (M/C) or citric acid/metal ratio (C/M). Finally, combustion conditions depend on the chemical nature of the formed reactive solution [45].

The main advantages of autocombustion are fast and simple processing (formation results in oxide mixture within a few seconds), rather cheap starting materials (usually nitrate salts), no special equipment, and low energy consumption. It is possible to prepare an enormous diversity of oxide materials, including magnetic metal oxides. One of disadvantages is that it is not always easy to define the optimal F/O ratio. The amount of fuel calculated strictly according to stoichiometric F/O for combustion frequently results in a product of undesired composition. In those cases, the processing may require an empirical approach leading to a larger number of experiments. On the other hand, since the combustion mode is in essence a thermal explosion, it is difficult to control the process. For many systems, the product yield is low. Nevertheless, autocombustion processing is useful for the magnetic metal oxide preparation including ferrites [46].

Shutka et al. [47] prepared ferrites with the spinel-type $NiFe_2O_4$ (NiF), $CoFe_2O_4$ (CoF), and hexagonal $BaFe_{12}O_{19}$ (BaF) structure type using citric acid as complexant and fuel. The compounds were synthesized starting from $Fe(NO)_3 \cdot 9H_2O$, Co $(NO)_3 \cdot 6H_2O$, $Ni(NO_3) \cdot 6H_2O$, $Ba(NO_3)_2$, and $C_6H_8O_7 \cdot H_2O$ as raw materials. The molar ratio M/C = 1:1. The neutralization of the solution was performed by adding NH_4OH to pH = 7. Then the solution was evaporated by intensive stirring and heating at 90°C for 1–4 h; highly viscous gel was formed and dried at 90°C. The autocombustion reaction was initiated at 250°C and the product was calcined at 800°C for BaF and CoF and at 1000°C for NiF. The corresponding ferrites were of a high level of purity and nanosized.

Mixed zinc ferrites with the general formula $M_x^{2+}Zn_{1-x}^{2+}Fe_2O_4$ (where M is a divalent metallic ion) are already established as technologically important materials due to valuable dielectric and magnetic properties and recently with respect to catalytic performance. Khorrami et al. [48] prepared nanosize $ZnFe_2O_4$ powders using Fe(NO3)3 · 9H2O, Zn(NO3)2 · 4H2O in molar ratio 1:2 and CS(NH2)2 as fuel. The foamy gel kept at 350°C spontaneously ignited to puffy, porous brown powders, afterward calcined at 750°C–1000°C for 4 h. The average particle size was 17–25 nm. Lorentzou et al. [49] prepared $Mn_xZn_{1-x}Fe_2O_4$ investigating the effect of F/O and M/C on the products' phase composition. For both glycine and citric acid, single-phase $Mn_xZn_{1-x}Fe_2O_4$ was formed, poorly or strongly crystallized depending on the ratios. In the case of glycine, for the ratio F/O 1/1, the reaction proceeds with a bright flame and violent expulsion of gases; formed dendrites are soft and friable, consisting of very well-crystallized ferrite phase. The higher ratios (2/1, 3/1) affected the change of combustion mode: combustion took place much slower, not in the explosion regime, and without a bright flame. The product was significantly less crystallized. Similar results were obtained for citric acid/oxidizer (metal) ratio (C/M = 1/1), that lead very easily to ignition and reaction propagation. The higher ratios (2/1, 3/1) induced slower combustion with formation of a much less foamed product and more sintered powder, whereas lower ratios (1/3) could not ignite the reaction, and it was necessary to sustain it with an external energy source.

Azadmanjiri et al. [50] prepared $NiFe_2O_4$ using $Fe(NO_3)_3 \cdot 9H_2O$, Ni $(NO_3)_2 \cdot 6H_2O$, $C_6H_8O_7 \cdot H_2O$, and NH_4OH with M/C = 1:1 at pH = 7, heated and stirred to transform into a xerogel. After ignition, dried gel was burned into powder, which was further calcined. Saturation magnetization at room temperature was 54 emu/g. Costa et al. [51] prepared NiZn ferrites with a nominal composition of $Ni_{0.5}Zn_{0.5}Fe_2O_4$ using urea as a fuel. The large quantity of developed gases inhibited particle aggregation and yielded soft powders suitable for dispersion and further use. Influence of calcination temperature on the morphology and magnetic properties of NiZn ferrite applied as an electromagnetic energy absorber was investigated. The powders calcined at 1200°C were suitable as narrow band frequency absorbers, while the powders calcined at 800°C and 1000°C were suitable for use as wide band absorbers. Sutka et al. [52] investigated the influence of Zn-ions on the $NiFe_2O_4$ p-type semiconductor gas response. The metal nitrates and citric acid with M/C = 1/1 and pH = 7 were stirred at 80°C until formation of a highly viscous gel, which was dried and then heated to 250°C to initiate a self-sustaining autocombustion reaction. The Zn-ion increased the gas-sensing properties, while the Ni-ion decreased them. Dzunuzovic et al. [53] prepared NiZn ferrites, with the general formula $Ni_{1-x}Zn_xFe_2O_4$ starting from metal nitrates and citric acid. Well-crystallized NiZn-ferrite was formed. The influence of the grain size on magnetic properties was studied. The saturation magnetization and remnant magnetization continuously increase up to $x = 0.3$ of Zn, and then decrease for more Zn. The values of 49 emu/g for NF, 79 emu/g for $x = 0.3$, and 76 emu/g for $x = 0.5$ indicate the effect of Zn ions. The presence of Zn promotes sintering, leads to the increase of grain size and has the influence on the coercivity of Zn-substituted NF materials.

Raval et al. [54] prepared nanosized cobalt ferrite starting with cobalt nitrate, iron nitrate, and citric acid in the deionized water in order to minimize the impurities in the final product. The sol was prepared under the molar ratio M/C = 1:1 and pH = 7. The $CoFe_2O_4$ ferrite was formed after self ignition and as-burnt powder was calcined at 950°C. Bahadur et al. [55] prepared single-domain barium ferrite. The citric acid was used as a fuel with ratio F/O or C/M = 1/1, 2/1, and 3/1. The ratio 2:1 gave the best results with respect to crystallinity, narrow size particle distribution (80–100 nm), and magnetic properties. Maximum magnetization and coercivity values at room temperature were 55 emu/g and 5000 Oe.

17.2.2 Physical methods

Physical methods are generally based on reaction in solid, gas, vapor, or plasma state. According to Tavakoli et al. [3] physical methods for the synthesis (magnetic) metal oxides are aerosol synthesis, gas phase deposition, pulsed laser ablation, laser induced pyrolysis, powder ball milling, and electron beam lithography. Coey et al. [2] pointed out that physical methods are based on thermal evaporation, electron beam evaporation, molecular beam epitaxy, and sputtering atomic layer deposition, which are mostly related to the preparation of films. However, for the magnetic metal oxides, especially for iron-based oxides and ferrites, the most used physical method is high-energy ball milling (mechanical activation). The important methods for the synthesis are also *solid-state reaction and vaporization*, as the transition between physical and chemical methods.

The vaporization method has been frequently used for the preparation of nanoparticles of the iron-based oxides, in which the target materials are vaporized by heat source and then rapidly condensed. The vaporization process can be indicated as a physical or chemical method, depending on the existence of a chemical reaction. If the resultant nanoparticles have the same composition as the target materials, they are prepared by physical vapor condensation (PVC). However, nanoparticles having a composition different from the target are usually obtained by chemical vapor condensation (CVC), because the chemical reaction occurs during the process. The CVC process has a benefit in controlling the composition, whereas the PVC products have higher purity. Of particular significance is the synthesis of nanopowders by thermal decomposition of metal-organic precursors using a focused laser beam, combustion flame, or plasma torch as heat source, and evaporation and condensation of volatile species in a reduced pressure environment. Disadvantages of these methods are low concentration of precursor in the carrier gas that minimizes the collision frequency between the clusters formed during the short residence time (less than 0.1 s), rapid expansion of the gas stream in the reactor tube, and rapid quenching of the nucleated clusters or nanoparticles after leaving the reactor tube. Lee et al. [56] studied the formation of iron oxide nanoparticles using pentacarbonyl iron vaporized in the condensation system. The rapid cooling allowed preparation of smaller iron oxide particles with narrower size distribution. The slow cooling caused coalescence of the particles inside the chamber because they could keep the resident heat energy even in the smoky state. Oh et al. [57] investigated the magnetic properties of iron oxide nanoparticles synthesized by the CVC process observing that the synthesized particles were nearly spherical for α-FeOOH, γ-FeOOH, and Fe_3O_4, but not for α-Fe_2O_3.

Laser-driven (or induced) pyrolysis of organometallic precursors allows the synthesis of nanoscale particles (2–20 nm) at rapid heating and cooling rates. Laser pyrolysis is based on the resonant interaction between laser photons and at least one of the gaseous species. The laser-induced pyrolysis offers the advantage of potentially clean process producing. The effects of the process conditions on the structural and magnetic properties of γ-Fe_2O_3 nanoparticles produced by laser pyrolysis have been studied, demonstrating that the size and crystallinity of the maghemite nanoparticles could be changed by the laser power and oxygen mole ratio in the carrier gas [58].

Solid-state processing is based on mechanically induced reaction (ball milling, e.g., mechanical activation or mechanical alloying) and solid-state reaction. Mechanical processing, in particular high-energy ball milling, is a convenient way to produce nanosize powders and is the most common method reported in the literature for the synthesis of numerous compounds. The mentioned method is a mechanical process that requires high energy. If the chemical reaction appears during the high-energy ball milling resulting in a new product, this process is known as the mechanochemical reaction. Compared to producing metal magnetic oxides through an unstable combustion reaction such as SHS, the redox reactions induced by reactive milling are much slower than those achieved during SHS. In spite of simplicity and efficiency of ball milling synthesis, there are some problems and limitations. The microstructure of the milling products is very sensitive to the grinding conditions and the product may be unpredictably followed by unwanted contamination from the milling media and atmosphere. Besides, obtaining small particles or new product may require extensive milling time. The milling time affects the size, size distribution, coercivity, magnetization, chemical composition, and morphology of magnetic metal oxides. Sepelak et al. [59] prepared Zn spinel ferrite by the high-energy milling of the ZnO/Fe_2O_3 mixture, and compared their properties with Zn ferrite synthesized by the conventional method (mechanical activation and thermal treatment). The powder mixture containing 66.34% α-Fe_2O_3 and 33.76% ZnO were milled in the planetary ball mill, the ball-to-powder weight ratio was 20:1, and milling time was 120 min. Both the high-energy milling of the Fe_2O_3/ZnO mixture and the conventionally prepared $ZnFe_2O_4$ results in the formation of nanocrystalline zinc ferrite. Druska et al. [60] investigated the formation of zinc ferrite by solid-state reaction using ball milling. The formation of the metastable inverse spinel structure was caused by mechanical activation or mechanochemical synthesis. In both cases, the inversion degree increases with increasing milling time.

Zak et al. [61] studied nickel ferrite prepared by a simple high-yielding and low-cost solid-state reaction route, using $NiSO_4 \cdot 6H_2O$, $Fe(NO_3)_3 \cdot 9H_2O$, NaOH, and NaCl as precursors. Solid precursors were mixed in the molar ratio of 1:2:8:10, and ground about 60 min. During the mixing process, exothermic reaction took place yielding a mixture of nickel and ferrite oxides. The calcination was performed at 700°C–800°C for 4 h. The formation of magnetic $NiFe_2O_4$ at 800°C was noticed.

17.2.3 Biological methods

According to Hasani et al. [17], protein-mediated, fungi-mediated, and bacteria-mediated synthesis routes belong to biological methods. Almost 70% of all biological methods for the preparation of magnetic metal oxides are related to protein-mediated processing and most of the published references are related to it. Bacteria-mediated methods make about 20% of the total number of published references dealing with biological methods.

In contrast to purely chemical procedures for the manufacture of magnetite particles, microbial reactions are characterized by their selectivity and precision for magnetite formation. In most bioprocesses it is assumed that the highly specific structures capable to drive highly specific interactions with the culture medium exist on the bacteria membrane (enzymes, proteins, etc.). Tavakoli et al. [3] pointed out that the use of bacteria as a novel biotechnology may facilitate the production of nanoparticles. A bacterially mediated electrochemical process was used to synthesize metal (Co, Cr, or Ni)-substituted magnetic powder employing iron (III)-reducing bacteria under anaerobic conditions. It was suggested that the bacteria might be viewed as a nonspecific source of electrons or as a catalyst supplying electrons for the crucial step of splitting the substrate (reduction of Fe (III)). In contrast to classical catalysts, however, the bacteria can extract some of the electrochemical energy in order to live, the electrons will be taken away from their surroundings, and energy is sufficient to precipitate the magnetite. It appears that the bacteria acts as an electrode that is substantially indifferent to the metal ions present in the surrounding medium [62]. The possible advantages of this method may include biologically facilitated production of magnetite that does not require the addition of exogenous electron carrier substances such as humid acids. However, for the sufficient understanding of the biological reactions and preparation of magnetic nanoparticles and materials, it is needed to be able to control them. New materials and understanding of their interaction with the environment (e.g., human body) may lead to better biocompatible nanomagnets.

Phumying et al. [63] propose the synthesis MFe_2O_4, with M = Ni, Co, Mn, Mg, and Zn based on an aloe vera plant-extracted solution hydrothermal route starting from metal acetylacetonates. This novel biosynthetic route is very simple and provides high-yield nanosize oxide materials with crystalline structure and acceptable properties. However, high calcination temperature is needed to convert the precursor into crystalline materials. The $ZnFe_2O_4$ obtained by this processing possesses plate-like structure of networked nanocrystalline particles. The specific saturation magnetization values at 10 kOe for $NiFe_2O_4$, $CoFe_2O_4$, $MnFe_2O_4$, $MgFe_2O_4$, and $ZnFe_2O$ were 31.9, 55.3, 52.4, 68.9, and 78.1 emu/g, respectively.

17.3 Synthesis of multiferroic materials

After an initial burst of interest, research in multiferroic materials remained static until early 2000. In 2003 the discovery of large ferroelectric polarization in epitaxial grown thin films restimulated activity in the field of multiferroics. In recent years, considerable attention has been drawn again because multiferroic materials have great potential for different applications in the new electronic devices, primarily for information storage devices such as multiple state memories (writing/reading/deleting data), sensors, spintronic devices, actuators, transducers, filters, attenuators, satellite communication, bubble devices, permanent magnets, and so on. MFs are formally defined as materials that exhibit more than one primary ferroic order parameter simultaneously. When two of the primary ferroic properties such as ferroelectric, ferromagnetic, or ferroelastic are united within one phase, it is known as single multiferroic. A composite of, for example, ferroelectric and ferromagnetic phase is called composite multiferroic [64,65].

$BiFeO_3$ (BFO) as a single multiferroic holds much potential, because magnetoelectric coupling occurs even at room temperature. The use of $BiFeO_3$ presents a few problems. One of the primary concerns regarding the synthesis of such single-phase multiferroic materials is to avoid formation of impurities. The impurities are caused by three different reasons: the evaporation of Bi component occurs easily due to low decomposition temperature of bismuth compounds, the chemical valence of Fe ion varies in an oxygen-deficient atmosphere and, according to the phase diagram, the thermodynamic area in which BFO can occur is very narrow. $Bi_2Fe_4O_9$ and $Bi_{25}FeO_{39}$ are the usual impurities. The wet-chemical route might be appropriate for the pure-phase synthesis of $BiFeO_3$. Kea et al. [66] proposed a simple chemical coprecipitation method using $Bi(NO_3)_3 \cdot 5H_2O$ and $Fe(NO_3)_3 \cdot 9H_2O$ powders dissolved in HNO_3. To precisely control the chemical precipitation process, the pH value was adjusted by ammonia. The production of impurities was avoided by controlling the processing parameters. The resultant BFO powders showed weak ferromagnetic behavior. Cai et al. [67] used the hydrothermal method, starting with Bi $(NO_3)_3 \cdot 5H_2O$, $Fe(NO_3)_3 \cdot 9H_2O$, HCl, PVA, and KOH. The addition of PVA stabilized the $Bi_{25}FeO_{40}$ phase. After the hydrothermal reaction, the autoclave was cooled and after filtering and washing the product was dried at 70°C. The phase evolution from sillenite $Bi_{25}FeO_{40}$ to perovskite $BiFeO_3$ was required. The prolongation of reaction time, temperature, and increase of KOH concentration may provide more $[Fe\ (OH)_4]^-$ ions that react with $Bi_{25}FeO_{40}$, resulting in BFO formation.

Lukovic Golic et al. [68] synthesized BFO powders by hydroevaporation and sonochemically assisted hydrothermal methods starting with $Bi(NO_3)_3 \cdot 5H_2O$, Fe $(NO_3)_3 \cdot 9H_2O$, and HNO_3. The solution was heated at 80°C under magnetic stirring until complete evaporation. The precursor powder obtained after drying at 90°C for 2 h was calcined at 620°C–770°C for 6 h and BFO was formed. Another method used was generally similar to the procedure applied by Cai et al., although an ultrasonic treatment was introduced before hydrothermal synthesis performed at 200°C on 0.5 MPa. However, BFO obtained by both methods contained small amounts of secondary phases, $Bi_2Fe_4O_9$ and $Bi_{25}FeO_{39}$. BFO showed antiferromagnetic behavior at room temperature, but at 5K it exhibited weak ferromagnetism. Hardy et al. [69] developed an aqueous solution-gel route based on aqueous Bi^{3+}- and Fe^{3+}-citrate solutions. A homogeneous, amorphous precursor gel was prepared by evaporation of the precursor solution. The addition of an excess of Bi^{3+} decreased the crystallization temperature, but also introduced a Bi_2O_3 secondary phases. Masoudpanah et al. [70] used the thermal decomposition method to investigate the effect of the ethylene glycol to metal nitrate molar ratio on the phase evolution, morphology, and magnetic properties. The starting materials were $Fe(NO_3)_3 \cdot 9H_2O$, $Bi(NO_3)_2 \cdot 5H_2O$, 1,2-ethanediol ($OH(CH_2)_2OH$), and HNO_3. The single phase $BiFeO_3$ has been synthesized by thermal decomposition of a glyoxylate complex achieved by the redox reaction between ethylene glycol and nitrate anions. The $BiFeO_3$ nanoparticles have quasispherical shape with the average particle size of 45 nm. With increasing of the ethylene glycol to nitrate ions molar ratio, the impurity $Bi_{25}FeO_{39}$ phase disappeared and the saturation magnetization increased from 1.3 to 3.2 emu/g. Furthermore, addition of H_2O_2 enhanced the photo-catalytic performance of BFO nanoparticles.

Das et al. [71] obtained BFO by microemulsion and sonochemical techniques in the temperature range between 400°C and 500°C. The starting materials in microemulsion processing included metal nitrates, NH_4OH, PEG 400, cationic surfactant, *iso*-octane, and butanol. In the sonochemical method, $Bi(NO_3)_3$ and $Fe(NO_3)_3$ solutions were mixed under sonication and the pH of the solution was kept at ~1. A small amount of decalin was added. Low volatility and low viscosity increase the energy transferred into the system. The phase formation temperature by microemulsion route is slightly lower compared to the temperature for sonochemically prepared powders. However, the sonochemical preparation route is relatively simple and does not require obnoxious and hazardous chemicals compared to the microemulsion route. Ilic et al. [72] prepared BFO precursor powders by two methods: autocombustion and soft chemical route. Both techniques are very simple, fast, and low cost. Chemicals used for synthesis were $Bi(NO_3)_3 \cdot 6H_2O$, $Fe(NO_3)_3 \cdot 9H_2O$, NH_4OH, citric acid, sucrose, urea, and tartaric acid as complexing agents and fuels. Soft-route synthesized ceramics possess smaller grains than autocombustion synthesized ceramics. The densities achieved by soft route were larger, but resistivity was rather low. BFO ceramics showed antiferromagnetic properties.

Nowadays multiferroic composites are among most attractive multifunctional materials. The ferroelectric-ferromagnetic composites, as two-phase multiferroic materials, are desired not only for the fundamental research of the magnetoelectric effect, but also for the potential applications in many electronic devices. The most widely studied systems correspond to Co or Ni ferrites, combined with lead zirconium titanate, lead niobium titanate, barium titanate (BT), or barium strontium titanate. Among them, the NiZn ferrites-$BaTiO_3$ systems (NZF-BT) are highly investigated because of high electrical resistivity, chemical stability, and excellent electromagnetic properties of the NiZn ferrites; and high permittivity, low dielectric loss, and high tunability of BaTiO3. Those composites have attracted much attention as a new class, expanding their use in other areas, such as drug delivery, heterogeneous catalysis, levitated railway system, magnetic refrigeration, microwave devices, antennas, and so on. To obtain the composite multiferroics, several routes for conventional material fabrication are being applied. Popular techniques within the multiferroic community are solid-state synthesis, hydrothermal synthesis, sol-gel processing, vacuum-based deposition, or other wet chemical synthesis methods. However, some types of multiferroic composites require specific processing conditions within more appropriate techniques. Consequently, they request methods for the synthesis of both components, ferroelectric and ferromagnetic.

A number of different methods were used for the preparation of composite multifferoic NiZn ferrites-$BaTiO_3$. Dzunuzovic et al. [73] prepared multiferroic composites NiZn ferrite—barium titanate from nanopowders obtained by autocombustion technique. It is a convenient method to obtain a pure phase of ferrites but for barium titanate it is more complex due to the possible appearance of secondary phases. Those phases, even in very low concentration, may later affect the change of the required properties of multiferroic composites. The multiferroic composite with the general formula $yNi_{1-x}Zn_xFe_2O_4-(1-y)BT$ was prepared by mixing NZF and BT powders in a liquid medium in the ball mill. The primary particle size was less than 100 nm for both, but the ceramic composites possessed differently shaped grains of around 1 μm in diameter. The hysteresis loop indicates the ferroelectric behavior of the composites with 70 and 90 mass amount of barium titanate phase (Fig. 17.3). Saturation magnetization and remnant magnetization continuously decrease with a barium titanate phase increase, and in comparison to pure nickel zinc ferrite (Fig. 17.4). These curves are typical for soft magnetic materials [74], since in these composites the only carrier of the magnetic properties is nickel zinc ferrite.

17.4 Summary

Various methods for the magnetic metal oxide synthesis based on the process media, including vapor, liquid, and solid-state processing routes are presented. The preparation methods are classified into physical, chemical, and biological methods. The physical methods are based on subdivision of bulk component, including mechanical crushing, pulverization, grinding, milling, mechanical activation of bulk

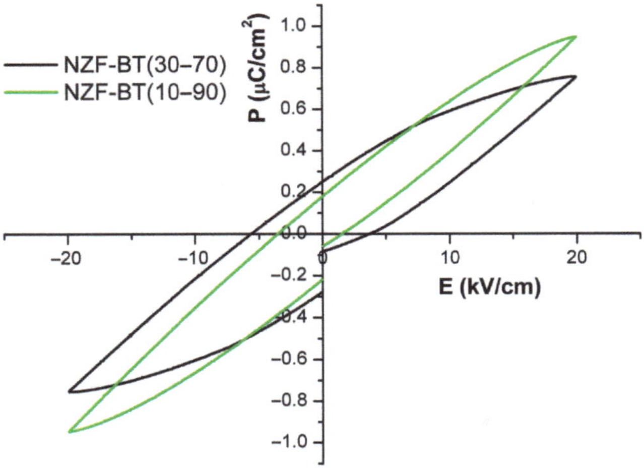

Figure 17.3 Ferroelectric hysteresis loops of NZF-BT composites.

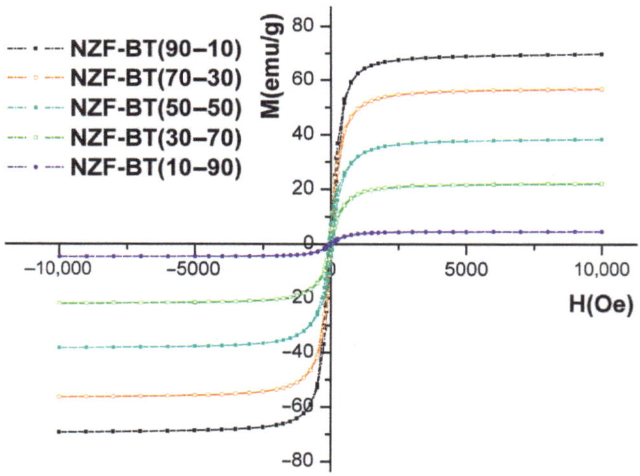

Figure 17.4 Magnetic properties of NZF-BT composites at room temperature.

component, and so on. The produced metal oxide particles are usually large in size and have wide size distribution. Solid-state reaction and mechanical activation are probably the most convenient methods for the preparation of larger amounts of product and belonging to both physical and chemical processing, they are often used in obtaining magnetic metal oxides. The chemical methods are based on precipitation or decomposition of precursors, leading to formation of new compounds usually with a narrow size distribution. Increasing interest in chemical synthesis, especially of nanoparticles is clearly indicated by the number of published references on this subject in recent years. The preparation of iron-based magnetic oxides is shown through various chemical processing, such as coprecipitation, sol-gel, hydrothermal, sonochemical, thermal decomposition, and solution autocombustion. The application of biological methods is still slightly neglected due to several serious factors that may influence them. Finally, attention is paid to multiferroic materials, both single and composite such as bismuth ferrites and composites based on ferrites-ferroelectric compounds.

References

[1] B. Cullity, C. Graham, Introduction to Magnetic Materials, second ed., IEEE Press, John Wiley & Sons, New Jersey, 2009.
[2] J. Coey, M. Viret, S. Von Molnar, Mixed valence manganites, Adv. Phys. 48 (2) (1999) 167–293.
[3] A. Tavakoli, M. Sohrabi, A. Kargari, A review of methods for synthesis of nanostructured metals with emphasis on iron compounds, Chem. Pap. 61 (3) (2007) 151–170.
[4] R. Lane, B. Craig, W. Babcock, Materials engineering with nature's building blocks, Newslett. Spring 6 (2002) 31–37.
[5] L. Liz-Marzan, P. Kamat, Nanoscale Materials, Kluwer Academic Publishers, Boston, 2003.
[6] C. Burda, X. Chen, R. Narayanan, M. El-Sayed, Chemistry and properties of nanocrystals of different shapes, Chem. Rev. 105 (4) (2005) 1025–1102.
[7] D. Huber, Synthesis, properties, and applications of iron nanoparticles, Small 1 (5) (2005) 482–501.
[8] M. Mohapatra, S. Anand, Synthesis and applications of nano-structured iron oxides/hydroxides – a review, Inter. J. Eng. Sci. Techn. 2 (8) (2010) 127–146.
[9] P. Tartaj, M. del Puerto Morales, S. Veintemillas-Verdaguer, T. Gonzalez-Carreno, C. Serna, The preparation of magnetic nanoparticles for applications in biomedicine, J. Phys. D Appl. Phys. 36 (2003) R182–R197.
[10] Z. Chen, L. Gao, Synthesis and magnetic properties of $CoFe_2O_4$ nanoparticles by using PEG as surfactant additive, Mater. Sci. Eng. B. 141 (2007) 82–86.
[11] Z. Liu, X. Li, Y. Leng, J. Lian, S. Liu, Z. Xiu, et al., Homogeneous precipitation synthesis and magnetic properties of cobalt ferrite nanoparticles, J. Nanomater. 2008 (2008) 1–4.
[12] K. Maaz, S. Karim, A. Mumtaz, S. Hasanain, J. Liu, J. Duan, Synthesis and magnetic characterization of nickel ferrite nanoparticles prepared by co-precipitation route, J. Magn. Magn. Mater. 321 (2009) 1838–1842.
[13] R. Langer, New methods of drug delivery, Science. 249 (1990) 1527–1533.
[14] J. Joubert, Magnetic microcomposites as vectors for bioactive agents: the state of art, An. Quim. 93 (1997) S70–S76.
[15] J. Lopez-Perez, M. Lopez-Quintela, J. Mira, J. Rivas, S. Charles, Advances in the preparation of magnetic nanoparticles by the microemulsion method, J. Phys. Chem. B. 101 (41) (1997) 8045–8047.
[16] I. Capek, Preparation of metal nanoparticles in water-in-oil (w/o) microemulsions, Adv. Colloid. Interface 110 (2004) 49–74.

[17] S. Hasany, I. Ahmed, J. Rajan, A. Rehman, Systematic review of the preparation techniques of iron oxide magnetic nanoparticles, Nanosci. Nanotech 2 (6) (2012) 148–158.
[18] G. Salazar-Alvarez, Synthesis, characterization and application of iron oxide nanoparticles, Doctoral Thesis, Stockholm, Sweden, 2004.
[19] Z. Dai, F. Meiser, H. Mohwald, Nanoengineering of iron oxide and iron oxide/silica hollow spheres by sequential layering combined with a sol-gel process, J. Colloid. Interface Sci. 28 (1) (2005) 298–300.
[20] L. Duraes, B. Costa, J. Vasques, J. Campos, A. Portugal, Phase investigation of as-prepared iron oxide/hydroxide produced by sol gel synthesis, Mater. Lett. 59 (7) (2005) 859–863.
[21] K. Kojima, M. Miyazaki, F. Mizukami, K. Maeda, Selective formation of spinel iron oxide in thin films by complexing agent-assisted sol-gel processing, J. Sol-Gel Sci. Techn. 8 (1997) 77–81.
[22] M. Raileanu, M. Crisan, C. Petrache, D. Crisan, A. Jitianu, M. Zaharescu, et al., Sol-gel Fe_xO_y-SiO_2 nanocomposites, Rom. J. Phys. 50 (5-6) (2005) 595–606.
[23] P. Kumar, P. Mishra, S. Kumar Sahu, Synthesis of Ni-Zn ferrites using low temperature sol-gel process, Inter. J. Sci. Engin. Res. 2 (8) (2011).
[24] D. Chen, X. He, Synthesis of nickel ferrite nanoparticles by sol-gel method, Mater. Res. Bull. 36 (2001) 1369–1377.
[25] J. Xu, H. Yang, W. Fu, K. Du, Y. Sui, J. Chen, et al., Preparation and magnetic properties of magnetite nanoparticles by sol–gel method, J. Magn. Magn. Mater. 309 (2007) 307–311.
[26] C. Xu, J. Lee, A. Teja, Continuous hydrothermal synthesis of lithium iron phosphate particles in subcritical and supercritical water, J. Supercrit. Fluid. 44 (1) (2008) 92–97.
[27] A. Lu, E. Salabas, F. Schuth, Magnetic nanoparticles: synthesis, protection, functionalization, and application, Angew. Chem. Int. Edit 46 (2007) 1222–1244.
[28] L. Cote, A. Teja, A. Wilkinson, Z. Zhang, Continuous hydrothermal synthesis and crystallization of $CoFe_2O_4$ nanoparticles, Fluid Phase Equilibr. 210 (2003) 307–317.
[29] S. Giri, S. Samanta, S. Maji, S. Ganguli, A. Bhaumik, Magnetic properties of α-Fe_2O_3 nanoparticle synthesized by a new hydrothermal method, J. Magn. Magn. Mater. 285 (2005) 296–302.
[30] V. Blanco-Gutierrez, E. Climent-Pascual, M. Torralvo-Fernandez, R. Saez-Puche, M. Fernandez-Diaz, Neutron diffraction study and superparamagnetic behavior of $ZnFe_2O_4$ nanoparticles obtained with different conditions, J. Solid State Chem. 184 (2011) 1608–1613.
[31] S. Yu, T. Fujino, M. Yoshimura, Hydrothermal synthesis of $ZnFe_2O_4$ ultrafine particles with high magnetization, J. Magn. Magn. Mater. 256 (2003) 420–424.
[32] R. Puche, M. Fernandez, V. Gutierrez, R. Gomez, V. Marquina, M. Marquina, et al., Ferrites nanoparticles MFe_2O_4 (M = Ni and Zn): hydrothermal synthesis and magnetic properties, Bol. Soc. Esp. Ceram. V 47 (3) (2008) 133–137.
[33] C. Hwang, N. Wang, Preparation and characteristics of ferrite catalysts for reduction of CO_2, Mater. Chem. Phys. 88 (2004) 258–263.
[34] P. Vlazan, I. Miron, P. Sfirloaga, Cobalt ferrite substituted with Mn: synthesis method, characterization and magnetic properties, Ceram. Inter. 41 (2015) 3760–3765.
[35] Q. Liu, J. Sun, H. Long, X. Sun, X. Zhong, Z. Xu, Hydrothermal synthesis of $CoFe_2O_4$ nanoplatelets and nanoparticles, Mater. Chem. Phys. 108 (2008) 269–273.
[36] J. Park, K. An, Y. Hwang, J. Park, H. Noh, J. Kim, et al., Ultra-large-scale syntheses of monodisperse nanocrystals, Nat. Mater. 3 (12) (2004) 891–895.

[37] Z. Li, Q. Sun, M. Gao, Preparation of water-soluble magnetite nanocrystals from hydrated ferric salts in 2-pyrrolidone: mechanism leading to Fe_3O_4, Angew. Chem. Int. Edit. 44 (2005) 123–126.
[38] F. Hu, L. Wei, Z. Zhou, Y. Ran, Z. Li, M. Gao, Preparation of biocompatible magnetite nanocrystals for in vivo magnetic resonance detection of cancer, Adv. Mater. 18 (19) (2006) 2553–2556.
[39] L. Lu, K. Hui-Zhong, M. Wenling, L. Huajie, W. Yuqiu, Surfactant-assisted synthesis of γ-Fe_2O_3 nanotubes and nanorods with shape-dependent magnetic properties, J. Phys. Chem. 110 (2006) 15218–15223.
[40] J. Bang, K. Suslick, Sonochemical synthesis of nanosized hollow hematite, J. Amer. Chem. Soc. 129 (2007) 2242–2243.
[41] R. Vijayakumar, Y. Koltypin, I. Felner, A. Gedanken, Sonochemical synthesis and characterization of pure nanometer-sized Fe_3O_4 particles, Mater. Proper. Microstruc. Process. 286 (1) (2000) 101–105.
[42] K. Patil, S. Aruna, S. Ekambaram, Combustion synthesis, Curr. Opin. Solid St. M 2 (1997) 158–165.
[43] T. Pine, X. Lu, R. Daniel, G. Mumm, J. Brouwer, Emission of pollutants from glycine–nitrate combustion synthesis processes, J. Amer. Ceram. Soc. 90 (2007) 3735–3740.
[44] S. Aruna, K. Rajam, Mixture of fuels approach for the solution combustion synthesis of Al_2O_3–ZrO_2 nanocomposite, Mater. Res. Bull. 39 (2004) 157–167.
[45] A. Mukasyan, P. Dinka, Novel approaches to solution–combustion synthesis of nanomaterials, Inter. J. Self-Prop. High-Temp. Synth. 16 (1) (2007) 23–35.
[46] K. Deshpande, A. Mukasyan, A. Varma, Direct synthesis of iron oxide nanopowders by combustion approach: reaction mechanism and properties, Chem. Mater. 16 (24) (2004) 4896–4904.
[47] A. Shutka, G. Mezinskis, A. Pludons, Preparation of dissimilarly structured ferrite compounds by sol–gel auto-combustion method, Chem. Techn 54 (1) (2010) 41–46.
[48] S. Khorrami, F. Gharib, G. Mahmoudzadeh, S. Sadat Sepehr, S. Sadat Madani, N. Naderfar, et al., Synthesis and characterization of nanocrystalline spinel zinc ferrite prepared by sol-gel auto-combustion technique, Inter. J. Nano Dimen. 1 (3) (2011) 221–224.
[49] S. Lorentzou, K. Karadimitra, C. Agrafiotis, A. Konstandopoulos, New routes for ferrite powders synthesis, PARTEC 2004.
[50] J. Azadmanjiri, S. Seyyed Ebrahimi, H. Salehani, Magnetic properties of nanosize $NiFe_2O_4$ particles synthesized by sol–gel auto combustion method, Ceram. Inter. 33 (2007) 1623–1625.
[51] A. Costa, E. Tortella, M. Morelli, R. Kiminami, Synthesis, microstructure and magnetic properties of Ni–Zn ferrites, J.Magn. Magn. Mater. 256 (2003) 174–182.
[52] A. Sutka, G. Mezinskis, A. Lusis, M. Stingaciu, Gas sensing properties of Zn-doped p-type nickel ferrite, Sensor. Actuat. B Chem. 171–172 (2012) 354–360.
[53] A. Dzunuzovic, N. Ilic, M. Vijatovic Petrovic, J. Bobic, B. Stojadinovic, Z. Dohcevic-Mitrovic, et al., Structure and properties of Ni–Zn ferrite obtained by auto-combustion method, J. Magn. Magn. Mater. 374 (2015) 245–251.
[54] A. Raval, N. Panchal, R. Jotania, Structural studies of co-spinel ferrite synthesized by an auto combustion method, J. Anal. Techniq. 1 (1) (2010) 1–2.
[55] D. Bahadur, S. Rajakumar, A. Kumar, Influence of fuel ratios on auto combustion synthesis of barium ferrite nano particles, J. Chem. Sci. 118 (1) (2006) 15–21.

[56] W. Chang, G. Skandan, S. Danforth, B. Kear, H. Hahn, Chemical vapor processing and applications for nanostructured ceramic powders and whiskers, NanoStructur. Mater. 4 (1994) 507–520.
[57] S. Veintemillas-Verdaguer, O. Bomati-Miguel, M. Morales, Effect of the process conditions on the structural and magnetic properties of γ-Fe_2O_3 nanoparticles produced by laser pyrolysis, Scripta Mater. 47 (2002) 589–593.
[58] R. Janot, D. Guerard, One step synthesis of maghemite nanometric powders by ball-milling, J. Alloy. Compd. 333 (2002) 302–307.
[59] V. Sepelak, S. Wisman, K. Becker, Magnetism of nanostructured mechanically activated and mechano synthesized spinel ferrites, J. Magn. Magn. Mater. 203 (1999) 135–137.
[60] P. Druska, U. Steinike, V. Sepelak, Surface structure of mechanically activated and of mechano synthesized zinc ferrite, J. Solid State Chem. 146 (1999) 13–21.
[61] T. Zak, V. Cosovic, A. Cosovic, B. David, N. Talijan, D. Zivkovic, Formation of magnetic microstructure of the nanosized $NiFe_2O_4$ synthesized via solid-state reaction, Sci. Sinter. 44 (2012) 103–112.
[62] Y. Roh, R. Lauf, A. McMillan, C. Zhang, C. Rawn, J. Bai, et al., Microbial synthesis and the characterization of metal-substituted magnetites, Solid State Commun. 118 (10) (2001) 529–534.
[63] S. Phumying, S. Labuayai, E. Swatsitang, V. Amornkitbamrung, S. Maensiri, Nanocrystalline spinel ferrite (MFe_2O_4, M = Ni, Co, Mn, Mg, Zn) powders prepared by a simple aloe vera plant-extracted solution hydrothermal route, Mater. Res. Bull. 48 (2013) 2060–2065.
[64] J. Scott, Prospects for ferroelectrics: 2012–2022, Hindawi Publishing Corporation, ISRN Mater. Sci. (2013), 24 pages.
[65] H. Palneedi, V. Annapureddy, S. Priya, J. Ryu, Status and perspectives of multiferroic magnetoelectric composite materials and application, Actuat 5 (2016) 9.
[66] H. Ke, W. Wang, Y. Wang, J. Xu, D. Jia, Z. Lu, et al., Factors controlling pure-phase multiferroic $BiFeO_3$ powders synthesized by chemical co-precipitation, J. Alloy. Compd. 509 (2011) 2192–2197.
[67] D. Cai, J. Li, T. Tong, D. Jin, S. Yu, J. Cheng, Phase evolution of bismuth ferrites in the process of hydrothermal reaction, Mater. Chem. Phys. 134 (1) (2012) 139–144.
[68] D. Lukovic Golic, A. Radojkovic, J. Cirkovic, A. Dapcevic, D. Pajic, N. Tasic, et al., Structural, ferroelectric and magnetic properties of $BiFeO_3$ synthesized by sonochemically assisted hydrothermal and hydro-evaporation chemical methods, J. Eur. Ceram. Soc. 36 (2016) 1623–1631.
[69] A. Hardy, D. Mondelaers, M. Van Bael, J. Mullens, L. Van Poucke, G. Vanhoyland, et al., Synthesis of $(Bi, La)_4Ti_{13}O_{12}$ by a new aqueous solution gel-route, J. Eur. Ceram. Soc. 24 (2004) 905–909.
[70] S. Masoudpanah, S. Mirkazemi, S. Shabani, P. Abadi, The effect of the ethylene glycol to metal nitrate molar ratio on the phase evolution, morphology and magnetic properties of single phase $BiFeO_3$ nanoparticles, Ceram. Inter 41 (8) (2015) 9642–9646.
[71] N. Das, R. Majumdar, A. Sen, H. Maiti, Nanosized bismuth ferrite powder prepared through sonochemical and microemulsion techniques, Mater. Lett. 61 (10) (2007) 2100–2104.
[72] N. Ilic, A. Dzunuzovic, J. Bobic, B. Stojadinovic, P. Hammer, M. Vijatovic Petrovic, et al., Structure and properties of chemically synthesized $BiFeO_3$. Influence of fuel and complexing agent, Ceram. Inter. 41 (2015) 69–77.

[73] A. Dzunuzovic, M. Vijatovic Petrovic, B. Stojadinovic, N. Ilic, J. Bobic, C. Foschini, et al., Multiferroic (NiZn)Fe_2O_4–$BaTiO_3$ composites prepared from nanopowders by auto-combustion method, Ceram. Inter. 41 (2015) 13189–13200.
[74] N. Spaldin, Magnetic Materials, Fundamentals and Applications, second ed., Cambridge University Press, Cambridge, 2011.

Ferrite-based composites for microwave absorbing applications 18

Ling Bing Kong[1,6], Lie Liu[2], Zhihong Yang[3], Sean Li[4], Tianshu Zhang[5] and Chuanhu Wang[6]
[1]Nanyang Technological University, Nanyang, Singapore, [2]General Test Systems Inc., Shenzhen, Guangdong, P.R. China, [3]Nanjing University of Aeronautics and Astronautics, Nanjing, P.R. China, [4]The University of New South Wales, Sydney, NSW, Australia, [5]Anhui Target Advanced Ceramics Technology Co. Ltd., Hefei, Anhui, P.R. China, [6]Bengbu University, Bengbu, Anhui, P. R. China

18.1 Introduction

Microwave absorbing capability of a material is usually represented by reflectivity as a function frequency [1]. In case of normal incidence, the reflectivity or reflection loss of the material means the power reflection of a plane wave reflected from an infinite slab that is backed by a metallic surface, which can be calculated directly from μ, ε, and t (thickness) of the materials that are experimentally measured [2]. In terms of practical applications, microwave absorbing materials should have a broad bandwidth, low *RL*, and small thickness or light weight. Theoretical maximum wavelength (or frequency) bandwidth is related to static permeability μ_0' of the materials, given by [3]

$$\Delta\lambda = \lambda_{\max} - \lambda_{\min} \leq 2\pi^2 \mu_0' t / \left| \ln(\rho_0) \right|, \tag{18.1}$$

where ρ_0 is reflectivity, t is thickness of the microwave absorbing material, $\lambda_{\max}$ and $\lambda_{\min}$ are the upper and lower wavelengths covering the bandwidth at the reflectivity ρ_0, respectively. Accordingly, a broad bandwidth absorption means a large $\Delta\lambda$ and a small thickness t, so that the microwave absorbing materials should have a large value of real permeability μ_0'.

At the same time, thickness t of microwave absorbing materials can be linked with imaginary permeability μ_0'', given by [4]

$$t = \frac{c}{2\pi\mu'' f}. \tag{18.2}$$

This implies that the thickness t is proportional to the reciprocal of μ_0''. Therefore, the thickness t of an absorbing material can be effectively reduced if a high value of imaginary permeability μ_0'' can be achieved.

Magnetic, Ferroelectric, and Multiferroic Metal Oxides. DOI: http://dx.doi.org/10.1016/B978-0-12-811180-2.00018-9

Additionally, an impedance matching between the microwave absorbing materials and free space, $\eta = \sqrt{\mu'/\varepsilon'}$, is also a determining factor to obtain a low reflection. Unfortunately, for most cases, the value of μ' microwave absorbing materials is usually lower than that of ε' at the microwave frequencies of interest. In this regard, achieving an impedance matching is still a challenge in developing microwave absorbing materials.

Electromagnetic absorbing materials are usually present as composites, which are fabricated by mixing active components (i.e., fillers or inclusions) and polymer matrix, with certain volume concentrations (p), designed according to desired requirements. Composites based on ferrite powders have formed a special group of microwave absorbing materials.

Ferrites can be classified into three groups, spinel, garnet, and barium (or hexagonal), depending on their crystal structures. Spinel ferrites have relatively high static permeability, together with low dispersion frequencies, so that they are suitable for applications at low frequencies, including radio frequency, very high frequency, and ultrahigh frequency. Garnet ferrites exhibited high gyromagnetic properties as well as very low magnetic and dielectric loss tangents. Therefore, they are usually used in reciprocal and nonreciprocal microwave devices. In comparison, hexagonal ferrites have been widely acknowledged to be candidates of high performance microwave absorbing materials.

Hexagonal ferrites are named because of their hexagonal structure [5–7]. For barium ferrites, their unit cells consist of three blocks: S blocks (spinel block) without barium ions, R blocks (hexagonal block) containing barium ions and two oxygen layers, and T blocks (hexagonal block) having barium ions and four oxygen layers. According to the numbers and arrangements of the three blocks, hexagonal ferrites are categorized into *M*-, *W*-, *Y*-, *Z*-, *X*-, and *U*-types.

Due to their relatively low hexagonal crystal symmetry, hexagonal ferrites have magnetocrystalline anisotropies of up to $K_1 \approx 10^6$ erg/cm^3, which is higher than those of spinel or garnet ferrites by 1–2 orders of magnitude. Therefore, hexagonal ferrites are more suitable for applications as microwave absorbing materials, since their resonance frequencies can be up to 100 GHz.

Depending on the easy magnetization directions, hexagonal ferrites have two types of anisotropy, *c*-axis and *c*-plane anisotropy, corresponding to the easy magnetization along the *c*-axis and in the *c*-plane. For *c*-axis anisotropy, the magnetic field, which is used to rotate the magnetization vectors from [001] to [00$\bar{1}$], is defined as the anisotropy field H_a. For *c*-plane anisotropy, there are six easy directions that are separated by 60 degrees equally. As a magnetic field is applied, the magnetization vectors can be rotated from one easy magnetization direction to another by going through the *c*-axis or rotating in the *c*-plane, corresponding to the magnetic fields of out-of-plane anisotropy field H_θ (about 10^4 Oe) and in-plane anisotropy field H_ϕ (tens to hundreds of Oe), respectively. The anisotropy fields H_a or H_θ and H_ϕ can be well linked to static real permeability μ_0' and resonance frequency f_R [8]. For *c*-axis anisotropy, there are:

$$f_R = \gamma H_a, \tag{18.3}$$

$$\mu_0' = \frac{2}{3}\frac{M_s}{H_a} + 1, \tag{18.4}$$

For c-plane anisotropy, there are:

$$f_R = \gamma\left(H_\theta H_\phi\right)^{1/2}, \tag{18.5}$$

$$\mu_0' = \frac{1}{2}\frac{M_s}{H_\phi} + 1, \tag{18.6}$$

Hexagonal ferrites with c-axis anisotropy are also known as hard magnetic materials because of their relatively large values of H_a. As seen in Eq. (18.3), both μ_0' and f_R are dependent on H_a, where it is impossible to have large μ_0' and high f_R simultaneously. Comparatively, hexagonal ferrites with c-plane anisotropy are called soft magnetic materials because of their very low values of H_ϕ. According to Eqs. (18.5) and (18.6), a large μ_0' can be obtained if H_ϕ is sufficiently low, whereas the resonance f_R is also related H_θ. Therefore, hexagonal ferrites with c-plane anisotropy are promising candidates as microwave absorbing materials, because large μ_0' and desired f_R could be achieved at the same time.

18.2 Theoretic considerations

High-frequency magnetic properties of a material are usually characterized by using complex permeability as a function of frequency (f); that is, permeability spectrum. At the low frequency band, μ' is nearly constant, while μ'' is almost zero. At the high-frequency band, there are domain wall resonance and relaxation. Therefore, μ' starts to decline, while μ'' begins to increase [9]. At the microwave band, natural resonance is dominant, which is usually accompanied by a resonant peak of μ'' [10].

18.2.1 Magnetic resonances

18.2.1.1 Domain wall resonance

Subject to an AC magnetic field, $h = h_0 e^{j\omega t}$, which is parallel to the plane of the materials, domain walls vibrate at the equilibrium position. Domain wall resonance usually leads to a relaxation type permeability spectrum, which is also called Debye type spectrum [11].

18.2.1.2 Natural resonance

Magnetization vector precesses about the static magnetic field. In this case, the energy is absorbed from the AC transverse field, when it is equal to the processional frequency, which is known as ferromagnetic resonance [12]. Ferromagnetic resonance without the presence of a magnetic field H_0 is generally known as natural resonance. In this case, the unsaturated magnetization vectors are located along the easy magnetization directions. For polycrystalline materials, the directions of crystal grains are randomly distributed.

18.2.1.3 Permeability spectra of natural resonance

Permeability spectra are determined by three parameters: (1) static real permeability μ_0, (2) intrinsic resonance frequency f_r, and (3) damping coefficient λ. μ_0 (or μ_0') is dependent on saturation magnetization M_s and anisotropy fields, H_a, H_θ, or H_ϕ.

Intrinsic resonant frequency (f_r)

Resonant frequency, f_r, is closely related to the anisotropy field and demagnetizing field H_d [10,13]. In case of single domain, if H_a or H_θ is along the z-axis, f_r is given by

$$f_r = \gamma\{[H_a+(N_x-N_z)M]\left[H_a+\left(N_y-N_z\right)M\right]\}^{1/2}, \tag{18.7}$$

for z-axis anisotropy, or

$$f_r = \gamma\{[H_\theta+(N_x-N_z)M]\left[H_\phi+\left(N_x-N_y\right)M\right]\}^{1/2}, \tag{18.8}$$

for c-plane anisotropy[14], where N_x, N_y, and N_z are the demagnetizing factors along x, y, and z crystallographic axes, respectively.

Damping coefficient (λ)

The resonant frequency obtained from permeability spectra, f_R, is usually the frequency at the maximum imaginary permeability, μ''_{max}, in the $\mu''-f$ spectra. In most cases, f_R and f_r have different values. For natural resonance, the maximum $\mu''(f)$ is given by [15]

$$\mu''_{max} = \frac{1}{2}\left(\mu'_0 - 1\right)\left(1+\frac{1}{\lambda^2}\right)^{1/2}, \tag{18.9}$$

at frequency

$$f_R = \left(\frac{1}{1+\lambda^2}\right)^{1/2} f_r. \tag{18.10}$$

Therefore, the smaller the value of λ, the larger the ratio $\mu''_{max}/(\lambda-1)$. Also, as $\lambda \to \infty$, $\mu''_{max}/(\mu_0'-1) = 1/2$. As a result, if the value of λ is sufficiently low, a large value of μ''_{max} can be achieved, even though μ_0 is low. Also, $f_R = f_r$ at $\lambda = 0$; otherwise, they can be significantly different from each other.

Magnetic dispersions

The type of dispersion of permeability spectra is determined by the value of λ. There are two types of spectra, resonance-like and relaxation-like dispersions, depending on the level of λ. Low λ results in resonance-like dispersion, while large λ leads to relaxation-like dispersion. Materials with relaxation-like dispersion have broader band absorption than those with resonance-like dispersion. Therefore, to develop wide bandwidth microwave absorbing materials, it is desirable to have relaxation-like dispersion permeability spectra.

18.2.2 Effects on magnetic properties of ferrite composites

18.2.2.1 Volume concentration

As stated earlier, ferrite composites are made of ferrite powders and polymer matrix. Therefore, high-frequency magnetic properties of the ferrite composites have a close relation to volume concentration of the ferrite particles, which is represented by p. Generally, static real permeability μ_0' of composites is much lower than that of their bulk counterpart, while they have a relatively higher resonant frequency f_R. Also, both μ_0' and f_R are not linearly varied as a function of p.

Different models, including Bruggeman and Maxwell Garnett mixing rules, magnetic circuit model [16–18], simulation electromagnetic energy, magnetic percolation model, and two-particle model, have been developed to describe the relationships between μ_0' and f_R and p of magnetic composites [19,20]. Take the two-particle model as an example, where an effective demagnetizing factor, N_{eff}, was introduced and two energies, the magnetocrystalline anisotropy energy and the demagnetizing energy between the two particles, were considered. Therefore, μ_0' and f_R can be derived from N_{eff}, while the correlation between f_R and free energy of the system, given by [20]

$$\mu_0' = \frac{p(\mu_b' - 1)}{N_{eff}(\mu_b' - 1) + 1} + 1, \tag{18.11}$$

$$f_R = f_{R,b}\left[1 + N_{eff}(\mu_b' - 1)\right]^{1/2}, \tag{18.12}$$

where μ_b' and $f_{R,b}$ are the static permeability and resonance frequency of the corresponding bulk material, $N_{eff} = \mathrm{F}(p)N_d$, with N_d to be the general demagnetizing factor determined by the shape of the particles and $\mathrm{F}(p)$ to be a correct function related to p. For spherical particles, $\mathrm{F}(p) = (1-p)$ and Eq. (18.11) becomes to be the same as the Maxwell–Garrett mixing law:

$$\mu_0' = \frac{p(\mu_b' - 1)}{(1 - p)(\mu_b' - 1)N_d + 1} + 1. \tag{18.13}$$

Multiplying Eq. (18.11) with the square of Eq. (18.12) yields

$$(\mu_b' - 1)f_R^2 = p(\mu_b' - 1)f_{R,b}^2 = pC. \qquad (18.14)$$

For composites, C is a constant that is related to μ_b' and $f_{R,b}$ of the corresponding bulk material. Eq. (18.14) is similar to Snoek's law [21], where the product of $(\mu_0' - 1)$ and f_R, instead of square of f_R, is a constant: $(\mu_0' - 1)f_R = C$.

18.2.2.2 Substitutions

Except Y-type, all hexagonal ferrites have c-axis anisotropy. In order to obtain high values of μ_0' and μ''_{max}, it is necessary to modify the hexagonal ferrites so that c-plane anisotropy is obtained, which is usually realized through ion substitution. For example, Co ion usually has a negative first-order magnetocrystalline anisotropy constant K_1, whereas the second magnetocrystalline anisotropy constant K_2 is nearly zero in all hexagonal ferrites without Co ion [22]. As a result, Co substitution can be used to increase c-plane anisotropy of hexagonal ferrites. In other words, the anisotropy of a hexagonal ferrite can be modified from c-axis to c-plane anisotropy through Co doping. Fig. 18.1 shows anisotropy type and anisotropy fields of the composites made with Z-type barium ferrite, $Ba_3Co_xZn_{2-x}Fe_{24}O_{41}$, as a function of Co concentration [23]. The ferrite had c-axis anisotropy without Co substitution, which was changed to c-plane anisotropy, as the Co concentration was $x \approx 0.6$. As the Co concentration was further increased from 0.8 to 2.0, the anisotropy field H_θ was increased from 3.5 to 12 kOe. At the same time, resonant frequency f_R of the composites was shifted from 1.6 to 3.0 GHz.

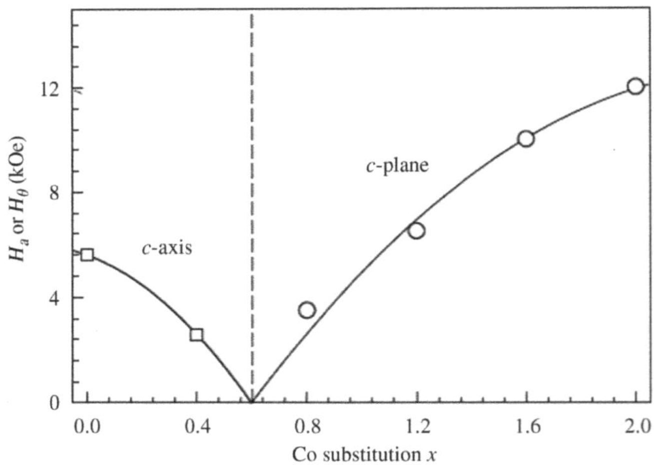

Figure 18.1 Anisotropy field H_a (open squares) or H_θ (open circles) for $Ba_3Co_xZn_{2-x}Fe_{24}O_{41}$ with Co substitutions.
Reproduced with permission from Z.W. Li, et al., Static and dynamic magnetic properties of CoZn substituted Z-type barium ferrite Ba3CoxZn2-xFe24O41 composites, J. Magn. Magn. Mater. 310 (1) (2007) 145–151, Copyright © 2007, Elsevier.

It has been demonstrated that nonmagnetic ions can also be used to modify the magnetocrystalline anisotropy of hexagonal ferrites. When using nonmagnetic ions for substitution, the variation in magnitude of anisotropy is closely related to distribution of the ions at the inequivalent sites [24,25]. For example, Al^{3+} and Ga^{3+} ions tended to occupy the 12*k* sites [26,27]. Therefore, if Al^{3+} and Ga^{3+} ions replaced Fe^{3+} ions, the *c*-axis anisotropy was enhanced, which led to an increase in their resonance frequency [26–28]. In comparison, complex ions, such as $Ti^{4+}Mn^{2+}$, $Zr^{4+}Zn^{2+}$, and $Sn^{4+}Zn^{2+}$, would reduce the degree of *c*-axis anisotropy, thus resulting in a downshift in f_R [29–33]. Generally, f_R of hexagonal ferrite ceramics and composites is proportional to the magnitude of H_a for *c*-axis anisotropy and that of the out-of-plane anisotropy field H_θ for *c*-plane anisotropy, respectively [34].

Distribution of the nonmagnetic ions at different sites is determined by the magnetocrystalline anisotropy and saturation magnetization (M_s). For instance, Zn ion preferred to occupy the spin-down sites of *M*- [32,35–40], *W*- [41–46], *Y*- [47–54], and Z-type hexagonal ferrites [15,23,55–58]. If a Zn ion is to substitute for a Fe^{3+} ion, the anisotropy fields, H_a or H_ϕ, were also decreased. Because an increase in Ms and a decrease in H_a or H_ϕ lead to an increase in μ_0', according to Eqs. (18.3–18.6), Zn^{2+} substitution can be employed to increase the value of μ_0'.

Additives

It was found that a low content of additive (e.g., <1.0 wt%) could be used to modify microstructure and magnetic domain characteristics of hexagonal ferrites so that certain special properties could be achieved, such as decrease in coercively H_c and increase in electrical resistivity. Various oxides have been employed as additives to optimize microstructures and magnetic properties of both spinel [59–66] and hexagonal ferrites [34].

18.2.2.3 Particle size and shape

As stated earlier, μ_0' is closely related to the reversible domain wall movement and magnetic moment rotation. It was found that the critical size of single domain particles for hexagonal ferrites is about 1 μm. In this case, nanoparticles usually have a single domain, while particles with sizes of ≥1 μm contained multiple domains [34]. As a result, the composite with nanoparticles of ferrite has no contribution by domain walls and thus smaller value of μ_0' (~2), while that made with microsized particles has both contributions from domain wall and the natural resonance. For meddle size particles, with an average size of 3 μm, some of them with sizes of <1 μm are single domain and the others with size of >1 μm are multidomain. Hence, a composite with these particles has a μ_0' of ~3. An example is shown in Fig. 18.2 [34].

According to the Maxwell–Garrett mixing law, Eq. (18.13), μ_0' of a composite also has a relationship with demagnetizing factor N_d, which is closely related to shapes of the ferrite particle [67]. Calculation results indicated that, for the plate with the normal parallel and perpendicular surface plane and bar or needle, the

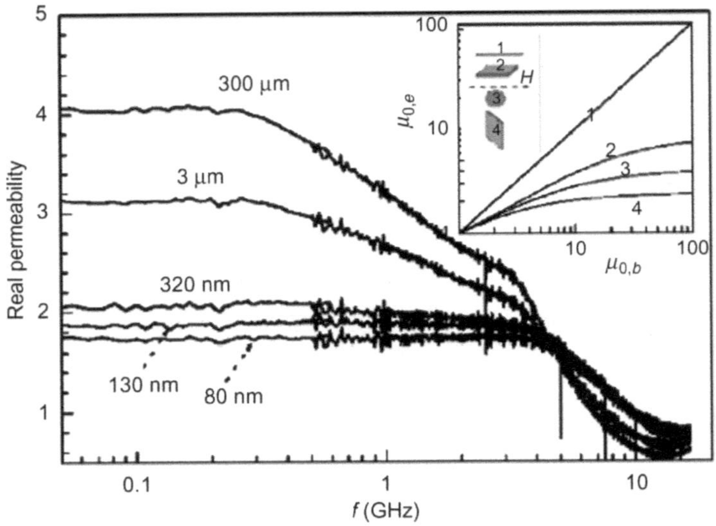

Figure 18.2 Effects of size and shape (inset) on real permeability of ferrite composites. Reproduced with permission from Z.W. Li, et al., High-frequency magnetic properties and attenuation characteristics for barium ferrite composites, J. Magn. Magn. Mater. 321 (7) (2009) 734–737, Copyright © 2009, Elsevier.

values of N_d are 1/3, 0, 1/2, and 0, respectively. Therefore, the static real permeability has an order of μ_0' (bar) > μ_0' (plate) > μ_0' (sphere) > μ_0' (plate), as seen as the inset in Fig. 18.2 [34]. At the same time, resonance frequency f_R has an order of f_R (bar) > f_R (spherical) > f_R (plate).

18.2.2.4 Damping and dispersion

μ''_{max} and f_R depend not only on μ_0' and intrinsic resonance frequency f_r, but also on the damping coefficient λ. For given values of μ_0' and f_r, the ratio of f_R/f_r is gradually increased with decreasing λ. The ratio of μ''_{max} to $(\mu_0'-1)$ is in the range of 0.5–0.7, as λ is changed from ∞ to 1. However, the ratios are increased quickly to ∞, as λ is reduced from 1 to 0. In summary, a small λ is necessary to ensure high value of μ''_{max} and high resonant frequency f_R. Depending on the magnitude of damping coefficient λ, composites can exhibit permeability spectra with either relaxation-like dispersion (relatively large λ) or resonance-like dispersion (relatively small λ).

Usually, composites based on hexagonal ferrite with c-plane anisotropy have relaxation-like permeability spectra. In this case, both μ_0' and μ''_{max} could have a relatively large value, so that the composites would have a high microwave absorption performance with a broad bandwidth. If ferrites with c-axis anisotropy are used to fabricate composites, resonance-like dispersion would be expected. Several special properties should be mentioned for the resonance-like dispersion ferrite composites. Upon approaching the resonance frequency, μ' is rapidly increased to the maximum value, which is much higher than static real permeability μ_0', while $\mu''(f)$ presents a narrow peak. Also, because of the small λ, the ratio of $\mu''_{max}/(\mu_0'-1)$ can be much higher than the typical value of 0.5 for permeability spectra with relaxation type dispersion. In addition, a sufficiently high f_R can be achieved, due to the presence of *c*-axis anisotropy.

18.3 Barium ferrite composites

Due to the interactions among the ferrite particles through the polymer matrix, magnetic properties of ferrite composites are different from those of their ceramic counterparts. The most pronounced differences include greatly decreased values of μ_0' and μ''_{max} and increased resonant frequencies. Properties and microwave absorbing performances of the composites based on different types of ferrites are discussed as follows.

18.3.1 M-type

M-type hexagonal ferrite, with a chemical formula of $BaFe_{12}O_{19}$, is hard magnetic material due to its *c*-axis anisotropy. It has a saturation magnetization M_s of 4.8 kGs, magnetocrystalline constant K_1 of 3.2×10^6 erg/cm^3, and anisotropy field H_a of 16.8 kOe [68]. Corresponding to its high *c*-axis anisotropy, *M*-type hexagonal ferrite has a static real permeability μ_0' to be only 1.3 and a high resonant frequency f_R of 45 GHz [69]. For practical applications, the resonance frequencies should be within the range of 1–18 GHz. To do this, it is necessary to modify the anisotropy from *c*-axis to *c*-plane or decrease the anisotropy field, H_a.

It is well known that Co^{2+}, Ru^{2+}, and Ir^{4+} ions could be used to modify hexagonal ferrites to achieve large *c*-plane anisotropy, through the substitution for Fe^{3+} ions in $BaFe_{12}O_{19}$ [70–72]. The substitutions are expected to bring out large permeability, μ_0' and μ''_{max} [35,73]. There would be a critical concentration of substitution to optimize the magnetic properties, which is different for different substituents. For example, the substituting concentrations should be $x \geq 1.2$ and $x \geq 0.4$ when using CoTi and CoRu to substitute Fe^{3+}, $BaFe_{12-2x}Co_xMe_xO_{19}$ (Me = Ti and Ru). In case of CoRu substitution, as the substitution level was increased from $x = 0.2$ to 0.4, although saturation magnetization M_s was almost not affected, coactivity H_c was decreased from 500 to 250 Oe [73], suggesting the decrease in anisotropy as a result of the substitution. As a consequence, the value of μ_0' was largely increased from 3 to 7, while the resonant frequency f_R was shifted from 14 to about 1 GHz. As the substitution level was further increased $x = 0.8$, μ_0' was decreased from 7 to 4, whereas f_R was increased from 1 GHz to 7 GHz.

Electromagnetic properties of the composites with 35 vol% ferrite powders were measured over the frequency range of 1–14 GHz. The real part of the complex permittivity was about 6.5, while the imaginary part was about 0.5 for all the samples.

The real and imaginary parts of the relative complex permeability, μ' and μ'', are shown in Fig. 18.3A and B, respectively [35].

Fig. 18.4 shows reflectivity curves of the composites based on the ferrite powders, estimated with the magnetic and dielectric properties. It was assumed that the samples were backed by a metal plate and thickness of the composites was 5 mm. The observed resonant peaks resulted in a low reflectivity for samples with $x = 0.8$ and $x = 1.2$, which could be understood in two aspects. On one hand, the thickness of a single layer isotropic material should be equal to quarter wavelength at the frequency of interest. Because of the magnetic resonance, the real part of the relative permeability was decreased with increasing frequency. The requirement of quarter wavelength could be realized within a frequency range near the magnetic resonant peak. In this case, a broadband low reflectivity was therefore achieved. On the other hand, more electromagnetic energy was dissipated because of the magnetic resonance, which was also beneficial to a low reflectivity.

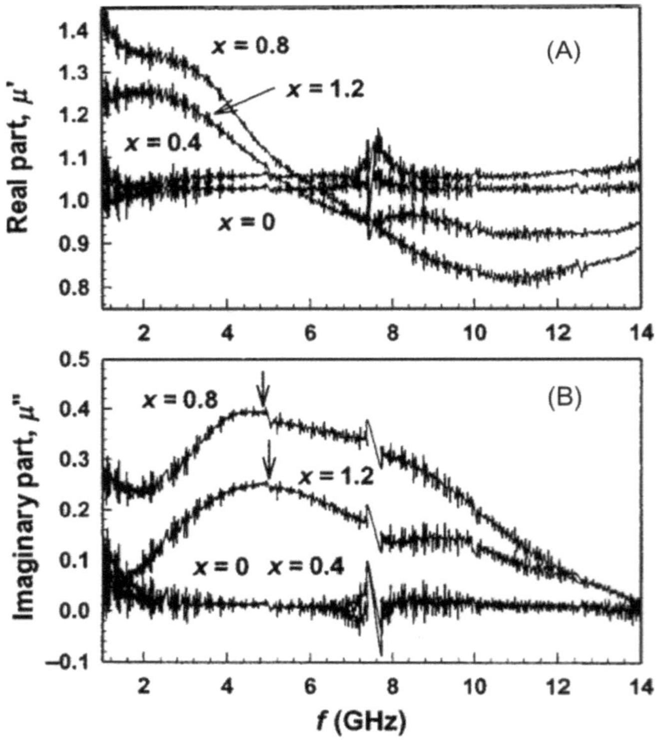

Figure 18.3 Complex permeability spectra of the composites based on the $BaFe_{12-2x}Co_xZr_xO_{19}$ powders: (A) real part and (B) imaginary part.
Reproduced with permission from Z.W. Li, L.F. Chen, and C.K. Ong, Studies of static and high-frequency magnetic properties for M-type ferrite BaFe12-2xCoxZrxO19, J. Appl. Phys. 92 (7) (2002) 3902–3907, Copyright © 2002, American Institute of Physics.

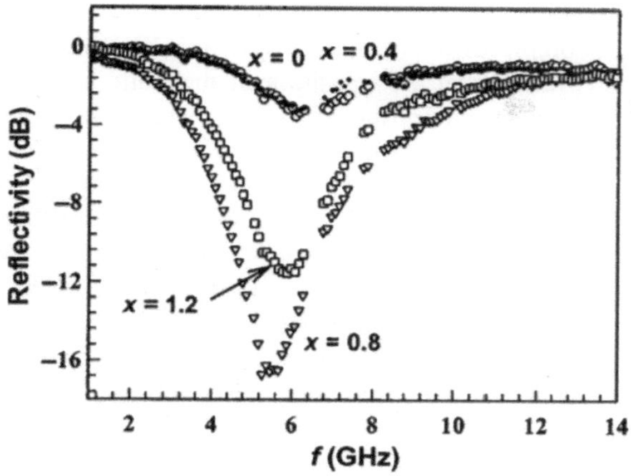

Figure 18.4 Reflectivity of the composites based on the $BaFe_{12-2x}Co_xZr_xO_{19}$ powders. Reproduced with permission from Z.W. Li, L.F. Chen, and C.K. Ong, Studies of static and high-frequency magnetic properties for M-type ferrite $BaFe_{12-2x}Co_xZr_xO_{19}$, J. Appl. Phys. 92 (7) (2002) 3902−3907, Copyright © 2002, American Institute of Physics.

It was reported that ferromagnetic resonant frequency f_R decreased from 46 GHz to 18 and 10 GHz, respectively, if *M*-type hexagonal ferrite was substituted with CoTi and Sc, at the substitution level of $x = 1.6$ [74]. High-frequency magnetic properties and EM absorbing performances of MnTi [75,76], ZnZr, and ZnSn [77] substituted *M*-type hexagonal ferrites and their composites, have been systematically studied. For instance, the composites based on $BaFe_{12-2x}Mn_xTi_xO_{19}$ with $x = 1.5$ and 2.0, had broad bandwidths over 14.1−20 GHz and 7.4−12 GHz, at thickness of 1.6 and 2.7 mm, respectively, at the reflectivity levels of $RL \leq 20$ dB.

18.3.2 W-type

Among all the hexagonal ferrites, *W*-type ferrites, with a chemical formula of $BaMe_2Fe_{16}O_{27}$ or Me_2W in short (Me = bivalent metallic ions), have the highest Curie temperatures T_c and the highest saturation magnetization M_s at room temperature, due to their highest ratio of magnetic ions to oxygen ions of 0.62:0.38. For instance, Fe_2W, Ni_2W and Zn_2W exhibited *c*-axis anisotropy, with relatively high *c*-axis anisotropy fields. Specifically, Ni_2W ferrite had resonant frequency, f_R, of 36 GHz, in a good agreement with the high anisotropy field of 12.5 kOe [78]. As stated earlier, Co substitution can be employed to change the anisotropy from *c*-axis to *c*-plane. According to ferromagnetic resonance [79] and magnetic measurement results of the aligned samples [80], Co_2W is *c*-plane anisotropy with an out-of-plane anisotropy, H_θ, to be 21.2 kOe. With Zn substitution, the value of μ_0' is able to increase. As a result, CoZn has been widely adopted to substitute Fe^{3+} so as to improve high-frequency magnetic properties of *W*-type barium ferrite.

High-frequency magnetic properties of $BaCo_xZn_{2-x}Fe_{16}O_{27}$ single crystals and ceramics have been comprehensively studied [80–82]. XRD patterns of the aligned composite samples revealed that the transition of magnetic anisotropy from c-axis to c-plane took place at the composite of $x = 0.7$, which was consistent with the dependence of coercivity H_c on x [80]. For W-type ferrite ceramics, as the level of Co substitution was increased from $x = 1$ to 2, the values of μ_0' and μ''_{max} were decreased from 14.5 to 4.0 and from 6.5 to 1.5, respectively. At the same time, resonant frequency f_R was increased from 0.7 to 2.5 GHz, [83]. For composites based on the ferrite powder with a volume concentration of $p = 0.35$, as the content Co was increased from 0.7 to 1.5, the values of μ_0' and μ''_{max} were decreased from 2.4 to 1.8 and 0.8 to 0.6, respectively. Correspondingly, resonant frequency f_R was significantly increased from 2.5 to 12 GHz. For those with a higher ferrite concentration of $p = 0.5$, the values of μ_0' were 3.6, 2.8, 2.0, and 1.5, where those of μ''_{max} were 1.3, 1.1, 0.8, and 0.3, for $x = 0.7$, 1.0, 1.5, and 2.0, respectively [82]. The resonant frequencies were 2.5, 6.0, 12.0, and 16.5 GHz.

Similarly, various oxides have been used to dope the hexagonal ferrites in order to increase the values of μ_0' and μ''_{max} of their composites [84–86]. If the undoped ferrite powders were used to make composites, the values of μ_0', μ''_{max}, and f_R were 3.2, 1.2, and 5.9 GHz, respectively. Once doped with oxides, both μ_0' and μ''_{max} of the composites were largely increased, which was accompanied with a downshift in resonant frequency. For example, the level of μ_0' was increased to 4.0–4.6, while the resonant frequency was lowered to the range of 2–3 GHz, if CaO, CuO, and MnO_2 were used as the dopants. It was found that the doping with IrO_2 and RuO_2 could lead to composites that exhibited the highest resonant frequencies, f_R, which were 4.6 and 4.2 GHz, respectively. This observation was readily ascribed to the large anisotropy of the ions in the ferrites. However, the doping with oxides resulted in an increase in permittivity ε' of the composites. This is an undesired outcome, because a high permittivity would lead to a poor impedance matching.

The effect of CoZn substitution, together with the doping of 1 wt% V_2O_5, on high frequency magnetic properties and microwave absorbing performance of the composites based on $BaCo_xZn_{2-x}Fe_{16}O_{27}$ with $x = 0$ to 2.0, has been reported in the open literature [87–89]. It was found that the ferrites with $x = 1.0$, 1.3, and 1.5 exhibited c-plane anisotropy, so that the corresponding composites displayed permeability spectra with relaxation-like dispersion. The values of μ_0' were decreased from 4.2 and 3.6 to 3.1, while the resonant frequency was upshifted to higher frequencies, from 4.0 and 6.5 to 8.0 GHz. The composites possessed promising microwave absorbing performances. Both lower limit and upper limit frequencies for $RL \leq -10$ dB were upshifted to higher frequencies. In this case, the lower limit f_{low} was increased from 3.8 and 5.0 to 6.3 GHz, whereas the upper limit f_{up} was raised from 13.5 and 16.2 to 16.5 GHz. The corresponding bandwidths W_R of 1:3.6, 1:3.2, and 1:2.6, for thicknesses of the composites to be 0.3, 0.26, and 0.22 cm, respectively [87,88].

The ferrites with $x = 0.5$, 0.6, and 0.65 had c-axis anisotropy and thus their composites possessed permeability spectra with resonance-like dispersion [89]. As a consequence, the values of μ_0' were quite low, in the range of 1.3–1.6.

Nevertheless, because the composites had very low values of damping coefficient, in the range of 0.2–0.3, a pretty high value of $\mu''_{max} \approx 1$ could be obtained. At the same time, a high ratio of $\mu''_{max}/(\mu_0'-1) = 2.7-1.6$ was observed. Also, due to the relatively large c-axis anisotropy, the resonant frequency f_R was high comparatively: 13, 9.8, and 6.4 GHz for the samples with compositions of $x = 0.5$, 0.6, and 0.65, respectively. Despite these, the composites could still have broad bandwidth, with very low reflectivity and small thickness. For example, bandwidths of the composites with $x = 0.5$, 0.6, and 0.65 for $RL \leq -20$ dB were over 12.5–16.5 GHz, 8.3–12.5 GHz, and 6.5–9.0 GHz, for the thicknesses of t=0.16, 0.22, and 0.28 cm, respectively. Therefore, the composites can be used for microwave absorption at X and Ku bands.

18.3.3 Y-type

Most Y-type hexagonal ferrites, $Ba_2Me_2Fe_{12}O_{22}$ or Me_2Y in short, demonstrate c-plane anisotropy, due mainly to their special crystal structure, in which $3 \times (ST)$ blocks are present without the presence of R block. The $2b$ or $2d$ sites in R block usually contribute to much higher c-axis magnetocrystalline anisotropy than other sites, due to their relatively high asymmetry. Therefore, due to the absence of R block, Y-type hexagonal ferrites generally exhibit c-plane anisotropy, even though there was no Co ion in them. As compared to other types of hexagonal ferrites, Me_2Y ferrites have relatively low M_s and low Curie temperature T_c. For example, T_c of Zn_2Y was as low as 130°C, while Ni_2Y has the highest T_c of 390°C [48,53,54]. However, because of its very low H_ϕ of about 1 Oe, high permeability still can be achieved in Zn_2Y.

High frequency and attenuation properties of Zn_2Y ferrites and their composites have been reported by various researchers [90–93]. Permeability spectra of single crystal Zn_2Y ferrite measured over 0.1–10 GHz revealed that the values of μ_0' and μ''_{max} were 13 and 12, respectively, while it had a resonant frequency of $f_R = 1.3$ GHz [92]. For $Ba_2Zn_{1.2-x}Co_xCu_{0.8}Fe_{12}O_{22}$ ferrites, as the substitution level of Co was increased from $x = 0$ to 0.2, the value of μ_0' decreased from 27 to 18. At the same time, the resonant frequency f_R was shifted to higher frequency, accordingly [94]. Fig. 18.5 shows real and imaginary parts of the permeability of the $Ba_2Zn_{1.2-x}Co_xCu_{0.8}Fe_{12}O_{22}$ ferrites, as a function of frequency. In comparison, the effect of substitution of Cu was as significant as that of Co, as seen in Fig. 18.6. Permeability curves of the two samples with $y = 0.2$ and 0.4 are almost the same.

For $Ba_2Co_{2-x}Zn_xFe_{12}O_{22}$ ferrites, as the substitution level of Zn was raised from $x = 0$ to 1.6, the values of μ_0' and μ''_{max} were increased from 4 to 15 and from 2 to 8, respectively, whereas the resonant frequency f_R was down-shifted from 1.5 to 0.5 GHz [83]. High frequency magnetic properties and microwave absorption behavior of the composites, with a volume fraction of $p = 0.43$, over the frequency range of 0.2–14 GHz, had been studied [90]. The values of μ_0' and μ''_{max} were in the ranges of 3–4 and 0.8–1.2, respectively. Multilayer structures designed by using the composites based on Mg_2Y or $(CoZn)_2Y$ ferrites exhibited broadband performance with $RL < -20$ dB, while the percentage bandwidths (W_P) were 43% and

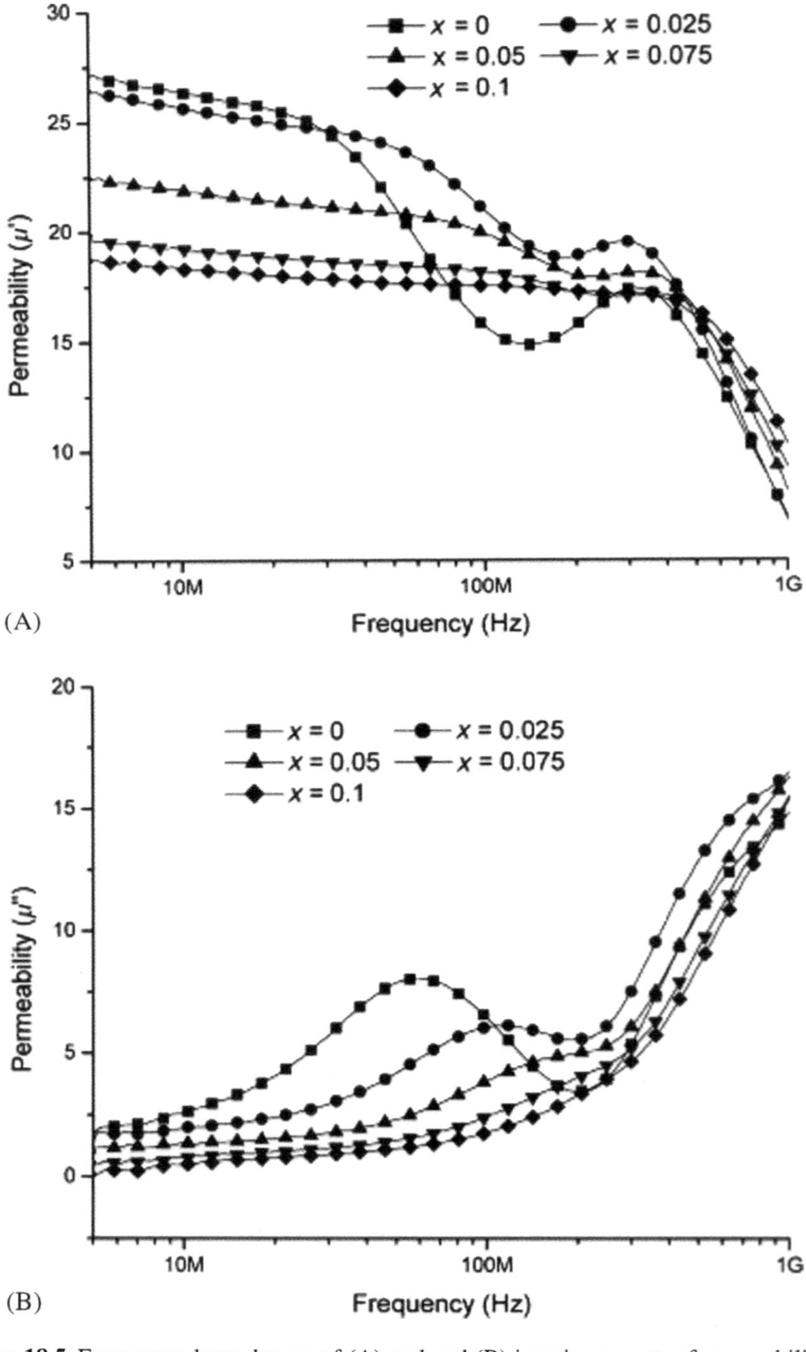

Figure 18.5 Frequency dependences of (A) real and (B) imaginary parts of permeability (μ' and μ'') of the $Ba_2Zn_{1.2-2x}Co_{2x}Cu_{0.4}Fe_{12}O_{22}$ with various contents of Co. Reproduced with permission from Y. Bai, et al., Effect of substitution on magnetization mechanism for Y-type hexagonal ferrite, Mater. Sci. Eng. B 103 (2) (2003) 115–117, Copyright © 2003, Elsevier.

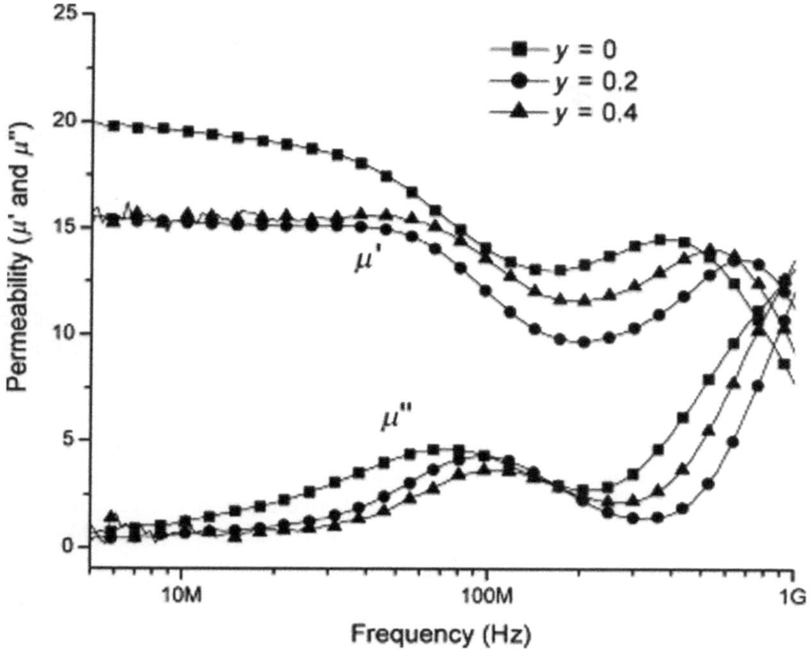

Figure 18.6 Frequency dependences of real and imaginary parts of permeability (μ' and μ'') of the $Ba_2Zn_{1.2-2y}Cu_{2y}Fe_{12}O_{22}$ samples with various contents of Cu. Reproduced with permission from Y. Bai, et al., Effect of substitution on magnetization mechanism for Y-type hexagonal ferrite, Mater. Sci. Eng. B 103 (2) (2003) 115–117, Copyright © 2003, Elsevier.

91%, for the composites of $(CoNi)_2Y$ and Mg_2Y, as the total thicknesses were 2.85 and 8.3 mm, respectively [91]. For ZnCuCo substituted *Y*-type ferrite composites with $p = 0.25$, the value of μ_0' was 4 and the resonant frequency was $f_R \approx 1$ GHz, at Zn substitution level of 1.2 [95].

It was found that the composites based on Cu substituted *Y*-type ferrite exhibited the most interesting microwave absorption performances [1]. Permeability spectrum of the $Ba_2CuZnFe_{12}O_{22}$ composites possessed both the natural and wall resonances, with almost the same weight. At the same time, the spectrum exhibited a typical relaxation-like dispersion, with μ_0' of 6.6 and μ''_{max} of 2.7 at 1.6 GHz due to the natural resonance and 2.5 at 0.5 GHz related to the wall resonance. There were two special characteristics, as compared with those of the W-type ferrite composites. On one hand, a broader bandwidth was achieved, because of the high value of μ_0' and μ''_{max}. For instance, the frequency band covered from 2.6 to 11.0 GHz for $RL \leq -10$ dB, so that a relative bandwidth W_R of 1:4.2 was achieved, as the composite had a thickness of $t = 0.35$ cm. On the other hand, the lower limit of frequency band was shifted to even lower frequency of 2.6 GHz, because of the lower resonant frequency f_R. As a result, this composite is a potential candidate for microwave absorption at S, C, and X bands. Moreover, for the composite based on $Ba_2CuZnFe_{12}O_{22}$ ferrite displayed a permeability spectrum with resonance-like dispersion due to the natural resonance, with $\mu_0' = 1-8$ and $\mu''_{max} = 1.0$, at $f_R = 9$ GHz. The frequency band for $RL \leq -15$ dB was as wide as 8.0–16.2 GHz, for a relatively small thickness of $t = 0.22$ cm, thus making it promising for microwave absorption at X and Ku bands.

18.3.4 Z-type

Z-type hexagonal ferrites are among the most attractive soft magnetic materials. Co_2Z ferrite, $Ba_3Co_2Fe_{24}O_{41}$, has relatively large permeability, with a saturation magnetization of 51 emu/g at room temperature and Curie temperature of 690 K. At room temperature, Co_2Z ferrite has *c*-plane anisotropy, with a large out-of-plane anisotropy field, H_θ, of 12 kOe and a low in-plane anisotropy field, H_ϕ, of about 0.12 kOe [22]. Static real permeability μ_0' of Co_2Z ferrite was in the range of 8–12, while its resonance frequency f_R was about 1.5 GHz [79,83,96].

The main task has been to enhance the high frequency properties of Co_2Z ferrites, to increase static real permeability μ_0', peak imaginary permeability μ''_{max}, and resonant frequency f_R. Substitution for Co^{2+} with other ions was among the most effective strategies to achieve this goal. For instance, substitution for Co^{2+} with Zn^{2+}could be used to increase the value of μ_0' up to 16, as the concentration of Zn was $x = 1.35$. If Fe^{2+} was employed to substitute Fe_2Z ferrite, the static permeability μ_0'could be as high as 21, after the ferrite was sintered at an oxygen partial pressure of 101.3 kPa [97]. Moreover, by using Cu^{2+} substitution, Z-type hexagonal ferrites with high properties could be fabricated at relatively low sintering temperatures [98,99].

Fig. 18.7 shows real permeability curves of the $Ba_3Co_{1.4}Cu_{0.6}Fe_{24}O_{41}$ ceramics sintered at different temperatures [98].

The value of μ_0' of Co_2Z could be largely increased by doping a small amount of SiO_2 [100]. Although the magnetization was only slightly increased from 0.268 to 0.287 T with increasing content of SiO_2 additive, electrical resistivities of the ferrite ceramics were significantly increased, with maximum resistivity of 10^5 Ω m, if the doping level of SiO_2 was >1.0 wt%. Fig. 18.8 shows complex permeability curves of the ceramics doped with different contents of SiO_2. The real permeability in the lower frequency region of <100 MHz was increased initially, as the doping concentration of SiO_2 was increased to 0.2 wt%. However, the value started to decline from 15 to 4, as the amount of SiO_2 was further increased. At the same time, as the SiO_2 content was increased, the natural resonance frequency was shifted toward higher frequency from 1 to 2.5 GHz.

Using substitution with $Zn^{2+}Ti^{4+}$ (without Co ion), Z-type ferrites could be made to have *c*-axis anisotropy. In this case, the value of μ_0' was reduced, while the temperature dependence of anisotropy field was positively set to the appropriate place modified [101,102]. Ni substitution has also been adopted to modify the properties of Z-type ferrites. For example, as the substitution level approached $x = 1$, the anisotropy of $Ba_3Ni_xCo_{2-x}Fe_{24}O_{41}$ began to transit from *c*-plane to *c*-axis. As the

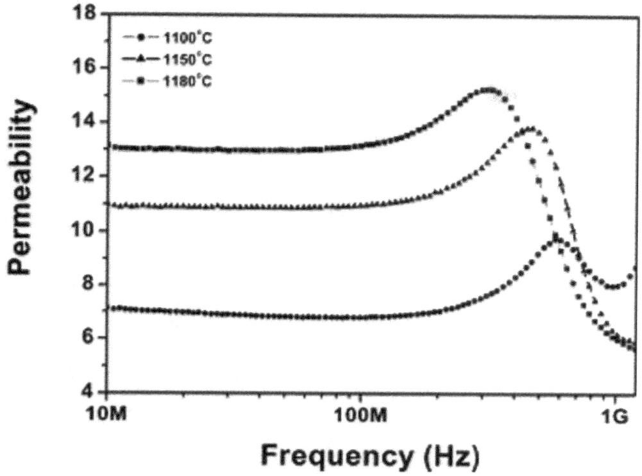

Figure 18.7 Frequency dependences of permeability of the $Ba_3Co_{1.4}Cu_{0.6}Fe_{24}O_{41}$ ceramics sintered at different temperatures.
Reproduced with permission from X.H. Wang, et al., Synthesis of Cu-modified Co_2Z hexaferrite with planar structure by a citrate precursor method, J. Magn. Magn. Mater. 234 (2) (2001) 255–260, Copyright © 2001, Elsevier.

substitution level of x was increased from 0.5 to 1.0, the value of μ_0' quickly decreased from 14 to 3.5 [103]. Representative complex permeability curves of the Ni-substituted Co_2Z hexagonal ferrite ceramics with different substitution levels are shown in Fig. 18.9.

One serious shortcoming of the CoZn substituted Z-type ferrite composites was the low resonant frequency f_R. For instance, the highest f_R was only 3.0 GHz as the maximum Co concentration was $x=2$. To further increase f_R, it was possible to substitute for Fe^{3+} with $Co^{2+}Ti^{4+}$ to form $Ba_3Co_{2+x}Ti_xFe_{24-2x}O_{41}$ [15]. As the substitution level of CoTi was increased from 0 to 0.5 and 1.0, the value of μ_0' was nearly unchanged. However, μ''_{max} was gradually increased from 1.2 to 1.6 and 1.8, while the resonant frequency f_R also raised from 3.0 to 3.5 and 4.5 GHz, respectively. The increase in μ''_{max} and f_R was directly related to the decrease in damping coefficients λ, from 1.44 to 1.22 and 0.88. In the corresponding $RL-f$ curves, although the lower limit f_{low} of the frequency bands for $RL \leq -10$ dB was almost not changed, the upper limit f_{up} was greatly shifted to higher frequencies. As a consequence, the relative bandwidth, W_R, was widened from 1:2.4 to 1:3.0 and 1:3.3, respectively, which was attributed mainly to the increased μ''_{max}. Meanwhile, the thickness of the composites corresponding to W_R was decreased from 0.35 to 0.30 and 0.25 cm, because of the increased resonant frequency f_R.

Similarly, doping with oxides has also been an effective way to improve magnetic properties of the Co_2Z ferrite composites. It was found that the most interesting dopant was TiO_2, which could increase all μ_0', μ''_{max}, and f_R. For example, with increasing doping content δ (wt%), the value of μ_0' increased

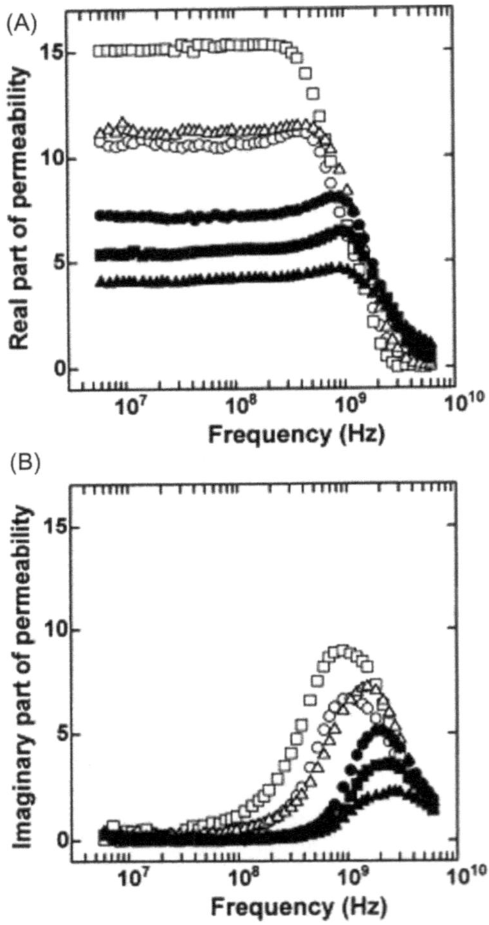

Figure 18.8 Complex permeability curves of ((A) real part and (B) imaginary part) of the Z-type ferrite $Ba_3Co_2Fe_{24}O_{41}$ ceramics with different contents of SiO_2: 0.0 (○), 0.2 (□), 0.5 (△), 1.0 (●), 1.5 (■), and 2.0 (▲).

Reproduced with permission from T. Nakamura, E. Hankui, Control of high-frequency permeability in polycrystalline (Ba,Co)-Z-type hexagonal ferrite, J. Magn. Magn. Mater. 257 (2–3) (2003) 158–164, Copyright © 2003, Elsevier.

from 12 for $\delta = 0$ to a maximum of 19 for $\delta = 1.0-1.5$ and then decreased to 16 for $\delta = 2.0$. The increase in $\mu_0'$was attributed to the increase in density of the ferrites caused by the doping. For the composites based on the CoZ powder doped with 1 wt% TiO_2, μ_0', μ''_{max}, and f_R were 4.0, 2.4, and 2.6 GHz, as compared to 3.6, 1.9, and 2.3 of the sample derived from the undoped Co_2Z ferrite powder. For the composites made with the 1 wt% TiO_2 doped Co_2Z ferrite, the frequency band for $RL \leq -210$ dB was from 3 to 12.5 GHz, corresponding to W_R

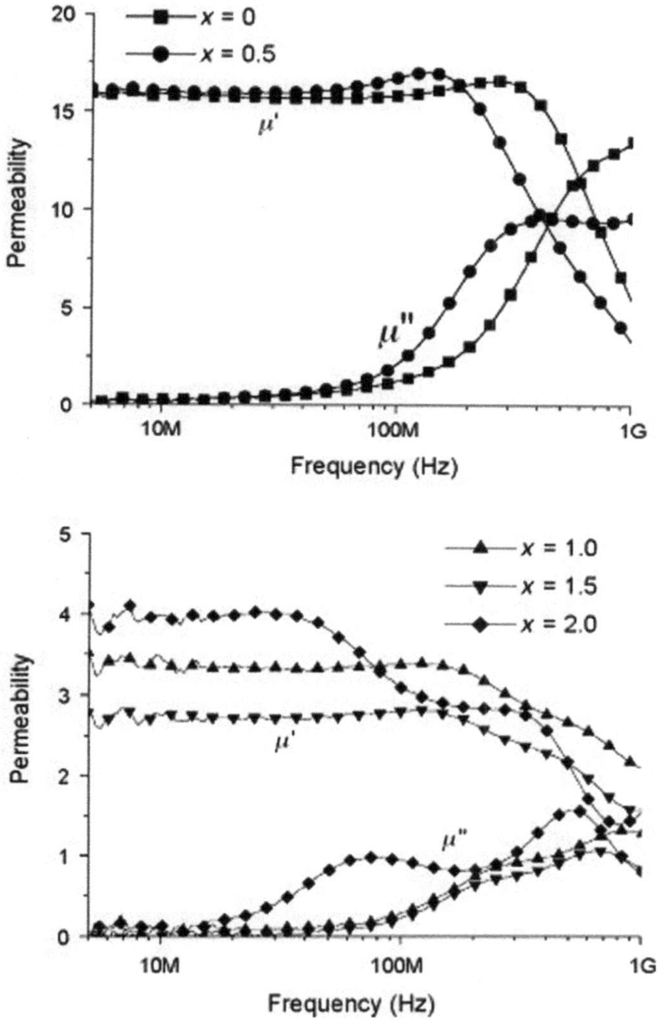

Figure 18.9 Complex permeability curves (μ' and μ'') of the Ni-substituted Co_2Z hexagonal ferrite ceramics with different substitution levels sintered at 1280°C. Reproduced with permission from T. Nakamura, E. Hankui, Control of high-frequency permeability in polycrystalline (Ba,Co)-Z-type hexagonal ferrite, J. Magn. Magn. Mater. 257 (2–3) (2003) 158–164, Copyright © 2003, Elsevier.

of 1:4.2, at a thickness of $t = 0.3$ cm. In comparison, the undoped Co_2Z ferrite composite had W_R of 2.4–3.1. Therefore, the TiO_2 doped Co_2Z composite could be a promising candidate as microwave absorbing materials with broad bandwidth and small thickness at S, C, and X bands.

18.4 Concluding remarks

Composites made with ferrites, especially hexagonal ferrites, have demonstrated most promising performances as candidates of microwave absorbing materials because of their advanced electromagnetic properties at microwave frequencies. Microwave absorption performance of a composite is closely related to its magnetic and dielectric properties, including resonant frequency, complex permeability, and permittivity. These properties of the composites are determined by various factors, such as concentration, magnetocrystalline structure, grain/particle size, size distribution, and shape of the ferrite fillers. Composites with desired microwave absorption performances can be designed by selecting composition of the ferrites, adopting appropriate processing parameters, and employing appropriate dopants. Currently, composites based on various hexagonal ferrite powders have been developed, with absorption capability covering different microwave frequency bands. Although theoretical studies have predicted that the performances of ferrite composites could be further improved by using ferrite fillers with special shapes/dimensions, such as plates, thin sheets, and fibers instead of the conventional spherical particles, further exploration is still necessary to prove them. Moreover, synthesis of ferrites with special shapes/dimensions and development of composites with them should be the new direction of this type of important functional material.

References

[1] L.B. Kong, et al., Recent progress in some composite materials and structures for specific electromagnetic applications, Intern. Mater. Rev. 58 (4) (2013) 203–259.
[2] E.F. Knott, J.F. Shaeffer, M.T. Tuley, Radar Cross Section, second ed., Artech House, Boston, MA, 1993.
[3] K.N. Rozanov, Ultimate thickness to bandwidth ratio of radar absorbers, IEEE Trans. Antennas Propag. 48 (8) (2000) 1230–1234.
[4] Z.W. Li, L.B. Kong, Z.H. Yang, *Study of matching characteristics for* $Ni_{0.97-x}Zn_xCo_{0.03}Fe_2O_4$ *spinel ferrites*, J. Appl. Phys. 105 (11) (2009) 113912.
[5] J. Smit, H.P.J. Wijn, Ferrites, Philips Technical Library, Eindhoven, 1959.
[6] H. Kojima, in: E.P. Wohlfarth (Ed.), Ferromagnetic Materials, North-Holland Publishing, Amsterdam, 1982, pp. 305–391.
[7] M. Sugimoto, in: E.P. Wohlfarth (Ed.), Ferromagnetic Materials, North-Holland Publishing, Amsterdam, 1982, pp. 393–440.
[8] J. Verwell, in: J. Smit (Ed.), Magnetic Properties of Materials, McGraw-Hill, New York, 1971.
[9] G.T. Rado, Magnetic spectra of ferrites, Rev. Mod. Phys. 25 (1) (1953) 81–89.
[10] C. Kittel, On the theory of ferromagnetic resonance absorption, Phys. Rev. 73 (2) (1948) 155–161.
[11] P.A. Miles, W.B. Westphal, A. Vonhippel, Dielectric spectroscopy of ferromagnetic semiconductors, Rev. Mod. Phys. 29 (3) (1957) 279–307.
[12] J.H.E. Grifiths, Anomalous high-frequency resistance of ferromagnetic metals, Nature 158 (1946) 670–671.

[13] H. Suhl, Ferromagnetic resonance in nickel ferrite between one kilomegacycle and 2 kilomegacycles, Phys. Rev. 97 (2) (1955) 555–557.
[14] Y.P. Wu, et al., Microstructural and high-frequency magnetic characteristics of W-type barium ferrites doped with V_2O_5, J. Appl. Phys. 97 (6) (2005) 063909.
[15] Z.W. Li, et al., $Co^{2+}Ti^{4+}$ substituted Z-type barium ferrite with enhanced imaginary permeability and resonance frequency, J. Appl. Phys. 99 (6) (2006) 063905.
[16] M.T. Johnson, E.G. Visser, A coherent model for the complex permeability in polycrystalline ferrites, IEEE Trans. Magn. 26 (5) (1990) 1987–1989.
[17] E.G. Visser, M.T. Johnson, A novel interpretation of the complex permeability in polycrystalline ferrites, J. Magn. Magn. Mater. 101 (1-3) (1991) 143–147.
[18] M.T. Johnson, et al., Microstructural dependence of the complex permeability in fine-grained MnZn ferrites, J. Magn. Magn. Mater. 116 (1-2) (1992) 169–176.
[19] Y. Konishi, H. Komori, The performance of absorbing materials with ferrite grains, Microw. Optic. Technol. Lett. 16 (3) (1997) 156–159.
[20] Z.W. Li, et al., Characteristics of effective permeability and resonance frequency for barium-ferrite/epoxy composites, J. Appl. Phys. 103 (7) (2008) 073901.
[21] J.L. Snoek, Dispersion and absorption in magnetic ferrites at frequencies above one Mc/s, Physica 14 (4) (1948) 207–217.
[22] G. Albanese, A. Deriu, S. Rinaldi, *Sublattice magnetization and anisotropy properties of $Ba_3Co_2Fe_{24}O_{41}$ hexagonal ferrite*, J. Phys. C-Solid State Phys. 9 (7) (1976) 1313–1323.
[23] Z.W. Li, et al., Static and dynamic magnetic properties of CoZn substituted Z-type barium ferrite $Ba_3Co_xZn_{2-x}Fe_{24}O_{41}$ composites, J. Magn. Magn. Mater. 310 (1) (2007) 145–151.
[24] N. Fuchikam, *Magnetic anisotropy of magnetoplumbite $BaFe_{12}O_{19}$*, J. Phys. Soc. Jpn. 20 (5) (1965) 760–769.
[25] Y. Xu, et al., Theory of the single ion magnetocrystalline anisotropy of 3D ions, Phys. Status Solidi B-Basic Res. 157 (2) (1990) 685–693.
[26] E.F. Bertaut, et al., *Substitution dans les hexaferrites de lion Fe^{3+} par Al^{3+}, Ga^{3+}, Cr^{3+}*, J. Phys. Radium 20 (2-3) (1959) 404–408.
[27] J.G. Rensen, J.A. Schulkes, J.S. van Wieringen, *Effet Mossvauer dans du $BaFe_{12}O_{19}$; subtitutions et mouvement thermique anisotrope*, J. Phys. Colloq. 32 (1971), C1-924-925.
[28] K. Haneda, H. Kojima, *Intrinsic coercivity of substituted $BaFe_{12}O_{19}$*, Jpn. J. Appl. Phys. 12 (3) (1973) 355–360.
[29] A.R. Corradi, et al., Effect of the cation occupancy on the magnetization reversal mechanism of hydrothermal barium ferrites, IEEE Trans. Magn. 24 (6) (1988) 2862–2864.
[30] X.Z. Zhou, et al., *Site preference for Co^{2+} and Ti^{4+} in Co-Ti substituted barium ferrite*, IEEE Trans. Magn. 27 (6) (1991) 4654–4656.
[31] C.L. Dube, et al., Dieletric and magnetic properties of Zn-Ti substituted M-type barium hexaferrite, Phys. Status Solidi A-Appl. Mater. Sci. 206 (11) (2009) 2627–2631.
[32] P. Wartewig, et al., Magnetic properties of Zn- and Ti-substituted barium hexaferrite, J. Magn. Magn. Mater. 192 (1) (1999) 83–99.
[33] Z.W. Li, et al., Site preference and magnetic properties for a perpendicular recording material: $BaFe_{12-x}Zn_{x/2}Zr_{x/2}O_{19}$ nanoparticles, Phys. Rev. B 62 (10) (2000) 6530–6537.
[34] Z.W. Li, et al., High-frequency magnetic properties and attenuation characteristics for barium ferrite composites, J. Magn. Magn. Mater. 321 (7) (2009) 734–737.
[35] Z.W. Li, L.F. Chen, C.K. Ong, Studies of static and high-frequency magnetic properties for M-type ferrite $BaFe_{12-2x}Co_xZr_xO_{19}$, J. Appl. Phys. 92 (7) (2002) 3902–3907.

[36] X. Obradors, et al., *Cation distribution and random spin canting in $LaZnFe_{11}O_{19}$*, J. Phys. C-Solid State Phys. 19 (33) (1986) 6605–6621.

[37] J. Li, et al., The transformation behavior of M-type barium ferrites due to Co-Ti substitution, J. Mater. Sci.-Mater. Electr. 26 (7) (2015) 4668–4674.

[38] J. Li, et al., Phase formation, magnetic properties and Raman spectra of Co-Ti co-substitution M-type barium ferrites, Appl. Phys. A-Mater. Sci. Process. 119 (2) (2015) 525–532.

[39] J. Li, et al., Co-Ti co-substitution of M-type hexagonal barium ferrite, Mater. Res. Exp. 2 (4) (2015) 046104.

[40] W.C. Li, et al., La and Co substituted M-type barium ferrites processed by sol-gel combustion synthesis, Mater. Res. Bull. 48 (11) (2013) 4449–4453.

[41] F. Leccabue, et al., Magnetic and morphological study of $BaZn_2Fe_{16}O_{27}$ hexagonal ferrite prepared by chemical coprecipitation method, J. Appl. Phys. 59 (6) (1986) 2114–2118.

[42] F. Aen, et al., Effect of holmium on the magnetic and electrical properties of barium based W-type hexagonal ferrites, Ceram. Int. 37 (6) (2011) 1725–1729.

[43] X.G. Huang, et al., Electromagnetic and microwave absorbing properties of W-type barium ferrite doped with Gd^{3+}, Rare Metals 30 (1) (2011) 44–48.

[44] Y.F. Wu, et al., *Pr^{3+}-substituted W-type barium ferrite: Preparation and electromagnetic properties*, J. Magn. Magn. Mater. 324 (4) (2012) 616–621.

[45] Z.H. Yang, Z.W. Li, Y.H. Yang, Structural and magnetic properties of plate-like W-type barium ferrites synthesized with a combination method of molten salt and sol-gel, Mater. Chem. Phys. 144 (3) (2014) 568–574.

[46] G. Albanese, M. Carbucicchio, G. Asti, *Spin-order and magnetic properties of $BaZn_2Fe_{16}O_{27}$ (Zn_2W) hexagonal ferrite*, Appl. Phys. 11 (1) (1976) 81–88.

[47] A. Deriu, et al., *Magnetic properties of the Cd substituted $(Cu,Zn)_2$-Y hexagonal ferrites*, J. Magn. Magn. Mater. 15-8 (JAN-) (1980) 1445–1446.

[48] A. Deriu, et al., Y-type hexagonal ferrites containing zinc, copper and cadmium - magnetic properties and cation distribution, J. Magn. Magn. Mater. 22 (3) (1981) 257–262.

[49] Mr Ahmad, et al., Temperature dependent structural and magnetic behavior of Y-type hexagonal ferrites synthesized by sol-gel autocombustion, J. Alloys Compd. 651 (2015) 749–755.

[50] I. Ali, et al., Effect of Eu-Ni substitution on electrical and dielectric properties of Co-Sr-Y-type hexagonal ferrite, Mater. Res. Bull. 49 (2014) 338–344.

[51] I. Ali, et al., Study of electrical and dielectric behavior of Tb^{3+} substituted Y-type hexagonal ferrite, J. Alloys Compd. 617 (2014) 863–868.

[52] Y. Bai, et al., Low-fired Y-type hexagonal ferrite for hyper frequency applications, J. Adv. Ceram. 1 (2) (2012) 100–109.

[53] C.A. Stergiou, G. Litsardakis, Y-type hexagonal ferrites for microwave absorber and antenna applications, J. Magn. Magn. Mater. 405 (2016) 54–61.

[54] F. Xu, et al., Characterization of a Y-type hexagonal ferrite-based frequency tunable microwave absorber, Int. J. Miner. Metal. Mater. 19 (5) (2012) 453–456.

[55] Z.W. Li, et al., Mossbauer spectra of CoZn-substituted Z-type barium ferrite $Ba_3Co_{2-x}Zn_xFe_{24}O_{41}$, Phys. Rev. B 72 (10) (2005) 104420.

[56] Z.W. Li, Z.H. Yang, L.B. Kong, Enhanced microwave magnetic and attenuation properties for Z-type barium ferrite composites with flaky fillers, J. Appl. Phys. 110 (6) (2011) 063907.

[57] T. Nakagawa, et al., Anomalous X-ray scattering study on site preference of Co and Fe ions in hexagonal Z-type barium ferrite, Jpn. J. Appl. Phys. Part 1-Reg. Pap. Brief Commun. Rev. Pap. 44 (2) (2005) 994–998.

[58] J.H. Shen, et al., Fabrication of Z-type barium ferrite/silica composites with enhanced microwave absorption, Sci. China-Technol. Sci. 57 (9) (2014) 1858–1864.
[59] G.C. Jain, et al., *Influence of V_2O_5 on the densification and the magnetic properties of Ni-Zn ferrite*, J. Magn. Magn. Mater. 14 (1) (1979) 80–86.
[60] P.K. Gallagher, et al., Preparation of Mn-Zn ferrites having a low tempeature coefficient of permeability, J. Am. Ceram. Soc. 66 (7) (1983) C110–C111.
[61] K.C. Chan, et al., *$Ni_{1-x}Co_xFe_{1.98}O_4$ ferrite ceramics with promising magneto-dielectric properties*, J. Am. Ceram. Soc. 91 (12) (2008) 3937–3942.
[62] L.B. Kong, et al., Electrical and magnetic properties of magnesium ferrite ceramics doped with Bi_2O_3, Acta Mater. 55 (19) (2007) 6561–6572.
[63] L.B. Kong, et al., Development of magneto-dielectric materials based on Li-ferrite ceramics - III. Complex relative permeability and magneto-dielectric properties, J. Alloy. Compd. 459 (1-2) (2008) 576–582.
[64] X.T. Liew, K.C. Chan, L.B. Kong, Magnetodielectric Ni ferrite ceramics with Bi_2O_3 additive for potential antenna miniaturizations, J. Mater. Res. 24 (2) (2009) 324–332.
[65] M.L.S. Teo, et al., Development of magneto-dielectric materials based on Li-ferrite ceramics - I. Densification behavior and microstructure development, J. Alloy. Compd. 459 (1-2) (2008) 557–566.
[66] M.L.S. Teo, et al., Development of magneto-dielectric materials based on Li-ferrite ceramics - II. DC resistivity and complex relative permittivity, J. Alloy. Compd. 459 (1-2) (2008) 567–575.
[67] G.Q. Lin, et al., Influence of demagnetizing field on the permeability of soft magnetic composites, J. Magn. Magn. Mater. 305 (2) (2006) 291–295.
[68] L. Jahn, H.G. Muller, Coercivity of hard ferrite single crystals, Phys. Status Solid. 35 (2) (1969) 723–730.
[69] R.C. Pullar, S.G. Appleton, A.K. Bhattacharya, The manufacture, characterisation and microwave properties of aligned M ferrite fibres, J. Magn. Magn. Mater. 186 (3) (1998) 326–332.
[70] D.J. Debitetto, Anisotropy fields in hexagonal ferrimagnetic oxides by ferrimagnetic resonance, J. Appl. Phys. 35 (12) (1964) 3482–3487.
[71] A. Tauber, R.O. Savage, J.A. Kohn, *Single-crystal ferroxdure, $BaFe_{12-2x}{}^{3+}Ir_x{}^{4+}Zn_x{}^{2+}O_{19}$, with strong planar anisotropy*, J. Appl. Phys. 34 (4) (1963) 1265–1267.
[72] J. Kreisel, et al., *An investigation of the magnetic anisotropy change in $BaFe_{12-2x}Ti_xCo_xO_{19}$ single crystals*, Journal of Magnetism and Magnetic Materials 224 (1) (2001) 17–29.
[73] H.S. Cho, S.S. Kim, M-hexaferrites with planar magnetic anisotropy and their application to high-frequency microwave absorbers, IEEE Trans. Magn. 35 (5) (1999) 3151–3153.
[74] I. Nedkov, A. Petkov, Microwave absorption in Sc-substituted and CoTi-substituted Ba hexaferrite powders, IEEE Trans. Magn. 26 (5) (1990) 1483–1484.
[75] S. Sugimoto, et al., Barium M-type ferrite as an electromagnetic microwave absorber in the GHz range, Mater. Trans. JIM 39 (10) (1998) 1080–1083.
[76] S. Sugimoto, et al., M-type ferrite composite as a microwave absorber with wide bandwidth in the GHz range, IEEE Trans. Magn. 35 (5) (1999) 3154–3156.
[77] T. Kagotani, et al., Enhancement of GHz electromagnetic wave absorption characteristics in aligned M-type barium ferrite $Ba_{1-x}La_xZn_xFe_{12-x-y}(Me_{0.5}Mn_{0.5})_yO_{19}$ ($x = 0.0$-0.5; $y = 1.9$-3.0, Me: Zr, Sn) by metal substitution, J. Magn. Magn. Mater. 272 (2004) E1813–E1815.
[78] Y.J. Kim, S.S. Kim, Microwave absorbing properties of Co-substituted Ni_2W hexaferrites in Ka-band frequencies (26.5-40 GHz), IEEE Trans. Magn. 38 (5) (2002) 3108–3110.

[79] R.A. Braden, I. Gordon, R.L. Harvey, Microwave properties of planar hexagonal ferrites, IEEE Trans. Magn. MAG2 (1) (1966) 43–47.

[80] Z.W. Li, L.F. Chen, C.K. Ong, *High-frequency magnetic properties of W-type barium-ferrite* $BaZn_{2-x}Co_xFe16O_{27}$ *composites*, J. Appl. Phys. 94 (9) (2003) 5918–5924.

[81] A. Paoluzi, et al., Magnetic, Mossvauer, and neutron-diffraction investigations of W-type hexaferrite $BaZn_{2-x}Co_xFe_{16}O_{27}$ single-crystals, J. Appl. Phys. 63 (10) (1988) 5074–5080.

[82] Z.W. Li, et al., Microwave attenuation properties of W-type barium ferrite $BaZn_{2-x}Co_xFe_{16}O_{27}$ composites, J. Appl. Phys. 96 (1) (2004) 534–539.

[83] T. Nakamura, K. Hatakeyama, Complex permeability of polycrystalline hexagonal ferrites, IEEE Trans. Magn. 36 (5) (2000) 3415–3417.

[84] Z.W. Li, Y.P. Wu, G.Q. Lin, Doping effect on complex permeability and permittivity for W-type barium ferrite composites, J. Appl. Phys. 102 (8) (2007) 083908.

[85] Z.W. Li, et al., Doping effect on microwave reflection characteristics for W-type ferrite/epoxy composites, J. Phys. D-Appl. Phys. 42 (9) (2009) 095007.

[86] Z.W. Li, et al., High-frequency properties and attenuation characteristics of WBa hexaferrite composites with doping of various oxides, IEEE Trans. Magn. 45 (2) (2009) 670–677.

[87] Z.W. Li, et al., The effect of V_2O_5 on high-frequency and attenuation properties for $BaCo_xZn_{2-x}Fe_{16}O_{27}$ composites, IEEE Trans. Magn. 42 (10) (2006) 3365–3367.

[88] Y.P. Wu, et al., Improved microwave magnetic and attenuation properties due to the dopant V_2O_5 in W-type barium ferrites, J. Phys. D-Appl. Phys. 39 (14) (2006) 2915–2919.

[89] Z.W. Li, Z.H. Yang, L.B. Kong, Resonancelike dispersion for W-type barium ferrite composites with c-axis anisotropy, J. Appl. Phys. 109 (3) (2011) 033916.

[90] H.J. Kwon, J.Y. Shin, J.H. Oh, The microwave absorbing and resonance phenomena of Y-type hexagonal ferrite microwave absorbers, J. Appl. Phys. 75 (10) (1994) 6109–6111.

[91] T. Inui, K. Koniski, K. Oda, Fabrications of broad-band RF-absorber composed of planar hexagonal ferrites, IEEE Trans. Magn. 35 (5) (1999) 3148–3150.

[92] M. Obol, C. Vittoria, Microwave permeability of Y-type hexaferrites in zero field, J. Appl. Phys. 94 (6) (2003) 4013–4017.

[93] M. Obol, C. Vittoria, Microwave permeability of Y-type hexaferrites in zero and low fields, J. Magn. Magn. Mater. 272 (2004) E1799–E1800.

[94] Y. Bai, et al., Effect of substitution on magnetization mechanism for Y-type hexagonal ferrite, Mater. Sci. Eng. B 103 (2) (2003) 115–117.

[95] H.J. Zhang, X. Yao, L.Y. Zhang, *The preparation and microwave properties of* $Ba_2Zn_zCo_{2-z}Fe_{12}O_{22}$ *hexaferrites*, J. Eur. Ceram. Soc. 22 (6) (2002) 835–840.

[96] I.G. Chen, S.H. Hsu, Y.H. Chang, *Preparation and magnetic properties of Ba-*Co_2Z *and Sr-*Zn_2Y *ferrites*, J. Appl. Phys. 87 (9) (2000) 6247–6249.

[97] T. Tachibana, et al., X-ray and neutron diffraction studies on iron-substituted Z-type hexagonal barium ferrite: $Ba_3Co_{2-x}Fe_{24+x}O_{41}$ (x = 0-0.6), J. Magn. Magn. Mater. 262 (2) (2003) 248–257.

[98] X.H. Wang, et al., *Synthesis of Cu-modified* Co_2Z *hexaferrite with planar structure by a citrate precursor method*, J. Magn. Magn. Mater. 234 (2) (2001) 255–260.

[99] X.H. Wang, et al., *Low-temperature sintering and high frequency properties of Cu-modified* Co_2Z *hexaferrite*, J. Eur. Ceram. Soc. 23 (5) (2003) 715–720.

[100] T. Nakamura, E. Hankui, Control of high-frequency permeability in polycrystalline (Ba,Co)-Z-type hexagonal ferrite, J. Magn. Magn. Mater. 257 (2-3) (2003) 158–164.

[101] P. Lubitz, F.J. Rachford, Z type Ba hexagonal ferrites with tailored microwave properties, J. Appl. Phys. 91 (10) (2002) 7613–7615.
[102] P. Lubitz, New substitutions in hexagonal ferrites to reduce anisotropy without using Co, J. Appl. Phys. 87 (9) (2000) 4978–4980.
[103] Y. Bai, et al., High frequency magnetic mechanism of Ni-substituted Co_2Z hexagonal ferrite, Mater. Res. Bull. 44 (4) (2009) 898–900.

Soft ferrite applications

19

Ljiljana Živanov[1], *Mirjana Damnjanović*[1], *Nelu Blaž*[1], *Andrea Marić*[1], *Milica Kisić*[1] *and Goran Radosavljević*[2]
[1]University of Novi Sad, Novi Sad, Serbia, [2]Vienna University of Technology, Vienna, Austria

Applications of magnetic materials in industry and consumer electronics have increased continuously. The ideal material is one with a large magnetic moment at room temperature, which is also an electrical insulator. Ferromagnetic metals and alloys have been widely exploited, but their low electrical resistivity (10^{-7} Ωm) is a large problem in radio-frequency (RF) applications because of the eddy-current losses. For this reason a number of magnetic oxides have become of great technical interest because of their high electrical resistivity ($1-10^{6}$ Ωm). Magnetic oxides show spontaneous magnetization, remanence, and other properties similar to ordinary ferromagnetic materials, but the spontaneous moment does not correspond to the value expected for full parallel alignment of the dipoles. In 1948 Néel put forward a theory for such materials; he suggested that they contain two sublattices (see Chapter 16: Metal Oxide Structure, Crystal Chemistry, and Magnetic Properties) in which the magnetizations are oppositely directed, but which give a net moment because the two sublattice moments are unequal. For this phenomenon he coined the word ferrimagnetism [1]. About the same time that Néel was developing his theory, Snoek was obtaining very interesting properties in a new class of oxide materials called ferrites that were very useful at high frequencies [2]. The arrangement of the ions of the crystal structure (see Chapter 16: Metal Oxide Structure, Crystal Chemistry, and Magnetic Properties) of the ferrite will play a most important role in determining the magnetic interactions, and therefore, the magnetic properties.

This chapter presents various kinds of applications of soft ferrite materials in magnetic components and sensors. Advanced integrated circuits need ferrite components to make them fully functional. Modern ferrite materials, size, and shape of components determine their applications. Development of components includes characterization of materials, design, and optimization of components, and finally, their implementation in electronic circuits. This chapter presents how components like electromagnetic interference (EMI) suppressors, inductors, microtransformers, or different types of sensors can be developed. For example, EMI suppressors are based on MnZn ferrites for suppression of noise in the lower frequency range (~MHz), while NiZn are used for suppression in the GHz range. Magnetic components can be manufactured in different technologies. Besides classical ceramic technology, low-temperature co-fired ceramic technology (LTCC) is efficient for microinductors, microtransformers, phase shifter, converters, and so on. Properties of ferrite materials make them useful for development of different kind of sensors, like detection of displacement, pressure, force, and so on.

Magnetic, Ferroelectric, and Multiferroic Metal Oxides. DOI: http://dx.doi.org/10.1016/B978-0-12-811180-2.00019-0

19.1 Characterization of ferrite material

Characterization of magnetic and electric properties of ferrite materials is very significant as these properties affect the operation of electronic devices. The intrinsic complex permittivity and permeability are the critical parameter for the optimization design, especially for high-frequency applications. The commonly used methods for magnetic material determination are the transmission/reflection method [3], equivalent circuit model method [4], and resonant method [5]. A simple experimental method to determine the initial complex permeability and permittivity is introduced by using the toroidal type of ferrite core. With the ferrite core installed in a shorted coaxial line, the fixture forms a one-turn coil of inductor or a shorted transmission line with distributed parameters. The relative permeability of the ferrite can then be extracted from S_{11} of the coaxial test holder with the aid of the vector network analyzer (VNA). Coaxial discontinuity structures are widely used as an element of microwave devices and commonly used in the permeability and/or permittivity measurement for materials. For permittivity and specific resistivity determination toroidal sample should be used (same sample can be used for permeability measurements). For permittivity measurements on sample surfaces, silver paste needs to be applied in order to have parallel plate capacitor structure. Capacitance of the sample can be measured by an impedance analyzer (IA) or LCZ meter (e.g., permittivity for commercial available NiZn ferrite material ESL 40012 was determined and presented in Fig. 19.1A).

It has been known that ferrites have a high permittivity and that this depends strongly on frequency. Permittivity, for presented measurement, slowly decreases from 6200 at 10 kHz to 5800 at 200 kHz. Above 200 kHz, permittivity falls rapidly from a value of 5800 to 490 and less.

The measurement of complex permeability will be discussed from two aspects: used frequency range and the applied measurement method. Low-frequency (10–1000 kHz) measurement can be performed by IA or LCZ meter and discrete turns of wire [6]. Examples of measurement results for NiZn ferrite material ESL40012 are presented in Fig. 19.1B.

The most commonly used method for high-frequency measurement of complex permeability is the coaxial line method [6]. For higher frequency measurements, VNA with a coaxial sample holder can be applied. The coaxial sample holder (with and without sample) is connected to the calibrated VNA via the 50 Ω coaxial cable. The reflection coefficient is measured, permitting the determination of the input impedance of the cell with sample. The equation for determination of the complex permeability is derived in [7]. The complex permeability is calculated from the difference between the impedance of the cell loaded with the toroidal sample and without it (e.g., complex permeability is calculated from measured *S*-parameters, and real and imaginary parts of ESL 40012 permeability are depicted in Fig. 19.2A).

Using the same method with a coaxial sample holder more measurement can be performed in the same frequency range, but on various temperatures (e.g.,

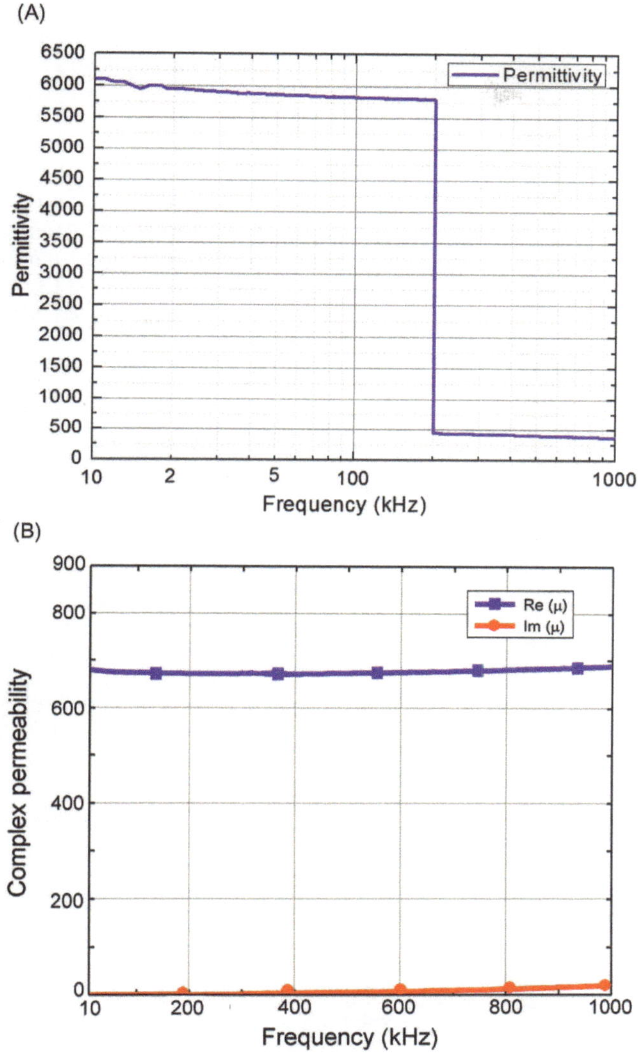

Figure 19.1 (A) Characteristic of permittivity versus frequency. (B) Real and imaginary part of relative permeability at low frequency.

Fig. 19.2A) variation of real and imaginary part of permeability spectra with temperature (at 27°C, 50°C, 70°C, 100°C, and 120°C) for ESL 40012 appears. For examined material it was found that the real part of permeability Re(μ) increases with rise of the temperature up to 70°C, and after 100°C the real part of permeability Re(μ) starts to decrease. The imaginary part of permeability Im(μ) slightly increases at 50°C but after 70°C it starts to decrease with the temperature increasing.

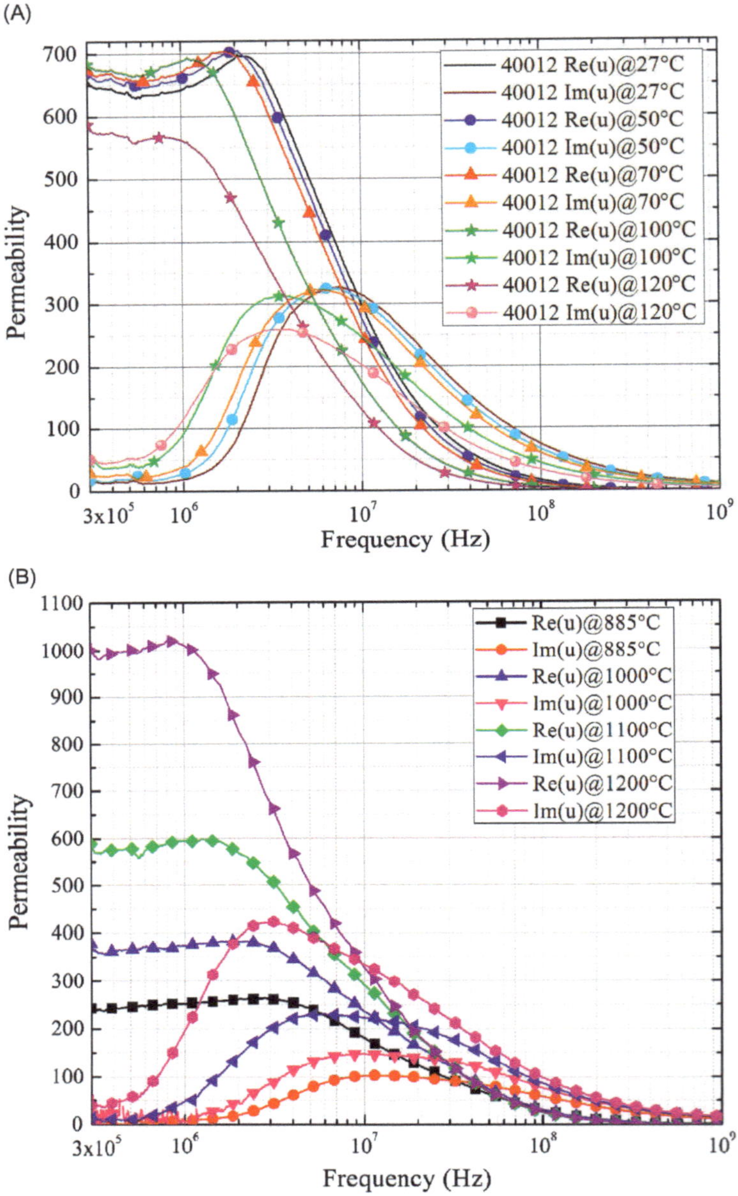

Figure 19.2 Complex permeability of LTCC sample (A) for ESL 40012 measured at various temperatures, and (B) for ESL 40011 at various sintering temperatures.

In addition, the presented measurement method was used to determine permeability of four ferrite LTCC samples (ESL 40011) sintered at various temperatures (Fig. 19.2B). Extracted values of the real part of complex permeability show a significant increase of permeability for higher peak sintering temperatures. The maximum of a real part permeability starting value of 263.19 for a sample sintered at $T_{max} = 885°C$ elevates to the value of 1022.09 for toroid sintered at $T_{max} = 1200°C$, exhibiting enhancement of almost four times. Maximum values of the imaginary part of complex permeability increase significantly (over four times) for samples sintered at elevated temperatures. However, this could be used as an advantage in certain applications (e.g., in EMI suppressors). More details about presented measurement methods and material characterization can be found in [7–10].

19.2 Passive ferrite components

In the days of modern integrated circuit technology and wireless industry, one of the key requirements is the availability of small, cost efficient, high performance, high power electronic circuit and components. Almost all of these modules demand the incorporation of high inductance and good performance passives with high packaging potential. Since all passive components (inductors, transformers, capacitors, etc.) occupy a large chip area, their fabrication on the top of the chip must be avoided. Moreover, in cases when high inductances need to be achieved the chip area becomes an even more important issue.

As a promising approach to answer the aforementioned requirements, implementation of the ceramic co-processing (surface-mount device (SMD) components) and LTCC components technology for the fabrication of microsized inductors and transformers with good performances has on more than one occasion been appreciated as an excellent solution [11–13]. It offers the possibility of embedding all cumbersome components inside the substrate, as well as different ways of design optimization in order to achieve better performance. The benefits of LTCC technology, such as the parallel manufacturing process with high yield and low cost, wide variety of commercially available materials for component fabrication, quick rounds in prototyping, environmental stability, compact structures, integration, and so on are well known.

19.2.1 Surface-mount device ferrite components

Surface-mount technology (SMT) was developed in the 1960s and became widely used in the late 1980s, thus replacing traditional through-hole technology. Space savings of 50% to 90% was achieved, but also reduced process costs, higher product quality, reduced handling costs, and higher reliability [14]. One of the leaders in introduction of this technology was IBM.

SMD is a miniature electronic component with two metal end caps, so it can be easily surface mounted on a printed circuit board (PCB). SMD components can be passive, active, or electromechanical. The most commonly used passive SMD components are inductors and capacitors.

The focus of this section is on ferrite SMD components. The size, performance, and reliability make these miniature inductors very attractive for a wide range of applications, such as EMI suppression in a universal series bus, low-voltage differential signaling, and in other high-speed digital interfaces incorporated in notebooks and personal computers, phones, and tablets. In addition, ferrite suppressors have been successfully employed for attenuating EMI in switching power supplies or electronic ignition systems. Ferrite components are efficient and cost effective for the prevention of—and protection against—spurious signals transmitted by conduction and radiation. Suppression components are offered in a number of ferrite materials, optimizing impedance over a wide range of frequencies. Depending on working frequency range, Nickel zinc (NiZn) or Manganese zinc (MnZn) ferrite materials are used [14]. NiZn materials have lower initial permeability than MnZn (up to 1000), but they maintain their permeability in a wider frequency range (over 100 MHz). Because of that, NiZn materials will not produce high impedance at low frequencies. They can be used for suppression of unwanted signals at frequencies greater than 20 MHz.

The inductor usually consists of a highly conductive layer embedded in a ferrite monolithic structure, which provides a good magnetic shielding and makes an inductor very suitable for high density mounting. There are three types of structure: wire-wound, multilayer, and film [15].They are made in EIA standard sizes: 0402, 0603, 0805, 1206, 1210, and 1812 (Fig. 19.3), and they have impedance between 6 and 2000 Ω at 100 MHz. Multilayer inductors are specially designed to reduce noise in low impedance circuits while keeping the signal free from distortion. They function as resistors at noise frequencies and suppress noise when they are configured in series with a noise-producing circuit [4].

In recent years, many authors studied high frequency characteristics of ferrite core and/or magnetic components such as ferrite inductors, based on the finite element method or physical analysis. Parasitic effects such as stray capacitance, self-resonance, and magnetic losses of ferrite play an important role in the design of such inductors. An understanding of the high-frequency parasitic and packaging

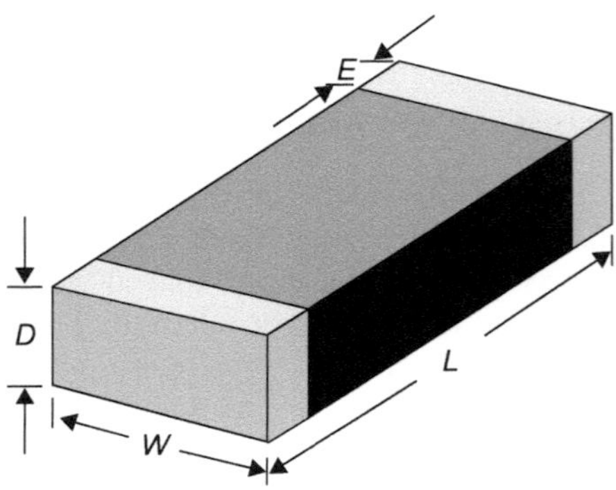

Figure 19.3 Label of mechanical dimensions of inductor SMD packages.

effects can be gained from equivalent circuit description of the inductor [4]. In Naishadham [16], necessary equations and calculations needed when designing a magnetic inductor or coupled inductor are presented.

An example of ferrite SMD components is presented in Fig. 19.4 [17]. The proposed structures have a conductive layer in the shape of zigzag and meander structures, with typical 0805 chip size (mechanical dimensions are $D \times L \times W = 1.1$ mm $\times$ 2.01 mm $\times$ 1.25 mm). A structure called a double coil has 1210 chip size (3.2 mm $\times$ 2.54 mm $\times$ 1.1 mm). The components were fabricated using the ceramic coprocessing technology. The conductive layer, platinum, $Pt = 10.6 \times 10^{-8}$ Ωm with the lower electric resistivity was chosen as the conductor to form the inductors. The width of the conductor is 150 mm and the thickness of the conductive layer is 10 mm. The inductor is embedded in the middle of a ferrite layer (Fig. 19.5). The ferrite materials used for components are F14

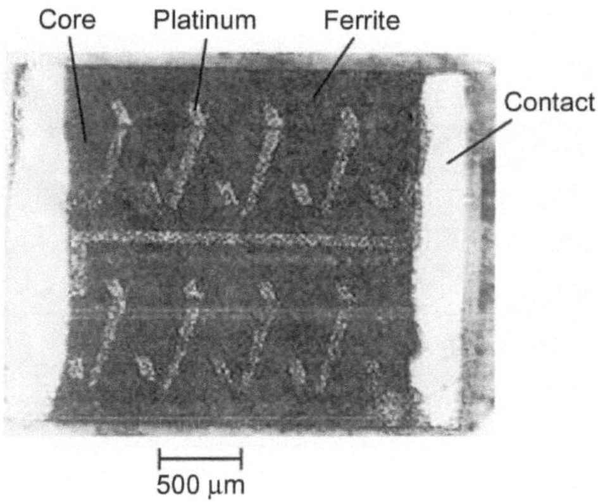

Figure 19.4 Cross-section of fabricated double coil with core and embedded in ferrite material, size 1210.

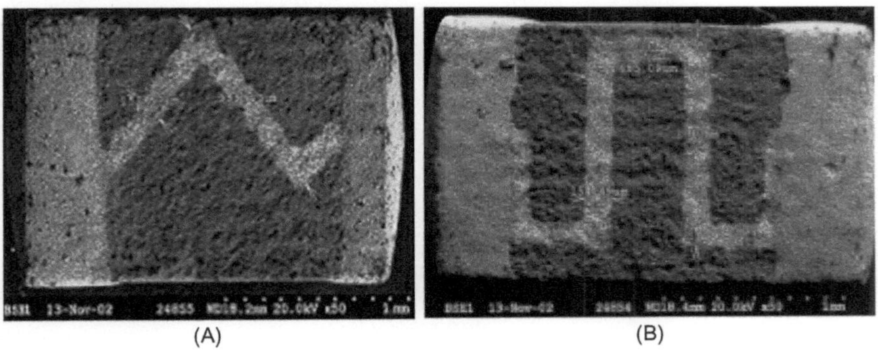

Figure 19.5 (A) Zigzag inductor and (B) meander inductor embedded in ferrite material, size 0805.

(low-permeability ferrite material $\mu_r = 220$) and F19 (high-permeability ferrite material $\mu_r = 1000$), from MMG Neosid Ltd. F14 ferrite has low loss factors at medium frequencies and high suppression impedance at high frequencies (over 100 MHz). F19 ferrite material is nickel–zinc ferrite, with initial permeability 1000 (while F14 has initial permeability 220). These ferrite materials find typical applications in EMI suppression components [17–19].

Comparison of measured and calculated values of inductance and total impedance Z for single and double coil in low permeability F14 material is depicted in Fig 19.6. At low frequencies (up to 3 MHz), losses in the suppressor are low. As frequency increases, series resistance R_S dominates and the suppressor behaves as a frequency-dependent resistor. This suppressor can be used for suppression of unwanted frequencies above 200 MHz. If the conductor's line is wider, for the

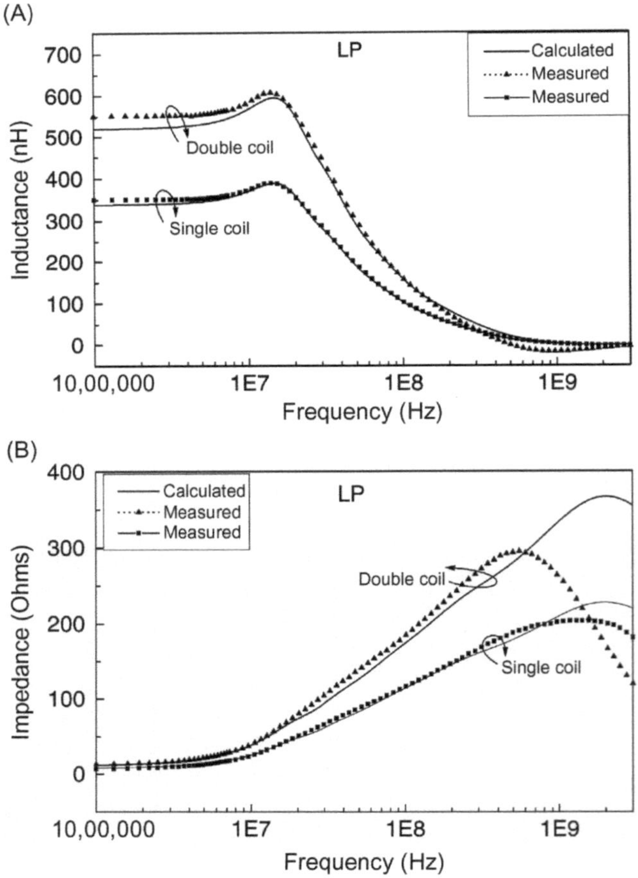

Figure 19.6 Frequency dependence of (A) the inductance and (B) the impedance for single and double coil in F14 ferrite.

same ferrite material, inductance L and DC resistance R_{DC} decrease. Because of that, series resistance R_S and reluctance X_L decrease, and hence, total impedance Z will be smaller.

On the other hand, if the conductive line is longer, then inductance L_0 is bigger (see zigzag and meander), and inductance L and impedance Z of the structure will be bigger. In Fig. 19.6, simulation and measurement results of inductance L and total impedance Z are compared.

19.2.2 LTCC ferrite components

In Marić et al. [20], presented planar type LTCC inductor is a one-port square spiral inductor, consisting of three 200 μm wide turns of 10 μm thick conductive segments embedded inside a combined ferrite-dielectric substrate stack. In order to meet the requirements of the LTCC technological process used for inductor fabrication, the ferrite stack is enclosed within two protective dielectric tapes from both sides (top and bottom). The corresponding 3D and cross-section representation of the ferrite inductor design is depicted in Fig. 19.7.

The ferrite stack was formed from four 70 μm thick ferrite ESL 40011 tapes, supported with two 100 μm thick ESL 41020 dielectric layers (dielectric constant of

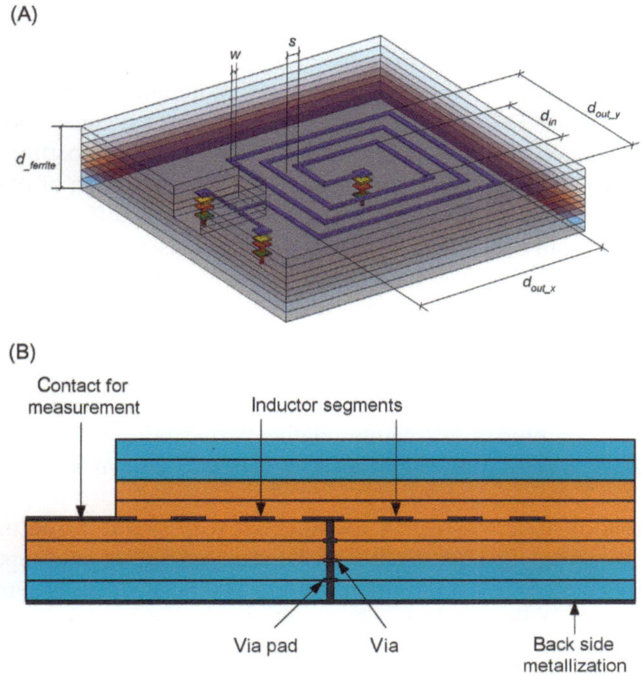

Figure 19.7 Schematic presentation of ferrite inductor stack: (A) 3D view and (B) cross-section.

7.5 at 1 MHz frequency) from the top and the bottom. The inductor was printed using silver ESL 903B paste on the top of the second ferrite tape, placing it in the middle of the stack. ESL 902 paste was implemented for via filling. The finalized inductor was characterized using standard RF measurement set-up (VNA, AC probe station).

It is evident from the obtained results (Fig. 19.8) that implementation of ferrite material substantially affects the inductance of the fabricated component. Inductance of 319.76 nH at Q-factor maximum frequency of 12.79 MHz is achieved with only a 240 μm thick ferrite stack. Quality factor maximum of a ferrite inductor is 11.35 and this value is relatively high, taking into consideration losses in ferrite material and very small thickness of used ferrite. Comparing inductance values of the ferrite inductor and same geometry inductor fabricated on a purely dielectric substrate increase of inductance over 11 times is achieved with implementation of ferrite material. This is very important in applications where high inductance values and high performances are required on a limited substrate area.

Marić et al. [21] considered that basic geometry of the 3D microinductor structure is a solenoid consisting of three interconnected half-turn circular windings positioned on the top of three adjacent layers of a dielectric stack formation. Line width of the solenoid conductive segment w is 500 μm, and the whole structure occupies $(4.00 \times 3.45)\ mm^2$ of the chip area. All inductors are embedded in purely dielectric ceramic substrate and Ceramtape GC and host different configurations of dielectric or ferrite core. Variations of dielectric core material are implemented in order to achieve wider operating frequency range and improve performance of the designed solenoid inductor. Ferrite cores are dices of commercially available ferrite (tapes ESL 40011 and ESL 40012) stacked in a different manner and inserted in dielectric foundation with intent to impact and tune inductor properties (inductance and quality factor) in fine manner. A 3D schematic representation of designed solenoid and cross-sections of all seven micro-inductors with different core material encircled with solenoid conductive segments [21] are presented in Fig. 19.9.

Investigated inductor designs incorporate an air core, cores from two different single ferrite materials with different permeability values, and three different cores realized as combination of these two ferrites. Inductors with cores of combined ferrites are designed with the purpose of finer inductance adjustment (inductance tuning) by means of material combination. Implemented ferrite materials are chosen due to their magnetic properties (mainly relative permeability value) to achieve different inductance values. Material properties are well known and determined in Blaž et al. [10].

A micrograph of the cross-section for solenoid with single ferrite material core (ESL 40011) presented in Fig. 19.10A reveal that individual core tapes have not integrated during sintering. Due to the relatively low peak sintering temperature of solenoids, ferrite cores are not fully cosintered and are prone to crack. Full formation of ferrite material grains appears at higher temperatures, which could not be applied in this case since they are above the silver melting point. This was one of the reasons for choosing solenoid geometry, which allows printouts of all conductive inductor segments purely on dielectric substrate and disenabling possible

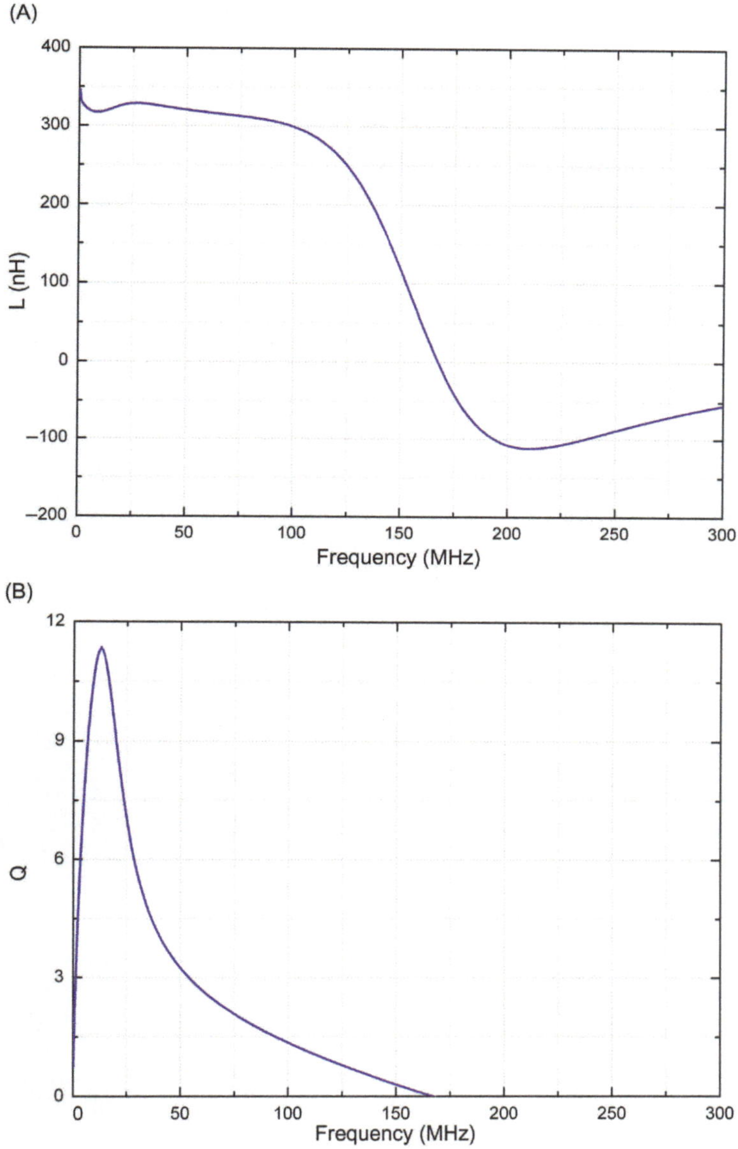

Figure 19.8 Measured results for ferrite inductor: (A) inductance and (B) Q-factor versus frequency.

abruptions of conductive paths. As can be seen in Fig. 19.10B, in the case of a solenoid with a 3:1 combination core most of dielectric-ferrite and ferrite 1 (ESL 40011) -ferrite 2 (ESL 40012) transition areas show no presence of delamination or air gaps, forming a good connection between core and dielectric surrounding.

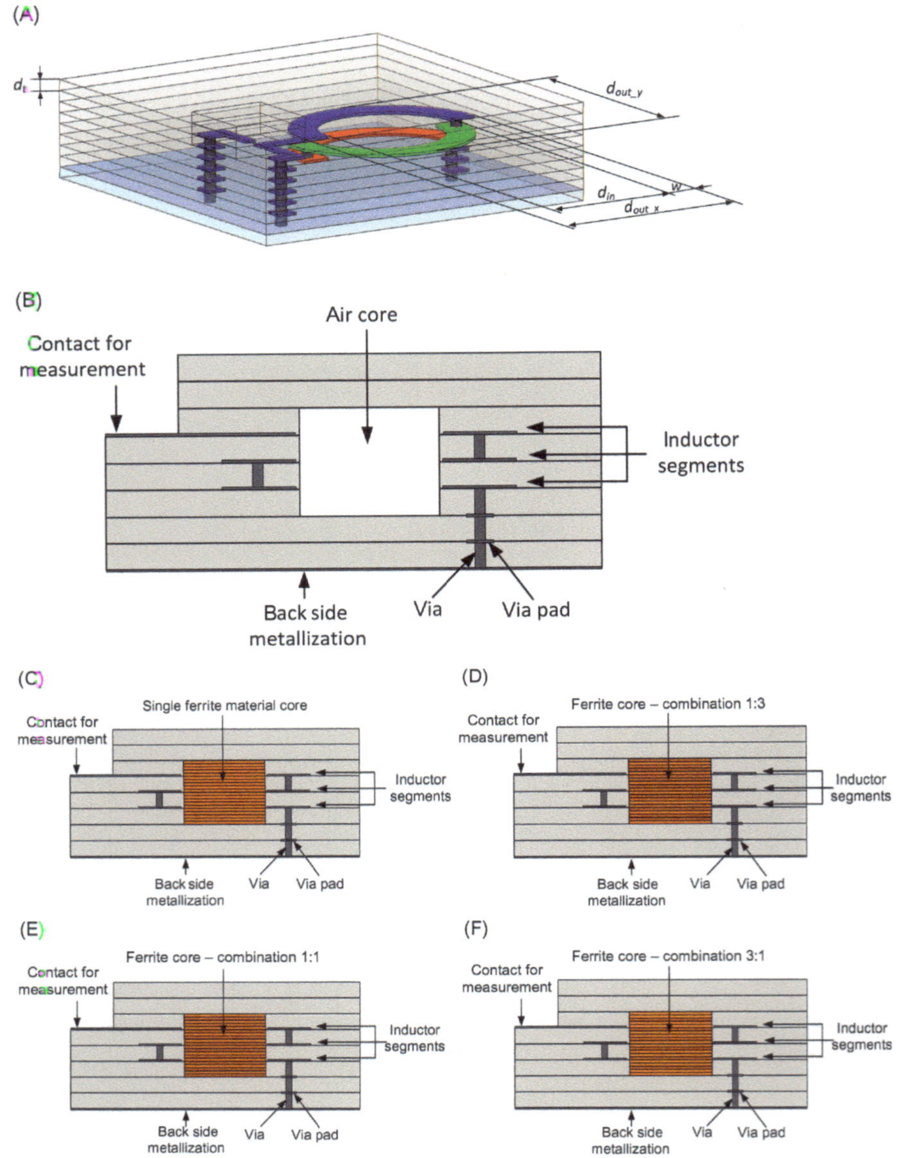

Figure 19.9 3D model (A), and schematic cross-sections of solenoids (B) air or dielectric, (C) single ferrite (ferrite 1 or ferrite 2), (D) 1:3 combination, (E) 1:1 combination, and (F) 3:1 combination.

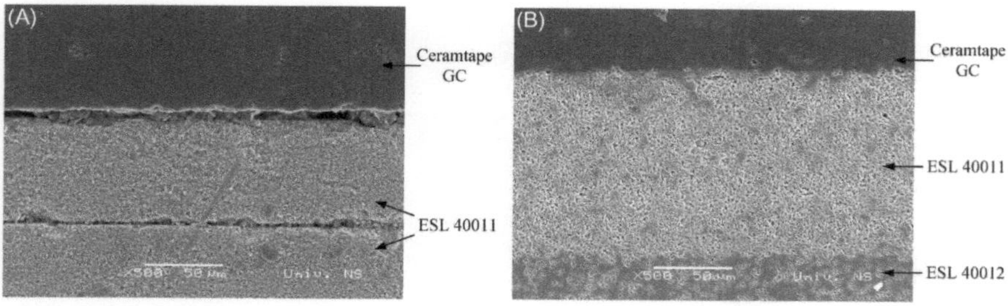

Figure 19.10 Micrographs of the cross-section for solenoid with (A) ESL 40011 core and (B) 3:1 combination core.

Extracted frequency dependences of solenoids inductance and quality factor are presented in Fig. 19.11.

Inductance for designs with ferrite cores increase approximately 35% and 60% for solenoids with single ferrite material cores is achieved. Significantly higher permeability of an ESL 40012 tape in comparison to an ESL 40011 tape leads to higher inductances of the solenoid. Achieved inductances at 100 MHz are 13.26 and 15.92 nH for solenoids with ESL 40011 and ESL 40012 cores, respectively. Finer inductance tuning is realized with cores fabricated as a combination of ferrite tapes. Raising the volume of ferrite with higher permeability in formed cores with ESL 40011: ESL 40012 ratios 3:1, 1:1, and 1:3, consequently leads to an inductance increase of approximately 2%–3%. An increase of almost 90% and an inductance value of 21.98 nH is achieved for an inductor with ESL 40012 core at a frequency of 1 GHz.

On the other hand, introduction of the air core improves inductor performance significantly. Effects are noticeable through up-growth of the quality factor by 40% and widening of the useful operating frequency range. A Q-factor maximum value of 109.15 at a frequency of 956.45 MHz and resonant frequency of 2.34 GHz is achieved with the air-core inductor. Quality factor maximums are lower if compared to corresponding values of solenoids with dielectric cores and their peaks are shifted toward lower frequencies. However, quality factor values are maintained at a rather high level (approximately 20 for single ferrite material core inductors and 40 for combined ferrite cores) due to relatively small thickness of used ferrites. Optimal core type and/or structure depend on the specifications of individual application.

An LTCC transformer is an interleaved planar transformer with three square 200 μm windings, occupying $(7.4 \times 7.4)\ mm^2$ of chip space [22]. Windings are sandwiched in the middle of two different LTCC stacks. One stack is a purely dielectric formation, while the other is a dielectric-ferrite combination. A transformer configuration embeds transformer windings between four ferrite LTCC tapes, enclosed with two supporting dielectric tapes from the top and bottom side. Cross-section schematic representations and a 3D model of transformer structure embedded in a ferrite dielectric combination is given in Fig. 19.12.

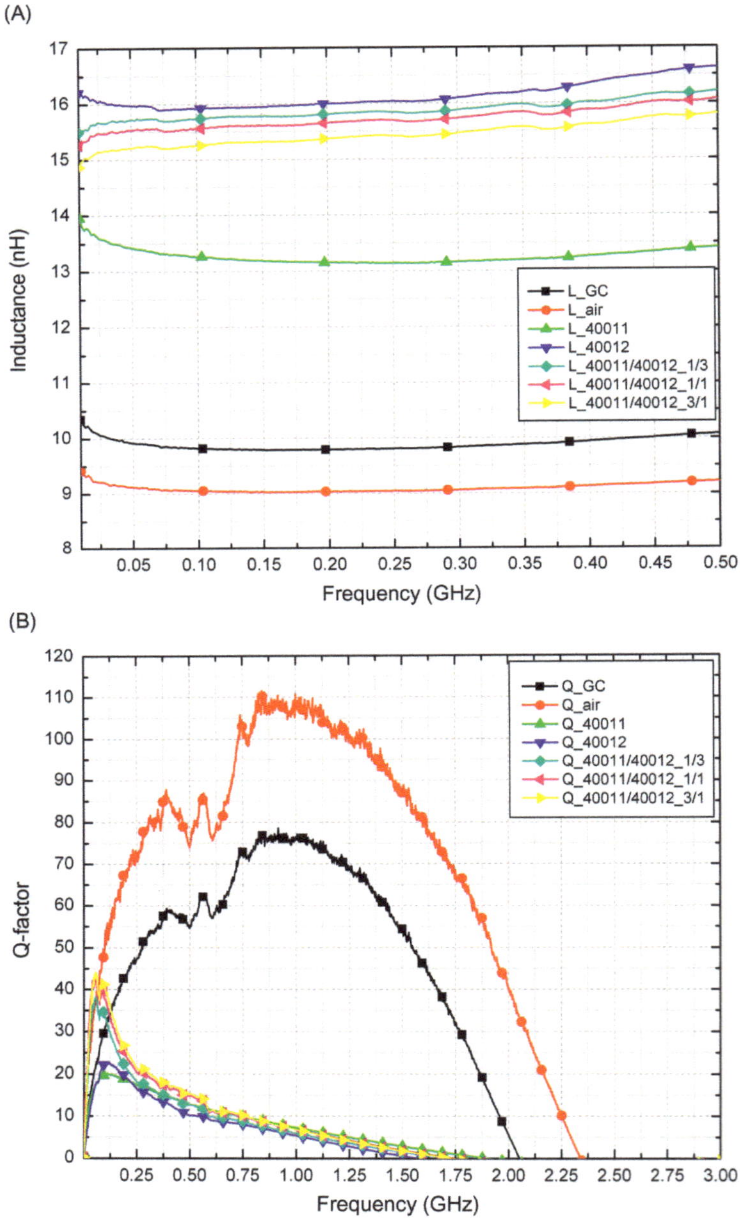

Figure 19.11 Measurement results of (A) inductance and (B) Q-factor frequency dependences.

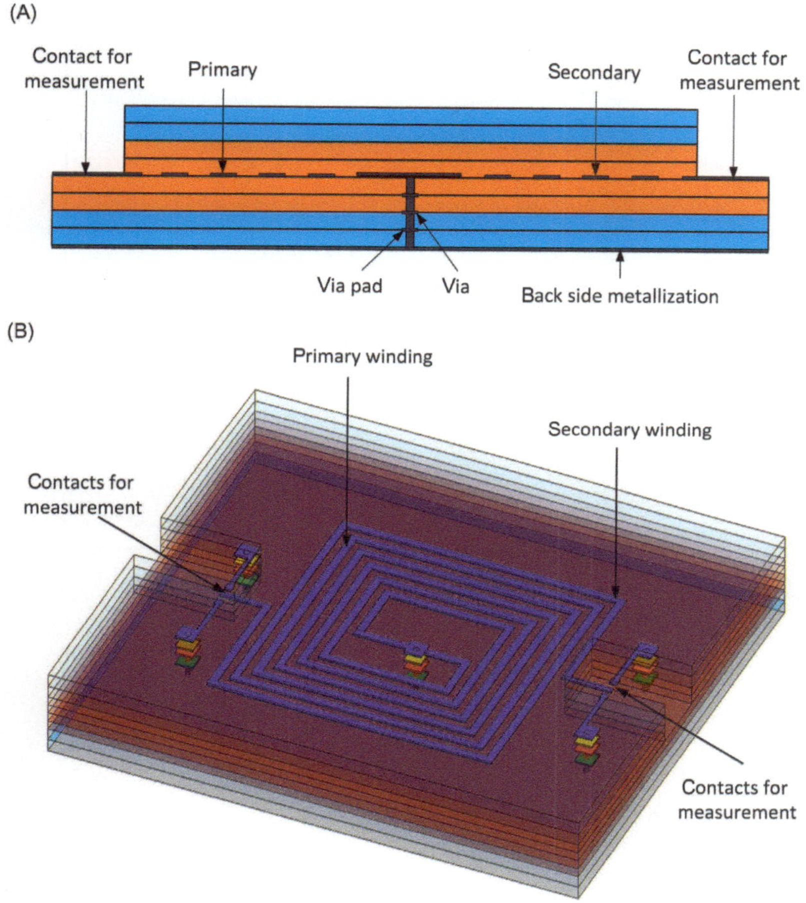

Figure 19.12 Schematic transformer presentation (A) cross-section and (B) 3D model.

Inductance and Q-factor of the primary and secondary windings were obtained from experimentally determined Y_{11} and Y_{22} parameters, respectively, while mutual inductance was calculated from Y_{12} values. Obtained dependences are given in Fig. 19.13A (inductance) and Fig. 19.13B (Q-factor).

At the frequency of 9.67 MHz (frequency of the ferrite transformer Q-factor maximum f_{Qmax}) inductance of the ferrite transformer exceeds 310 nH, while inductance of the corresponding all dielectric transformer is around 43 nH at the frequency of the dielectric transformer Q-factor maximum. At the same time, good performance of the transformer is maintained at lower frequencies. Maximum values of primary and secondary Q-factor (Q_{P_max} and Q_{S_max}) of the ferrite transformer decreases not more than 10% compared to the values obtained for the dielectric transformer. Encapsulation of the transformer inside the LTCC stack represents another advantage, allowing surface-mount of other components and diminishment of the size of the chip that carries passive components.

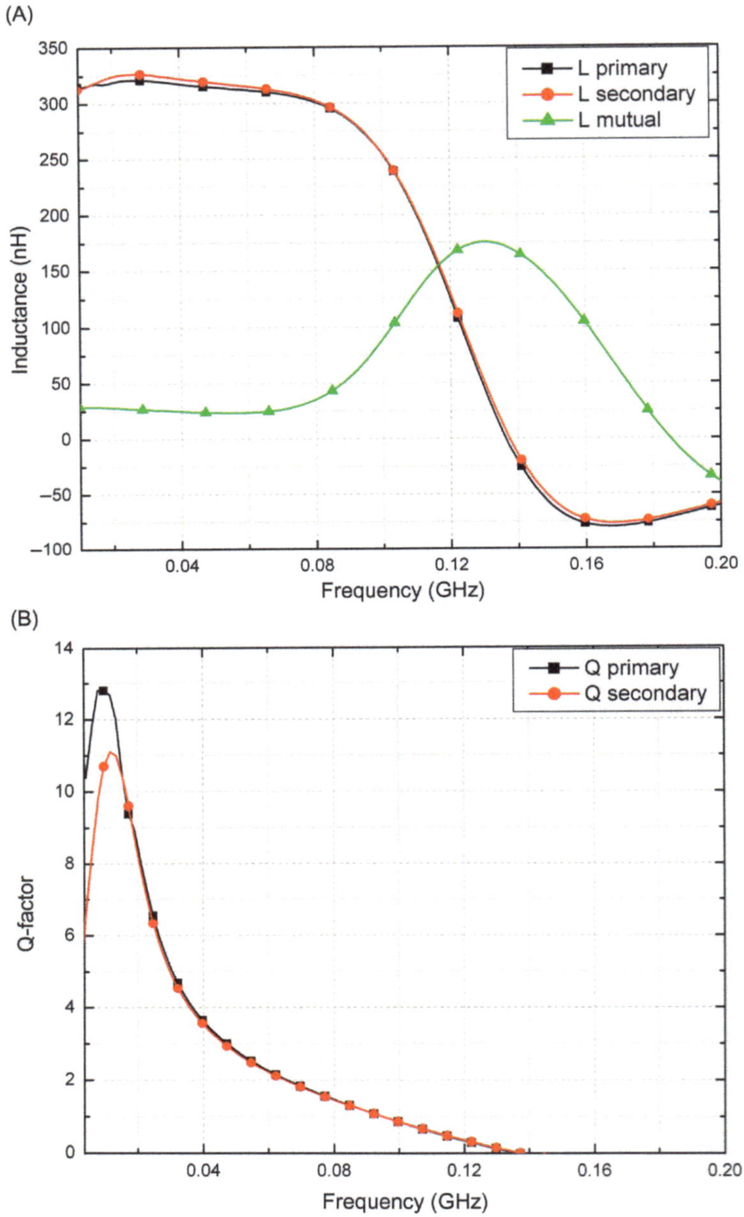

Figure 19.13 Measured results: inductance versus frequency (A) and Q-factor versus frequency (B) for ferrite transformer.

The multilayer LTCC ferrite inductors can be fabricated as the magnetic substrate in an integrated point-of-load converters working at high frequency. The magnetic substrate has active components at the top. The mixed laminate with alternating layers of ESL 40010 and ESL 40012 in a 1:1 ratio presents the highest permeability and the lowest core loss density among all examined samples when DC bias is above 1000 A/m. The integrated point-of-load converter was demonstrated to work at high efficiency with a power density as high as 1000 W/in^3 [23].

Kagita et al. [24] present the design of an electrically tunable ferrite phase shifter and a tunable notch filter. These components are made electrically tunable by winding the bias coils around the ferrite substrate. They operate in the partially magnetized state at magnetic fields <100 Oe. This biasing mechanism eliminates the use of an external biasing magnet for tuning. The phase shifter occupies a total volume of about 2 cm^3 including the bias circuitry and the test fixture. For a biased current of 700 mA, measured differential phase shift of 215 degrees having a figure of merit of 48°/dB is achieved at 10.6 GHz. Tunable reflection-type notch filter is designed using the same method of biasing. Center frequency of the notch is observed at 12.35 GHz. Tunability of 3.5% is achieved by varying the bias current from 50 to 450 mA.

19.3 Ferrite sensors

Ferrite materials have been extensively used for realization of different kinds of magnetic sensors for many decades. Numerous sensors consisting of various types and forms of ferrites are fabricated for specific sensing applications. The self-resonant frequency modulation of an integrated coil caused by the relative displacement of a ferrite core attached to a silicone rubber membrane is used for fabrication of passive pressure sensors [25]. The integration of the parallel capacitor is avoided using the coil and its self-capacitance, which is unnecessary for connecting the separate capacitor in parallel with the coil. The sensor consists of an inductor (outer dimensions of 3×3 mm^2 and 30 turns) and cylindrical ferrite core (0.95 mm diameter and 0.5 mm height). The obtained sensitivity is 9.6 kHz/kPa. A similar structure is used as a wireless intraocular pressure sensor [26]. A NiZn ferrite is attached on the bottom flexible membrane. By applying pressure on the sensor the ferrite gets closer to the cooper coil and causes the increase in the inductance of the inductor. The obtained change of the sensor's resonant frequency is 0.4 MHz in the pressure range 0–70 mmHg. A cylinder-shaped ferrite core is implemented in an orthogonal fluxgate sensor for the magnetic treatment facility for submarines and surface vessels [27]. The ovality meter is realized by the ferrite core and the specially designed inductor for the measurement of ovality in a pipeline in a plant site or in any industrial establishment [28]. The ferrite core can be used for the realization of a linear variable differential transducer (LVDT) for measuring and detecting the level of fluid. In addition to the step-down transformer, transducer coils, a floating object (plastic ball), and a multimeter, a ferrite core is the main part of the sensor, whose displacement determines the flux linkage, induced electromotive force, as well as the sensitivity of the LVDT [29]. An interesting application of ferrite

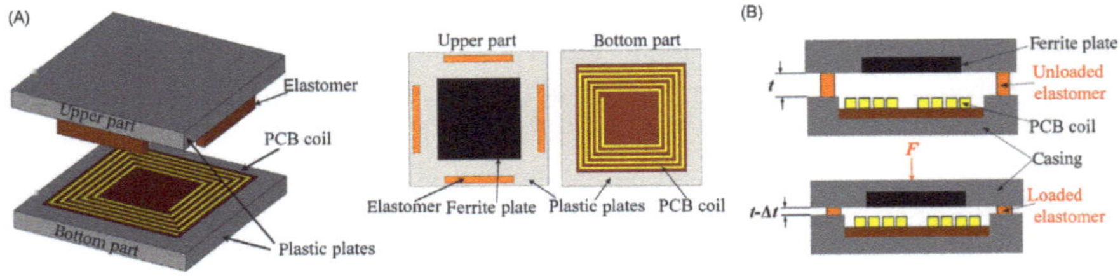

Figure 19.14 The view of the sensor model: (A) exploded 3D and views from above the upper and the bottom parts of the sensor, (B) the cross-section of the sensor and its operational principle.

material is in an instrument for inspection in a nondestructive manner of underground oil and gas pipelines. This instrument is passed through them to capture the data about corrosion of pipes and to report critical corrosion spots to pipeline operators. Instead of using a highly sensitive hall sensor, ferrite cores are used as a back yoke for enhancing sensitivity in caliper tools for pipeline inspection [30].

The force sensor can be miniaturized using a soft ferrite as low-cost material. The force sensor consists of a single coil and a commercially available elastomer with a ferrite plate is developed [31]. The exploded 3D view of the sensor and the cross-section with its operational principle are shown in Fig. 19.14. The PCB coil is designed as a planar square spiral type with outer dimensions of 15×15 mm^2. Four prismatic pieces of elastomer are placed and aligned with the four sides of the coil in order to enable the sensitivity on the force. The used ferrite plate has a great permeability and it covered the entire surface of the coil. The coil and the ferrite plate are embedded in a protective plastic casing.

Changes of the sensor's resonant frequency for variable values of applied force are wirelessly done with the phase-dip technique, using an external coil antenna. By applying force to the sensor, the elastomer deforms and the ferrite plate approaches the coil, causing changes to the inductance of the coil. The resonant frequency of the antenna–sensor system shifts toward lower frequencies. The characteristic of the sensor can provide information about the force applied on the sensor as seen in Fig. 19.15. The similar sensor is integrated with a suitable mechanical device and used as a seat occupancy detector with adjustable weight threshold [32].

As a displacement sensor the structure of the coil and the ferrite disk are used for realization of displacement measurement. In Kisić et al. [33] sensors for normal displacement measurements are developed using the heterogeneous integration process of traditional technologies (PCB and LTCC) with a flexible polyimide foil as membrane with a ferrite disk attached to it. The coil is designed as a square spiral type with outer dimensions of 19 mm. With the displacement of the polyimide foil, the ferrite disk gets closer to the coil, causing an increase in its inductance and a decrease of the resonant frequency of the system (coil, ferrite disk, and antenna). The dimensions of the used ferrite disk's different radii ($r = 6$ mm and $r = 9.5$ mm) define the sensitivity of the realized sensor. The obtained characteristics of systems with two ferrite disks of different size are shown in Fig. 19.16. The sensor with the smaller ferrite disk has a

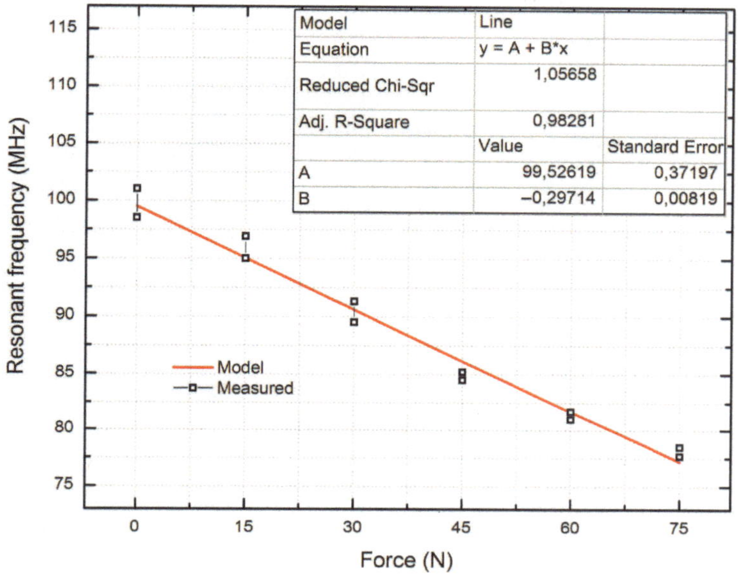

Figure 19.15 Resonant frequency versus force.

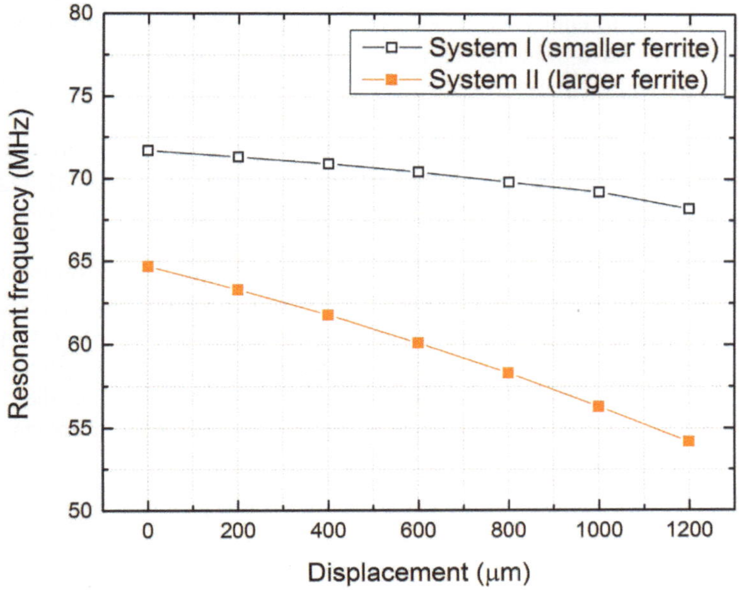

Figure 19.16 Resonant frequency characteristics of systems with two different ferrite disk radii.

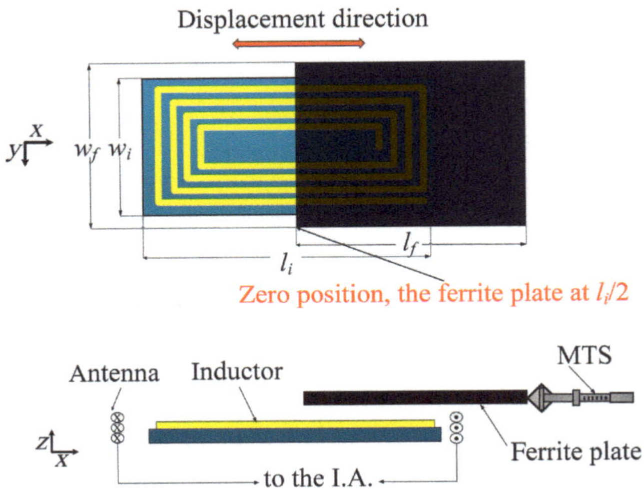

Figure 19.17 Inductive sensor and its sensing mechanism.

smaller inductance value, and consequently resonant frequency of the System I (sensor coil, smaller ferrite disk, and antenna) is higher compared to the resonant frequency of the System II (sensor coil, larger ferrite disk, and antenna). Using sensor structures with the same outer dimensions but larger ferrites yields a greater change in the resonant frequency and the sensitivity of the system for the same amount of displacement. Therefore, in order to enhance the performance of the sensor, the dimensions of the coil and the ferrite have to be appropriately selected. The same structure with the coil and the ferrite disk on a polyimide membrane can be used for pressure measurements, as presented in Kisić et al. [34].

A tangential displacement sensor consists of one stationary inductor, and a movable ferrite plate relative to the inductor is presented in Kisić et al. [35]. Fig. 19.17 illustrates the design concept of inductive tangential displacement sensor and its working mechanism.

A PCB inductor is a rectangular spiral type with outer dimensions of $l_i = 17.5$ mm and $w_i = 9.5$ mm. The inductor is fixed and a moving object (such as a manual translation stage) moves the ferrite plate in one x-axis. At zero position the ferrite plate overlaps half of the area of the inductor. As the overlapping area between the inductor and the ferrite increases, the inductance of the inductor increases too; consequently, the resonant frequency of the system decreases. The characteristic of the system is shown in Fig. 19.18.

19.4 Conclusion

It is evident from presented results that implementation of ferrite materials substantially affects inductance and quality factors of fabricated components. Comparing

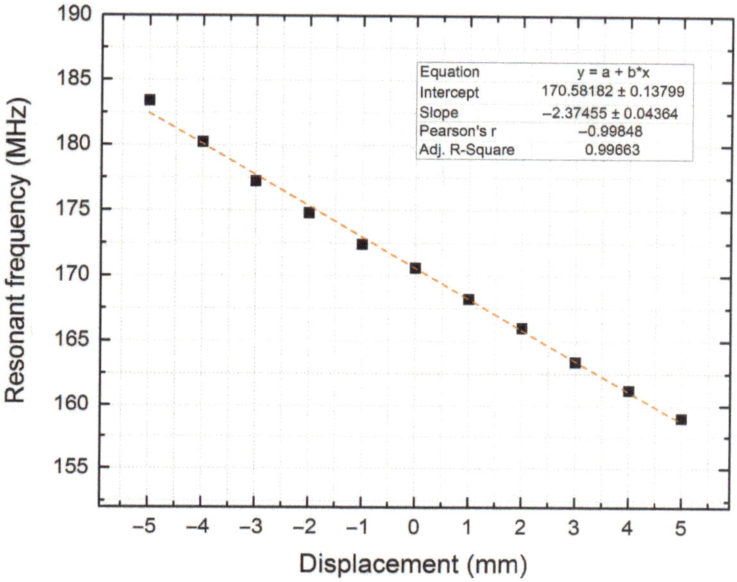

Figure 19.18 Resonant frequency versus displacement.

inductance values of the ferrite component and the same geometry of components fabricated on a purely dielectric substrate, a significant increase of inductance is achieved with implementation of ferrite material. This is very important in applications where high inductance values and high performances are required on a limited substrate area.

The multilayer LTCC ferrite inductors can be fabricated as the magnetic substrate in an integrated point-of-load converter working at high frequency. The magnetic substrate has active components at the top. Using soft ferrite materials enables the work of an electrically tunable ferrite phase shifter and a tunable notch filter. These components are made electrically tunable by winding the bias coils around the ferrite substrate. This biasing mechanism eliminates the use of an external biasing magnet for tuning.

Ferrite materials are extensively applied for realization of different kinds of magnetic sensors. Various types and forms of ferrites are used for specific sensing applications. The sensors can be miniaturized with improved efficiency and functionality using soft ferrite materials as low-cost materials.

References

[1] B.I. Bleaney, B. Bleaney, Electtricity and Magnetism, Oxford university press, 1976.
[2] A. Goldman, Modern Ferrite Technology, Springer, 2006.
[3] B. Tellini, R. Giannetti, S. Lizón-Martínez, Sensorless measurement technique for characterization of magnetic material under nonperiodic conditions, IEEE Trans. Instrum. Meas. 57 (7) (2008) 1465–1469.

[4] Q. Yu, T. Holmes, K. Naishadham, RF equivalent circuit modelling of ferrite core inductors and characterization of core materials, IEEE Trans. Electromagn. Compat. 44 (1) (2002) 258–262.
[5] C.F. Foo, D.M. Zhang, A resonant method to construct core loss of magnetic materials using impedance analyzer, IEEE Power Electronics Specialists Conference, Fukuoka, Japan, May 22–22, 1998, PESC 98 Record 29th Ann. IEEE Power Electronics Specialists Conference vol. 2, (1998) 1997–2002.
[6] R. Dosoudil, E. Ušak, V. Olah, Computer controlled system for complex permeability measurement in the frequency range of 5Hz–1GHz, J. Elect. Eng. 57 (8/S) (2006) 105–109.
[7] N. Blaz, A. Maric, G. Radosavljevic, Lj Zivanov, G. Stojanovic, I. Atassi, et al., Modeling and characterization of frequency and temperature variation of complex permeability of ferrite ltcc material, Prog. Electromagn. Res. B 23 (2010) 131–146.
[8] N. Blaž, A. Marić, G. Radosavljević, N. Mitrović, I. Atassi, W. Smetana, et al., Determination of electric and magnetic properties of commercial LTCC soft ferrite material, J. Microelectron. Electron. Packag. 8 (1) (2011) 1–9.
[9] A. Marić, N. Blaž, G. Radosavljević, S. Toškov, L. Živanov, Affect of peak sintering temperature variation on LTCC ferrite tape complex permeability, Proceedings of the 37th International Spring Seminar on Electronics Technology, ISSE, May 7–11, 2014, Dresden, Germany, (2014) 18–19.
[10] N. Blaž, A. Marić, I. Atassi, G. Radosavljević, L. Živanov, H. Homolka, et al., Complex permeability changes of ferritic LTCC samples with variation of sintering temperatures, IEEE Trans. Magn. 48 (4) (2012) 1563–1566.
[11] K. Liu, T.W. Law, P.L. Cheng, I.T. Chong, D.C.C. Lam, Low temperature processing and electrical properties of embedded ferrite-cored inductor, Proceedings of the 52nd Electronic Components and Technology Conference, May 28–31, 2002, San Diego California, USA, (2002) 490–494.
[12] A. Shamim,J. Roy, N. Hojjat, R. AbouElasoued, D. Baillargeat, Microwave and Magnetostatic Characterization of Ferrite LTCC for Tunable and Reconfigurable SiP Applications, Proceedings of the IEEE MTT-S International Microwave Symposium, June 3–8, 2007, Honolulu, Hawaii, (2007) 691–694.
[13] A. Roesler, J. Schare, S. Glass, K. Ewsuk, G. Slama, D. Abel, et al., Planar LTCC transformers for high-voltage flyback converters, IEEE Trans. Comp. Packag. Technol. 33 (2) (2010) 359–372.
[14] P. Williams (Ed.), Status of the Technology Industry Activities and Action Plan, Surface Mount Council, 1999.
[15] Murata Manufacturing Co., Ltd., Chip Inductors, 2013. Available at: www.murata.com.
[16] K. Naishadham, A rigorous experimental characterization of ferrite inductors for RF noise suppression, 1999 IEEE Radio and Wireless Conference, RAWCON 99, August 1–4, 1999, Denver, 271–274.
[17] M. Damnjanović, Lj Živanov, G. Stojanović, V. Desnica, Comparison of different structures of ferrite EMI suppressors, Microelectron. Int. 3 (3) (2006) 42–48.
[18] M. Damnjanović, G. Stojanović, V. Desnica, Lj Živanov, R. Raghavendra, P. Bellew, et al., Analysis, design and characterization of ferrite EMI suppressors, IEEE Trans. Magn. 42 (2) (2006) 270–277.
[19] M. Damnjanović, L. Živanov, G. Stojanović, A. Menićanin, Influence of conductive layer geometry on maximal impedance frequency shift of zig-zag ferrite EMI suppressor, IEEE Trans. Magn. 46 (6) (2010) 1303–1306.

[20] A. Marić, G. Radosavljević, N. Blaž, W. Smetana, L.J. Živanov, Embedded ferrite LTCC inductors, Proceedings of the EightInternational CICMT Conference and Exhibition 2012, IMAPS/ACerS, April 16–19, 2012, Erfurt, Germany, 2012, 000388–000393.
[21] A. Marić, G. Radosavljević, N. Blaž, L. Živanov, Fine tuning of 3D LTCC inductor properties using combination of different ferrite and dielectric tapes, Int. J. Appl. Ceram. Technol. 12 (5) (2015) 1034–1044.
[22] A. Marić, G. Radosavljević, N. Blaž, L. Živanov, High quality spiral transformer embedded in LTCC substrate, Proceedings of the 36th International Spring Seminar on Electronics Technology ISSE 2013, (May 2013), Alba Iulia, Romania, 436–440.
[23] W. Zhang, M. Mu, D. Hou, Y. Su, Q. Li, F. Lee, Characterization of low temperature sintered ferrite laminates for high frequency point-of-load (POL) converters, IEEE Trans. Magn. 49 (11) (2013) 5454–5463.
[24] S. Kagita, A. Basu, S. Koul, Characterization of LTCC-based ferrite tape in x-band and its application to electrically tunable phase shifter and notch filter, IEEE Trans. Magn. 53 (1) (2017) 1–8.
[25] A. Baldi, W. Choi, B. Ziaie, A self-resonant frequency-modulated micromachined passive pressure transensor, IEEE Sensors J. 3 (6) (2003) 728–733.
[26] B. Kang, H. Hwang, S.H. Lee, J.Y. Kang, J. Hu Park, C. Seo, et al., A wireless intraocular pressure sensor with variable inductance using a ferrite material, J. Semicond. Technol. Sci. 13 (4) (2013) 355–360.
[27] Y. Kim, B. Yoon, Y.H. Kim, C.S. Yang, K.H. Shin, Wide range orthogonal fluxgate sensor fabricated with cylinder-shaped ferrite core, IEEE Trans. Magn. 47 (10) (2011) 3744–3747.
[28] S. Ramesh, M.P. Rajiniganth, P. Sahoo, Deployment of inductance-based pulsating sensor toward development of measurement technique for ovality in pipe, IEEE Trans. Instrum. Meas. 65 (6) (2016) 1443–1449.
[29] U.K. Muhammad, S. Umar, Sensitivity determination of linear variable differential transducer in fluid level detection techniques, Int. J. Mod. Eng. Sci. 2 (2) (2013) 73–83.
[30] R. Keshwani, S. Bhattacharya, Ferrite cores as back yoke for enhancing hall sensor sensitivity in caliper toolfor pipeline inspection, Sens. Rev. 29 (4) (2009) 355–360.
[31] M. Kisic, N. Blaz, K. Babkovic, A. Maric, G. Radosavljevic, L. Zivanov, et al., Passive wireless sensor for force measurements, IEEE Trans. Magn. 51 (1) (2015) 1–4.
[32] M. Kisic, N. Blaz, K. Babkovic, L. Zivanov, M. Damnjanovic, Detection of seat occupancy using a wireless inductive sensor, IEEE Trans. Magn. 53 (4) (2017) 1–4.
[33] M. Kisić, N. Blaž, K. Babković, A. Marić, G. Radosavljević, L. Živanov, et al., Performance analysis of a flexible polyimide based device for displacement sensing, Factauniversitatis Series: Electron. Energ. 28 (2) (2015) 287–296.
[34] M. Kisić, N. Blaž, A. Marić, G. Radosavljević, L. Živanov, M. Damnjanović, Influence of coil design on sensing performance of pressure sensor with polyimide membrane, Proceedings of the 37th International Spring Seminar on Electronics Technology, ISSE, May, 7–11, 2014, Dresden, Germany, (2014) 421–426.
[35] M. Kisic, N. Blaz, K. Babkovic, L. Zivanov, M. Damnjanovic, Inductive tangential displacement sensor, Proceedings of the 39th International Spring Seminar on Electronics Technology, ISSE, May 18–22, 2016 Pilsen, Czech Republic, (2016) 433–438.

Biomedical applications

20

Drago M. Djordjevic[1], *Sasa T. Cirkovic*[1] *and Dusanka S. Mandic*[2]
[1]University of Belgrade, Belgrade, Serbia, [2]MADU Clinic, Belgrade, Serbia

20.1 Introduction

Natural magnetic particles can be found everywhere. Metals and metal oxides are the primary magnetic materials. Magnetic particles and nanoparticles (1–100 nm) can be divided into magnetic metal oxides, pure magnetic metals, and magnetic composites [1]. Magnetic metal oxide particles (biological, artificial) are the most represented magnetic particles on Earth.

Biological magnetic oxides include primarily enormous amounts of sediments in the water and land that contain particles of metal oxides in Earth's material. Only small amounts of metal iron oxide particles are formed from magnetosomes, which represent special organelles in bacteria [2]. Throughout nature, microorganisms are known to synthesize iron oxide nanoparticles (IONPs) through intra- and extracellular mineralization processes [3].

Iron oxide particles are present in a wide range of biological species. An integral part of natural minerals are iron (Fe) oxides: magnetite (Fe_3O_4), hematite (α-Fe_2O_3), and maghemite (γ-Fe_2O_3) [4], as well as greigite (Fe_3S_4) and various ferrites [$MO \cdot Fe_2O_3$, where M = cobalt (Co), nickel (Ni), magnesium (Mg), manganese (Mn), zinc (Zn), and O = oxide] that are crystallized by living organisms under ambient conditions [5]. The most common iron storage protein ferritin $(FeOOH)n$, which contains magnetic nanoparticles, is present in almost every cell of plants and animals including humans [6]. The human brain contains over 10^8 magnetic nanoparticles of Fe_3O_4–γ-Fe_2O_3 per gram of tissue [6].

Magnetism is a minor sense that has been used for navigation from magnetotaxic bacteria, insects, fishes, and birds to mammals that is realized by magnetic particles [7]. All of these organisms contain an aggregate of magnetic particles mainly attached to nerve cells [8]. Sensitivity to electrical field is another minor sense in some sharks, rays, dolphins, duck-billed platypuses, and salt and fresh water fish [7].

Biomineralization in living organisms leads to inorganic–organic nanocomposite formation of hybrid nanomaterials [5]. The formation of inorganic crystals in the process of biomineralization is genetically controlled by organic biomolecules [9], and presumably incorporated in the genome of higher organisms, including humans [10]. Interactions between metal-containing particles, biopolymers (proteins, nucleic acids, polysaccharides), and cells are crucial in enzymatic catalysis, biomineralization, and many biological processes [11].

Artificial magnetic oxides are synthesized on the bases of biogenic magnetic particles with a tendency to use the properties of these particles not only for material sciences but also for natural sciences. According to all major stages of transformation

Magnetic, Ferroelectric, and Multiferroic Metal Oxides. DOI: http://dx.doi.org/10.1016/B978-0-12-811180-2.00020-7

of metal atoms (particle size > metal atom > cluster > nanoparticle > colloid > bulk metal) [11], metal oxides are synthesized in size, shape, and properties suitable for use in biomedical purpose. Especially important is synthesis and use of magnetic nanoparticles or nanostructures, as well as their interactions with biological magnetic materials and biopolymers [12–14].

20.2 Biomedical applications of magnetic oxides

Among a wide range of the magnetic nanomaterials used in biomedicine are nanoparticles of magnetic metals, simple and complex magnetic oxides, and magnetic alloys or composites. In particular, the widespread applications have iron, cobalt, nickel, chromium (e.g., $MgCr_{0.9}Fe_{11}O_4$), titanium alloys, and iron oxide (ferrite) (e.g., barium (Ba) ferrite ($BaFe_{12}O_{19}$), $CoFe_2O_4$, and $ZnO/ZnFe_2O_4$) [1]. Iron oxides have multiple applications in various fields of biomedicine. Lesser significance used other magnetic oxides. Lanthanides metals and oxides have more and more usage in many fields of biomedicine.

In biomedicine, mainly used are magnetic alloys and polymer nanocomposites, which represent magnetic nanoparticles in a polymer inorganic or organic matrix [3,6]. The most magnetic nanoparticles for biomedical applications are comprised of iron oxide-based nanocrystalline structures of magnetite and/or maghemite encased or coated with organic stabilizers and organic or inorganic coatings (metal or nonmetal) [15] in a core-shell, matrix, and core-shell-core structures [16]. However, novel nanoparticles with advanced magnetic properties have different compositions. Bimagnetic materials include a combination of IONPs and other magnetic inorganic materials in a core-shell architecture [3]. They involve transition metal ferrite (MFe_2O_4) nanoparticles, where M is a divalent metal ion (M = Mn, Co, Zn, Ni, Fe, Cu, etc.) [3], while others are metal alloys such as iron cobalt (FeCo) [6,15], iron platinum (FePt) [3,6,15], iron silver (FeAg), or iron gold (FeAu) nanoparticles [1,4].

Biomedical applications of magnetic oxide nanomaterials, especially nanoparticles, can be classified according to their application inside or outside the body (in vivo, in vitro). In medicine, nanoparticles are usually used as components of (1) magnetic liposomes (liposomes with combination of ferromagnetic iron oxide and other magnetic or paramagnetic ions) [17]; and (2) ferrofluids: (a) surfacted (iron oxide coated with shell of surfactant molecule) and (b) ionic-cationic or anionic (iron oxide coated with electric shell) [14].

Some of the main possible applications of magnetic nanoparticles in biomedicine and health care are summarized in Table 20.1 [18–41].

20.2.1 Magnetic oxide sensors

Sensors are devices that detect and measure a physical and chemical stimulus. The sensors primarily consist of sensing elements (e.g., different biorecognition

Table 20.1 Biomedical and health care applications of magnetic nanoparticles [18–41]

	Applications of magnetic nanoparticles
Biomedical	Information technology [18] (e.g., ferrites for microwave 1–100 GHz applications [19]), printed electronics [4], nanoelectronics and magnetooptics [20], nanoelectronics, and nanoelectromechanical systems [21]. Magnetic/thermoresponsive technology [22]. Magnetic nanoparticles technology [23]. Medical instrumentation [24], machine tool positioning, and for rotation and speed sensing in devices [25]. Nanocoatings [23,24]. Magnetic macro- and nanosensors, and consumer electronics [25]. Biosensors (electrochemical sensors, immunosensors, optical (surface plasmon resonance) biosensors, DNA or RNA sensors) [3]. Flexible electronics (flexible magnetic field sensors, biomedical flexible appliances, and artificial devices like robotics) [26]. Magnetic particle (particularly coated with liposomes [4]—magnetic liposomes targeted liposomal delivery systems) [17] for molecular biology investigations [6], fluorescent biological labels [17,21,23], biorecognition (antibodies, aptamers, or antibiotics) [27], specific capture of targeted biomolecules [3], drug delivery and gene delivery [1,17,21,23]. Biomedical imaging [23]: contrast agents (MRI, etc.) [1,6,17,23,28], contrast enhancement [17,28], color imaging [28]. Cell sorting [28]. Drug discovery [23]. Biotechnology [21,29]. Agriculture, food, and cosmetics [30]. In vitro application: diagnostic (selection and separation, magnetorelaxometry and magneto-optical relaxation [31]; bioseparation (DNA, antibody, protein, gene, enzyme, cell, virus, and bacteria) [32], and magnetic based biosensors [1,14,16,32]; catalysis applications [33]; magnetic immunoassay [23]). Bioanalytics (analytical target signal, fluorescence techniques [33], imaging) [17], probing of DNA structure [17], biodetection and selectively separation of biomaterials (cells, proteins, genes, and pathogens) [3,14,17,23]. Blood purification [14]. Models to predict the in vivo behavior of nanoparticles for drug delivery [34]. In vivo targeted drug delivery (including control by superconducting magnetic force [14,16,35]), magnetic contrast agents in MRI [14], and hyperthermia [14,16] or thermoablation agents [32]; MRI of enzyme activity, tracking of labeled dendritic cells and macrophages cells, monitoring stem cell migration [36]. Cellular therapy (e.g., cell labeling [1,33], cell separation and handling of cells, purifying cell population, transfection (magnetofection), tissue repair [1,4,33], diseases of the musculoskeletal system, severe inflammation, disability, and pain) [4], gene therapy [6,23], tumor destruction via heating (magnetic hyperthermia) [1,3,17,23]. Immunomagnetic methods in cell biology and cell separation, and in pure medical application [1,6,17]. Engineering [37], genetic and tissue engineering [1,23,38].

(*Continued*)

Table 20.1 (Continued)

	Applications of magnetic nanoparticles
Health care	Magnetic cooling and heating systems [28]. Smartphone-based medical diagnosis system [39]. Diagnostic imaging (contrast enhancement in MRI, magnetic particle imaging) [2,14,23]; multimode nanoprobes (fluorescent magnetic, etc.) [1,21]. Theranostics [32]. Therapeutic in vivo applications of magnetic carriers, target specific drug delivery [1,3,6,14,21,23] (including painkillers, cancer gene therapy [17,21,23], magnetic hyperthermia [14], and detoxification) [30]; MRI contrast agents [1,3,23] and delivery of therapeutic radioisotopes [17]; therapeutic targets in (cancer and tumor) chemotherapy and radiotherapy (including nanoneedles) [1]; nanoscale biosensors and imaging [1,4,32]; nanocoatings on surfaces [24] including implants [40]; nanocarrier for vaccination and antimicrobial activities [4,17,29,30]; medical implant infection control [40]. Solid lipid nanoparticles in drug delivery and research; nanophotothermolysis for the treatment of cancer, hepatitis B virus, and many other viruses; delivering antigens for a particular disease into the blood stream; and preventing aging of the skin [4]. Biomechanical applications [41]. Regenerative medicine [1,14,21,30,38].

Abbreviations: *DNA*, deoxyribonucleic acid; *MRI*, magnetic resonance imaging; *RNA*, ribonucleic acid.

elements) and a range of transducers (e.g., electrochemical, magnetic, optical, and acoustic) to translate the stimulus (recognition) into physical (e.g., electrical) signals [42]. They are generally classified as (1) electric, magnetic, electromagnetic wave sensors; (2) heat, temperature sensors; (3) optical sensors; (4) physical (mechanical displacement) sensors; and (5) chemical sensors [42].

Magnetic field sensors are sensors that are used to transform magnetic or magnetically coded information into electrical signals (magnetic transducers). Magnetic nanoparticles, clusters, and partially thin films are used as magnetic field sensors. Magnetic sensors are mainly based on the same working principle: the change of magnetic moment, change of temperature, or magnetoelasticity that create the medium in which they are immersed [43]. In biomedical applications, the medium can be a fluid such as blood, the cerebrospinal fluid, or a culture medium or organic tissue.

Natural biological sensors (biosensors) are found in all biological species in the form of magnetic particles such as ferromagnetic iron oxides (magnetite, maghemite) and sulphides (greigite) [5]. Magnetic particles are the basis of the navigation system of living beings responsible for their behavior and orientation in relation to the Earth's magnetic field lines, including seasonal migrations over long distances [7,44]. The human brain and other organs [10] also contain biogenic magnetite [6,44], which is important for the interactions with the external magnetic/electromagnetic fields (MFs/EMFs) [12,13].

The behavior of the magnetic materials that absorb or emit heat in magnetizing or demagnetizing processes is called magnetocaloric effect (MCE) [45]. Many nanoparticles of biologic origin take part in sustainability of thermodynamic balance in organisms like magnetite, maghemite, and greigite, because MCE is inherent physical property of all magnetic materials [46].

Artificial magnetic field sensors are devices that detect or measure the magnetic field in a medium. They are measuring the magnitude of the field or the vector component of the field. Magnetic nanoparticles, clusters, and partially thin films are used as magnetic field sensors [47,48]. Sensors mostly serve for other applications; for example, magnetoresistive sensors are mainly used as reading heads in the computer industry for biomedical purposes, superconducting quantum interference device (SQUID) sensors for biomagnetic applications, inductive sensors on resonance principle for medical applications [49], and magnetocaloric sensors for temperature changes in the medium [50]. In addition to nuclear magnetic resonance (NMR), detection methods of homogeneous biosensing cover inductive coils, Hall sensors, magnetoresistive sensors, SQUID sensors, and optical sensors [51].

Caloric sensors are sensors that register changes in environment temperature, and then switch on specific thermodynamic triggers for reversible decrease or increase of temperature. They are working on the principle of reversible thermal effects, which are activated in solids by magnetic, electric, or stress fields (magnetocaloric, electrocaloric, and elastocaloric effect) [50]. Magnetocaloric sensors are made of magnetocaloric materials, which show reversible temperature changes in response to the changes of applied magnetic field [46,52]. All magnetic materials express MCE, but the intensity depends on the characteristics of each material [53].

Magnetic field sensors are generally divided into two large groups: vector sensors and scalar sensors. Vector sensors are magnetometers specialized in measuring low magnetic fields (<1 mT) and gaussmeters specialized in measuring high magnetic fields (>1 mT) [54]. Magnetometer sensors are (1) SQUID; (2) magnetoresistive; (3) fiber-optic; (4) fluxgate; (5) inductive [54]; (6) magnetoimpedance; (7) resonance; (8) cholographic; and (9) others (spintronic sensors, sensors prepared by using a microelectromechanical system technology, etc.) [49]. Gaussmeter's sensors are (1) Hall; (2) magnetoresistive; (3) magnetodiode; (4) magnetotransistors [54]; and (5) others. Scalar sensors include proton precession and optically pumped sensors [54].

Almost all types of sensors can be used in biomedicine, but five types of sensors dominate in biomedical applications: Hall sensors (high magnetic fields), magnetoresistive sensors (medium magnetic fields), flux-gate sensors (small magnetic fields), SQUID sensors, and magnetoimpedance sensors [55] (very small magnetic fields) [49]. From an application point of view, most of them represent a separate group consisting of biosensors: (1) electrochemical biosensors (label-based and label-free enzyme sensors, DNA sensors, and nanomaterial-based labels vs nanomaterial-modified sensors) [56]; (2) optical biosensors (surface plasmon resonance and fluorescence from evanescent wave excitation); and (3) acoustic biosensors [57]. Particularly important are biosensors for noninvasive detections and measurements of biological parameters in media other than blood [58], nanosized biosensors or nanobiosensors (1–100 nm) [59], and wearable sensors for biomedical applications [42].

A complete new generation of flexible and wearable devices was developed based on sensors, which are completely autonomous with ultrathin and flexible modules navigation, body tracking, and relative position monitoring systems [26]. Such appliances are required for soft robotics; functional medical implants; and epidermal, imperceptible, and transient electronics [26]. These applications include real-time monitoring of artificial joints or valves of the heart to diagnose early stages of dysfunctions [26].

Modern refrigeration technology is highly efficient technology friendly to the environment, based on use of MCE or electrocaloric effect [60]. It is used in all fields of biomedicine, from research activities to numerous useful applications. Its vital application can be prenatal (sperm banks), throughout lifespan (storing food, biomedicine), and after end of life (storing human organs, tissues, or whole bodies). Without refrigeration technology many activities or procedures in biology or medicine would not be possible, like cold storing of sampled materials (biomaterial, tissues, organs), organ transplantation, cryosurgery, normal functioning of magnetic resonance imaging (MRI), or comfortable working conditions in laboratories, operating rooms, or intensive care units.

Stimuli-responsive polymers are macromolecules mainly composed of magnetic metallopolymer nanocomposites (mostly ferromagnetic metal particles) that change their properties in response to a various stimuli: (1) small change in temperature (thermally responsive polymers); (2) electric or magnetic field (electroresponsive polymers, light/photoresponsive polymers); (3) pH (pH-sensitive polymers); (4) molecular stimuli; (5) mechanical stress; (6) combinations of different stimuli; and (7) others [14]. These "smart" materials are called smart polymers [14] and are used as high-density information media, materials for permanent magnets, magnetic cooling systems, and for production of magnetic sensors [28] as well as in medicine as microfluidic devices, pulsatile drug release systems, bioadhesion mediators, and motors/actuators [14]. Materials made of thermosensitive polymers, which can perform local cooling, are increasingly applied in control of transfer and release of drugs, hormones, and biomedical substances (e.g., antiinflammatory and antiproliferative agents, glues, and others from the surface of the implant) to the specified locations in the body, as in control of state of implant coating [52]. Magnetic materials with large MCE values make magnetic hyperthermia, causing local heating, which is used for cancer treatment [52]. Such materials could be used in any situation where there is necessity for common hypothermia, as in the case of treating patients with super-refractory convulsive status epilepticus (resistant to general anesthesia), by induced hypothermia (32°C–34°C for 24 h) [61].

The future creation of all types of sensors of flexible and wearable sensor networks is associated with wearable sensing or wearable computing goals to create body sensor network systems and electronic textiles as a particular class of wireless sensor networks for continuous healthcare services (e.g., medical monitoring and communication in emergency situations) [42].

20.2.2 Magnetic oxide labels

Detection of biomarkers in vitro and in vivo by magnetic particle labels is a central focus of biomedical research and applications. There are two main types of labeling: *direct labeling* (by nanoparticles and chemical agents), and *indirect labeling* (by genetic incorporation of reporter genes) [62].

Direct labeling mostly use contrast agents in form of nanoparticles and radionuclides; rarely use chemical agents which have larger dimensions than nanoparticles. Direct labeling refers to nonspecific internalization of agent (contrast medium or fluorescent probe), which can be visualized (by radiotracer) in special cell population that is examined, or on specific linking of agents to target molecules, which can be short term and followed radiologically (e.g., paramagnetic nanoparticles for MRI) [63].

Direct labeling is exogenous labeling with MRI-based contrast agents or radioisotopes for positron emission computed tomography (PET)/single-photon emission computed tomography (SPECT) imaging [62,64]. Magnetic nanoparticles convenient for MRI include (1) ultra-small (<50 nm) and superparamagnetic (>50 nm) iron oxide nanoparticle (USPIONP and SPIONP)-based T_2 contrast agents (i.e., ferucarbotran, ferumoxytol, ferumoxtran-10, ferumoxsil, feruglose, and very small superparamagnetic iron oxide particles (VSOP) coated with citrate (VSOP-C184) for USPIONs vs ferumoxides, ferucarbotran for SPIONs) [62,64–66]; and (2) paramagnetic gadolinium (Gd)-filled microcapsules and liposomes, perfluorocarbon nanoparticle- or manganese (Mn) particle-based T_1 contrast agents [62,64,66].

Magnetic nanoparticles serve as labels for monitoring biological binding reactions, which can be measured by magnetorelaxometry and magnetooptic relaxation devices [31]. Larger particles like liposomes are very convenient flexible carriers for contrast agents, whose phospholipid surface can be modified with various molecules such as polyethylene glycol, antibodies, metal chelates, and radioisotopes (positron emission and single photon radionuclides) [67].

Radionuclide labeling includes use of adequate (MRI, PET/SPECT) radionuclides (e.g., gallium-67 (^{67}Ga), indium-111 (^{111}In), technetium-99m (^{99m}Tc), oxygen-15 (^{15}O)) [62].

According to radiolabeling methods, there are two categories of radiolabeled IONPs: (1) chelator-based radiolabeled IONPs (e.g., ^{99m}Tc-labeled IONPs, ^{111}In- and iodine-125 (^{125}I)-labeled IONPs, copper-64 (^{64}Cu)- and gallium-68 (^{68}Ga)-labeled IONPs); and (2) chelator-free radiolabeled IONPs (e.g., fluorine-18 (^{18}F)- and carbon-11 (^{11}C)-labeled IONPs, arsenic [*As (* = 71, 72, 74, 76)]- and germanium-69 (^{69}Ge)-labeled IONPs, ^{64}Cu- and zirconium-89 (^{89}Zr)-labeled IONPs) [68].

Nanoparticles are used as important radionuclide carriers because of their physicochemical properties, enormous possibility for surface modification (functionalization), great linking capacity, biological stability, and compatibility [69]. Surface modification is carried out with inorganic and organic coatings for biomedical application. After labeling with radionuclide functionalized nanostructures, nanoparticles can serve as candidates for molecular imaging (MRI, PET, or SPECT) [64,69].

Indirect labeling consider using genetically modified cells to express reporter genes by coding proteins which generate imaging signals, often under interaction with molecular sample or substrate which the cells take [62]. Reporter genes inserted in genome express surface proteins, channels, storage proteins, receptors, or enzymes, which are detectable or (which) bind detectable samples [62], visible with fluorescence or bioluminescence imaging (e.g., luciferase activity) [63].

Indirect labeling is genetic labeling, which refers to stable insertion of reporter genes whose expression correlates with molecular processes during examination, or is specific for a given cell population, with possible radiological long-term proliferation follow-up [63]. However, insertion of a reporter gene, whose expression is examined in vivo as a biomarker of tumor cells, limits use of this method in preclinical studies because of genetic modification of the cell population [63].

Cell labeling is performed adequately by the techniques used for in vivo cell imaging. For in vivo cell imaging with MRI usually paramagnetic nanoparticles of iron oxide are used [63,64]; gadolinium chelates or manganese chelates are used, and also used for radionuclide labeling (e.g., ^{99m}Tc, ^{111}In, ^{64}Cu) [63]. Radionuclide labeled substances are used as contrast agents by binding radionuclides through a chelating agent to targeted components, including antibodies, antibody fragments, peptides, proteins, oligonucleotides, or receptor ligands [69].

Cell labeling and monitoring enables insight in a normal and pathological state of cells, tissues, and organs; making diagnoses, prescribing therapy; and following its course. Direct labeling helps the assessment of an illness's course and response to therapy [63]. This labeling also helps in vivo investigation of the role of health and ill cells in mechanisms of genesis and development of pathological process, stopping it with adequate therapy, and follow-up treatment efficiency [63]. Indirect genetic labeling, on the other hand, gives more detailed, intracellular insight, but only in ex vivo active processes in cells [63]. Together, direct and indirect labeling enable complex analysis of the whole state of an organism, necessary for follow-up of the regression of an illness's course, treatment efficiency, and eventual therapy correction.

20.2.3 Magnetic oxide contrast agents

Magnetic contrast agents (media) are pharmaceutical substances that are used to enhance contrast between structures or fluids in the body for the purpose of biomedical imaging. With support of contrast agents the differences between healthy and ill tissue can be determined, in order to the assess state of functionality of organs or the bloodstream after insertion of mediums. At a cellular level, molecular imaging is defined as "in vivo visualization, characterization and measurement of biological processes at the molecular and cellular levels in humans and other living systems" [70]. According to the chemical composition, magnetic contrast agents can be divided into nonorganic, organic, and mixed, or in natural and artificial agents.

The contrast media in biomedical imaging is classified according to the use of imaging modalities. In general, molecular imaging modalities include optical bioluminescence (optical fluorescence or fluorescence imaging), ultrasound, magnetic resonance techniques, and tomographic techniques [68].

Magnetic contrast agents are used in many classical biomedical techniques:

1. NMR and related techniques (MRI, functional MRI (fMRI), NMR spectroscopy (MRS), magnetic resonance spectroscopic imaging (MRSI))
2. X-ray computed tomography (CT)
3. Γ-ray positron emission tomography (PET)
4. Single photon emission computed tomography (SPECT)
5. Others techniques [69].

NMR and similar techniques (MRI, MRS, and MRSI) are relatively harmless, because they emit nonionizing radiation, as opposed to CT, PET, and SPECT, which emit ionizing radiation.

All contrast agents can be used for research, diagnostic, or therapeutic purposes. Some contrast agents can also be radiopharmaceutical agents, and can be used alone or as part of a diagnostic system.

20.2.3.1 Magnetic contrast agents of nuclear magnetic resonance imaging and related techniques

The MRI is a noninvasive imaging method that is used for function and structure visualization of live biological systems from molecules, cells, tissues, and organs to whole organisms for the purpose of research, diagnostics, and therapy or their special combination (theranostics) [71]. The fMRI is a variant of MRI, which provides real-time images of the metabolic activity or response of the human brain areas to external stimuli using, for example, blood-oxygen-level dependent contrast [72].

For biomedical purpose two groups of magnetic contrast agents apply: (1) metal-based contrast agents; and (2) chemical exchange saturation transfer (CEST) contrast agents [73].

Metal-based contrast agents can be alkaline earth metal-based, transition metal-based (e.g., iron oxide, manganese complex, and copper complex) [74], and lanthanide-based complex (e.g., gadolinium (Gd), dysprosium (Dy), and ytterbium (Yb) complex) [75]. CEST MRI contrast agents can be exogenous and endogenous [73].

Contrast agents that are used in MRI (paramagnetic, superparamagnetic, and ferromagnetic) are divided into four major groups: (1) two classical (positive (bright signal) T_1-weighted and negative (dark signal) T_2-weighted [64,76]) and one additional T_2 exchange relaxation (T_{2ex}) [77]; and (2) two new groups (two CEST: diamagnetic CEST (diaCEST) and paramagnetic CEST (paraCEST; lanthanide is the source of paramagnetism)) and one hyperpolarized molecules (e.g., carbon-13 (^{13}C) MRI or hyperpolarized ^{13}C MRS/I (MR spectroscopy/spectroscopic imaging)) [77].

Complementary combinations of various types of materials (e.g., fluorescent molecules and radioisotopes) allow nanoparticle dual imaging as well as development of the new types of contrast agents that are used in emerging imaging technologies. Magnetic nanoparticles for T_1-T_2 MRI, MRI-optical, and MRI−PET/SPECT dual-modal imaging can serve as nontraditional multimode imaging agents such as (1) magnetic particle imaging (MPI-MRI dual-modal imaging); (2) magneto-motive ultrasound (US) imaging (MMUS)-US dual-modal imaging; and (3) magneto-photoacoustic imaging (MPA)-US dual-modal imaging [78].

Magnetic MRI contrast agents can be divided into (1) extracellular contrast agents; (2) blood pool contrast agents (intravascular agents); (3) targeted contrast agents; (4) responsive contrast agents; and (5) CEST and paraCEST MRI contrast agents [76].

20.2.3.2 Magnetic contrast agents of X-ray computed tomography

X-ray CT magnetic contrast agents are often applied in biomedicine. Most frequently used contrast-generating elements are iodine (I), gold (Au), and bismuth (Bi); less commonly used are barium (Ba), bromine (Br), tantalum (Ta), platinum (Pt), ytterbium (Yb), yttrium (Y), gadolinium (Gd), wolfram (W), dysprosium (Dy), and lutecium (Lu) [67].

Lanthanide-based contrast agents are the most frequently used in CT, especially Gd chelates. Ytterbium-based nanoparticulate contrast agents usually also include Ba-, Bi-, Au-, Pt-, and Ta-based nanoparticles [79].

Metal-based contrast agents are less frequently used than lanthanides like copolymeric nanoparticles of iron oxides, zinc, copper [80], Au, and silver (Ag) [79,80]. Dextran-coated bismuth–iron oxide nanohybrid contrast agents are also synthesized for in vivo CT and MR imaging [81].

Other metallic contrast agents include in vitro and in vivo use of basic elements like Ba (e.g., $BaSO_4$), Bi (e.g., Bi_2S_3), Ta (e.g., Ta_2O_5) [79,80], and Pt (e.g., FePt) [82].

Various combinations of elements are used for the dual-modal imaging, for example, core/shell Fe_3O_4/TaO_x nanoparticles as a bifunctional agent for CT and MRI [83].

20.2.3.3 Magnetic contrasts of positron emission tomography

A large number of classical and new (nanostructured and others) contrast agents are used in PET for preclinical and clinical purposes. The most commonly used PET radioisotopes (radionuclides) are ^{11}C, nitrogen-13 (^{13}N), ^{15}O, ^{18}F, ^{64}Cu, ^{68}Ga, rubidium-82 (^{82}Rb), and holmium-166 (^{166}Ho) [84]; less commonly used are arsenic-72 (^{72}As) and ^{89}Zr [85].

Magnetic nanoparticles, especially IONPs, are in development as MRI contrast agents, but also in iron oxide-based PET/MR dual applications [86] for detection of many disorders and diseases. Contrast agents for NMR and MRI can be achieved by labeling a variety of PET isotopes (e.g., ^{11}C, ^{18}F, ^{64}Cu, ^{68}Ga, ^{69}Ge, ^{72}As, and ^{89}Zr) and SPECT isotopes (e.g., ^{99m}Tc, iodine-123 (^{123}I) or ^{125}I, and ^{111}In) to water-soluble IONPs [68].

Radionuclide (e.g., ^{99m}Tc, ^{111}In, ^{64}Cu) cell labeling is used ex vivo for labeling tumor cells, lymphocytes, dendritic cells, and monocytes in PET and SPECT examinations [63]. Iodine-124 (^{124}I) radiolabeled IONPs are developed for optical/PET/MR modality tumor imaging [86]. ^{89}Zr radiolabeled nanoparticles are explored for PET oncological imaging and indium-120 (^{120}In) for colon carcinoma imaging [86]. Gold nanoshell with ^{64}Cu radiolabeling [86] and SPIONs with multiple radioisotopes (e.g., ^{64}Cu, ^{68}Ga, ^{89}Zr, and ^{69}Ge) [66] were used for multimodality theranostics [66,86].

20.2.3.4 Magnetic contrasts of single photon emission computed tomography

SPECT is a nuclear biomedicine technology that deals with preclinical and clinical monitoring of the distribution, kinetics, and targeting of radiolabeled organic and nonorganic molecules into tomographic (two- or three-dimensional) images. SPECT is very similar to PET in its use of radioactive tracer and detection of γ-rays, because it emits only a single γ-ray during decay of a radioisotope that is measured directly, unlike PET, which records pairs of γ-rays that are emitted and measured indirectly [84].

Target-specific SPECT radioisotopes can be divided into two main parts: a targeting biomolecule (e.g., peptide, antibody fragment, vitamin) and a γ-radiation emitting radionuclide (e.g., ^{99m}Tc, ^{111}In) [87]. Most radionuclides consist of metallic radioisotopes. In successive processes of conjugation, radiolabeling, and target binding, bifunctional chelator links to the targeting biomolecule and radiometal that form a target-specific radiobioconjugate that binds to the targeting structure [87].

The most commonly used radioisotopes for SPECT imaging are ^{99m}Tc, ^{111}In, ^{67}Ga, and ^{123}I [87]. The most commonly used therapy/SPECT isotopes are lutetium-177 (^{177}Lu), rhenium-186 (^{186}Re), rhenium-188 (^{188}Re), copper-67 (^{67}Cu), and iodine-131 (^{131}I) [86]. Less commonly used isotopes are ^{125}I, tin-117m (^{117m}Sn), and radiolanthanides gadolinium-159 (^{159}Gd), samarium-153 (^{153}Sm), ^{166}Ho, thallium-201 (^{201}Tl), and itrium-90 (^{90}Y).

Among the metallic radioisotopes, SPECT imaging ^{99m}Tc is the most widely applied radioisotope of diagnostic agents [66,87]. Among all nonmetallics, the most used isotope is ^{123}I; less used are ^{125}I and ^{131}I [87].

The SPIONs have been labeled with multiple radioisotopes for both SPECT and PET (e.g., ^{99m}Tc, ^{125}I, ^{111}In, and ^{131}I) [66]. The USPIONs are conjugated by arginylglycylaspartic acid (RGD) peptides and labeled by ^{99m}Tc (^{99m}Tc-USPION-RGD) in order to be successively used for molecular imaging as a targeting probe for SPECT imaging, dual-contrast (T_1/T_2-weighted) MRI, and dual-modality (MRI/SPECT) imaging of tumor angiogenesis [64].

The SPECT radionuclides may be used alone or in combination with other radionuclides, depending on the nature of the examination.

The SPECT tracer media are used for (1) biomolecule labeling [88]; (2) biodistribution [89]; (3) bioassay [88]; (4) tissue imaging [90]; (5) diagnostic applications [87,88]; (6) therapeutic applications [87] (including targeted radiotherapy [84] and radioimmunotherapy [87]); and (7) theranostics [32,66].

20.2.4 Magnetic oxide diagnostics and therapy applications

Medical application of magnetic oxides in human and veterinary medicine is widespread in numerous medical fields.

In human medicine, magnetic oxides are used in many fields: (1) microbiology (e.g., magnetic IONPs can stimulate or inhibit microbial growth [29]); (2) dentistry (e.g., dental repair and dental implants [1], secure dentures, or prostheses) [14,91,92]; (3) orthopedy (e.g., bone repair and bone implants) [1,93]; (4) cardiology (e.g., activation of artificial heart valves by permanent magnets [94], self-controlled nanovalves [14]); (5) neurology (e.g., provide communication and mobility [95,96]); (6) surgery (e.g., navigating within the brain [14]); (7) gynecology (e.g., magnetic intrauterine devices for contraception [15]); (8) oncology (e.g., magnetic fluid (nanoparticles) hyperthermia–iron oxide (SPIONs and others ferrofluids) and transition metal nanoparticles [97], tumor destruction and detoxification [3,17,30]); (9) radiology (diagnostic imaging [1,2,6,64], deliver therapeutic radioisotopes [17] and therapeutic targets in chemotherapy [1,4]); and (10) physical medicine and rehabilitation (e.g., treatment of various diseases [98]; aid feeble muscles in the eyelids or bladder [94], guiding catheters through the body with permanent magnets [14,94]).

In veterinary medicine, magnetic oxides are used much less frequently than in human medicine. Their use is mainly for applications of contrast agents in veterinary radiology, therapeutic agents in veterinary oncology (e.g., electrochemotherapy and gene therapy), [99] and veterinary physical medicine and rehabilitation (e.g., healing a horse's injury) [100].

Magnetic oxides are used in human and veterinary medicine for diagnostic and therapeutic purposes, including assistive technology as a special type of therapy.

20.2.4.1 Medical diagnostics

Medical diagnostics use a multitude of devices, appliances, handy tools, and other equipment that contain magnetic oxide materials.

1. **Electrodiagnostics**

   Electrodiagnostics is based on the registration of bioelectric fields in certain areas of the body and cover electrodiagnostic techniques and electrodiagnostic testing for research and clinical purposes. Registration is performed by metallic or nonmetallic electrodes, both noninvasive (sensor patches) and invasive (needles).

   Electrodiagnostic techniques include many procedures such as electrocardiography, electroencephalography, electrogastrography, spirometry, audiometry, and electrodermal activity (measurement of changes in skin conductance).

   Electrodiagnostic testing measures electrical activity of muscles and nerves and typically consists of needle electromyography and electroneurography (or nerve conduction studies for motor and sensory nerves), including long latency reflex testing (F-waves) [101]. Somatosensory evoked potentials, magnetic stimulation motor evoked potentials, and autonomic testing are rarely used [101].

2. **Magnetodiagnostics**
   Magnetic field diagnostics is based on the registration of biomagnetic fields for research and clinical purposes such as magnetoencephalography, magnetospinography, magnetoneurography, magnetoretinography, magnetocardiography, magnetomyography, magnetic impurity detection (magnetopneumography), magnetic susceptometry (liver iron measurement), and magnetogastrography [102].
   Magnetic trackers are used for biological detection (e.g., pathogens or other targeted biomolecules such as DNA, RNA, antibodies, and metabolites), determine the position of medical tools inside the body (e.g., endoscope, colonoscope, and biopsy needle), and to observe biomechanical motions [103]. An external magnetic field can exert on the particle the force at range of 10^{-12} to 10^{-9} newtons, which represent the typical levels that cells experience in vivo [23].

20.2.4.2 Medical therapy

Magnetic oxides are widely used in various fields of medical therapy, especially in physical medicine and rehabilitation as electrotherapy, magnetotherapy, and assistive technology.

1. **Electrotherapy versus electrophysical agents**
   Electrotherapy includes delivery and therapeutic application of electrical and possibly also electromagnetic energies emitted by various sources [104]. Electrophysical agents have the wider concept involving ultrasound, light, vibration, and various heat therapies [104]. Together they are divided into three main subgroups: electrical stimulation, thermal modalities, and nonthermal modalities [104].
2. **Magnetotherapy**
   Magnetotherapy involves therapeutic application of MF/EMF modalities emitted by various sources: from the simplest such as magnetophores and elastic magnetic stripes (made of iron oxides [98] or lanthanides) to the most complex such as MF/EMF-emitting devices (magnetic devices) [13]. Therapeutic modalities of MFs/EMFs can be classified into the following groups: (1) static/permanent MFs; (2) low-frequency sine waves; (3) pulsed EMFs; (4) pulsed radio-frequency fields; (5) magnetic/electric stimulation; (6) millimeter waves; and (7) electroporation and nanoelectroporation [13].
3. **Assistive technology**
   Assistive technology devices are used in physical medicine and rehabilitation for communication disorders, mobility impairments, ergonomics and prevention of secondary injuries, hearing impairments, visual impairments, and cognitive/learning disabilities [105].

20.2.5 Other magnetic oxide techniques and applications

In biomedicine, besides the already mentioned techniques, mainly magnetic separation and magnetic tweezers are in use.

20.2.5.1 Magnetic separation

Magnetic separation is the technique for the extraction of magnetic particles from nonmagnetic materials. For bioseparation processes, the magnetic nanoparticles usually are coated with various polymeric materials [1]. By using magnetic particles, the magnetic separation method is used for sensitive and reliable capture of specific proteins, genetic material, and other biomolecules [17]. The magnetic separation is the technique of sorting specific biological targets based on magnetically labeled surfaces and is used in a wide range of applications from in vitro diagnostics to cell-based therapies in vivo. Possible therapy includes extracorporeal magnetic separation-based blood purification due to rapid and selective removal of disease-causing compounds from whole blood [106].

The magnetic separation uses the magnetic field gradient-induced magnetophoresis, which can be either toward (positive magnetophoresis) or away (negative magnetophoresis) from the magnetic source due to separate particles and cells based on the difference in physical properties [107]. Positive magnetophoresis refers to magnetic particles or magnetically tagged cells, and negative magnetophoresis to diamagnetic particles and cells in magnetic fluids [107].

20.2.5.2 Magnetic tweezers

Magnetic tweezers are techniques of biological micromanipulators that use the magnetic force on superparamagnetic or ferromagnetic particles performed by gradients in an applied magnetic field [72,94]. They belong to single-molecule techniques that allow research of the behavior of individual biological molecules in solution in real time [108]. These are molecule techniques for determination of functions of various enzymes at a molecular level by exerting forces on individual proteins [109] that are used to monitor the mechanical properties of nucleic acids and their interactions with the proteins [108].

An important feature of magnetic tweezers is their ability to apply torque while maintaining a constant stretching force [108] and their ability to quickly and accurately twist macromolecules [110]. Magnetic tweezers offer an appropriate tool for handling two or more DNAs [110]. This can be easily achieved by increasing the amount of DNA during its attachment to magnetic particles [110]. NonDNA applications of magnetic tweezers include a number of other noninvasive applications such as protein folding, rotating protein assemblies, cellular cortical stiffness, cellular mechanotransduction processes, and intracellular transport [110].

References

[1] X. Li, J. Wei, K.E. Aifantis, Y. Fan, Q. Feng, F.-Z. Cui, et al., Current investigations into magnetic nanoparticles for biomedical applications, J. Biomed. Mater. Res. Part A 104A (2016) 1285−1296.

[2] W. Andrä, U. Häfeli, R. Hergt, R. Misri, Application of magnetic particles in medicine and biology, in: H. Kronmüller, S. Parkin (Eds.), Handbook of Magnetism and Advanced Magnetic Materials, vol. 4, Novel Materials, John Wiley & Sons, Ltd, Chichester, 2007, pp. 1−33.

[3] W. Wu, C.Z. Jiang, V.A.L. Roy, Designed synthesis and surface engineering strategies of magnetic iron oxide nanoparticles for biomedical applications, Nanoscale (2016). Available from: http://dx.doi.org/10.1039/c6nr07542h.

[4] A. Ali, H. Zafar, M. Zia, I. Ul Haq, A.R. Phull, J.S. Ali, et al., Synthesis, characterization, applications, and challenges of iron oxide nanoparticles, Nanotechnol. Sci. Appl. 9 (2016) 49–67.
[5] A.D. Pomogailo, V.N. Kestelman, Chemical methods of metal–polymer nanocomposite production, Metallopolymer Nanocomposites, Springer-Verlag Berlin, 2005, pp. 135–236.
[6] S.P. Gubin, Introduction, in: S.P. Gubin (Ed.), Magnetic Nanoparticles, Wiley-VCH Verlag GmbH & Co. KGaA, Weinheim, 2009, pp. 1–23.
[7] T.A. Waigh, The senses, The Physics of Living Processes: A Mesoscopic Approach, first ed., John Wiley & Sons, Ltd, Chichester, 2014, pp. 541–554.
[8] R.K. Hobbie, B.J. Roth, Biomagnetism, Intermediate Physics for Medicine and Biology, fifth ed., Springer International Publishing Switzerland, Cham, 2015, pp. 213–238.
[9] A.D. Pomogailo, V.N. Kestelman, Nanobiocomposites, Metallopolymer Nanocomposites, Springer-Verlag, Berlin, 2005, pp. 377–422.
[10] O. Strbak, P. Kopcansky, I. Frollo, Biogenic magnetite in humans and new magnetic resonance hazard questions, Meas. Sci. Rev. 11 (3) (2011) 85–91.
[11] A.D. Pomogailo, V.N. Kestelman, Nanoparticles in materials chemistry and in the natural sciences (introduction), Metallopolymer Nanocomposites, Springer-Verlag, Berlin, 2005, pp. 3–24.
[12] D.M. Djordjevich, S.R. De Luka, I.D. Milovanovich, S. Janković, S. Stefanović, S. Vesković-Moračanin, et al., Hematological parameters' changes in mice subchronically exposed to static magnetic fields of different orientations, Ecotoxicol. Environ. Saf. 81 (2012) 98–105.
[13] M.S. Markov, Benefit and hazard of electromagnetic fields, in: M.S. Markov (Ed.), Electromagnetic Fields in Biology and Medicine, Taylor & Francis Group, LLC, Boca Raton, 2015, pp. 15–28.
[14] S.F. Medeiros, A.M. Santos, H. Fessi, A. Elaissari, Stimuli-responsive magnetic particles for biomedical applications, Int. J. Pharm 403 (2011) 139–161.
[15] C. Bárcena, A.K. Sra, J. Gao, Applications of magnetic nanoparticles in biomedicine, in: J.P. Liu, E. Fullerton, O. Gutfleisch, D.J. Sellmyer (Eds.), Nanoscale Magnetic Materials and Applications, Springer Science+Business Media, LLC, Springer Dordrecht, 2009, pp. 591–626.
[16] M.R. Ghazanfari, M. Kashefi, S.F. Shams, M.R. Jaafari, Perspective of Fe_3O_4 nanoparticles role in biomedical applications, Biochem. Res. Int. 2016 (2016), Article ID 7840161. http://dx.doi.org/10.1155/2016/7840161.
[17] V.N. Nikiforov, E. Yu Filinova, Biomedical Applications of Magnetic Nanoparticles, in: S.P. Gubin (Ed.), Magnetic Nanoparticles, Wiley-VCH Verlag GmbH & Co. KGaA, Weinheim, 2009, pp. 393–455.
[18] S. Mørup, M.F. Hansen, Superparamagnetic particles, in: H. Kronmüller, S. Parkin (Eds.), Handbook of Magnetism and Advanced Magnetic Materials, vol. 4, Novel Materials, John Wiley & Sons, Ltd, Chichester, 2007, pp. 1–18.
[19] S. Tumanski, Magnetic materials, Handbook of Magnetic Measurements, CRC Press/Taylor & Francis Group, Boca Raton, 2011, pp. 117–158.
[20] G.B. Khomutov, Y.A. Koksharov, Organized ensembles of magnetic nanoparticles: preparation, structure, and properties, in: S.P. Gubin (Ed.), Magnetic Nanoparticles, Wiley-VCH Verlag GmbH & Co. KGaA, Weinheim, 2009, pp. 117–195.
[21] A. Masotti, A. Caporali, Preparation of magnetic carbon nanotubes (Mag-CNTs) for biomedical and biotechnological applications, Int. J. Mol. Sci. 14 (2013) 24619–24642.

[22] L. Chen, X. Wang, W. Lu, X. Wua, J. Li, Molecular imprinting: perspectives and applications, Chem. Soc. Rev. (2016). Available from: http://dx.doi.org/10.1039/c6cs00061d.
[23] A. Aseri, S.K. Garg, A. Nayak, S.K. Trivedi, J. Ahsan, Magnetic nanoparticles: magnetic nano-technology using biomedical applications and future prospects, Int. J. Pharm. Sci. Rev. Res. 31 (2) (2015) 119−131, 20.
[24] D.V. Louzguine-Luzgin, Bulk metallic glasses and glassy/crystalline materials, in: A. Zhukov (Ed.), Novel Functional Magnetic Materials: Fundamentals and Applications, Springer International Publishing Switzerland, 2016, pp. 396−440.
[25] S.A. Solin, L.R. Ram-Mohan, Geometry-driven magnetoresistance, in: H. Kronmüller, S. Parkin (Eds.), Handbook of Magnetism and Advanced Magnetic Materials, volume 5, Spintronics and Magnetoelectronics, John Wiley & Sons, Ltd, Chichester, 2007, pp. 1−21.
[26] M. Melzer, J.I. Mönch, D. Makarov, Y. Zabila, G.S. Cañón Bermúdez, D. Karnaushenko, et al., Wearable magnetic field sensors for flexible electronics, Adv. Mater. 27 (2015) 1274−1280.
[27] C. Wang, J. Wang, P. Li, Z. Rong, X. Jia, Q. Ma, et al., Sonochemical synthesis of highly branched flower-like Fe_3O_4@SiO_2@Ag microcomposites and their application as versatile SERS substrates, Nanoscale (2016). Available from: http://dx.doi.org/10.1039/c6nr07295j.
[28] G.I. Dzhardimalieva, A.D. Pomogailo, A.S. Rozenberg, M. Leonowicz, Magnetic metallopolymer nanocomposites: preparation and properties, in: S.P. Gubin (Ed.), Magnetic Nanoparticles, Wiley-VCH Verlag GmbH & Co. KGaA, Weinheim, 2009, pp. 59−85.
[29] R. Dinali, A. Ebrahiminezhad, M. Manley-Harris, Y. Ghasemi, A. Berenjian, Iron oxide nanoparticles in modern microbiology and biotechnology, Crit. Rev. Microbiol. (2016). Available from: http://dx.doi.org/10.1080/1040841X.2016.1267708.
[30] K.S. Siddiqi, A. Ur Rahman, Tajuddin, A. Husen, Biogenic fabrication of iron/iron oxide nanoparticles and their application, Nanoscale Res. Lett. 11 (2016) 498.
[31] L. Trahms, Biomedical applications of magnetic nanoparticles, in: S. Odenbach (Ed.), Colloidal Magnetic Fluids, Basics, Development and Application of Ferrofluids, Springer, Berlin, 2009, pp. 327−358.
[32] W. Wu, Z. Wu, T. Yu, C. Jiang, Woo-Sik Kim, Recent progress on magnetic iron oxide nanoparticles: synthesis, surface functional strategies and biomedical applications, Sci. Technol. Adv. Mater. 16 (2015) 023501, 43pp.
[33] A. Akbarzadeh, M. Samiei, S. Davaran, Magnetic nanoparticles: preparation, physical properties, and applications in biomedicine, Nanoscale Res. Lett. 7 (2012) 144.
[34] P. Chetprayoon, M. Matsusaki, U. Yokoyama, T. Tejima, Y. Ishikawa, M. Akashi, Use of three-dimensional arterial models to predict the in vivo behavior of nanoparticles for drug delivery, Angew. Chem. Int. Ed. 55 (2016) 4461−4466.
[35] M. Muralidhar, M. Murakami, Superconductivity in oxides: from processing to applications, in: M. Lorenz, M.S. Ramachandra Rao, T. Venkatesan, E. Fortunato, P. Barquinha, R. Branquinho, et al., (Eds.), The 2016 oxide electronic materials and oxide interfaces roadmap, *J. Phys. D: Appl. Phys.* **49**, 2016, 433001 (53pp): pp. 23−24.
[36] M. Colombo, S. Carregal-Romero, M.F. Casula, L. Gutiérrez, M.P. Morales, I.B. Böhm, et al., Biological applications of magnetic nanoparticles, Chem. Soc. Rev. 41 (2012) 4306−4334.
[37] V.A. L'vov, V.A. Chernenko, J.M. Barandiaran, Magnetic shape memory materials with improved functional properties: scientific aspects, in: A. Zhukov (Ed.), Novel

Functional Magnetic Materials: Fundamentals and Applications, Springer International Publishing Switzerland, 2016, pp. 1–40.

[38] B.D. Plouffe, S.K. Murthy, L.H. Lewis, Fundamentals and application of magnetic particles in cell isolation and enrichment, Rep. Prog. Phys. 78 (1) (2015) 016601.

[39] T. Takamura, P.J. Ko, J. Sharma, R. Yukino, S. Ishizawa, A. Sandhu, Magnetic-particle-sensing based diagnostic protocols and applications, Sensors 15 (2015) 12983–12998.

[40] J. Coffel, E. Nuxoll, Magnetic nanoparticle/polymer composites for medical implant infection control, J. Mater. Chem. B (2015). Available from: http://dx.doi.org/10.1039/c5tb01540e.

[41] R. Riemer, A. Shapiro, Biomechanical energy harvesting from human motion: theory, state of the art, design guidelines, and future directions, J. Neuroeng. Rehabil. 8 (2011) 22.

[42] C. Crean, C. McGeough, R. O'Kennedy, Wearable biosensors for medical applications, in: S. Higson (Ed.), Biosensors for Medical Applications, Woodhead Publishing Limited, Cambridge, 2012, pp. 301–330.

[43] G. Rivero, M. Multigner, J. Spottorno, Magnetic sensors for biomedical applications, in: K. Kuang (Ed.), Magnetic Sensors - Principles and Applications, InTech, Rijeka, 2012, pp. 125–150.

[44] V.A.M. Brabers, Ferrimagnetic insulators, in: H. Kronmüller, S. Parkin (Eds.), Handbook of Magnetism and Advanced Magnetic Materials, vol. 4, Novel Materials, John Wiley & Sons, Ltd, Chichester, 2007, pp. 1–19.

[45] J.R. Sun, B.G. Shen, F.X. Hu, Magnetocaloric effect and materials, in: J.P. Liu, E. Fullerton, O. Gutfleisch, D.J. Sellmyer (Eds.), Nanoscale Magnetic Materials and Applications, Springer Science+Business Media, LLC, Springer Dordrecht, 2009, pp. 441–483.

[46] V.K. Pecharsky, K.A. Gschneidner Jr., Magnetocaloric materials, in: H. Kronmüller, S. Parkin (Eds.), Handbook of Magnetism and Advanced Magnetic Materials, volume 4, Novel Materials, John Wiley & Sons, Ltd, Chichester, 2007, pp. 1–15.

[47] V. Kuncser, P. Palade, A. Kuncser, S. Greculeasa, G. Schinteie, Engineering magnetic properties of nanostructures via size effects and interphase interactions, in: V. Kuncser, L. Miu (Eds.), Size Effects in Nanostructures, Basics and Applications, Springer-Verlag, Berlin, 2014, pp. 169–237.

[48] M.D. Glinchuk, A.V. Ragulya, V.A. Stephanovich, Ferroics, Nanoferroics, Springer Science+Business Media, Dordrecht, 2013, pp. 1–32.

[49] S. Tumanski, Modern magnetic field sensors – a review, Prz. Elektrotechniczny. R. 89 (2013) NR 10/2013.

[50] V.K. Pecharsky, J. Cui, D.D. Johnson, (Magneto)caloric refrigeration: is there light at the end of the tunnel? Phil. Trans. R. Soc. A 374 (2016) 20150305.

[51] A.-Y. Chang, M.S.-C. Lu, A CMOS magnetic microbead- based capacitive biosensor array with on-chip electromagnetic manipulation, Biosens. Bioelectron. 45 (2013) 6–12.

[52] A.M. Tishin, Y.I. Spichkin, V.I. Zverev, P.W. Egolf, A review and new perspectives for the magnetocaloric effect: new materials and local heating and cooling inside the human body, Int. J. Refrig. (2016), http://dx.doi.org/10.1016/j.ijrefrig.2016.04.020.

[53] J.Q. Zhao, C.L. Zhang, Y.G. Nie, H.F. Shi, E.J. Ye, Z.D. Han, et al., Tunable magnetostructural phase transition and magnetocaloric effect in $Mn_{1-x}Ni_{1-x}Co_{2x}Si_{1-x}Ge_x$ system, J. Alloys Compd. 698 (2017) 7–12.

[54] U.C. Inan, Design and Implementation of Magnetic Field Sensors for Biomedical Applications, A Thesis Submitted to The Graduate School of Natural and Applied

Sciences of Middle East Technical University, Middle East Technical University, Department of Electrical and Electronics Engineering, Ankara, 2015. http://etd.lib.metu.edu.tr/upload/12618588/index.pdf

[55] K. Mohri, T. Uchiyama, L.V. Panina, M. Yamamoto, K. Bushida, Recent advances of amorphous wire CMOS IC magneto-impedance sensors: innovative high-performance micromagnetic sensor chip, J. Sens. 2015 (2015), Article ID 718069. http://dx.doi.org/10.1155/2015/718069.

[56] A.J. Veloso, X.R. Cheng, K. Kerman, Electrochemical biosensors for medical applications, in: S. Higson (Ed.), Biosensors for Medical Applications, Woodhead Publishing Limited, Cambridge, 2012, pp. 3–40.

[57] C. Parsajoo, J.-M. Kauffmann, M. Elkaoutit, Biosensors for drug testing and discovery, in: S. Higson (Ed.), Biosensors for Medical Applications, Woodhead Publishing Limited, Cambridge, 2012, pp. 233–262.

[58] L. Micheli, D. Moscone, G. Palleschi, Biosensors for non-invasive measurements, in: S. Higson (Ed.), Biosensors for Medical Applications, Woodhead Publishing Limited, Cambridge, 2012, pp. 263–300.

[59] J.C. Claussen, J. Shi, C.S. Rout, M.S. Artiles, M.M. Roushar, M.C. Stensberg, et al., Nano-sized biosensors for medical applications, in: S. Higson (Ed.), Biosensors for Medical Applications, Woodhead Publishing Limited, Cambridge, 2012, pp. 65–102.

[60] M.A. Hamad, Magnetocaloric effect in Sr_2FeMoO_6/Ag composites, Process. Appl. Ceram. 9 (1) (2015) 11–15.

[61] S. Legriel, V. Lemiale, M. Schenck, J. Chelly, V. Laurent, F. Daviaud, et al., Hypothermia for neuroprotection in convulsive status epilepticus, for the HYBERNATUS Study Group N. Engl. J. Med. 375 (2016) 2457–2467.

[62] A.V. Naumova, M. Modo, A. Moore, C.E. Murry, J.A. Frank, Clinical imaging in regenerative medicine, Nat. Biotechnol. 32 (8) (2014) 804–818.

[63] L. Ottobrini, C. Martelli, G. Lucignani, Imaging cell trafficking in cancer research, in: X. Chen (Ed.), Molecular Imaging Probes for Cancer Research, World Scientific Publishing, Co., Pte. Ltd., Singapore, 2012, pp. 905–948.

[64] S. Xue, C. Zhang, Y. Yang, L. Zhang, D. Cheng, J. Zhang, et al., ^{99m}Tc-labeled iron oxide nanoparticles for dual-contrast (T_1/T_2) magnetic resonance and dual-modality imaging of tumor angiogenesis, J. Biomed. Nanotechnol. 11 (2015) 1027–1037.

[65] D. Couto, M. Freitas, F. Carvalho, E. Fernandes, Iron oxide nanoparticles: an insight into their biomedical applications, Curr. Med. Chem. 22 (2015) 1808–1828.

[66] S. Goel, C.G. England, F. Chen, W. Cai, Positron emission tomography and nanotechnology: a dynamic duo for cancer theranostics, Adv. Drug Deliv. Rev. (2016). Available from: http://dx.doi.org/10.1016/j.addr.2016.08.001.

[67] N. Lee, S.H. Choi, T. Hyeon, Nano-sized CT contrast agents, Adv. Mater. 25 (2013) 2641–2660.

[68] F. Ai, C.A. Ferreira, F. Chen, W. Cai, Engineering of radio labeled iron oxide nanoparticles for dual modality imaging, WIREs Nanomed. Nanobiotechnol. 8 (2016) 619–630.

[69] W.-K. Rhim, M. Kim, K.L. Hartman, K.W. Kang, J.-M. Nam, Radionuclide-labeled nanostructures for in vivo imaging of cancer, Nano Converg 2 (2015) 10. Available from: http://dx.doi.org/10.1186/s40580-014-0041-3.

[70] D.A. Mankoff, A definition of molecular imaging, J. Nucl. Med. 48 (2007) 18N–21N.

[71] O.L. Gobbo, K. Sjaastad, M.W. Radomski, Y. Volkov, A. Prina-Mello, Magnetic nanoparticles in cancer theranostics, Theranostics 5 (11) (2015) 1249–1263.

[72] T.A. Waigh, Experimental techniques, The Physics of Living Processes: A Mesoscopic Approach, first ed., John Wiley & Sons, Ltd, Chichester, 2014, pp. 357–433.

[73] J. Kim, Y. Wu, Y. Guo, H. Zheng, P.Z. Sun, A review of optimization and quantification techniques for chemical exchange saturation transfer (CEST) MRI toward sensitive in vivo imaging, Contrast Media Mol. Imaging 10 (3) (2015) 163−178.
[74] D. Pan, S.D. Caruthers, A. Senpan, C. Yalaz, A.J. Stacy, G. Hu, et al., Synthesis of nanoQ, a copper-based contrast agent, for high-resolution magnetic resonance imaging characterization of human thrombus, J. Am. Chem. Soc. 133 (24) (2011) 9168−9171.
[75] H.U. Rashid, K. Yu, J. Zhou, Lanthanide(III) chelates as MRI contrast agents: a brief description, J. Struct. Chem. 54 (1) (2013) 223−249.
[76] G.P. Yan, L. Robinson, P. Hogg, Magnetic resonance imaging contrast agents: overview and perspectives, Radiography 13 (2007) e5−e19.
[77] S. Sinharay, M.D. Pagel, Advances in magnetic resonance imaging contrast agents for biomarker detection, Annu. Rev. Anal. Chem. 9 (2016) 95−115.
[78] T.-H. Shin, Y. Choi, S. Kim, J. Cheon, Recent advances in magnetic nanoparticle-based multi-modal imaging, Chem. Soc. Rev. 44 (2015) 4501−4516.
[79] Y. Liu, K. Ai, L. Lu, Nanoparticulate X-ray computed tomography contrast agents: from design validation to in vivo applications, Acc. Chem. Res. 45 (10) (2012) 1817−1827.
[80] H. Lusic, M.W. Grinstaff, X-ray computed tomography contrast agents, Chem. Rev. 113 (3) (2013). Available from: http://dx.doi.org/10.1021/cr200358s.
[81] P.C. Naha, A.A. Zaki, E. Hecht, M. Chorny, P. Chhour, E. Blankemeyer, et al., Dextran coated bismuth−iron oxide nanohybrid contrast agents for computed tomography and magnetic resonance imaging, J. Mater. Chem. B (2014). Available from: http://dx.doi.org/10.1039/c4tb01159g.
[82] R.M. Taylor, T.C. Monson, R.R. Gullapalli, Influence of carbon chain length on the synthesis and yield of fatty amine-coated iron-platinum nanoparticles, Nanoscale Res. Lett 9 (2014) 306, http://www.nanoscalereslett.com/content/9/1/306.
[83] S. Sheng-Nan, W. Chao, Z. Zan-Zan, H. Yang-Long, S.S. Venkatraman, X. Zhi-Chuan, Magnetic iron oxide nanoparticles: synthesis and surface coating techniques for biomedical applications, Chin. Phys. B. 23 (3) (2014) 037503. Available from: http://dx.doi.org/10.1088/1674-1056/23/3/037503.
[84] F.-M. Lu, Z. Yuan, PET/SPECT molecular imaging in clinical neuroscience: recent advances in the investigation of CNS diseases, Quant Imaging Med. Surg. 5 (3) (2015) 433−447.
[85] X. Sun, W. Cai, X. Chen, Positron emission tomography imaging using radiolabeled inorganic nanomaterials, Acc. Chem. Res. 48 (2) (2015) 286−294.
[86] Y. Liu, M.J. Welch, Nanoparticles labeled with positron emitting nuclides: advantages, methods, and applications, Bioconjug. Chem. 23 (4) (2012) 671−682.
[87] C. Müller, R. Schibli, Single photon emission computed tomography tracer, in: O. Schober, B. Riemann (Eds.), Molecular Imaging in Oncology, Springer-Verlag Berlin, 2013, pp. 65−105.
[88] S.A. Albu, S.A. Al-Karmi, A. Vito, J.P.K. Dzandzi, A. Zlitni, D. Beckford-Vera, et al., I-tetrazines and inverse-electron-demand Diels − Alder chemistry: a convenient radioiodination strategy for biomolecule labeling, screening, and biodistribution studies, Bioconjug. Chem. 27 (1) (2016) 207−216.
[89] M. Bączyk, Radioisotope therapy of bone metastases, Nucl. Med. Rev. 14 (2) (2011) 96−104.
[90] T.A. Henderson, The diagnosis and evaluation of dementia and mild cognitive impairment with emphasis on SPECT perfusion neuroimaging, CNS Spectr. 17 (2012) 176−206.

[91] P. Ceruti, S.R. Bryant, J.-H. Lee, M.I. MacEntee, Magnet-retained implant-supported overdentures: review and 1-year clinical report, J. Can. Dent. Assoc. 76 (2010) a52.
[92] R. Raghavan, M. Ramzi, R. Kumar P, P.A. Shajahan, J. Usman, S. Balakrishnan, Magnets in complete dentures, Int. J. Oral Health Dent. 1 (3) (2015) 133–137.
[93] M. Mazaheri, N. Eslahi, F. Ordikhani, E. Tamjid, A. Simchi, Nanomedicine applications in orthopedic medicine: state of the art, Int. J. Nanomed. 10 (2015) 6039–6054.
[94] J.M.D. Coey, Special topics, Magnetism and Magnetic Materials, Cambridge University Press, Cambridge, 2009, pp. 542–579.
[95] S.T. Carmichael, Emergent properties of neural repair: elemental biology to therapeutic concepts, Ann. Neurol. 79 (2016) 895–906.
[96] Y. Suzuki, Mental health services and related factors in health care of traumatic brain injury survivors, Cogent Med 4 (2017) 1273568, http://dx.doi.org/10.1080/2331205X.2016.1273568.
[97] P. Kandesar, P. Chikode, S. Sabale, Perspective of magnetic fluid hyperthermia (MFH) for the treatment of tumor, J. Tumor Res. 2 (1) (2016) 111.
[98] A.Ž. Ilić, S. Ćirković, D.M. Djordjevic, S.R. De Luka, I.D. Milovanovich, A.M. Trbovich, et al., Analytical description of two-dimensional magnetic arrays suitable for biomedical applications, IEEE Trans. Magn. 49 (12) (2013) 5656–5663.
[99] M. Cemazar, T. Kotnik, G. Sersa, D. Miklavcic, Electroporation for electrochemotherapy and gene therapy, in: M.S. Markov (Ed.), Electromagnetic Fields in Biology and Medicine, Taylor & Francis Group, LLC, Boca Raton, 2015, pp. 395–413.
[100] R. Parker, M.S. Markov, Electromagnetic fields in the treatment of tendon injury in human and veterinarian medicine, in: M.S. Markov (Ed.), Electromagnetic Fields in Biology and Medicine, Taylor & Francis Group, LLC, Boca Raton, 2015, pp. 433–452.
[101] T. Schelle, Electrodiagnostic testing of the peripheral nerves, in: G. Penkert, J. Böhm, T. Schelle (Eds.), Focal Peripheral Neuropathies: Imaging, Neurological, and Neurosurgical Approaches, Springer-Verlag, Berlin, 2015, pp. 25–56.
[102] S. Tumanski, Magnetic field measurements and their applications, Handbook of Magnetic Measurements, CRC Press/ Taylor & Francis Group, Boca Raton, 2011, pp. 335–382.
[103] H.-X. Peng, F. Qin, M.-H. Phan, Giant magnetoimpedance sensors and their applications, Ferromagnetic Microwire Composites: From Sensors to Microwave Applications, Springer International Publishing Switzerland, 2016, pp. 99–117.
[104] T. Watson, Electrotherapy, in: S.B. Porter (Ed.), Tidy's Physiotherapy, fifteenth ed., Elsevier Ltd, Edinburgh, 2013, pp. 417–455.
[105] C. Bodine, Computer assistive devices and environmental controls, in: D.X. Cifu (Ed.), Braddom's Physical Medicine and Rehabilitation, fifth ed., Elsevier, Inc, Philadelphia, 2016, pp. 407–423.
[106] I.K. Herrmann, A.A. Schlegel, R. Graf, W.J. Stark, B. Beck-Schimmer, Magnetic separation-based blood purification: a promising new approach for the removal of disease-causing compounds? J. Nanobiotechnol. 13 (2015) 49.
[107] Y. Zhou, X. Xuan, Diamagnetic particle separation by shape in ferrofluids, Appl. Phys. Lett. 109 (2016) 102405.
[108] J. Lipfert, M. Lee, O. Ordu, J.W.J. Kerssemakers, N.H. Dekker, Magnetic tweezers for the measurement of twist and torque, J. Vis. Exp. 87 (2014) e51503.
[109] A.S. Adhikari, J. Chai, A.R. Dunn, Multiplexed single-molecule force proteolysis measurements using magnetic tweezers, J. Vis. Exp. 65 (2012) e3520.
[110] R. Sarkar, V.V. Rybenkov, A guide to magnetic tweezers and their applications, Front. Phys. 4 (2016) 48.

Section V

Multiferroics: Fundamentals

Ferroelectric perovskite–spinel ferrite ceramics

21

Cristina E. Ciomaga[1] *and Liliana Mitoseriu*[2]
[1]Research Department, Faculty of Physics, Dielectrics, Ferroelectrics & Multiferroics Group, Al. I. Cuza Univ. of Iasi, 11 Carol I Bv., 700506, Iasi, Romania, [2]Faculty of Physics, Dielectrics, Ferroelectrics & Multiferroics Group, Al. I. Cuza Univ. of Iasi, 11 Carol I Bv., 700506, Iasi, Romania

21.1 Introduction

Multiferroics are a class of multifunctional materials characterized by the coexistence of at least two ferroic orders (ferroelectric, antiferromagnetic, or ferroelastic) in the same phase [1]. Research in the field of multiferroics dates back to the 1950s, but in the last 10 years, key discoveries in theory, synthesis, and characterization techniques have led to a renewed interest concerning the study and applications of such materials [2,3]. The high interest in multiferroic magnetoelectrics (MEs) can be explained from two points of view: (1) the coexistence of ferroelectricity and ferromagnetism in the same structure, which is highly desired but rarely found in single-phase materials; and (2) the coupling between the magnetic and ferroelectric orders, which might induce novel functionalities, such as a ME effect not present in either state alone. The ME effect represents the appearance of an electric polarization upon applying a magnetic field and/or the appearance of a magnetization under the electric field. Secondary ME coupling may refer in general to a change of a magnetic property (e.g., magnetization, M(H) loop, magnetic permeability, etc.) upon the application of an electric field or vice versa.

Since multiferroic character is rare in single-phase materials or is present at very low temperatures, artificially produced composites with magnetic- and ferroelectric-coupled phases were proposed. In such structures, due to the intimate mechanical coupling between the two ferroic components, by applying an electric field, a deformation in the ferropiezoelectric phase will be induced, which will be transmitted to the magnetostrictive component, thus giving rise to a change of its magnetic properties. Therefore, the applied electric field causes a change in the magnetic state of such composite structure. The control of the magnetic properties through electric fields and vice versa giving rise to a multiple state memory element, the bitunable character (modification of permittivity and permeability by electric and magnetic fields) in such structures, are examples showing the high potential of artificial ME multiferroics for technological applications. Additionally, their investigation also may pose interesting problems from the point of view of fundamental research and moreover, the study of multiferroic composites contributed to the development of a few neighboring research areas, such as developing of innovative complex fabrication processes to generate controlled micro/

Magnetic, Ferroelectric, and Multiferroic Metal Oxides. DOI: http://dx.doi.org/10.1016/B978-0-12-811180-2.00021-9

nanostructures and phase connectivity to maximize the ME response, the discovery of new paths for ferroelectric switching, understanding the ME coupling mechanisms in specific structures, and developing realistic models to describe them [1,4].

After the first artificially ME composites (the quaternary system Fe-Co-Ti-Ba-O) prepared in the Philips laboratory [5], pioneering experiments have been later performed on a large variety of ferroelectric−ferrite composites consisting of ferroelectrics as $BaTiO_3$ (BTO), $PbZr_{1-x}Ti_xO_3$ (PZT), $PbTiO_3$ (PT), polyvinylidenefluoride (PVDF), with ferromagnetic materials such as $CoFe_2O_4$ (CFO), $Tb_{1-x}Dy_xFe_2$ (Terfenol-D), $NiFe_2O_4$ (NFO), $CuFe_2O_4$ (CuFO), $LaMnO_3$ (LMO), and so on. Their study pointed out a few conceptual conditions for obtaining the ME response in ferroelectric-magnetic composites: (1) the individual phases should be in chemical equilibrium, (2) a good structural compatibility between the magnetic and ferroelectric phases in order to ensure a good mechanical coupling between them, (3) high values of magnetostriction coefficient of piezomagnetic phase and of the piezoelectric coefficient of the ferro/piezoelectric phase in order to acquire a high ME coupling, (4) no lack of the accumulated charges through the low-resistivity piezomagnetic/magnetostrictive phase in order to keep an overall dielectric character of the structure, and (5) a deterministic strategy for poling, with reproducible effects at the same applied field inputs [6−12].

In the last years, a number of excellent reviews of the topic have synthesized the main results concerning fundamental aspects, innovative methods for designing, preparation, characterization, theoretical modeling [13−16], and applications of ME composites [2,3,17−22]. Important attention was paid to the investigation of epitaxial ferromagnetic/ferroelectric heterostructures, multilayers, superlattices, and 3D artificially grown composites for potential applications as multistate field effect transistors, magnetic sensors, and tunneling junctions [23]. The ME composites are already used as prototype devices and integrated systems based on the strong ME response for producing ME transducers, sensors, actuators, and microwave active/passive ME components, and so on [24].

In these investigations, the most commonly used piezoelectric ceramic was the perovskite lead zirconium titanate ($PbZr_xTi_{1-x}O_3$; PZT), known for its superior piezoelectric properties, in combination with a relatively high Curie temperature. PZT presents maximized dielectric, piezoelectric, and pyroelectric properties at the morphotropic phase boundary (MPB) due to the easy reorientation of its ferroelectric domains during the poling process. PZT with a composition at MPB ($x = 0.52$) is usually doped with donor or acceptor ions in order to obtain so-called soft or hard piezoelectric properties. Soft PZT is realized by doping with donor ions such as La^{3+} (for Pb^{2+} at A-site) and Nb^{5+}. As result, this substitution leads to the creation of A-site cationic vacancies in the perovskite lattice and therefore, the domain walls can be moved easier (easy poling/depoling), which results in a lower coercitive field, and improved piezoelectric properties, higher permittivity, but also larger dielectric losses [25]. Among the soft PZTs, Nb-doped PZT (PZTN) was most studied [25−27], and due to its characteristics, it was commonly used for applications requiring good piezoelectric properties as actuators, sensors, and generators. Hard PZTs are produced by doping with acceptor ions such as K^+, Na^+ (for A-site), or Fe^{3+}, Ni^{2+}, and others (for B-site), creating oxygen vacancies in the lattice [28].

These ceramics are more difficult to be poled and present high coercive field, lower piezoelectric coefficients, but also lower dielectric losses, which make them more suitable for applications at higher frequencies.

As magnetic counterpart in combination with perovskite ferroelectrics, mostly spinel ferrites were employed due to a high structural compatibility between the perovskite (size of about ~4 Å) and spinel (size of about ~8 Å) unit cells. The spinel ferrites, represented by the chemical formula $MeFe_2O_4$ (Me is a divalent metal ion like Co, Ni, Zn, Mn, Cu, Mg, etc.), are magnetic oxides usually with dielectric or semiconducting character, which are of great technological importance by virtue of their interesting electrical and magnetic properties [29,30]. Their crystalline structure has two interstitial sites, namely tetrahedral (A) and octahedral (B) sites in which cations with different valence and ionic radii can be incorporated [31]. The ferrites present two extreme types of spinel structure: normal spinel and inverse spinel, but in practice mixed spinel structures are also observed. In normal spinel structure, A-sites are occupied by Me^{2+} ions and the B-sites are occupied by Fe^{3+} cations. In the inverse spinel structures, Fe^{3+} cations are distributed between the A- and B-sites and Me^{2+} cations are placed in the B-sites. The mixed spinel structure has the cations distributed in a combined mode between the normal and inverse structure. For a long time, ferrites have been used in various electromagnetic applications and electronic devices, such as transformer cores, antenna rods, memory chips, high density magnetic recording media, permanent magnets, transducers, activators, microwave absorbing materials, and computer technology [32–35]. After 1990, ferrites have been investigated from the viewpoint of global environmental protection [36], such as for wastewater treatment process, hybridization process for mixing solar and fossil energies and solar energy conversion into hydrogen energy, as well as for biomedical applications like magnetically guided drug delivery, hyperthermia, peptide synthesis, detoxification, magnetic resonance imaging (MRI), catalyst, humidity and gas sensors, magnetic fluids, among others [37,38]. For being employed in ME structures, the main requirements are good magnetostrictive/piezomagnetic character and high resistivity with low dielectric losses.

To produce bulk ME materials, a large number of combinations of magnetostrictive ferrites with piezoferroelectric were chosen and different synthesis and processing methods were employed, in order to result in controlled phase interconnectivity and valuable functional properties suitable for new electromagnetic devices. Literature presents a large number of publications and a few reviews in which the most promising combinations for a high ME coefficient are comparatively shown [2,17,39–42]. However, irrespective of the material choice and of the advanced processing methods used to produce a good phase interconnectivity, the obtained ME coefficient in bulk sintered composites is quite low, in the range between 10 and 130 mV/cmOe [18,42–44], which is far below ones resulting in laminar structures (ME coefficient up to 3000 mV/cmOe) [18,45–47]. Apart from the initial technological interest in increasing the ME coupling for sensing applications, the complex investigation of such bulk ME systems still present a high interest for understanding the relationship preparation/composition–micro/nanostructures–functional properties in such complex materials and for exploring other potential applications. Thus, the aim of the present work is to show as example the results of the investigations of two ME composite ceramics based on Nb-PZT ferroelectric with Mn or Ni ferrites, produced by two different synthesis methods, to discuss the effect of composition and phase connectivity and to understand the role of the intrinsic/extrinsic contributions to the functional properties.

21.2 Ceramic composites of Nb-doped $Pb(Zr,Ti)O_3$ with $MnFe_2O_4$

The first spinel–perovskite composite system analyzed here has the formula $x MnFe_2O_4$-$(1-x)Pb_{0.988}(Zr_{0.52}Ti_{0.48})_{0.976}Nb_{0.024}O_3$ (xMF-$(1-x)$PZTN). The synthesis method for obtaining this system was described in detail in Ciomaga et al. [48]. MF is a ferrite with either normal or inverse spinel structure, according to the distribution of the divalent and trivalent cations filling the tetrahedral and octahedral sites (denoted as A- and B-sites, respectively) in the *fcc* sublattice of oxygen. Due to the fact that MF is a ferrimagnetic material whose interaction between the magnetic ions at A- and B-sites is inferred as being antiferrimagnetic [49,50], it presents a total spin magnetic moment of 5 μ_B per formula unit. Unlike the other ferrites, this one doesn't present any change in the total spin magnetic moment between the normal and inverse spinel structure, due to the fact that there are only very small changes in the values of the spin moments of the transition metal ions, which are compensated by changes in the induced oxygen spin moments. The magnetic susceptibility of MF is higher than for other ferrites such as Fe_3O_4, $CoFe_2O_4$, and $NiFe_2O_4$ [51,52].

The mixed PZTN and MF with different compositions of $x = 10$, 25, and 35 wt% $MnFe_2O_4$ composite powders were sintered at 1100°C/1 h. The structural and microstructural investigations of composite samples were realized by X-ray diffraction (XRD) and scanning electron microscopy (SEM) analysis (Fig. 21.1). The XRD patterns confirmed the formation of PZTN perovskite with tetragonal structure and MF with cubic spinel structure. The intensity of (110) perovskite peak decreases and the broadening increases, while intensity of (311) ferrite peak increases with increase in ferrite content in the composites. The calculations based on the Debye-Sherrer relation have revealed that pure PZTN has crystallite size in the range of 18–25 nm, as also confirmed by transmission electron microscopy (TEM) analysis [48]. With addition of ferrite in the ceramic composite, the proportion of the PZTN and MF phases determines a decreasing of crystallite size around 10 nm for the composition with 35% MF in the PZTN matrix. The pure ferrite presents a crystallite size around 43 nm [48]. The backscattered SEM images for pure PZTN and for composition with 35% of MF confirm the good dispersion of the ferrite phase into the ferroelectric matrix, the existence of two different phases in the PZTN–MF samples (light regions are PZTN, dark regions are MF), and the presence of some pores resulted from the mechanical mixture. The grain size of the

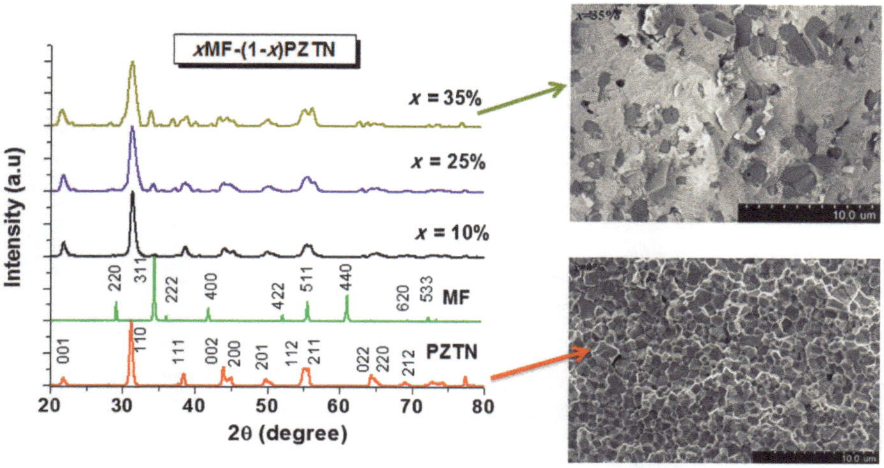

Figure 21.1 X-ray diffraction pattern of *x*MF-(1-*x*)PZTN composites for different concentrations of Mn ferrite. Scanning electron micrographs of fractured ceramic surfaces for composition with $x = 0\%$ and 35%.

PZTN component in the pure and in composite ceramics has values of about 1–2 μm, while the ferrite phase presents distinctly grain size of ~1.5–2.5 μm. The relative density of the sintered ceramics shows a decrease of density from 94% in pure PZTN to 86% for composition with $x = 35\%$ MF. In the ME composites, a very important feature for the functional properties is the fact that the magnetic phase has to be well isolated by the high resistivity PZTN phase and the samples to presents good homogeneity and density.

The dielectric characteristics in the *x*MF-(1-*x*)PZTN ceramic composites were investigated in the range of frequency of 100 Hz–1 MHz and at different temperatures. The variation of dielectric constant with frequency for different temperatures is shown in Fig. 21.2A–D. It can be clearly observed that all the composites present temperature-dependent multiple relaxations in the investigated frequency range. The dielectric permittivity contains multiple contributions as a combined effect of dipolar, electronic, ionic, and interfacial polarizations, besides the lattice intrinsic one. For low frequency, the dipolar and interfacial polarizations contribute significantly to the dielectric permittivity, while at higher frequency only the electronic polarization became significant. The dielectric behavior in these composites can also be understood on the basis of the polarization mechanism in ferrites, since conduction beyond phase percolation limits is due to the ferrite component. Similar results reported in literature showed that in ferrites, the ac conduction is due to the exchange of electrons between $Fe^{2+} \leftrightarrow Fe^{3+}$, accompanied by changes of the oxygen local stoichiometry, which produce local displacements of the electrons and of oxygen vacancies (polaron hopping) in the direction of alternative applied electric field and induce a polarization in the ferrite phase. Thus, the exchange mechanism

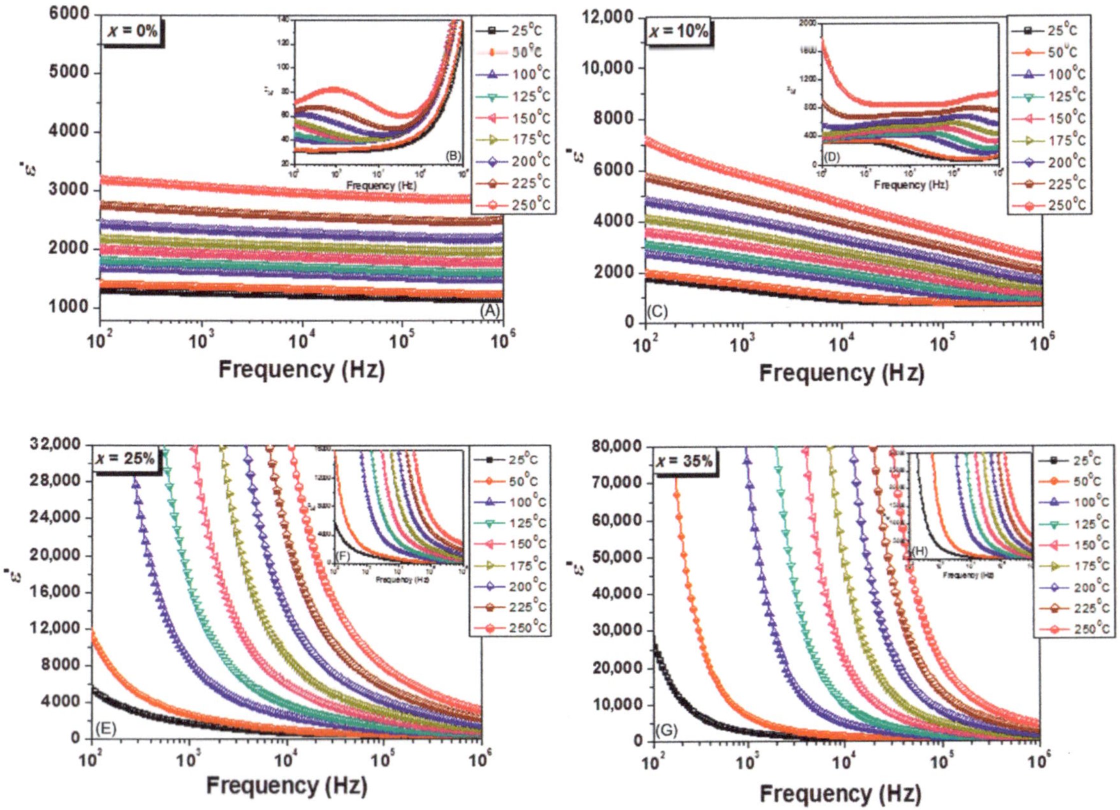

Figure 21.2 Dielectric permittivity versus frequency of xMF-(1-x)PZTN ceramic composites with $x = 0\%$, 10%, 25%, and 35% of ferrite: (A, C, E, G) real, and (B, D, F, H) different temperatures.

of charges among the ions situated at crystallographically equivalent sites determines the transport properties such as electrical conductivity and dielectric dispersion of ferrites and of ferrite-based composites [52,53].

The imaginary part of permittivity (ε') with frequency for different temperature (inset Fig. 21.2D, F, H) clearly shows the dielectric and conductivity relaxation characteristics: (1) at low-frequency dispersion (below 1 kHz), xMF-(1-x)PZTN ceramic composites present Maxwell-Wagner space charge mechanism [54,55], most likely due to the presence of local electrical inhomogeneity within the samples; and (2) a Debye relaxation (10^2-10^4 Hz range), which decreases and shifts toward higher frequency when increasing the ferrite addition (Fig. 21.2B and D). The observed Maxwell-Wagner relaxation is associated with uncompensated surface charges at the interfaces between dissimilar phases (perovskite−spinel) inside the composite ceramics. As discussed in detail in Mitoseriu et al. [55], such extrinsic phenomena cause the peculiar dielectric properties observed in all the ME composites reported in the literature.

At room temperature, the dielectric constant (ε) of these composites decreases with increase of the ferrite addition, as a consequence of the sum property. The sum property represents an enhancement of the effects already present in constituent phases being proportional to the weight/volume fractions of the parent phases. The composition with $x = 0\%$ has the highest value of $\varepsilon' \sim 1240$. With the addition of Mn ferrite, the permittivity decreases down to $\varepsilon' \sim 400$ for $x = 35\%$, at $f = 10$ kHz. However, the dielectric constant of the composites is not determined intrinsically only by the values of permittivity of the two component phases and by their volume ratio. The dielectric losses above unity indicate that the influence of the low resistivity MF phase dominates the electrical properties in radiofrequency range, as it was detailed in Ciomaga et al. [48]. Extrinsic effects related to possible small doping at the interfaces that create small compositional variations between grain boundaries and bulk properties, the interfaces between the dissimilar materials where uncompensated charges are usually located, as well as the presence of uncontrolled volume charge defects as oxygen vacancies, are at the origin of extrinsic contributions to the final dielectric and conduction properties in such composites. These extrinsic contributions usually generate dielectric and conductive relaxation effects, as observed for the present xMF-(1-x)PZTN composites and also reported by many authors in similar ferroelectric−ferromagnetic systems [11,55−58].

The observed relaxation and conduction processes can be described by using the dielectric modulus formalism [56]. Thus, the frequency dependences of the imaginary part of dielectric modulus (M'') for different temperatures in the range of 25°C−250°C are shown in Fig. 21.3A−D). It can be observed that the PZTN sample presents dispersion phenomena below 10 kHz. With addition of magnetic phase in the composite system, $M''(f)$ attains two maxima in the lower and higher frequency regions, which have the tendency to shift both toward higher frequency with increasing temperature (Fig. 21.3A−D). These peaks indicate the transition from long-range to short-range mobility with increasing frequency. At the low frequency (of the peak) the ions are capable of moving a long distance (i.e., performing successful hopping from one site to the neighboring site), but for the high

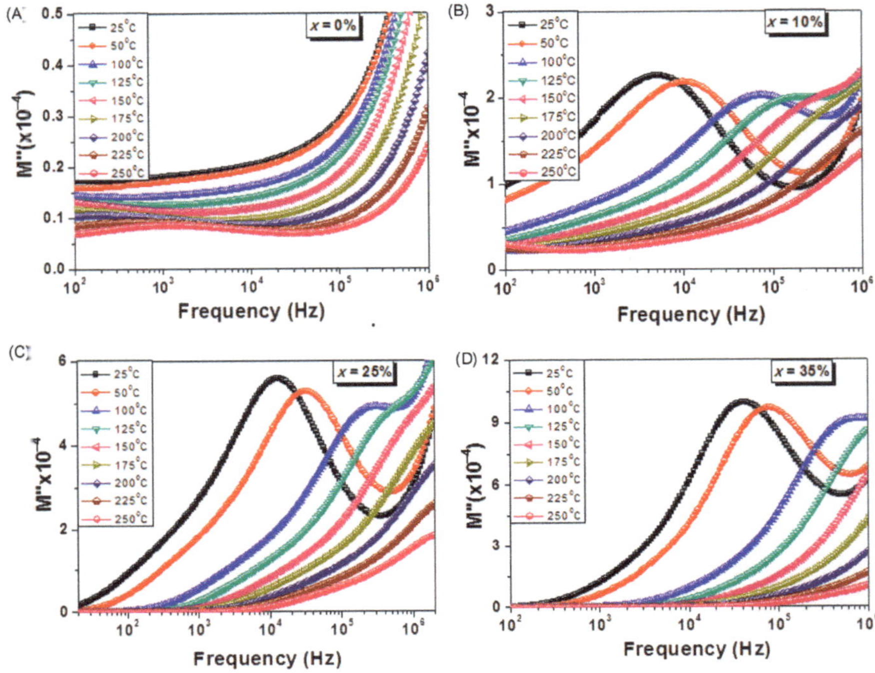

Figure 21.3 Frequency dependence of the imaginary part of the dielectric modulus in *x*MF-(1-*x*)PZTN ceramic composites with (A) $x = 0$, (B) $x = 10$, (C) $x = 25$, (D) $x = 35$ wt%, at different temperatures.

frequency side of peak, the ions are spatially confined to their potential wells and they can execute only localized motions within the well [59]. The appearance of two peaks in M'' versus frequency plots again confirms the presence of both grain and grain boundary effect in the material. The frequency dependence of dielectric modulus indicates that the peak relaxation corresponds to the plot of the imaginary part of permittivity versus frequency, indicating dielectric relaxation processes for all the compositions.

The dielectric relaxations observed in the *x*MF-(1-*x*)PZTN ceramic composites are thermally activated phenomena whose characteristic activation energy was computed by the maximum of the imaginary part of the dielectric modulus using the Arrhenius law:

$$\tau = \tau_0 \exp\left(-\frac{E_a}{k_B T}\right) \tag{21.1}$$

where $\tau = 1/2\pi f_m$ is the relaxation time related to the frequency f_m corresponding to the relaxation peak of M'', τ_0 is a preexponential factor, k_B is the Boltzmann constant, and E_a is the activation energy. Thus, the evolution with frequency and temperature of the characteristic relaxation time of each composition determined whenever possible in the available frequency and temperature range indicates the corresponding activation energies. It appears that with increase of MF content (x) in these ME composite ceramics, the activation energy (E_a) presents close values of 0.39 eV for $x = 10\%$ and 0.41 eV for samples with $x = 25\%$ and 35%. The obtained activation energies are values typically observed in complex oxides and they are in the range of those reported in other similar composites [48,60–62]. These values are assigned to uncompensated charges, fluctuations of the magnetic ion oxidation states between ferrous and ferric ions $Fe^{2+} \leftrightarrow Fe^{3+} + e^{-}$, which contribute to local electron displacements in the direction of the applied external field and to different levels of local oxygen stoichiometry within the ceramic volume. Similarly, the $Mn^{3+} \leftrightarrow Mn^{2+} + e^{+}$ reaction gives a number of holes (e^{+}) in the octahedral sites that ensure local displacements in the opposite direction with respect to the external applied field. These displacements are responsible for the polarizability and macroscopic dielectric properties of the ferrite-based composite ceramics [63,64].

In Fig. 21.4 the frequency dependence of AC conductivity at room temperature for xMF-(1-x)PZTN ceramic composites is represented. While for pure PZTN ($x = 0$) and for the composite with $x = 10\%$ the low-frequency conductivity still decreases when reducing frequency, showing that a very low DC conductivity below 10^{-8} S/m is expected, an increase of low frequency conductivity with 3–4 orders in magnitude results for $x = 25\%$ and $x = 35\%$. Therefore, the composites with a high content of MF show a stabilized low-frequency conductivity (horizontal plateau at low frequencies), which allow the precise determination of their dc conductivity of 10^{-5} S/m for $x = 25\%$ and 4×10^{-5} S/m for $x = 35\%$, respectively. The jump of conductivity with a few orders in magnitude between $x = 10\%$ and $x = 25\%$ is related to the proximity of percolation around $x = 30\%$. Above this composition, a transition from dielectric to semiconductor character as derived from the properties of ferrite is likely. The compositional differences in the AC conductivity

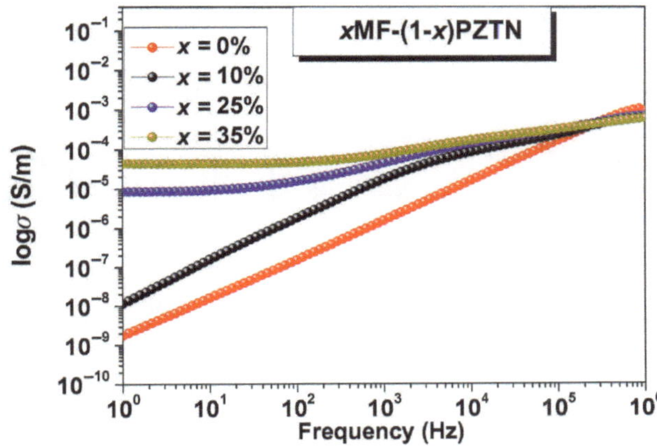

Figure 21.4 Variation of AC conductivity with frequency for xMF-(1-x)PZTN ceramic composites, at room temperature.

disappear at high frequencies, when the conductivity for all the compositions tends to saturate to a value of about 6×10^{-4} S/m(1 MHz).

To check the magnetic properties of these composites the magnetization as a function of magnetic field (hysteresis loops *M-H*) at three different temperatures have been measured and represented in Fig. 21.5A–D. The composites show hysteretic behavior at all the temperatures, with an evident small value of coercitivity, at room temperature ($T = 300$K). A decreasing value of the coercitive field (H_c) is observed when increasing temperature, from 0.1 (at $T = 15$K) to 0.043 T (at $T = 300$K), indicating a soft ferrimagnetic behavior of the magnetic component of the composite ceramics (Fig. 21.5A and D). The magnetic hysteresis loops presented in Fig. 21.5A–D show an increase of saturation magnetization with increasing temperature from 15K–300K for all the compositions. In this case, the existence of the ferromagnetic behavior can be observed even at 15K, where all the samples present a more rectangular hysteresis loop (Fig. 21.5D). The coercivity is generally a measure of the magnetocrystalline anisotropy for the magnetic phase in the composite system and this behavior may originate from the exchange anisotropy due to the spin disorder at the grain surface at low temperatures [65]. As it can be noted, at low temperature the composites reach their saturation for a higher magnetic field ($H \sim 1.5$ T), than at room temperature ($H \sim 0.5$ T). This behavior can be

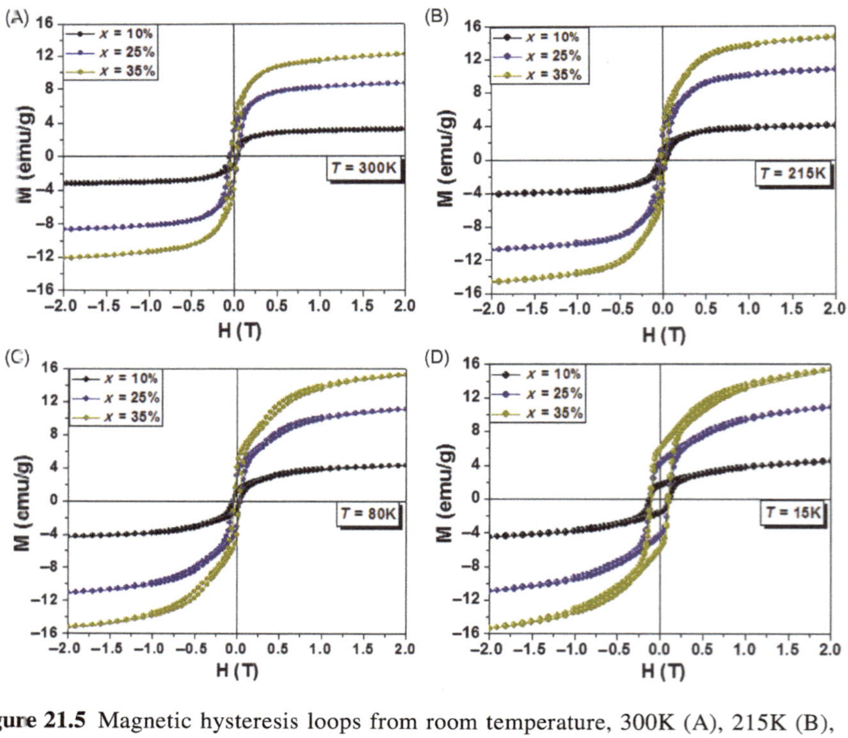

Figure 21.5 Magnetic hysteresis loops from room temperature, 300K (A), 215K (B), 80K(C), and 15K (D), for all *x*MF-(1-*x*)PZTN ceramic composites.

attributed to the presence of a spin disorder surface layer, which requires a larger field to be saturated. This is much better observed in the composition with higher addition of magnetic phase [66]. An increase of the remnant magnetization with addition of magnetic phase ($x = 35\%$ MF) from ~5.08 emu/g, at $T = 300$K, to ~6.43 emu/g at $T = 15$K, is observed. As expected, by combining ferrites with ferroelectrics, mostly ferrite-derived magnetic properties were obtained, as result of a sum property [55].

In order to test the properties in microwave range, the permeability and permittivity of xMF-(1-x)PZTN composites have been measured in GHz range in order to search for interesting properties for applications as microwave devices with magnetic tunable properties. The experimental results and the numerical simulation using a high frequency structure simulator have shown that, in the range of 2–6 GHz, the composition with $x = 35\%$ can be used for designing a magnetic tunable microwave band stop filter device. From Fig. 21.6, it was found that the central stop-band frequency of the designed filter can be tuned with 18 MHz by changing the magnetic field from 0 to 2 kOe; more details are given in Ciomaga et al. [48].

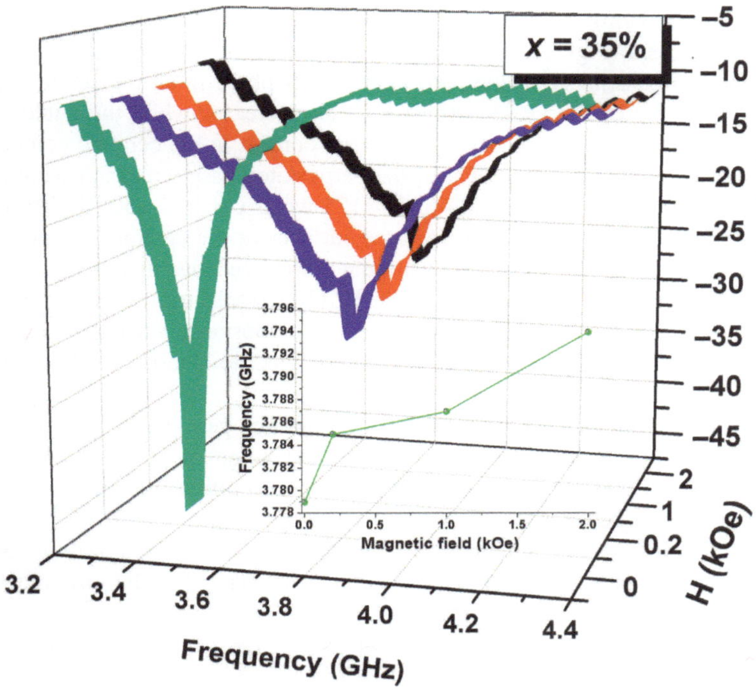

Figure 21.6 Measured transmission coefficient for the magnetically tunable microwave filter at different magnetic fields.

21.3 Nb-doped Pb(Zr,Ti)O_3-ferrite composites prepared by in situ sol-gel combustion method

In our previous works, studies concerning various ferroelectric-ferrite composite ceramics prepared by solid state reaction were presented [11,55]. By comparison with other literature data and our own works, it was demonstrated that a better mixing of parent phases with a high degree of homogeneity of the microstructures could be provided by using in situ methods for producing the composites. Therefore, a reduction of the dielectric losses in such ceramic composites was expected, as due to a better microstructural control. In the following, we will present the results of investigation of a ceramic composite based on Nb-doped Pb(Zr,Ti)O_3 with different compositions of $NiFe_2O_4$ ferrites. These ceramic composites were prepared in situ by using a combination of solid state reaction method for producing the PZTN powders further used as template for producing the ferrite by a sol-gel combustion technique [67–69]. The sol-gel combustion method is one of the wet chemical methods, which ensured a high degree of homogeneity in the composites, reducing the atomic interdiffusion during the calcination and sintering step and better control of the microstructural characteristics. The $Pb_{0.988}(Zr_{0.52}Ti_{0.48})_{0.976}Nb_{0.024}O_3$ (PZTN) powder prepared by conventional mixed oxides and calcinated at 800°C/4 h was used as a template. The ferrite ($NiFe_2O_4$) formation directly onto PZTN powder template was realized by sol-gel combustion, following the procedure described in detail in Iordan et al. [68].

The XRD analysis (Fig. 21.7) indicates the presence of inverse spinel $NiFe_2O_4$ structure and the maintenance of the perovskite PZTN phase in all the ceramic composites. The intensity of the (101) perovskite peak decreases while the cubic (311) ferrite peak increases with the increasing of ferrite content in the ceramic composite. A reduction of tetragonal distortion c/a of the perovskite phase with increasing addition of NF was observed (from 1.0244 to 1.01175 when $x = 10\%-30\%$) [70]. From the backscattered SEM images of the fractured surfaces of the ceramic composites it can be observed that all the compositions present a homogeneous microstructure with a good dispersion of the NF crystals into the PZTN ferroelectric matrix. Therefore, the use of sol-gel method for in situ preparation of composites contributes to a better degree of connectivity of the magnetic and ferroelectric phases in the ceramic composites.

The experimental dielectric characteristics (real part of permittivity and dielectric losses) versus frequency at room temperature in the range of 100 Hz–1 MHz are shown in Fig. 21.8A and B. The overall decrease of permittivity in xNF-(1-x)PZTN ceramic composites when increasing the ferrite amount is attributed to the low permittivity characteristic to the NF addition. However, the decrease is not monotonous at all the frequencies and some relaxation anomalies are typical in the low-frequency range, as often observed in such composites and attributed to interfaces or Maxwell-Wagner effects [54,55,71–73]. At higher frequencies, the permittivity reduction is regular and monotonous, as shown by the inset of Fig. 21.8A (at 10^5 Hz).

The observed composition dependence of permittivity can be attributed to the superposition of two effects: modification of piezoelectric PZTN material with NF

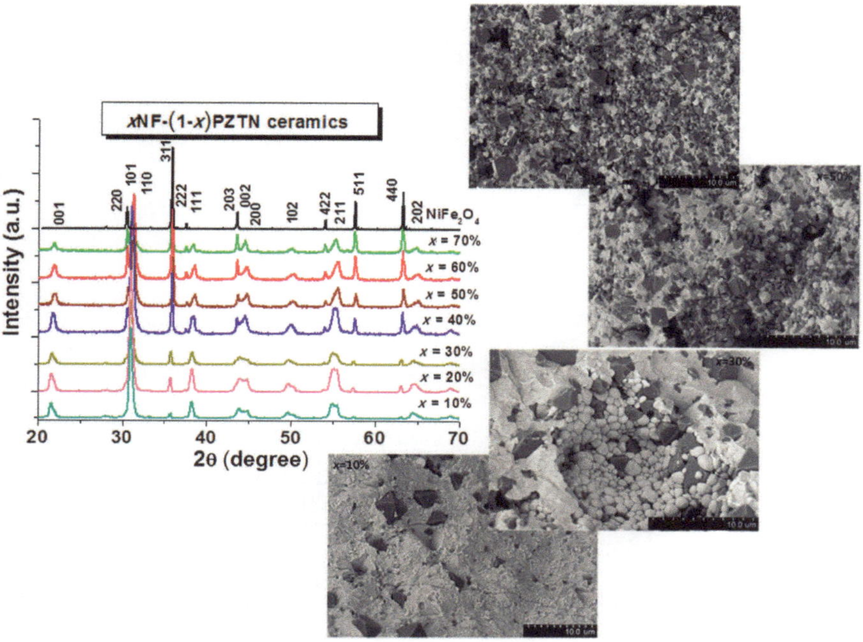

Figure 21.7 X-ray diffraction patterns of xNF-(1-x)PZTN ceramic composites with $x = 10$; 20; 30; 40; 50; 60; 70; 100 wt% and the scanning electron microscopy images for $x = 10\%$, 30%, 50%, and 70% of NF.

composition (small doping at interfaces) and change in the grain size of the magnetostrictive NF material at different compositions. As mentioned, at lower frequency, the samples with higher NF concentration (up to 30%) present dispersion and after that, permittivity reaches a constant value with further increasing of frequency. The dispersion at low frequencies suggests a contribution of space charge polarization or grain boundary phenomena related to the possible reciprocal doping at the interfaces between dissimilar phases. However, the low-frequency relaxations are strongly reduced by comparison with ones observed for the xMF-(1-x)PZTN prepared by simple mixing and solid state method (Fig. 21.2). We can infer that the use of in situ sol-gel processing may contribute to the reduction of space charge phenomena via a better microstructural control. As the content of NF increases beyond the percolation limit, there is a sharp fall of resistivity, derived from the electrical properties of ferrite phase, as previously explained. The dielectric losses of the composites (Fig. 21.8(B)) increase with an increase in the ferrite concentration, mainly due to the interfacial polarization, at frequencies lower than 10 kHz. As shown in the study reported in Ciomaga et al. [67] the multiple relaxation phenomena in such composite systems are thermally activated phenomena determined by the electrical inhomogeneity of the ceramic composites.

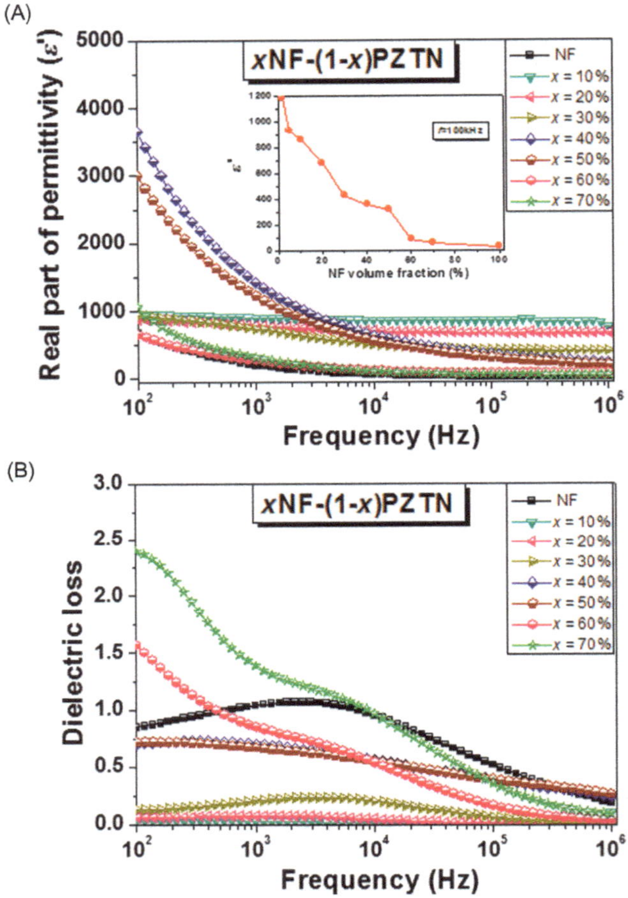

Figure 21.8 Frequency dependences of (A) real part of permittivity and (B) dielectric losses of *x*NF-(1-*x*)PZTN ceramic composites with $x = 10; 20; 30; 40; 50; 60; 70; 100$ wt%.

Thus, by using an in situ synthesis method for a better mixing of the ferroelectric and magnetic components, an improvement of the microstructural characteristics (homogeneity, dispersion, lack of cracks) could be obtained. However, even for the *x*NF-(1-*x*)PZTN ceramic composites with relative good dielectric properties at room temperature ($\tan \delta < 10\%$ for $f = 1$ kHz), some activated relaxation and conductivity are still present.

The variation of AC conductivity with frequency at room temperature shown in Fig. 21.9A indicates that the increase in the ferrite content generally results in an increase of conductivity, but the variation is not fully monotonous. The extrapolated DC conductivity (Fig. 21.9B) shows that the composites shift from a dielectric character for lower NF addition (conductivity below 10^{-8} S/m) to a semiconductive one, as proved by the sudden increase with two orders in magnitude at $x = 0.20$,

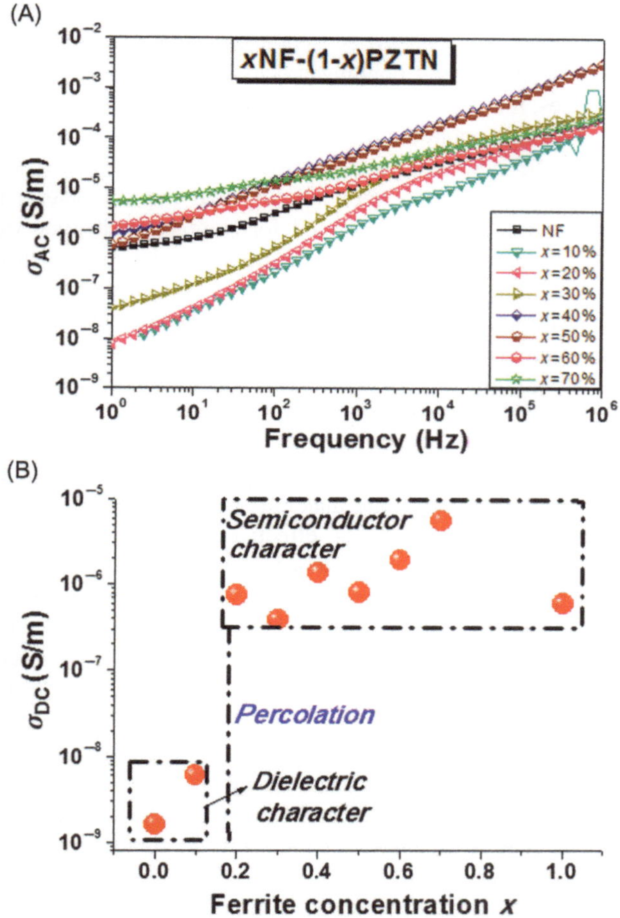

Figure 21.9 (A) Frequency dependence of the AC conductivity of xNF-(1-x)PZTN ceramic composites, at room temperature; (B) Extrapolated room temperature DC conductivity versus ferrite concentration.

where percolation most probably takes place. All the other compositions show DC conductivities in the range of 10^{-6}–10^{-5} S/m with slightly nonmonotonous variation versus the ferrite addition, but with a dominant semiconductor character derived from the electrical properties of ferrites. The conduction mechanism observed in these types of composites with $x \geq 0.20$ can be explained again by polaron hopping among the localized state as usually present in ferrites [74]. Hopping conduction is favored in ionic lattices in which the same kind of cation is found in two different oxidation states. In the conduction process an important role is playing by the hopping of 3d electrons among Fe^{2+} and Fe^{3+} as well as between Ni^{2+} and Ni^{3+} in the case of $NiFe_2O_4$ ferrite.

The increasing AC conductivity with increase of frequency indicates mixed (small and large) polaron conduction. At lower frequencies, the grain boundaries are more active and the role of the polaron hopping process is minor. As the frequency of the applied field increases, the conductive phase becomes more active, thereby increasing the importance of mixed polaron hopping conduction [75]. As mentioned before, with increase of the NF addition, the composite ceramics present a semiconducting behavior rather than an insulating one, showing that the percolation is reached at lower composition ($x = 20\%$) than expected for samples with good dispersion and (0–3) connectivity (expected at $x \sim 33\%$). In similar composites, such an effect was explained by the fact that ferrite grains may form some higher conductivity chains and therefore the electrical resistivity of these composites is reduced significantly due to the parallel connectivity between ferrite and ferroelectric grains [76,77]. It can be concluded that although in situ preparation generated a superior uniformity and better dispersion of the two phases, still the electrical properties are not fully controlled to be dielectric in a large compositional range.

To describe the evolution of permittivity and permeability with composition and for predicting and designing composite materials with desired properties, a few theoretical approaches were developed, as effective medium approximation (EMA) [78,79] and more recently, finite element method (FEM) calculations [16,80]. The most common EMA models for predicting the dielectric properties in composites are (1) Maxwell-Garnett approximation (MG), corresponding to isolated spherical inclusions embedded in a continuous background matrix [81]; (2) Bruggeman approximation (BG), considering that the properties of matrix or host are the same as that of the effective medium itself and that assumes a symmetry between inclusion and background phases [82]; (3) Lichtenecker mixing formula (LI), which is derived by differential analysis and is considered appropriate for inhomogeneous microstructures [83]. Due to the fact that experimentally different types of microstructures are generated and the difficulty to control the phase interconnectivity and degree of percolation in real ceramic composites when the volume filler is changing, the applicability of the EMA approach is quite limited to small compositional ranges. Thus, the FEM approach proved to be extremely useful because it allowed calculation of the electric field distribution inside the composite and the effective dielectric constant for realistic microstructures with combined types of interconnectivities [80].

By comparison with EMA approximations, FEM takes into consideration the shape of composites and the local permittivity, and performs a more rigorous calculation of the local potential without supplementary approximations. The effective permittivity as a function of the volume fraction of inclusions and at various fields can be calculated using this method in all the compositional range and for various types of phase interconnectivities. For any type of approximation, it is important to consider the intrinsic value of the dielectric parameters of the two phases (if available), since permittivity and conductivity of the two phases and mainly of the composite are frequency dependent and are strongly affected by extrinsic phenomena, as we observed and described previously, thus making a relevant comparison of the experimental data with model predictions difficult.

For example, in the case of our discussed ferrite–ferroelectric composites, the experimental values of permittivity were compared with ones predicted by EMA and FEM calculations (Fig. 21.10). As already discussed, the dielectric properties in such composites, mostly at low frequencies, are strongly affected by extrinsic phenomena and by conductivity. For this reason, the comparison between experimental data and the theoretical simulations was performed for permittivity determined at the frequency of 10^8 Hz, where the extrinsic contributions to the dielectric constant are strongly reduced. From Fig. 21.10, it can be observed that not all the types of approaches could describe the experimental permittivity of the investigated ferroelectric–ferrite composites in a large compositional range: (1) for the MF-PZTN composites, prepared by mixing and classical ceramic method, the experimental data better fit with Lichtenecker's approximation and FEM results; (2) for NF-PZTN composites obtained by sol-gel technique, the predictions of MG and BG approximations are close to the experimental values for small addition of NF magnetic phase (NF magnetic inclusions in PZTN ferroelectric matrix), while for low PZTN amounts (PZTN islands inside the NF matrix), the best agreement, in overall composition range, was provided by the values derived by FEM approach and LI equation [16]. These results can be explained in terms of their specific microstructural characteristics. The best results for both types of composites in a large compositional range were provided by the FEM calculations because realistic virtual microstructures have been simulated and local field distributions have been computed for the estimation of the effective permittivity, while the LI model considers that microstructures present randomly dispersed grains with different grain size and shapes of the constituent phases, which for certain compositions are quite realistic hypotheses as well.

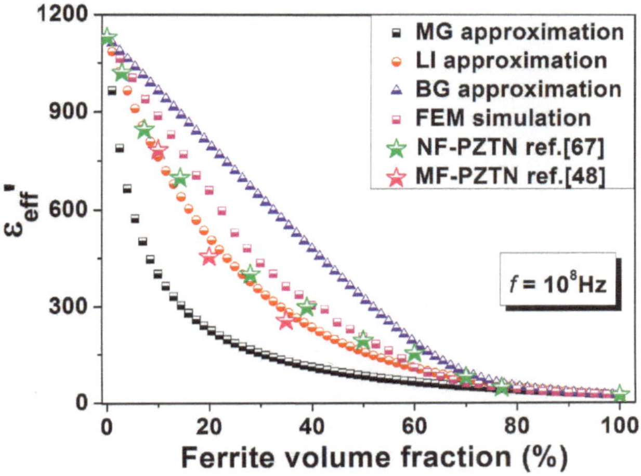

Figure 21.10 Comparison between the simulated and experimental permittivity of various ferroelectric–ferrite ceramic composites, at the frequency of 10^8 Hz.

A comparison of the saturation magnetization (M_s) in the two types of composites with various compositions discussed in this chapter is shown in Fig. 21.11. The magnetic properties of *x*NF-(1-*x*)PZTN ceramics with $x = 10$; 20; 30; 40; 50; 60; 70; 100 wt% described in detail in Ciomaga et al. [67] and ones of the *x*MF-(1-*x*)PZTN system have shown that magnetic characteristics are derived from the ferrite's properties, which are sensitive to composition and preparation methods. The increase in saturation magnetization with increasing ferrite content provides an indication that the spontaneous magnetization originates from the unbalanced antiparallel spins (ferromagnetic character). In the investigated diphase ferroelectric–ferrite composites, it is important to show that the intrinsic properties of the ferroelectric and magnetic phases are maintained, without major changes to the interaction mechanisms due to the presence of secondary phases, especially the magnetic properties. Since the presence of nonferromagnetic PZTN phase does not cause intimate changes of the magnetic interactions, as expected the magnetic character increases with increasing the addition of the magnetic phase. The magnetic characteristics of ME composites predominantly depend on the magnetic behavior of the parent ferrite phases (cation distribution, site preference energies, covalence bonds, and the molecular field) and rarely are the result of extrinsic phenomena, as in case of the dielectric character. As other data from literature revealed, our data also indicated a magnetic character in composite, which is a consequence of the sum property, a dilution of the magnetic character when increasing the ferroelectric PZTN amount [72,84–87]. Magnetic properties in such composites may also be described either by FEM approximations or by FEM calculations, but literature data does not report such detailed calculations, mostly because the interest was mainly dedicated to the dielectric properties in such ME composites.

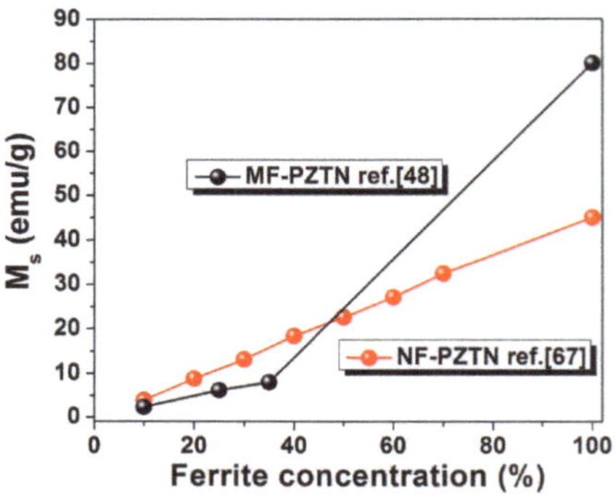

Figure 21.11 Saturation magnetization as a function of ferrite composition in PZTN-ferrite ceramic composites.

21.4 Conclusions

The studies have showed that ferroelectric–ferrite ME composites with relative good dielectric properties mostly at high frequencies and ferromagnetic character at room temperature can be produced either by simple mixing or by more sophisticated wet chemical methods. If the sintering temperatures are not too high, pure diphase composites can be obtained in ceramic form. The microstructures (in terms of large-scale homogeneity, lack of aggregates, large pores or cracks) and phase interconnectivity toward the desired (0–3) one can be better controlled by using wet or combined methods with respect to the simple ceramic mixing methods. In such composites, the functional properties are determined by the ones of the parent phases, phase interconnectivity, composition, and also by the interfaces separating dissimilar phases, which may contain regions with reciprocal doping, small compositional variations, and uncompensated charges. While the magnetic properties are mainly the result of a sum response (i.e., a reduction of the magnetic character with an increase in the ferroelectric addition is always noticed), the electrical properties (permittivity, conductivity) are influenced by extrinsic phenomena rather than by the electrical properties of the parent phases. These extrinsic contributions are mainly related to the microstructures, compositional and structural heterogeneity of the constituent phases and of the interfaces, inhomogeneity of the local oxygen stoichiometry mainly within the ferrite phase, and the presence of electron exchange between Fe^{2+} and Fe^{3+} or of other transition ions present in the magnetic phase. These contributions manifest by multiple relaxations in radiofrequency range and high dielectric losses. However, by using chemical methods, for example the in situ sol-gel synthesis method, a better mixing of the ferroelectric and magnetic components and improved homogeneity contribute to the reduction of dielectric losses and of the space charge phenomena in the composites. Even the improved microstructures did not result in stronger ME coefficients in bulk ceramic composites, which remain far lower than for layered ferrite–ferroelectric counterparts. Instead, the exploration of various ferroelectric–ferrite combinations provided knowledge and methods to produce composites with desired dielectric and magnetic properties at room temperature mainly at high frequencies, which can be tuned to some extent by electric and magnetic fields and provide electromagnetic properties. This is interesting for some innovative applications, such as materials for low-loss miniaturized antennas or tunable filters, and further stimulated the search for metamaterial-like character in ferroelectric–ferrite ceramic composites.

Acknowledgments

A part of this work was financially supported by Romanian UEFISCDI project no. PN-II-PT-PCCA-2013-4-1119. The group of Dr. Carmen Galassi, from ISTEC-CNR, Faenza, Italy is acknowledged for the collaboration on the topic of ME composites. Dr. Mirela Airimioaei from University Alexandru Ioan Cuza of Iasi is acknowledged for the preparation of ferrites.

References

[1] H. Schmid, Multi-ferroicmagnetoelectrics, Ferroelectrics 162 (1994) 317–338.
[2] M. Fiebig, T. Lottermoser, D. Meier, M. Trassin, The evolution of multiferroics, Nat. Rev. Mater 1 (2016) 1–14.
[3] H. Palneedi, V. Annapureddy, S. Priya, J. Ryu, Status and perspectives of multiferroic-magnetoelectric composite materials and applications, Actuators 5 (9) (2016) 1–31.
[4] S. Dong, J.-M. Liu, S.-W. Cheong, Z. Ren, Multiferroic materials and magnetoelectric physics: symmetry, entanglement, excitation, and topology, Adv. Phys. 64 (2015) 519–626.
[5] J. Van Suchetelene, Product properties: a new application of composite materials, Philips Res. Rep. 27 (1972) 28–37.
[6] A.M.J.G. Van Run, D.R. Terrell, J.H. Scholing, An in situ grown eutectic magnetoelectric composite material, J. Mater Sci. 9 (1974) 1710–1714.
[7] J. Van den Boomgaard, A.M.J.G. Van Run, J. Van Suchetelene, Magnetoelectricity in piezoelectric–magnetostrictive composite, Ferroelectrics 10 (1976) 295–298.
[8] R.P. Mahajan, K.K. Patankar, M.B. Kothale, S.A. Patil, Conductivity, dielectric behaviour and magnetoelectric effect in copper ferrite-barium titanate composites, Bull. Mater. Sci. 23 (2000) 273–279.
[9] Y.R. Dai, P. Bao, J.S. Zhu, J.G. Wan, H.M. Shen, J.M. Liu, Internal friction study on $CuFe_2O_4/PbZr_{0.53}Ti_{0.47}O_3$ composites, J. Appl. Phys. 96 (2004) 5687–5690.
[10] J. Zhai, N. Cai, Z. Shi, Y. Lin, C.-W. Nan, Magnetic-dielectric properties of $NiFe_2O_4$/PZT particulate composites, J. Phys. D: Appl. Phys. 37 (2004) 823–827.
[11] L. Mitoseriu, V. Buscaglia, M. Viviani, M.T. Buscaglia, I. Pallecchi, C. Harnagea, et al., $BaTiO_3$–$(Ni_{0.5}Zn_{0.5})Fe_2O_4$ ceramic composites with ferroelectric and magnetic properties, J. Eur. Ceram. Soc. 27 (2007) 4379–4382.
[12] A. Shan, X. Wu, J. Lu, C. Chen, R. Wang, Phase formations and magnetic properties of single crystal nickel ferrite ($NiFe_2O_4$) with different morphologies, Cryst. Eng. Commun. 17 (2015) 1603–1607.
[13] C.W. Nan, Magnetoelectric effect in composites of piezoelectric and piezomagnetic phases, Phys. Rev. B 50 (1994) 6082–6088.
[14] B. Sareni, L. Krahenbuhl, A. Beroual, C. Brosseau, Complex effective permittivity of a lossy composite material, J. Appl. Phys. 80 (1996) 4560–4565.
[15] T.M. Peng, R.T. Hsu, J.H. Jean, Low-fire processing and properties of ferrite + dielectric ceramic composite, J. Amer. Ceram. Soc. 89 (2006) 2822–2827.
[16] C.E. Ciomaga, C.S. Olariu, L. Padurariu, A.V. Sandu, C. Galassi, L. Mitoseriu, Low field permittivity of ferroelectric-ferrite ceramic composites: experiment and modeling, J. Appl. Phys. 112 (2012) 094103–094107.
[17] N.A. Spaldin, M. Fiebig, The renaissance of magnetoelectricmultiferroics, Science 309 (2005) 391–392.
[18] C.W. Nan, M.I. Bichurin, S. Dong, D. Viehland, G. Srinivasan, Multiferroicmagnetoelectric composites: historical perspective, status, and future directions, J. Appl. Phys. 103 (2008) 031101.
[19] G. Lawes, G. Srinivasan, Introduction to magnetoelectric coupling and multiferroic films, J. Phys. D: Appl. Phys. 44 (2011) 243001.
[20] M. Bichurin, V. Petrov, S. Priya, A. Bhalla, Multiferroicmagnetoelectric composites and their applications, Adv. Condens. Matter Phys. 2012 (2012) 1–3, Article ID 129794.

[21] R.A. Islam, S. Priya, Progress in dual (piezoelectric-magnetostrictive) phase magnetoelectric sintered composites, Adv. Condens. Matter Phys. 2012 (2012) 1–29, Article ID 320612.
[22] A.S. Tatarenko, M.I. Bichurin, Microwave magnetoelectric devices, Adv. Condens. Matter Phys. 2012 (2012) 1–10, Article ID 286562.
[23] E.Y. Tsymbal, A. Gruverman, V. Garcia, M. Bibes, A. Barthelemy, Ferroelectric and multiferroic tunnel junctions, MRS Bull. 37 (2012) 138–143.
[24] Y.J. Wang, J.F. Li, D. Viehland, Magnetoelectrics for magnetic sensor applications: status, challenges and perspectives, Mater. Tod. 17 (2014) 269–275.
[25] V.D. Kugel, L.E. Cross, Behavior of soft piezoelectric ceramics under high sinusoidal electric fields, J. Appl. Phys. 84 (5) (1998) 2815–2830.
[26] W. Wersing, Applications of piezoelectric materials: an introductory review, in: N. Setter (Ed.), Piezoelectric Materials in Devices, EPFL, Lausanne, 2002, pp. 29–66.
[27] J. Ryu, J.-J. Choi, H.-E. Kim, Effect of heating rate on the sintering behavior and the piezoelectric properties of lead zirconatetitanate ceramics, J. Am. Ceram. Soc. 84 (2001) 902–904.
[28] D. Damjanovic, M. Demartin, H.S. Shulman, M. Testorf, N. Setter, Instabilities in the piezoelectric properties of ferroelectric ceramics, Sens. Actuator A: Phys. 53 (1996) 353–360.
[29] L. Jaswal, B. Singh, Ferrite materials: a chronological review, J. Integr. Sci. Technol. 2 (2) (2014) 69–71.
[30] A. Hajalilou, S. AmriMazlan, A review on preparation techniques for synthesis of nanocrystalline soft magnetic ferrites and investigation on the effects of microstructure features on magnetic properties, Appl. Phys. A 122 (2016) 680–695.
[31] J. Smit, H.P.J. Wijn, Ferrites, Philips Technical Library, Eindhoven, The Netherlands, 1959, pp. 1–369.
[32] S.M. Chavan, M.K. Babrekar, S.S. More, K.M. Jadhav, Structural and optical properties of nanocrystalline Ni–Zn ferrite thin films, J. Alloys Compd. 507 (2010) 21–25.
[33] H. Waqus, A.H. Quresghi, Influence of pH on nanosized Mn-Zn ferrite synthesized by sol-gel auto combustion process, J. Therm. Analy. Calori. 98 (2009) 355–360.
[34] T. Nakamura, T. Miyamoto, Y. Yamada, Complex permeability spectra of polycrystalline Li–Zn ferrite and application to EM-wave absorber, J. Magn. Magn. Mater. 256 (2003) 340–347.
[35] M. Sugimoto, The past, present, and future of ferrites, J. Am. Ceram. Soc. 82 (1999) 269–280.
[36] Y. Tamaura, Ferrites for global environmental protection technology, J. Magn. Soc. Jpn. S1 (1998) 396–399.
[37] X. Chu, B. Cheng, J. Hu, H. Qin, M. Jiang, Semiconducting gas sensor for ethanol based on $LaMg_xFe_{1-x}O_3$ nanocrystals, Sens. Actuat. B 129 (2008) 53–58.
[38] C.N.R. Rao, A.K. Cheetam, Science and technology of nanomaterials: current status and future prospects, J. Mater. Chem. 11 (2001) 2887–2894.
[39] M. Fiebig, Revival of the magnetoelectric effect, J. Phys. D: Appl. Phys. 38 (2005) R123–R152.
[40] W. Eerenstein, N.D. Mathur, J.F. Scott, Multiferroic and magnetoelectric materials, Nature 442 (2006) 759–764.
[41] S. Priya, R. Islam, S. Dong, D. Viehland, Recent advancements in magnetoelectric particulate and laminate composites, J. Electroceram. 19 (2007) 147–164.
[42] N. Ortega, Ashok Kumar, J.F. Scottand, R.S. Katiyar, Multifunctional magnetoelectric materials for device applications, J. Phys.: Condens. Matter. 27 (2015) 504002, 23pp.

[43] J. Ryu, Ch-W. Baek, N.-K. Oh, G. Han, J.-W. Kim, B.-D. Hahn, et al., Effect of Microstructure on magnetoelectric properties of $0.9Pb(Zr_{0.52}Ti_{0.48})O_3$-$0.1Pb(Zn_{1/3}Nb_{2/3})O_3$ and $Ni_{0.8}Zn_{0.2}Fe_2O_4$ particulate composites, Jpn. J. App. Phys. 50 (2011) 111501.

[44] J.Y. Zhai, N. Cai, Z. Shi, Y.H. Lin, C.W. Nan, Coupled magnetodielectric properties of laminated $PbZr_{0.53}Ti_{0.47}O_3/NiFe_2O_4$ ceramics, J. Appl. Phys. 95 (2004) 5686–5690.

[45] G. Srinivasan, E.T. Rasmussen, J. Gallegos, R. Srinivasan, Yu.I. Bokhan, V.Ms Laletin, Magnetoelectric bilayer and multilayer structures of magnetostrictive and piezoelectric oxides, Phys. Rev. B 64 (2001), 214408-6.

[46] N. Cai, J. Zhai, C.W. Nan, Y. Lin, Z. Shi, Dielectric, ferroelectric, magnetic, and magnetoelectric properties of multiferroic laminated composites, Phys. Rev. B 68 (2003) 224103–224107.

[47] Y. Lin, N. Cai, J. Zhai, G. Liu, C.-W. Nan, Large high-frequency magnetoelectric response in laminated composites of piezoelectric ceramics, rare-earth iron alloys and polymer, Appl. Phys. B 84 (2004) 3516–3518.

[48] C.E. Ciomaga, O.G. Avadanei, I. Dumitru, M. Airimioaei, S. Tascu, F. Tufescu, et al., Engineering magnetoelectric composites towards application as tunable microwave filter, J. Phys. D: Appl. Phys. 49 (2016) 125002, pp9.

[49] A.J. Heeger, T.W. Houston, Nuclear magnetic resonance in ferrimagnetic $MnFe_2O_4$, Phys. Rev. 135 (1964) A661–A671.

[50] G.A. Sawatzky, F. Van der Woude, A.H. Morrish, Mössbauer study of several ferrimagnetic spinels, Phys. Rev. 187 (1969) 747–757.

[51] M. Pénicaud, B. Siberchicot, C.B. Sommers, J. Kübler, Calculated electronic band structure and magnetic moments of ferrites, J. Magn. Magn. Mater. 103 (1992) 212–220.

[52] Z. Szotek, W.M. Temmerman, D. Kodderitzsch, A. Svane, L. Petit, H. Winter, Electronic structures of normal and inverse spinel ferrites from first principles, Phys. Rev. B 74 (2006), 174431-12.

[53] B. Viswanathan, V.R.K. Murthy, Ferrite Materials: Science and Technology, Springer, Narosa Publishing House, New Delhi, 1990, pp. 1–198.

[54] Z. Yu, C. Ang, Maxwell–Wagner polarization in ceramic composites $BaTiO_3$–$(Ni_{0.3}Zn_{0.7})Fe_{2.1}O_4$, J. Appl. Phys. 91 (2002) 794–797.

[55] L. Mitoseriu, V. Buscaglia, Intrinsic/extrinsic interplay contributions to the functional properties of ferroelectric-magnetic composites, Phase Trans. 79 (2006) 1095–1121.
L. Mitoseriu, I. Pallecchi, V. Buscaglia, A. Testino, C.E. Ciomaga, A. Stancu, Magnetic properties of the $BaTiO_3$–$(Ni,Zn)Fe_2O_4$ multiferroic composites, J. Magn. Magn. Mater. 316 (2007) e603–e606.

[56] A. Molak, M. Paluch, S. Pawlus, J. Klimontko, Z. Ujma, I. Gruszka, Electric modulus approach to the analysis of electric relaxation in highly conducting $(Na_{0.75}Bi_{0.25})(Mn_{0.25}Nb_{0.75})O_3$ ceramics, J. Phys. D. Appl. Phys. 38 (2005) 1450–1460.

[57] L.P. Curecheriu, M.T. Buscaglia, V. Buscaglia, L. Mitoseriu, P. Postolache, A. Ianculescu, et al., Functional properties of $BaTiO_3$–$Ni_{0.5}Zn_{0.5}Fe_2O_4$ magnetoelectric ceramics prepared from powders with core-shell structure, J. Appl. Phys. 107 (2010) 104106–104111.

[58] R.S. Devan, S.B. Deshpande, B.K. Chougule, Ferroelectric and ferromagnetic properties of $(x)BaTiO_3 + (1 - x)Ni_{0.94}Co_{0.01}Cu_{0.05}Fe_2O_4$ composite, J. Phys. D: Appl. Phys. 40 (2007) 1864–1868.

[59] D.K. Pradhan, R.N.P. Choudhary, C. Rinaldi, R.S. Katiyar, Effect of Mn substitution on electrical and magnetic properties of $Bi_{0.9}La_{0.}1FeO_3$, J. Appl. Phys. 106 (2009) 024102.

[60] G. Montanari, L. Costa, S. Albonetti, C. Galassi, Nb-doped PZT material by sol-gel combustion, J. Sol-Gel Sci. Technol. 36 (2005) 203−211.
[61] L. Curecheriu, P. Postolache, M.T. Buscaglia, V. Buscaglia, A. Ianculescu, L. Mitoseriu, Novel magnetoelectric ceramic composites by control of the interface reactions in Fe_2O_3@$BaTiO_3$ core-shell structures, J. Appl. Phys. 116 (2014) 084102.
[62] S.A. Lokare, D.R. Patil, R.S. Devan, S.S. Chougule, Y.D. Kolekar, B.K. Chougule, Electrical conduction, dielectric behavior and magnetoelectric effect in (x)$BaTiO_3$ + (1-x) $Ni_{0.94}Co_{0.01}Mn_{0.05}Fe_2O_4$ ME composites, Mat. Res. Bull. 43 (2008) 326−332.
[63] M. Airimioaei, C.E. Ciomaga, N. Apostolescu, L. Leontie, A.R. Iordana, L. Mitoseriu, et al., Synthesis and functional properties of the $Ni_{1-x}Mn_xFe_2O_4$ ferrites, J. Alloy Compd. 509 (2011) 8065−8072.
[64] Z.V. Mocanu, M. Airimioaei, C.E. Ciomaga, L. Curecheriu, F. Tudorache, S. Tascu, et al., Investigation of the functional properties of $Mg_xNi_{1-x}Fe_2O_4$ ceramics, J. Mater. Sci. 49 (2014) 3276−3286.
[65] Y. Köseoğlu, Structural, magnetic, electrical and dielectric properties of $Mn_xNi_{1-x}Fe_2O_4$ spinel nanoferrites prepared by PEG assisted hydrothermal method, Ceram. Int. 39 (2013) 4221−4230.
[66] H. Jacobsen, K. Lefmann, E. Brok, C. Frandsen, Temperature dependence of the magnetization of canted spin structures, J. Magn. Magn. Mater. 324 (19) (2012) 3218−3222.
[67] C.E. Ciomaga, A.M. Neagu, M.V. Pop, M. Airimioaei, S. Tascu, G. Schileo, et al., Ferroelectric and dielectric properties of ferrite-ferroelectric ceramic composites, J. Appl. Phys. 113 (2013) 074103.
[68] A.R. Iordan, M. Airimioaiei, M.N. Palamaru, C. Galassi, A.V. Sandu, C.E. Ciomaga, et al., In situ preparation of $CoFe_2O_4$-$Pb(ZrTi)O_3$ multiferroic composites by gel-combustion technique, J. Eur. Ceram. Soc. 29 (2009) 2807−2813.
[69] D. Wu, W. Gong, H. Deng, M. Li, Magnetoelectric composite ceramics of nickel ferrite and lead zirconatetitanate via in situ processing, J. Phys. D: Appl. Phys. 40 (2007) 5002−5005.
[70] C.E. Ciomaga, M. Airimioaei, V. Nica, L.M. Hrib, O.F. Caltun, A.R. Iordan, et al., Preparation and magnetoelectric properties of $NiFe_2O_4$-PZT composites obtained in-situ by gel-combustion method, J. Eur. Ceram. Soc. 32 (12) (2012) 3325−3337.
[71] B.K. Bammannavar, L.R. Naik, R.K. Kotnala, Study of electrical properties and the magnetoelectric effect in $Ni_{0.2}Co_{0.2}Fe_2O_4$−$PbZr_{0.8}Ti_{0.2}O_3$ particulate composites, Smart Mater. Struct. 20 (2011) 045005−045012.
[72] H. Zhang, C.-L. Mak, Impedance spectroscopic characterization of fine-grained magnetoelectric $Pb(Zr_{0.53}Ti_{0.47})O_3$−$(Ni_{0.5}Zn_{0.5})Fe_2O_4$ ceramic composites, J. Alloys Compd. 513 (2012) 165−171.
[73] M. Bichurin, V. Petrov, Modeling of magnetoelectric effects in composites, Springer Ser. Mater. Sci. 201 (2014), Chapter 3.
[74] C.M. Kanamadi, G.S.R. Raju, H.K. Yang, B.C. Choi, J.H. Jeong, Conduction mechanism and magnetic properties of (x)$Ni_{0.8}Cu_{0.2}Fe_2O_4$ + (1 − x)$Ba_{0.8}Pb_{0.2}Ti_{0.8}Zr_{0.2}O_3$ multiferroics, J. Alloys Compd. 479 (2009) 807−811.
[75] R.C. Kambale, P.A. Shaikh, C.H. Bhosale, K.Y. Rajpure, Y.D. Kolekar, Dielectric properties and complex impedance spectroscopy studies of mixed Ni−Co ferrites, Smart Mater. Struct. 18 (2009) 085014.
[76] A.S. Fawzi, A.D. Sheikh, V.L. Mathe, Effect of magnetostrictive phase on structural, dielectric and electrical properties of $NiFe_2O_4$ + $Pb_{0.93}La_{0.07}(Zr_{0.6}Ti_{0.4})O_3$ composites, Solid State 11 (2009) 1979−1984.

[77] J. Ryu, S. Priya, K. Uchino, H.-E. Kim, Magnetoelectric effect in composites of magnetostrictive and piezoelectric materials, J. Electroceram 8 (2002) 107–119.
[78] J.V. Mantese, A.L. Micheli, D.F. Dungan, R.G. Geyer, J. Baker-Jarvis, J. Grosvenor, Applicability of effective medium theory to ferroelectric/ferrimagnetic composites with composition and frequency dependent complex permittivities and permeabilities, J. Appl. Phys. 79 (3) (1996) 1655–1660.
[79] A. Sihvola, Electromagnetic mixing formulas and applications, Electromagnetic Waves Series 47, The Institution of Electrical Engineers, London, 1999, pp. 1–296.
A. Sihvola, Self-consistency aspects of dielectric mixing theories, IEEE Trans. Geosci. Remote Sensing 27 (4) (1989) 403–415.
[80] L. Padurariu, L.P. Curecheriu, L. Mitoseriu, Nonlinear dielectric properties of paraelectric-dielectric composites described by a 3D Finite Element Method based on Landau-Devonshire theory, Acta Mater. 103 (2016) 724–734.
[81] J.C. Maxwell-Garnett, Colours in Metal Glasses and Metal Films, Trans. of the Royal Society CCIII, London, 1904, pp. 385–420.
J.C. Maxwell, A Treatise on Electricity and Magnetism, Dover, New York, 1954.
[82] D.A.G. Bruggeman, Annalen derPhysik (Leipzig) 24 (1935) 636–664.
[83] A.V. Goncharenko, V.Z. Lozovski, E.F. Venger, Lichtenecker's equation: applicability and limitations, Opt. Commun. 174 (2000) 19–32.
[84] H.K. Liu, J.H. Huang, C.W. Hsieh, H.-T. Tu, Piezoelectric and magnetic properties of PZT-$(Ni_{0.284}Zn_{0.549}Cu_{0.183})Fe_{1.984}O_4$ composites, J. Mater. Sci. 40 (2005) 1979–1985.
[85] S.S. Chougule, D.R. Patil, B.K. Chougule, Electrical conduction and magnetoelectric effect in ferroelectric rich $(x)Ni_{0.9}Zn_{0.1}Fe_2O_4 + (1\text{-}x)PZT$ ME composites, J. Alloy Compd. 452 (2008) 205–209.
[86] K.C. Verma, R.K. Kotnala, Nanostructural and lattice contributions to multiferroism in $NiFe_2O_4/BaTiO_3$, Mater. Chem. Phys. 174 (2016) 120–128.
[87] A.S. Dzunuzovic, M.M. VijatovicPetrovic, B.S. Stojadinovic, N.I. Ilica, J.D. Bobic, C. R. Foschin, et al., Multiferroic $(Ni,Zn)Fe_2O_4–BaTiO_3$ composites prepared from nanopowders by auto-combustion method, Ceram. Int. 41 (2015) 13189–13200.

Single-phase, composite and laminate multiferroics

22

Cristina Pascual-Gonzalez[1], *Giorgio Schileo*[2] *and Antonio Feteira*[1]
[1]Christian Doppler Lab for Advanced Ferroic Oxides, Materials and Engineering Research Institute, Sheffield Hallam University, Sheffield, United Kingdom, [2]Department of Materials Science and Engineering, The University of Sheffield, Sheffield, United Kingdom

22.1 Introduction

Over the past decade, magnetoelectric multiferroics (MMs) have occupied a prevalent research position in the field of metal oxides. MMs are materials that exhibit simultaneously ferroelectric and magnetic order, and coupling between these two order parameters i.e. magnetization, M, can be manipulated by an electric field, E, and similarly the polarization, P, by a magnetic field, H, as schematically illustrated in Fig. 22.1 [1]

This phenomenon commonly referred to as the magnetoelectric (ME) effect has been envisaged as a promising route to develop new electronic technologies such as uncooled magnetic field sensors, nonvolatile memories, and ferroelectric photovoltaics, just to name a few. An intrinsic ME behavior can occur in a single-phase compound, as schematically represented in the left column of Fig. 22.1. Unfortunately, to date no single-phase MM material was discovered that meets the room-temperature ME requirements for memory applications. The most severe weaknesses of single-phase MMs are linked to their operational temperatures (most only operate well below room temperature), but also to the magnitudes of the order parameters. In the film form, their switchable polarization is extremely low, usually in the order of magnitude of nC/cm^2. Basically, such values are one order of magnitude lower than the desirable minimum of $\sim 1\ \mu C/cm^2$ required by a computer memory sense amplifier to discriminate between 1 and 0 states [2]. Multiferroics (MFs) that exhibit reasonably high operating temperatures tend to be weak ferromagnets or antiferromagnets, however strong ferromagnets are desirable, for the ease of READ/WRITE operations. Finally, the magnetic fields currently required to switch the polarization of single-phase MFs are orders of magnitude larger than the mT required for a magnetic random access memory (MRAM). Hence, the deployment of a multiferroic RAM is not envisaged in the near future; however, there are other equally important devices with less stringent requirements that may be developed.

The lack of single-phase MMs can be intuitively understood from the now classic Venn diagram (see Fig. 22.2), created in 2006 by Eerenstein and coauthors [3]. It immediately shows that chances for coexistence of magnetism (spin order) and ferroelectricity (electric dipole order) in a single-phase material may be limited.

Magnetic, Ferroelectric, and Multiferroic Metal Oxides. DOI: http://dx.doi.org/10.1016/B978-0-12-811180-2.00022-0

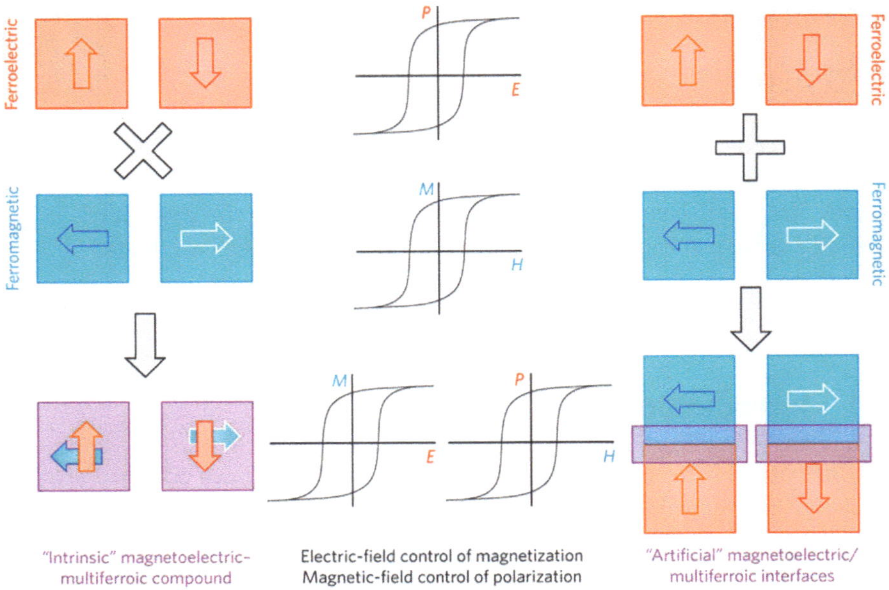

Figure 22.1 Ferromagnetic and ferroelectric orders may coexist and influence each other through a magnetoelectric coupling.
With permission from M. Bibes, Nanoferronics is a winning combination. Nat. Mater. 2012. 11 (5) (2012) 354–357.

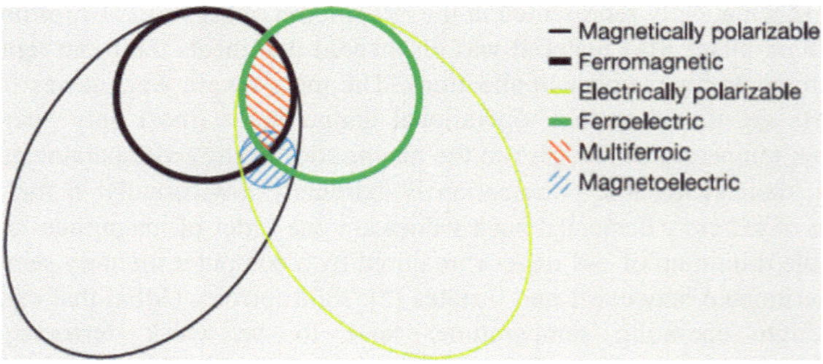

Figure 22.2 Relationship between multiferroic and magnetoelectric materials.
With permission from W. Eerenstein, N. Mathur, J. Scott, Multiferroic and magnetoelectric materials. Nature 442 (7104) (2006) 759–765.

Moreover, MMs result from the intersection of two phenomena, multiferroicity and magnetoelectricity.

Already in 2000, Hill [4] postulated that the conditions conducive to the simultaneous occurrence of ferroelectricity and ferromagnetism in single-phase transition metal oxide perovskites are mutually exclusive, as discussed later. For this reason researchers have explored different microscopic conditions that may allow the two orders to coexist in a single-phase material. Nevertheless, this is only part of the challenge. Actually, achieving an efficient ME coupling between the two orders appears to be considerably more challenging than their mere coexistence. Indeed, most single-phase MMs exhibit relatively low coupling coefficients, α, ranging between 10^{-11} and 10^{-10} s/m.

In order to overcome this limitation, many researchers have been investigating MM composites. Basically, in MM composites the ME effect results from the product of the magnetostrictive response (i.e., a magnetic/mechanical effect) of a magnetostrictive phase and the piezoelectric response (i.e., a mechanical/electrical effect) of a piezoelectric phase, given either as:

$$ME_H \text{ effect} = \frac{\text{magnetic}}{\text{mechanical}} \times \frac{\text{mechanical}}{\text{electric}}$$

or

$$ME_E \text{ effect} = \frac{\text{electric}}{\text{mechanical}} \times \frac{\text{mechanical}}{\text{magnetic}}$$

The two ferroic phases can be arranged according to various geometries, as described by the phase connectivity concept introduced by Newnham et al. [5]. Hence, the structure of two-phase composites can, for example, be referred to as 0–3, where one-phase particles (denoted as 0) are embedded in a matrix (denoted as 3). Multilayer materials, with alternating layers of dissimilar phases, are referred to as 2–2-type composites, whereas the 1–3 notation is used to refer to tubes, pillars, and other elongated structures embedded into a matrix. In any of these geometries, the ME coupling is achieved via mechanical strain, therefore a good contact between the surfaces of the two phases is essential to transfer the strain and achieve an useful ME response. Composites exhibit higher coupling coefficients than single-phase materials, with values ranging from 10^{-8} to 10^{-6} s/m. Often, the ME response of MM composites is measured using capacitor-type devices, where either the charge or the induced voltage is measured as a function of the applied magnetic field. Hence, the ME voltage coefficient, α_{ME} is defined as $\alpha_{ME} = \frac{E}{H} \cong \frac{\alpha}{\varepsilon_0 \varepsilon_r}$, where E is the electric field and H is the magnetic field, whereas α is the ME coupling coefficient, ε_0 is the permittivity of free space, and ε_r is the relative permittivity. α_{ME} is often expressed in units of V/cm Oe.

Although fundamental research into MMs can be traced back to pioneering works from the 1960s, an upsurge of interest in these materials occurred in the mid-1990s and early 2000s. This renaissance was accompanied by the publication of reviews dealing with the state-of-art back then. Excellent reviews of the general study of magnetoelectricity were published in 1994 by Schmid [6], and later in 2005 by Fiebig [7] and in 2006 by Eerenstein et al. [3]. Also in 2006, Khomskii [8] elucidated about the different ways of combining magnetism and ferroelectricity. In 2008, Schmid [9] provided a detailed classification of possible symmetries allowing for multiferroicity and magnetoelectricity. Other important reviews dealing with more contemporary findings such as the discovery of a new class of spin-driven improper ferroelectrics appeared subsequently. The reader is referred to those influential reviews in order to gather more specific and in-depth details on those particular topics. The purpose of this chapter is to discuss the present status and recent progress in the development of MM materials and their applications in devices. An overview of the pioneering works will be provided, but emphasis will be placed on the latest major contributions. Hence, the first part of this chapter will be devoted to single-phase MMs. Mechanisms of multiferroicity are covered alongside emerging concepts to achieve room-temperature MMs. The second part of this chapter covers MM composites consisting of physically separated magnetic and electrically ordered phases, which are arranged in different geometries.

22.2 Single-phase multiferroics

Historically, the ME effect was first experimentally demonstrated in Cr_2O_3 in the 1961 by Astrov. Interestingly, this compound is neither ferroelectric nor ferromagnetic. Actually, Cr_2O_3 is antiferroelectric and antiferromagnetic, with a Néel Temperature of 307 K and a linear ME coupling coefficient, α, of 4.17×10^{-12} s/m (or ME voltage coefficient, $\alpha_{ME} \sim 28$ mV/cm Oe). The transformation from α to α_{ME} requires the knowledge of the relative permittivity of the material, and it can be done as shown by Vaz et al. [10]. Although Cr_2O_3 is historically significant, technologically it is less attractive than other MFs, because it is considerably easier to address (READ/WRITE) materials exhibiting simultaneously ferroelectricity and ferromagnetism.

Pioneering works to combine both ferromagnetic and ferroelectric properties in one system were carried out in the 1960s by Smolenskii and Venevtsev's groups in the former USSR. Those works were essentially focused on oxide perovskites and revolved around the replacement of some of B cations in a known ferroelectric by magnetic cations. However, in these mixed perovskites coupling between ferroelectric and magnetic subsystems is not very strong. Following empirical observations that conventional ferroelectric perovskites, such as $BaTiO_3$, contain transition metal ions with an electronic configuration d^0 (i.e., an empty *d*-shell), it was postulated that "d^0-ness" was a necessary but not sufficient condition for the occurrence of ferroelectricity in this family of perovskites. Nevertheless, this condition seems to be at odds with requirements for the occurrence of magnetic ordering, which involve partially filled inner shells (*d*- or *f*-levels of transition metals or rare-earth ions). Obviously, the absence of *d* electrons creating localized magnetic moments precludes magnetic ordering of any type, either ferro-, ferri-, or antiferromagnetic. In 2000, Hill [4] explored the fundamental physics behind the scarcity of materials exhibiting coexistence of ferromagnetism and ferroelectricity. Based on *ab initio* calculations, the d^0-ness condition for ferroelectricity was qualitatively explained. Hence, the empty *d*-levels of transition metals ions such as Ti^{4+} in $BaTiO_3$

establish strong covalent bonds with the surrounding oxygens, but a particularly stronger bond with one favored oxygen leads to off-centering of Ti^{4+} within the oxygen octahedra. Somehow the presence of d electrons in magnetic transition metals suppresses this process, preventing ferroelectricity. This discussion was implicitly limited to transition-metal oxides, particularly to perovskites, and has been now a long matter of debate. Fortunately, this d^0-ness rule can be broken, providing that details of chemistry can lead to stable crystal structures. This prospect of alternative routes for MFs prompted researchers to explore different crystal chemistries, which resulted in the discovery of a series of mechanisms by which MF behavior can be attained, as summarized in Table 22.1.

In general, there are two classes of MFs, as indicated in Table 22.1. In principle, their microscopic origin of magnetism is basically the same and is linked to the presence of localized electrons, mostly in partially filled *d* or *f* shells of transition-metal and/or rare-earth ions. In contrast, there are several different microscopic mechanisms giving rise to ferroelectricity, as briefly described later and schematically represented in Fig. 22.3. For a detailed description of each mechanism the reader is referred to works by Fiebig et al. [11] and Tokura et al. [12].

In $BiFeO_3$, the polarization arises from the electronic lone-pair (s^2 orbital) of A-site Bi^{3+}, whereas magnetization emerges from the B-site Fe^{3+}. $BiFeO_3$ is the prototypical example of a so-called Type I MFs [8,13,14], in which ferroelectricity and magnetism have different sources and appear largely independent of each other, though there is some coupling between them. This route for multiferroicity prompted researchers to investigate other perovskites with potential A-site ferroelectricity and B-site magnetism such as $Bi(Cr,Fe)O_3$ [15] and $BiMnO_3$ [16,17]. Because the ferroelectric and magnetic orders arise from two different subsystems, the coupling between them is weak.

In the last decade, thousands of research papers were devoted to $BiFeO_3$. This huge interest in $BiFeO_3$ was essentially motivated by the 2003 work by Wang et al. [18], who showed thin films to exhibit an unexpectedly large remnant polarization (15 times larger than their bulk counterparts) and very large ferromagnetism of $\sim$1.0 Bohr magneton (mB) per unit cell. Catalan and Scott [13] subsequently have reviewed and summarized in detail both the basic physics and unresolved issues in $BiFeO_3$. Although $BiFeO_3$ emerged as a potential candidate for room-temperature multiferroic single-phase devices, its magnetization is too small for many applications, and its dielectric loss and leakage current too large. In the past decade, thousands of works dealt with processing of $BiFeO_3$, in particular chemical doping was employed to improve its electrical resistivity.

There are other Bi-containing ferroelectrics, such as the Aurivillius phases, represented by the general formulae $(Bi_2O_2)(A_{n-1}B_nO_{3n+1})$ (where A is a large 12-coordinated cation, and B is a small 6-coordinated cation). In practice, their structure can be described by alternating layers of $[Bi_2O_2]^{2+}$ and pseudo-perovskite blocks, with perovskite layers that are n octahedral layers in thickness. Aurivillius have invariably high ferroelectric Curie temperatures (>600°C) and large in-plane spontaneous polarizations. Some researchers have been exploiting chemical substitution with magnetic ions as a route to create room-temperature MMs. Actually,

Table 22.1 General mechanisms for multiferroics

Classes	Mechanism of inversion symmetry breaking	Description	Examples	
Type I	Lone-pair effects	Lone pairs of electrons on the A-site of perovskites distort the geometry of the BO_3 leading to ferroelectricity.	$BiFeO_3$, $BiMnO_3$	Proper
	Structural transition "Geometric ferroelectrics"	Rotations of oxygen polyhedra accompanied by long-range dipole–dipole interactions can induce a stable ferroelectric state.	h-$YMnO_3$ $HoMnO_3$ $Ca_3Mn_2O_7$	Improper
	Charge-ordering electronic ferroelectrics	Some noncentrosymmetric ion arrangements can induce ferroelectricity in magnetic materials.	Fe_3O_4 $LuFe_2O_4$ $Pr_{1-x}Ca_xMnO_3$	
Type II	Spin-driven magnetic ferroelectrics	Magnetic long-range order in which magnetic dipole arrangement lacks reflection symmetry can induce ferroelectricity.	o-$TbMnO_3$, $DyMnO_3$, $TbMn_2O_5$ $Ni_3V_2O_8$ Ca_3CoMnO_6	

Adapted from M. Fiebig, et al., The evolution of multiferroics, Nat. Rev. Mater. 1 (8) (2016) and Y. Tokura, S. Seki, N. Nagaosa, Multiferroics of spin origin, Rep. Prog. Phys. 77 (7) (2014).

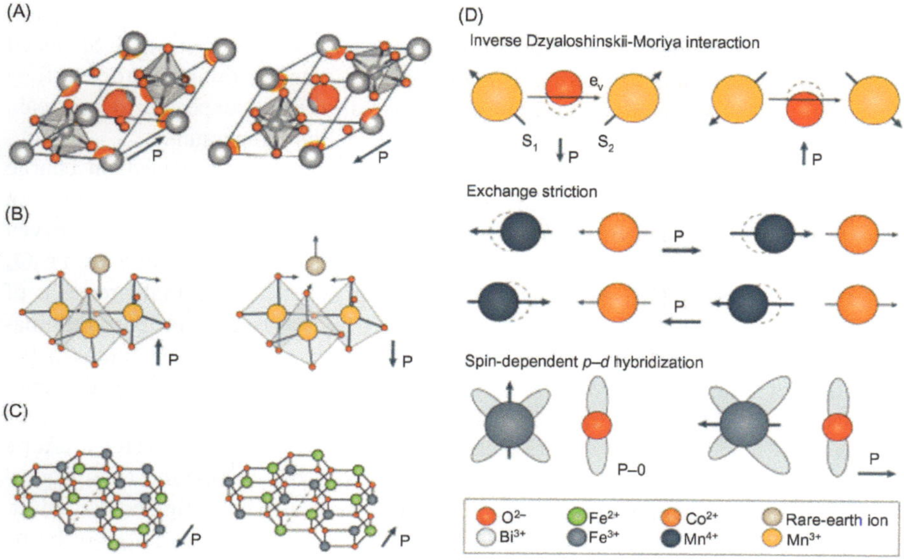

Figure 22.3 Mechanisms for multiferroicity. (A) Lone-pair mechanism; (B) geometric ferroelectricity; (C) charge ordering; (D) spin-driven mechanisms.
With permission from M. Fiebig, et al., The evolution of multiferroics. Nat. Rev. Mater. 1(8) (2016).

ferroelectricity and ferromagnetism above room temperature were reported for cobalt-substituted, four-layered $Bi_5Ti_3Fe_{0.5}Co_{0.5}O_{15}$ [19] and $Bi_5Ti_3Fe_{0.7}Co_{0.3}O_{15}$, and five-layer $Bi_6Ti_{2.8}Fe_{1.52}Mn_{0.68}O_{18}$ [20] phases. Later, it was shown that $Bi_5Ti_3Fe_{0.7}Co_{0.3}O_{15}$ is not ferromagnetic, and its magnetic response was actually affected by the presence of Fe- and/or Co-rich secondary magnetic spinels [21]. In 2016, Li et al. [22] claimed room temperature ME coupling in the Aurivillius $Bi_{4.25}La_{0.75}Ti_3Fe_{0.5}Co_{0.5}O_{15}$ phase.

A second subclass of Type I MFs that violates the d^0-ness rule is the so-called geometrically driven ferroelectrics, which are mainly represented by hexagonal manganites $RMnO_3$ (R = Y or small rare-earths) [23,24]. Often these compounds are referred to as hexagonal perovskites; however, they are structurally dissimilar from typical perovskites. Indeed, in these compounds Mn^{3+} are fivefold coordinated (instead of sixfold as in perovskites) and are located at the center of O_5 trigonal biprisms, whereas R ions are sevenfold coordinated. The origin of ferroelectricity is not linked to displacements of the Mn or R ions, but to the tilting of the rigid MnO_5 trigonal biprisms, leading to a loss of inversion symmetry and to ferroelectricity [25–28]. In 2011, Benedek and Fennie [29] introduced the term hybrid improper ferroelectricity, which describes a polarization state induced by a complex distortion pattern consisting of more than one octahedron rotation mode. Hence, hybrid improper ferroelectricity has been theoretically predicted in ordered perovskites and Ruddlesden-Popper compounds ($Ca_3Ti_2O_7$, $Ca_3Mn_2O_7$, and (Ca/Sr/

$Ba_3(Sn/Zr/Ge)_2O_7$). In 2015, Oh et al. [30] demonstrated experimentally room-temperature switchable polarization in bulk crystals of $Ca_3Ti_2O_7$ and Sr-doped $Ca_3Ti_2O_7$. Again, in these materials it is evident that magnetism and ferroelectricity have independent origins, therefore coupling between them is expected to be weak. Usually these materials exhibit high ferroelectric transition temperatures.

A third subclass of Type I MFs violating the d^0-ness rule is based on charge ordering. Basically, certain noncentrosymmetric charge-ordering arrangements can cause ferroelectricity. den Brink and Khomskii [31] discussed multiferroicity driven by different forms of charge ordering. In particular, iron-oxides such as Fe_3O_4 [32,33] and $LuFe_2O_4$ [34] exhibit charge ordering due to the formal presence of both $Fe^{2+}(d^6)$ and $Fe^{3+}(d^5)$. Ferroelectricity in $LuFe_2O_4$ arises from the combination of two factors: (1) the bilayer character of the crystal structure, and (2) the frustration of charge ordering, which results in the formation of charged planes [34]. Despite some authors reporting that the polarization in $LuFe_2O_4$ can be as high as 24 $\mu C/cm^2$ [34] (which is comparable to the polarization of $BaTiO_3$), after a decade of research the occurrence of ferroelectricity in this compound is still questioned. Mixed manganites such as $Pr_{1-x}Ca_xMnO_3$ were also investigated for multiferroicity [35,36], however they received less attention. Currently, charge-ordering as a route for ferroelectricity remains at conceptual stages.

It is worth noting that in Type I MFs, the ferroelectric phase transition typically appears at higher temperatures than the magnetic one. For example, $BiMnO_3$ shows a ferroelectric transition at $T_{FE} \sim 800K$ and a ferromagnetic transition at $T_{FM} \sim 110K$ [37]. $YMnO_3$ shows a ferroelectric transition at $T_{FE} \sim 914K$ and an antiferromagnetic transition at $T_{AFM} \sim 80K$ [28].

In 2003, a major experimental development concerned the discovery of a novel class of MFs, Type II, in which magnetic ordering induces ferroelectricity. In these materials the formation of conventional Néel ground states for antiferromagnetically coupled spins is geometrically frustrated. Kimura et al. [38] discovered this phenomenon in $TbMnO_3$ and Hur et al. [39] found a similar effect in $TbMn_2O_5$. Those are the most emblematic representatives of the so-called Type-II MFs, where ferroelectricity is present only in the magnetically ordered state. Delafossite systems such as $CuMO_2$ (where M = Fe or Cr) [40], $Ni_3V_2O_8$ [41,42], $MnWO_4$ [43], and Ca_3CoMnO_6 [44] also fall into the magnetic-order category.

The physics of spin-driven ferroelectricity are complex and mainly explained by three mechanisms: spin current (often referred to as inverse Dzyaloshinskii–Moriya interaction) [45], exchange striction, and spin dependent p-d hybridization. These mechanisms are schematically represented in Fig. 22.3, and the reader is referred to the works of Fiebig et al. [11] and Tokura et al. [12] to gather more insight into each mechanism.

Since ferroelectricity is caused by magnetism this implies a strong coupling between the two. Disappointingly, these materials exhibit much smaller polarizations, usually around 10^{-2} $\mu C/cm^2$, and also low magnetic transition temperatures (<40K). In 2012, Johnson et al. [46] reported $CaMn_7O_{12}$ to exhibit the largest spin spiral-driven polarization 0.3 $\mu C/cm^2$, which evolves below 90K. Nevertheless, these discoveries have been instrumental in triggering the quest for other ways to achieve coexistence of ferroelectricity and magnetism in oxides.

The pioneering idea of alloying ferroelectrics with magnetic ions has been recently revisited and led to the successful discovery of new MFs with a ME response at room temperature. Moreover, it was observed that the ME response becomes particularly pronounced near phase boundaries. Solid solution phases between $PbZr_{0.5}Ti_{0.5}O_3$ (PZT) and $PbFe_{0.5}Ta_{0.5}O_3$ (PFT) were reported to show both ferroelectric and ferromagnetic/ferrimagnetic hysteresis loops at room temperature. This observation was rather surprising because PFT possesses a Néel temperature of ~ 150K, and its magnetic susceptibility shows characteristics of an antiferromagnet. Nevertheless, ferromagnetic hysteresis loops can be observed in PFT up to 300K due to partially ferromagnetic spin clustering and partly due to Dzyaloshinskii–Moriya anisotropic-exchange canting. This discovery led researchers to explore other solid solutions between PZT and $PbFe_{0.5}Nb_{0.5}O_3$ (PFN) and $PbFe_{0.66}W_{0.33}O_3$ (PFW). The Venn diagram in Fig. 22.4, produced by Scott [2], shows the overlap of perovskites obtainable as single-phase MFs from $PbZr_{0.5}Ti_{0.5}O_3$ and $Pb(Fe,M)O_3$ solid solutions.

In order to unequivocally demonstrate room temperature ME coupling in the $(PbZr_{0.53}Ti_{0.47}O_3)_{(1-x)}$ $(PbFe_{0.5}Ta_{0.5}O_3)_x$ (PZT-PFT) solid solution, Evans et al. [47] monitored changes to ferroelectric domain patterns by magnetic fields, as illustrated in Fig. 22.5.

It is remarkable that a magnetic field applied perpendicular to a PZT-PFT lamella led to switching of the ferroelectric domains, to a level comparable to that produced by an electric field. This work clearly showed significant room-temperature ME coupling, estimated as high as $\sim 1 \times 10^{-7}$ s/m.

Intensive research activity is being pursued towards the discovery of new single-phase multiferroic materials and the design of new engineered materials with strong ME coupling.

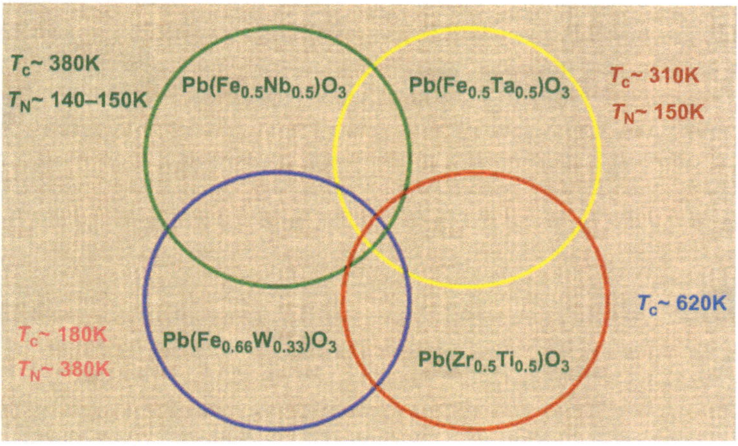

Figure 22.4 Venn diagramm showing the overlap of perovskites obtainable as single-phase multiferroics from $PbZr_{0.5}Ti_{0.5}O_3$ and $Pb(Fe,M)O_3$ solid solutions.
With permission from J.F. Scott, Room-temperature multiferroic magnetoelectrics. NPG Asia Mater. (2013) 5.

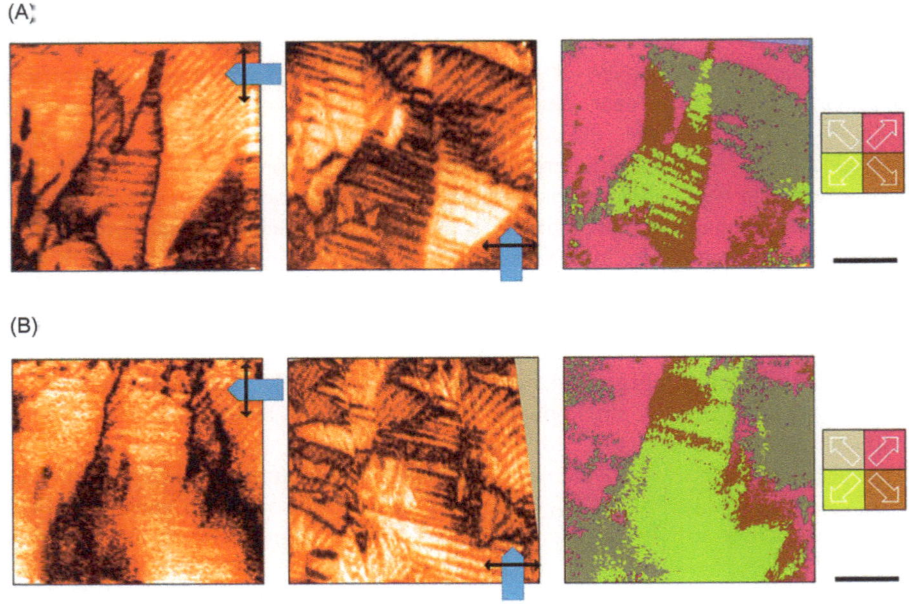

Figure 22.5 Ferroelectric domain configuration modification in single-phase PZT-PFT, upon the application of magnetic field of 1.8 T at 295K. (A) Before magnetic field and (B) after 18 kOe field.
With permission from J.F. Scott, Room-temperature multiferroic magnetoelectrics. NPG Asia Mater. (2013) 5.

A strategy for development of multiferroicity in materials with the tetragonal tungsten bronze structure was recently discussed [48]. This involves A- and B-site substitutions in the starting compound $Ba_6GaNb_9O_{30}$. Large B-site cations and local strain variations created by isovalent A-site substitutions enhance dipole stability but lead to a dilute, weakly correlated dipolar response and canonical relaxor behavior. The incorporation of small lanthanide cations in the A-site compensated with A-site vacancies induces dipole ordering and ferroelectricity. Moreover, the unfilled 4*f* shells from the lanthanides introduce magnetic moments and the possibility of MF behavior.

The coexistence of ferroelectricity and a magnetically ordered state in the complex perovskite $Pb(Fe_{0.5}Ti_{0.25}W_{0.25})O_3$ and observation of a possible coupling between them was observed near room temperature [49]. An admixture of a ferroelectric phase (P4mm space group) (34%) and a paraelectric phase ($Pm\overline{3}m$ space group) (66%) was determined from room-temperature neutron powder diffraction data. The three B cations (Fe, Ti, W) are distributed at random at the octahedral sites in these two phases and Pb is shifted away from the center of the cubic (sub)cell. The appearance of magnetic interactions above room temperature was linked to the existence of

Fe-rich islands, where the strong Fe^{3+}-O-Fe^{3+}superexchange interactions govern the magnetic behavior.

A concept based on MF clusters was advanced in 2009 by the Christian Doppler Lab for Advanced Ferroic Oxides and demonstrated in 2016 by Henrichs et al. [50]. This concept was demonstrated for the single-phase relaxor ferroelectric $(BiFe_{0.9}Co_{0.1}O_3)0.4-(Bi_{0.5}K_{0.5}TiO_3)0.6$, where polar nanoregions (PNR) transform into static-PNR below the freezing temperature of 300°C and simultaneously enable congruent MF clusters to appear from inherent strongly magnetic $Bi(Fe,Co)O_3$ rich regions. This material exhibits a Néel temperature $T_N = 670K$ and an exceptional room-temperature ME coupling coefficient, $\alpha \sim 1.0 \times 10^{-5}$ s/m, comparable to values measured for composites. Another concept to combine ferroelectricity and ferromagnetism at room temperature in bulk perovskite oxides was proposed in 2016 by Mandal et al. [51], through the construction of a percolating network of magnetic ions with strong superexchange interactions within a structural scaffold with a morphotropic phase boundary. In principle, engineered structure enhances polarization switching and permits canting of the ordered magnetic moments. This concept was exemplified in $0.85BiTi_{0.125}Fe_{0.75}Mg_{0.125}O_3-0.15CaTiO_3$, with a saturated magnetization of 0.0097 μB per Fe, a linear ME susceptibility of 0.19×10^{-12} s/m, and a polarization of 66 μC/cm^2 at room temperature.

$Bi_{0.68}Pb_{0.32}Fe_{0.655}Mn_{0.025}Ti_{0.32}O_3$ was presented as novel perovskite oxide, which was chemically designed to present a MF MPB showing distinctive room-temperature ME responses by means of these two mechanisms: field-driven phase changes and transverse lattice softening [52]. Novel MF $(Pb_{1\text{-}3x/2}Nd_x)(Ti_{0.98\text{-}y}Fe_yMn_{0.02})O_3$ ceramics with coexisting ferroelectricity and ferromagnetism at ambient temperature were reported by Craciun et al. [53]. The presence of both Fe^{3+} (spin 5/2) and Ti^{3+} (spin 1/2) magnetic ions in the B-site drives the occurrence of magnetic behavior from low temperature to above room temperature via superexchange Fe^{3+}-O-Fe^{3+}, Fe^{3+}-O-Ti^{3+} and Ti^{3+}-O-Ti^{3+} interactions.

Previously, researchers explored MM responses in other crystal structures. For example, in 2010, Kitagawa et al. [54] also reported a room-temperature ME effect in z-type hexaferrite $Sr_3Co_2Fe_{24}O_{41}$. Nevertheless, symmetry arguments point toward electric rather than true ferroelectric behavior. Interestingly, $Sr_3Co_2Fe_{24}O_{41}$ can operate at room temperature, and unlike most MEs does not require a prohibitively high switching magnetic field.

The langasite $Ba_3NbFe_3Si_2O_{14}$ was also found to be a geometrically frustrated MF, where the Fe^{3+} spins appear in isolated triangular units stacked along the c-direction. $Ba_3NbFe_3Si_2O_{14}$ exhibits a magnetic ordering temperature of $T_N = 26K$, below which spontaneous electric polarization is also observed. Coupling between those two other parameters was also reported [55]. Recently, spin-induced ferroelectricity and ME coupling was reported for Mn_3TeO_6 [56]. Two types of incommensurate spin structures coexist in this material, cycloidal and helical. This may lead to new ways of manipulating the ferroelectric polarization in a MF material. Table 22.2 lists MFs based on the perovskite structure alongside other oxides.

Table 22.2 **Multiferroics based on the perovskite structure and related structures**

Compounds	Magnetic ion	Multiferroic mechanism	Ferroelectric polarization ($\mu C/cm^2$)	Ferroelectric transition temperature (K)	Magnetic transition temperature (K)	References
Relaxors **Characteristics: Ferroelectric (FE), antiferromagnetic (anti-FM), and weak ferromagnetism (FM)**						
$Pb(Fe_{0.5}Ta_{0.5})O_3$ PFT	Fe^{3+}	Pb-lone pair and Ta^{5+} induce ferroelectricity Fe^{3+} induces magnetism	1.4 at RT ~10 at 200 K	220–250	Two possible MT 180 55	[57,58]
$Pb(Fe_{0.5}Nb_{0.5})O_3$ PFN	Fe^{3+}	Pb-lone pair and Nb^{5+} induce ferroelectricity Fe^{3+} induces magnetism	~8	379–393	143–170	[57,59]
$Pb(Fe_{2/3}W_{1/3})O_3$ PFW	Fe^{3+}	Pb-lone pair and W^{6+} induce ferroelectricity Fe^{3+} induces magnetism	14.3	343	180	[60,61]
Bismuth compounds **Characteristics: Weakly anti-FM and FE.**						
$BiFeO_3$	Fe^{3+}	Lone-pair at A-site	8.9	1083	653	[62,63]
$BiMnO_3$	Mn^{3+}	Lone-pair at A-site	–	750–770	105 (FMT)	[62]

(Continued)

Table 22.2 (Continued)

Rare-earth manganites: $RMnO_3$ **Characteristics: Anti-FM, and FE. The magnetoelectric coupling results from weak spin-orbit interactions.**						
$YMnO_3$	Mn^{3+}	Geometric ferroelectricity	5.5	1270	80	[64]
$YbMnO_3$	Mn^{3+}	Geometric ferroelectricity	–	995	80	[65]
$HoMnO_3$	Mn^{3+}	Geometric ferroelectricity	5.6	875	76	[66]
$ErMnO_3$	Mn^{3+}	Geometric ferroelectricity	–	600	77	[67]
Solid solutions **Characteristics: Solid solutions of relaxor with PZT for enhancing FE.**						
$PbFe_{0.5}Ta_{0.5}O_3$-$PbZr_xTi_{1-x}O_3$ PFT-PZT	Fe^{3+}	Ionic displacements	25 at RT 40 at 550 K	475–520	Two possible MT associated with PFT	[47,68]
$PbFe_{0.5}Nb_{0.5}O_3$-$PbZr_xTi_{1-x}O_3$ PFN-PZT	Fe^{3+}	Ionic displacements	20–30	255	–	[68,69]
$PbFe_{2/3}W_{1/3}O_3$-$PbZr_xTi_{1-x}O_3$ PFW-PZT	Fe^{3+}	Ionic displacements	~22	500K	–	[70]

22.3 Magnetoelectric multiferroic composites

MM composites emerged as an alternative to overcome the fundamental difficulties of creating efficient single-phase materials able to operate at room temperature. Bulk artificial MM composites are often fabricated using ferromagnetic magnetostrictive and ferroelectric piezoelectric components assembled either in granular or layered arrangements. In theory, coupling in MM composites can be up to 10^8 times larger than in their single-phase counterparts. Intrinsic coupling in single-phase MMs is achieved by direct polarization–magnetization interactions, whereas in MM composites coupling occurs extrinsically by any of the three following mechanisms: (1) strain, (2) charge carrier, and (3) spin exchange. The strain-mediated ME coupling mechanism is now well established, whereas the other two mechanisms are still in infancy stages. Hence, the remainder of this chapter will be devoted to MM composites based on strain-mediated coupling, starting from the pioneering works from the 1970s until the most recent works.

The strain-mediated ME effect in composites is a product tensor property and arises from the elastic coupling between magnetostrictive and piezoelectric components. A magnetic field applied to a MM composite will induce strain in the magnetic component through the magnetostrictive effect. This strain is simultaneously transferred to the piezoelectric component leading to an electric displacement via the piezoelectric effect. A converse effect also exists, when an electric field induces strain in the ferroelectric component through the inverse piezoelectric effect. This strain is simultaneously transferred to the magnetic component giving rise to a magnetization change or domain reorientation via the piezomagnetic effect. Hence, a significant ME response requires ferroelectric materials with large piezoelectric coefficients, and magnetic materials with large magnetostriction coefficients. Tables 22.3 and 22.4 summarize the properties of piezoelectric and magnetic materials of choice for the fabrication of MM composites. The preferred piezoelectric ceramic oxides have been morphotropic $Pb(Zr,Ti)O_3$-based compositions, because they offer a good combination between piezoelectric coefficients and electromechanical coupling coefficients, as shown in Table 22.3. $Pb(Zn_{1/3}Nb_{2/3})O_3$-$PbTiO_3$ (PZN-PT) and $Pb(Mg_{1/3}Nb_{2/3})O_3$-$PbTiO_3$ (PMN-PT) single

Table 22.3 Properties of piezoelectric materials of choice for the fabrication of magnetoelectric multiferroic composites

	$BaTiO_3$	Soft PZT-5H	Hard PZT-4	PZN-PT	PMN-PT	PVDF
d_{31} (pC/N)	−90	−175	−109	−1100	700	16
d_{33} (pC/N)	190	400	300	2200	2000	−33
ε_r	1500–2000	1750	1350	7200	5000	10
T_c (°C)	120–130	360	320	160	80	130
k_{33}	0.63	0.72	0.68	0.94	0.9	0.19

Table 22.4 Properties of magnetic materials of choice for the fabrication of magnetoelectric multiferroic composites

	$Y_3Fe_5O_{12}$	$NiFe_2O_4$	$CoFe_2O_4$	Terfenol-D $Tb_xDy_{1-x}Fe_2$	Galfenol Fe-Ga	Metglas 2605
λ (ppm)	3	27	150–400	1400	200–400	40
μ_r		20		6–10	20	>40,000
T_c (°C)	285	535	400	357	700	395

crystals exhibit even better electromechanical properties, however inferior mechanical properties and lower Curie temperatures, which may limit their high-temperature applications. Polyvinylidene difluoride (PVDF) is a polymeric ferroelectric that offers some advantages in terms of elastic constants, but also has a low melting point.

Ferrites are the most common oxides employed in the manufacture of MM composites; however, the most promising magnetostrictive materials are metallic alloys. Indeed, $Tb_xDy_{1-x}Fe_2$, often referred to as Terfenol-D, exhibits the largest magnetostriction of any known material, as shown in Table 22.4, however it is very brittle. $CoFe_2O_4$ is often chosen as the magnetic phase because it possesses the highest magnetostrictive coefficient among oxides ($\sim$150–400 ppm), depending on sintering temperature, grain size, and other factors [71].

The ME response of a composite also relies on the chemical compatibility and the elastic coupling between their components, therefore phase connectivity and interfacial bonding play a critical role in MM composite performance. Nevertheless, composite structures afford considerable flexibility for tailoring the ME response, through a choice of phase properties, volume fraction, shape, and connectivity. Indeed, MM composites can have a variety of connectivity schemes, but the most common, as schematically illustrated in Fig. 22.6, are the (1) 0–3 type particulate composites where the magnetic/piezoelectric particles are embedded in a matrix of piezoelectric/magnetic phases, (2) 2–2 laminates consisting of alternating piezoelectric and magnetic phases, and (3) 1–3 type fibers of one phase embedded on the matrix of the other phase.

Good mechanical bonding at the interface between the composite phases is essential to achieve efficient strain transfer, and is characterized by the interface coupling factor, k. The variation of the ME coupling coefficient with the volume fraction, f, of the magnetic phase in 2–2 type $Pb(Zr,Ti)O_3$-$CoFe_2O_4$ composites in function of the interface coupling factor, k, was calculated by Bichurin et al. [73]. Fig. 22.7 shows that the maximum ME to occur is $f = 0.4$, but also decreases dramatically with poorer interface bonding.

Calculations by Nan [74] have also shown that 0–3 composites of BaTiO-$CoFe_2O_4$ should display a stronger ME coupling than the corresponding 2–2 structure, as schematically illustrated in Fig. 22.8.

However, experimental values are actually much lower than expected because of (1) high porosity and/or (2) high electrical conductivity due to the magnetic phase, which creates conduction paths across the bulk of the material, preventing poling

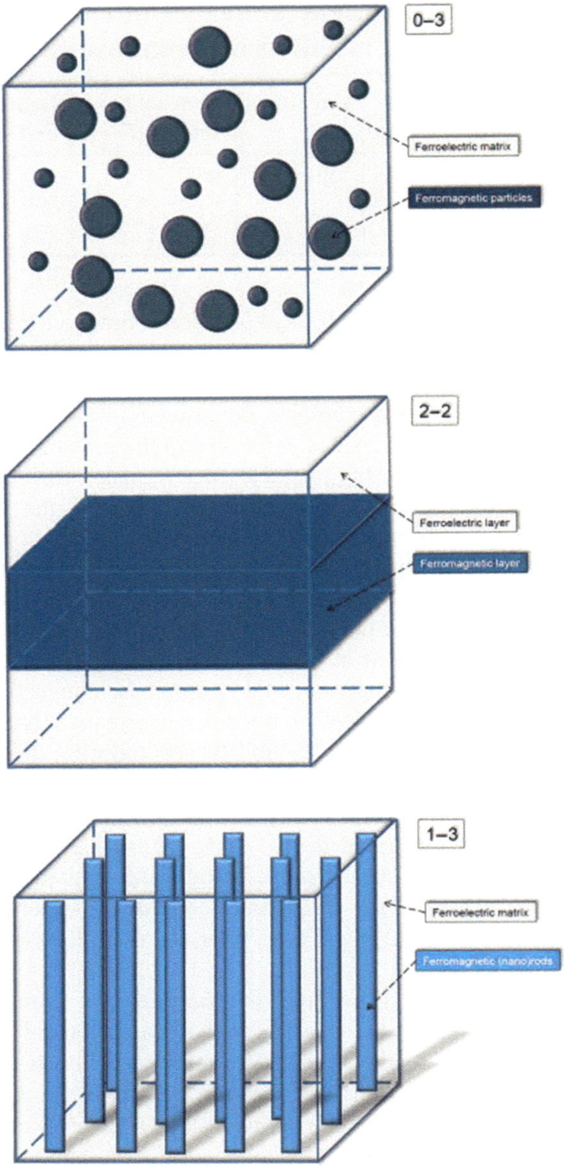

Figure 22.6 Schematic representation of 0–3, 2–2, and 1–3 type connectivities in composite materials.

With permission from G. Schileo, Recent developments in ceramic multiferroic composites based on core/shell and other heterostructures obtained by sol-gel routes, Prog. Solid State Ch. 41 (4) (2013) 87–98.

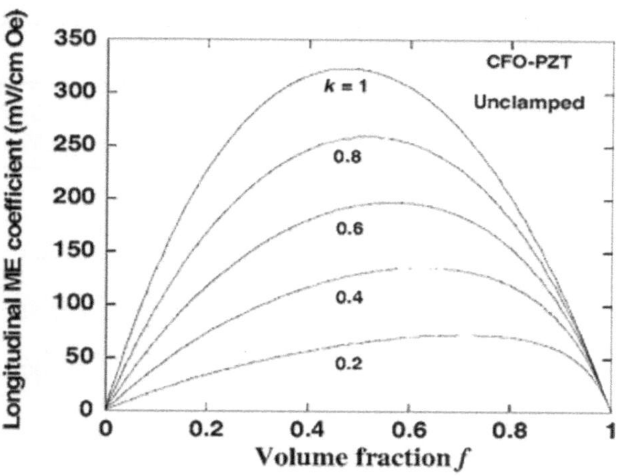

Figure 22.7 Simulated dependence of the longitudinal ME voltage coefficient for Pb(Zr,Ti) O_3-$CoFe_2O_4$ composites in function of volume fraction and the interface coupling factor, *k*. With permission from M.I. Bichurin, V.M. Petrov, G. Srinivasan, Theory of low-frequency magnetoelectric effects in ferromagnetic-ferroelectric layered composites, J. Appl. Phys. 92 (12) (2002) 7681–7683.

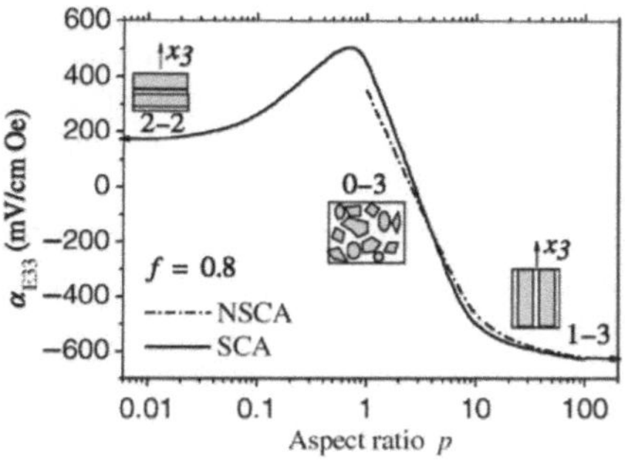

Figure 22.8 Effect of phase connectivity on the magnetoelectric response of $BaTiO_3$-$CoFe_2O_4$ composites.
With permission from C.W. Nan, Magnetoelectric effect in composites of piezoelectric and piezomagnetic phases, Phys. Rev. B 50 (9) (1994) 6082–6088.

and charge storage. By confining the more conductive magnetic layers between insulating ferroelectric layers, 2–2 multilayer structures avoid the latter issue. Hence, in theory, large fractions of magnetic phase can be employed as long as they are isolated between two ferroelectric layers.

Table 22.5, compiled by Eerenstein and coauthors [3], summarizes the state-of-art in the field of MMs as of 2006.

Hereafter, a more comprehensive description of the development of particulate and laminate MM composites will be presented and the latest advances will be

Table 22.5 Examples of magnetoelectric multiferroic composites and their magnetoelectric coupling coefficients

Arrangement	Materials	Coupling coefficient (mV/cm Oe)
Particulate	$BaTiO_3$ and $CoFe_2O_4$	50
Particulate	Terfenol-D and $Pb(Zr,Ti)O_3$ in polymer matrix	42
Laminate	Terfenol-D in polymer matrix/$Pb(Zr,Ti)O_3$ polymer matrix	3000
Laminate	Terfenol-D/$Pb(Zr,Ti)O_3$	4800
Laminate	$La_{0.7}Sr_{0.3}MnO_3$/$Pb(Zr,Ti)O_3$	60
Laminate	$NiFe_2O_4$/$Pb(Zr,Ti)O_3$	1400

contrasted with those in Table 22.5. The reader will notice that many major advances were mainly due to processing improvements, and to the introduction of a few new combinations of ferroelectric and magnetic phases.

22.3.1 Particulate composites

The first work devoted to MM composites was conducted at Philips Laboratories in 1972, through the unidirectional solidification of a eutectic composition of the quaternary Co–Fe–Ba–Ti–O system. A ME coupling coefficient of ~130 mV/cm Oe was reported for these composites, which was larger than the value previously measured from any single-phase MM material. For example, single-phase Cr_2O_3 was known to exhibit a ME coupling of ~28mV/cm Oe. This promising result prompted the investigation of other methods of processing $CoFe_2O_4$-$BaTiO_3$ particulate composites, including simple solid-state reaction methods, sol-gel processes, spark plasma sintering, and the synthesis of so-called core-shell-type nanostructures [75]. The variability of ME coupling coefficients obtained for $CoFe_2O_4$-$BaTiO_3$ particulate composites was astonishing. Researchers reported values ranging from 0.52 to 35 mV/cm Oe [76] composites prepared by conventional sintering to values as large as 2540 mV/cm Oe for composites prepared by a so-called one-pot method [77].

This variation on the ME coefficients showed that processing plays a key role in the development of high-performance MM composites. In addition, the nature of the magnetic and ferroelectric phases and their relative chemical stability were also important parameters considered during the development of particulate MM composites. Hence, several different permutations in terms of materials chosen to fabricate particulate composites have been investigated, as discussed later. The main purpose was the evaluation of the ultimate impact on the ME coupling coefficients. One of the first permutations that investigators explored was the impact of replacing $BaTiO_3$ by other ferroelectric materials such as $Pb(Zr,Ti)O_3$, $BiFeO_3$, $Sr_{0.5}Ba_{0.5}Nb_2O_6$, $Ba_{0.8}Ca_{0.2}TiO_3$, and PVDF. With respect to the magnetic phase, other ferrites such as $NiFe_2O_4$, $CuFe_2O_4$, $LaMnO_3$, and $Y_3Fe_5O_{12}$ were employed to fabricate particulate

MM composites. The suitability of other nonoxide magnetic phases such as Terfenol-D ($Tb_{1-x}Dy_xFe_2$) to fabricate particulate composites was also explored.

Hence, several composites based on $BaTiO_3$ with spinel ferrites other than $CoFe_2O_4$ were the focus of many investigations. Actually, many led to composites with inferior properties than those composites, for example composites encompassing $CuFe_2O_4$ and $BaTiO_3$ [78], exhibit a maximum ME coupling of 520 μV/cm Oe, composites of $BaTiO_3$ (BT) and $Y_3Fe_5O_{12}$ (YIG) [79] exhibit a maximum ME coupling of 24–60 μV/cm Oe. BT-YIG ceramics exhibited both a good densification and distribution of the YIG particles within the BT matrix, as illustrated in Fig. 22.9. Hence, the lower ME coefficients for YIG–BT composites in comparison with the well-studied $CoFe_2O_4$–BT composites (~130 mV/cm Oe) was then ascribed to the lower magnetostriction coefficient of YIG (~3 ppm) in comparison with $CoFe_2O_4$(~165 ppm), as shown in Table 22.4. Subsequently, Wang et al. showed 0–3 type $BiY_2Fe_5O_{12}$–BT composites to exhibit an ME as high as 0.24 mV/cm Oe.

Composites based on $CoFe_2O_4$ combined ferroelectric phases other than $BaTiO_3$ have also received considerable attention. For example, a ME coupling coefficient

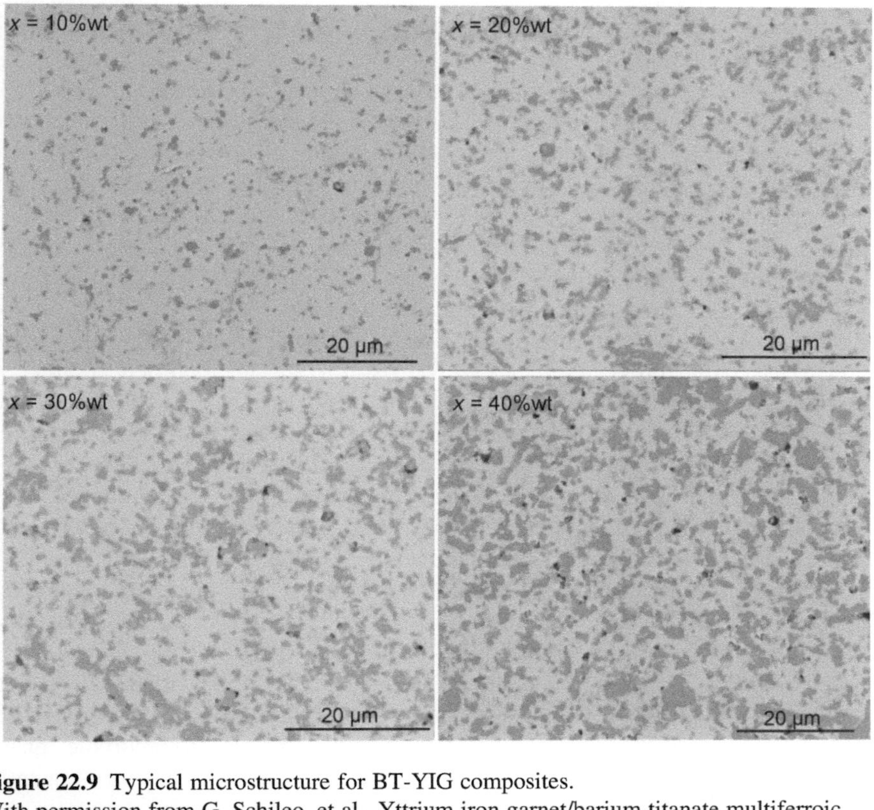

Figure 22.9 Typical microstructure for BT-YIG composites.
With permission from G. Schileo, et al., Yttrium iron garnet/barium titanate multiferroic composites, J. Amer. Ceram. Soc. 99 (5) (2016) 1609–1614.

of 24.8 mV/cm Oe was reported for composites consisting of 0.4 $CoFe_2O_4$ and 0.6 $Sr_{0.5}Ba_{0.5}Nb_2O_6$ [80]. $CoFe_2O_4$−$BiFeO_3$ nanocomposites were shown to exhibit a ME coupling coefficient as large as 285 mV/cm Oe [81].

The use of morphotropic $Pb(Zr_{0.52}Ti_{0.48})O_3$ (PZT) in the fabrication of MM particulate composites incorporating magnetic phases such as $CoFe_2O_4$, $NiFe_2O_4$ also became an active area of research. For example, the ME coupling for PZT−$NiFe_2O_4$ particulate composites with various PZT volume fractions was studied by Zeng et al. [82], who reported maximum ME voltage coefficient resonance values of 6700 mV/cm Oe. An even larger value was reported for the same system for composites prepared by Wu et al. [83]. These fabricated the composites using a simple in situ processing method based on a sol-gel, followed by conventional sintering. The in situ grown magnetic $NiFe_2O_4$ grains were well dispersed in the PZT matrix. 0.35 $NiFe_2O_4$/0.65PZT composites exhibited a remarkably large ME response of ~28,500 mV/cm Oe at a resonance frequency of ~287 kHz.

Chemical doping of well-known spinel ferrites was also pursued to evaluate the impact on the ME coupling coefficients of MM particulate composites. For example, in 2009, Ramanaa et al. [84] reported a relatively high ME coupling coefficient for particulate composites consisting of piezomagnetic phase $Ni_{0.83}Co_{0.15}Cu_{0.02}Fe_{1.9}O_{4-\delta}$ (NCCF) and the piezoelectric phase $Pb(Zr_{0.52}Ti_{0.48})O_3$ (PZT). In particular, the composite consisting of 50% NCCF and 50% PZT exhibited the largest ME coupling coefficient of ~3150 mV/cm Oe.

The impact induced by chemical doping of the ferroelectric phases was also explored by some researchers. For example, Miclea et al. [85] investigated the ME properties' particulate composites based on $CoFe_2O_4$ (CFO) and $Pb_{0.95}Sr_{0.05}Ni_{0.06}Zr_{0.49}Ti_{0.40}O_3$, a soft PZT composition. They reported a maximum ME coefficient of ~70 mV/cm Oe for composites containing 40% CFO and 60% PZT, in a DC magnetic field of 3 kOe at a frequency of 1 KHz of the ac magnetic field. A maximum ME voltage coefficient ~86 mV/cm Oe was measured for CFO-$Pb_{0.7}Ca_{0.3}TiO_3$ composites with 80% ferrite content [86].

Lead-free 3−0 ME composites have been studied using three types of piezoelectric materials based on $K_{0.5}Na_{0.5}NbO_3$, $Na_{0.5}Bi_{0.5}TiO_3$, and modified $BaTiO_3$. A rather low ME coupling coefficient ~0.5 mV/cm Oe was reported to 65NBT-35 CFO composites [87]. However, a more promising result was obtained for lead-free $Ni_{0.83}Co_{0.15}Cu_{0.02}Fe_{1.9}O_{4-\delta}$-$Na_{0.5}Bi_{0.5}TiO_3$ (NCCF-NBT) composites fabricated by spark plasma sintering in conjunction with high-energy ball milling. Indeed, 0.5 NCCF−0.5 NBT composites show enhanced ME properties with a ME coupling coefficient of ~663 mV/cm Oe measured under a magnetic field of 5 kOe [88]. Investigations of magnetodielectric (MD) effects in lead-free $CoFe_2O_4$/$Na_{0.5}K_{0.5}NbO_3$ (CFO/KNN) MF bilayers was carried by He et al. [89]. Recently, composites encompassing lead-free $0.5Ba(Zr_{0.2}Ti_{0.8})O_3$-$0.5(Ba_{0.7}Ca_{0.3})TiO_3$ (BCZT) and $CoFe_2O_4$ have been fabricated [90]. Composite exhibits larger ME effects than their $BaTiO_3$−$CoFe_2O_4$ counterparts with similar content of the ferrite phase. Basically, BCZT has a composition close to the polymorphic phase boundary between the orthorhombic and tetragonal phases, and consequently exhibits a large piezoelectric coefficient ~600 pC/N.

Core-shell structure particulate composites have been also investigated by several researchers [75,91,92]. Recent developments in ceramic MF composites based on core/shell and other heterostructures obtained by sol-gel routes have been reviewed by Schileo [72].

22.3.2 Laminate composites

Laminate MM composites were developed to circumvent issues of low electrical resistance associated with particulate composites. Another immediate advantage of laminate composites is associated with the ability to achieve enhanced poling of the piezoelectric component because of the low resistivity of the magnetic component. They also afford better control over the mechanical coupling compared to the particulate composites, as reflected by the observation of larger ME couplings, both at low frequencies and at resonance.

Laminate MM composites can be prepared in a variety different shapes, geometries, and sizes, varying from circular to square-shaped and rings [93]. In terms of stacking, they can range from simple bilayered to multilayered structures. Several processes have been employed for the fabrication of laminate composites. For example, composites can be prepared by bonding bulk magnetostrictive and piezoelectric phases together with adhesive such as conductive silver epoxy, but often composites fabricated by this simple method exhibit much lower ME coupling coefficients than predicted, due to the weak interfacial mechanical coupling. Alternatively, cosintering of magnetostrictive and piezoelectric layers has been envisaged as an avenue to improve the interfacial mechanical coupling. Despite the potential enhancement of the ME coupling coefficients, several issues need to be taken into account to ensure an effective cosinterability. First, the densification temperatures and the shrinkage rates need to be similar for both phases. Second, any chemical reaction and diffusion near the interfaces may create severe problems, leading to degradation of the ME properties, as observed in particulate composites. Several 2–2 laminate composites consisting of alternating magnetic and piezoelectric ceramic layers have been fabricated by tape casting followed by cosintering. This relatively simple fabrication process affords the possibility of mass industrial production.

Guzdek [94] has successfully prepared composites by tape casting based on ferromagnetic ferrite $Ni_{0.3}Zn_{0.62}Cu_{0.08}Fe_2O_4$ (NZCF) as the magnetostrictive component and ferroelectric relaxor $Pb(Fe_{0.5}Nb_{0.5})O_3$(PFN) as the ferroelectric component. This laminate structure consisted of seven NZCF and eight PFN layers. The maximum values of ME coefficients achieved by these multilayered composites are about 50 mV/cm Oe. Hao et al. [95] also fabricated by tape casting multilayer structures encompassing nine alternated layers of $BaTiO_3$ and $CoFe_2O_4$, as shown in Fig. 22.10. Nevertheless, these composites exhibited a disappointingly low ME coefficient of 8.1 μV/cm Oe in comparison with values measured for their particulate composites, which can reach 50 mV/cm Oe.

The low ME coefficient was attributed to interfacial defects, which reduced dramatically the interfacial coupling coefficient, k.

The preparation of multilayer structures by uniaxial pressing has been conducted by some authors. In this case, powders of magnetic and ferroelectric phases are

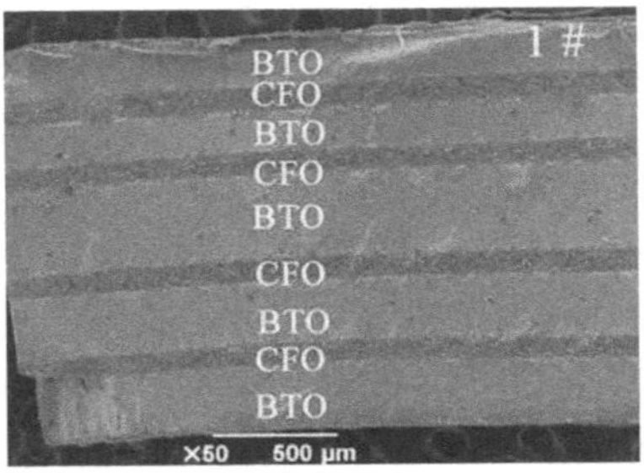

Figure 22.10 $BaTiO_3/CoFe_2O_4$ ceramic multilayer.
With permission from L.B. Hao, et al., Multiferroic properties of multilayered $BaTiO_3$-$CoFe_2O_4$ composites via tape casting method, J. Mater. Sci. 48 (1) (2013) 178–185.

layered alternately inside a mold and subsequently pressed. This green body is then fired at high temperatures in order to simultaneously provide a strong bonding between the layers and increase the mechanical strength of the multilayered structure. For example, Babu et al. [96] used this process to prepare structures encompassing ferrimagnetic $Ni_{0.93}Co_{0.02}Mn_{0.05}Fe_{1.95}O_4$ (NCMF) confined between piezoelectric PZT. In their study they investigated the ME output of NCMF/PZT/NCMF laminate composites with different thickness ratios as a function of the dc bias field. They observed an enhancement of the ME coupling coefficients as the thickness ratio between NCMF and PZT was increasing and reported a ME coefficient of 182 mV/cm Oe for structures where the PZT layer was roughly one-third thinner than the NCMF layers.

Zhou et al. [97] also fabricated 2–2-type $PbZr_{0.52}Ti_{0.48}O_3$-$CoFe_2O_4$ composites employing a conventional uniaxial pressing technique followed by cosintering. They observed interdiffusion of the elements between different layers occurred, which modified material properties. Moreover, the dielectric response was dominated by the relative thickness of the different layers, due to the much higher dielectric permittivity of the $PbZr_{0.52}Ti_{0.48}O_3$ layers in comparison with the CFO layer. The ME behavior in this PZT/CFO/PZT structured composite was found to be strongly dependent on the relative thickness of the CFO layer.

Recent legislation concerning the use of hazardous lead in the fabrication of electrical and electronic components prompted the investigation of composites where the piezoelectric phase was based on one of the $Bi_{0.5}Na_{0.5}TiO_3$ or $K_{0.5}Na_{0.5}NbO_3$ ceramics. For example, Yang et al. [98] prepared $0.65BaTiO_3$-$0.35Bi_{0.5}Na_{0.5}TiO_3/CoFe_2O_4$ both particulate and laminated composites. Higher resistivity and lower leakage current was observed in the laminated composites, making electric poling more effective and thus enhancing the ME coupling. The maximum ME coefficient of the laminate composites reaches up to about 113.10 mV/cm Oe, which is 10 times as large as that of the particulate composites (10.08 mV/cm Oe).

Among all-oxide 2:2 type composites the largest ME coupling has been reported for bilayers and multilayers of $NiFe_2O_4$ with PZT, ranging from 460 mV/cm Oe in bilayers to 1500 mV/cm Oe for multilayers [99].

In 2001, a milestone in the development of MM bulk composites was the use of giant magnetostrictive rare-earth iron alloy $Tb_{1-x}Dy_xFe_2$ (Terfenol-D), and ME coupling coefficients as high as 4800 mV/cm Oe were achieved for laminated composites encompassing Terfenol-D and $Pb(Zr,Ti)O_3$. Mori and Wuttig [100] studied the elastic coupling in 2:2 type composites made from Terfenol-D bonded to PVDF using epoxy and observed a ME coefficient of 1430 mV/cm Oe. Subsequent improvements were achieved for 2:2 type composites using Terfenol-D combined with ferroelectric phases such as PMN-PT. Nevertheless, to date the largest ME

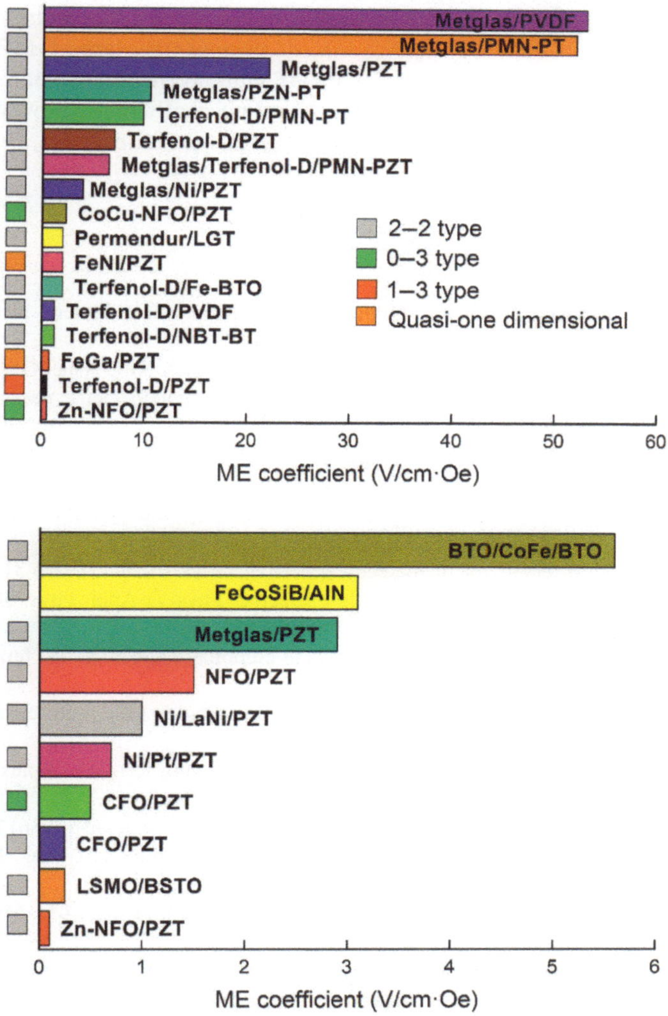

Figure 22.11 Reported values of off-resonance magnetoelectric voltage coefficients for various material systems.

coefficient has been reported for Metglas/PVDF composites. Fig. 22.11 summarizes the ME coupling coefficients achievable for different combinations of ferroelectric and magnetic phases in different geometries. For details on each of the composites listed in Fig. 22.11, refer to Palneedi et al. [101] and the references herewith.

22.4 Final remarks

In spite of many efforts to find intrinsic MM single-phase materials, the small ME coupling coefficients achieved so far have motivated the design of heterogeneous systems, where coupling is artificially engineered between ferroelectric and ferromagnetic components. Among heterogeneous systems the largest ME coupling coefficient has been reported for Metglas-PVDF multilayered composites. All-oxide composites exhibit significantly lower ME coupling coefficients, however processing can still play an important role in improving their performance.

References

[1] M. Bibes, Nanoferronics is a winning combination, Nat. Mater. 11 (5) (2012) 354–357.
[2] J.F. Scott, Room-temperature multiferroic magnetoelectrics, NPG Asia Mater. (2013) 5.
[3] W. Eerenstein, N. Mathur, J. Scott, Multiferroic and magnetoelectric materials, Nature 442 (7104) (2006) 759–765.
[4] N.A. Hill, Why are there so few magnetic ferroelectrics? J. Phys. Chem. B 104 (29) (2000) 6694–6709.
[5] R.E. Newnham, D.P. Skinner, L.E. Cross, Connectivity and piezoelectric-pyroelectric composites, Mater. Res. Bull. 13 (5) (1978) 525–536.
[6] H. Schmid, Multi-ferroic magnetoelectrics, Ferroelectrics 162 (1) (1994) 317–338.
[7] M. Fiebig, Revival of the magnetoelectric effect, J. Phys. D-Appl. Phys. 38 (8) (2005) R123–R152.
[8] D.I. Khomskii, Multiferroics: different ways to combine magnetism and ferroelectricity, J. Magn. Magn. Mater. 306 (1) (2006) 1–8.
[9] H. Schmid, Some symmetry aspects of ferroics and single phase multiferroics, J. Phys. Conden. Matter. 20 (43) (2008) 24.
[10] C.A.F. Vaz, et al., Magnetoelectric coupling effects in multiferroic complex oxide composite structures, Adv. Mater. 22 (26-27) (2010) 2900–2918.
[11] M. Fiebig, et al., The evolution of multiferroics, Nat. Rev. Mater. 1 (2016) 8.
[12] Y. Tokura, S. Seki, N. Nagaosa, Multiferroics of spin origin, Rep. Prog. Phys. 77 (7) (2014).
[13] G. Catalan, J.F. Scott, Physics and applications of bismuth ferrite, Adv. Mater. 21 (24) (2009) 2463–2485.
[14] T. Zhao, et al., Electrical control of antiferromagnetic domains in multiferroic $BiFeO_3$ films at room temperature, Nat. Mater. 5 (10) (2006) 823–829.
[15] R. Nechache, et al., Growth, structure, and properties of epitaxial thin films of first-principles predicted multiferroic Bi_2FeCrO_6, Appl. Phys. Lett. 89 (10) (2006) 3.

[16] R. Seshadri, N.A. Hill, Visualizing the role of Bi 6s "lone pairs" in the off-center distortion in ferromagnetic $BiMnO_3$, Chem. Mater. 13 (9) (2001) 2892–2899.

[17] N.A. Hill, K.M. Rabe, First-principles investigation of ferromagnetism and ferroelectricity in bismuth manganite, Phys. Rev. B 59 (13) (1999) 8759–8769.

[18] J. Wang, et al., Epitaxial $BiFeO_3$ multiferroic thin film heterostructures, Science 299 (5613) (2003) 1719–1722.

[19] X.Y. Mao, et al., Multiferroic properties of layer-structured $Bi_5Fe_{0.5}Co_{0.5}Ti_3O_{15}$ ceramics, Appl. Phys. Lett. 95 (8) (2009) 3.

[20] L. Keeney, et al., Magnetic field-induced ferroelectric switching in multiferroic aurivillius phase thin films at room temperature, J. Amer. Ceram. Soc. 96 (8) (2013) 2339–2357.

[21] L. Keeney, et al., Room temperature ferroelectric and magnetic investigations and detailed phase analysis of Aurivillius phase $Bi_5Ti_3Fe_{0.7}Co_{0.3}O_{15}$ thin films, J. Appl. Phys. 112 (5) (2012) 10.

[22] Z. Li, et al., Room temperature magnetoelectric coupling in intrinsic multiferroic Aurivillius phase textured ceramics, Dalton Trans. 45 (36) (2016) 14049–14052.

[23] T. Goto, et al., Ferroelectricity and giant magnetocapacitance in perovskite rare-earth manganites, Phys. Rev. Lett. 92 (25) (2004) 4.

[24] M. Fiebig, et al., Observation of coupled magnetic and electric domains, Nature 419 (6909) (2002) 818–820.

[25] B.B. Van Aken, et al., The origin of ferroelectricity in magnetoelectric $YMnO_3$, Nat. Mater. 3 (3) (2004) 164–170.

[26] M. Lilienblum, et al., Ferroelectricity in the multiferroic hexagonal manganites, Nat. Phys. 11 (12) (2015) 1070.

[27] H. Das, et al., Bulk magnetoelectricity in the hexagonal manganites and ferrites, Nat. Commun. (2014) 5.

[28] Z.J. Huang, et al., Coupling between the ferroelectric and antiferromagnetic orders in $YMnO_3$, Phys. Rev. B 56 (5) (1997) 2623–2626.

[29] N.A. Benedek, C.J. Fennie, Hybrid improper ferroelectricity: a mechanism for controllable polarization-magnetization coupling, Phys. Rev. Lett. 106 (10) (2011) 4.

[30] Y. Oh, et al., Experimental demonstration of hybrid improper ferroelectricity and the presence of abundant charged walls in $(Ca, Sr)_3Ti_2O_7$ crystals, Nat. Mater. 14 (4) (2015) 407–413.

[31] J. van den Brink, D.I. Khomskii, Multiferroicity due to charge ordering, J. Phys.-Conden. Matter. 20 (2008) 43.

[32] K. Kato, et al., Ferrimagnetic ferroelectricity of Fe_3O_4, J. Magn. Magn. Mater. 31-4 (FEB) (1983) 783–784.

[33] K. Yamauchi, T. Fukushima, S. Picozzi, Ferroelectricity in multiferroic magnetite Fe_3O_4 driven by noncentrosymmetric Fe^{2+}/Fe^{3+} charge-ordering: first-principles study, Phys. Rev. B 79 (21) (2009) 4.

[34] N. Ikeda, et al., Ferroelectricity from iron valence ordering in the charge-frustrated system $LuFe_2O_4$, Nature 436 (7054) (2005) 1136–1138.

[35] D.V. Efremov, J. Van den Brink, D.I. Khomskii, Bond-versus site-centred ordering and possible ferroelectricity in manganites, Nat. Mater. 3 (12) (2004) 853–856.

[36] M. Fiebig, et al., Ultrafast magnetization dynamics of antiferromagnetic compounds, J. Phys. D-Appl. Phys. 41 (2008) 16.

[37] T. Kimura, et al., Magnetocapacitance effect in multiferroic $BiMnO_3$, Phys. Rev. B. 67 (18) (2003) 4.

[38] T. Kimura, et al., Magnetic control of ferroelectric polarization, Nature 426 (6962) (2003) 55–58.

[39] N. Hur, et al., Electric polarization reversal and memory in a multiferroic material induced by magnetic fields, Nature 429 (6990) (2004) 392–395.

[40] T. Kimura, J.C. Lashley, A.P. Ramirez, Inversion-symmetry breaking in the noncollinear magnetic phase of the triangular-lattice antiferromagnet $CuFeO_2$, Phys. Rev. B 73 (22) (2006) 4.

[41] A.B. Harris, et al., Towards a microscopic model of magnetoelectric interactions in $Ni_3V_2O_8$, Phys. Rev. B 73 (2006) 18.

[42] G. Lawes, et al., Magnetically driven ferroelectric order in $Ni_3V_2O_8$, Phys. Rev. Lett. 95 (2005) 8.

[43] O. Heyer, et al., A new multiferroic material: $MnWO_4$, J. Phys. Conden. Matter 18 (39) (2006) L471–L475.

[44] Y.J. Choi, et al., Ferroelectricity in an Ising chain magnet, Phys. Rev. Lett. 100 (4) (2008) 4.

[45] I.A. Sergienko, E. Dagotto, Role of the Dzyaloshinskii-Moriya interaction in multiferroic perovskites, Phys. Rev. B 73 (9) (2006) 5.

[46] R.D. Johnson, et al., Giant improper ferroelectricity in the ferroaxial magnet $CaMn_7O_{12}$, Phys. Rev. Lett. 108 (6) (2012) 4.

[47] D. Evans, et al., Magnetic switching of ferroelectric domains at room temperature in multiferroic PZTFT, Nat. Commun. (2013) 4.

[48] A. Rotaru, et al., Towards novel multiferroic and magnetoelectric materials: dipole stability in tetragonal tungsten bronzes, Philos. T. R. Soc. A 372 (2009) (2014) 13.

[49] S.A. Larregola, et al., Novel near-room-temperature Type I multiferroic: Pb $(Fe_{0.5}Ti_{0.25}W_{0.25})O_3$ with coexistence of ferroelectricity and weak ferromagnetism, Chem. Mater. 24 (14) (2012) 2664–2672.

[50] L.F. Henrichs, et al., Multiferroic clusters: a new perspective for relaxor-type room-temperature multiferroics, Adv. Funct. Mater. 26 (13) (2016) 2111–2121.

[51] P. Mandal, et al., Controlling phase assemblage in a complex multi-cation system: phase-pure room temperature multiferroic $(1\text{-}x)BiTi_{(1-y)/2}Fe_yMg_{(1-y)/2}O_3\text{-}xCaTiO_3$, Adv. Funct. Mater. 26 (15) (2016) 2523–2531.

[52] C. Fernandez-Posada, et al., A novel perovskite oxide chemically designed to show multiferroic phase boundary with room-temperature magnetoelectricity, Nat. Commun. (2016) 7.

[53] F. Craciun, et al., Novel multiferroic $(Pb_{1-x}Nd_{3x/2x})(Ti_{0.98-y}Fe_yMn_{0.02})O_3$ ceramics with coexisting ferroelectricity and ferromagnetism at ambient temperature, Mater. Design. 110 (2016) 693–704.

[54] Y. Kitagawa, et al., Low-field magnetoelectric effect at room temperature, Nat. Mater. 9 (10) (2010) 797–802.

[55] H.D. Zhou, et al., $Ba_3NbFe_3Si_2O_{14}$: a new multiferroic with a 2D triangular Fe^{3+} motif, Chem. Mater. 21 (1) (2009) 156–159.

[56] L. Zhao, et al., Mn_3TeO_6 - a new multiferroic material with two magnetic substructures, Phys. Stat. Sol. Rap. Res. Lett. 9 (12) (2015) 730–734.

[57] A. Falqui, et al., Low- temperature magnetic behavior of perovskite compounds $PbFe_{1/2}Ta_{1/2}O_3$ and $PbFe_{1/2}Nb_{1/2}O_3$, J. Phys. Chem. B 109 (48) (2005) 22967.

[58] R. Martinez, et al., Nanoscale ordering and multiferroic behavior in $Pb(Fe_{1/2}Ta_{1/2})O_3$ (vol 82, 134104, 2010). Phys. Rev. B 82 (2010) 13.

[59] M.H. Lente, et al., Nature of the magnetoelectric coupling in multiferroic Pb $(Fe_{1/2}Nb_{1/2})O_3$ ceramics, Phys. Rev. B - Conden. Matter. Mater. Phys. 78 (2008) 5.

[60] L. Mitoseriu, et al., Magnetic properties of $PbFe_{2/3}W_{1/3}O_3\text{-}PbTiO_3$ solid solutions, Appl. Phys. Lett. 83 (26) (2003) 5509–5511.

[61] M. Kempa, et al., Bulk dielectric and magnetic properties of PFWPZT ceramics: absence of magnetically switched-off polarization, J. Phys. Conden. Matter 22 (44) (2010) 445902.
[62] W. Prellier, M.P. Singh, P. Murugavel, The single- phase multiferroic oxides: from bulk to thin film, J. Phys. Conden. Matter. (2005) R803−R832.
[63] D. Varshney, et al., Structural, magnetic and dielectric properties of Pr-modified $BiFeO_3$ multiferroic, J. Alloy. Compd. 584 (2014) 232−239.
[64] S.C. Abrahams, Ferroelectricity and structure in the $YMnO_3$ family, Acta Crystallogr. Sect. B Struct. Sci. 57 (2001) 485−490.
[65] T.-C. Han, et al., Grain size effect on magnetic and dielectric properties of hexagonal $YbMnO_3$ nanoparticles, J. Appl. Phys. 117 (2015) 17.
[66] P. Murugavel, et al., Physical properties of multiferroic hexagonal $HoMnO_3$ thin films, Appl. Phys. Lett. 90 (2007) 14.
[67] A. Ahlawat, et al., Effect of Fe doping on the magnetic ordering temperature of $ErMnO_3$, Appl. Phys. Lett. 106 (2015) 25.
[68] D. Sanchez, et al., Room-temperature single phase multiferroic magnetoelectrics: $Pb(Fe,M)_x(Zr, Ti)_{(1-x)}O_3$, J. Appl. Phys. 113 (2013) 7.
[69] N. Kumar, A. Ghosh, R. Choudhary, Electrical behavior of $Pb(Zr_{0.52}Ti_{0.48})_{0.5}(Fe_{0.5}Nb_{0.5})_{0.5}O_3$ ceramics, Mater. Chem. Phys. 130 (1-2) (2011) 381−386.
[70] A. Kumar, et al., Magnetic control of large room- temperature polarization, J. Phys. Conden. Matter 21 (38) (2009) 382204.
[71] P.N. Anantharamaiah, P.A. Joy, Enhancing the strain sensitivity of $CoFe_2O_4$ at low magnetic fields without affecting the magnetostriction coefficient by substitution of small amounts of Mg for Fe, Phys. Chem. Chem. Phys. 18 (15) (2016) 10516−10527.
[72] G. Schileo, Recent developments in ceramic multiferroic composites based on core/shell and other heterostructures obtained by sol-gel routes, Prog. Solid State Ch. 41 (4) (2013) 87−98.
[73] M.I. Bichurin, V.M. Petrov, G. Srinivasan, Theory of low-frequency magnetoelectric effects in ferromagnetic-ferroelectric layered composites, J. Appl. Phys. 92 (12) (2002) 7681−7683.
[74] C.W. Nan, Magnetoelectric effect in composites of piezoelectric and piezomagnetic phases, Phys. Rev. B 50 (9) (1994) 6082−6088.
[75] V. Corral-Flores, D. Bueno-Baques, R.F. Ziolo, Synthesis and characterization of novel $CoFe_2O_4$-$BaTiO_3$ multiferroic core-shell-type nanostructures, Acta Mater. 58 (3) (2010) 764−769.
[76] H.B. Yang, G. Zhang, Y. Lin, Enhanced magnetoelectric properties of the laminated $BaTiO_3/CoFe_2O_4$ composites, J. Alloy. Compd. 644 (2015) 390−397.
[77] S.Q. Ren, et al., $BaTiO_3/CoFe_2O_4$ particulate composites with large high frequency magnetoelectric response, J. Mater. Sci. 40 (16) (2005) 4375−4378.
[78] R.P. Mahajan, et al., Conductivity, dielectric behaviour and magnetoelectric effect in copper ferrite-barium titanate composites, Bull. Mater. Sci. 23 (4) (2000) 273−279.
[79] G. Schileo, et al., Yttrium iron garnet/barium titanate multiferroic composites, J. Amer. Ceram. Soc. 99 (5) (2016) 1609−1614.
[80] X.M. Chen, et al., Dielectric and magnetoelectric characterization of $CoFe_2O_4$/$Sr_{0.5}Ba_{0.5}Nb_2O_6$ composites, J. Appl. Phys. 96 (11) (2004) 6520−6522.
[81] X.M. Liu, S.Y. Fu, C.J. Huang, Synthesis and magnetic characterization of novel $CoFe_2O_4$-$BiFeO_3$ nanocomposites, Mater. Sci. Engin. B-Solid State Mater. Adv. Techn. 121 (3) (2005) 255−260.

[82] M. Zeng, et al., Resonance magnetoelectric effect in bulk composites of lead zirconate titanate and nickel ferrite, J. Appl. Phys. 95 (12) (2004) 8069–8073.
[83] D.D. Wu, et al., Magnetoelectric composite ceramics of nickel ferrite and lead zirconate titanate via in situ processing, J. Phys. D-Appl. Phys. 40 (16) (2007) 5002–5005.
[84] M.V. Ramanaa, et al., Enhanced mangnetoelectric voltage in multiferroic particulate $Ni_{0.83}Co_{0.15}Cu_{0.02}Fe_{1.9}O_{4-\delta}/PbZr_{0.52}Ti_{0.48}O_3$ composites - dielectric, piezoelectric and magnetic properties, Curr. Appl. Phys. 9 (5) (2009) 1134–1139.
[85] C. Miclea, et al., Magnetoelectric properties of multiferroic cobalt ferrite and soft piezoelectric PZT particulate composites, J. Optoelectron. Adv. M. 12 (2) (2010) 272–276.
[86] A. Sharma, R.K. Kotnala, N.S. Negi, Observation of multiferroic properties and magnetoelectric effect in (x)$CoFe_2O_4$ -(1-x)$Pb_{0.7}Ca_{0.3}TiO_3$ composites, J. Alloy. Compd. 582 (2014) 628–634.
[87] R.V. Krishnaiah, et al., Effect of $CoFe_2O_4$ mole percentage on multiferroic and magnetoelectric properties of $Na_{0.5}Bi_{0.5}TiO_3/CoFe_2O_4$ particulate composites, Ceram. Inter. 40 (6) (2014) 7799–7804.
[88] V.R. Mudinepalli, S.H. Song, B.S. Murty, Enhanced magnetoelectric properties in lead-free $Ni_{0.83}Co_{0.15}Cu_{0.02}Fe_{1.9}O_4$-delta -$Na_{0.5}Bi_{0.5}TiO_3$ composites by spark plasma sintering, Scripta Mater. 82 (2014) 9–12.
[89] S.M. He, et al., Magnetodielectric effect in lead-free multiferroic $CoFe_2O_4$/$K_{0.5}Na_{0.5}NbO_3$ bilayers, Mater. Lett. 89 (2012) 159–162.
[90] M. Naveed-Ul-Haq, et al., A new (Ba,Ca)(Ti,Zr)O_3 based multiferroic composite with large magnetoelectric effect, Sci. Rep. 6 (2016) 10.
[91] A. Sakanas, et al., Microstructural influence on the broadband dielectric properties of $BaTiO_3$-$Ni_{0.5}Zn_{0.5}Fe_2O_4$ core-shell dcomposites: experiment and modeling, J. Appl. Phys. 118 (2015) 17.
[92] L.P. Curecheriu, et al., Functional properties of $BaTiO_3$-$Ni_{0.5}Zn_{0.5}Fe_2O_4$ magnetoelectric ceramics prepared from powders with core-shell structure, J. Appl. Phys. 107 (2010) 10.
[93] L. Li, Y.Q. Lin, X.M. Chen, $CoFe_2O_4/Pb(Zr_{0.52}Ti_{0.48})O_3$ disk-ring magnetoelectric composite structures, J. Appl. Phys. 102 (6) (2007) 4.
[94] P. Guzdek, The magnetostrictive and magnetoelectric characterization of $Ni_{0.3}Zn_{0.62}Cu_{0.08}Fe_2O_4$-Pb(FeNb)$O_3$ laminated composite, J. Magn. Magn. Mater. 349 (2014) 219–223.
[95] L.B. Hao, et al., Multiferroic properties of multilayered $BaTiO_3$-$CoFe_2O_4$ composites via tape casting method, J. Mater. Sci. 48 (1) (2013) 178–185.
[96] S.N. Babu, et al., Magnetoelectric properties in NCMF/PZT particulate and bulk laminate composites, J. Phys. D-Appl. Phys. 41 (16) (2008) 6.
[97] J.P. Zhou, et al., Dielectric, magnetic, and magnetoelectric properties of laminated $PbZr_{0.52}Ti_{0.48}O_3/CoFe_2O_4$ composite ceramics, J. Appl. Phys. 100 (9) (2006) 6.
[98] H.B. Yang, et al., Simultaneous enhancement of electrical and magnetoelectric effects in $BaTiO_3$-$Bi_{0.5}Na_{0.5}TiO_3/CoFe_2O_4$ laminate composites, J. Alloy. Compd. 646 (2015) 1104–1108.
[99] G. Srinivasan, et al., Magnetoelectric bilayer and multilayer structures of magnetostrictive and piezoelectric oxides, Phys. Rev. B 64 (21) (2001) 6.
[100] K. Mori, M. Wuttig, Magnetoelectric coupling in terfenol-D/polyvinylidenedifluoride composites, Appl. Phys. Lett. 81 (1) (2002) 100–101.
[101] H. Palneedi, et al., Status and perspectives of multiferroic magnetoelectric composite materials and applications, Actuators 5 (1) (2016) 31.

Section VI

Multiferroic Metal Oxides: Properties and Applications

Single and heterostructure multiferroic thin films

23

Antoine Barbier
Service de Physique de l'Etat Condensé, SPEC, CEA, CNRS, UMR 3680, Université Paris-Saclay, CEA Saclay, 91191 Gif-sur-Yvette Cedex, France

23.1 Introduction

Multiferroic systems combining at least two ferroic orders (e.g., ferroelectric, (anti)-ferromagnetic, ferroelastic, etc.) motivate an intense research activity, driven by fundamental physics interests trying to understand the coupling between ferroic orders, as well as numerous potential applications in genuine multifunctional devices, such as sensors, transducers, memories, and spintronics. Although the basic effect was discovered more than a century ago [1] the potential application in novel heterostructures has revived the interest in such composites in recent years [2–4]. The electric field control of magnetism in the solid state is a very interesting concept in condensed matter physics. Defining and tailoring new functional multiferroic devices is as a hot topic in modern spintronics, which has been reviewed recently in references [5–9]. In the current industrial context new devices with higher performance, environment friendly production and/or recycling processes, and lower energy consumption, are thoroughly sought. Many studies focus on electron spin-based electronics (spintronics [10]) in thin-film systems with ferromagnetic layers because of their huge technological potential in a variety of domains including memory cells and sensors. Large research efforts in the field are devoted to the reduction of storage device sizes (i.e., the increase of the storage density) and cross-field or eventually all-electric device control. As a matter of fact, multifunctional and/or multiple field addressable memory cell elements are ideally suited to fulfill the requirements for increasing data storage density and limit the use of natural resources. The required materials imply additional properties to the magnetization carried by the typical layers used in spintronics. Such a class of material has been thoroughly studied in the second half of twentieth century and is called multiferroics after Schmid [2]. It was inferred since the 1970s that there may exist systems in which two types of ordering (ferro)magnetism (i.e., the spontaneous ordering of orbital and spin magnetic moments) and ferroelectricity (i.e., the spontaneous ordering of electric dipole moments) can coexist in some material in the absence of external fields. Moreover strong cross-coupling may exist in such solids (i.e., the appearance of magnetization M in an electric field E, or the inverse effect of electric polarization P generated by the application of magnetic field H), properties opening fully new avenues in spintronics if implementable in device structures. Unfortunately, single-phase multiferroics with ordering temperatures above 300K

Magnetic, Ferroelectric, and Multiferroic Metal Oxides. DOI: http://dx.doi.org/10.1016/B978-0-12-811180-2.00023-2

are rare (if any exist) since cooperative cation off-centering and magnetic ordering are antagonist properties (in bulk compounds). The nonavailability of single-phase compounds combining robust electric and magnetic polarizations appeared as a major and limiting issue and the research activity slowed down by the 1970s.

The concomitant need for spintronic devices to develop multifunctional properties and the huge improvements in oxide thin films elaboration using various methods in the end of the last century allowed an impressive revival of the field especially by the use of novel thin-film multiferroic structures. In 2003 the successful growth of several new thin-film multiferroic compounds were reported. In particular, $BiFeO_3$ thin films, one of the most popular multiferroics, could be elaborated [11]. Fortunately, in the thin-film form its multiferroic properties are greatly enhanced with respect to the bulk, in which form the compound exhibits fairly weak multiferroic behavior. Although $BiFeO_3$ is rather a ferroelectric canted antiferromagnet, it revived the research field because of the potential use of magnetic interfacial exchange coupling that could lead to new spintronic devices. A second major experimental breakthrough occurred at the same moment: the discovery of a novel class of multiferroics, in which magnetism causes ferroelectricity rather than just the independent coexistence of both long-range orders. This phenomenon was reported in $TbMnO_3$ [12] and $TbMn_2O_5$ [13]. In addition, by this time theoretical concepts were developed, allowing an understanding of why the coexistence of magnetism and ferroelectricity is very seldom [14,15].

Although multiferroicity corresponds in principle to the occurrence, in the same phase, of one or more of the four primary ferroic properties—ferroelectricity, ferromagnetism, ferrotoroidicity, and ferroelasticity—nowadays what most people mean by multiferroic mostly applies to the coexistence of magnetism and ferroelectricity likely due to the large expected applications of this particular class of material. When dealing with thin films, even if magnetoelectric coupling effects are sought, we should often include in this category a third type of ordering: spontaneous deformation, which leads to ferroelasticity. Indeed, lattice deformations, strain relaxation via elastic or plastic deformations are intrinsic to lattice mismatched growth of heterostructures as long as the growth is epitaxial. As a matter of fact, the thin-film form brings this way an additional degree of freedom that appeared to be highly beneficial for the multiferroic coupling properties in most systems studied to date.

The literature already contains a large number of reviews on multiferroic thin films [3,4,15–23]. We will discuss more specifically here the interplay between thin films and multiferroics. In view of the recent prolific literature on multiferroics, with an accent on magnetoelectrics, significant illustrative examples were chosen on criteria that could appear as subjective and review papers were prioritized. As a matter of fact, all relevant thin-film studies could not be cited within the available space and the author apologizes to the authors whose relevant work is not cited here. However, most pertinent review references are provided so that interested readers will be able to go beyond.

First we will provide some key concepts regarding epitaxial thin-film growth and discuss their pertinence within the field under discussion here as well as the different routes allowing for elaborating thin-film multiferroics. Next we will examine to what extend thin films are compatible (or not) with multiferroic properties in view of the present understanding of the physics behind these behaviors.

23.2 Elements of thin-film growth: Thin films versus multiferroics

In this section we will discuss some mechanisms intrinsic to thin-film growth as well as their implication when applied to multiferroic thin films.

23.2.1 Growth stress in thin films

Thin-film growth is a very mature research field and a great variety of growth methods are available going from heavy ultra-high vacuum systems to chemical deposition. In general, a film or coating is a region at the surface that has different composition and properties than the under layer. Usually, the film is formed by deposition of a certain material on top of a different one and the body coated by the film is called substrate. Thin-film growth occurs typically from thermodynamically equilibrium conditions that may even be limited by the atomic species kinetics. It results in the possibility to grow new atomic arrangements with genuine properties that have no counterpart in the bulk. Moreover the reduced dimensionality allows exploiting quantum mechanics effects otherwise not realizable. It results that the thin-film terminology should be limited to films with thicknesses small enough that the physical properties of the layer are affected in some way.

The spectrum of goals being pursued by thin films utilization appears to be amazingly broad. They are technologically important in a large variety of fields, much broader than the multiferroic films under discussion here; they are widely used in optical coatings, corrosion protection, semiconductors, and spintronics devices. Many of these applications involve sophisticated growth processes. Thin films are also important within physical science itself since they allow exploring the differences between three-dimensional and (quasi-)two-dimensional states of matter. Modifications with respect to film thickness may concern, for example, thin metal conductivity or chemical reactivity. The nucleation and growth of thin films has been reviewed in a number of excellent review papers from general points of view [24,25] as well as for multiferroics in particular [17,20]. The overall understanding of the epitaxial growth is nowadays advanced enough to allow detailed modeling and simulation [26].

Multiferroics properties can appear in a large variety of materials. Therefore, many material fabrication techniques are being applied. Popular techniques within the multiferroic community are solid-state synthesis, hydrothermal synthesis, sol-gel processing, vacuum-based deposition, and floating zone. However some types of multiferroics require specific processing conditions: vacuum-based deposition allows efficient two-dimensional layered structures in the form of heterostructures or multilayers while high pressure solid state synthesis allows stabilizing metastable or distorted structures. The most popular multiferroic magnetoelectric layers consist of insulating ferroelectric oxides. Growth methods like pulsed laser deposition and molecular beam epitaxy (MBE) are well suited to grow single crystalline layers of oxides especially when using molecular or better atomic oxygen residual atmospheres during the growth.

Growing epitaxial ferroelectric layers raises specific questions that will be discussed hereafter. From a rough macroscopic point of view, during growth an intrinsic stress builds up within the thin film-substrate compounds, mainly caused by atomic mismatch at the interface between substrate and deposit. Additionally, thermal stresses resulting from the difference between deposition and room temperature should be considered as well as other effects, for example, like ion bombardment coming from the deposition process itself when using sputtering techniques. Thin films usually show some anisotropy of their structure due to the fact that the growth direction is very particular. In case of polycrystalline films evolutionary selection takes place selecting the grains having the highest growth velocity. This macroscopic anisotropy may have a distinct influence on the mechanical behavior of the films.

Thin films are usually under mechanical stress in lateral directions because in the direction perpendicular to the surface the material can move freely. This is the so-called biaxial stress state. The sources of stress in the film may arise from the lattice constants being matched by force to the substrate parameters, but also from cooling down or recrystallization. The effects of stress will result in elastic deformation, which is a very important parameter in the context of multiferroic growth, as well as plastic deformation taking the form of the inclusion of edge dislocation in the films after a critical thickness has been reached. The stress state is a very important parameter in the film properties modifications and for multiferroics in particular.

The problem of intrinsic stresses was historically addressed within the so-called thin plate simplification, which consists of neglecting all stress components pointing in the direction of the plate's normal axis. In those cases, where a 2D film biaxial stress is considered, the original Stoney formulae [27] have to be modified. When measuring thin films deposited on plate-like substrates, the biaxial deformation has to be taken into account by using the biaxial modulus, $E_{b,S} = E_S/(1-\nu_S)$, of the substrate rather than the Young's modulus, E_S, alone, leading thus to the following stress value formula:

$$\sigma_S = -\frac{E_S}{(1-\nu_S)}\frac{h_S^2}{6h_f}\left(\frac{1}{R}-\frac{1}{R_0}\right) \tag{23.1}$$

with R and R_0 the curvature radii of the substrate after and before deposition, h_S the thickness of the substrate, h_f the thickness of the film, and ν_S—Poisson's ratio of the substrate. Typical Poisson's coefficient for solid materials is in the 0.2–0.3 range yielding a modification in calculated values around 30% as compared to the initial 1D approach. Since the layer is free to move along the surface normal, z, the biaxial stress only acts in the x–y plane. It results that the deformation parallel to the stress direction is smaller and the strain perpendicular to the stress direction is bigger than

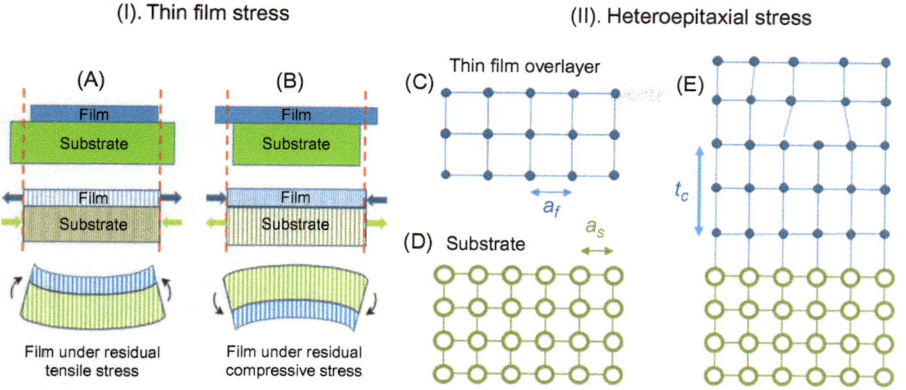

Figure 23.1 Stress in thin films. I. Thin-film stress: Equilibrium shape of a nonrigid film + substrate system under (A) a tensile stress and (B) a compressive stress. II. Heteroepitaxial stress: structural modifications for an adsorbate (C) and substrate layer (D) within an epitaxial overlayer structure (E). The strain relaxation is elastic until a critical layer thickness, t_C, where plastic relaxation starts to occurs via the introduction of heteroepitaxy edge dislocations.
Adapted from M. Ohring, Materials Science of Thin Films, Copyright (2002), Published by Academic Press, Chapter 12, with permission from Elsevier [28].

in the uniaxial case. If the film would be deposited on a nonrigid substrate that could follow the film stress we could distinguish specific deformations related to tensile or compressive stress as depicted in Fig. 23.1A and B.

Besides the macroscopic stress build up, heteroepitaxy (i.e., the growth of a material B on a different material A) leads intrinsically to epitaxial stress as soon as crystal symmetry, lattice constant, and/or chemical bonds do not match perfectly. At the beginning of the growth, the film is forced to adopt the substrate lattice constants (i.e., pseudomorphic growth), which is connected with the formation of stresses in the film (Fig. 23.1, II). Assuming enough symmetry, the misfit strain ε_{mf} can be written as:

$$\varepsilon_{xx} = \varepsilon_{yy} = \frac{a_s - a_f}{a_f} = \varepsilon_{mf} \tag{23.2}$$

The epitaxial strain results in a misfit stress, σ_{mf}, in the film. For single crystal films the anisotropic elasticity is adapted to calculate them as fit stress. If we take the example of a cubic crystal overlayer structure with its {001} planes parallel to the {001} planes of a cubic substrate we can write Hook's law as follows:

$$\begin{bmatrix} \sigma_{xx} \\ \sigma_{yy} \\ \sigma_{zz} \end{bmatrix} = \begin{bmatrix} C_{11} & C_{12} & C_{12} \\ C_{12} & C_{11} & C_{12} \\ C_{12} & C_{12} & C_{11} \end{bmatrix} \begin{bmatrix} \varepsilon_{xx} \\ \varepsilon_{yy} \\ \varepsilon_{zz} \end{bmatrix} \quad \begin{matrix} \overline{e}_x \| [100] \\ \overline{e}_y \| [010] \\ \overline{e}_z \| [001] \end{matrix} \tag{23.3}$$

Since the film experiences in plane stress we have $\sigma_z = 0$. The misfit stress in the film plane parallel to the substrate is thus given by:

$$\sigma_{mf} = \left[C_{11} + C_{12} - \frac{2C_{12}^2}{C_{11}}\right]\varepsilon_{mf}, \text{ and } \varepsilon_{z.z} = -\frac{2C_{12}}{C_{11}}\varepsilon_{mf}. \quad (23.4)$$

For films with $\{111\}$ planes parallel to the substrate we would have:

$$\sigma_{mf} = \frac{6C_{44}(C_{11} + 2C_{12})}{C_{11} + 2C_{12} + 4C_{44}}\varepsilon_{mf}, \text{ and } \varepsilon_{z.z} = -\frac{2C_{11} + 4C_{12} - 4C_{44}}{C_{11} + 2C_{12} + 4C_{44}}\varepsilon_{mf}. \quad (23.5)$$

The driving force for film relaxation increases with strain and film thickness. After a certain critical thickness, t_C, a further increase of the elastic strain energy is prevented by the formation of edge dislocations (i.e., misfit dislocations) (Fig. 23.1E). When the film thickness increases, it becomes energetically favorable for misfit dislocation at the interface between film and substrate to reduce the stress in the film. The critical thickness can be derived from the minimization of the total strain energy that builds up with thickness. This concept is of more important impact however for the growth of semiconductors than for oxide layers that are the major part of the ferroelectric layers involved in multiferroics.

Usual growth conditions of multiferroic oxide layers involve relatively high elaboration temperatures that result in additional thermal stress due to differences in linear thermal expansion coefficients between the adsorbate and substrate. Additionally, in the case of 3D growth we should note that coalescence of islands tends to cause tensile stress which then gradually reduces with increasing film thickness. The growth method has also an impact to consider on the strain state of the film since films grown at low particle energy have tensile stress while under high energy particle impact films with compressive stress are formed. Thus for a given material, especially for ferroelectric ones, the growth method may lead to serious differences in ferroelectric response.

23.2.2 Thin-film configuration

From a microscopic point of view it is generally accepted that there are three possible modes of crystal growth on surfaces. In the island, or Volmer–Weber mode, small clusters are nucleated directly on the substrate surface before growing into islands of the condensed phase. This happens when the deposited atoms or molecules are more strongly bound to each other than to the substrate. This mode is very likely to occur when growing metals on insulators or for any situation where the adsorbate surface energy is fairly larger than the substrate's one. The layer by layer, or Frank–van der Merwe mode, consists of growing each layer fully before starting the next one. In this mode where the atoms are more strongly bound to the substrate than to each other, the atoms condense in full layers. This growth mode is observed for adsorbed gases, in some metal-metal systems, and in semiconductor growth on semiconductors. The layer plus island, or Stranski–Krastanov, growth mode is an intermediate case; after forming a finite number of full monolayers, subsequent layer growth takes the form of island growth. Simple thermodynamic considerations allow describing the shape of the islands formed in terms of surface and interfacial free energies. The equilibrium form of crystals of type A condensed on a surface of type B can be described in terms of the surface free energies, γ_A and γ_B, and the interfacial free energy γ^* between A and B. The equilibrium form can be calculated using the Wulff theorem [29]. In consequence, island (3D) growth or layer (2D) growth is favored as $\gamma_A + \gamma^*$ is larger or smaller than γ_B, respectively. The growth mode of nanoparticles can be modeled with high accuracy [30].

Many multiferroic systems crystallize in oxide lattices that can be seen as an oxygen atom skeleton in which smaller metal cations take place within tetrahedral and/or octahedral sites; for example, a perovskite structure. On the one hand, the lattice parameter in oxides is mainly governed by the big oxygen atom skeleton resulting in most oxides exhibiting close lattice parameters within a given crystalline structure. On the other hand, because of the presence of ionic species, the surface energy in oxides depends fairly more on the polarity of the surface than on any other considerations: a low surface energy crystalline orientation will promote quite easily the same low surface energy surface for the growing layers. These observations are very encouraging in the perspective of epitaxial thin-film growth of oxide layers deposited on other oxides. Although growth engineering is a very time-consuming task requiring many trial-and-error samples, it is usually possible to find parameter windows for a given epitaxial systems that enables the growth in the form that is expected. Depending on the application this can be either 2D or 3D. Because of this intrinsically favorable growth situation we may build composite multiferroic films with a large variety of structures and morphologies in addition to the case of a simple single-phase layer; some typical examples are illustrated in Fig. 23.2.

23.2.3 Strain engineering in multiferroic thin films

It is worthy to note that the macroscopic bending of a thin film-substrate compound depends mostly on the intrinsic mechanical parameters of the involved materials.

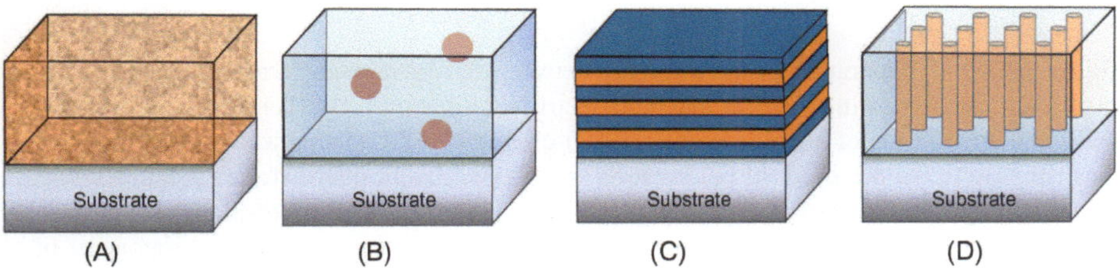

Figure 23.2 Schematic illustration of four types of multiferroic composite films: (A) single-phase layer, (B) particulate composite, (C) laminate composite, and (D) fiber/rod composite films.

Dealing with ferroelectric and piezoelectric layers raises specific questions here since the epitaxial growth of a thin layer intrinsically applies a residual stress that will heavily influence the ferroelectric behavior of the films. Which means that in the case of thin-film multiferroics, a third type of ordering should always be considered as derived from the epitaxial stress state of the film, which leads to ferroelasticity. The deformations experienced by the multiferroic layer, as a result of the growth stress, is likely coupled to the electric polarization, P, and the magnetization, M, via piezoelectric and magnetostrictive coefficients respectively as depicted in Fig. 23.3A. In studies that concentrated specifically on the magnetoelectric effect the reverse argumentation has sometimes been used; that is, an indirect magnetoelectric coupling transmitted through strain: electric polarization leading to local deformation that via ferroelastic coupling leads to a structural stress resulting in magnetization modification via magnetostriction and vice versa. Growth stress is inherent to thin-film elaboration as described earlier, leading necessarily to the coexistence of ferroelasticity with other ferroic orders. Direct and indirect coupling mechanisms coexist, and disentangling their relative weight in the occurrence of multiferroic behaviors, can be a very complex and tricky task.

The stress mechanisms occurring during epitaxial growth are in fact quite favorable if the grown layer is ferroelectric. Ferroelectric materials demonstrate a spontaneous nonzero polarization even without an applied electric field below a well-defined Curie temperature T_C above which they are paraelectric. Their spontaneous polarization can be reversed by a suitably strong applied electric field in the opposite direction within hysteresis loop behavior. The internal electric dipoles are

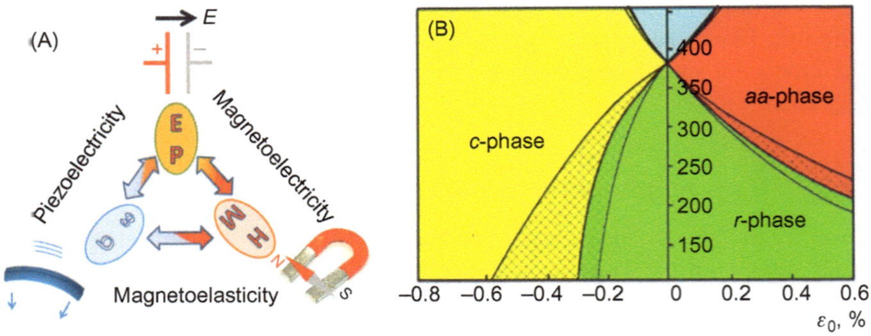

Figure 23.3 (A) Illustration of the interplay between ferromagnetic, ferroelectric, and ferroelastic properties in thin-film magnetoelectric multiferroics. (B) Temperature–misfit strain phase diagrams of a single-domain (001)-oriented $BaTiO_3$ using a free-energy expansion up to the 6th order in polarization. ε_0, biaxial parent misfit strain; T, temperature. The phases are denoted as *(1)* the *c*-phase (yellow), polarization is out-of-plane; *(2)* the *aa*-phase (red), where polarization is in-plane; *(3)* *r*-phase (green), where all components of polarization are nonzero; and *(4)* paraelectric phase (blue), polarization is 0.
Reprinted figure with permission from A. Kvasov, A. Tagantsev, Role of high-order electromechanical coupling terms in thermodynamics of ferroelectric thin films, Phys. Rev. B 87 (2013) 184101, Copyright © (2013) by the American Physical Society.

coupled to the material lattice; changes to the lattice will in turn change the strength of the dipoles (i.e., the spontaneous polarization) that will accordingly lead to a change in the surface charge. Whether the ferroelectric phase transition is displacive (such as for $BaTiO_3$), order-disorder (such as for $NaNO_2$), or a mixture of both, a lattice stress is a great stimuli allowing tuning the ferroelectric state. Indeed for most conventional ferroelectrics the electric polarization state is driven by the chemical bonding, which results in an off-center position of the transition metal ions in the lattice that requires by the way noncentrosymmetric space groups. As we described previously, epitaxial stress results in an anisotropic out-of-surface plane relaxation linked to the in-plane lattice constants misfit strain through the elasticity constants of the growing film. In other words, growing ferroelectric layers on a substrate with slightly different lattice parameters may lead to a more tetragonal lattice that, if oriented adequately with respect to electric polarization, can be very favorable and even greatly improve its ferroelectric properties.

This has nicely been evidenced from theoretical and experimental approaches in the case of the archetypical $BaTiO_3$ layers. Choi et al. [32] demonstrated that by growing $BaTiO_3$ layers on substrates imposing increasing lattice misfit stresses, the ferroelectric−paraelectric transition Curie temperature T_C can be increased by up to 500°C when the growth is realized on $DyScO_3$ substrates. This compares well with the T_C enhancement predicted from thermodynamic analysis for a $BaTiO_3$ thin film under biaxial strain as derived from the Landau thermodynamic theories. Choi et al. stressed that since the driving force for film relaxation increases with strain and film thickness and since the critical thickness at which dislocations begin to form varies approximately inversely with lattice mismatch [33], the possible T_C improvement will depend on the expected film thickness; lower lattice mismatch situations will favor larger possible thicknesses.

The strain appearing in the film not only shifts T_C but can also change the properties of the material or even can induce room temperature ferroelectricity in a non-ferroelectric material [34]. The basic theoretical framework used for explaining and predicting equilibrium states and phase transitions in bulk materials is given by the Landau−Ginsburg−Devonshire thermodynamic theory. A modified thermodynamic potential G derived by an appropriate transformation has been introduced to take into account the mechanical interactions occurring at the interface between the ferroelectric layer and the substrate. The minimization of this potential yields the equilibrium state of the film as a function of the temperature and misfit strain. In the case of $BaTiO_3$ the potential had to be extended up to sixth order in polarization to lead to reliable predictions and as a matter of fact to a more complex strain-temperature scheme than previously considered, keeping the main features [31] (Fig. 23.3B).

Epitaxial growth of ferroelectric layers, within the framework of multiferroic materials, opens thus the possibility of using strain engineering as a strategy to explore the special behavior of ferroelectric and multiferroic materials that may be built this way. Superior multiferroic properties, well above the bulk characteristics, can reasonably be expected when considering thin films. However, the close interplay between ferroelasticity, ferro- and/or piezoelectricity, and magnetostriction may quite easily mingle the interpretation of the actual mechanisms leading to multiferroic couplings. In summary, thin-film growth elaboration brings to multiferroics nearly by essence the required additional degree of freedom to allow tuning their properties to the goal defined by device needs.

23.3 Pertinence of multiferroic thin films: Multiferroics versus thin films

In this part we will consider the implication of multiferroicity with respect to thin-film growth. We will examine to what extent multiferroic properties are compatible or not with thin-film materials or to what extent thin-film elaboration may prevent or not the use of some aspects of multiferroics.

23.3.1 Thin films and spintronics application of multiferroics

For multiferroic composite systems, the presence of a strong magnetoelectric coupling enables the development of new electronic devices where the spin of the electron is controlled by electric (instead of magnetic) fields; conversely, magnetic fields can also be transduced directly to electromotive forces [35–38]. Such functionalities are appealing to spintronic devices in particular, where both charge and spin degrees of freedom are used to carry information [39–44]. Beyond the simple manipulation of the magnetization by a magnetic field, several new approaches appeared based on the use of an electrical field including the evidence of domain wall motion in ferromagnetic layers [45,46], high enough electric fields able to modify the coercive field or magnetization of nanostructures [47,48] and the magnetoresistance change in tunnel junctions using a ferroelectric barrier as a function of the electric polarization [49,50]. Such devices may also overcome the impending scaling limits in CMOS technology [51]. More generally, there is also a pressing need for smarter and more energy-efficient devices suitable for a broad range of applications ranging from electronics, to sensors and transducers, to medicine, and to energy harvesting and storage. Hence, besides the large body of academic works, the expected high added-value devices that may be designed using multiferroics has motivated an intense industrial research effort and an ever-increasing number of patents are taken [52]. These applications are expected to open a new high added-value industrial era and should thus drain large-scale societal advances.

Magnetoelectric multiferroics are expected to bring new functionalities to spintronic devices making them intrinsically multifunctional and allowing new genuine device applications. The magnetoelectric effect is defined as an induced electric polarization (P) of a material with an applied magnetic field (H). The converse effect is an induced magnetization (M) with an external electric field (E). This means that the electric polarization/magnetization can be modified by the application of a magnetic or electric field on the materials via the magnetoelectric coupling. The induction of polarization by a magnetic field or magnetization by an electric field is known as the linear magnetoelectric effect. It can be shown that the square of the magnetoelectric response coefficient is limited by the product of the electric and magnetic susceptibilities. This means that the magnetoelectric effect can only be large in ferroelectric and/or ferromagnetic materials [53].

The goal of spintronic devices is to exploit, using a spin current, the spin as well as the charge of the electrons that pass through them while conventional electronics ignored simply the electron spin state. Exploiting spintronics properties requires, however, confinement in at least one space direction, which derives directly from quantum mechanical concepts and more generally because length and time scales determine the physics of a system. Quantum mechanical basic principles imply that confinement of electrons by lowering dimensions affects the electronic states that may span from discrete energy levels for quantum dots to a continuum for bulk materials (Fig. 23.4). The electronic state dictates the transport properties of the resulting material and its macroscopic properties.

Tuning the material configuration to lower dimensions like rods or dots not only produces a closer interplay between ferroic phases for multiferroics (Fig. 23.2B–D), but also implies electronic state quantification modifications that by themselves may modify the resulting composite film properties. These effects will come in addition to the growth stress effects discussed in Section 23.2. Finite size effects will start to affect the material properties when at least one of its dimensions becomes close to the length scale of the physical property of interest. This size may not necessarily be very small; for example, optical mirror effects depending on the light wave length occur for multilayer film thicknesses in the range of

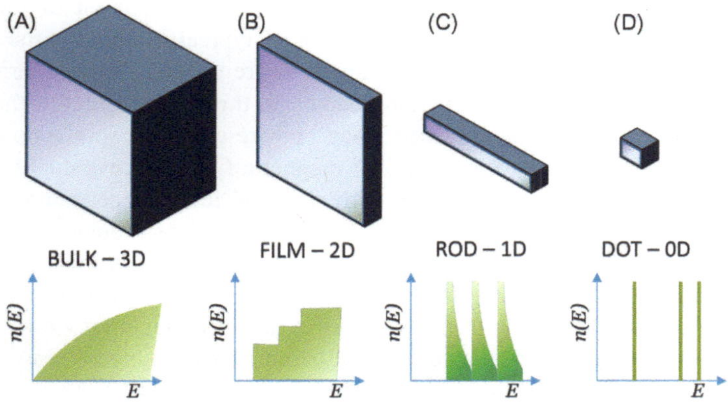

$n(E) = a\sqrt{E}$	$= \sum a_{am}\theta(\varepsilon - \varepsilon_m)$	$= \sum a_{m,n} \frac{1}{\sqrt{\varepsilon - \varepsilon_m - \varepsilon_n}}$	$= \sum_{m,n,l} \delta(\varepsilon - \varepsilon_{m,n,l})$

Figure 23.4 Top panel: Extended systems for 3 (A), 2 (B), 1 (C), and 0 (D) dimensions. Middle panel: Illustration of the density of states $n(E)$ for each material configuration. Bottom panel: Expressions for the corresponding density of states. The summation runs over the discrete eigenvalue spectra.

hundreds of nanometers. To the contrary, the typical thickness range of ferroelectric layers used in ferroelectric random access memory cell (FRAM) elements is one order of magnitude smaller and has reliable functioning conditions within this thickness range because finite size effects are already negligible for this physical mechanism. Such capacitor memory cells store information as +P or −P polarization orientation in a ferroelectric layer and have two distinct states, which can be read by an applied voltage. They combine fast access time and nonvolatility. Unfortunately, they are limited by the need for a destructive read and reset operation and elaborated mainly using lead zirconate titanate thus incorporating toxic lead.

Magnetic interactions are based on the exchange interaction, which acts in the nanometer range before decaying exponentially. As a consequence, typical magnetoresistance-based devices [10] are built using magnetic and spacer layers in a very few 10 nm range. The most popular resulting sensors are based on a three-layer sandwich configuration called spin valve [54], where two ferromagnetic layers, a hard reference layer and a soft sensing layer, are separated by a nonmagnetic layer that may be metallic for the thicker less sensitive devices or insulating for the thinner more sensitive ones. Insulating spacers require smaller thickness because the physical phenomenon is based on electron tunneling that requires overlapping wave functions, reducing furthermore the required thicknesses to several atomic spacing. The spin valve resistance state directly depends on the relative magnetization orientation of the two ferromagnetic layers, an effect that can be exploited in magnetic field sensors and magnetic random access memory cells (MRAM) that are based on an array of spin-valves. MRAM present the great advantage of nondestructive magnetic read operation. Because of the limited mean free path of electrons and tunneling distance, spintronics devices consist of thin-film devices with at least one reduced dimension. Another typical example is the lateral spin transistor [55] in which spin-polarized electrons are injected (from a ferromagnetic source) into a narrow semiconductor channel. In this semiconductor, the electron spins can only move in the confined plane or wire and their spin state can be switched between up and down by an applied magnetic field, and eventually by the gate voltage, determining the output spin current in the ferromagnetic drain material. The output voltage can be modulated so as to make the spin device work with a spin degree of freedom. Describing extensively all possible spintronics devices is beyond the scope of this chapter. The interested reader may find more examples and a larger discussion about spintronics devices in dedicated recent reviews [56].

In view of the previously discussed concepts, multiferroic films within the field of spintronics will necessarily have thicknesses limited to several atomic planes. Anyhow, multiferroic composite materials are expected to be of great importance in the drive toward creating faster, smaller, and more energy-efficient devices and in particular for information and communications technologies. Two distinct concepts have been identified as potential technological breakthroughs: (1) switching of the ferroelectric state with a magnetic field and vice versa (Fig. 23.5A), which leads to the ability to manipulate and control magnetic properties by the application of an electric field, for information storage and processing; such a technology would remove the need to apply electrical currents that create heat dissipation and stray

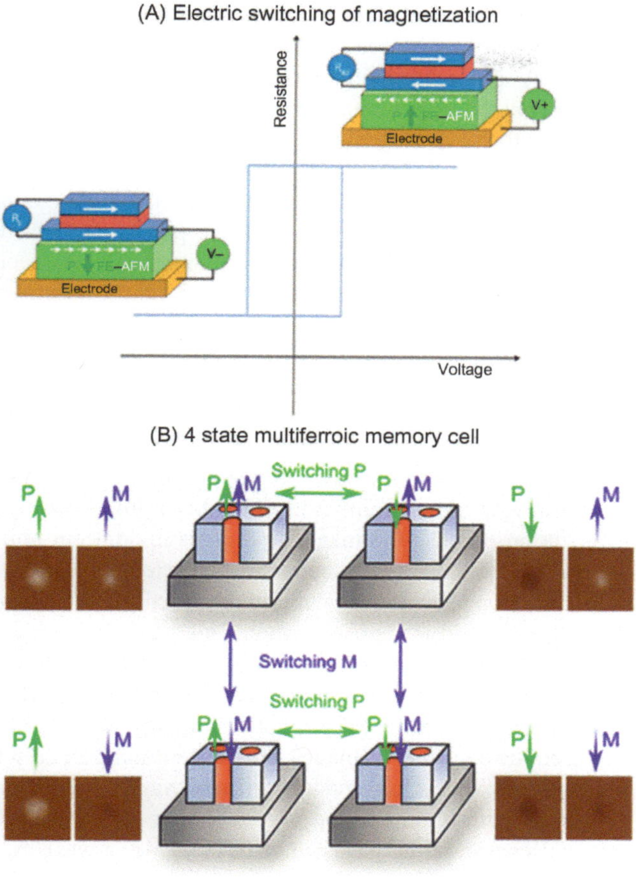

Figure 23.5 (A) Illustration of the electric switching of the magnetization using multiferroic layers. A tunnel spin valve (blue-red layers) is grown on a multiferroic layer (green layer). The binary information is stored by the magnetization direction of the bottom ferromagnetic layer in contact with the ferroelectric layer. If the magnetization of the bottom ferromagnetic layer is coupled to the spins in the multiferroic (small white arrows) and if the magnetoelectric coupling is strong enough, reversing the ferroelectric polarization P in the multiferroic changes the magnetic configuration in the trilayer and in consequence the resistance state of the top spin valve. (B) Illustration of a four-state multiferroic memory cell. Practical realization obtained using Mn-doped $BaTiO_3$ nanorods elaborated by dip-pen nanolithography grown on a Nb-doped $SrTiO_3$ substrate. PFM and MFM images are used to determine the polarization and magnetization directions, respectively. Independent switching is obtained because of negligible magnetoelectric coupling.
(A) Reprinted by permission from Macmillan Publishers Ltd: M. Bibes, A. Barthélémy, Nat. Mater. 7 (2008) 425–426, Copyright © (2008); (B) Reprinted with permission from (Jong Yeog Son et al., ACS Nano 7(2013) 5522–5529 [75]), Copyright (2013) American Chemical Society.

magnetic fields; and (2) four-state memory devices combining the best qualities of FRAMs and MRAMs: fast low-power (without current) electrical write operation, and nondestructive magnetic read operation (Fig. 23.5B). A four (or even higher)-state logic would permit an exponentially increased computing capacity without increasing significantly the necessary materials.

The cross-coupling between polarization and magnetization vectors could be evidenced at cryogenic temperatures for manganite multiferroic compounds $HoMnO_3$ [59] and $TbMnO_3$ [12] as a proof of principle of the potential effect. However for device application, strong cross-coupling at room temperature has been considered to be necessary. A first remarkable step in the quest of the electrical control at room temperature of magnetization has been the observation of the multiferroic and magnetoelectric properties of single crystals of bismuth ferrite ($BiFeO_3$) that contain just one electrical and magnetic domain [60]. $BiFeO_3$ is a ferroelectric canted antiferromagnet and is usually also considered as multiferroic. It is the most widely considered compound because both polarization and magnetic orders coexist at room temperature especially because most systems (such as the manganites) exhibit multiferroicity only at low temperatures [13,61]. This ferroelectric compound has no overall magnetization because the magnetic dipoles in adjacent unit cells cancel out. Below $T_C \sim 1100$K, bulk $BiFeO_3$ becomes ferroelectric with a polarization value of $\sim 100°C/cm^2$ along the $[111]_C$ cubic-like directions. The structure is best described by the rhombohedral R3c space group. This one allows antiphase octahedral tilts and ionic displacements from the centrosymmetric positions along the same polar $[111]_C$ direction. Below the Néel temperature $T_N \sim 640$K bulk $BiFeO_3$ exhibits a G-type antiferromagnetic order: the Fe magnetic moments are coupled ferromagnetically within the pseudocubic $\{111\}_C$ planes and antiferromagnetically between first neighbors. A long-range spin cycloid modulation resulting in a rotation of the spin axis through the crystal with a long period of ~ 62 nm is superimposed on the antiferromagnetic order. Because of the presence of this cycloid, a weak magnetoelectric coupling arises from a quadratic term. Destroying the cycloid using high magnetic fields leads to a small canting of the spin moments through the oxygen octahedra tilts and gives rise to a weak magnetization producing a linear magnetoelectric coupling. At low temperature two reorientation transitions were reported at 200K and 140K, respectively [62]. The magnetoelectric properties of $BiFeO_3$ were reviewed recently [63], with emphasis to spintronics [64], strain effects [65], and pressure–temperature phase diagram and polarity [66]. In its thin-film form it is a typical single-phase multiferroic system corresponding to the situation depicted in Fig. 23.2A.

Lebeugle et al. [60] have shown that electrically induced magnetic switching also occurs in bulk $BiFeO_3$. They reported the electrically induced conversion of a single antiferromagnetic domain into a multidomain state. This evidence lets expect the possible voltage driven (low-power) magnetic writing on a MRAM spin valve cell that could be grown on top of $BiFeO_3$ as depicted in Fig. 23.5A. The process could be described as follows: an electric field would magnetoelectrically switch the antiferromagnetic order of a $BiFeO_3$ layer, which would, in turn, switch the magnetization of an overlying thin ferromagnetic film grown on top of it, via magnetic exchange coupling that occurs between ferromagnetic and antiferromagnetic layers [67]. Even when considering antiferromagnetic single crystal substrates [68], the effect can be tuned using a magnetic field cooling through the Néel temperature, likely due to different surface and bulk ordering temperatures [69]. A first important step was the report of a 1:1 correlation between antiferromagnetic and ferroelectric domain structure of a 600 nm thick $BiFeO_3$ film using PFM to evidence the ferroelectric domain structure and linear X-ray dichroism XPEEM [70] imaging to evidence the antiferromagnetic one [71] at room temperature. The described electric control of magnetization inferred from the single crystal studies received further confirmation in a study where the magnetic domains in an exchange-biased ferromagnetic film of $Co_{0.9}Fe_{0.1}$ were controlled by the application of a lateral electric field to an underlying layer of epitaxial $BiFeO_3$ [72]. The authors could evidence a perfect correlation of the ferroelectric domains as written and evidenced by PFM measurements and the overlayer ferromagnetic domain structure as evidenced by XMCD-XPEEM measurements realized at the Co-L_3 absorption edge. After demonstrating the possibility to manipulate the ferroelectric domain structure (Fig. 23.6C–E) by an applied electric field, within a device specially prepared to

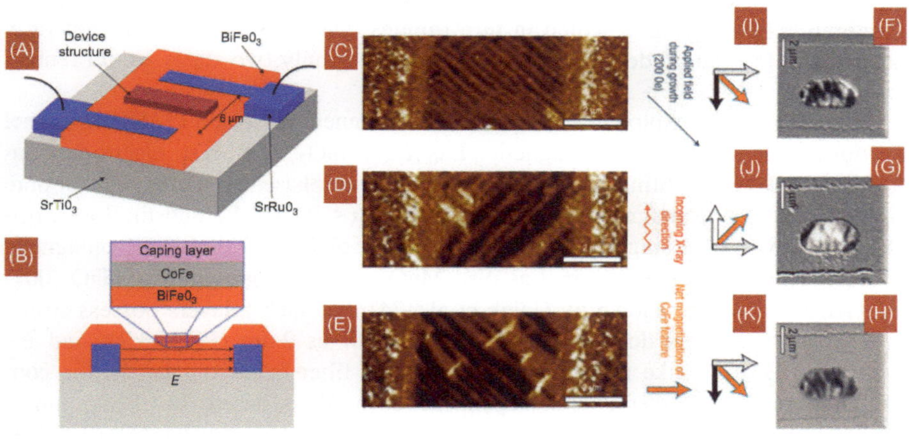

Figure 23.6 Three-dimensional (A) and cross-sectional (B) schematic diagrams of the coplanar epitaxial electrode device enabling controlled ferroelectric switching. C–E. In-plane PFM images showing the ferroelectric domain structure for a device in the as-grown state (C), after the first electrical switch (D), and after the second electrical switch (E). F–H. XMCD–PEEM images taken at the Co L-edge revealing the ferromagnetic domain structure of the CoFe features in such a coplanar electrode device structure in the as-grown state (F), after the first electrical switch (G), and after the second electrical switch (H). I–K. Schematic descriptions of the observed magnetic contrast (gray, black, and white) in the corresponding XMCD–PEEM images, respectively. Application of an electric field is found to rotate the net magnetization of the structures by 90 degrees. The direction of the applied growth field and the incoming X-ray direction are labeled as well.
Reprinted by permission from Macmillan Publishers Ltd: Y.-H. Chu, L.W. Martin, M.B. Holcomb, M. Gajek, S.-J. Han, Q. He et al., Electric-field control of local ferromagnetism using a magnetoelectric multiferroic, Nat. Mater. 7 (2008) 478–482, Copyright © (2008).

do so (Fig. 23.6A and B), they observed the magnetization switching resulting from this manipulation using XMCD-XPEEM taking advantage of the synchrotron radiation circular polarization sensitivity to the orientation of the magnetization in the overlayer (Fig. 23.6F–K).

Because of the epitaxial stress and of the ferroelastic nature of $BiFeO_3$ it is however difficult to determine, as pointed out in Section 23.2, if the magnetic domain switching has been due solely to the magnetoelectric coupling in the $BiFeO_3$ and exchange coupling between the two materials or if the effect is also due to, or assisted by, strain. Specifically for $BiFeO_3$ thin films it has been demonstrated that substrate vicinality allows tuning the domain configuration of the overlayer and that the ferroelectric nature of the film remains stable down to thicknesses as small as 2 nm [73].

Concerning the second exciting topic, namely the setup of four state memory cells, considerable progress has also been made in recent years. In order to make a multiple-state magnetoelectric memory, we must be able to access the four combined states formed by electric polarization $\pm P$ and magnetization $\pm M$ (see Fig. 23.5B). This is not just a matter of symmetry, but also of multiferroic coupling. Switch from (+P, + M) to (+P,−M) by applying an electric field requires a significant P–M coupling, which in turn implies that the four states are not fully independent [58] and indeed for a too high coupling only two states would remain accessible.

Gajek et al. [74] exploited the large tunnel magnetoresistance (TMR) in tunnel junctions that have a ferromagnetic $La_{0.1}Bi_{0.9}MnO_3$ (LBMO) electrode. La-substituted $BiMnO_3$ films stabilize the perovskite phase as compared to pure $BiMnO_3$ and is ferromagnetic at low temperature (<100K). La substitution facilitates the growth of a single-phase perovskite film of LBMO that is ferromagnetic below 95K when deposited on $La_{2/3}Sr_{1/3}MnO_3$ (LSMO) grown on $SrTiO_3(001)$ [64]. Also it could be shown by Gajek et al. [74] that such bilayers possess strong ferroelectric behaviors down to thicknesses as low as 2 nm (Fig. 23.7A and B). Moreover LBMO is like tunnel barriers with spin-filter behavior, that is, the conduction through the layer is spin dependent. Spin filters based on LBMO tunnel barriers with LSMO and Au electrodes have been fabricated and show TMR values up to 90% corresponding to spin-filtering efficiencies of about 35%. In addition the ferroelectric character of the tunnel barrier produces an electroresistance effect (TER) influenced by the electric polarization P. The combination of these two effects—magnetoresistance (due to the different magnetization directions in both films) plus electroresistance (which corresponds to a modulation of the barrier height)—yields a four-state resistive memory element (Fig. 23.7B and C) with specific resistance values interpreted as related to the relative orientation of the polarization and magnetization in the tunnel barrier. Hence, the resistance state of the junctions is encoded by the two order parameters, M and P, in the tunnel barrier.

This example sets a proof of principle for the use of multiferroics showing strong M-P coupling in four-state memory cells. Additional ferroic orders could increase the number of possible states of a single memory cell. But transposition to room temperature requires the discovery of a room-temperature ferroelectric and

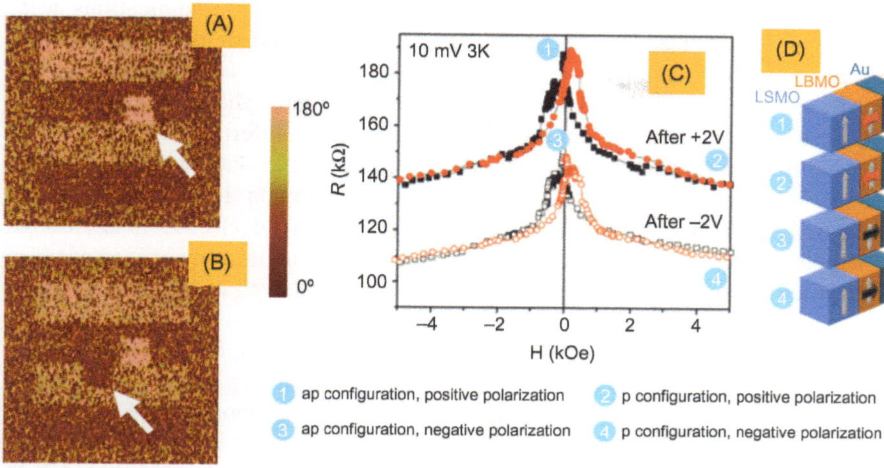

Figure 23.7 Thin film $La_{0.1}Bi_{0.9}MnO_3$ (LSMO) properties. PFM phase images of a 2-nm LBMO film grown on an LSMO electrode, after writing first four voltage stripes and then a first (A) and a second (B) 1 μm^2 squares at opposite voltage. (C) TMR curves of LSMO/LBMO (2 nm)/Au tunnel junction measured at 10 mV after applying either a positive (2 V) or negative (−2 V) voltage across the junction. (D) Sketches of the magnetic (white arrows) and electric (black and red arrows) configurations proposed to correspond to the four resistance states.

Reprinted by permission from Macmillan Publishers Ltd: M. Gajek, M. Bibes, S. Fusil, K. Bouzehouane, J. Fontcuberta, A. Barthélémy, et al., Tunnel junctions with multiferroic barriers, Nat. Mater. 6 (2007) 296, Copyright© (2007).

ferromagnetic multiferroic material or compound, which is a very difficult task as we will see in Section 23.3.

More recently Jong Yeog Son et al. [75] showed the possible four states encoding using vertically aligned 2 at% Mn-doped $BaTiO_3$ nanorods elaborated by dip-pen lithography on a Nb-doped $SrTiO_3$ substrate realized using a precursor solution filled into a removable nanopore template (30 nm rod diameters on an array with a 100 nm step size). Such a system is close to the material configuration depicted in Fig. 23.2D. Although antiferromagnetic long-range order should be marginally more stable than ferromagnetic ordering, the nanorods were found slightly ferromagnetic with negligible coupling to the electric polarization. Consequently independent polarization, P, and magnetization, M, could be realized in the vertical nanorods in which P and M were found always parallel to the *c*-axis direction, defining a four-state system, the single-domain magnetic state allowed easy observations in MFM (Fig. 23.5B). The practical manipulation of the states of such a system with negligible P-M coupling remains quite questionable however.

Magnetoelectric films may find new applications depending on our ability to master the growth conditions of the films; poor epitaxial growth often degraded their properties compared to bulk materials. For example, tunable devices for magnetic information storage, processing, and microwave communications were proposed provided the ferromagnetic film has large magnetostriction like in Galfenol ($Fe_{81}Ga_{19}$) and narrow microwave resonance line widths [76]. Another example is the recent prediction based on model calculation that a magnetoelectric tunneling electroresistance effect could be expected in multiferroic tunnel junctions consisting of ferromagnetic electrodes and magnetoelectric antiferromagnetic barriers. The resistance of the junction is predicted to depend on the antiferromagnetic order parameter in the barrier that may vary in an applied electric field through magnetoelectric coupling. Single-phase magnetoelectric antiferromagnet barriers in tunnel junction may be considered within the realization of nonvolatile memory cells [77].

For the particular application of multiferroic materials to spintronic devices, the limited length scales involved in the mechanisms underlying magnetic exchange imposes thin-film structures limited to a small few tens of nanometers. For application of multiferroics to other fields or for nonmagnetoelectric coupling this condition can be fairly more relaxed. Magnetoelectric multiferroic likely corresponds to the lower thickness limit. However, the successful implementation in actual devices will require additional challenging steps to be solved including defining reliable strategies to manipulate the ferroelectric state, even at a macroscopic scale, and the implementation into semiconducting devices [78].

23.3.2 Availability and classification

23.3.2.1 Availability

Multiferroic single-phase materials, exhibiting at least one of the four primary ferroic orders (ferroelectric, ferromagnetic, and ferrotoroidic vectors) are intrinsically rare because their crystal structures have to obey strict symmetry rules [5]. Spontaneous polarization, magnetization, and toroidization (i.e., spontaneous currents) are characterized by a vector and a single Heesch–Shubnikov point group while spontaneous strain is described by a symmetric second rank tensor. The point groups compatible with each ferroic order have to correspond to a particular symmetry operation depending on the order parameter.

- Spatial inversion symmetry breaking is a necessary condition for ferroelectricity. This is obvious if we remember that an electrical dipole results from a splitting in the barycenter of positive and negative charges in the unit cell. Therefore, if we apply a spatial inversion operator to the cell, positive and negative barycenters will change places and the electrical polarization reverses. Thus only noncentrosymmetric point groups can exhibit ferroelectricity. It can be shown that a total of 31 point groups permit a spontaneous polarization (invariant under time reversal).
- Magnetic moments are well described as pseudovectors corresponding to a charge in a closed orbit, which is immune to changes produced by a spatial inversion operation. To the contrary, time reversal would make the charge move in the opposite direction and hence switch the magnetic moment. Therefore when time reversal symmetry is preserved, the total magnetic moment is unchanged and when it is broken switchable magnetic

ordering can occur and defines the compatible point groups. A total of 31 groups permit a spontaneous magnetization (invariant under space inversion).

- The ferrotoroidic current density is characterized by a time-odd polar vector, invariant under the product of space and time reversal. It is also called magnetoelectric vector, and indeed crystals with a spontaneous toroidal moment are necessarily associated with the linear magnetoelectric effect. Again a total of 31 point groups permit spontaneous currents (invariant under the product of time and space reversal).
- To allow ferroelasticity to occur simultaneously in addition to the three vectorial primary ferroics, a change of crystal class has to take place.

From the symmetry analysis it appears that only 13 point groups break both spatial inversion and time reversal symmetries and allow both ferroelectric and ferromagnetic properties in the same phase (intersection of the two sets of groups). This number reduces to 9 groups able to allow all three types of vectors (intersection of all 3 sets of groups) [5]. This still represents a large amount of possible materials and compounds, nickel iodine boracite, $Ni_3B_7O_{13}I$, having been the first material, in which (weak) ferromagnetism and ferroelectricity were shown to occur in the same phase already in the 1960s.

Although many studies concentrate on magnetoelectric multiferroics because of their huge technological potential, this configuration suffers additional limitations dictated by chemical bonding considerations. In most conventional ferroelectrics including transition metals, the chemical bonding, which results in an off-center position of the transition metal ions creating electric polarization, requires a formally empty *d*-electron configuration while, to the contrary, ferromagnetism requires unpaired *d*-electrons. Correlatively, for perovskites with ABO_3 formulae, partially filled *d*-shells on the B-sites reduce the tendency of perovskites to display ferroelectricity. Then, the coexistence of the two properties seems incompatible from the local chemistry point of view. Therefore, more sophisticated ferroelectric driving mechanisms must be present for electric and magnetic ferroic order to occur simultaneously. For perovskites lone-pair hybridization may overcome this problem for a few compounds like $BiFeO_3$, $BiMnO_3$, and $PbVO_3$ where the A-site cation drives the displacement via hybridization of a lone-pair 6*s* electron with a neighboring O 2*p* orbital and partially filled *d*-shell on the B-site contributes to the magnetism.

Thin films will have to obey the same materials science rules, however with several major advantages that may overcome the difficulty of the very limited number of intrinsic multiferroics: (1) the growth results in out-of-equilibrium layer configurations that may stabilize phases not available in the bulk phase diagram, (2) a higher impact of ferroelasticity is gained from the growth stress, and (3) the possibility to engineer a close proximity of distinct ferroic phases embedded in a composite material that is called artificial multiferroics [6]. A noticeable example is the self-assembly of $CoFe_2O_4$ nanopillars in a $BaTiO_3$ layer, with an average spacing of 20−30 nm, for which temperature-dependent magnetic measurements evidenced the coupling between the two order parameters [79]. $CoFe_2O_4$ is a robust ferrimagnetic spinel, ideally inverse, ferrite with a high Curie temperature and a magnetization that depends on its spinel inversion fraction [80]. It is a very well-suited ferrimagnet to be combined with a robust ferroelectric because of sizeable ferroelastic cross-coupling enabling cross-correlations at room temperature. In recent experiments [81], fully epitaxial $CoFe_2O_4/BaTiO_3$ bilayers were grown on 1 at% doped $SrTiO_3$ (001) substrates by atomic oxygen assisted MBE, with variable $CoFe_2O_4$ overlayer thickness. The crystalline structure has been subsequently investigated by grazing incidence X-ray diffraction that is a very well-suited technique to investigate oxide layer structure. It allows great surface sensitivity and is insensitive to charge build-up [82]. Mapping the reciprocal space in regions corresponding to asymmetric Bragg reflections of all layers allows the determination of the in and out-of-plane lattice parameters. For a $BaTiO_3$ layer thickness of 7 nm, Fig. 23.9 shows reciprocal space maps for two $CoFe_2O_4$ layer thicknesses. The position shifts of the Bragg peaks show obvious changes on both the $BaTiO_3$ and $CoFe_2O_4$ lattice parameters. And indeed, reporting the lattice evolution for increasing $CoFe_2O_4$ bilayer thicknesses evidence a structural relaxation of both layers. Due to epitaxial stress, $CoFe_2O_4$ evolves from an artificial tetragonal structure to its cubic form while the $BaTiO_3$ layer sees its substrate clamping strength reduced and relaxes toward more bulk-like c/a ratios. This example evidences cross-correlated strain effects that can be used to strain engineer composite multiferroic thin-film systems, a route that makes great promise to overcome the scarcity of room temperature single-phase multiferroics. (Fig. 23.8).

Many of the popular multiferroics are transition metal oxides with perovskite ABO_3 crystal structure and we will discuss them in more detail. They have been subdivided into two classes as introduced by Khomskii [83].

23.3.2.2 Type I multiferroic perovskites

In type I systems, the magnetic and ferroelectric orders are established on different sublattices of the material. They are weakly coupled, but enough to allow the electric field to drive magnetization changes. They typically exhibit high ferroelectric and lower magnetic ordering temperatures. Typical examples are $BiFeO_3$, $YMnO_3$, and $BiMnO_3$. In the bulk form, $BiMnO_3$ exhibits a ferroelectric−paraelectric transition at 750K and ferromagnetic-to-paramagnetic transition at 105K. $BiFeO_3$ exhibits an antiferromagnetic long-range ordering with Néel temperature 653K and a ferroelectric one up to 1083K. $YMnO_3$ in bulk form exhibits ferroelectric as well as antiferromagnetic behavior. When deposited in thin-film form, due to the beneficial epitaxial strain, most of these single-phase multiferroic materials exhibit somewhat different material characteristics as compared to their bulk counterpart.

In a simplified view, type I compounds can be described as ferroelectrics that happen to be (antiferro) magnetic. Because the distortions that lead to ferroelectricity in perovskites seem to be inhibited, as soon as *d*-shells, necessary for magnetism, begin to fill, the antiferromagnetic long-range ordering is far more favorable. Besides single-phase multiferroics of type I, progress in composite heterostructures of multiferroics have been made allowing the realization of artificial systems in thin-film form.

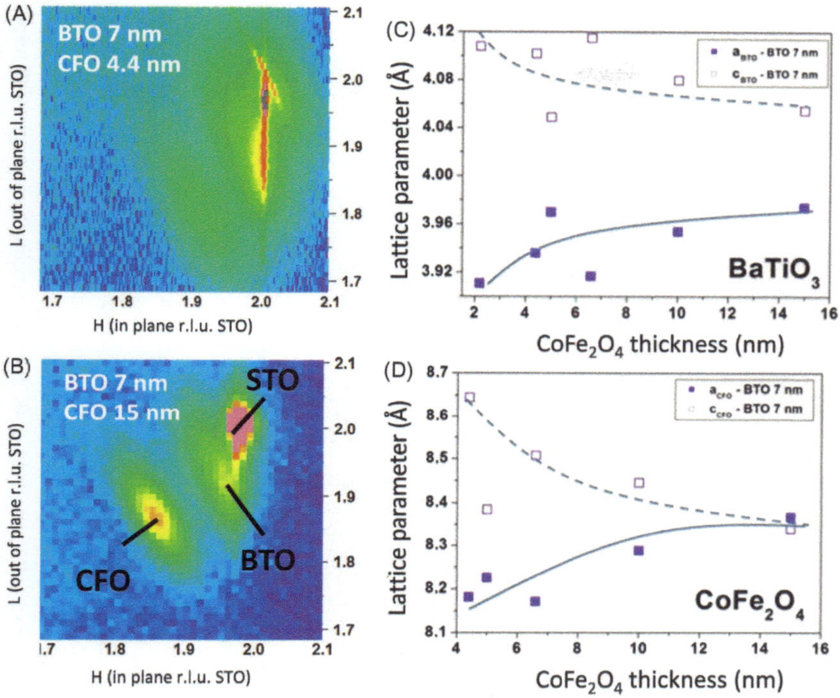

Figure 23.8 Structural relaxations of $CoFe_2O_4$ (CFO)/$BaTiO_3$ (BTO)/Nb 1 at% Nb-$SrTiO_3$(001) (STO) bilayers as a function of the CFO thickness. Grazing incidence reciprocal space maps around the 220 Bragg peak for 7 nm thick BTO layers covered by (A) 4.4 nm and (B) 15 nm of CFO. The units are given in the reciprocal lattice units (r.l.u.) of STO. Evolution of the in-plane (full square) and out-of-plane (open square) of the BTO (C) and CFO (D) lattice parameters with respect to the CFO overlayer thickness.

Doping a ferroelectric material is a possible route to achieve multiferroic compounds that will belong to this class of compounds. $BaTiO_3$ is an excellent candidate since it can be grown with high crystalline quality on $SrTiO_3$ substrates (that may be rendered conducting by Nb doping), although with some BaO surface excess when growing it with atomic oxygen assisted MBE [84]. It undergoes complex phase changes at low temperature but is mostly ferroelectric at room temperature, a property fairly enhanced for such thin films. Doping with 3*d* transition metal ions led to a number of divergent reports and mostly evidenced a dominant role of the preparation conditions and sometimes substrate magnetic contamination. Doped $BaTiO_3$ layers including Fe and $NiFe_2$ doping were thoroughly studied [85]. While pure Fe doping appeared detrimental to the structural quality and led to a loss in lattice anisotropy as compared to $NiFe_2$ doping, X-ray absorption spectroscopy on Fe and Ni L-edges evidenced predominant Fe^{2+} configuration at low doping levels before Fe^{3+}ions become prevalent (Fig. 23.9A and B). Importantly, no cation in its metallic state could be detected. The doping scheme could be understood in terms of preferential substitution

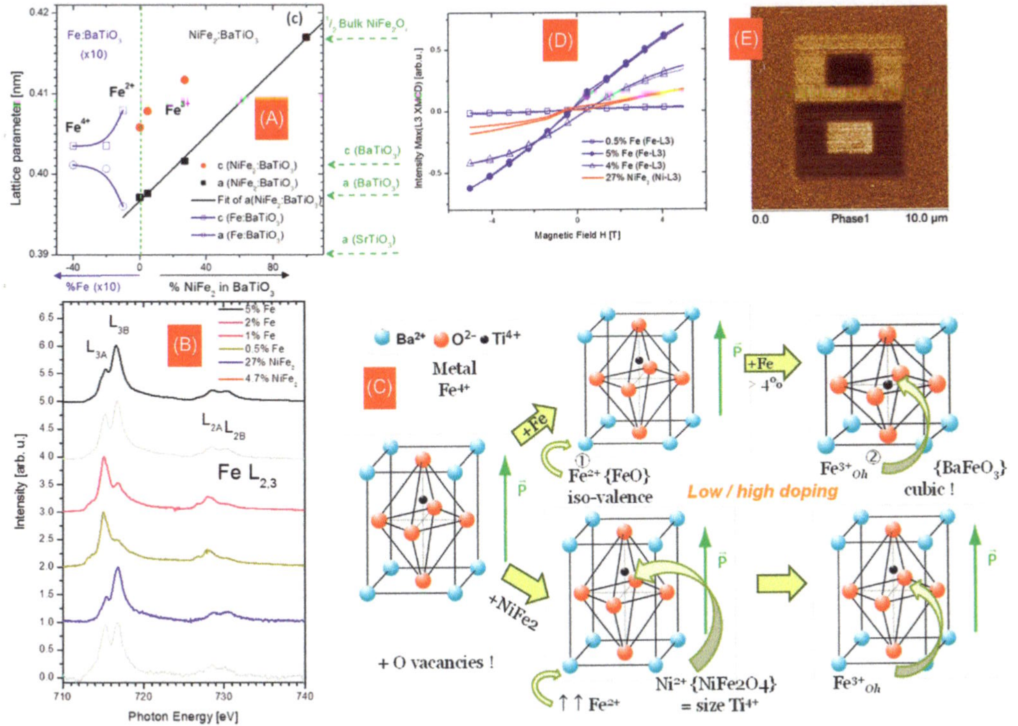

Figure 23.9 Doping of MBE grown epitaxial $BaTiO_3$/Nb(1%)-$SrTiO_3$(001) layers. (A) In- and out-of-surface plane lattice parameters observed for pure and doped $BaTiO_3$(001) layers. The abscissa axis reports the dopant contents. The bulk lattice parameters (right arrows) at room temperature for the $SrTiO_3$ substrate, tetragonal $BaTiO_3$, and half of the parameter of $NiFe_2O_4$ are indicated. (B) X-ray absorption spectra of Fe and Ni in doped $BaTiO_3$ layers. (C) Summary of the dopant substitution sites in Fe and $NiFe_2$ doped $BaTiO_3$ layers with respect to the doping level and dopant nature. (D) Examples of magnetization loops realized using the L_3 absorption edge maximum XMCD signal while looping the magnetic field. The Fe edge L_3 evolution is shown for purely Fe-doped samples and the Ni-L_3 edge for the highly doped $NiFe_2$ sample. The samples were held at 4K and the glancing X-ray angle with respect to the surface was 30 degrees (i.e., grazing incidence). (E) PFM phase image (using a 4 V drive amplitude) of a written pattern on a 20 nm thick 4 at% Fe-doped $BaTiO_3$ film. The pattern was realized using the PFM tip using DC voltages of +6 V and −6 V for the bright and dark regions, respectively. (A, B, D, E). Reprinted figure with permission from A. Barbier et al., Phys. Rev. B 91 (2015) 035417 [85]. Copyright © (2015) by the American Physical Society.

sites as a function of the doping level and site/ion type (Fig. 23.9C). All films were found antiferromagnetic without hysteresis loop opening or remanent magnetization but with very strong ferroelectric behaviors (Fig. 23.9D and E).

A last family of compounds often described as improper geometric ferroelectrics possess a high temperature structural phase transition driving the appearance of ferroelectricity. Hexagonal rare-earth manganites (h-$RMnO_3$ with R = Ho-Lu, Y), show such structural phase transitions providing the necessary symmetry lowering by tilting of the MnO_5 bipyramids. The induced polarization is low and the multiferroic phase is entered only at cryogenic temperatures when antiferromagnetic order due to spin frustration arises [86]. $YMnO_3$ was one of the earliest thin-film multiferroics and is both a ferroelectric and an A-type antiferromagnet.

Overall, for type I multiferroics, even if some debates are still open, the antiferromagnetic spin ordering that may be exploited via exchange coupling to produce multiferroic composite layers appears as the largely most favorable situation.

23.3.2.3 Type II multiferroic perovskites

In type II systems the magnetic order breaks the inversion symmetry of the material, driving a secondary ferroelectric phase transition in which the ferroelectric polarization is coupled to the magnetic structure. Typical type II multiferroics include rare-earth manganites such as R-MnO_3 with R = [Gd,Tb,Dy] and $HoMn_2O_5$ [23]. The ferroelectric ordering is driven by magnetism and the ordering temperatures are usually identical and in the cryogenic range. Although their incorporation in practical devices is questionable they provide excellent test situations for a better understanding of multiferroics or proof-of-principle devices operating at low temperature. They exhibit typically low net magnetization and low polarization.

The R-MnO_3 series of perovskite manganites, where R = [Gd,Tb,Dy], belong to the orthorhombic (space group Pb nm). The Mn^{3+} ions have an electronic configuration of $3t_{2g}e_g$ with the nearest neighbor t_{2g} electrons having antiferromagnetic couplings and the e_g electrons having ferromagnetic coupling. The competition between these couplings gives rise to various Mn magnetic orderings, ranging from the antiferromagnetism in $LaMnO_3$ to a complex incommensurate ordering in $TbMnO_3$. At low temperature the spontaneous ferroelectric polarization can be switched by the application of a strong magnetic field. The ordering wave-vector of $TbMnO_3$ was found along the *b*-axis and incommensurate with the incommensurability changing with temperature.

A wide range of hexagonal R-MnO_3 (R = [Nd, Ho, Tm, Lu, Yb, Tb, Dy, Gd, Sm]) materials have been grown in form of thin films [54]. Unfortunately, none of these compounds exhibit both room temperature ferroelectricity and magnetism. However, these systems serve as a model thin-film systems.

23.4 Conclusion

The possibility to interact with a device using several external fields or to modify a long-range ordering normally linked to another field via cross-coupling opens large avenues for new devices with additional or novel functionalities. In multiferroic materials, the coupling interaction between the different order parameters can be used to produce additional functionalities in devices, such as a magnetoelectric effect. Strong electron correlations drive the coupling of magnetic and electric order parameters in intrinsic multiferroics. A large amount of fundamental physics work still has to be done in this framework to provide a solid understanding of the physics behind magnetoelectric coupling. Artificial multiferroics structures are multifunctional by essence and likely belong to the elemental building blocks of the next generation of spintronic devices. For each additional addressable ferroic order that can be incorporated in the elemental structure the storage capacity doubles, without the expense of additional matter. This does not only render the device highly cost effective but will also contribute to a better usage of the natural resources.

Within spintronics applications the thin-film configuration is mostly mandatory to create the necessary confinement required for most spin-dependent interactions. From the growth stress point of view thin-film configurations are very favorable and tend to improve the ferroelectric properties of the films. It is noticeable, and seems quite general, that the ferroelectric long-range ordering remains very robust down to spintronic device thicknesses and can thus be implemented even as in the form of tunnel barriers. Single-phase multiferroics being seldom by essence, especially at room temperature, rather antiferromagnetic, new approaches need to be developed. Thin-film technologies are now mature enough to produce perfectly well-controlled interfaces and composite films that allow promoting close coexistence of ferromagnetic and ferroelectric phases. This opens large avenues of new possibilities of combination in terms of materials, compounds, and devices.

Some great challenges are still open for the future in the field, from both very fundamental approaches as well as from more application-dedicated ones. They include a better understanding of the interplay between ferroelasticity and ferromagnetic-ferroelectric coupling that is a very difficult task because, as discussed here, the order parameters appear as highly intricate. A second considerable challenge is the implementation in practical room-temperature working devices of the concept that could be demonstrated at the laboratory level. This task will probably also impose the successful implementation into semiconducting devices [78] that appears very challenging.

References

[1] P. Curie, Sur la symétrie dans les phénomènes physiques, symétrie d'un champ électrique et d'un champ magnétique, J. Phys. 3 (1894) 393.

[2] H. Schmid, Introduction to the proceedings of the 2nd international conference on magnetoelectric interaction phenomena in crystals, MEIPIC-2, Ferroelectrics 161 (1994) 1.

[3] M. Fiebig, Revival of the magnetoelectric effect, J. Phys. D: Appl. Phys 38 (2005) R123.

[4] W. Eerenstein, N.D. Mathur, J.F. Scott, Multiferroic and magnetoelectricmaterials, Nature 442 (2006) 759.

[5] H. Schmid, Some symmetry aspects of ferroics and single phase multiferroics, J. Phys.: Condens. Matter 20 (2008) 434201.

[6] C.A.F. Vaz, Electric field control of magnetism in multiferroic heterostructures, J. Phys.: Condens. Matter 24 (2012) 333201.
[7] J. Ma, J. Hu, Z. Li, C.-W. Nan, Recent progress in multiferroic magnetoelectric composites: from bulk to thin films, Adv. Mater. 23 (2011) 1062–1087.
[8] C.A.F. Vaz, J. Hoffman, C.H. Ahn, R. Ramesh, Magnetoelectric coupling effects in multiferroic complex oxide composite structures, Adv. Mater. 22 (2010) 2900–2918.
[9] H.S. Virk, W. Kleemann, Ferroics and multiferroics, Solid State Phenomena 189 (2012) (2012) 1662–9787.
[10] A. Fert, Origin, development, and future of spintronics (nobel lecture), Angew. Chem. Int. Ed. 45 (2008) 5956.
[11] J. Wang, J.B. Neaton, H. Zheng, V. Nagarajan, S.B. Ogale, B. Liu, et al., Epitaxial $BiFeO_3$ multiferroic thin film heterostructures, Science 299 (2003) 1719.
[12] T. Kimura, T. Goto, H. Shintani, K. Ishizaka, Y. Arima, Y. Tokura, Magnetic control of ferroelectric polarization, Nature 426 (2003) 55–58.
[13] N. Hur, S. Park, P.A. Sharma, J.S. Ahn, S. Guha, S.-W. Cheong, Electric polarization reversal and memory in a multiferroic material induced by magnetic fields, Nature 429 (2004) 392–395.
[14] N.A. Hill, Why are there so few magnetic ferroelectrics? J. Phys. Chem. B 104 (2000) 6694.
[15] D.I. Khomskii, Multiferroics: different ways to combine magnetism and ferroelectricity, J. Magn. Magn.Mater. 306 (2006) 1.
[16] S.W. Cheong, M.V. Mostovoy, Multiferroics: a magnetic twist for ferroelectricity, Nat. Mater. 6 (2007) 13.
[17] R. Ramesh, N.A. Spaldin, Multiferroics: progress and prospects in thin films, Nat. Mater. 6 (2007) 21.
[18] Special issue, J. Phys. Condens.Matter 20, 434201–434220 (2008).
[19] S. Roy, S.B. Majumder, Recent advances in multiferroic thin films and composites, JALCOM 538 (2012) 153–159.
[20] R.C. Kambale, D.-Y. Jeong, J. Ryu, Current status of magnetoelectric composite thin/thick films, Adv.Cond.Mat.Phys. 2012 (2012) 824643.
[21] O.G. Udalov, N.M. Chtchelkatchev, I.S. Beloborodov, Electric field control of magnetic properties and magneto-transport in composite multiferroics, J. Phys.: Condens. Matter 27 (2015) 186001.
[22] N. Ortega, A. Kumar, J.F. Scott, R.S. Katiyar, Multifunctional magnetoelectric materials for device applications, J. Phys.: Condens. Matter 27 (2015) 504002.
[23] W. Ratcliff, J.W. Lynn, V. Kiryukhin, P. Jain, M.R. Fitzsimmons, Magnetic structures and dynamics of multiferroic systems obtained with neutron scattering, Quant. Mater. 1 (2016) 16003.
[24] J.A. Venables, G.D.T. Spiller, M. Hanbücken, Rep. Prog. Phys. 47 (1984) 399–459.
[25] N. Kaiser, Review of the fundamentals of thin-film growth, Appl. Opt. 41 (2002) 3053.
[26] Rl.E. Caflisch, Growth, structure and pattern formation for thin films, J. Sci.Comput. 37 (2008) 3–17.
[27] G.G. Stoney, The Tension of Metallic Films Deposited by Electrolysis, Proc, A82, Royal Soc, London, 1909, p. 172.
[28] M. Ohring, In Materials Science of Thin Films, Copyright, Published by Academic Press, 2002
[29] R. Kern, G. Le Lay, J.J. Metoisin, in: E. Kaldis (Ed.), Current Topics in Materials Science, 3, North-Holland, Amsterdam, 1979, pp. 139–419.
[30] A. Barbier, Generalized model for interface description, Surf. Sci. 406 (1998) 69–89.

[31] A. Kvasov, A. Tagantsev, Role of high-order electromechanical coupling terms in thermodynamics of ferroelectric thin films, Phys. Rev. B 87 (2013) 184101.

[32] K.J. Choi, M. Biegalski, Y.L. Li, A. Sharan, J. Schubert, R. Uecker, et al., Enhancement of ferroelectricity in strained $BaTiO_3$ thin films, Science 306 (2004) 1005.

[33] W.D. Nix, Mechanical properties of thin films, Metall. Trans. A 20 (1989) 2217.

[34] J. Haeni, P. Irvin, W. Chang, R. Uecker, P. Reiche, Y.L. Li, et al., Room-temperature ferroelectricity in strained $SrTiO_3$, Nature 430 (2004) 758.

[35] R.M. Hornreich, The magnetoelectric effect: materials, physical aspects, and applications, IEEE Trans. Magn. 8 (1972) 584.

[36] J.F. Scott, Magnetoelectric memories—a disruptive technology? Chem. Phys. Chem. 10 (2009) 1761.

[37] M.I. Bichurin, V.M. Petrov and S.Priya, Magnetoelectric multiferroic composites Ferroelectrics—Physical Effects (2011) p. 277, (Ed.) M.Lallart (Rijeka: InTech).

[38] J.F. Scott, Applications of magnetoelectrics, J. Mater. Chem. 22 (2012) 4567.

[39] G.A. Prinz, Magnetoelectronics, Science 282 (1998) 1660.

[40] S.A. Wolf, D.D. Awschalom, R.A. Buhrman, J.M. Daughton, S. von Molnár, M.L. Roukes, et al., Spintronics: a spin-based electronics vision for the future, Science 294 (2001) 1488.

[41] Igor Žutić, Jaroslav Fabian, S. Das Sarma, Spintronics: fundamentals and applications, Rev. Mod. Phys. 76 (2004) 323.

[42] H. Munekata, I. Zutic, J. Fabian, S. Tarucha, M. Stopa, S. Sasaki, et al., in: S. Maekawa (Ed.), Concepts in Spin Electronics, Oxford University Press, 2006.

[43] S.D. Bader, S.S.P. Parkin, Spintronics, Annu. Rev. Condens. Matter Phys. 1 (2010) 71.

[44] T. Taniyama, E. Wada, M. Itoh, M. Yamaguchi, Electrical and optical spin injection in ferromagnet/semiconductor heterostructures, NPG Asia Mater. 3 (2011) 65.

[45] E.-J. Guo, R. Roth, A. Herklotz, D. Hesse, K. Dörr, Ferroelectric 180° domain wall motion controlled by biaxial strain, Adv. Mater. 27 (2015) 1615−1618.

[46] L.J. McGilly, P. Yudin, L. Feigl, A.K. Tagantsev, N. Setter, Controlling domain wall motion in ferroelectric thin films, Nat. Nanotechnol. 10 (2015) 145−150.

[47] M. Weisheit, S. Fähler, A. Marty, Y. Souche, C. Poinsignon, D. Givord, Electric field-induced modification of magnetism in thin-film ferromagnets, Science 315 (2007) 349.

[48] T. Maruyama, Y. Shiota, T. Nozaki, K. Ohta, N. Toda, M. Mizuguchi, et al., Large voltage-induced magnetic anisotropy change in a few atomic layers of iron, Nat. Nanotechnol. 4 (2009) 158.

[49] C.-G. Duan, S.S. Jaswal, E.Y. Tsymbal, Predicted magnetoelectric effect in Fe/$BaTiO_3$ multilayers: ferroelectric control of magnetism, Phys. Rev. Lett. 97 (2006) 047201.

[50] S. Valencia, A. Crassous, L. Bocher, V. Garcia, X. Moya, R.O. Cherifi, et al., Interface-induced room-temperature multiferroicity in $BaTiO_3$, Nat. Mater. 10 (2011) 753.

[51] Y. Kim, C. Pham, J.P. Chang, Potentials and challenges of integration for complex metal oxides in CMOS devices and beyond, J. Phys. D: Appl. Phys. 48 (2015) 063001.

[52] C. Lekha, V. Sudarsanan, S.S. Nair, Recent advances in multiferroics and magnetoelectrics based devices: a review of patents, Recent Pat. Mater. Sci. 7 (2014) 90−98.

[53] L.W. Martin, S.P. Crane, Y.-H. Chu, M.B. Holcomb, M. Gajek, M. Huijben, et al., Multiferroics and magnetoelectrics: thin films and nanostructures, J. Phys.: Condens. Mat. 20 (2008) 434220.

[54] B. Dieny, V.S. Speriosu, S.S. Parkin, B.A. Gurney, D.R. Wilhoit, D. Mauri, Giant magnetoresistive in soft ferromagnetic multilayers, Phys. Rev. B 43 (1991) 1297 (R)-1300.

[55] S. Datta, B. Das, Electronic analog of the electro-optic modulator, Appl. Phys. Lett. 56 (1990) 665–667.
[56] V.K. Joshi, Spintronics: a contemporary review of emerging electronics devices, Eng. Sci. Techn. 19 (2016) 1503–1513.
[57] M. Bibes, A. Barthélémy, Nat. Mater. 7 (2008) 425–426.
[58] M. Fiebig, Th Lottermoser, M.K. Kneip, M. Bayer, Correlations between magnetic and electrical orderings in multiferroic manganites, J. Appl. Phys. 99 (2006) 08E302.
[59] T. Lottermoser, T. Lonkai, U. Amann, D. Hohlwein, J. Ihringer, M. Fiebig, Magnetic phase control by an electric field, Nature 430 (2004) 541–544.
[60] D. Lebeugle, D. Colson, A. Forget, M. Viret, A.M. Batailleand, A. Gukasov, Electric-field-induced spin flop in $BiFeO_3$ single crystals at room temperature, Phys. Rev. Lett. 100 (2008) 227602.
[61] Z.J. Huang, Y. Cao, Y.Y. Sun, Y.Y. Xue, C.W. Chu, Coupling between the ferroelectric and antiferromagnetic orders in $YMnO_3$, Phys. Rev. B 56 (1997) 2623.
[62] J.F. Scott, Magnetic phases of bismuth ferrite, J. Magn. Magn. Mater. 321 (2009) 1689–1691.
[63] G. Catalan, J.F. Scott, Physics and applications of bismuth ferrite, Adv. Mater. 21 (2009) 2463–2485.
[64] M. Bibes, J.E. Villegas, A. Barthélémy, Ultrathin oxide films and interfaces for electronics and spintronics, Adv. Phys. 60 (2011) 5–84.
[65] Y. Yang, I.C. Infante, B. Dkhil, L. Bellaiche, Strain effects on multiferroic $BiFeO_3$ films, C. R. Phys. 16 (2015) 193–203.
[66] M. Guennoua, M. Viret, J. Kreisel, Bismuth-based perovskites as multiferroics, C. R. Phys. 16 (2015) 182–192.
[67] J. Nogués, I.K. Schuller, Exchange bias, J. Magn. Magn. Mater. 192 (1999) 203–232.
[68] C. Mocuta, A. Barbier, S. Lafaye, P. Bayle-Guillemaud, M. Panabière, Single-crystalline model spin valves using single-crystalline NiO(111) substrates, Phys. Rev. B 68 (2003) 014416.
[69] A. Barbier, C. Mocuta, W. Neubeck, M. Mulazzi, F. Yakhou, K. Chesnel, et al., Surface and bulk spin ordering of antiferromagnetic materials: NiO(111), Phys. Rev. Lett. 93 (2004) 257208.
[70] E. Bauer, Surface Microscopy With Low Energy Electrons, Published by Springer-Verlag, New York, 2014.
[71] T. Zhao, A. Scholl, F. Zavaliche, K. Lee, M. Barry, A. Doran, et al., Electrical control of antiferromagnetic domains inmultiferroic $BiFeO_3$ films at room temperature, Nat. Mater. 5 (2006) 823.
[72] Y.-H. Chu, L.W. Martin, M.B. Holcomb, M. Gajek, S.-J. Han, Q. He, et al., Electric-field control of local ferromagnetism using a magnetoelectric multiferroic, Nat. Mater. 7 (2008) 478–482.
[73] Y.-H. Chu, L.W. Martin, M.B. Holcomb, R. Ramesh, Controlling magnetism with multiferroics, Mater. Today 10 (2007) 16.
[74] M. Gajek, M. Bibes, S. Fusil, K. Bouzehouane, J. Fontcuberta, A. Barthélémy, et al., Tunnel junctions with multiferroic barriers, Nat. Mater. 6 (2007) 296.
[75] J.Y. Son, J.-H. Lee, S. Song, Y.-H. Shin, H.M. Jang, Four-states multiferroic memory embodied using Mn-doped $BaTiO_3$ nanorods, ACS Nano 7 (2013) 5522–5529.
[76] D.E. Parkes, L.R. Shelford, P. Wadley, V. Holý, M. Wang, A.T. Hindmarch, et al., Magnetostrictive thin films for microwave spintronics, Sci. Rep. 3 (2013) 2220.

[77] P. Merodio, A. Kalitsov, M. Chshiev, J. Velev, Electric field control of the resistance of multiferroic tunnel junctions with magnetoelectric antiferromagnetic barriers, Phys. Rev. Appl. 5 (2016) 064006.
[78] Reji Thomas, J.F. Scott, DwarkaN. Bose, RamS. Katiyar, Multiferroic thin-film integration onto semiconductor devices, J. Phys.: Condens. Matter 22 (2010) 423201.
[79] H. Zheng, J. Wang, S.E. Lofland, Z. Ma, L. Mohaddes-Ardabili, T. Zhao, et al., Multiferroic $BaTiO_3$-$CoFe_2O_4$ nanostructures, Science 303 (2004) 661–663.
[80] T. Aghavnian, J.-B. Moussy, D. Stanescu, R. Belkhou, N. Jedrecy, H. Magnan, et al., Determination of the cation site distribution of the spinel in multiferroic $CoFe_2O_4$/$BaTiO_3$ layers by X-ray photoelectron spectroscopy, J. Electron. Spectrosc. Relat. Phenom 202 (2015) 16–21.
[81] T. Aghavnian, PhD – Université Paris-Saclay (2016).
[82] A. Barbier, C. Mocuta, G. Renaud, in: H.S. Nalwa (Ed.), Handbook of Thin Film Materials - Chapter 11, In *situ* Synchrotron Structural Studies of the Growth of Oxides and Metals, Published by Academic Press, 2002, pp. 1–67.
[83] Daniel Khomskii, Classifying multiferroics: mechanisms and effects, Physics 2 (2009) 20.
[84] A. Barbier, C. Mocuta, D. Stanescu, P. Jegou, N. Jedrecy, H. Magnan, Surface composition of $BaTiO_3$/$SrTiO_3$(001) films grown by atomic oxygen plasma assisted molecular beam epitaxy, J. Appl. Phys. 112 (2012) 114116.
[85] A. Barbier, T. Aghavnian, V. Badjeck, C. Mocuta, D. Stanescu, H. Magnan, et al., Antiferromagnetic long-range spin ordering in Fe- and $NiFe_2$-doped $BaTiO_3$ multiferroic layers, Phys. Rev. B 91 (2015) 035417.
[86] F. Yen, C. De la Cruz, B. Lorenz, E. Galstyan, Y.Y. Sun, M. Gospodinov, et al., Magnetic phase diagrams of multiferroic hexagonal $RMnO_3$ (R = Er, Yb, Tm, and Ho), J. Mater. Res. 22 (2007) 2163–2173.

$BiFeO_3$ ceramics and thick films: Processing issues and electromechanical properties

24

Tadej Rojac[1], *Evgeniya Khomyakova*[1], *Julian Walker*[2], *Hana Ursic*[1] *and Andreja Bencan*[1]
[1]Jozef Stefan Institute, Ljubljana, Slovenia,
[2]The Pennsylvania State University, State College, PA, United States

24.1 Processing issues

Since early studies in 1960s it has been clear that processing of $BiFeO_3$ is a nontrivial task. A number of these early papers reported difficulties in preparing phase-pure $BiFeO_3$ [1–3], which were accredited to the experimentally observed high-temperature decomposition of $BiFeO_3$ into secondary phases, identified as Bi_2O_3, Fe_2O_3, and $Bi_2Fe_4O_9$ [1,3,4]. Only recently, however, the question as to why these secondary phases form has been addressed in more detail [5]. This chapter begins by presenting the multiple origins of the secondary-phase formation, which will be explained by analyzing how these phases distribute inside a $BiFeO_3$ ceramic grain matrix.

Although several authors reported on the metastability of polycrystalline and single-crystal $BiFeO_3$ [6–8], it was not until the key studies by Selbach et al. [9] and Valant et al. [10] that explanations were provided for the intriguing decomposition first observed almost 60 years ago [1–4]. Using a combination of experimental and theoretical analysis, Selbach et al. [9] showed that, according to equilibrium reaction (1), $BiFeO_3$ tends to decompose to Bi-rich and Fe-rich secondary phases (often referred to in the literature as the sillenite and mullite phases, respectively):

$$BiFeO_3 \leftarrow\ \rightarrow 1/49\ Bi_{25}FeO_{39} + 12/49\ Bi_2Fe_4O_9 \tag{24.1}$$

The secondary phases are weakly but more thermodynamically stable than $BiFeO_3$ at temperatures between approximately 447°C and 767°C (reaction (1) spontaneously shifted to the right). Outside this temperature interval the free energy of reaction (1) reverses sign, shifting the reaction toward $BiFeO_3$ formation (reaction (1) spontaneously shifted to the left) [9].

Perhaps the most critical feature of the solid-state synthesis of $BiFeO_3$ was found by Valant et al. [10] when the role of impurities in the stabilization of the $Bi_{25}FeO_{39}$ and $Bi_2Fe_4O_9$ phases was questioned. It was found that impurities that are soluble in these two phases (examples are SiO_2 and Al_2O_3) have a crucial impact on the equilibrium phase composition. The presence of such impurities,

Magnetic, Ferroelectric, and Multiferroic Metal Oxides. DOI: http://dx.doi.org/10.1016/B978-0-12-811180-2.00024-4

even if in small concentrations (<1 wt%), may stabilize large amounts (several tens of vol%) of secondary phases, leading to the compositional degradation of $BiFeO_3$.

One method for solving this problem is to avoid impurities by using ultrapure starting powders and carefully performing the synthesis steps to prevent contamination [10]. An alternative method is to incorporate a third component into the $Bi_2O_3-Fe_2O_3$ system that is highly soluble in the perovskite, for example, Sm_2O_3. This was shown to have beneficial effects in terms of reducing the susceptibility of $BiFeO_3$ to phase degradation by Si impurities [11].

The intrinsic thermodynamic instability of $BiFeO_3$ [9] and the strong effect of impurities on destabilization of $BiFeO_3$ [10] can explain why the two secondary phases often appear in physical contact, sharing interfaces in sintered $BiFeO_3$. This is illustrated in Fig. 24.1A, where a large $Bi_2Fe_4O_9$ crystal appears to embed smaller sillenite inclusions. The appearance of two phases with large differences in their chemical composition in physical contact would be normally unexpected as the chemical gradient should drive the reaction at elevated temperatures. Owing to the

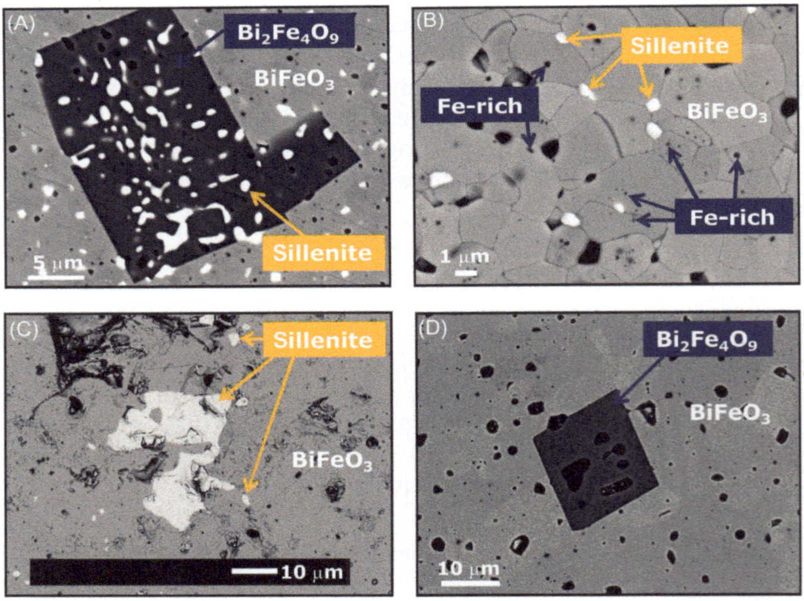

Figure 24.1 Type and distribution of secondary phases in $BiFeO_3$ ceramics. Back-scattered scanning electron microscopy micrographs illustrating four different examples of the often observed morphology and distributions of the Bi-rich (sillenite) and Fe-rich ($Bi_2Fe_4O_9$ or Fe_2O_3) secondary phases in sintered $BiFeO_3$ ceramics. The sintering conditions for the four cases were (A) 770°C for 6 h in air, (B) 760°C for 6 h in air, (C) 820°C for 20 h with the pellet immersed in a packing powder, (D) same as (C) with additional postannealing of the pellet at 840°C for 10 h in air (without packing powder). The secondary phases are indicated with arrows in respective images.

particular thermodynamics of $BiFeO_3$ [9,10], however, the two secondary phases can be stabilized and do not react even if in physical contact.

In contrast to the phase distribution shown in Fig. 24.1A, Fig. 24.1B presents a different situation where the sillenite phase resides at the grain boundaries and sub-μm Fe-rich inclusions are observed in the interior of the $BiFeO_3$ grains. Note that the two secondary phases are not in physical contact; instead they are separated by $BiFeO_3$. Bernardo et al. [12] explained this phase distribution as kinetically stabilized phase coexistence where the diffusion of bismuth through the $BiFeO_3$ grain was insufficient to complete the reaction with Fe_2O_3, leaving unreacted Fe-rich grain cores. The kinetics of the $Bi_2O_3-Fe_2O_3$ reaction is thus as important as the thermodynamics of $BiFeO_3$ for stabilizing the secondary phases [5].

The sillenite phase is the first phase to appear during the reaction of a $Bi_2O_3-Fe_2O_3$ 1:1 powder mixture [7,12,13]. As a result, sintering the ceramic at temperatures exceeding the peritectic decomposition temperature of the sillenite phase ($T_p \sim 790°C$ [6]) can be problematic. Incongruent melting of sillenite can lead to liquid-phase segregation [14], resulting in large sillenite inclusions (tens of μm) that grow at the expense of smaller inclusions (Fig. 24.1C). By separating Bi-rich from Fe-rich reaction counterparts, this segregation stops the reaction, leading to multiphase ceramics.

Sillenite segregation was identified by annealing ceramics at 820°C in the presence of a packing powder (Fig. 24.1C; see also [14]), which is commonly used to control excessive loss of the volatile Bi_2O_3 from the ceramic. In the absence of the packing powder and at temperature >820°C, however, significant loss of Bi_2O_3 may occur and the resulting Bi_2O_3 deficiency leads to the formation of large $Bi_2Fe_4O_9$ crystals, as shown in Fig. 24.1D (note that here the sillenite is not present, which is different from the case shown in Fig. 24.1A). We also note that the volatility of Bi_2O_3 at temperatures higher than T_p of the sillenite ($\sim$790°C)should be enhanced due to the presence of the liquid phase, particularly if the sintering atmosphere is not controlled (for example, by sintering in the absence of a packing powder) [5,10,14].

The thermodynamic and kinetic issues so far discussed provide a number of critical factors that need to be addressed during the synthesis of $BiFeO_3$ ceramics: (1) the diffusion distances (determined by homogeneity and particle size of the reaction mixture), (2) the impurity level and its control, (3) the reaction and sintering temperatures (with respect to the $BiFeO_3$ instability ($\sim$447°C–767°C), T_p of the sillenite phase ($\sim$790°C), and volatility of Bi_2O_3 ($\geq$820°C)), and (4) the sintering atmosphere (e.g., by using packing powder). For this reason a combination of ultrapure (>99.99%) starting powders and reaction sintering of a mechanochemically activated $Bi_2O_3-Fe_2O_3$ mixture has recently been shown to be efficient in producing ceramics with a minimum amount of secondary phases (<1%) [5], although a large body of work has been published on $BiFeO_3$ using a wide range of different synthesis methods.

A potential application of $BiFeO_3$ is in piezoelectric microelectromechanical systems (piezo MEMS). This technology requires integration of the perovskite onto substrates using thick-film deposition, for example, by the widely used screen-printing method [15,16]. During integration, however, problems additional to those relating to bulk ceramic processing (Fig. 24.1) may arise.

The integration of $BiFeO_3$ thick films (with typical thicknesses of several tens of μm) onto polycrystalline Al_2O_3 and noble metal substrates by screen printing have been comprehensively investigated by Khomyakova et al. [17–19]. Fig. 24.2 summarizes their results by presenting a compilation of back-scattered–scanning electron microscopy cross-section images of a sintered $BiFeO_3$ thick film deposited on an Al_2O_3 substrate with an Au electrode (Fig. 24.2A). The reaction and mass-transport processes, which were determined to be critical for the sintering of the $BiFeO_3$ thick films, are summarized schematically in Fig. 24.2B.

Reactive sintering of the deposited Bi_2O_3–Fe_2O_3 mixture at 750°C resulted in a number of different observations. First, the distribution of secondary phases across the $BiFeO_3$ film thickness was uneven: μm-sized sillenite inclusions were observed in the interior of the film, while bigger $Bi_2Fe_4O_9$ crystals were concentrated at the film/electrode interface (Fig. 24.2A). Second, an additional Cu- and Bi-containing layer was observed underneath the $Bi_2Fe_4O_9$ crystals (see also upper inset of Fig. 24.2A). Copper, which was found as the major impurity in the Au electrode (noted in Fig. 24.2A and B), most likely reacted with bismuth oxide from the $BiFeO_3$ film, possibly creating a Bi_2CuO_4 phase reported in the Bi_2O_3–CuO phase diagram [20]. This reaction is expected to cause local bismuth deficiency in the film near the electrode layer, resulting in the formation of the $Bi_2Fe_4O_9$ crystals (see Fig. 24.2B) [17].

Third, a Bi-rich phase was found in between the grains of the polycrystalline Al_2O_3 substrate (Fig. 24.2A, see also bottom inset). It was shown that significant

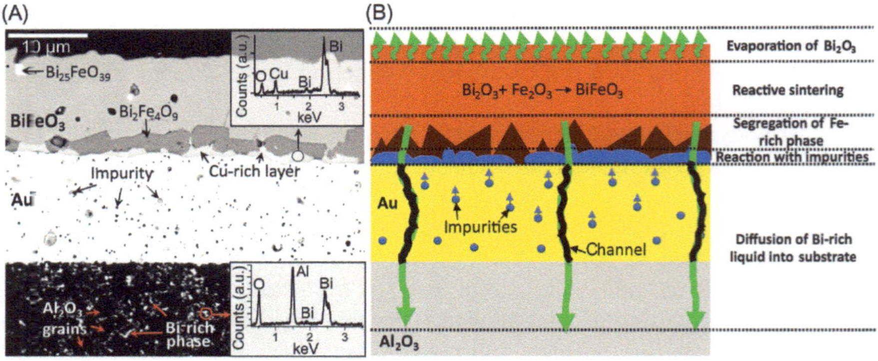

Figure 24.2 Integration of $BiFeO_3$ thick films by screen printing on Al_2O_3 substrate. (A) Compilation of back-scattered–scanning electron microscopy cross-section images of a $BiFeO_3$ thick film screen printed on an Au/Al_2O_3 substrate and sintered at 750°C for 4 h. The different phases and impurities observed in the layers are indicated by arrows. Insets show energy-dispersive X-ray spectra acquired from indicated circled areas. (B) Schematic illustration of the different reaction and mass-transport processes taking place during sintering of screen-printed $BiFeO_3$ thick films.

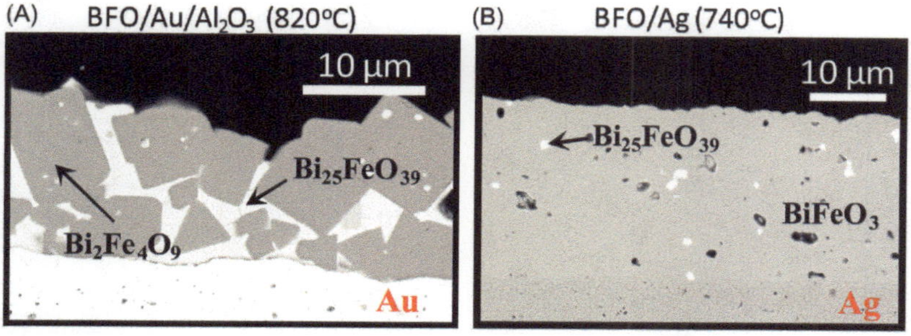

Figure 24.3 Optimization of thick-film processing parameters toward minimization of secondary phases in $BiFeO_3$ thick films. Cross-section of screen-printed $BiFeO_3$ thick films deposited on (A) Au/Al_2O_3 substrate (820°C) and (B) highly pure Ag substrate (740°C). The identified phases are indicated with arrows.

diffusion of bismuth from the $BiFeO_3$ film occurred at temperatures higher than T_p of the sillenite phase ($\sim$790°C) [17]. This suggests that bismuth penetrates into the substrate as a Bi-rich liquid phase, most likely by diffusion through pores or channels present in the Au layer (see Fig. 24.2B), similarly as proposed for the Pb-rich liquid phase in $(Pb,La)(Zr,Ti)O_3$ thick films [21,22].

Finally, evaporation/sublimation of Bi_2O_3, which is expected to be extensive in films due to the large surface-to-volume ratio, was suggested to occur at the sintering temperature of 820°C (Fig. 24.2B; see [17]).

Identification of the thick-film integration problems, summarized in Fig. 24.2B, allows minimizing the concentration of secondary phases through process design, particularly sintering temperature and type of electrode/substrate (including purity and porosity). To demonstrate this point, we illustrate in Fig. 24.3 two extreme cases [17,18]. Almost complete decomposition of a $BiFeO_3$ thick film was observed after sintering the film at 820°C on an Au/Al_2O_3 substrate (Fig. 24.3A). In contrast, a $BiFeO_3$ film on a highly pure (99.9%) Ag substrate sintered at a lower temperature of 740°C resulted in small amounts of secondary phases (Fig. 24.3B). These results confirm the key role of the sintering temperature and electrode/substrate purity in the preparation of high-quality $BiFeO_3$ thick films.

24.2 Polarization switching, piezoelectricity, and local electrical conductivity

The ferroelectric/ferroelastic domain switching in $BiFeO_3$ ceramics resembles that of so-called hard ferroelectric materials [14]. Two of the common features of this class of materials are pinched polarization–electric-field hystereses loops and aging behavior [23–28], both of which were observed in $BiFeO_3$ ceramics [14,29]. An example of a pinched loop of $BiFeO_3$ is displayed in Fig. 24.4A (as-sintered curve).

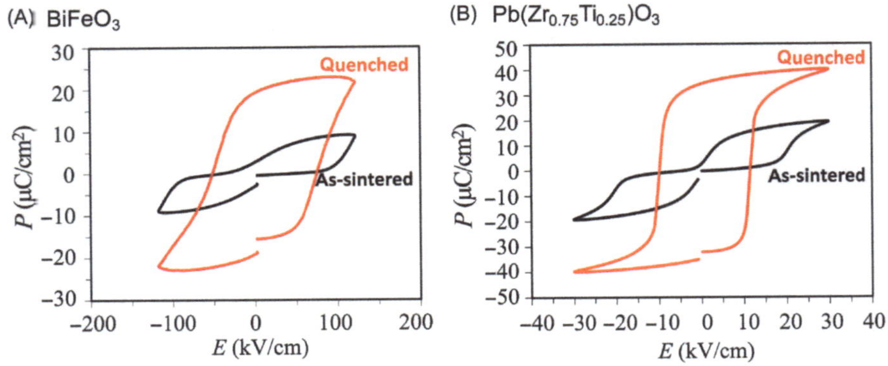

Figure 24.4 Polarization switching behavior in $BiFeO_3$ ceramics. Polarization–electric-field hysteresis loops of as-sintered and quenched ceramics of (A) $BiFeO_3$ and, for comparison, (B) $Pb(Zr_{0.75}Ti_{0.25})O_3$. In both cases the de-aging was performed by quenching the ceramics in water from above the respective Curie temperatures of the two rhombohedral perovskites; that is, from 900°C in the case of $BiFeO_3$ ($T_C = 825°C$) and from 600°C in the case of Pb $(Zr_{0.75}Ti_{0.25})O_3$ ($T_C = 320°C$).

In acceptor-doped hard ferroelectrics, such as Mn- or Fe-doped $Pb(Zr,Ti)O_3$ (PZT) [24,27] and $BaTiO_3$ [25,30], the origin of the domain-wall pinning effect, which is responsible for the pinched loops and aging, is related to the presence of acceptor–oxygen-vacancy defect complexes (an example is $Fe'_{zr,Ti}-V_O^{\bullet\bullet}$ complex in Fe-doped PZT [31–33]). Through oxygen vacancy hopping, these complexes tend to align along the direction of spontaneous polarization (P_s) inside domains, providing electric and elastic pinning forces for the domain-wall motion [23–26,28,34]. The alignment of oxygen-vacancy-related complexes in PZT and $BaTiO_3$ was experimentally proven using electron paramagnetic resonance analyses [33,35–38].

As in Fe-doped PZT, the loop-pinching behavior in undoped PZT ceramics (Fig. 24.4B, as-sintered curve) was also explained in terms of defect-induced pinning effects, this time associated with $V''_{Pb}-V_O^{\bullet\bullet}$ complexes [32] that form as a result of PbO loss during sintering [39]. The qualitative similarity of the pinched loops of $BiFeO_3$ and PZT (as-sintered curves in Fig. 24.4) likely indicates a common defect-related pinning mechanism [14]. Consistent with this hypothesis is the opening (depinching) of the hystereses loops observed when the two materials are water-quenched to room temperature from above their respective Curie temperatures (quenched curves in Fig. 24.4). This behavior is interpreted in terms of order-disorder defect transition [14,27], meaning that a disordered defect arrangement, which develops in the paraelectric phase because of the absence of P_S, is considered to become frozen at room temperature owing to the rapid cooling. In contrast to the ordered defect state in aged samples (pinched loops in Fig. 24.4), this metastable, disordered defect configuration in de-aged samples releases the domain walls from the pinning, hence the loops open up (open loops in Fig. 24.4).

Recent studies using atomic-resolution microscopy provide further insights into the domain-wall pinning in $BiFeO_3$ by revealing the presence of bismuth vacancies and Fe^{4+} defects accumulated in domain-wall regions [40]. These charged defects are consistent with the defect-chemistry model developed for $BiFeO_3$ and might play a dominant role in the observed pinning effect (Fig. 24.4A). In particular, the accumulation of electron holes (associated with Fe^{4+} states), whose electrical mobility is much higher than that of bismuth vacancies, suggested *p*-type hopping conduction localized at the domain walls. This domain-wall conduction was first observed in $BiFeO_3$ thin films [41] and later confirmed in polycrystalline $BiFeO_3$ ceramics [42] using conductive atomic-force microscopy.

It was suggested that the domain-wall conductivity in $BiFeO_3$ ceramics strongly affects its macroscopic piezoelectric response through two interrelated mechanisms [42]. First, a dynamic interaction is expected between the accumulated defects and domain walls, meaning that displacements of domain walls may be coupled to the

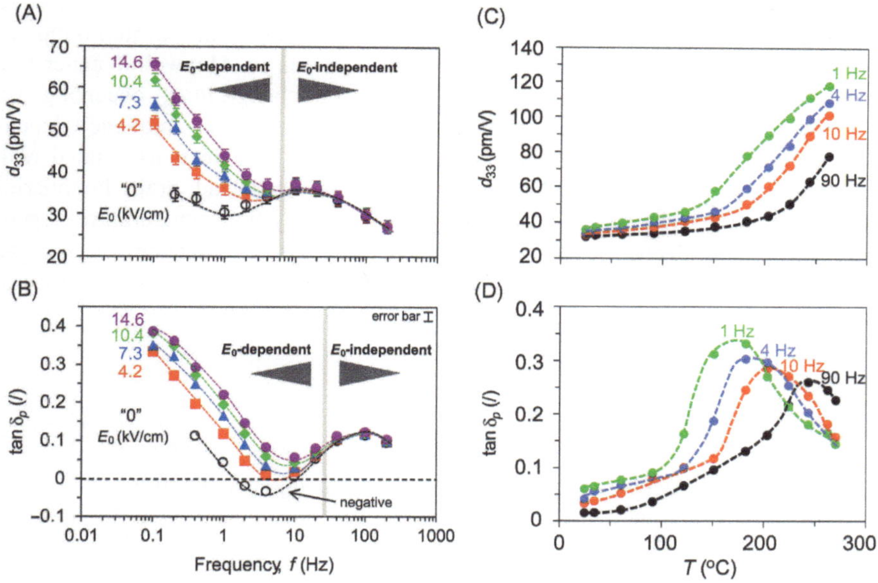

Figure 24.5 Converse longitudinal piezoelectric response of $BiFeO_3$ ceramics. (A) Piezoelectric coefficient (d_{33}) and (B) piezoelectric phase angle (tan δ_p) measured at room temperature as a function of driving field frequency (f) and driving field amplitude (E_0; indicated on the graphs). (C) d_{33} and (D) tan δ_p measured as a function of temperature at different driving field frequencies (indicated on the graphs). In panels (A) and (B) the field-dependent and field-independent dispersion regimes are indicated by a thick line. The open-symbol curves in panels (A) and (B) represent the linear contribution to the piezoelectric response (calculated for zero electric field), exhibiting a negative tan δ_p indicated with an arrow in panel (B). Panels (A) and (B) are reprinted from [42], with the permission of John Wiley & Sons, Inc. Panels (C) and (D) are reprinted from [44], with the permission of AIP Publishing.

p-type hopping conductivity. Second, the presence of the local conductivity itself (also that at grain boundaries [42]) may lead to piezoelectric Maxwell-Wagner relaxations (the piezoelectric analogue of the more common dielectric Maxwell-Wagner effect), similarly as observed in heterogeneous ferroelectrics [43].

These hypothetical mechanisms have been shown to be consistent with the converse piezoelectric response of $BiFeO_3$ measured at room (Fig. 24.5A and B; [42]) and elevated temperatures (Fig. 24.5C and D; [44]). At room temperature, strong dispersions in d_{33} (Fig. 24.5A) and piezoelectric phase angle (tan δ_p; Fig. 24.5B) were observed at low frequencies (<10 Hz), both of which are dependent on the driving electric-field amplitude. This low-frequency nonlinearity and hysteresis suggest irreversible domain-wall displacements that occur only at low frequencies probably because the *p*-type hopping conductivity allows depinning and motion of the walls. In addition, by extracting the linear (field-independent) contribution from the piezoelectric data shown in Fig. 24.5A and B, a negative tan δ_p emerged (Fig. 24.5B, open symbols), which was predicted and experimentally observed in systems dominated by the Maxwell-Wagner effect [43,45–48]. These results are consistent with the piezoelectric behavior of $BiFeO_3$ measured at elevated temperatures (Fig. 24.5C and D), which shows typical features of Maxwell-Wagner-like relaxation: a step-like increase in $d_{33}(T)$ with an abrupt change in d_{33} occurring at a frequency-dependent temperature point (Fig. 24.5C), coupled to a frequency dependent peak in tan δ_p (T) (Fig. 24.5D). Maxwell-Wagner-like effects associated with local conductive paths in $BiFeO_3$ (such as domain walls and grain boundaries [42,44]) thus appear to dominate the complex frequency-field-temperature dependence of the piezoelectric response. The results demonstrate that the microscopic electrical conductivity of $BiFeO_3$ should be important in controlling the stability of its macroscopic piezoelectric response.

Acknowledgments

This work was supported by the Slovenian Research Agency (program P2-0105).

References

[1] V.S. Filipev, N.P. Smolyaninov, E.G. Fesenko, I.N. Belyaev, Synthesis of $BiFeO_3$ and determination of the unit cell, Kristallografiya 5 (1960) 958–959.

[2] S.A. Fedulov, Yu.N. Venevtsev, G.S. Zhdanov, E.G. Smazhevskaya, High-temperature X-ray and thermal-analysis results for bismuth ferrite, Kristallografiya 6 (1961) 795–796.

[3] G.D. Achenbach, W.J. James, R. Gerson, Preparation of single-phase polycrystalline $BiFeO_3$, J. Am. Ceram. Soc. 50 (1967) 437.

[4] S.M. Skinner, Magnetically ordered ferroelectric materials, IEEE Trans. Parts Mater. Packag. PMP-6, (1970) 68–90.

[5] T. Rojac, A. Bencan, B. Malic, G. Tutuncu, J.L. Jones, J.E. Daniels, et al., $BiFeO_3$ ceramics: processing, electrical, and electromechanical properties, J. Am. Ceram. Soc. 97 (2014) 1993–2011.

[6] R. Palai, R.S. Katiyar, H. Schmid, P. Tissot, S.J. Clark, J. Robertson, et al., β phase and γ-β metal-insulator transition in multiferroic $BiFeO_3$, Phys. Rev. B 77 (2008), 014110-1-11.

[7] M.I. Morozov, N.A. Lomanova, V.V. Gusarov, Specific features of $BiFeO_3$ formation in a mixture of bismuth (III) and iron (III) oxides, Zh. Obshch. Khim. 73 (2003) 1772–1776.

[8] T.T. Carvalho, P.B. Tavares, Synthesis and thermodynamic stability of multiferroic $BiFeO_3$, Mater. Lett. 62 (2008) 3984–3986.

[9] S.M. Selbach, M.-A. Einarsrud, T. Grande, On the thermodynamic stability of $BiFeO_3$, Chem. Mater. 21 (2009) 169–173.

[10] M. Valant, A.-K. Axelsson, N. Alford, Peculiarities of a solid-state synthesis of multiferroic polycrystalline $BiFeO_3$, Chem. Mater. 19 (2007) 5431–5436.

[11] J. Walker, B. Budic, P. Bryant, V. Kurusingal, Ch.C. Sorrell, A. Bencan, et al., Robust polarization and strain behaviour of Sm-modified $BiFeO_3$ piezoelectric ceramics, IEEE Trans. UFFC 62 (2015) 83–87.

[12] M.S. Bernardo, T. Jardiel, M. Peiteado, A.C. Caballero, M. Villegas, Reaction pathways in the solid state synthesis of multiferroic $BiFeO_3$, J. Eur. Ceram. Soc. 31 (2011) 3047–3053.

[13] M. Thrall, R. Freer, C. Martin, F. Azough, B. Patterson, R.J. Cernik, An in situ study of the formation of multiferroic bismuth ferrite using high resolution synchrotron X-ray powder diffraction, J. Eur. Ceram. Soc. 28 (2008) 2567–2572.

[14] T. Rojac, M. Kosec, B. Budic, N. Setter, D. Damjanovic, Strong ferroelectric domain-wall pinning in $BiFeO_3$ ceramics, J. Appl. Phys. 108 (2010), 074107-1-8.

[15] N.M. White, J.D. Turner, Thick-film sensors: past, present and future, Meas. Sci. Technol. 8 (1997) 1–20.

[16] D. Belavic, M. Hrovat, J. Holc, M. Santo Zarnik, M. Kosec, M. Pavlin, The application of thick-film technology in C-MEMS, J. Electroceram. 19 (2007) 363–368.

[17] E. Khomyakova, J. Pavlic, M. Makarovic, H. Ursic, J. Walker, T. Rojac, et al., Integration of $BiFeO_3$ thick films onto ceramic and metal substrates by screen printing, J. Eur. Ceram. Soc. 35 (2015) 4163–4171.

[18] E. Khomyakova, J. Pavlic, M. Makarovic, H. Ursic, J. Walker, V.Ya Shur, et al., Screen-printed $BiFeO_3$ thick films on noble metal foils, Ferroelectrics 496 (2016) 196–203.

[19] E. Khomyakova, M. Sadl, H. Ursic, J. Daniels, B. Malic, A. Bencan, et al., Self-poling of $BiFeO_3$ thick films, ACS Appl. Mater. Inter. 8 (2016) 19626–19634.

[20] B.G. Kakhan, V.B. Lazarev, I.S. Shaplygin, Study of the subsolidus part of phase diagrams of Bi_2O_3-MO (M = Ni, Cu, or Pd) binary systems, Zh. Neorg. Khim. 24 (1979) 1663–1668.

[21] M. Kosec, B. Malic, J. Holc, M. Hrovat, M. Mandeljc, A. Bencan, et al., Interface reactions among electrodes, substrates and $Pb(Zr,Ti)O_3$-based films, Acta Chim. Slov. 48 (2001) 51–62.

[22] M. Kosec, J. Holc, B. Malic, V. Bobnar, Processing of high performance lead lanthanum zirconate titanate thick films, J. Eur. Ceram. Soc. 19 (1999) 949–954.

[23] G.H. Jonker, Nature of aging in ferroelectric ceramics, J. Am. Ceram. Soc. 55 (1972) 57–58.

[24] K. Carl, K.H. Hardtl, Electrical after-effects in $Pb(Ti,Zr)O_3$ ceramics, Ferroelectrics 17 (1978) 473–486.

[25] P.V. Lambeck, G.H. Jonker, The nature of domain stabilization in ferroelectric perovskites, J. Phys. Chem. Solids 47 (1986) 453–461.

[26] U. Robels, G. Arlt, Domain wall clamping in ferroelectrics by orientation of defects, J. Appl. Phys. 73 (1993) 3454–3460.

[27] M.I. Morozov, D. Damjanovic, Hardening-softening transition in Fe-doped $Pb(Zr,Ti)O_3$ ceramics and evolution of the third harmonic of the polarization response, J. Appl. Phys. 104 (2008), 034107-1-8.

[28] Y.A. Genenko, J. Glaum, M.J. Hoffmann, K. Albe, Mechanisms of aging and fatigue in ferroelectrics, Mat. Sci. Eng. B 192 (2015) 52–82.

[29] G.L. Yuan, Y. Yang, S.W. Or, Aging-induced double ferroelectric hysteresis loops in $BiFeO_3$ multiferroic ceramic, Appl. Phys. Lett. 91 (2007), 122907-1-3.

[30] L. Zhang, X. Ren, Aging behavior in single-domain Mn-doped $BaTiO_3$ crystals: implication for a unified microscopic explanation of ferroelectric aging, Phys. Rev. B 73 (2006), 094121-1-6.

[31] P. Jakes, E. Erdem, R.-A. Eichel, L. Jin, D. Damjanovic, Position of defects with respect to domain walls in Fe^{3+}-doped $Pb[Zr_{0.52}Ti_{0.48}]O_3$ piezoelectric ceramics, Appl. Phys. Lett. 98 (2011), 072907-1-3.

[32] A. Chandrasekaran, D. Damjanovic, N. Setter, N. Marzari, Defect ordering and defect–domain-wall interactions in $PbTiO_3$: a first-principles study, Phys. Rev. B 88 (2013), 214116-1-7.

[33] R.-A. Eichel, Structural and dynamic properties of oxygen vacancies in perovskite oxides – analysis of defect chemistry by modern multi-frequency and pulsed EPR techniques, Phys. Chem. Chem. Phys. 13 (2011) 368–384.

[34] X. Ren, Large electric-field-induced strain in ferroelectric crystals by point-defect-mediated reversible domain switching, Nat. Mater. 3 (2004) 91–94.

[35] W.L. Warren, D. Dimos, G.E. Pike, K. Vanheusden, Alignment of defect dipoles in polycrystalline ferroelectrics, Appl. Phys. Lett. 67 (1995) 1689–1691.

[36] W.L. Warren, G.E. Pike, K. Vanheusden, D. Dimos, B.A. Tuttle, J. Robertson, Defect-dipole alignment and tetragonal strain in ferroelectrics, J. Appl. Phys. 79 (1996) 9250–9257.

[37] W.L. Warren, K. Vanheusden, D. Dimos, G.E. Pike, B.A. Tuttle, Oxygen vacancy motion in perovskite oxides, J. Am. Ceram. Soc. 79 (1996) 536–538.

[38] L. Zhang, E. Erdem, X. Ren, R.-A. Eichel, Reorientation of $(Mn''_{Ti} - V_{O}^{\bullet\bullet})^{\times}$ defect dipoles in acceptor-modified $BaTiO_3$ single crystals: an electron paramagnetic resonance study, Appl. Phys. Lett. 93 (2008), 202901-1-3.

[39] T. Rojac, S. Drnovsek, A. Bencan, B. Malic, D. Damjanovic, Role of charged defects on the electrical and electromechanical properties of rhombohedral $Pb(Zr,Ti)O_3$ with oxygen octahedra tilts, Phys. Rev. B 93 (2016), 014102-1-11.

[40] T. Rojac, A. Bencan, G. Drazic, N. Sakamoto, H. Ursic, B. Jancar, et al., Domain-wall conduction in ferroelectric $BiFeO_3$ controlled by accumulation of charged defects, Nat. Mater. 16 (2016) 322–327.

[41] J. Seidel, L.W. Martin, Q. Hel, Q. Zhan, Y.-H. Chu, A. Rother, et al., Conduction at domain walls in oxide multiferroics, Nat. Mater. 8 (2009) 229–234.

[42] T. Rojac, H. Ursic, A. Bencan, B. Malic, D. Damjanovic, Mobile domain walls as a bridge between nanoscale conductivity and macroscopic electromechanical response, Adv. Funct. Mater. 25 (2015) 2099–2108.

[43] D. Damjanovic, M. DemartinMaeder, P. Duran Martin, C. Voisard, N. Setter, Maxwell–Wagner piezoelectric relaxation in ferroelectric heterostructures, J. Appl. Phys. 90 (2001) 5708–5712.

[44] T. Rojac, M. Makarovic, J. Walker, H. Ursic, D. Damjanovic, T. Kos, Piezoelectric response of $BiFeO_3$ ceramics at elevated temperatures, Appl. Phys. Lett. 109 (2016), 042904-1-4.
[45] T. Furukawa, K. Ishida, E. Fukada, Piezoelectric properties in the composite systems of polymers and PZT ceramics, J. Appl. Phys. 50 (1979) 4904–4912.
[46] H. Ueda, E. Fukada, F.E. Karasz, Piezoelectricity in three-phase systems: effect of the boundary phase, J. Appl. Phys. 60 (1986) 2672–2677.
[47] A.V. Turik, G.S. Radchenko, Maxwell–Wagner relaxation in piezoactive media, J. Phys. D: Appl. Phys. 35 (2002) 1188–1192.
[48] A.V. Turik, A.I. Chernobabov, G.S. Radchenko, S.A. Turik, Giant piezoelectric and dielectric enhancement in disordered heterogeneous systems, Phys. Solid State 46 (2004) 2213–2216.

Properties of single multiferroics: Complex transition metal oxides

25

Nikola I. Ilic and Biljana D. Stojanovic
Institute for Multidisciplinary Research, University of Belgrade, Belgrade, Serbia

25.1 Introduction

In 1994, Schmid blended the term multiferroic to describe a special class of materials [1], and in following years multiferroics were introduced and widely recognized in material science. The search for the philosopher's stone of new era was initiated. Today's definition of single multiferroic material implies existence of at least two primary ferroic properties (primarily (anti)ferromagnetic and ferroelectric, but also ferroelastic and ferrotoroidic) in one phase. Those properties imply spontaneous magnetization, polarization, strain, and alignment of magnetic vortices. Secondary and higher order ferroics are not necessarily multiferroic despite being polarizable, magnetizable, and strainable, or manifesting alignment of toroidal moments in electric and magnetic fields and under mechanical stress, although literature sometimes equalize magnetoelectrics (with secondary ferroic properties) with multiferroics [1,2]. Possible direct coupling between different properties through coupling in crystalline structure, magnetic and electric domains, and spin interactions in multiferroic materials offer a good way toward proper and efficient management of electric and magnetic fields [3]. The interest in the field of single multiferroics is obvious and legitimate.

Single multiferroic (MF) materials are very rare (at temperatures above 10K), and that is the main reason why they haven't attracted the attention of materials scientists earlier. The cause of such scarcity is different origin of properties existing in those materials (for example, the need of empty *d* orbitals in ferroelectrics and the need for *d* electrons in ferromagnets). Therefore, either ferroelectricity or ferromagnetism of MFs must not be of the usual kind, which requires many structural, symmetric, and compositional requirements (explained in Chapter 22: Single-Phase, Composite, and Laminate Multiferroics) [4]. Composite MFs, on the other hand, could be theoretically made of any two or more materials possessing a primary ferroic property, so they are ubiquitous and various, with a higher degree of connection between different properties, but with limitation of specific properties to contact surfaces between different phases [5]. Because of that, there are two main directions in which MF materials research is heading today:

1. Obtaining single MFs with stronger interconnection among primary ferroic properties. It is very important to achieve stability of properties in wider temperature ranges, especially at room temperature (since most MF materials are multiferroic only at very low temperatures).
2. Obtaining composite MFs with more homogenous multiferroic properties.

Magnetic, Ferroelectric, and Multiferroic Metal Oxides. DOI: http://dx.doi.org/10.1016/B978-0-12-811180-2.00025-6

25.2 Classification of single multiferroics

There are only two confirmed naturally occurring single MF crystals: congolite ($Fe_3B_7O_{13}Cl$) and chambersite ($Mn_3B_7O_{13}Cl$) [6]. The first compounds synthesized with the idea to be both ferromagnetic (FM) and ferroelectric (FE) were Pb($Fe_{0.5}Nb_{0.5}$)O_3, Pb($Fe_{0.5}Ta_{0.5}$)O_3 and Pb($Fe_{0.66}W_{0.33}$)O_3, but they exhibited only antiferromagnetic (AFM) and FE behavior with very small coupling or without it. The first successfully obtained FM FE was $Ni_3B_7O_{13}I$, although at low temperatures (below 61K) [1]. Even today, FM MFs count a few materials. Up-to-date engineered MFs could be classified into several main material classes [1,6–8]:

1. Perovskite ferrites, manganites and chromites of Bi, Pb and rare earths (RE): (Pb,Bi,RE)(Fe,Mn,Cr)O_3
2. Double perovskites with transition metal ions (TM) of d^4-d^6 and d^0 electron configurations: $(Pb,Bi,RE)_2(TM'TM'')O_6$
3. Double transition metal oxides: $CoCr_2O_4$, $Ni_3V_2O_8$, $MnWO_4$, $LiCu_2O_2$, $CuFeO_2$, $CuCrO_2$, $AgCrO_2$
4. Mixed manganese oxides: $REMn_2O_5$
5. Transition metal compounds with complex structure like different types of alkaline earth (AE) hexaferrites, Aurivillius phase bismuth iron titanate ($Bi_5Ti_3FeO_{15}$), and borates: RE$(Al,Ga,Sc,Fe,Cr,Mn,Co,Ni)_3(BO_3)_4$
6. Halides (boracites: $(Cr,Mn,Fe,Co,Ni,Cu)_3B_7O_{13}(F,Cl,Br,I)$; and fluorides: Ba(Mn,Ni,Co)F_4) and chalcogenides ($BaFe_2Se_3$)
7. Organics

Multiferroicity is theoretically expected but still not confirmed in several metal oxide material classes like some Pb perovskites [9], high pressure TM titanates [10], $AE_3TM_2O_7$ [8], tetragonal tungsten bronzes ($A_6B_{10}C_4O_{30}$; A, B, and C are metal ions arranged in different positions) [11], and melilite ($AE_2(V,Ni,Co,Mn,Cu)(Ge,Si)_2O_7$) structures [12]. Going at very low temperatures (below 10K), many other metal oxides were reported multiferroic [1], but from the application point of view, they are not interesting. But there are materials like bismuth-ferrite, and Aurivillius structures with MF properties present at close-to-room temperatures, and they are attracting lot of attention these days.

In this chapter, attention is directed to transition metal oxide MFs. Probably the clearest way to distinguish between different groups is by the cause of magnetic or ferroelectric properties. The nature of magnetism and ferroelectricity usually prevents single MFs to be ferroelectrics of typical type [1]. Instead of usual d^0cation decentralization, electrical polarization in transition metal oxides could be [3,13,14]:

1. A-site driven.
2. Geometrically driven.
3. Charge ordering (CO) driven.
4. Spin driven (often equalized with type II MFs).

But there are also those with a combination of different driving forces causing ferroelectricity, and those with a complex and poorly understood nature of multiferroicity. According to the degree of interconnection between magnetism and ferroelectricity, MF materials are sometimes also classified into two groups. Type I MFs have different causes of ferroelectricity and ferromagnetism, which appear mostly independently within them, from different reasons (A-site driven, geometrically driven, and charge ordering driven ferroelectricity). In type II MFs, there is a specific magnetic ordering (arising in temperature often called $T_{lock\text{-}in}$) where magnetic field can cause changes in symmetry through spin interactions, inducing ferroelectricity. That means that magnetic and electric domains and domain walls in type I MFs occur independently, while in type II MFs they coincide [3]. Type I MFs are more numerous, and have significantly higher values of electrical polarization than type II MFs, which have larger magnetoelectric coupling [15]. Temperature range within some type I and type II MFs are multiferroic is indicated in Table 25.1. Theoretically, other mechanisms of multiferroicity like ferromagnetic order induced by electric field were predicted ($LiNbO_3$, $FeTiO_3$) [3].

Review of structure and properties of some of the most investigated single MFs (except bismuth-ferrite, which is covered in previous chapter 24) will be covered in this chapter. Substituted (doped) systems and nanomaterials will not be covered, but it is worth mentioning that modifications like these may induce many improvements in properties of MF materials.

Since it is possible that compounds of the same class or structure possess different origin of properties, it could be useful to make the review of MF materials according to the cause of their specific properties.

25.3 A-site driven ferroelectricity multiferroics

As mentioned previously, because of the different origins of magnetic and ferroelectric properties, MFs must not be (anti)ferromagnets and ferroelectrics of typical type. One of the most common ways for this in ABO_3 perovskites is ferroelectricity caused not by hybridization and decentralization of d^0 B-cations, but by hybridization between $2p$ orbitals of oxygen and $6p$ orbitals of A ions and localization of $6s^2$ electrons on A-site ions. All A-site driven MFs contain either Bi^{3+} or Pb^{2+} as A-site ions providing ferroelectricity. The most promising and for sure the most studied of them is $BiFeO_3$, which was presented in detail in previous chapters. Materials of this group have significantly higher polarization than other multiferroics [3].

Bismuth manganite, $BiMnO_3$, crystallizes in a distorted perovskite structure, probably thanks to Jahn-Teller (JT) active Mn^{3+} ion. Because of JT distortion, $BiMnO_3$ was believed to be ferroelectric. It is further expected that d^4 state of Mn^{3+} ions would give AFM symmetry through superexchange interactions, but $BiMnO_3$ exhibits peculiar behavior, being ferromagnetic (under 114K) [15]. Of all perovskite manganites, $BiMnO_3$ was the only one thought to be both ferroelectric and ferromagnetic, which would make it the most promising single MF material. But the situation is more complicated. Current opinions and experimental results are highly conflicted. Some authors claim that $BiMnO_3$ have centrosymmetric space group and could not be ferroelectric, and, therefore is not multiferroic, although

Table 25.1 **Specific ferroelectric Curie and (anti)ferromagnetic Curie (Néel) temperatures of selected single MF ceramic materials and saturated values of polarization and magnetization achieved in them under various conditions**

Material	Ferroelectric T_C or $T_{lock-in}$, K	(Anti)ferromagnetic T_C or T_N^*, K	Type of ferroelectricity	P_S, μC/cm^2	M_S, emu/g	References
Type I multiferroics						
$BiFeO_3$	1103	643*	A-site driven	60–100	1×10^{-1}	[16]
$BiCoO_3$	>600	470*	A-site driven	120	–	[15,17]
$BiMnO_3$	? (760–800)	99–114	? (A-site driven)	? (1×10^{-1})	70	[15,16,18]
$BiCrO_3$	? (440)	109	? (A-site driven)	–	1	[15]
$PbVO_3$	?	45*	A-site driven	–	2×10^{-6}	[9]
$Bi_2(TM'TM'')O_6$	>293	100–800	A-site driven	5	28	[7]
$Pb(TM'_{1-x}TM''_x)O_3$	180–387	77–350*	A-site driven	–	2	[16,19,20]
$RE_2(TM'TM'')O_6$	>293	226–270	CO driven	–	66	[7,21,22]
$Ca_{0.4}Pr_{0.6}MnO_3$	235	160*	CO driven	–	–	[23]
Fe_3O_4	120	860	CO driven	–	–	[23]
$REFe_2O_4$	330–350	240–250	CO driven	24	20	[23,24]
h-$REMnO_3$	500–1050	50–130*	Geometrically driven	6	4×10^{-1}	[8,25–27]
h-$REFeO_3$	470–563	115–155*	Geometrically driven	6.5	–	[8]
$RECrO_3$	440–516	113–140*	Local off-centering	–	1	[8,28]
$Sr_{0.5}Ba_{0.5}MnO_3$	400	200*	Strain caused JT off-centering	4.5	–	[8]
$FeTiO_3$	>293	110*	d^0	–	–	[10]
$(Bi,RE)_5Ti_3FeO_{15}$	800–1077	80*/497	Bi displacement/A-site driven/d^0	8	500	[29,30]

Type II multiferroics						
o-$REMnO_3$	18–30	42–45*	Inverse DMI	1.8×10^{-1}	1	[8,31,32]
$MnWO_4$	12.5	13.5*	Inverse DMI	6×10^{-3}	–	[23]
$CoCr_2O_4$	26	93	Inverse DMI	2×10^{-4}	1	[33,34]
$(Ba,Sr)_2TM_2Fe_{12}O_{22}$	130	326–553*	Inverse DMI	1.5×10^{-2}	50	[33,35,36]
$SrFe_{12}O_{19}$	?	729*	Inverse DMI?	40	40	[37]
$LiCu_2O_2$	23	?	Inverse DMI	8×10^{-4}	–	[35]
CuO	230	230*	Inverse DMI	1.6×10^{-2}	–	[8]
$RECrO_3$	120–250	120–300*	Magnetostriction	8×10^{-1}	4	[38]
Ca_3CoMnO_6	16	16	Magnetostriction	9×10^{-3}	–	[23]
$REMn_2O_5$	38–41	42–45*	Magnetostriction	4×10^{-2}	–	[8,18]
$CuCrO_2$	23.6	24.2*	Spin chirality	1.6×10^{-2}	–	[39]
$Cu_3Nb_2O_8$	24	26*	Spin chirality?	1.8×10^{-3}	–	[40]
$CaMn_7O_{12}$	90	90	Inverse DMI and magnetostriction/spin chirality?	2.9×10^{-1}	–	[41]

Significantly higher values were reported for thin films but they are not presented here unless no data for ceramics was found (as for $Bi_2(TM'TM'')O_6$). $RECrO_3$ is among type I multiferroics for smaller rare-earth ions (Y, Sc, Ho-Lu), and among type II multiferroics for larger rare-earth ions (Ce-Tb).

room temperature ferroelectric $BiMnO_3$ was experimentally obtained under structural strains in thin films and doped ceramics or in single crystals under very high pressure (with remnant polarization of up to 16 μC/cm^2). Nonsaturated FE loops with P_S of 0.1 μC/cm^2 were measured even in bulk form [15,17]. Attempts to explain the bulk polarization of centrosymmetric lattice was with variations in stoichiometry leading to formation of some polar phases and with hidden AFM interactions disturbing the symmetry [15]. Other authors reported distorted monoclinic or triclinic structure with A-site driven ferroelectricity. In favor of the latter goes the fact that large degree of covalency in Bi-O bond that would exist in A-site driven ferroelectric may also reduce the overlap of 3*d* orbitals of Mn ions and 2*p* orbitals of oxygen, reducing the strength of AFM superexchange and enabling ferromagnetism [4].

It is obvious that nature of $BiMnO_3$ is under question, but with very large magnetization of up to 70 emu/g and potential for ferroelectricity, it is the second most studied single MF material [15].

Many other bismuth perovskite compounds are among A-site driven ferroelectrics, but only $BiCoO_3$ exhibit multiferroic properties [15]. It has C-type AFM structure, with ferromagnetically arranged antiferromagnetic planes. Six *d* electrons of Co^{3+} ions give rise to tetragonality ($c/a = 1.27$), so very large values of polarization were theoretically expected (180 μC/cm^2). However, low resistivity usually prevents ferroelectric hysteresis loop to be measured. $BiCrO_3$ could also be MF, but so far ferroelectric properties were observed only in thin films [15].

$PbVO_3$ has two-dimensional AFM structure stable under 43K. Electronic configuration of V^{4+} ions (one *d* electron situated in lowest energy d_{xy} orbital) stabilizes large tetragonal distortion (as well as 2D AFM structure), and $6s^2$ lone pairs on Pb^{2+} ions face the *c*-axis direction. The ferroelectricity is expected to be large, but neither the polarization nor the Curie temperature were successfully measured yet [9]. $PbNiO_3$ has similar potential.

Several double and mixed perovskites described with a chemical formula of (Bi, Pb,RE)$_2$(TM′TM″)O$_6$ were among the first reported multiferroic materials (mixed perovskites labels materials with randomly distributed various ratios of two or more transition metals, while the term double stands for orderly distributed TM′ and TM″) [19,21].

It was thought that if one of the present transition metals in perovskites has empty d electronic states, the regular, d^0 type ferroelectricity will be aroused. However, that is probably not the case. Ordered or disordered B-sites, (anti)ferromagnetism occurs from TM(d^n)-O-TM(d^n) superexchange and other magnetic transitions at lower temperatures originate from TM(d^n)-O-TM(d^0)-O-TM(d^n) superexchange, while ferroelectricity is caused by A-site localized $6s^2$ electronic pairs and not d^0 TMs [19,20]. In Bi-based double perovskites, ferro(i)magnetic to paramagnetic phase transitions occurs at 140K (Bi_2NiMnO_6), 440K (Bi_2FeMnO_6), and even as high as 800K (Bi_2CoMnO_6), which was tried to be explained by the monoclinic tilt and possible interaction between the local magnetization and $6s^2$ pairs of electrons [42]. The saturation polarization ($\sim 10^0$ μC/cm^2) is smaller than usual for A-site driven ferroelectricity [7], and magnetoelectric coupling is weak [16]. Still, it is not certain that those all are intrinsic properties, since they were reported for epitaxial thin films, where strain can strongly influence the properties [8].

Even if A-site ions are not Bi or Pb, this class of materials may be ferroelectric when both the transition metal ions possess *d* electrons (usually Mn, Co, and Ni), but their size and electronegativity differ enough to introduce some degree of ordering in arrangement of different ions. The ordering of ions with different valences (i.e., charge ordering) in double perovskites changes the symmetry and introduces electrical polarization, which may be weak, but is strongly coupled with the magnetic order parameters [21]. The magnetic transition in RE double perovskites occurs at 226K–270K and in some cases indicates ferromagnetism with very large saturation magnetization of the order of magnitude of 10^1 emu/g and large coercive field (0.2-10 kOe) [19,21,22].

25.4 Geometrically driven ferroelectricity multiferroics

Another way of ferroelectric arrangement could be tilting in the structure because of large mismatching in sizes of ions. Ions move in order to pack closer, making imbalance and arising electrical polarization.

The small size of A-site ions (smaller than ideal according to Goldschmidt tolerance factor) in h-$REMnO_3$ (RE = Sc, Y, In, lanthanides from Ho to Lu) make changes in their structure. The *d*-levels of Mn^{3+} ions are split in two doublets and one singlet, and four *d* electrons occupy only lower energy doublets without formation of orbital degeneracy, and without JT effect (which is present in orthorhombic manganites) [16]. RE^{3+} ions are coordinated with 7 and not 12 O^{2-} ions, as it is in perovskites, and Mn^{3+} ions are coordinated with not 6, but 5 O^{2-} ions forming the hexahedron (trigonal bipyramid) around each of them. The structural changes cause tilting (Glazer tilts) of MnO_5 hexahedra (connected in two-dimensional layers with noncollinear 120 degrees AFM structure), which don't disturb the center of symmetry, but shifts the RE ions intermittently up and down along the *c*-axis, making a difference in length of RE-O bonds and forming chains of antiparallel dipole moments [25,43]. They exist up to very high temperatures, so all h-$REMnO_3$ have ferroelectric Curie temperature higher than usual ferroelectrics—more than 900K for most of them [16,26]. The polarization is lower than most d^0, A-type, and CO-driven ferroelectrics, but still significant—of the order of 10^0 $\mu C/cm^2$ [25].

It is interesting that h-$REMnO_3$ become polar below 1200K–1300K (T_P), but ferroelectricity is macroscopically present only at somewhat lower temperatures, T_{FE} (T_C in Table 25.1), and processes leading to the difference are still not understood. The most probable case is that between T_P and T_{FE}, hexagonal manganites are antiferroelectric, and at the lower temperature they are not usual ferroelectrics, but rather canted triangular antiferroelectrics, exhibiting ferroelectricity in a similar way as it is with weak ferromagnetism in canted antiferromagnets [27].

The tilting also results in frustrated in-plane and interplane exchange paths, making strong coupling between magnetic, electric, and elastic properties.

However, linear magnetoelectric effect is forbidden in this hexagonal symmetry [26]. Anyway, linear magnetoelectric effect was observed in some h-$REMnO_3$ and it was explained with possible mutual clamping of FE and AFM domains/domain walls [18]. Higher order, magnetodielectric coupling is large, especially in case of manganites of lanthanides, and at a magnetic phase transition, dielectric constant abruptly increases. Below T_N, there are more magnetic transitions, in which not only spins of *d* electrons in Mn, but also f electrons in RE ions contribute to total magnetic moment and magnetodielectric coupling [44].

Hexagonal RE ferrites too exhibit geometrically driven ferroelectricity and the polarization of same order, with week ferromagnetism in the AFM phase due to Dzyaloshinskii–Moriya interactions (DMI) [45,46]. Similarly to manganites, h-$REFeO_3$ express two ferroelectric transitions, only at lower temperatures. They are not stable, so are usually synthesized in the form of epitaxial thin films [8].

25.5 Charge ordering driven ferroelectricity multiferroics

In crystals that contain ions with different valences and/or have specific structure, simultaneous inequivalences in strength (and length) of M′–O and M″–O bonds break the inversion center, causing ferroelectricity. This is charge ordering driven (also called electronic) ferroelectricity, and it is common in compounds containing transition metal ions, especially mixed valence ions like Mn^{3+} and Mn^{4+}, Fe^{2+} and Fe^{3+}, etc. (it is important to point out that the charge localization is never complete, so the real valence states are not integer numbers, and the charge difference between different ions is smaller; e.g., $Mn^{3.4+}$ and $Mn^{3.6+}$ [16]), but sometimes it is possible that O^{2-} ions play the role of ion with different valence (like in orthorhombic $REMnO_3$, where magnetic fields induce charge ordering by moving Mn and oxygen sublattices in opposite directions) [23]. Ordered existence of differently charged ions at various distances arranges local dipole moments unevenly, making the system ferroelectric. This kind of ferroelectricity can result in very large polarization (similar to typical d^0 ferroelectrics). Containing TM ions with different number of *d* electrons, those materials are usually ferrimagnetic with large magnetization and, therefore, have open potential to be used as MFs.

There are several ways of introducing mixed valences. Thanks to electron holes in Ca^{2+} doped RE manganites, Mn^{3+} and Mn^{4+} ions are present. Both, site- and bond-centered ordering take place in some doping degree, and for that composition, ferroelectricity is possible [23]. RE double perovskites (already mentioned in Section 25.2) probably have (at some extent) the same cause of ferroelectricity.

The situation is very similar in inverse spinel Fe_3O_4. Fe^{3+} and Fe^{2+} ions are in the equal number present in octahedral positions (and other half of Fe^{3+} is in tetrahedral ones). Below the Verwey transition temperature (120K), beside the alternation of differently charged Fe ions, there is the alternation in Fe-O bone length in the [110] direction.

$REFe_2O_4$ (for small RE ions) have completely different structure with pairs of $(FeO_2)^{5-}/(FeO_2)^{4-}$ layers. The pairs consist of triangular lattices, wherein one contains 2:1 ratio of Fe^{2+}:Fe^{3+} and the other 1:2. Layers in a pair are closer than neighboring pairs, so lot of local dipole moments don't completely annul, and the system is ferroelectric with high polarization of the order of 10^1 μC/cm^2 [23]. Since the ferrimagnetic and ferroelectric orders originate from the same Fe^{3+}-Fe^{2+} relations, larger coupling than normal for type I MFs is present [24].

A different kind of CO-driven ferroelectricity is present in *E*-type antiferromagnets, but since they are type II MFs, this will be discussed later.

25.6 Type I multiferroics with complex or unknown origin of ferroelectricity

Many of the multiferroic materials cannot be completely described with any of the exposed causes of ferroelectricity. Electrical polarization in them can be a consequence of a combination of usual driving forces, but also of some other phenomenon, not described yet.

Chromites of smaller RE ions (Y, Ho-Lu) were also reported multiferroic with low predicted polarization of 10^{-1} μC/cm^2 order of magnitude below 472K and weak ferromagnetism, probably canted antiferromagnetism below 140K. Larger RE chromites haven't exhibited this paraelectric-to-ferroelectric transition. The nature of ferroelectricity is still unknown, but it probably isn't the geometrical factor, since structural research indicated that at the transition, noncentrosymmetric monoclinic structure becomes locally more stable than the high temperature, orthorhombic one. And the A-site ion is the one that is off-centered, so $RECrO_3$ could be considered as another kind of A-site driven ferroelectric. The chromite of Bi has very similar Curie and Néel temperatures, but it is probable that its ferroelectricity is of usual A-site driven, lone $6s^2$ electronic pair type [28,47].

However, there are also reports that RE chromites exhibit type II multiferroicity with relatively high (for type II MF) polarization of 10^{-1} μC/cm^2 with (probably) magnetostriction origin [38].

$FeTiO_3$ in its high pressure form (the $LiNbO_3$ phase) is worth mentioning as very rare ferroelectric with DMIs inducing weak ferromagnetism (similarly as in $BiFeO_3$, though the origin of ferroelectricity is different) of the order of 10^{-1} emu/g [10].

Alkaline earth (AE) manganites were also reported to be ferroelectric if strain or chemical pressure is introduced by substrate in thin films or by AE ion doping, thanks to the second order JT distortion. The polarization is significant (10^0 μC/cm^2) and depends on AFM ordering through magnetostriction, although it is not caused by it [8].

Some of the Aurivillius layered compounds (usually containing bismuth) are among the complex MF structures. They can be described by the general formula $Bi_{n-1}Bi_2TM_nO_{3n+3}$. The structure will not be explained here in detail, but it's enough to know that *n* perovskite $(Bi_3TM_4O_{13})^{2-}$ layers are intermittently arranged with tetrahedral $(Bi_2O_2)^{2+}$ layers of bismuth oxide. $Bi_5Ti_3FeO_{15}$ is the most studied one. Ferroelectric Curie temperature could be very high (up to 1077K), while both AFM ($T_N = 80K$) and FM ($T_N = 497K$) Aurivillius Bi materials were studied [29,30]. Ferromagnetic $Bi_4NdTi_3Fe_{0.7}O_{15}$ exhibited M_S of the order of almost 500 emu/g ($M_R = 190$ emu/g, $H_C = 300$ Oe), and P_S of 8 $\mu C/cm^2$ ($P_R = 4\ \mu C/cm^2$, $E_C = 50$ kV/cm). Magnetoelectric coupling (maybe through strain) is higher than usual for type I MFs, which could be seen on improved P_R and E_C under magnetic field [30]. The real cause of polarization is not known, but some authors believe that the slight displacement of Bi^{3+} ions along *a*-axes gives the main contribution [48]. However, it should not be forgotten that $6s^2$ electrons of Bi^{3+} are present as well as d^0 Ti^{4+} ions, giving their contribution to ferroelectricity.

25.7 Magnetically driven ferroelectrics: Type II multiferroics

There are MF materials with ferroelectricity existing only in the presence of a magnetic field, and only in a particular type of magnetic ordering. Those materials are called type II MFs. The important difference between MFs of types I and II is the nature of domain walls. Ferroelectric and (anti)ferromagnetic domain walls may coincide in type I MFs, and in some degree that will be the case, but they must coincide in MFs of type II (Fig. 25.1A), arousing magnetoelectric coupling and making the magnetic and electric domains to coincide as well [3]. The coupling between magnetic and ferroelectric orders is much stronger than in type I MFs, but the ferroelectric polarization is very weak (several orders of magnitude weaker than in type I MFs) [14]. It is usually not real coupling, since it can go in only one direction—from magnetism to ferroelectricity. Several phenomena may be the cause of ferroelectricity in type II MFs:

1. Spin-orbit interactions in cycloidal and transverse conical types of spiral spin structures (inverse effect of DMI).
2. Spin-lattice interaction (magnetostriction) in materials with inequivalent magnetic ions, where both charge and magnetic spin structure are symmetric, but with different inversion center, so under magnetic field, the symmetry is disturbed and ionic polarization arises (though total polarization is very small due to electronic polarization of opposite direction formed as a result of ionic polarization) [3,23,33].
3. Variation in metal-ligand hybridization and the phenomenon of spin chirality in screw-like spiral spin structure of low symmetry crystals [35].

25.7.1 Inverse Dzyaloshinskii–Moriya interaction-driven ferroelectrics

Superexchange interactions in neighboring spin pairs make the TM-O-TM angle change. In this way, structure of the systems with noncollinear AMF spin order will be changed, and, therefore, polarization will be induced in a direction perpendicular

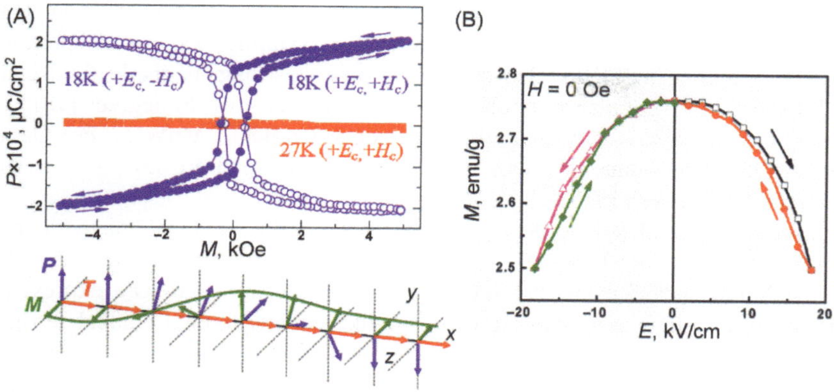

Figure 25.1 Examples of coupled properties in type II multiferroics: (A) Influence of magnetic field on electrical polarization in inverse DMI $CoCr_2O_4$ at 27K and at 18K with scheme of rotating directions of possible clamping of spin and polarization orders in coinciding domain walls with intact toroidal moment, *T*. (B) The change of magnetization in sample premagnetized at 5 kOe as a function of a square wave electric field of 18.5 kV/cm for $SrCo_2Ti_2Fe_8O_{19}$ at room temperature.
(A) Reprinted with permission from Y. Yamasaki, S. Miyasaka, Y. Kaneko, J.-P. He, T. Arima, Y. Tokura, Magnetic reversal of the ferroelectric polarization in a multiferroic spinel oxide, Phys. Rev. Lett. 96 (2006) 207204, Copyright 2006 by the American Physical Society (doi:10.1103/PhysRevLett.96.207204) and (B) Reprinted by permission from Macmillan Publishers Ltd: Scientific Reports L. Wang, D. Wang, Q. Cao, Y. Zheng, H. Xuan, J. Gao, et al., Electric control of magnetism at room temperature, Sci. Rep. 2 (2012) 223, advanced online publication, 13.01.2012. (doi:10.1038/srep00223).

to the plane of neighboring spins. The polarization will exist only for the spin orders where the spin-bearing axis is not the symmetry axis for their rotation; that is, in cycloidal and transverse conical orders [33].

In orthorhombic RE manganites, o-$REMnO_3$, the Mn *d*-levels are split in one triplet and one doublet, resulting in usual MnO_6 octahedrons connected through the three-dimensional crystalline network, which prevents tilting to cause ferroelectricity like in h-$REMnO_3$ [25]. Larger RE ions (lanthanides from La to Dy, but also Y and Ho-Lu in high pressure form) have orthorhombic centrosymmetric nonpolar structure with antiferromagnetism of A-type with antiferromagnetically arranged ferromagnetic planes (RE = La-Gd) or E-type (spins are ordered in ↑↑↓↓ manner) for smaller RE ions (Y, Ho-Lu), with transition sinusoidal order for RE = Tb, Dy [31]. The difference in magnetic structure is a consequence of stronger distortion in lanthanides of small RE ions [50]. In larger RE manganites, distortion creates deviation from perfect ordering of antiparallel magnetic moments, and the resulting cycloidal spin structure and inverse DMI introduce polarization. It is worth mentioning that not only the spiral magnetic structure of material, but also 180 degrees domain walls of Néel type may induce electrical polarization, since there is a spin half-rotation in them, and that is especially expressed in thin films [18].

Dielectric constant increases up to 500% at magnetic transition that induces ferroelectricity, implying colossal magnetodielectric coupling. At low temperatures, when RE magnetic moments also come into play, a strong magnetic field causes spin-flop transition, but also rotation of polarization vector by 90 degrees [18].

Similar is the behavior of many other systems: Cr_2BeO_4, $MnWO_4$, $FeVO_4$, and $LiCu_2O_2$ [35]. The values of polarization and magnetization are low, as it is usual for inverse DMI-driven MFs.

Transverse conical spin order is present in $CoCr_2O_4$. Typical magnetic field-polarization dependence below $T_{lock\text{-}in}$ of this material is present in Fig. 25.1A [34].

Some Y-type hexaferrites like $Ba_2Mg_2Fe_{12}O_{22}$ and $Sr_2Zn_2Fe_{12}O_{22}$ express type II multiferroicity thanks to slanting of screw-like or longitudinal conical ferrimagnetic spin structures (which normally occur along *c*-axes and alternating Fe layers) in the magnetic field, which make them partially transversal and susceptible to inverse DMI [33]. At higher fields, the spin structure changes to noncollinear ferrimagnetic through an unknown transitional phase, followed with an increase in magnetization (10^1 emu/g), but also in polarization, which reaches up to 1.5×10^{-3} $\mu C/cm^2$ (quite high for type II MF), before becoming collinear ferrimagnetic and paraelectric above 2 T. The magnetoelectric coupling is very strong, even at close-to-room temperatures, and low magnetic fields (~ 100 Oe) although the electrical resistivity becoming too low prevents ferroelectric properties above 130K [35].

The situation is similar in Z-type ($Sr_3Co_2Fe_{24}O_{41}$) and, with much smaller magnetically induced polarization, in U-type hexaferrites ($Sr_4Co_2Fe_{36}O_{60}$) [35]. M-type hexaferrite $SrFe_{12}O_{19}$ (boron doped), on the other hand, was reported with much-improved properties, suitable more to a hybrid of type I and type II MFs. It doesn't need a magnetic field to exhibit ferroelectricity, but the polarization is still probably caused by transversal conical spin structure stable even without the field. Very large values of T_N (729K), M_S (40 emu/g), and P_S (40 $\mu C/cm^2$), and excellent magnetoelectric coupling, were measured [37]. M-hexaferrites even exhibited possible effect of electric field on magnetization (Fig. 25.1B) [37,49], which gives them huge application potential. However, the mechanism for this kind of type II multiferroicity was not proposed.

Very interesting is the behavior of CuO. It shows two AFM transitions: paramagnetic to noncollinear spiral spin antiferromagnetic at 230K, and then to collinear antiferromagnetic at 213K. Between those two temperatures, CuO is ferroelectric with small polarization (10^{-2} $\mu C/cm^2$), but the temperature region in which magnetically induced ferroelectricity exists is exceptionally high thanks to strong superexchange [8].

25.7.2 Magnetostriction-driven ferroelectrics

Although being studied for long time, mixed manganese oxides, $REMn_2O_5$ (RE = Pr-Lu, Bi, Y) were recognized as MFs very recently. They contain two types of manganese ions, Mn^{3+} and Mn^{4+}, which occupy different crystallographic positions. Mn^{4+} ions are in oxygen octahedra, while Mn^{3+} ions are placed at the bases of oxygen square pyramids [44]. If watched from the direction parallel to the *c*-axis, the pyramids, *p*, and edge sharing pairs of octahedra, 2o, form vertex (v) and edge (e) connected loops in a manner-p(e)p(v)2o(v)p(v)2o-. Exchange prefers antiparallel ordering of neighboring spins in loops (helixes in 3D), but that is not sustainable within the loop of an odd number of Mn^{3+}/Mn^{4+} ions, so frustration takes place. In distorted Mn^{3+}-$2Mn^{4+}$ neighboring triplets, Mn^{3+} ions shift to or away from the other two, depending on their spins being more antiparallel or parallel, respectively. As explained in Section 25.4, the change of bonds of differently charged ions can induce unbalanced dipole moments. The net polarization is induced along the *b*-axis. There are several magnetic transitions, and the one at 38K−41K arises ferroelectricity and makes a sharp rise in value of the dielectric constant, indicating large magnetodielectric coupling, similarly to $REMnO_3$. Other contributions to polarization come into play at lower temperature magnetic transitions, reducing and even reversibly reorienting electric dipoles [18].

Under 16K, Ca_3CoMnO_6 develops E-type of AFM spin order (↑↑↓↓) in quasi-one-dimensional-O-Co-O-Mn-O- chains. The Co-Mn distance shortens or elongates depending on parallel or antiparallel spins in the pair, and displacements of differently charged ions induces polarization in the direction of the chain [8].

The case is similar to already mentioned smaller RE manganites in metastable orthorhombic phase. Their E-type AFM structure is two-dimensional, and although all the cations carry the same charge, bond bending and displacement of O^{2-} ions produce the polarization [8]. The highest theoretically predicted values reach up to 6 μC/cm^2 for o-$HoMnO_3$, although measured ones are 10^3 times lower [50]. The difference could be a consequence of competition between predicted and some other orders in real material, the quality of samples, but also the polarization of an electronic cloud, which can cancel out the ionic polarization [8]. Orthorhombic RE chromites and at very low temperatures (<10K), RE ferrites too exhibit this kind of type II MF property (explained by RE-Fe/Cr exchange caused magnetostriction) with 10^{-1} μC/cm^2 order of polarization and quite high magnetization (10^0 emu/g) [8]. Nickelates of smaller rare earth ions, $RENiO_3$ exhibit spontaneous charge disproportionation of Ni^{3+} into Ni^{2+} and Ni^{4+} below metal-insulator transition, so they are candidates for this type of MF [23].

25.7.3 Proper screw spin systems

Even the proper-screw noncollinear spin orders can produce polarization in some materials. Delafossite compounds, AMO_2, ($A = Cu^+$, Ag^+; $M = Cr^{3+}$, Fe^{3+}) crystallize in centrosymmetric space group and they have a proper screw-spin structure that does not allow inverse DMI, but polarization is still possible in low symmetry crystals like triclinic and monoclinic thanks to d-p hybridization [35]. In delafossite structures, triangular (quasi-2D) sublattice of each element with 120 degrees magnetic order in the plane perpendicular to the triangular layers breaks the inversion center, enabling the system to be polar [33]. And the direction of polarization can be reversed by change in vector spin chirality [35]. Once induced, it can be reversed by electric field. Polarization reaches 10^{-2} μC/cm^2, and electric coercive field strength increases with increasing the strength of the magnetic field [39].

A variation of the proper screw-spin structure is possibly present in triclinic $Cu_3Nb_2O_8$. Coupling the chiral component of the magnetic structure with the crystal structure is believed to be the cause of the electrical polarization [40].

The ferroelectricity mechanisms are not completely unambiguous, so some type II MFs with a mixture of different causes of polarization may exist. The class of so-called quadruple perovskites, like $CaMn_7O_{12}$ or $LaMn_3Cr_4O_{12}$, contains both non-collinear and collinear spin orders and it is believed that inverse DMI and magnetostriction in them may exist simultaneously, but other mechanisms like spin chirality were also proposed [8,41]. Magnetically induced polarization in those materials could be several orders of magnitude larger than in other type II MFs [41].

There are also materials that are not multiferroic by definition, but their properties are close to it. Cobaltites of type (Y,Ca)$BaCo_4O_7$ are antiferrimagnetic below 70K−110K, and have polar orthorhombic space group, and strong magneto(di)electric coupling is present, but the polarization is not switchable, neither by electric nor magnetic fields [51].

25.8 Conclusion

Some of the representative single multiferroic materials were presented very briefly in this chapter. The uprising field of multiferroics needs much more space, probably a chapter for each class, to systematically present all specific issues concerning those materials: origin and nature of interconnection between specific electric and magnetic properties and structure, overcoming synthesis problems, improvement of modification methods (compositional as well as nano-, micro-, and macrostructural). The main, but also the hardest task is to understand all the knowledge acquired worldwide and use it efficiently in finding the proper application.

References

[1] H. Schmid, Multi-ferroic magnetoelectrics, Ferroelectrics 162 (1994) 317−338.

[2] H. Schmid, Some symmetry aspects of ferroics and single phase multiferroics, J. Phys.: Condens. Matter 20 (2008) 434201.

[3] M. Fiebig, T. Lottermoser, D. Meier, M. Trassin, The evolution of multiferroics, Nat. Rev. Mater. 1 (2016) 16046.

[4] Nicola A. Hill, Why are there so few magnetic ferroelectrics? J. Phys. Chem. B104 (2000) 6694−6709.

[5] W. Eerenstein, N.D. Mathur, J.F. Scott, Multiferroic and magnetoelectric materials, Nature 442 (2006) 759−765.

[6] H. Zheng, Growth and characterization of multiferroic $BaTiO_3$-$CoFe_2O_4$ thin film nanostructures, (Ph.D. thesis), University Of Maryland, College Park, 2004.

[7] Yi Du, Multiferroic transition metal oxides: structural, magnetic, ferroelectric and thermal properties (Ph.D. thesis), Institute for Superconducting and Electronic Materials, University of Wollongong, NSW, Australia, 2011.

[8] S. Dong, J.-M. Liu, S.-W. Cheong, Z. Ren, Multiferroic materials and magnetoelectric physics: symmetry, entanglement, excitation, and topology, Adv. Phys. 64 (5-6) (2015) 519–626.

[9] K. Oka, I. Yamada, M. Azuma, S. Takeshita, K.H. Satoh, A. Koda, et al., Magnetic ground-state of perovskite $PbVO_3$ with large tetragonal distortion, Inorg. Chem. 47 (2008) 7355–7359.

[10] T. Varga, A. Kumar, E. Vlahos, S. Denev, M. Park, S. Hong, et al., Coexistence of weak ferromagnetism and ferroelectricityin the high pressure $LiNbO_3$-type phase of $FeTiO_3$, Phys. Rev. Lett. 103 (2009) 047601.

[11] A. Rotaru, A.J. Miller, D.C. Arnold, F.D. Morrison, Towards novel multiferroic and magnetoelectric materials: dipole stability in tetragonaltungsten bronzes, Phil. Trans. R. Soc. A372 (2014) 20120451.

[12] P. Barone, K. Yamauchi, S. Picozzi, Jahn-Teller distortions as a novel source of multiferroicity, Phys. Rev. B 92 (2015) 014116.

[13] L.W. Martin, D.G. Schlom, Advanced synthesis techniques and routes to new single-phase multiferroics, Curr. Opin. Solid St. M 16 (2012) 199–215.

[14] D. Khomskii, Classifying multiferroics: mechanisms and effects, Physics 2 (2009) 20.

[15] A.A. Belik, Polar and nonpolar phases of $BiMO_3$: a review, J. Solid State Chem. 195 (2012) 32–40.

[16] D.I. Khomskii, Multiferroics: different ways to combine magnetism and ferroelectricity, J. Magn. Magn. Mater. 306 (2006) 1–8.

[17] M. Guennou, M. Viret, J. Kreisel, Bismuth-based perovskites as multiferroics, C. R. Phys. 16 (2015) 182–192.

[18] S.-W. Cheong, M. Mostovoy, Multiferroics: a magnetic twist for ferroelectricity, Nat. Mater. 6 (2007) 13–20.

[19] J.Y. Moon, M.K. Kim, H.Y. Choi, S.H. Oh, Y. Jo, N. Lee, et al., Direct observation of magnetodielectric effect in type-I multiferroic $PbFe_{0.5}Ti_{0.25}W_{0.25}O_3$, Curr. Appl. Phys. 15 (2015) 1545–1548.

[20] S.A. Ivanov, P. Nordblad, R. Tellgren, H. Rundlof, G. André, F. Bourée, Neutron powder diffraction and magnetic study of perovskites $Pb(Mn_{1/2}Nb_{1/2})O_3$ and $Pb(Mn_{1/4}Fe_{1/4}Nb_{1/2})O_3$, Mater. Res. Bull. 43 (2008) 3074–3087.

[21] M.P. Singh, K.D. Truong, S. Jandl, P. Fournier, Multiferroic double perovskites: opportunities, issues, and challenges, J. Appl. Phys. 107 (2010) 09D917.

[22] R.I. Dass, J.B. Goodenough, Multiple magnetic phases of $La_2CoMnO_{6-\delta}$ ($0 \leq \delta \leq 0.05$), Phys. Rev. B 67 (2003) 014401.

[23] J. van den Brink, D.I. Khomskii, Multiferroicity due to charge ordering, J. Phys.: Condens. Matter 20 (2008) 434217.

[24] N. Ikeda, Ferroelectric properties of triangularcharge-frustrated $LuFe_2O_4$, J. Phys.: Condens. Matter 20 (2008) 434218.

[25] C.-Y. Ren, Atomic, electronic, and ferroelectric properties of manganite $RMnO_3$ (R = Ho, Er,Tm,Lu) in hexagonal and orthorhombic phases, Phys. Rev. B 79 (2009) 125113.

[26] Timothy James Stevenson, Magnetic and electric properties of bismuth ferrite lead titanate ceramics (Ph.D. thesis), Institute for Materials Research, School of Process, Environmental and Materials Engineering, The University of Leeds, 2010.

[27] Th. Lonkai, D.G. Tomuta, U. Amann, J. Ihringer, R.W.A. Hendrikx, D.M. Többens, et al., Development of the high-temperature phase of hexagonal manganites, Phys. Rev. B 69 (2004) 134108.

[28] J.R. Sahu, C.R. Serrao, N. Ray, C. Umesh, V. Waghmare, C.N.R. Rao, Rare earth chromites: a new family of multiferroics, J. Mater. Chem. 17 (2007) 42–44.

[29] P.P. Jiang, Z.H. Duan, L.P. Xu, X.L. Zhang, Y.W. Li, Z.G. Hu, et al., Phase transformation in multiferroic $Bi_5Ti_3FeO_{15}$ ceramicsby temperature-dependent ellipsometric and Raman spectra: an interband electronic transition evidence, J. Appl. Phys. 115 (2014) 083101.
[30] X. Chen, J. Xiao, J. Yao, Z. Kang, F. Yang, X. Zeng, Room temperature magnetoelectric coupling study in multiferroic $Bi_4NdTi_3Fe_{0.7}Ni_{0.3}O_{15}$ prepared by a multicalcination procedure, Ceram. Int. 40 (2014) 6815–6819.
[31] B. Lorenz, Y.Q. Wang, Y.Y. Sun, C.W. Chu, Large magnetodielectric effects in orthorhombic $HoMnO_3$ and $YMnO_3$, Phys. Rev. B 70 (2004) 212412.
[32] T. Kimura, Y. Tokura, Magnetoelectric phase control in amagnetic system showing cycloidal/conical spin order, J. Phys.: Condens. Matter 20 (2008) 434204.
[33] Y. Tokura, S. Seki, Multiferroics with spiral spin orders, Adv. Mater. 22 (2010) 1554–1565.
[34] Y. Yamasaki, S. Miyasaka, Y. Kaneko, J.-P. He, T. Arima, Y. Tokura, Magnetic reversal of the ferroelectric polarization in a multiferroic spinel oxide, Phys. Rev. Lett. 96 (2006) 207204.
[35] T. Kimura, Magnetoelectric hexaferrites, Annu. Rev. Condens. Matter Phys. 3 (2012) 93–110.
[36] R.B. Jotania, H.S. Virk, Y-type hexaferrites: structural, dielectric and magnetic properties, Solid State Phenom. 189 (2012) 209–232.
[37] V.G. Kostishyn, L.V. Panina, L.V. Kozhitov, A.V. Timofeev, A.N. Kovalev, Synthesis and multiferroic properties of M-type $SrFe_{12}O_{19}$ hexaferrite ceramics, J. Alloy Compd. 645 (2015) 297–300.
[38] B. Rajeswaran, D.I. Khomskii, A.K. Zvezdin, C.N.R. Rao, A. Sundaresan, Field-induced polar order at the Néel temperature of chromium in rare-earth orthochromites: interplay of rare-earth and Cr magnetism, Phys. Rev. B 86 (2012) 214409.
[39] K. Kimura, H. Nakamura, S. Kimura, M. Hagiwara, T. Kimura, Tuning ferroelectric polarization reversal by electric and magnetic fields in $CuCrO_2$, Phys. Rev. Lett. 103 (2009) 107201.
[40] R.D. Johnson, S. Nair, L.C. Chapon, A. Bombardi, C. Vecchini, D. Prabhakaran, et al., $Cu_3Nb_2O_8$: a multiferroic with chiral coupling to the crystal structure, Phys. Rev. Lett. 107 (2011) 137205.
[41] R.D. Johnson, L.C. Chapon, D.D. Khalyavin, P. Manuel, P.G. Radaelli, C. Martin, Giant improper ferroelectricity in the ferroaxial magnet $CaMn_7O_{12}$, Phys. Rev. Lett. 108 (2012) 067201.
[42] M.P. Singh, K.D. Truong, P. Fournier, P. Rauwel, E. Rauwel, L.P. Carignan, et al., Anomalously large ferromagnetic Curie temperature of epitaxial Bi_2CoMnO_6 thin films, Appl. Phys. Lett. 92 (2008) 112505.
[43] N.A. Spaldin, Analogies and differences between ferroelectrics and ferromagnets, in: K.M. Rabe, C.H. Ahn, J.-M. Triscone (Eds.), Physics of Ferroelectrics, Springer-Verlag Berlin Heidelberg, 2007, pp. 175–218.
[44] W. Prellier, M.P. Singh, P. Murugavel, The single-phase multiferroic oxides: from bulk to thin film, J. Phys.: Condens. Matter 17 (2005) R803–R832.
[45] I.E. Dzyaloshinskii, Thermodynamic theory of "weak" ferromagnetism in antiferromagnetic substances, Sov. Phys. Jetp 5 (1957) 1259–1272.
[46] T. Moriya, Anisotropic superexchange interaction and weak ferromagnetism, Phys. Rev. 120 (1960) 91–98.
[47] C.R. Serrao, A.K. Kundu, S.B. Krupanidhi, U.V. Waghmare, C.N.R. Rao, Biferroic YCrO3, Phys. Rev. B 72 (2005) 220101.

[48] C.H. Hervoches, A. Snedden, R. Riggs, S.H. Kilcoyne, P. Manuel, P. Lightfoot, Structural behavior of the four-layer Aurivillius-phase ferroelectrics $SrBi_4Ti_4O_{15}$ and $Bi_5Ti_3FeO_{15}$, J. Solid State Chem. 164 (2002) 280–291.
[49] L. Wang, D. Wang, Q. Cao, Y. Zheng, H. Xuan, J. Gao, et al., Electric control of magnetism at room temperature, Sci. Rep. 2 (2012) 223.
[50] S. Picozzi, K. Yamauchi, I.A. Sergienko, C. Sen, B. Sanyal, E. Dagotto, Microscopic mechanisms for improper ferroelectricity in multiferroic perovskites: a theoretical review, J. Phys.: Condens. Matter 20 (2008) 434208.
[51] K.R.S. Preethi Meher, C. Martin, V. Caignaert, F. Damay, A. Maignan, Multiferroics and magnetoelectrics: a comparison between some chromites and cobaltites, Chem. Mater. 26 (2014) 830–836.

Bulk composite multiferroics: $BaTiO_3$-ferrites

26

Mirjana M. Vijatovic Petrovic, Jelena D. Bobic and Biljana D. Stojanovic
Institute for Multidisciplinary Research, University of Belgrade, Belgrade, Serbia

26.1 Preparation procedures of bulk multiferroics

In recent years, composite materials consisted of two or more ferroic orders, such as ferroelectric, ferromagnetic, or ferroelastic, in which coupling between these phases can produce a magnetoelectric effect has been drawing significant attention [1]. It is well known that structural, magnetic, and electrical properties of ferroelectric as well as ferrite phases are strongly dependent on composition, preparation method, microstructure, and so on. It is very important to find an optimal ratio between these phases in order to obtain a composite with desired properties. Even though there are lot of publications referring to multiferroic composites, the preparation problems of these materials are still present and need to be solved. It is very important for preparation parameters to be monitored in order to obtain high homogeneity, high density, high structural compatibility of the components, no secondary phases and so on.

First, pure ferroelectric and ferrite phases should be prepared by a selected method. In a great number of papers, the most commonly used preparation procedure for ferroelectric and ferrite phases has been the convectional solid-state route. Detailed procedures for barium titanate (BT) and ferrite preparation by a solid-state reaction can be found elsewhere Although the solid-state route has many advantages, chemical syntheses are being recognized as methods that give much better properties of the resulting powders. Therefore, constituents of multiferroic composites often are being prepared by a sol-gel method, autocombustion method, hydrothermal of polymeric precursors method, among others [2–7].

The simplest way for multiferroic (MF) preparation is simple mixing of pure ferroelectric and ferrite powders in a desired mass ratio. These powders are being homogenized in a ball mill up to 24 h [4] or using agate mortar and pestle. Highly pure ethanol or isopropyl alcohols were frequently used as a medium for the homogenization. Thus, obtained precursor powders were pressed into pallets and sintered in the temperature range from 1000°C–1200°C, depending on the composition. Optimization of the sintering process (sintering temperature, time, and heating/cooling rate) is highly important and still quite difficult, since it is significant to obtain systems without secondary phases, with high density, clean interfaces, and so on. Even though it is very simple, a lot of disadvantages can appear in this preparation route. The difference in thermal expansion coefficient and/or the lattice mismatch between the two phases can lead to microcracks and low density of the final

Magnetic, Ferroelectric, and Multiferroic Metal Oxides. DOI: http://dx.doi.org/10.1016/B978-0-12-811180-2.00026-8

ceramics [8]. If the sintering temperature is too high for both phases, secondary phases can be found at the interfaces and can affect properties of composite material. During the optimization of sintering process for the $Ni_xZn_{(1-x)}Fe_2O_4$ (0.3, 0.5, 0.7)-$BaTiO_3$ (further in the text, NZF-BT) in the mass ratio 50-50 prepared by this method, the chosen temperature was 1200°C [9]. The formation of both BT and ferrite phases can be notices, as well as the peaks of secondary phases, which is identified as the $BaFe_{12}O_{19}$ phase. Literature data [10,11] have shown that the appearance of this secondary phase is quite often due to high sintering temperature. The formation of the $BaFe_{12}O_{19}$ phase causes structure and microstructure change and further affects electrical and magnetic properties of a final multiferroic composite. Since the sintering temperature used, T_s, is quite low for achieving good dielectric properties of BT ceramics, the resulting multiferroic material will possibly have lower permittivity than theoretically expected. Since ferrites are highly conductive, there is a possible problem with percolation that can make electrical poling impossible. Further optimization of sintering temperature has shown that 1170°C is the highest temperature at which no secondary phases appear. After selecting the sintering temperature it is necessary to find a holding time that will provide proper homogeneous microstructure and good density of ceramics.

Another method for the MF preparation is the wet route, where Fe-Ni-Zn (and/or other metals) nitric salts are coprecipitated in a NaOH solution, in which $BaTiO_3$ (or other ferroelectric) powder is previously dispersed [10]. Further treatment consists of drying, calcinations, pressing, and sintering in the optimized process for each composition. Regarding properties of obtained MFs, Testino et al. [12] compared these two preparation routes. More homogeneous microstructure with good dispersion of BT and NZF grains and small pores were found in the ceramics prepared by the second method, when the formation of NZF hard aggregates is strongly inhibited. The authors also noticed that composites prepared by the second method have presented much better dielectric and magnetic properties.

During the years of research an innovative wet chemistry route combined with solid state in order to obtain composite materials with a core-shell structure. Fig. 26.1 represents a preparation route in which ferrite particles in the form of nitric or chloride salts previously dissolved in water have been added into a suspension prepared by mixing Ti (IV) solution (usually Ti-isopropoxide, Ti-isobutoxide)

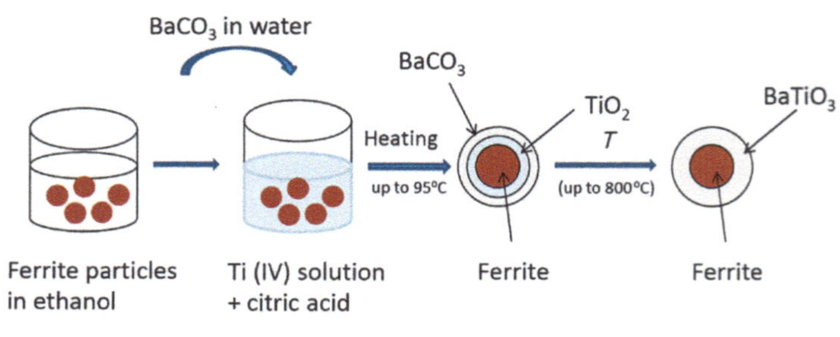

Figure 26.1 Preparation procedure of multiferroics with core-shell structure.

with Ba ions in the form of carbonates or acetates in distilled water. Vigorous stirring and heating enabled the formation of the core-shell particles. The ferromagnetic phase provides the core of the composite material being surrounded by a shell of insulating ferroelectric phase. It is also possible to prepare ferroelectric particles coated with magnetic phases but it makes composite material highly conductive and as it was mentioned earlier that will affect the poling process and formation of short circuits [8].

Therefore, the preparation procedure should be modified in order to obtain desired properties. To date, much literature data about core-shell structure preparations have been released [13]. The most investigated are $CoFeO_3$@$BaTiO_3$ [14–16], $CaFeO_3$@$BaTiO_3$ [17], and $NiFe_2O_4$@$BaTiO_3$ [18], among others. The main problem in the formation of core-shell structure composites is that every ferrite particle should be covered with a ferroelectric shell. However, there is a lack of appropriate TEM and HRTEM micrographs of these structures that will prove the real existence of true core-shell composites.

26.2 Ferroelectric-dependent electrical properties of the multiferroics

BT is a well-known ferroelectric material with excellent electrical properties (high dielectric constant, low tangent losses, high piezoelectric coefficient). When being mixed with ferro/ferrimagnetic phases in the multiferroic composites these properties are being compromised. Dielectric constant starts to decrease in comparison with pure BT ferroelectric material. Since BT has the capability to host ions of different sizes in its lattice, this problem could be solved by doping with various dopants. Literature data [19–20] have shown that addition of ions such as La^{3+}, Sb^{3+}, Sm^{3+}, Dy^{3+}, Gd^{3+}, and so on could enhance overall dielectric constant of BT and therefore the dielectric properties of multiferroic material (see Chapter 2: Perovskite and Aurivillius: Types of Ferroelectric Metal Oxides).

Dielectric anomalies that correspond to three structural transitions (cubic to tetragonal—$T_{C\text{-}T}$; tetragonal to orthorhombic—$T_{T\text{-}O}$; orthorhombic to rhombohedral —$T_{O\text{-}R}$) could be detected in the ceramics on the curve typical for classical ferroelectric BT material prepared by the Pechini method (Figure 26.2D). Sharp permittivity maximum, T_C at around 123°C and 13°C for $T_{T\text{-}O}$ and −73°C for $T_{O\text{-}R}$ can be noticed. The dielectric permittivity value at room temperature was around 2000 and at the Curie temperature was 6700; the dielectric losses tangent was around 0.02 [21].

BT powder prepared by a polymeric precursor method and nickel ferrite (NF) prepared by an autocombustion method were constituent phases of multiferroic materials prepared by a simple mixed route [4]. The composite ceramics were BT-NF composites at several frequencies: (1) BT-NF/80-20, (2) BT-NF/70-30, and (3) BT-NF/50-50 and sintered at 1170°C for 4 h. The dielectric properties of obtained ceramic composite are presented in Fig. 26.2A–C and show rather complex

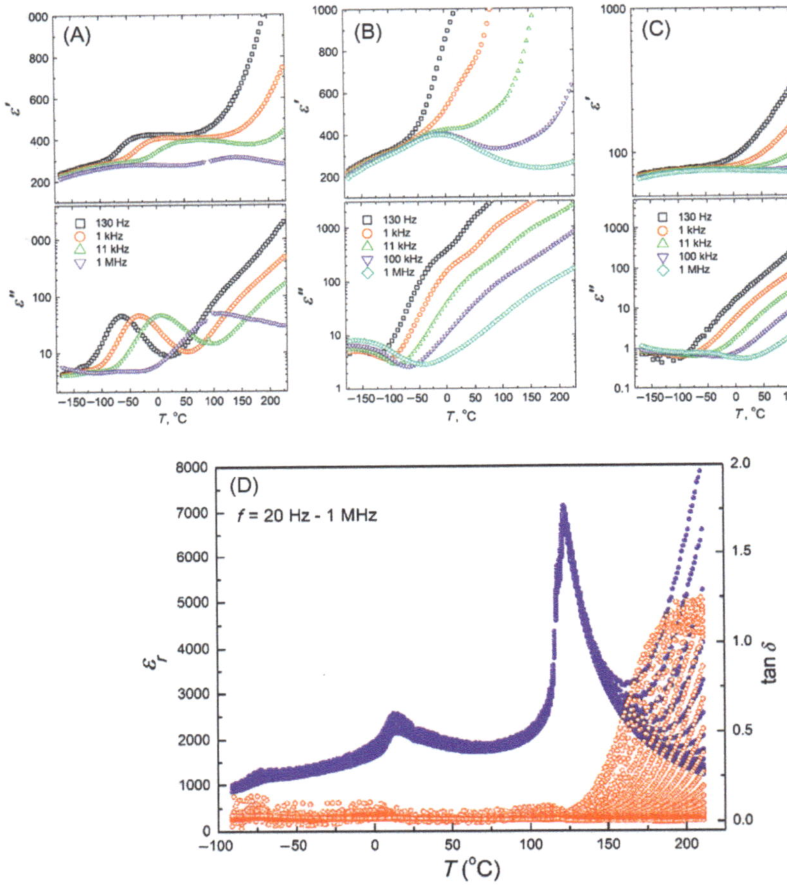

Figure 26.2 The temperature dependence of dielectric permittivity of BT-NF composites at several frequencies: (A) BT-NF/80-20, (B) BT-NF/70-30, (C) BT-NF/50-50 and (D) pure BT. Reprinted from R. Grigalaitis, M.M.Vijatović Petrović, J.D. Bobić, A. Dzunuzovic, R. Sobiestianskas, A. Brilingas, et al., Dielectric and magnetic properties of BaTiO3–NiFe2O4 multiferroic composites. Ceram. Int. 40 (2013) 6165–6170, Copyright (2014) with permission from Elsevier.

behavior. In comparison with dielectric measurements for pure BT given in Figure 13.3.2d, the shape of the characteristic curve was significantly changed. Three different regions of dispersion can be easily identified in the composite BT-NF/80-20. At high temperatures and low frequencies the dielectric spectra are governed by conductivity effects, which are seen as a steep increase of the dielectric permittivity curves. The main feature of this dispersion is the maximum seen in both the real and the imaginary parts of the dielectric permittivity, which shift to the lower temperatures with the decrease of frequency [4].

For the BT-NF/70-30 composition around 0°C and frequencies above 100 kHz one broad peak can be distinguished. In the BT-NF/50-50 ceramics we can also notice a very broad peak at slightly lower temperature, hidden by the high conductivity of the ceramics. The evident changes in dielectric dispersion can be also explained by a defect chemistry.

It is well known that different types of changes can be expected, depending on the type and concentration of dopants. Some dopants shift BT transition temperatures or induce broadening of the T_C curve and many of them cause diffuseness of the ferroelectric phase transition [22]. Merging of characteristic BT phase transitions into one broad peak is also possible, and is often called a pinching effect [23]. The degree of diffuseness of the transition could show a continuous change from classical ferroelectric to diffuse ferroelectric or to relaxor-like behavior. Literature data have shown that a partial substitution of Ba or Ti ions with dopants such as Zn, Y, or Sb causes the formation of diffuse transition and Sn, Ce, Zr, Bi, or Hf cause the appearance of ferroelectric relaxor behavior [24]. Therefore, it is quite possible that some amount of Fe^{3+} acted as a dopant in the BT lattice in the investigated composites. It cannot be seen in the XRD because the concentration is very low, but BT can be affected by these impurities. Fe^{3+} most likely replaces Ti^{4+} in the BT lattice and Ti−O-Ti linkages and also the ferroelectric properties can be destroyed. Thus, it can affect the movement and broadening of the dielectric permittivity maximum and induce the relaxor-like behavior in the composite materials [25].

On the other hand, during the years of investigation of ferroelectric materials, a close connection between ceramics microstructure and its dielectric properties is established. We have to point out that sintering temperature for MF preparation is quite low for BT ceramics and therefore grains formed in the multiferroics (are much rounded) differ from these in pure BT [21,26].

Even though the microstructure shows homogeneous distribution of ferroelectric and ferrite phases in the composite material, the densities of these composite ceramics are much lower (~85%). Thus, it was expected for dielectric properties to be diluted and the permittivity of obtained MFs is rather low and goes up to 400 at room temperature. Lokare et al. [27] connected the dielectric constant decrease with increase in ferrite content with Verway type of electron exchange interaction in ferrite on one hand and the polarization being an inherent property of ferroelectric on the other. Conduction in ferrites due to electron hopping also causes local displacement of charges in the direction of the external applied field, influencing dielectric polarization in ferrites.

Dielectric permittivity increases at higher temperatures possibly due to activation of conductivity mechanisms at these temperatures, causing also high dielectric losses (see Fig. 26.2). This behavior was also found by other authors investigating various compositions [2,7,23,27,28]. The problem is attributed to space charge effects, Maxwell−Wagner relaxations at the interfaces of the ferroelectric-magnetic phases, or even to other defect mechanisms, particularly those activated at low frequencies and high temperatures.

Frequency dependence of dielectric constants show quite similar behavior in different types of composite materials [2,6,28,29]. The high dielectric constant at lower frequencies is explained on the basis of space charge polarization and attributed to the fact that ferroelectric regions are surrounded by nonferroelectric regions just as that in the case of relaxor ferroelectric materials [2] and the heterogeneity of the samples. It can be also attributed to the changes in valence state of cations Fe^{3+}/Fe^{2+}, Co^{2+}/Co^{3+}, Cu^{2+}/Cu^{3+}, and Ni^{3+}/Ni^{2+}. Others attributed these phenomena to interfacial polarization present at the interface of two phases (Maxwell–Wagner). This is interpreted through the concept of hopping of charge carriers, which takes place in the ferrite phase. It is a consequence of the occupation of Ni^{2+} ions on the octahedral sites and Fe^{3+} on the tetrahedral sites. The apparent high dielectric constant is due to high losses. The profile of quick decrease of the dielectric loss with frequency clearly shows the polarization mechanism in ferrites, which is equivalent to a conduction process and mainly due to simultaneous presence of Fe^{2+} and Fe^{3+} on crystallographically equivalent sites. The hopping of the electrons in the neighboring sites causes the conduction and hence dielectric loss. This hopping follows the electric field at lower frequency, which causes dispersion in the dielectric constant [30,31].

In recent years, barium strontium titanate has been widely studied due to very high dielectric permittivity [6,32] and possibility to modify dielectric behavior not only by adjusting Ba/Sr ratio but by doping with different dopants. Therefore, this material has often been used as a ferroelectric part of multiferroic composite materials. In the work of Pahuja et al. [6] composite materials consisted of ferroelectric phase $Ba_{0.95-1.5x}Sr_{0.05}R_xTiO_3$ (where $x = 0.00$, 0.01, 0.02, 0.03; and R are rare earths Dy, Gd, Sm) (BST) prepared by a solid-state reaction and ferrite, $Ni_{0.8}Co_{0.2}Fe_2O_4$ by a ball milling, were prepared by a mixing route and then sintered at 1200°C for 6 h. The authors detected the formation of a two-phase system with no traces of secondary phases in pure as well as in rare-earth–doped BST ceramics. It was concluded that the dielectric properties were mainly affected by microstructure and phase composition. Some of the essential features given by careful investigation of dielectric properties are given as follows: (1) dielectric constant of both the pure and composite materials decreases with increase in frequency following Maxwell–Wagner relaxation mechanism, (2) composite samples possess lower dielectric constant as compared to corresponding pure samples, (3) addition of rare earths significantly increased the dielectric constant of BST and the same result has also been observed in composite samples that possess high value of dielectric constant as compared to BST composite. Also, T_C shifting and broadening of peaks due to BST doping with rare earths in pure and in composite materials as well, was noticed. The same preparation methods and also the best microstructures and the increase of ferroelectric phase content does not automatically lead to a reduction of losses. Complete study of the conduction mechanisms for each composition, correlated with optimum processing parameters, is necessary in order to understand the dielectric properties and the mechanisms of losses in the system [12].

Polarization-electric field hysteresis loops for composite material prepared from $(1 - x)$ $BaTiO_3$—$xNi_{0.5}Zn_{0.5}Fe_2O_4$ where $x = 5\%$, 10%, and 15% have shown that Ps decreases with increase of ferrite phase in the composite. In comparison with pure BT, Ps is much lower and as expected, the addition of ferrite phase dilute ferroelectric properties [3].

On the other hand, the value of coercive field increases with the increase of ferrite concentration. Also, the values of H_c of composites are higher than for pure BT. The possible explanation could be the existence of hindered and pinned domain wall motion of ferroelectric regions due to ferrite phase in the material. Similar dependence of ferroelectric properties and composition are obtained for different types of composites [33–35]. Pahuja et al. [6] have presented the investigation of composites prepared using barium strontium titanate doped with different dopants in various concentrations and nickel cobalt ferrite. They have given the conclusions that the ferroelectric properties are highly dependent on a number of oxygen vacancies exhibiting in the ceramics. These oxygen vacancies inhibit the movement of domains causing the domain wall pinning. Net result is increase in the value of coercive field and decrease in remnant polarization.

26.3 Ferrite-dependent magnetic properties of multiferroics

In terms of their magnetic properties ferrites can be classified into soft and hard ferrites, which refer to their low or high magnetic coercivity. Ferrites having a low coercivity are called soft ferrites. The low coercivity means that the material's magnetization can easily reverse direction without dissipating much energy (hysteresis losses), while high resistivity prevents eddy currents in the core, another source of energy loss [36]. Literature data have shown that the most often used soft ferrite materials to be a constituent of multiferroic composites are nickel-zinc ferrite (NZF with the formula $Ni_xZn_{(1-x)}Fe_2O_4$) and manganese-zinc ferrite (MZF with the formula $Mn_xZn_{(1-x)}Fe_2O_4$). NZF ferrites exhibit higher resistivity than MZF, and therefore they are more suitable for frequencies above 1 MHz.

On the other hand, ferrites that have a high coercivity and high remanence after magnetization are called hard ferrites. The high coercivity means the materials are very resistant to becoming demagnetized, an essential characteristic for a permanent magnet. They also have high magnetic permeability. These so-called ceramic magnets are cheap, and are widely used in household products such as refrigerator magnets. The most commonly used hard ferrite in multiferroics is cobalt ferrite, $CoFe_2O_4$ ($CoO \cdot Fe_2O_3$), which can be used in some media for magnetic recording.

In order to ensure the applicability of ferrites as a component in multiferroic devices, it is necessary to pay special attention to processing methods and properties of ferrites that are important for its functionality in multiferroic materials. How the concentration of Ni and Zn in the $Ni_xZn_{(1-x)}Fe_2O_4$ (0.0, 0.3, 0.5, 0.7, 1.0) can affect structure and magnetic properties and which composition is the best choice for usage in the multiferroics will be presented. These NZF compositions can be prepared by different types of methods, from solid state to chemical methods such as sol-gel, co-precipitation, citrate precursors, sonochemical process, among others [37].

$Ni_xZn_{(1-x)}Fe_2O_4$ (0.0, 0.3, 0.5, 0.7, 1.0) materials were prepared by an autocombustion method starting from nitric salts of appropriate metals [37]. Autocombustion synthesis represents an attractive method for ferrite preparation due to its short reaction time and low costs compared to the conventional methods. The purpose of the presented study was to investigate the effect of Zn content on the morphological, structural, magnetic, and electrical properties of nickel ferrite nanopowders and ceramics, having in mind possible use of ferrites as a component in the multiferroic devices [37]. Powders obtained from the autocombustion method were calcined at 1000°C for 1 h. The formation of pure nickel and nickel zinc ferrite phases with cubic spinel structure was detected in each powder. It was also noticed that calculated crystallite size has shown an increasing trend with Zn^{2+} content, indicating the influence of larger Zn^{2+} ions (45 nm) in comparison with Ni^{2+} (38 nm). In order to investigate structural, electrical, and magnetic properties of these compositions, the ceramics were prepared by sintering at 1250°C for 4 h. The authors noticed the formation of pure ferrite phases in all compositions. It was suggested that larger Zn^{2+} ions displaced the smaller Fe^{3+} ions (0.67 Å) from the tetrahedral to octahedral sites causing the elongation of the NZF crystal lattice [37]. The increase of zinc content led to the increase of the average grain size. It was observed that the shape of the grains in the case of ZF is more pyramidal compared to polygonal grains of NZF.

SEM images indicated that with increase of molar fraction of Zn, the density increases. It was assumed that the increase in density may be a consequence of Zn^{2+} concentration that slightly promotes the grain growth and improves densification, which was also observed in the work of Shirsath et al. [38]. Magnetic measurements showed that magnetization was rising with increasing of Zn content up to 30%, and followed by a further decrease. The resistance was found to decrease with the increase in zinc content up to NZF (50−50) and then increase. For all obtained ceramic samples grain boundary resistance has the highest contribution to the total resistance. From this research it was concluded that NZF (70−30) and NZF (50−50) possess the highest values of magnetization and the lowest value of coercive field, indicating that this material may be a good candidate as a part of multiferroic composites.

Doping of nickel ferrite with various metals such as zinc, chromium, copper, and manganese are usually used to improve some of their electric and magnetic properties [7,27]. NF substituted with cobalt and cooper was selected as a ferrite phase presuming that Ni^{2+} and Cu^{2+}, being Jahn−Teller ions, have high mechanical coupling leading to high magnetostriction [2]. Van Uitert has shown that a very large increase in room temperature resistivity of nickel ferrite can be achieved by substituting 1% or 2% of copper or cobalt ions [2]. The Jahn−Teller like Mn^{2+}, Ni^{2+} has a high coupling coefficient, and its corresponding contribution to magnetostriction is high [27].

Literature data have shown that $Ni_{0.5}Zn_{0.5}Fe_2O_4$ is the most frequently used composition as a part of multiferroic materials [10,12,18,28]. The authors have used different processing methods and sintering parameters in order to obtain the best densities and ceramics without secondary phases.

As it is very important to choose good ferrite material it is also necessary to find the best mass ratio between ferroelectric and ferro/ferrimagnetic phases. In order to prepare materials with desired properties for specific applications we can vary the mass ratio between the constituent components. Grigalaitis et al. [4] presented a study of composite materials prepared from NF and BT powders (prepared by chemical methods) and processed by a mixing route. Sintering was performed at 1170°C for 4 h and no secondary phases were detected. The presence of non-magnetic perovskite phase was expected to influence the magnetic properties of composites.

The hysteresis loops show the presence of the ordered magnetic structure derived from unbalanced antiparallel spins between Fe^{3+} ions at tetrahedral sites and Ni^{2+} ions at octahedral sites as in the pure NF material [4]. Compared to pure NF, magnetization in composite materials is reduced. As the ferrite concentration increases, saturation of the magnetization moment increases. The coercivity H_c (Oe) increases with BT concentration in obtained multiferroic materials.

Pure $Ni_{0.5}Zn_{0.5}Fe_2O_4$ prepared by the autocombustion method has shown the characteristics of soft magnetic material with a well-defined magnetic hysteresis loop [37]. Saturation of the magnetization occurs at around 70 emu/g, which is higher than 48 emu/g found for pure NF ceramics prepared by the same type of synthesis. The coercive field value is 0.574 Oe, which is much lower than 14 Oe obtained for NF ceramics. On the other hand, NZF prepared by the coprecipitation method has shown higher values of saturation magnetization, ~130 emu/g, indicating the influence of the method used on magnetic properties of ferrite [12].

As already mentioned, this type of soft ferrite material is the most commonly used as a constituent phase of MFs. The literature data of composites xBT-(1-x) NZF prepared using NZF prepared by the coprecipitation method have given similar information. Hysteresis loops for all compositions show the presence of an ordered magnetic structure and clearly observed dilution of magnetic properties caused by a ferroelectric BT. Small values of coercivity and saturation fields were also evident [12]. The work of Upadhyay et al. [3] also proved the same trend of changes of the magnetic properties in the compositions with a lower amount of ferrite phase ($x = 5\%$, 10%, and 15%). Similar behavior was found in the study of $0.5BaTiO_3$-$0.5MgFe_3O_4$ by Ravindar et al. [39]. Soft magnetic properties and saturation with increasing magnetic fields were detected in all samples.

Cobalt ferrite as a hard ferrite is the most frequently used of the multiferroic composites. The advantage of cobalt ferrite among other hard spinel ferrites as a part of MFs is its high chemical stability, large magnetooptic effect, and high magnetostriction coefficient [40]. It is not very useful for high frequency application due to its very low permeability [41–42]. All properties, including the magnetostrictive coefficient, could be modified by a small change made in the structure and in preparation procedures. Bhame et al. [43] have shown that the sintering regime highly affected the microstructure and magnetostriction of $CoFe_3O_4$. Ceramics sintered at 1100°C had the highest magnetostriction, even though these samples were very porous but with uniform and small grain size. The ceramics sintered at 1200°C possessed much higher density, ~95%, with small grain size but the magnetostriction was evidently reduced. The investigation of the composites $0.5BaTiO_3 + 0.5CoFe_2O_4$ prepared by a wet chemical route have shown the magnetostriction about 30 ppm, which is less than half of magnetostriction that shows pure $CoFe_2O_4$ (~130 and ~70 ppm for parallel and perpendicular measurements, respectively). This indicates that the part of the strain that produces ferrite phase is adsorbed due to different elastic constants of constituents as well as of actual microstructure [15].

The substitution of one of the metallic ions with another or with a combination that ensures electrical neutrality significantly influences the electric, magnetic, and magnetostrictive properties of cobalt ferrite [42,44,45]. The substitution of Fe with Mn and Cr increases stress sensitivity and decreases saturation magnetization compared to stoichiometric cobalt ferrite [45].

Cr substitution also decreased the Curie temperature and the maximum magnitude of magnetostriction. It was proved by a Mössbauer measurement that Cr^{3+} ions substitute into the octahedral sites with stronger preference than Mn^{3+} and displace the Co^{2+} ions to tetrahedral sites more effectively than Mn^{3+} ions do. For small amounts of substituent, both Cr-substituted and Mn-substituted cobalt ferrites have maximum strain derivatives $d\lambda/dH_{max}$ greater than that of pure cobalt ferrite, and therefore show promise for magnetoelastic sensing applications and better properties in the composite multiferroic materials [41].

The systems $BaTiO_3$-$CoFe_2O_4$ and $BaTiO_3$-$CoMn_{0.2}Fe_{1.8}O_4$ were prepared by mixing route and sintering for 1150°C/1 h in the study of Hrib et al. [45]. In both systems, only pure ferroelectric and ferrite phases were found without the phase percolation. It was noticed that when the iron is substituted by manganese ions coercive field deceases and maximum magnetization slightly increases. These authors connected this small increase of magnetization with the oxidation state of the manganese ions. From spectroscopic measurements, it was determined that Mn^{3+} ions have strong site preference for octahedral sites, while Mn^{2+} have no particular preference for either coordination. If the amount of Mn^{2+} ions from octahedral sites exceeds that of Mn^{3+}, a large increase in the magnetization is expected, whereas if the amount of Mn^{3+} ions is larger, the increase in the magnetic moment will be smaller. On the other hand, the manganese substitution for iron ions decreases the coercive field. Mossbauer studies performed on these systems [46] have shown that the substitution of iron with manganese ions causes the displacement of cobalt ions from octahedral sites. The large anisotropy in cobalt ferrite is due to the presence of cobalt ions in the octahedral sites; their displacement decreases the anisotropy and, consequently, the value of H_C. A high magnetostriction coefficient is desirable to achieve a high magnetoelectric signal. In the study of these authors, the substitution of iron with a small amount of manganese did not show any significant alteration of magnetostrictive properties.

References

[1] H. Palneedi, V. Annapureddy, S. Priya, J. Ryu, Status and perspectives of multiferroic magnetoelectric composite materials and application, Actuators 5 (2016) 9.

[2] R.S. Devan, B.K. Chougule, Magnetic properties and dielectric behavior in ferrite/ferroelectric particulate composites, Physica B 393 (2007) 161–166.

[3] S.K. Upadhyay, V.R. Reddy, N. Lakshmi, Study of (1 − x) $BaTiO_{3-x}$ $Ni_{0.5}Zn_{0.5}Fe_2O_4$ (x = 5, 10 and 15%) magneto-electricceramic composites, J. Asian Ceram. Soc. 1 (2013) 346−350.

[4] R. Grigalaitis, M.M. Vijatović Petrović, J.D. Bobić, A. Dzunuzovic, R. Sobiestianskas, A. Brilingas, et al., Dielectric and magnetic properties of $BaTiO_3$−$NiFe_2O_4$ multiferroic composites, Ceram. Int. 40 (2013) 6165−6170.

[5] M.M. Vijatović, J.D. Bobić, B.D. Stojanović, History and challenges of barium titanate: Part I, Sci. Sinter. 40 (2) (2008) 155−165.

[6] P. Pahuja, R.K. Kotnala, R.P. Tandon, Effect of rare earth substitution on properties of barium strontium titanate ceramics and its multiferroic composite with nickel cobalt ferrite, J. Alloy Compd. 617 (2014) 140−148.

[7] R.J. Pandya, U.S. Joshi, O.F. Caltun, Microstructural and electrical properties of barium strontium titanate and nickel zinc ferrite composites, Proc. Mater. Sci. 10 (2015) 168−175.

[8] G. Schileo, Recent developments in ceramic multiferroic composites based on core/shell and other heterostructures obtained by sol-gel routes, Prog. Solid State Ch. 41 (2013) 87−98.

[9] A.S. Dzunuzovic, M.M. Vijatovic Petrovic, B.S. Stojadinovic, N.I. Ilic, J.D. Bobic, C.R. Foschini, et al., Multiferroic $(NiZn)Fe_2O_4$−$BaTiO_3$ composites prepared from nanopowders by auto-combustion method, Ceram. Int. 41 (2015) 13189−13200.

[10] L. Mitoseriu, V. Buscaglia, M. Viviani, M.T. Buscaglia, I. Pallecchi, C. Harnagea, et al., $BaTiO_3$− $(Ni_{0.5}Zn_{0.5})Fe_2O_4$ ceramic composites with ferroelectric and magnetic properties, J. Eur. Ceram. Soc. 27 (2007) 4379−4382.

[11] K. Verma, S. Sharma, Impedance spectroscopy and dielectric behavior in barium strontium titanate-nickel zinc ferrite composites, Phys. Status Solidi B 249 (2012) 209−216.

[12] A. Testino, L. Mitoseriu, V. Buscaglia, M.T. Buscaglia, I. Pallecchi, A.S. Albuquerque, et al., Preparation of multiferroic composites of $BaTiO_3$−$Ni_{0.5}Zn_{0.5}Fe_2O_4$ ceramics, J. Eur. Ceram. Soc. 26 (2006) 3031−3036.

[13] V.V. Srdi, B. Moji, M. Nikoli, S. Ognjanović, Recent progress on synthesis of ceramics core/shell nanostructures, Process. Appl. Ceram. 7 (2) (2013) 45−62.

[14] M. Yu, J. Hu, J. Liu, S. Li, Electromagnetic properties of multiferroic magnetoelectric $BaTiO_3$−$Co_xFe_{3-x}O_4$ core−shell particles obtained by homogeneous coprecipitation, J. Magn. Magn. Mater. 326 (2013) 31−34.

[15] V.G. Duong, R. Turtelli, R. Groessinger, Magnetoelectric properties of $CoFe_2O_4$−$BaTiO_3$ core-shell structure composite studied by a magnetic pulse method, J. Magn. Magn. Mater. 322 (2010) 1581−1584.

[16] A. Chaudhuri, K. Mandal, Large magnetoelectric properties in $CoFe_2O_4$: $BaTiO_3$ core−shell nanocomposites, J. Magn. Magn. Mater. 377 (2015) 441−445.

[17] T. Woldu, B. Raneesh, B. Krishna Hazra, S. Srinath, P. Saravanan, M.V. Ramana Reddy, et al., A comparative study on structural, dielectric and multiferroic properties of $CaFe_2O_4$/$BaTiO_3$ core-shell and mixed composites, J. Alloy Compd. 691 (2017) 644−652.

[18] J.-P. Zhou, L. Lv, Q. Liu, Y.-X. Zhang, P. Liu, Hydrothermal synthesis and properties of $NiFe_2O_4$@$BaTiO_3$ composites with well-matched interface, Sci. Technol. Adv. Mat. 13 (2012) 045001.

[19] M.M. Vijatovic Petrovic, J.D. Bobic, J. Banys, B.D. Stojanovic, Electrical properties of antimony doped barium titanate ceramics, Mater. Res. Bull. 48 (2013) 3766−3772.

[20] M.M. Vijatović Petrović, J.D. Bobić, T. Ramoška, J. Banys, B.D. Stojanović, Electrical properties of lanthanum doped barium titanate ceramics, Mater. Charact. 62 (10) (2011) 1000−1006.

[21] M.M. Vijatovic Petrovic, J.D. Bobic, A.M. Radojkovic, J. Banys, B.D. Stojanovic, Improvement of barium titanate properties induced by attrition milling, Ceram. Int. 38 (2012) 5347–5354.

[22] N.M. Burange, S.S. Chougule, D.R. Patil, R.S. Devan, Y.D. Kolekar, B.K. Chougule, Studies on structural, electrical and magnetic properties of $y(Ni_{0.5}Zn_{0.3}Co_{0.2}Fe_2O_4)$ + (1-y)($BaTiO_3$) composites, J. Alloy Compd. 479 (1–2) (2009) 569–573.

[23] C.E. Ciomaga, M. Airimioaei, V. Nica, L.M. Hrib, O.F. Caltun, A.R. Iordan, et al., Preparation and magnetoelectric properties of $NiFe_2O_4$–PZT composites obtained in-situ by gel-combustion method, J. Eur. Ceram. Soc. 32 (12) (2012) 3325–3337.

[24] J. Azadmanjiri, S.A. Seyyed, Ebrahimi, H.K. Salehani, Magnetic properties of nanosize $NiFe_2O_4$ particles synthesized by sol-gel autocombustion method, Ceram. Int. 33 (2007) 1623–1625.

[25] R. Grigalaitis, M.M. Vijatovic Petrovic, D. Baltrunas, K. Mazeika, B.D. Stojanovic, J. Banys, Broadband dielectric and Mössbauer studies of $BaTiO_3$–$NiFe_2O_4$ composite multiferroics, J. Mater. Sci.-Mater. El. 26 (2015) 9727–9734.

[26] M.M. Vijatovic Petrovic, J.D. Bobic, H. Ursic, J. Banys, B.D. Stojanovic, The electrical properties of chemically obtained barium titanate improved by attrition milling, J. Sol-Gel Sci. Techn. 67 (2013) 267–272.

[27] S.A. Lokare, D.R. Patil, R.S. Devan, S.S. Chougule, Y.D. Kolekar, B.K. Chougule, Electrical conduction, dielectric behavior and magnetic effect in $xBaTiO_3$-(1-x) $Ni_{0.94}Co_{0.01}Mn_{0.05}Fe_2O_4$ ME composites, Mater. Res. Bull. 43 (2008) 326–332.

[28] A. Sakanas, R. Grigalaitis, J. Banys, L. Mitoseriu, V. Buscaglia, P. Nanni, Broadband dielectric spectroscopy of $BaTiO_3$–$Ni_{0.5}Zn_{0.5}Fe_2O_4$ composite ceramics, J. Alloy Compd. 602 (2014) 241–247.

[29] S. Khader, T. Sankarappa, Dielectric, magnetic and ferroelectric studies in (x) $Mn_{0.5}Zn_{0.5}Fe_2O_4$ + (1-x)$BaTiO_3$ magnetoelectric nano-composites, Mater. Today: Proc. 3 (2016) 2358–2365.

[30] N. Adhlakna, K.L. Yadav, Study of structural, dielectric and magnetic behavior of $Ni_{0.75}Zn_{0.25}Fe_2O_4$-$Ba(Ti_{0.85}Zr_{0.15})O_3$ composites, Smart Mater. Struct. 21 (2012) 115021.

[31] S.A. Khader, T. Sankarappa, T. Sujatha, J.S. Ashwajeet, R. Ramanna, Structural and dielectric studies on magneto electric nano-composites, Mater. Today: Proc. 2 (2015) 4334–4343.

[32] A. Kumar Yadav, C.R. Gautam, P. Singh, Effect of donor and acceptor dopants on crystallization, microstructural and dielectric behaviors of barium strontium titanate glass ceramics, J. Alloy Compd. 672 (2016) 52–58.

[33] Y. Wang, Y. Wang, W. Rao, M. Wang, G. Li, Y. Li, et al., Dielectric, ferromagnetic and ferroelectric properties of the (1-x)$Ba_{0.8}Sr_{0.2}TiO_3$–$xCoFe_2O_4$ multiferroic particulate ceramic composites, J. Mater. Sci.-Mater. El. 23 (2012) 1064–1071.

[34] C. Ciomaga, A. Neagu, M. Pop, M. Airimioaei, S. Tascu, G. Schileo, et al., Ferroelectric and dielectric properties of ferrite-ferroelectric ceramic composites, J. Appl. Phys. 113 (2013) 074103.

[35] L. Zhang, J. Zhai, W. Mo, X. Yao, Electric and magnetic properties of (x)$CoFe_2O_4$-(1-x)$BaTiO_3$ thick film prepared by electrophoretic deposition technique, Solid State Sci. 13 (2011) 321–325.

[36] https://en.wikipedia.org/wiki/Ferrite_(magnet).

[37] A.S. Džunuzović, N.I. Ilić, M.M. Vijatović Petrović, J.D. Bobić, B. Stojadinović, Z. Dohčević-Mitrović, et al., Structure and properties of Ni–Zn ferrite obtained by auto-combustion method, J. Magn. Magn. Mater. 374 (2015) 245–251.

[38] S.E. Shirsath, S.M. Patange, R.H. Kadam, M.L. Mane, K.M. Jadhav, Structure refinement, cation site location, spectral and elastic properties of Zn^{2+} substituted $NiFe_2O_4$, J. Mol. Struct. 1024 (2012) 77–83.
[39] R. Tadi, Y.I. Kim, K.S. Ryu, C.G. Kim, Effect of magnetic field on the dielectric properties of multiferroic composites, J. Korean Phys. Soc. 61 (2012) 1545–1549.
[40] M. Etier, C. Schmitz-Antoniak, S. Salamon, H. Trivedi, Y. Gao, A. Nazrabi, et al., Magnetoelectric coupling on multiferroic cobalt ferrite–barium titanate ceramic composites with different connectivity schemes, Acta Mater. 90 (2015) 1–9.
[41] S.J. Lee, ChesterC.H. Lo, P.N. Matlage, S.H. Song, Y. Melikhov, Magnetic and magnetoelastic properties of Cr substituted cobalt ferrite, J. Appl. Phys. 102 (2007) 073910.
[42] P.A. Shaikh, R.C. Kambale, A.V. Rao, Y.D. Kolekar, Studies on structural and electrical properties of $Co_{1-x}Ni_xFe_{1.9}Mn_{0.1}O_4$ ferrite, J. Alloy Compd. 482 (2009) 276–282.
[43] S.D. Bhame, P.A. Joy, Effect of sintering conditions and microstructure on the magnetostrictive properties of cobalt ferrite, J. Amer. Ceram. Soc. 91 (2008) 1976–1980.
[44] L.V. Leonel, J.B. Silva, A.S. Albuquerque, J.D. Ardisson, W.A.A. Macedo, N.D.S. Mohallem, Structural and Mössbauer investigation on barium titanate–cobalt ferrite composites, J. Phys. Chem. Solids 73 (2012) 1362–1371.
[45] L.M. Hrib, O.F. Caltun, Effects of the chemical composition of the magnetostrictive phase on the dielectric and magnetoelectric properties of cobalt ferrite–barium titanate composites, J. Alloy. Compd. 509 (2011) 6644–6648.
[46] K. Krieble, T. Schaeffer, J.A. Paulsen, A.P. Ring, C.C.H. Lo, J.E. Snyder, Mössbauer spectroscopy investigation of Mn-substituted Co-ferrite $CoMn_xFe_{2-x}O_4$, J. Appl. Phys. 97 (10) (2005) F101.

Complex composites: Polymer matrix-ferroics or multiferroics

27

Biljana D. Stojanovic, Adis S. Dzunuzovic, Nikola I. Ilic and Mirjana M. Vijatovic Petrovic
Institute for Multidisciplinary Research, University of Belgrade, Belgrade, Serbia

The emerging field of electronics integration and low dimensionality places new physical requirements on electronic components. Over the past several decades, the continuous growth of the microelectronics industry has relied on silicon-based devices. The requirements for faster, cheaper, and more energy-efficient electronics have fueled the continued progress of integrated circuits. The aggressive reductions in the critical dimensions of electronic devices and components have led to enormous increases in speed, functionality, and the possibility for various applications, but require different configuration and more complex composite architecture design. Besides efforts to improve the performance of electronic devices that are based on hard and brittle substrates, a new challenge in electronics is recently the flexibility to survive the mechanical deformation of material. Flexible electronics, which can be bent, rolled, and stretched into arbitrary shapes, would significantly expand the applications of modern electronic devices in the areas of computation, communications, displays, sensing, and energy harvesting and storage devices. There have been tremendous efforts to develop electronic devices based on mechanically flexible substrates. The first new concept was developed in 1978 by Skinner et al. [1] and in 1986 by Newnham [2], where the manner in which the individual phases are self-connected is described.

As there are no truly stretchable electronics, the most successful approach is to use flexible electronics conceived as a composite combination of highly flexible polymers with functional components. A composite material or shorter composite is a material made from two or more constituent materials with specific physical or chemical properties that, when combined, produce a material with characteristics different from those of the individual components. The new materials are preferred for many reasons: they may be stronger, lighter, or less expensive when compared to traditional materials or may have new or advanced behavior. Recently, the composites are involved in sensing, actuation, computation, and communication.

There are two main categories of constituent materials: matrix and active components. Many different polymers are available as matrix materials due to their capability of being molded or shaped by the application of heat and pressure. The plasticity, often found in combination with low density, low electrical conductivity, transparency, and toughness, allows their use in a great variety of products. Various active components can be used. Ferroics and multiferroics are very attractive materials for the next generation computer memories, smart sensors, high frequency

Magnetic, Ferroelectric, and Multiferroic Metal Oxides. DOI: http://dx.doi.org/10.1016/B978-0-12-811180-2.00027-X

microelectronic devices, electric field-controlled magnetic data storage, highly sensitive field sensors, transducers, actuators, energy harvesting, among others. Ferroic (ferroelectrics, ferromagnetics, and/or ferroelastics) and multiferroic magnetoelectric (ME; simultaneously exhibit ferroelectricity and ferromagnetism) materials have recently stimulated a sharply increasing scientific interest and significant technological promise in the novel multifunctional devices [3]. The coupling in ME between the constituent ferroic orders due to spontaneous electrical polarization in a magnetic field and an induced magnetization in an electrical field offers multifunctional properties and enhanced functionalities absent in each of the individual phases.

Currently, there are two possible approaches in the implementation and design of the flexible composites for new electronic devices. The first method is based on the combination of the inorganic material (rigid functional component) with polymer, whereby the functional component is applied onto the polymer flexible support by process of printing (screen-printing or ink-jet printing) or gas deposition or deposition from the solution. The supporting organic materials should ensure the transparency and flexibility, and need to be chemically inert. It is usually based on polyimide such as Kapton or poly(dimethyl)siloxane (PDMS). Recently, various devices/circuits on flexible and/or transparent substrates are designed, such as field effect transistors, organic light emitter diodes, transparent monitors, transparent mobile phone design, antennas, and GPS assisted driving [4]. Skin-mounted electronics (epidermal electronics) based on that concept is also the subject of huge interest due to possible important applications in biomedical devices [4]. The second approach to design flexible electronics is more recent and essentially represents an integration of inorganic and organic materials into a unique complex composite. In that design, the inorganic materials are the functional component with electric, magnetic, or electromagnetic properties, and the organic component is a polymer that should possess some of the properties such as conductivity, piezoelectricity, ferroelectricity, and/or dielectric properties.

In recent years, there has been a significant research effort to design flexible composite of ferroic materials. Unlike ceramic-based ME composites, polymer-based ME composites provide the advantages of being flexible, versatile, lightweight, low cost, able to conform to complicated shapes, and typically involve a low-temperature fabrication process [5]. However, the evident problem in the design and fabrication is that the active component (ferroic or multiferroic) uses inorganic structures that are rigid, and that the matrix is an organic component, which is flexible. For flexible electronics based on the composites mentioned earlier, the need is to achieve the compatibility between active inorganic and matrix organic phases, which is rather difficult because they are, by definition, incompatible related to the processing temperature. Namely, the preparation of oxides, in which the functional properties mainly exist in their crystalline state, requires the high temperature to crystallize (600°C−700°C), unlike organics, in which the plastic deformations, melting, and decomposition take place above 500°C. To overcome issues related to inorganic and organic material properties, it is necessary to solve many problems and raise the level of knowledge in many fields, such as chemistry, physics, modeling, design, materials sciences, and fabrication. Reducing the synthesis temperature for inorganic materials and production of new micro/nanostructures is necessary for development of new synthesis methods, which enable the preparation of inorganic products at lower temperatures. Numerical methods considering the system's geometry or percolation probabilities, understanding relationship between components in composite (e.g., active component-polymer) and development of new methods for producing the inorganic products at lower temperatures are an important tool to design such composites.

Although the magnetoelectric response has been observed as an intrinsic effect in single-phase multiferroic materials, the ME coefficients are usually too weak to be technologically viable [6]. The difficulties associated with uniting electrical and magnetic ordering in a single phase have been circumvented by forming two-phase multiferroic composites of piezoelectric and magnetostrictive components that can be electromagnetically coupled by stress mediation. It was pointed that for high ME voltage coefficient in composite, a suitable combination of piezoelectric (with high piezoelectric voltage coefficient) and magnetostrictive (with large piezomagnetic coefficient) components, synthesized in optimized conditions with homogenous microstructure, is required. For example, ferrite/piezoelectric ceramic composites, such as $CoFe_2O_4/BaTiO_3$ or $NiFe_2O_4$/PZT, have been found to exhibit much higher ME coupling than the single-phase multiferroics. ME effect in such composites is known as an extrinsic product property, resulting from the interaction of the ferrite and piezoelectric constituents [7].

As previously mentioned, compared with traditional ceramics, polymer-based multiferroic materials have an advantage of easy processing, mechanical flexibility, lightweight, and the ability to conform to complicated shapes in addition to their potentially higher ME responses. By developing poly(vinylidene fluoride) (PVDF)-metglas composites and multilayered PVDF/Terfenol-D 2−2 composites, the importance of laminate configuration and high-permeability magnetostrictive layers in the composites was highlighted in the recent published data [8−14].

The study of polymer-based ferroics and multiferroics is mostly centered on two-phase materials. The potentials of flexible polymer-magnetostrictive composite materials, synthesized by economic chemical-based processing routes, has been outlined by several authors. Polymer-ferroelectric nanocomposites are currently of interest as high permittivity materials for electronic applications such as embedded capacitors, multilayer capacitors, high-energy-density capacitors, gate insulators, and transistors. These nanocomposites can have high permittivity and dielectric strength and they can be processed on large and flexible substrates at low temperatures and ambient pressures. However, the high surface energy of dielectric nanoparticles usually leads to agglomeration and phase separation from the polymer matrix, resulting in high defect density and poor process ability of the films. Kim et al. [8] prepared high-energy-density nanocomposites based on phosphonic acid modified barium titanate and poly(vinylidene-fluoride-*co*-hexafluoropropylene) − P (VDF-HFP). Nanocomposites were prepared by ball-milling in *N,N* dimethyl formamide adding P(VDF-HFP). Thin films fabricated by spin-coating possess energy density of 3.2 J/cm^3 for volume fraction of nanoparticles of 50%. Guo et al. [9]

prepared the $Ni_{0.5}Zn_{0.5}Fe_2O_4$/PVDF composite films with low dielectric loss, and giant magnetodielectric coupling effect. Composites like $La_{0.7}Sr_{0.3}MnO_3$/PVDF and Fe_3O_4/PVDF possess multiferroic properties with significant ME coupling.

Most known ferrite/$BaTiO_3$ or ferrite/PZT composites have some limitations due to chemical reaction, interdiffusion between the ferroelectric (FE) and magnetostrictive phases, and different thermal expansion, which complicate their application [10]. To simplify processing and avoid the problem of interdiffusion, the polymer-based composite materials may be prepared by filling polymers with magnetic, dielectric, and FE components with lower processing temperatures [11]. PVDF, among the most important polymers due to its FE and piezoelectric properties, is a semicrystalline polymer exhibiting at least four crystalline phases, α, β, γ, and δ. Among these, polar β-phase is the most desirable phase with high permittivity or FE value. PVDF is easy to process and can be cast into any desirable shape by simple methods. Another well-known ferroelectric polymer is poly(vinylidene fluoride trifluoroethylene)—P(VDF-TrFE)—possessing high remnant polarization value, short switching time ($\sim$1 s), and low process temperature (120°C$\sim$140°C).

To make flexible ferroic or multiferroic composites, Roy et al. [12] synthesized the ferromagnetic $CoFe_2O_4$ (CFO) spinel powders further dispersed into ferroelectric PVDF matrix. The polymer-spinel composite was cast using different weight fractions of the spinel component. Gupta et al. [13] prepared three-phased composite $BaTiO_3$/$CoFe_2O_4$/PVDF with large magnetoelectric response, as a result of the effective mechanical interaction between CFO and BT through the PVDF matrix. Processing temperatures for composite preparation are low, ensuring no chemical reaction or interdiffusion between phases. Gupta et al. [14] prepared $Co_{0.6}Zn_{0.4}Fe_{1.7}Mn_{0.3}O_4$/PZT/PVDF composite with a high ME voltage coefficient of 122 mV/cm and $(x)BaTiO_3$-$(1\text{-}x)Co_{0.6}Zn_{0.4}Fe_{1.7}Mn_{0.3}O_4$/PVDF composites, noting the dielectric anomaly close to the ferroelectric to paraelectric transition temperature of the BT [15]. Martins et al. [16] reported incorporation of ferrite nanoparticles (35–55 nm) of $CoFe_2O_4$ or $Ni_{0.5}Zn_{0.5}Fe_2O_4$ into P(VDF-TrFE) matrix. In both cases, the nanocomposites exhibited improved ferroelectric, ferromagnetic, and ME properties, strongly dependent on the ferrite filler concentrations. Alsaнar et al. [5] investigated a nanocomposite based on ferroelectric copolymer P (VDF-TrFE) and ferromagnetic iron nanowires embedded inside it, exhibiting improved ME coupling compared to nanoparticle-based nanocomposites.

Texeira et al. [17] obtained $NaNbO_3$ nanoparticles and $Na_2Nb_2O_6 \cdot nH_2O$ fibers by microwave hydrothermal synthesis. Composite was prepared with α-PVDF in powder form mixed in a mortar with pure $NaNbO_3$ particles. In addition, the $NaNbO_3$ fiber-like particles coated with PAni (polyaniline) were mixed with PVDF until a homogeneous mixture was formed. The ceramic particles covered with PAni possess higher electrical conductivity. The hysteresis loop, dielectric constant, and conductivity confirmed that the properties of flexible and rather transparent composite depend on particle size and its morphology and the $NaNbO_3$/PVDF ratio. The anisotropy of ceramic particles improves the composite characteristics, such as piezoelectric coefficient d_{33}, remnant polarization, conductivity, and dielectric constant values.

Graz et al. [18] demonstrated a monolithically integrated bifunctional front plane element based on a composite foil of piezoelectric lead titanate nanoparticles embedded in a P(VDF-TrFE) matrix that ensures the mechanical flexibility to the sensor front-plane. A bifunctional sensory cell is made of two (one pressure-sensitive, and one temperature-sensitive) subcells that are only piezoelectric or only pyroelectric (Fig. 27.1A–C). The flexible ceramic polymer sensor is laminated on a flexible transistor (A) and the pyroelectric subelements have a parallel orientation of the polarization in the ceramic nanoparticles and in the ferroelectric polymer matrix, while the piezoelectric subelements have an antiparallel orientation of the polarization. Equivalent circuit diagram of the sensing circuit is presented in (B). The pyroelectric and piezoelectric sensors are modeled with the same circuit (C).

Krause et al. [19] reported $PbTiO_3$/P(VDF-TrFE) nanocomposites for flexible skin tailored to exhibit exclusively piezoelectric or pyroelectric responses depending on the applied poling process. The composites consist of 70% P(VDF-TrFE) and 30% $PbTiO_3$ nanoparticles. The copolymer films serve as front-plane in a flexible electronic surface, based on amorphous silicon or organic pentacene field effect transistors. Lee et al. [20] indicated that P(VDF-TrFE)-$BaTiO_3$ capacitors fabricated on the PEN substrate showed good electrical properties without significant degradation. In recent years, the ferroelectric-gate thin film transistors (Fe-TFTs) on the

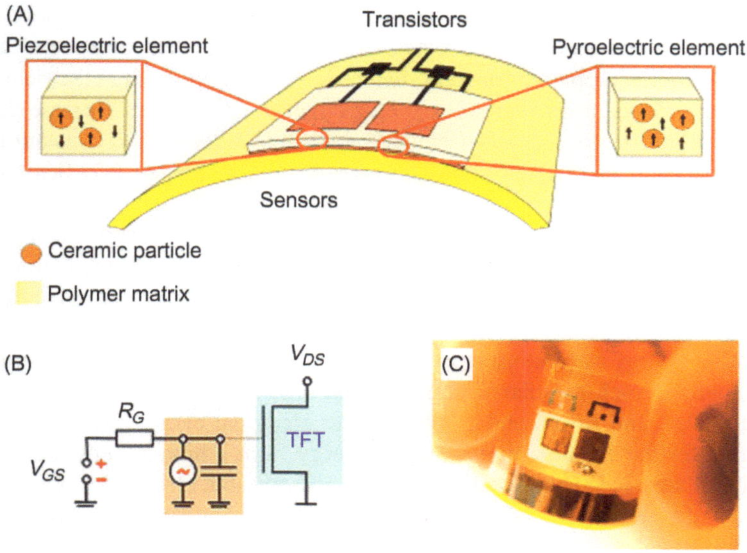

Figure 27.1 Bifunctional sensor array: (A) scheme of the bifunctional sensor array, (B) equivalent diagram of the sensing circuit, (C) a two-element sensor prototype.
Reprinted from I. Graz, M. Krause, S. Bauer-Gogonea, S. Bauer, S. Lacour, B. Ploss, et al., Flexible active-matrix cells with selectively poled bifunctional polymer-ceramic nanocomposite for pressure and temperature sensing skin, J. Appl. Phys. 106 (2009) 034503, with the permission of AIP Publishing.

flexible substrate have attracted much attention for the embeddable nonvolatile memories, radio frequency identification tags, sensor arrays, and displays.

Building flexible transparent energy conversion and storage units plays a key role in fabrication of new devices. Energy harvesting using mechanical energy sources (e.g., human walking, transportation movement, and even sound waves) is a promising candidate for generation of energy with improved accessibility, eco-compatibility, and body sensor networks. The general physical process for energy conversion is based on the generation, separation, and flow of charge. These steps were accomplished in piezoelectric nanogenerators (NGs) by employing the piezoelectric potential created in piezoelectric materials under strain. Recently, Wu et al. [21] specified few problems: the piezoelectric ceramics need to be calcined to form a ferroelectric phase, flexible substrates cannot endure high calcining temperature, and manipulating piezoelectrics could be difficult due to their brittleness and shrinkage. Jung et al. [22] demonstrate the device based on $NaNbO_3$ nanowire-PDMS polymer. $NaNbO_3$ nanowires were grown by hydrothermal method at low temperature and poled at room temperature. Due to the merit of ferroelectricity (piezoelectricity) and the demerit of random alignment, a composite was formed in a volume ratio of 1:100 to permeate NGs with flexible component. The piezoelectric device consists of four layers. With this configuration, an output voltage of 3.2 V and current density of 16 nA/cm^2 was obtained (Fig. 27.2). Jung et al. [23] presented a flexible piezoelectric NG composed of randomly aligned $KNbO_3$ nanorods in PDMS, sandwiched by Au/Cr-coated polymer substrates. A thin poly(methyl methacrylate) layer was deposited between the $KNbO_3$/PDMS composite and the Au/Cr electrode to completely prevent dielectric breakdown during electrical poling and to reduce leakage current during excessive straining. The flexible $KNbO_3$/PDMS composite shows a nearly frequency-independent dielectric constant ($\sim$3.2) and low dielectric loss ($<$0.006), which makes it suitable for use as a flexible high-output nanogenerator and high-k capacitor. Jeong et al. [24] demonstrated the lead-free NG for high-output energy harvesting using alkaline niobate-based particles and copper nanorods filler distributed in a PDMS matrix prepared by a simple spin-casting method. It successfully converted mechanical energy to electricity up to 12 V and 1.2 μA.

Generally, energy harvesting is a process by which energy can be captured from external sources, such as solar power, thermal energy, vibrational energy, electromagnetic waves, and so on, and converted into electrical energy. The development of a method for the efficient harvesting of energy from a human body could enable extraordinary advances in biomedical devices and portable electronics. Of particular interest is developing nanopiezoelectrics in flexible form for wearable or implantable applications. Qi and McAlpine [25] presented an overview of the opportunities, progresses, and challenges in the field of nanopiezoelectrics based on highly efficient nanoassemblies, such as piezoelectric fiber composites, polymer-based generators, nanowire generators, and piezoelectric ribbons printed from thin films onto flexible and stretchable substrates. Kim et al. [26], Swallow et al. [27], and Venkatragavaraj et al. [28] indicated piezoelectric materials that can be applied in films, fibers, or composite form, such as PMT, PMT/PT, PZT, PZT/PVDF, PZT/

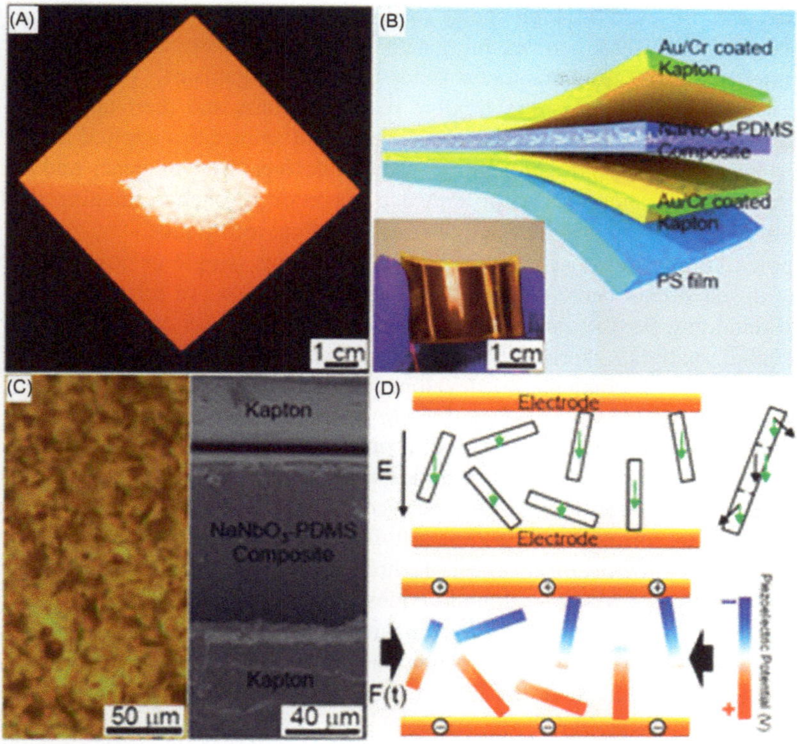

Figure 27.2 Piezoelectric device: (A) $NaNbO_3$ nanowires, (B) piezoelectric device scheme, (C) surface and cross-section views on the device, (D) scheme of the piezoelectric power generation mechanism.
Reprinted with permission from J. Jung, M. Lee, J. Hong, Y. Ding, C. Chen, L. Chou, et al., Lead-free NaNbO3 nanowires for a high output piezoelectric nanogenerator, J. Am. Chem. Soc. 5(12) (2011) 10041–10046. Copyright 2011 American Chemical Society.

fiber composite, fashioned at a variety of length scales, from bulk materials to films and microscale cantilever actuators. The last mentioned can be fabricated via processes for microelectromechanical systems such as a miniature machine that has both mechanical and electronic components. Piezoelectric PZT/PVDF composites have been located inside the soles of shoes yielding power outputs up to 10 mW from walking or implanted inside the rib cage in dogs, generating 20 mW of power during respiration. An application of piezoelectrics to biomechanical power generation has already started.

Recently, it was shown that the magnetic materials on flexible substrates can have a very significant application. Flexible magnetic devices are very attractive in noncontact actuators and microwave devices because of their stretchable, biocompatible, lightweight, portable, and low-cost properties. Spintronic devices with giant magnetoresistance (GMR) or tunneling magnetoresistance (TMR) multilayered structures can be used in high-speed read/write heads in disk memory devices. Recently, GMR or TMR sensors fabricated on flexible substrates have attracted a lot of interest in detecting magnetic fields in living organisms. Yiwei et al. [29] reviewed recent advances in the flexible magnetic films and their applications in magnetic sensors, actuators, microwave devices, and spintronics applied in biomedical devices. Magnetic particles can be used to deliver drug or gene, and help to detect proteins, nucleic acids, or to enhance magnetic resonance imaging contrast, all of which are very important for modern biomedical techniques.

Complex composites based on polymer-multiferroics (single or composite) were recently investigated. There are very few publications that are related to the latest trend of obtaining flexible composites based on a combination of inorganic–organic functional materials (ferroics or multiferroics) posted on organic flexible and transparent support. Prabhakaran developed highly flexible $BiFeO_3$/PVDF nanocomposites, fabricating by a solvent casting method. However, the research is still in progress and no detail results about properties of flexible composites [30]. Stojanovic et al. [31] developed complex composites based on polymer–single multiferroic ($BiFeO_3$-BF) and polymer–composite multiferroic (composite of $BaTiO_3$-BT with (Ni,Zn)Fe_2O_4-NZF). All functional inorganic components were prepared by autocombustion synthesis technique (Chapter 17: Review of Methods for the Preparation of Magnetic Metal Oxides). Obtained materials were further calcined to obtain ferroelectric (BT), ferromagnetic (NZF), or multiferroic (BF) powders. PVDF was used as an organic component. The main steps of the preparation are presented in Fig. 27.3. In the first step of processing the powders of inorganic and organic components were well mixed (Fig. 27.3A). Further, the mixed powders were set onto flexible and transparent component Kapton (Fig. 27.3B). It is important to point out that for the formation of flexible composite it is necessary to use hot pressing as presented in Fig. 27.3D. The flexible composites are shown in Fig. 27.3E (based on BF/PVDF composite) and in Fig. 27.3F (BT/NZF/PVDF composite). Microstructure, ferroelectric, and magnetic properties of multiferroics (single or composites) differ from microstructure of complex composites. Ferroelectric

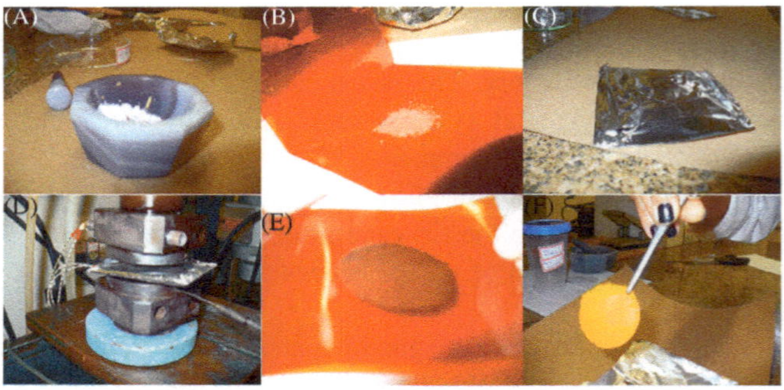

Figure 27.3 Preparation of complex composites: (A) mixing powders in mortar (multiferroics with PVDF), (B) deposition of powder on Kapton foil, (C) sample in aluminum foil, (D) hot-pressing, (E) and (F) flexible composite polymer multiferroics.

properties are more expressed for flexible composite than for $BiFeO_3$ ceramics, which is probably connected with decrease of conductivity. The difference for multiferroic composite ferrite/BT and complex composite ferrite/BT/PVDF are less expressed because in both cases barium titanate, independently of the presence of polymer, contributes to ferroelectric properties, and ferrites to magnetic properties (Fig. 27.4).

27.1 Summary

Polymer–metal oxide nanocomposites are currently of considerable interest for electronic applications as embedded multilayer and high-energy-density capacitors, gate insulators, transistors, nanogenerators, energy harvesters, and so on. Recently, energy conversion and storage units play a key role in realizing new devices based on the ferroelectrics and multiferroics. Energy harvesting using mechanical energy sources (human walking, transportation movement, and even sound waves) from external sources (solar power, thermal and vibrational energy, electromagnetic waves, etc.) is a promising candidate for energy generation with improved accessibility and eco-compatibility. Among various energy harvesting technologies, there has been significant interest in the area of vibrational energy based on piezoelectric and magnetic harvesters. The development of a method for efficiently harvesting energy from the human body could enable extraordinary advances in biomedical devices and portable electronics. Of particular interest is developing nanopiezoelectrics as promising new materials for biomechanical energy nanogenerators that are highly efficient and, with flexible form, highly suitable for wearable or implantable applications. Transparent nanogenerators and self-powered pressure sensors are recently of special interest. Flexible multiferroic structure opens the new field in the application of composites: multiferroic active component on polymeric support or multiferroic-polymer complex composites.

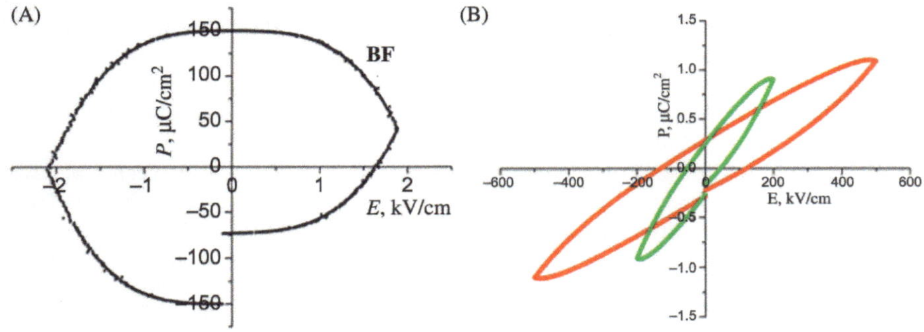

Figure 27.4 Ferroelectric hysteresis loops: (A) bismuth ferrite ceramics; (B) complex composite: PVDF/BFO (red) and PVDF/NZF/BT (green).

References

[1] D. Skinner, R. Newnham, L. Cross, Flexible composite transducers, Mater. Res. Bull. 13 (6) (1978) 599–607.

[2] R. Newnham, Composite electroceramics, Ferroelectrics 68 (1) (1986) 1–32.

[3] J. Jin, S. Lu, C. Chanthad, Q. Zhang, M. Haque, Q. Wang, Multiferroic polymer composites with greatly enhanced magnetoelectric effect under a low magnetic bias, Adv. Mater. 23 (2011) 3853–3858.

[4] K. Nomura, H. Ohta, A. Takagi, T. Kamiya, M. Hirano, H. Hosono, Nature 432 (2004) 488–492.

[5] M. Alnassar, A. Alfadhel, Y. Ivanov, J. Kosel, Magnetoelectric polymer nanocomposite for flexible electronics, J. Appl. Phys. 117 (2015) 17D711.

[6] J. Ma, J. Hu, Z. Li, C. Nan, Recent progress in multiferroic magnetoelectric composites: from bulk to thin films, Adv. Mater. 23 (2011) 1062–1087.

[7] Y. Lin, N. Cai, J. Zhai, G. Liu, C.W. Nan, Giant magnetoelectric effect in multiferroic laminated composites, Phys. Rev. B 72 (2005), 012405-012405-4.

[8] P. Kim, N. Doss, J. Tillotson, P. Hotchkiss, M. Pan, S. Marder, et al., High energy density nanocomposites based on surface-modified $BaTiO_3$ and a ferroelectric polymer, Am. Chem. Soc. 3 (9) (2009) 2581–2592.

[9] Y. Guo, Y. Liu, J. Wang, R. Withers, H. Chen, L. Jin, P. Smith, Giant magnetodielectric effect in $0-3$ $Ni_{0.5}Zn_{0.5}Fe_2O_4$-poly (vinylidene-fluoride) nanocomposite films, J. Phys. Chem. C 114 (2010) 13861–13866.

[10] J. Ma, Z. Shi, C. Nan, Magnetoelectric properties of composites of single $Pb(Zr,Ti)O_3$ rods and terfenol-D/epoxy with a single-period of 1-3-type structure, Adv. Mater. 19 (2007) 2571–2573.

[11] C. Thirmal, C. Nayek, P. Murugavel, V. Subramanian, Magnetic, dielectric and magnetodielectric properties of PVDF-$La_{0.7}Sr_{0.3}MnO_3$ polymer nanocomposite film, AIP. Adv. 3 (2013), 112109-112109-8.

[12] S. Roy, B. Biswas, S. Majumder, Investigations on flexible multiferroic composites, Mesosc. Nanosc. Macrosc. Mater. 1063 (2008) 276–289.

[13] A. Gupta, R. Chatterjee, Magnetic, dielectric, magnetoelectric, and microstructural studies demonstrating improved magnetoelectric sensitivity in three-phase $BaTiO_3-CoFe_2O_4$–poly(vinylidene-fluoride) composite, J. Appl. Phys. 106 (2009) 024110.

[14] A. Gupta, R. Chatterjee, Study of dielectric and magnetic properties of $PbZr_{0.52}Ti_{0.48}O_3-Mn_{0.3}Co_{0.6}Zn_{0.4}Fe_{1.7}O_4$ composite, J. Magn. Magn. Mater. 322 (2010) 1020–1025.

[15] A. Gupta, A. Huang, S. Shannigrahi, R. Chatterjee, Improved magnetoelectric coupling in Mn and Zn doped $CoFe_2O_4-PbZr_{0.52}Ti_{0.48}O_3$ particulate composite, Appl. Phys. Lett. 98 (2011) 112901.

[16] P. Martins, A. Lasheras, J. Gutierrez, J. Barandiaran, I. Orue, S. Lanceros-Mendez, Optimizing piezoelectric and magnetoelectric response on $CoFe_2O_4$/P(VDF-TrFE) nanocomposites, J. Phys. D: Appl. Phys. 44 (2011), 495303-495303-7.

[17] G. Teixeira, R. Ciola, W. Sakamoto, M. Zaghete, Ferroelectric materials - synthesis and characterization, Electrical and Electronic Engineering, Chapter 3, 2015.

[18] I. Graz, M. Krause, S. Bauer-Gogonea, S. Bauer, S. Lacour, B. Ploss, et al., Flexible active-matrix cells with selectively poled bifunctional polymer-ceramic nanocomposite for pressure and temperature sensing skin, J. Appl. Phys. 106 (2009) 034503.

[19] M. Krause, N. Gaar, R. Schwodiauer, S. Bauer-Gogonea, S. Bauer, ERC Project "Soft-Map", Toy brick tensile tester, Radial stretching system DOI: 10.1109/ise.2008.4814070.
[20] G. Lee, E. Tokumitsu, S. Yoon, Y. Fujisaki, J. Yoon, H. Ishiwara, The flexible non-volatile memory devices using oxide semiconductors and ferroelectric polymer poly (vinylidene fluoride-trifluoroethylene), Appl. Phys. Lett. 99 (2011), 012901-012901-3.
[21] W. Wu, S. Bai, M. Yuan, Y. Qin, Z. Wang, T. Jing, Lead zirconate titanate nanowire textile nanogenerator for wearable energy-harvesting and self-powered devices, J. Am. Chem. Soc. (2012), A-E.
[22] J. Jung, M. Lee, J. Hong, Y. Ding, C. Chen, L. Chou, et al., Lead-free $NaNbO_3$ nanowires for a high output piezoelectric nanogenerator, J. Am. Chem. Soc. 5 (12) (2011) 10041–10046.
[23] J. Jung, C. Chen, B. Yun, N. Lee, Y. Zhou, W. Jo, et al., Lead-free $KNbO_3$ ferroelectric nanorod based flexible nanogenerators and capacitors, Nanotechnology 23 (2012), 375401-375401-6.
[24] C. Jeong, K. Park, J. Ryu, G. Hwang, K. Lee, Large-area and flexible lead-free nanocomposite generator using alkaline niobate particles and metal nanorod filler, Adv. Funct. Mater. 24 (2014) 2620–2629.
[25] Y. Qi, M. McAlpine, Nanotechnology-enabled flexible and biocompatible energy harvesting, Ener. Env. Sci. 3 (9) (2010) 1275–1285.
[26] H. Kim, Y. Tadesse, S. Priya, Piezoelectric energy harvesting, Ener. Harv. Techn. New York, 2008.
[27] L. Swallow, J. Luo, E. Siores, I. Patel, D. Dodds, A piezoelectric fibre composite based energy harvesting device for potential wearable applications, Smart. Mater. Struct. 17 (2008), 025017-025017-7.
[28] E. Venkatragavaraj, B. Satish, P. Vinod, M. Vijaya, Piezoelectric properties of ferroelectric PZT-polymer composites, J. Phys. D: Appl. Phys. 34 (2001) 487–492.
[29] L. Yi-Wei, Z. Qing-Feng, L. Run-Wei, Fabrication, properties, and applications of flexible magnetic films, Chin. Phys. B 22 (12) (2013), 127502-127502-14.
[30] T. Prabhakaran, J. Hemalatha, Highly flexible poly (vinyldine fluoride)/bismuth iron oxide multiferroic polymer nanocomposites, AIP Conference Proceedings, 1447 (2012) 1309.
[31] B.D. Stojanovic, A.S. Dzunuzovic, N.I. Ilic, G. Texeira, M.A. Zaghete, CAPES Report, IQ-UNESP, Araraquara, Brazil, 2015, unpublished data.

Ferroelectric, ferromagnetic, and multiferroic heterostructures for possible applications as tunnel junctions

28

Shalini Kumari[1,2], *Dhiren K. Pradhan*[1,3], *Ram S. Katiyar*[1] *and Ashok Kumar*[4]
[1]University of Puerto Rico, Puerto Rico, United States, [2]West Virginia University, Morgantown, WV, United States, [3]Carnegie Institution for Science, Washington, DC, United States, [4]CSIR-National Physical Laboratory, New Delhi, India

28.1 Introduction

The ferroelectric fraternity is looking for the so-called universal memory, one that combines fast read, fast write, nonvolatile, unlimited endurance, with low-power, scalability, and is cheap to make. In the late 1980s and 1990s, ferroelectric random access memory (FeRAM) was considered one of the most suitable candidates for universal memory. It received more attention because it combines nonvolatility, fast write process, and low power consumption; however, it has limited commercialization due to cost-effective integration with established Si-based CMOS technologies and manufacturing difficulties to scale-up memory density of microelectronics industries [1,2]. Since 2010, multiferroic materials are the mainstream research topics between both ferroelectric and magnetic research communities. These magnetoelectric multiferroic systems exhibit more than two ferroic orders (such as ferroelectric, ferromagnetic, ferroelastic, or ferrotoroidic) and having coupling between them has recently attracted significant interest in novel multifunctional devices [3–6]. Among these materials, special attention has been dedicated to artificially engineered multiferroics, which combine ferroelectric and ferromagnetic constituents in two-phase heterostructures [7,8]. Due to the interplay between magnetic, electric, and transport properties, these multiferroic heterostructures reveal new physics and could be used to design novel multifunctional devices such as multiferroic tunnel junctions (MFTJs). It has the capability to control electron and spin tunneling employing ferroelectric and ferromagnetic polarizations of the MFTJs component. In MFTJs, a multiferroic ultrathin film serves as a tunneling barrier across two normal metal or magnetic-metal electrodes. It is also possible to design

Magnetic, Ferroelectric, and Multiferroic Metal Oxides. DOI: http://dx.doi.org/10.1016/B978-0-12-811180-2.00028-1

a MFTJ heterostructure with an ultrathin layer of ferroelectric materials sandwiched between two ferromagnetic electrodes. However, if the top electrodes are only normal metal then it forms ferroelectric tunnel junctions (FTJs). The crucial property of a MFTJ is the coexistence of the tunnel magnetoresistance (TMR) and tunnel electroresistance (TER) effects [9,10].

As the demands for faster, smaller, and nonvolatile electronics increase enormously, the limit of semiconductor devices have been pushed to ever smaller dimensions. However, the challenge of power dissipation places a limit on future device scaling. Size effects may also post constraints on device miniaturization. These have led to the emergent need, development, and investigation of new concepts for information processing and storage. For the next generation of devices, multifunctional structures have been envisioned in which functionality can be manipulated by several external independent controls, and can operate with low power. One of the easy approaches to multifunctional devices is to combine ferromagnetic (FM)-based functionality with ferroelectric (FE)-based functionality. It can be utilized as memory cells with multiple resistance states or to combine logic and memory operations. It may also be able to achieve cross-control of the spin and charge degrees of freedom. These offer significant opportunities for new device concepts. Coexistence of several order parameters in multiferroic materials brings about novel physical phenomena and offers possibilities for new device applications. Of particular interest is the existence of a cross coupling between the magnetic and electric orders, termed magnetoelectric (ME) coupling. This coupling enables the control of the FE polarization by a magnetic field and the manipulation of magnetization by an electric field. Hence coexistence of several order parameters and the ME coupling in such materials can be exploited for novel memory application [11–14]. As FE polarization and magnetization are used to encode binary information in FeRAMs and magnetic random access memories (MRAMs), respectively, the coexistence of magnetization and polarization in a multiferroic material allows the realization of four-state logics in a single device. Magnetoelectric random access memory's (MeRAM's) key advantage over existing technologies is that it requires extraordinary low operating voltage with very high-density microelectronic elements, high-speed reading and writing times, and nonvolatility, as shown in Fig. 28.1 [15,16].

The basic operation of such MeRAMs combines the magnetoelectric coupling with the interfacial exchange coupling between a multiferroic and a ferromagnet to switch the magnetization of the ferromagnetic layer by using a voltage and vice versa. Due to the advancement in thin-film technology, it became possible to fabricate composite multiferroics by artificially making FEs and ferromagnets in nanoscale heterostructures. All of these structures exhibit magnetic and electric properties and provide a unique opportunity to exploit several functionalities that could yield a new device concept, such as MFTJs. Furthermore, conceptually new multilevel memory and logic devices are possible with MFTJs. A typical tunnel junction consists of two metal layers separated by a thin barrier layer (tunnel barrier). However, the tunneling probability becomes significant only for ultrathin (<10 nm thin) barriers.

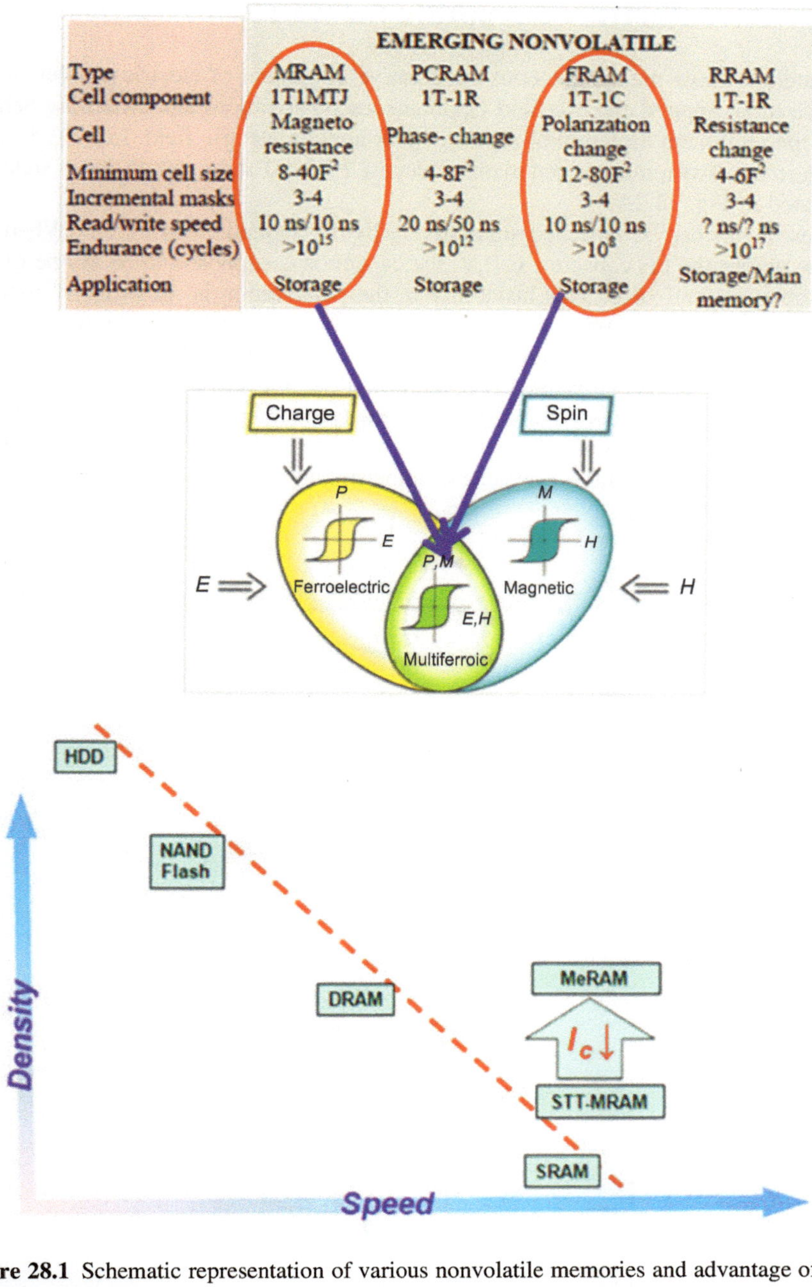

Figure 28.1 Schematic representation of various nonvolatile memories and advantage of MeRAM over other existing memories. Density and speed relation for various memory devices.

28.2 Ferroelectric nonvolatile memories

FE materials show the spontaneous cooperative alignment of electric dipoles in the absence of an applied electric field (spontaneous polarization) and switching behavior of the alignment direction of dipoles under applied electric field. In a FE capacitor, there are two remanent polarization states, P_r for 0 state and P_r for 1 state, as illustrated in Fig. 28.2A.

These states can be switched alternatively by the applied electric field. When the voltage across the FE capacitor is 0 V, the capacitor is supposed to be at one of the two stable states: 1 or 0. This bistability in the polarization is the property used to store data in a FeRAM memory. There are four kinds of cell configurations differing for the number of transistors and capacitors adapted to the memory cells: the 2T-2C cell, the 1T-2C cell, the 1T-1C cell, and the 1 T cell (1 T is also known as FeFET cell). The 1T-1C cell with a high density of integration is the most developed cell. This cell consists of a single FE capacitor that is connected to the drive line (PL) at one end and via an access transistor to the bit line (BL) at the other ends (see Fig. 28.2B).

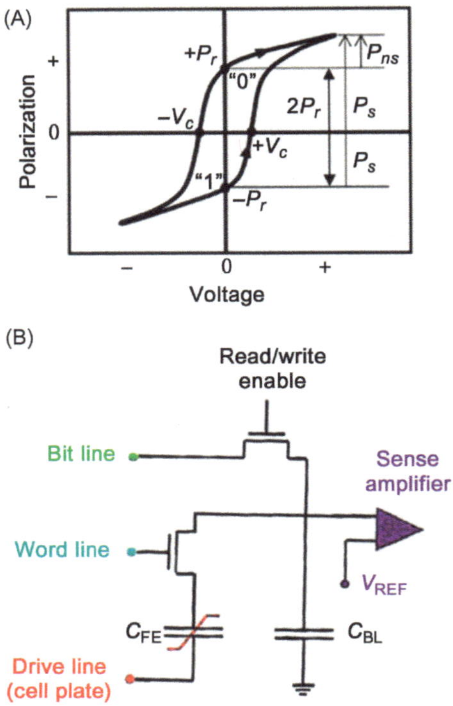

Figure 28.2 (A) Schematic hysteresis loop of a ferroelectric capacitor, (B) 1T-1C configuration for a FeRAM cell.

Voltage pulses are used to write and read the digital information in FeRAM. In FeRAM like in dynamic random access memory for read-out operation, the memory cell requires each bit accessed to be rewritten in a refresh function. But FeRAM has inexhaustible write endurance (100 trillion write/read cycles); however, a few reliability issues such as fatigue, imprint, depolarization, and decrease of switching charges with decrease in device dimension must be addressed in any further downscaling to fabricate FE-based devices with memory functionality.

28.3 Ferroelectric tunnel junctions

A junction consisting of an ultrathin insulating layer between two conducting electrodes is known as a tunnel junction. However, the junction made of a FE tunnel barrier sandwiched between two electrodes are called FTJs. The important characteristic of bulk FE materials is the spontaneous electrical polarization, which can be switched between at least two stable orientations by applying an external electric field. Switching the polarization direction of FE barrier produces a notable change in the resistance of the junction known as the TER effect. FTJs unite the phenomena of quantum-mechanical tunneling and switchable spontaneous polarization of a few nanometer-thick FE films into novel device functionality. The switching of resistance in a FTJ is closely linked to the orientation of FE polarization and hence is different and unique from other types of resistive switching known for other oxide materials. Moreover, the readout current densities are much larger than in FE diodes, thereby opening the path to possible applications as high-density FE memories with nondestructive readout. In 1971 Esaki first proposed the concept of a polar switch involving a switchable FE material in thin-film form [11] as there were no experimental techniques and capabilities available at that time to fabricate an ultrathin film of FE material to serve as a tunneling barrier layer. It was also assumed that the critical thickness for the existence of ferroelectricity in thin films was much higher than the thickness necessary for tunneling to take place. The experimental investigations of the existence of ferroelectricity in thin films of a nanometer thick opened up exciting expectations for FTJs. One of the most important considerations in fabricating FTJs is that the barrier thickness must be greater than the minimum coherence length to obtain stable FE polarization. For many years this was thought to fail. Until the work of Gregg's group, most of the researchers believed that ferroelectricity disappeared due to depolarization fields for films much thinner than 100 nm [12]. Since the tunneling of electrons in oxide films thicker than 8 nm is negligible, this appeared to imply that FE tunnel junctions were impossible. However, we now know that the minimum thickness for stable ferroelectricity in a FE oxide is $\sim$2.4 nm [17], hence giving rise to the possibility of fabricating FTJs. It only aroused considerable research interest recently, largely because only in recent years experiments have shown that ferroelectricity can persist down to the nanometer scale. This makes it possible to use FEs as tunneling barriers. The concept of the FTJ is illustrated in Fig. 28.3 [18], which shows a simplified band

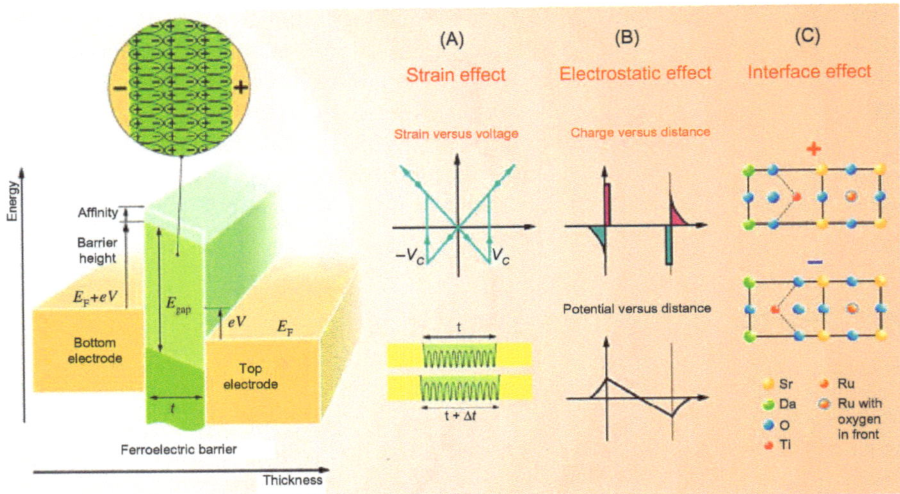

Figure 28.3 Schematics of a ferroelectric (FE) tunnel junction consisting of two electrodes separated by a nm-thick FE barrier layer. The FE nature of the barrier changes the transmission probability in three ways (A, B, C).

diagram of the metal-FE-metal heterostructure. Due to the reversible electric polarization, the current-voltage characteristics of FTJ are expected to be different from those of a convention metal-insulator-metal junction.

The electric polarization (P) in the FE barrier, which can be switched by an applied electric field, may have a pronounced effect on the conductance of a FTJ. The reversal of polarization from $+P$ to $-P$ in the FE alters the switching of polarization charges at the barrier/electrode interfaces, changing the position of the ions in the FE unit cell and modifying the lattice strain inside the barrier. The TER effect of an FTJ could at least be obtained through the three effects shown in Fig. 28.3. These are as follows: (1) the incomplete screening of the FE bound charge leads to an electrostatic potential that superimposes the contact potential in the tunnel junction; (2) the displacement of the B-site atom affects the atomic orbital hybridizations at the interface, which makes the transmission probability different for the two opposite polarization orientations; and (C) due to piezoelectricity of a FE barrier an applied voltage produces a strain that changes transport characteristics of the barrier, such as the barrier width and the attenuation constant. Using first-principles density-functional calculations, Burton and Tsymbal [19] predicted that a giant TER effect could be obtained by exploiting a ME interaction between a FE $BaTiO_3$ (BTO) tunneling barrier and a magnetic $La_{1-x}Sr_xMnO_3$ electrode. These results were confirmed experimentally by Yin et al. [20] in 2013. They fabricated $La_{0.7}Sr_{0.3}MnO_3$ (50 nm)/BTO (3 nm)/$La_{0.5}Ca_{0.5}MnO_3$ (0.4–2 nm)/$La_{0.7}Sr_{0.3}MnO_3$ (30 nm) planar tunnel junctions where the thin $La_{0.5}Ca_{0.5}MnO_3$ interlayer next to the FE BTO barrier provides the magnetoelectrically active response. They demonstrated a significant enhancement of TER due to a ferroelectrically induced phase transition at a magnetic complex oxide interface.

28.4 Critical thickness for the existence of ferroelectricity

The ferroelectricity is a bulk as well as interface property of materials due to capacitive device configuration. Due to many interface problems in thin films, initially, researchers are expected to sustain a spontaneous electric polarization only above some critical thickness. The basic understanding of size effects in FEs is complicated since the cooperative phenomenon of polarization strongly depends on the order parameters such as compositions, strain across the interface, charge distribution, and depolarization field. These extrinsic effects strongly influence the intrinsic polarization, which in turn provides false information [13]. The critical thickness for ferroelectricity has been discussed since the early 1970s. It was believed for a long time that the critical size for ferroelectricity is of a few tens of nm. However a crucial condition to realize a FTJ is the existence of ferroelectricity in nm-thick barrier film, which means FTJ devices require the fabrication of ultra-thin film retaining pronounced FE properties at a thickness of only a few unit cells. However, to achieve this length scale reduction, to characterize and understand the scaling of ferroelectricity, it is essential to have synthesis, characterization, and theoretical analysis at the nanoscale. However, due to development in the synthesis process, it has now become possible to make heterostructures in which single-crystalline perovskite–oxide films of thickness down to 1 to 2 lattice constants can be epitaxially matched at atomically sharp interfaces [21]. Typically these films are prepared with advanced thin-film vapor deposition techniques such as molecular beam epitaxy (MBE), pulsed laser deposition (PLD), chemical vapor deposition (CVD), and atomic layer deposition (ALD). Another important point to consider is the appropriate choice of the substrate materials that permits control of the lattice mismatch between the FE film and the substrate, allowing the film's growth coherently. PLD technique is one of the most versatile methods to grow high-quality films. Fig. 28.4 shows high-quality ultrathin FE film can be grown by PLD technique, with sharp interfaces, defect-free layer-by-layer epitaxial growth

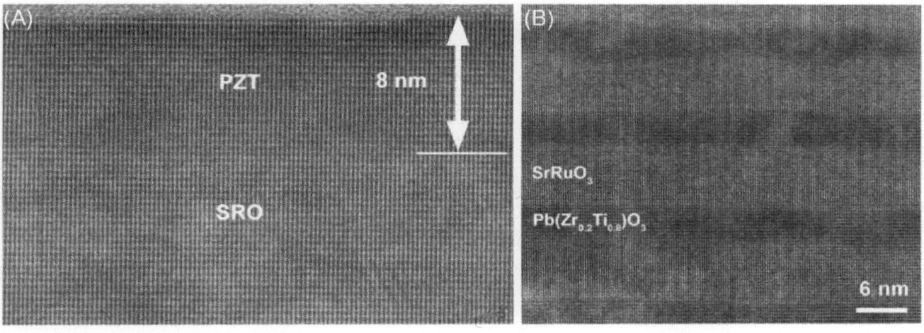

Figure 28.4 Cross-section high-resolution transmission electron microscopy micrograph of (A) 8 nm PZT20/80/SRO film and (B) PZT20/80/SRO superlattice with 4 nm thin PZT20/80 layers and 6 nm thin SRO layers, coherently grown on STO(100) [22].

with no misfit dislocations [22]. Additionally, several works had found stable polarization in ultrathin FE films. Tybell et al. [23] have demonstrated the presence of a stable polarization in a 4-nm thick epitaxial film of perovskite FE Pb $(Zr_{0.2}Ti_{0.8})O_3$.

Experimental evidence for ferroelectricity in ultrathin epitaxial $PbTiO_3$ films was found in Argonne National Laboratory where structural investigations by synchrotron radiation demonstrated FE properties in these films down to 1.2 nm thickness [13]. Also, ferroelectricity has been observed in monomolecular layers of FE polymers in the form of PVDF and its copolymers with TrFE [24]. Since the capability to produce high-quality ultrathin FE films is available, novel applications can be realized by incorporating such films into heterostructures with other materials, producing FTJ and MFTJ structures. Systematic studies of the thickness effect on FE/dielectric properties in thin films have been done [25–31] to know the response of this material (i.e., PZT) when the thickness is decreasing to nanometer scale. Barrionuevo et al. [31] reported in the case of PZT a drastic decrease of remanent polarization (P_r) and increase of coercive field (E_c) when the thickness of PZT is reduced to few nm level (10 nm). This effect was explained by the effect of the passive layer (also called low dielectric or dead layer) originating at the FE–electrode interface. The room temperature polar switching and tunneling behavior in $PbZr_{0.52}Ti_{0.48}O_3$ (PZT) ultrathin films with 3–7 nm of thickness, sandwiched between platinum metal and FM $La_{0.67}Sr_{0.33}MnO_3$ (LSMO) layers are shown in Fig. 28.5.

This junction also shows magnetic field dependent tunnel current switching in Pt/PZT/LSMO heterostructure. The capacitance versus voltage graphs show butterfly loops above the coercive field ($> \pm 3$ V) of PZT for a small probe area ($\sim 16\ \mu m^2$). The effect of FE switching was observed in current density versus voltage curves with a large variation in high-resistance/low-resistance (HRS/LRS) ratio (2:1 to 100:1, for different samples). However, these effects were more prominent

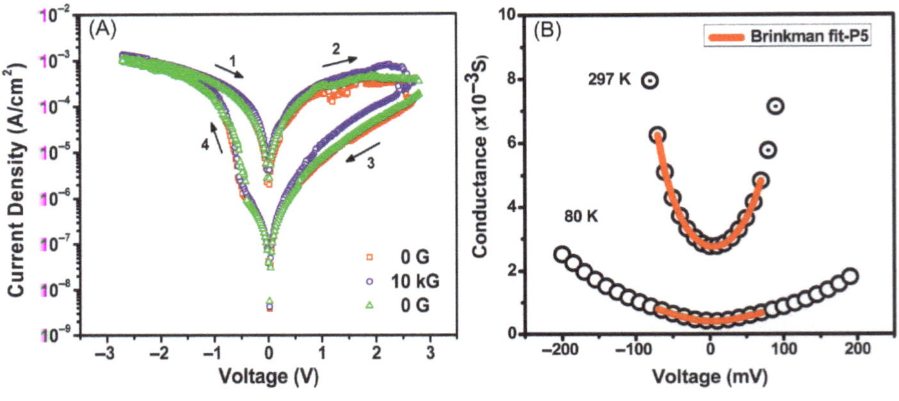

Figure 28.5 (A) Current density versus voltage *J*(*V*) curves of the sample showing the effect of applied 10 kG magnetic field. (B) Brinkman's model fitting of the low-resistance conductance-voltage curves for PZT-5 nm [31].

in the presence of an in-plane external magnetic field. Fig. 28.5A shows the effect of magnetic field along the in-plane direction; it was observed that when 10 kG was applied to the junction during *J*(*V*) measurements, the resistance is decreased only in the positive voltage range (controlled by the LSMO electrode), while no significant change was observed in the negative voltage range (controlled by the Pt electrode). This is reasonable, taking into consideration that LSMO becomes more metallic in its FM state. For this junction, the resistance switching between LRS and HRS became sharper, and the HRS/LRS ratio values at zero bias were ~60 at 0 G and increased to 110 with 10 kG. The conductance is fitted with Brinkman's model, and the parabolic conductance upon bias voltage implies electron tunneling governs the transport (see Fig. 28.4B). Similar behavior in P_r and E_c was observed by other researchers when the thickness of PZT ranged between 100 nm and 4 μm [32,33]. An early study of coercive fields verified the Kay–Dunn Law [34] and showed that E_c increases rapidly in FEs with decreasing thickness; however, the Kay–Dunn Law is semiempirical [35].

28.5 Magnetic tunnel junctions

Extensive research has been carried out on the design and development of magnetic tunnel junctions (MTJs) due to the ever-increasing demand for compact and high-density hard disk drive storage capacities. MTJ devices are based on the basic physical phenomenon called magnetoresistive effect at room temperature. MTJs are the most widely used spintronic devices in computer read heads, sensors, and memories [36,37]. A MTJ consists of two layers of magnetic metal with different coercive field, separated by an ultrathin layer of insulator, with a thickness of about few nm. The insulating layer is so thin that electrons can tunnel through the barrier if a bias voltage is applied between the two metal electrodes. In the MTJ, the tunneling current depends on the relative orientation of the magnetizations of the two magnetic layers, which can be changed by an applied magnetic field. This phenomenon is known as tunneling magnetoresistance (TMR), which is a consequence of spin-dependent tunneling. The parallel and antiparallel magnetization alignments of the two electrodes represent two logic states with different resistances [14,38]. When electrons are tunneling between two magnetic metals, the magnitude of the tunneling current depends on the relative orientation of the magnetization of both electrodes. This can be established from a few elementary arguments: (1) the tunneling current is, in first order, proportional to the product of the electrode density of states (DOS) at the Fermi level; (2) the ground-state energy bands in the vicinity of the Fermi level are shifted in energy, yielding separate majority and minority bands for electrons with opposite spins in ferromagnetic materials; and (3) assuming spin conservation for the tunneling electrons, there are two parallel currents of spin-up and spin-down character. As a result of these aspects, the current between electrodes with the same magnetization direction should be higher than those with opposite magnetization as can be seen from Fig. 28.6.

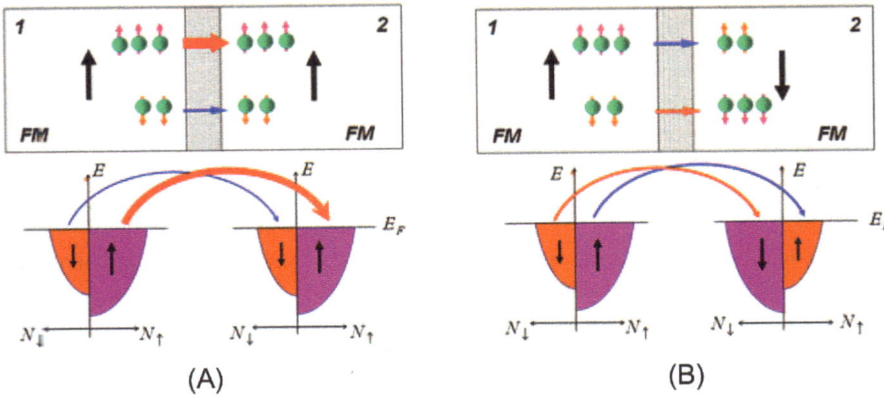

Figure 28.6 Schematic diagram showing the mechanisms of tunnel magnetoresistance for both resistance states [14].

Using this simple model, the change in resistance between antiparallel and parallel magnetization (normalized to the parallel resistance) can be given by [39–41]:

$$TMR = \frac{2P_1P_2}{1 - P_1P_2} \quad \text{with} \quad P_{1,2} = \frac{N_{1,2}{}^{\text{maj}} - N_{1,2}{}^{\text{min}}}{N_{1,2}{}^{\text{maj}} + N_{1,2}{}^{\text{min}}} \tag{28.1}$$

where P_1 and P_2 are the so-called tunneling spin polarizations determined by the relative difference in DOS at the Fermi level [14].

It is critical to realize that not all electrons present at the Fermi level can efficiently tunnel through the barrier and that this simple equation is not able to explain the physics behind some observations in MTJs. In many cases, the spherically symmetrics-like electrons, which have a much lower DOS at the Fermi level, dominantly tunnel through the barrier, and the interface between the insulating tunnel barrier and the magnetic electrodes play an important role. This expression clearly shows the presence of a magnetoresistance effect and the relevance of the magnetic nature for the spin polarization of the tunneling electrons. Moreover, it demonstrates that so-called half-metallic metals with only one of the two spin species available at the Fermi level may, in principle, induce infinitely high TMR. The direction of spins/magnetization of the electrodes determined the magnitude of TMR, also the ability to tune the spin as desired for various logics states may play an important role for design in MRAM. It can be achieved by several methods, which include the use of intrinsic differences in magnetic hysteresis and coercive field of the ferromagnetic materials, exchange biasing with antiferromagnetic thin films, and antiferromagnetic interlayer coupling across nonmagnetic metallic films. TMR ratios well above 100% have been shown in epitaxial (001)-oriented Fe/MgO/Fe junctions and sputtered CoFe/MgO/CoFe by Yuasa et al. [42] and Parkin et al. [43], respectively. Since then, the TMR of junctions involving MgO as a barrier has steadily been improved, in particular by using the ternary CoFeB alloy as the ferromagnetic electrode [42,44]. It has been shown that high-quality MgO can be adequately stabilized between the as-grown, amorphous CoFeB electrodes, which, after annealing at temperatures up to almost 400°C, attain their required crystal structure. Due to this giant magnetoresistance, there is an ongoing effort not only in explaining the large tunneling spin polarization of CoFeB [45], but also in further implementing CoFeB for novel spintronics devices and industrial applications [46].

28.6 Multiferroic tunnel junctions

MFTJs have emerged as an important research area in recent years for exploring ME materials with high ME coupling at room temperature as the tunnel layer. The MFTJ consists of two ferromagnetic electrodes separated by a FE layer or by a single-phase multiferroic layer with strong coupling between magnetic and FE order parameters. One major advantage with the multiferroic tunnel layer compared to FTJ or MTJ is the ability to create various logic states up to (16) with proper tuning of various order parameters. Nowadays MFTJ is receiving much attention due to their potential applications in information and spintronics devices. The ME coupling in MFTJs allows the electric control of magnetic polarization and the magnetic control of electric polarization, which makes them very promising for the application in nondestructive RAMs with fast read/write speed and high memory density [20,32,47]. Equivalently, MFTJ may be considered a particular type of FTJ that has ferromagnetic metal electrodes. The key property of a MFTJ is the coexistence of the TMR and TER effects. Therefore, MFTJ represents at least four state resistance devices where resistance can be switched both by electric and magnetic fields as seen in Fig. 28.7.

Recent theoretical and experimental studies have explored the coexistence of TMR (the magnetization aligned/disaligned in the metallic electrodes) and TER (dependence with the polarization state of the barrier) properties in TJs, using polar barrier (i.e., FTJs/MFTJs) [31,48]. These systems can produce more than four resistance state systems; this is a novel tunnel junction with potential applications in future technology. Density functional calculations also reveal the key functional properties of MFTJs, predicting the coexistence of TMR and TER effects. With the miniaturized and multifunctional developments of devices, the actual application of MFTJs faces a number of challenges. The first challenge is that the decreasing thickness of the FE/MF barrier layer in a MFTJ will degrade FE polarization if the quality and processing techniques are not up to the atomic level perfection. Thus, it is necessary to enhance the FE stability of a FE/MF barrier layer to ensure the reliable performance of devices based on MFTJs. Although the potential impact of these technologies is tremendous, some scientific issues remain to be resolved to produce commercially viable devices. Fundamental issues that need to be addressed for further advances in the field include the stability of FE ordering in ultrathin films and its relationship to electrical and mechanical boundary conditions. Due to the electronic and transport properties of MFTJs being inherently related to the FE polarization, the fundamental mechanisms of FE switching and the effect of structural and interfacial defects on electronic transport properties are also of critical importance.

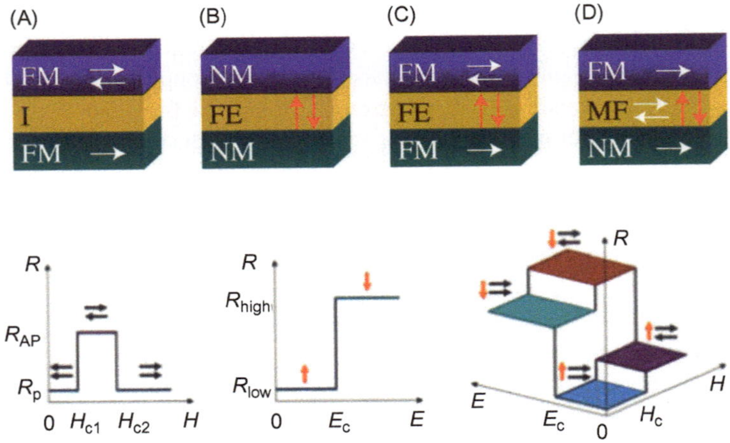

Figure 28.7 Schematic view of the different types of tunnel junctions: (A) magnetic tunnel junction (MTJ), (B) ferroelectric tunnel junction (FTJ), and (C,D) multiferroic tunnel junction (MFTJ). Ferromagnetic (FM), ferroelectric (FE), normal metal (NM), and insulating (I) layers are indicated where appropriate. Bottom panels show the resistance response of these junctions to magnetic (*H*) and electric (*E*) fields. Horizontal and vertical arrows indicate orientations of magnetization and electric polarization, respectively [9].

As a result of the pronounced scaling effect in FEs, transition to heterostructures with nanoscale lateral dimensions will likely entail significant changes in their properties, so that the issues of FE and transport behavior in MFTJs may need to be readdressed at the appropriate length scale. The magnetic properties of ferromagnetic materials employed in MFTJs are critical for the operation of these devices. The essential role is played by interfaces where a magnetically dead layer may occur due to either nonstoichiometry or defects formed at the interfaces, suppressing the transport spin-polarization and TMR. Magnetization switching behavior of the magnetic thin films in a MFTJ is also important and interesting due to a possible effect of the FE polarization on interface magnetic anisotropy and coercivity of the film. The interrelationship between ferroelectricity of the barrier layer and ferromagnetism of the electrodes through an interface ME effect may also control other functional properties in MFTJs, such as TMR and TER, and needs to be investigated in depth for future multistate logics. Although meeting these challenges is not easy the prediction and recent demonstration of giant resistive switching effects and the control of tunneling spin polarization are a strong testament to the fact that the field of MFTJs has an exciting future.

28.7 Multiferroic heterostructure-based tunnel junctions

Hybrid complex oxide heterostructures with controlled geometry and dimensionality offer a pioneering platform to introduce and explore competing for physical phenomena in novel functional materials. Artificially engineered epitaxial heterostructures enable unique functionalities that cannot be realized with the individual constituents [49,50]. The ability to fabricate well-defined heterostructures on a nanoscale has not only led to advanced technology such as higher electronic device speeds, a broad range of interfacial functionality, and band gap engineering of materials, but also has revealed fundamentally new phenomena. Rising density requirements in the microelectronics and magnetic-storage industries promote the construction of devices that function reproducibly at ever smaller dimensions. Nanometer-scale control of material properties has already enabled technologies that exploit electron spin and the discreteness of electronic charge. In recent years, the extensive interest in perovskites shifted from bulk ceramics and single crystals to thin films and heterostructures due to advances in deposition techniques. Several synthesis techniques based on layer-by-layer epitaxial growth have been developed, which involve sequential exposure of the growing components. Artificially engineered perovskite-based heterostructures offer unique opportunities for combining materials with various ferroic orders such as ferroelectricity/antiferroelectricity and ferromagnetism/antiferromagnetism at the nanoscale.

Various heterostructures with different parent components that are paraelectric, FE, antiferroelectric, FM, and multiferroic, have been studied experimentally and theoretically in detail and shown enhancement of functional properties as compared to parent components, which can emerge from interaction of charge, spin, orbital, and strain interactions developed at the interfaces between the layers. The fabrication of complex oxide heterostructures with single-unit-cell, control, and atomically sharp interfaces may lead new pathways to collective-ordering phenomena in condensed matter physics. Stacking different perovskite materials in epitaxial superlattice (SL) structures not only allows their properties to be combined but, sometimes, also completely new phenomena to be induced. Epitaxial strain in thin films and SLs might play an important role in enhancing functional properties required for multifunctional device applications and can even induce new ferroic properties, which can also be modified by controlling the layer thickness and varying the modulation period (Λ), composition, deposition conditions, and substrate, as well as altering the lattice parameters by doping [51]. Conclusively, in oxide SLs where the SL modulation period is below a characteristic length-scale, the nanostructure might show unique and enhanced properties. The multiple interfaces that exist in these structures leads to the carrier confinement effect, magnetic interaction, and lattice strain effect at the interface, which plays an important role in determining the physical properties of the SL structures. So, the ordering of objects on a nanoscale is recently a basic and required goal within the nanoscience and technology community that has been used in the synthesis of nanostructures.

Recently, a tremendous amount of research has been focused on multiferroic heterostructure that combines both ferroic order ferroelectricity and magnetism in heterostructure and the possibility of creating new functionalities in these structures. Multiferroic heterostructure is receiving huge attention in the science community due to the search for strong and enhanced coupling between these two properties, and these structures have been fueled by the dream of magnetically readable and electrically writable data storage. The enhanced multifunctional properties of these heterostructures are essential for dielectric and FE device applications such as nonvolatile FE memories, dynamic RAM, tunable microwave devices, and tunnel junctions [52–54]. Multiferroic heterostructure-based tunnel junctions refer to MTJs with FE barriers or the FTJs with ferromagnetic electrodes. MFTJ is expected to be characterized by the TMR and TER simultaneously and exhibits four resistance states in a single memory unit cell, which inevitably receives much attention as the next generation of information storages. Velev et al. [47] theoretically predicted that an FM/FE/FM MFTJ, such as $SrRuO_3/BaTiO_3/SrRuO_3$ (SRO/BTO/SRO) heterostructure, has four resistance states. Garcia et al. [55] experimentally studied the MFTJs of Fe (or Co)/$BaTiO_3/La_{0.7}Sr_{0.3}MnO_3$ on the $NdGaO_3$ substrate with the tunneling barrier BTO having a thickness of 1 nm and demonstrated the control of spin polarization by electrically reversing the FE polarization. Pantel et al. [32] fabricated Co/$PbZr_{0.2}Ti_{0.8}O_3/La_{0.7}Sr_{0.3}MnO_3$ MFTJs and studied that the magnitudes and the signs of the TMR could be changed by FE switching due to the ME coupling at such a multiferroic interface. Li's group observed four resistance states and the exchange bias effect in $La_{0.6}Sr_{0.4}MnO_3/BiFeO_3/La_{0.6}Sr_{0.4}MnO_3$ MFTJs [56]. This implies that the multistate memory devices with an enhanced capacity of information storage would be achieved using MFTJs, which could be operated by magnetic and electric fields [56,57].

28.8 Tunneling electroresistance for the realization of nondestructive ferroelectric polarization readout

The achievement of room-temperature stable polarization and switching in ultrathin films is very challenging. Therefore, piezoresponse force microscopy (PFM) is a reliable technique to investigate robust ferroelectricity in MFTJ. PFM has emerged as a powerful and nondestructive tool for experimental investigations of FE and multiferroic materials since its invention. PFM technique gives the conclusive evidence of the existence of ferroelectricity [58,59]. It also allows visualization of the FE domain structures with nanometer spatial resolution in the imaging mode. By applying sufficient external electrical bias through the conductive scanning probe microscopy (SPM) tip local polarization switching can be induced, and it also can be extended to the creation of the tailored domain structures. The acquisition of the piezoresponse signals during polarization switching allows measuring of the local piezo hysteresis loops, which can be used to get the information of the switching process in the vicinity of the tip. Using the PFM technique, it is possible to avoid the undesirable leakage current dominated by tunneling at very low thicknesses and read the polarization state without destroying it. SPM techniques allow the localization of an electric field within nanometer-scale regions with the simultaneous probing of polarization by PFM and tunneling current by conducting atomic force microscopy (C-AFM). SPM measurements can be done in two basic geometries: with the conductive

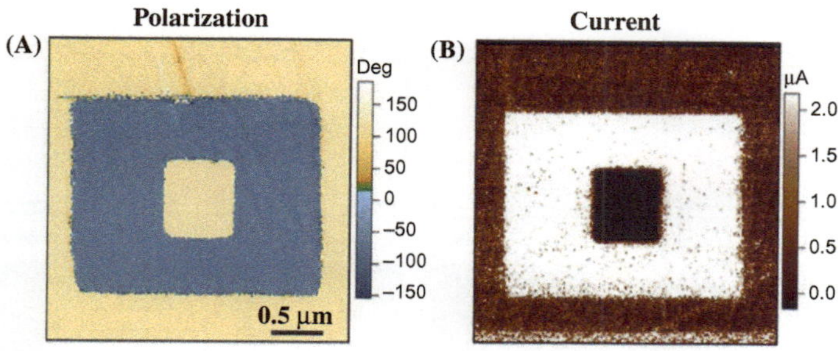

Figure 28.8 (A) Piezoresponse force microscopy image of a polarization switching pattern recorded by scanning probe microscopy with the tip under 3.0 V (blue region corresponds to polarization switched downward; yellow region represents upward polarization in the as-grown film). (B) Tunneling current mapping acquired in the same region after polarization switching. Bright contrast indicating higher current is observed in the regions with polarization switched downward direction [60].

tip used as a top electrode in contact with the surface of the FE/multiferroic layers or with the tip contacting a deposited top electrode.

Gruverman et al. [60] have demonstrated the reproducible tunneling electroresistance effect on nanometer-thick epitaxial $BaTiO_3$ single-crystalline thin films on $SrRuO_3$ bottom electrodes using a set of SPM techniques shown in Fig. 28.8. To probe the nanoscale switching behavior and transport properties, they used epitaxial FE ultrathin films of $BaTiO_3$ grown on atomically smooth TiO_2 terminated (001) $SrTiO_3$ substrates with single crystalline $SrRuO_3$ electrodes. The polarization-dependent TER effect was monitored and controlled by the electric polarization and correlating local FE switching with the local conductivity. The combination of SPM techniques for nanoscale polarization detection (PFM) and spatially resolved local conductance measurements (C-AFM) were utilized. By polarization reversal, TER changes by almost two orders of magnitude on a lateral scale of about 20 nm. Retention of polarization was not affected by conductance measurements, and thus allowing multiple nondestructive polarization readouts and opening a possibility for application as noncharge-based logical switches in nonvolatile memory devices.

28.9 Advantages of band excitation over single frequency excitation piezoresponse force microscopy

Fig. 28.9A shows the measurement sequence of a single-frequency (SF) PFM. In SF PFM single-frequency voltage waveform is applied to the surface of the sample via cantilever tip. The surface of the sample expands due to the piezoelectric effect, and this response gets registered by a lock-in amplifier as a time-varied deflection signal of a particular amplitude and phase shift between the excitation and response.

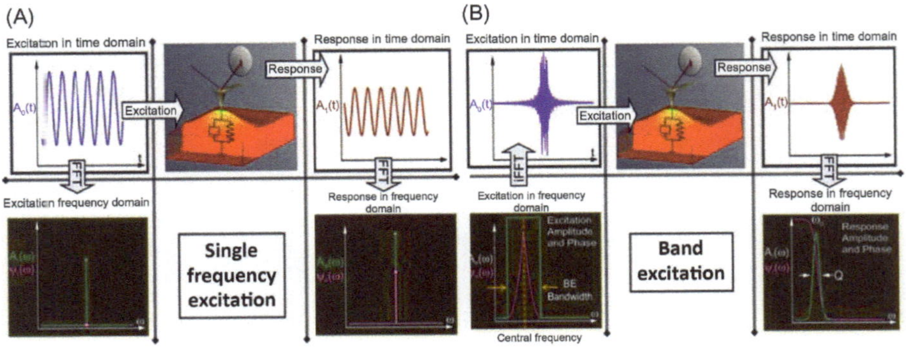

Figure 28.9 (A) Single frequency excitation piezoresponse force microscopy (PFM); (B) band excitation PFM [62,63].

In the wide frequency domain, both excitation and response look like a single vertical line [61,62].

In band excitation PFM, the band excitation signal in the frequency domain is a rounded boxcar function of a certain bandwidth with a particular central frequency, which is chosen to encompass a resonance of the tip (Fig. 28.9B). Once the excitation signal is selected in the frequency domain, the inverse Fourier transform can be used to convert it to the time domain, which can be applied to the surface of the sample via an arbitrary waveform generator. The response signal (cantilever deflection) is recorded as a function of time simultaneously during excitation, registered, and converted to the frequency domain via fast Fourier transform (FFT). Data processing is performed by fitting with a suitable equation derived from a physical model of the system instead of lock in. The most simple physical model to fit the cantilever motion under excitation signal is the damped simple harmonic oscillator (SHO) motion [63–65].

One of the first studies of the polarization effect on conductivity using SPM involved combined detection of conduction changes in a domain-patterned Pb(Zr, Ti)O_3 (PZT) film. The large film thickness in these studies (69 nm) prohibited direct tunneling through the FE layer, and the observed conduction modulation was explained by band bending at the metal-FE interface due to the charge injection into the PZT film. Maksymovych et al. [66] have demonstrated a highly reproducible control of local electron transport through a FE oxide via its spontaneous polarization. Electrons were injected from the tip of an atomic force microscope into a thin film of lead-zirconate-titanate ($Pb(Zr_{0.2}Ti_{0.8})O_3$), in the regime of electron tunneling assisted by a high electric field (Fowler-Nordheim tunneling). The tunneling current exhibits a pronounced hysteresis with abrupt switching events that coincide, within the experimental resolution, with the local switching of FE polarization. The large spontaneous polarization of the PZT film results in up to 500-fold amplification of the tunneling current upon FE switching. The magnitude of the effect is subject to electrostatic control via FE switching, suggesting possible applications in ultrahigh-density data storage and spintronics.

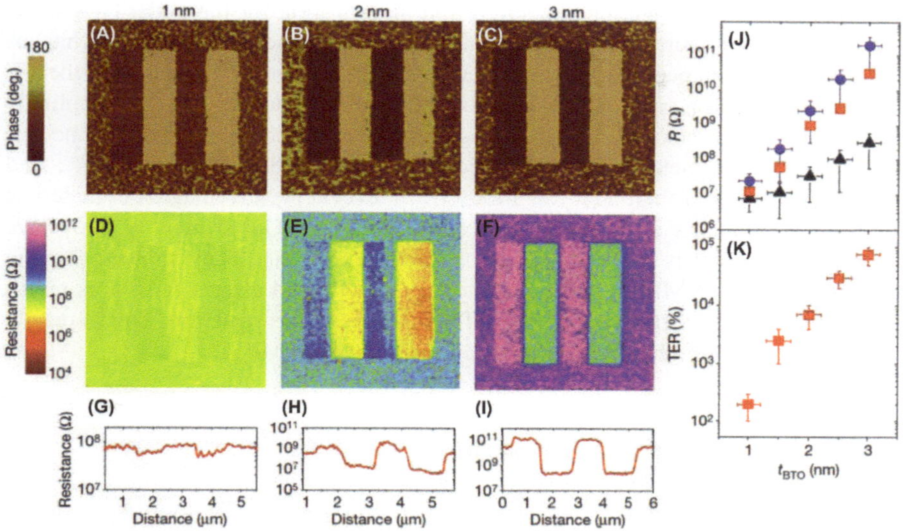

Figure 28.10 Direct evidence for ferroelectricity-related giant TER with ultrathin strained BTO films. (A–F) Parallel PFM phase image (A–C) and CTAFM resistance mapping (D–F) of four written ferroelectric stripes ($1 \times 4\ \mu m^2$) for BTO films 1, 2, and 3 nm thick, respectively. (G–I) Corresponding resistance profiles of the poled area. (J) Thickness dependence of the resistance (R) of unpoled (red squares), and positively (black triangles) and negatively (blue circles) poled, regions. An exponential increase with t_{BTO} is found, as expected for direct tunnelling. (K) The TER also increases exponentially with increase in BTO thickness. In (J and K) vertical error bars correspond to the standard deviation of the resistance distribution over selected areas (and corresponding TER variations), and the horizontal error bars to the thickness uncertainty from growth calibrations.

Garcia et al. [67] have used PFM to show robust ferroelectricity down to 1 nm in highly strained $BaTiO_3$ films and conductive-tip atomic force microscopy to demonstrate resistive readout of the polarization state, and its influence on the tunnel current can be seen from Fig. 28.10. The resulting electroresistance effect scales exponentially with FE film thickness, reaching 75,000% at 3 nm. They exploited the tunneling at these very low thicknesses to read the polarization state without destroying it. They demonstrated scalability down to 70 nm, corresponding to potential densities of >16 Gbit in^2. The observed results showed the great importance for high-density data storage, as efficient nondestructive resistive readout might greatly simplify FeRAM architectures and reading protocol. A MFTJ model consists of a FE $BaTiO_3$ sandwiched between two ferromagnetic $SrRuO_3$ electrodes, and have been studied by Velev et al. [47]. The TMR effect is the consequence of wave-function symmetry conservation across the epitaxial $SrRuO_3/BaTiO_3$ interfaces.

Garcia et al. [68] fabricated MFTJs of a LSMO electrode and a $BaTiO_3$ FE tunnel barrier (1–3 nm), with a Fe or Co counter-electrode. At 4K, a large negative

TMR was observed, reflecting a negative spin polarization for the Fe/$BaTiO_3$ interface. By applying short voltage pulses of ± 1 V, they observed reversible changes of the tunnel resistance linked to the variation of the barrier height with the FE polarization direction, with a TER of about 30%. More interestingly, the amplitude of the TMR was also found to strongly depend on the direction of the FE polarization. Both metallic-ferromagnetic LSMO ($a = 3.875$ Å, $\alpha = 89.74°$, and $T_c = \sim 310$K) and Cobalt ($T_c = \sim 1400$K) are excellent ferromagnetic materials useful for high-quality multiferroic tunnel junctions [69]. Quindeau et al. [70] have observed that cobalt is a necessary factor for the tunneling. They showed that the FE–FM interface in MFTJs is responsible for the observed effects as well as reproducing the coupling between TER and TMR. These heterostructures can be easily fabricated by PLD to obtain a practical room temperature MFTJ.

28.10 Summary and outlook

The FE, magnetic, and MFTJs promise exciting future technological applications as nonvolatile RAM memory logics and devices. Recently, tremendous research activities have been aimed at investigating the next generation of memory/logic devices that would overcome scaling limitations of conventional semiconductor technology based on charge storage. The electronic and transport properties of FTJs and MFTJs are intrinsically related to the FE polarization, coupling among various ferroics order parameters; mechanisms of FE switching and the effect of structure and interface on electronic transport properties are also of equal importance. For many reasons, the effect of scaling on FE polarization in heterostructures with different dimensions will likely show significant changes in their functional properties. These issues of FE and transport characteristics in FTJs and MFTJs must be readdressed at the appropriate length scale. In the case of MFTJs, the relationship between TMR and TER and manipulating them in a controlled manner needs to be investigated in depth. This report provides glimpses of state of the art development in the direction of FE and ME memory elements with simplified architectures, higher densities, and faster operation with further exploration of the interplay between quantum tunneling and ferroelectricity at the nanoscale.

References

[1] J.F. Scott, Ferroelectric Memories, Springer, Berlin, (2000).
[2] O. Auciello, C.M. Foster, R. Ramesh, Annu. Rev. Mater. Sci. 28 (1998) 501.
[3] A. Smolenskii, I.E. Chups, Sov. Phys. Usp. 25 (1982) 475.
[4] W. Eerenstein, N.D. Mathur, J.F. Scott, Nature 442 (2006) 759–765.
[5] H. Schmid, Ferroelectrics 162 (1994) 317.
[6] N.A. Spaldin, M. Fiebig, Science 309 (2005) 391.
[7] N. Ortega, P. Bhattacharya, R.S. Katiyar, P. Dutta, A. Manivannan, M.S. Seehra, et al., J Appl. Phys. 100 (2006) 126105.

[8] N. Ortega, Ashok Kumar, P. Bhattacharya, S.B. Majumder, R.S. Katiyar, Phys. Rev. B. 77 (2008) 014111.
[9] E.Y. Tsymbal, A. Gruverman, V. Garcia, M. Bibes, A. Barthélémy, MRS Bull. 37 (2012).
[10] André Chanthbouala, Arnaud Crassous, Vincent Garcia, Karim Bouzehouane, Stéphane Fusil, Xavier Moya, et al., Mathur, manuel bibes, and Agnès Barthélémy, Nat. Nanotechnol. 7 (2012) 101.
[11] L. Esaki, R.B. Laibowitz, P.J. Stiles, IBM Tech. Discl. Bull. 13 (1971) 2161.
[12] A. Lookman, R.M. Bowman, J.M. Gregg, J. Kut, S. Rios, M. Dawber, et al., J. Appl. Phys. 96 (2004) 555.
[13] D.D. Fong, G.B. Stephenson, S.K. Streiffer, J.A. Eastman, O. Auciello, P.H. Fuoss, et al., Science 304 (2004) 1650.
[14] Jian-Gang (Jimmy) Zhu, and Chando Park, Mater. Today, 9 (2006).
[15] D. Khomskii, Physics 2 (2009) 20.
[16] D. SeokJeong, R. Thomas, R.S. Katiyar, J.F. Scott, H. Kohlstedt, A. Petraru, et al., Rep. Prog. Phys. 75 (2012) 076502.
[17] J. Junquera, P. Ghosez, Nature 422 (2003) 506.
[18] E.Y. Tsymbal, H. Kohlstedt, Science 313 (2006).
[19] J.D. Burton, E.Y. Tsymbal, Phys. Rev. B 80 (2009) 174406.
[20] Y.W. Yin, J.D. Burton, Y.-M. Kim, A.Y. Borisevich, S.J. Pennycook, S.M. Yang, et al., Nat. Mater. 12 (2013) 397.
[21] A. Ohtomo, D.A. Muller, J.L. Grazul, H.Y. Hwang, Nature 419 (2002) 378.
[22] I. Vrejoiu, M. Alexe, D. Hesse, U. Gösele, Adv. Funct. Mater. 18 (2008) 3892.
[23] T. Tybell, C.H. Ahn, J.-M. Triscone, Appl. Phys. Lett. 75 (1999) 856.
[24] A.V. Bune, V.M. Fridkin, S. Ducharme, L.M. Blinov, S.P. Palto, A.V. Sorokin, et al., Nature 391 (1998) 874.
[25] T. Haccart, E. Cattan, D. Remiens, Semicond. Phys. Quant. Electron. Optoelectron 5 (2002) 78.
[26] J. Pérez de la Cruz, E. Joanni, P.M. Vilarinho, A.L. Kholkin, J. Appl. Phys. 108 (2010) 114106.
[27] H. Fijisawa, S. Nakashima, M. Shimizu, H. Niu, Proceedings of the 11th International Symposium on Applications of Ferroelectrics (ISAF), 77 (1998).
[28] C.-R. Cho, W.-J. Lee, B.-G. Yu, B.-W. Kim, J. Appl. Phys. 86 (1999) 2700.
[29] A.K. Tagantsev, M. Landivar, E. Colla, N. Setter, J. Appl. Phys. 78 (1995) 2623.
[30] L. Ran, J. Gen-Shan, L. Bin, Z. Quan-Liang, Z. De-Qing, Y. Jie, et al., Chin. Phys. Lett. 29 (2012) 058101.
[31] D. Barrionuevo, Le Zhang, N. Ortega, A. Sokolov, A. Kumar, Pankaj Misra, et al., Nanotechnology 25 (2014) 495203.
[32] D. Pantel, S. Goetze, D. Hesse, M. Alexe, Nat. Mater. 11 (2012) 289−293.
[33] J.F. Scott, Nat. Mater. 6 (2007) 256−257.
[34] J.F. Scott, L. Kammerdiner, M. Parris, S. Traynor, V. Ottenbacher, A. Shawabkeh, et al., J. Appl. Phys. 64 (1988) 787.
[35] H.F. Kay, J.W. Dunn, Phil. Mag. 7 (1962) 2027.
[36] M.N. Baibich, J.M. Broto, A. Fert, F. Nguyen Van Dau, F. Petroff, P. Etienne, et al., Phys. Rev. Lett. 61 (1988) 2472−2475.
[37] I. Žutić, J. Fabian, S.Das Sarma, Rev. Mod. Phys. 76 (2004) 323−410.
[38] S.S.P. Parkin, C. Kaiser, A. Panchula, P.M. Rice, B. Hughes, M. Samant, et al., Nat. Mat. 3 (12) (2004) 862−867.
[39] M. Julliere, Phys. Lett. A 54 (1975) 225.

[40] R.J. Soulen Jr., J.M. Byers, M.S. Osofsky, B. Nadgorny, T. Ambrose, S.F. Cheng, et al., Science 282 (1998) 85.
[41] D.J. Monsma, S.S.P. Parkin, Appl. Phys. Lett. 77 (2000) 720.
[42] S. Yuasa, T. Nagahama, A. Fukushima, Y. Suzuki, K. Ando, Nat. Mater. 3 (2004) 868–871.
[43] S.S.P. Parkin, C. Kaiser, A. Panchula, P.M. Rice, B. Hughes, M. Samant, et al., Nat. Mater. 3 (2004) 862–867.
[44] D.D. Djayaprawira, K. Tsunekawa, M. Nagai, H. Maehara, S. Yamagata, N. Watanabe, et al., Appl. Phys. Lett. 86 (3) (2005) 092502.
[45] P.V. Paluskar, J.J. Attema, G.A. de Wijs, S. Fiddy, S. Snoeck, J.T. Kohlhepp, et al., Phys. Rev. Lett. 100 (4) (2008) 057205.
[46] C. Chappert, A. Fert, F. Nguyen, Nat. Mater. 6 (2007) 813–823.
[47] J.P. Velev, C.G. Duan, J.D. Burton, A. Smogunov, M.K. Niranjan, E. Tosatti, et al., Nano Lett. 9 (2009) 427–432.
[48] Y.-W. Yin, M. Raju, W.-J. Hu, X.-J. Weng, K. Zou, J. Zhu, et al., Front. Phys. 7 (4) (2012) 380–385.
[49] S. Singh, J.T. Haraldsen, J. Xiong, E.M. Choi, P. Lu, D. Yi, et al., Phys. Rev. Lett. 113 (2014) 047204.
[50] T.D.N. Ngo, J.-W. Chang, K. Lee, S. Han, J.S. Lee, Y.H. Kim, et al., Nat. Commun. 6 (2015) 8035.
[51] N. Driza, S. Blanco-Canosa, M. Bakr, S. Soltan, M. Khalid, L. Mustafa, et al., Nat. Mater. 11 (2012) 675–681.
[52] W. Huang, S. Yang, X. Li, J. Materiomics 1 (2015) 263–284.
[53] E.Y. Tsymbal, A. Gruverman, V. Garcia, M. Bibes, A. Barthelemy, MRS Bull. 37 (2012) 138–143.
[54] V. Garcia, M. Bibes, Nat. Commun. 5 (2014) 4289.
[55] V. Garcia, M. Bibes, L. Bocher, S. Valencia, F. Kronast, A. Crassous, et al., Science 327 (2010) 1106–1110.
[56] Y.K. Liu, Y.W. Yin, S.N. Dong, S.W. Yang, T. Jiang, X.G. Li, Appl. Phys. Lett. 104 (2014) 043507.
[57] Y.W. Yin, W.C. Huang, Y.K. Liu, S.W. Yang, S.N. Dong, J. Tao, et al., Adv Electron Mater. 1 (2015) 1500183.
[58] O. Kolosov, A. Gruverman, J. Hatano, K. Takahashi, H. Tokumoto, Phys. Rev. Lett. 74 (1995) 4309.
[59] D.A. Bonnell, S.V. Kalinin, A.L. Kholkin, A. Gruverman, MRS Bull. 34 (2009) 648.
[60] A. Gruverman, D. Wu, H. Lu, Y. Wang, H.W. Jang, C.M. Folkman, et al., Nano Lett. 9 (2009) 3539.
[61] S.V. Kalinin, A. Gruverman, Scanning Probe Microscopy: Electrical and Electromechanical Phenomena at the Nanoscale, Springer Science & Business Media, 2007.
[62] N. Balke, P. Maksymovych, S. Jesse, A. Herklotz, A. Tselev, C.B. Eom, et al., ACS Nano 9 (2015) 6484–6492.
[63] S. Jesse, R.K. Vasudevan, L. Collins, E. Strelcov, M.B. Okatan, A. Belianinov, et al., Annu. Rev. Phys. Chem. 65 (2014) 519–536.
& Stephen Jesse, SergeiV. Kalinin, J. Phys. D: Appl. Phys. 44 (2011) 464006.
[64] R.K. Vasudevan, S. Zhang, J. Ding, M.B. Okatan, S. Jesse, S.V. Kalinin, et al., Appl. Phys. Lett. 106 (2015) 222901.
[65] R.K. Vasudevan, S. Zhang, M.B. Okatan, S. Jesse, S.V. Kalinin, N. Bassiri Gharb, J. Appl. Phys. 118 (2015) 072003.
[66] P. Maksymovych, S. Jesse, P. Yu, R. Ramesh, A.P. Baddorf, S.V. Kalinin, Science 1421 (2009).

[67] V. Garcia, S. Fusil, K. Bouzehouane, S. Enouz-Vedrenne, N.D. Mathur, A. Barthélémy, et al., Nature 460 (2009).
[68] V. Garcia, M. Bibes, L. Bocher, S. Valencia, F. Kronast, S. Enouz-Vedrenne, et al., Science 327 (2010) 1106.
[69] L.C. Phillips, F. Maccherozzi, X. Moya, M. Ghidini, W. Yan, J. Soussi, et al., J. Magn. Magn. Mater. 355 (2014) 331–333.
[70] A. Quindeau, I. Fina, X. Marti, G. Apachitei, P. Ferrer, C. Nicklin, et al., Sci. Rep. 5 (2015) 9749.

Index

Note: Page numbers followed by "*f*" and "*t*" refer to figures and tables, respectively.

B

F

Q